Density Functional Theory

Theoretical and Mathematical Physics

The series founded in 1975 and formerly (until 2005) entitled *Texts and Monographs in Physics* (TMP) publishes high-level monographs in theoretical and mathematical physics. The change of title to *Theoretical and Mathematical Physics* (TMP) signals that the series is a suitable publication platform for both the mathematical and the theoretical physicist. The wider scope of the series is reflected by the composition of the editorial board, comprising both physicists and mathematicians.

The books, written in a didactic style and containing a certain amount of elementary background material, bridge the gap between advanced textbooks and research monographs. They can thus serve as basis for advanced studies, not only for lectures and seminars at graduate level, but also for scientists entering a field of research.

For further volumes:
http://www.springer.com/series/720

Eberhard Engel · Reiner M. Dreizler

Density Functional Theory

An Advanced Course

Springer

apl. Prof. Dr. Eberhard Engel
Center for Scientific Computing
and Institute of Theoretical Physics
Johann Wolfgang Goethe-Universität
Frankfurt/Main
Max-von-Laue-Str. 1
60438 Frankfurt am Main
Germany
engel@th.physik.uni-frankfurt.de

Prof. Dr. Reiner M. Dreizler
Institute of Theoretical Physics
Dept. of Physics
Johann Wolfgang Goethe-Universität
Frankfurt/Main
Max-von-Laue-Str. 1
60438 Frankfurt am Main
Germany
dreizler@th.physik.uni-frankfurt.de

ISSN 1864-5879 e-ISSN 1864-5887
ISBN 978-3-642-14089-1 e-ISBN 978-3-642-14090-7
DOI 10.1007/978-3-642-14090-7
Springer Heidelberg Dordrecht London New York

Library of Congress Control Number: 2010933963

Printed on acid-free paper

Springer is part of Springer Science+Business Media (www.springer.com)

Preface

This book emerged from a course on density functional theory (DFT), first given at the University of Munich more than a decade ago. The course was based on the classic texts by Dreizler and Gross (Springer, 1990) and by Parr and Yang (Oxford University Press, 1989). More recent topics of that time, such as time-dependent DFT or orbital-dependent functionals, were added to the material covered by the two books. However, already at that time restriction to the most relevant and/or most illustrative statements on a particular aspect of DFT was necessary, in order to keep the length of the course under control. When the course was later given again at the University of Frankfurt it soon turned out to be impossible to integrate the exploding number of new results, concerning both the formalism as well as important applications, into the course: So, even a selection of the branches of DFT covered in the course was unavoidable.

The present text reflects this, admittedly subjective, choice of topics: it concentrates on the basics of the most widely used variants of DFT. This implies a thorough discussion of the corresponding existence theorems and effective single-particle equations as well as of the key approximations utilized in implementations. Ground state DFT (on the nonrelativistic level) is addressed in Chaps. 2–6. Chapter 2 introduces the fundamental Hohenberg-Kohn theorem and its extensions to spin-, current- and current-spin-density functional theory, together with some basic notions such as v-representability. The resulting Kohn-Sham equations are collected in Chap. 3. This chapter also includes a discussion of the relation between the Kohn-Sham wavefunctions and eigenvalues and the true many-body wavefunctions and energies. Chapter 4 is devoted to a detailed exposition of the currently available approximations for the exchange-correlation functional, based on two exact representations of this quantity. The most important virial relations valid for density functionals are summarized in Chap. 5. The discussion of the exchange-correlation functional is then resumed in Chap. 6, in which the concept of orbital-dependent functionals is introduced. This chapter also serves as a demonstration of the first-principles character of DFT, in that it shows that the true exchange-correlation energies and potentials can be systematically approached by use of orbital-dependent functionals. On the other hand, the discussion of the existence theorem, of the basic formalism and of standard approximations is concentrated in a single chapter in the case of time-dependent DFT (Chap. 7). The same statement applies to relativistic DFT which is presented in Chap. 8. An extended review of the problem of renormalization (in the Appendices F–I) serves as background information for relativistic DFT.

Throughout the text formal statements are complemented by selected quantitative results, which primarily aim at an illustration of the strengths and weaknesses of a particular approach or functional. However, no attempt is made to review the full range of present-day DFT applications, not even their boundaries will be marked out. In order to stay within reasonable bounds, the discussion also omits a few topics which have recently attracted much interest, such as DFT for superconducting or hadronic systems. An overview of the topics not covered explicitly in this book is provided by Chap. 9 (*Further Reading*) which offers a substantial collection of pertinent papers together with some comments.

The students in the courses indicated above had quite diverse backgrounds, ranging from mineralogy to biochemistry. So, the courses had to be rather self-contained, requiring neither the audience's familiarity with standard many-body theory nor extensive experience with the quantum theory of solids. Again, the book reflects this fact: the reader will find that all concepts of many-body theory which are indispensable for the discussion of DFT, such as the single-particle Green's function or response functions, are introduced step by step, rather than just used. The same applies to some basic notions of solid state theory, as, for instance, the Fermi surface. In fact, even the language of second quantization is introduced systematically in an Appendix. When starting with this Appendix, reading this book should require little more than a strong background in elementary quantum mechanics (at least, if one accepts some of the more advanced relations of Chap. 4 without going through their derivations[1]).

As is clear from these remarks, this book does not target only one particular scientific community. On the other hand, the material is easily restricted to the needs of a more specialized course. Many of the advanced chapters require little more than knowledge of the most elementary parts of the introductory chapters. Several redundancies help to support this modular structure. It should therefore be possible to find one's own way through the material. Although detailed recommendations have obvious problems, some suggestions for selected reading are made in the following table:

[1] An alternative, low level entry point to the discussion of the exchange-correlation energy functional is provided by Appendix D, in which the local density approximation for the exchange functional is derived explicitly without using concepts from many-body theory. With the background of this appendix it should be possible to continue with the more advanced results and explicit functionals of Chap. 4.

Section	Bias solid state physics		Bias chemistry, atomic physics	
	Basic	Advanced	Basic	Advanced
1	Complete		Complete	
2.1	Complete	Complete	Complete	Complete
2.2	Optional	Complete	Optional	Complete
2.3	1.+2. paragraph	Complete	1.+2. paragraph	Complete
2.4		Complete		Complete
2.5	Complete	Complete	Complete	Complete
2.6		Optional		Optional
2.7	Complete	Complete	Complete	Complete
3.1	Complete	Complete	Complete	Complete
3.2,3.3		Complete		Complete
3.4		Optional		Optional
3.5	Complete	Complete	Complete	Complete
3.6	Complete	Complete	1st paragraph	3.6.1
3.7		Optional		Optional
4.1	Complete	Complete	Complete	Complete
4.2	Only results	Complete		Only results
4.3	Complete	Complete	App.D+4.3.2–6	Complete
4.4	Only results	Complete		Only results
4.5	4.5.4+4.5.5	Complete	4.5.4+4.5.5	Complete
4.6				
4.7		Optional		Optional
4.8	Optional	Complete		Only results
4.9	Complete	Complete		
5.1–5.4	Complete	Complete	Complete	Complete
6.1	Complete	Complete	Complete	Complete
6.2	6.2.1/2/5/6	6.2.1–6		6.2.1/2/5/6
6.3	Complete	Complete		Complete
6.4		6.4.1+6.4.2		6.4.1
6.5		6.5.4	6.5.4	6.5.4
6.6		Complete		Complete
6.7				
7.1–7.6	Complete	Complete	Complete	Complete
8.1		Complete		Complete
8.2–8.5		Optional		
8.6		Complete		Complete
8.7		Complete		Optional
8.8		Optional		Optional

Depending on the background of the reader, Appendix B (on second quantization) might have to be included between Chaps. 1 and 2.

At various points explicit derivations of important results are given, rather than just summaries of the results. These derivations might not be suitable for presentation in a course, depending on its scope and audience. However, there is always an option to skip the details and restrict oneself to the essentials. Sometimes, such details are distinguished from the main text by use of a smaller font, in other cases the details have been relegated to Appendices. At some points the reader will find suggestions which portion of the text may be skipped, if one wants to focus on results only.

It is a pleasure for us to acknowledge the help and support that we received when writing this book. This book has benefited enormously from extensive discussions with Dr. D. Ködderitzsch. His comments and suggestions helped in particular to improve the readability of this text for less experienced readers. Many thanks go to the two referees of our manuscript (unknown to us). Their reports encouraged us to include a number of additional topics in the text (some suggested by the referees, others which we felt appropriate), although this led to an increase of its length, way beyond initial plans. Equally important, however, was the additional time which the referees comments have given us: this allowed us to straighten out some paragraphs in the initial manuscript. We would also like to thank Dr. S. Varga and H. Engel for their careful proof-reading and many suggestions for improvements. We are grateful to M. Hellgren and U. von Barth for making their atomic RPA potentials available to us. Last, but not least, our thanks go to Dr. Ch. Caron from Springer Verlag, who supervised the production of this book. His support, flexibility and patience were instrumental in making this book what it is.

Frankfurt am Main,
March 2010

Eberhard Engel
Reiner M. Dreizler

Contents

Acronyms

ALDA	adiabatic local density approximation
AF	antiferromagnetic
ASA	atomic sphere approximation
B86	Becke 86 GGA for exchange energy
B88	Becke 88 GGA for exchange energy
B3LYP	hybrid constructed on basis of Becke-Lee-Yang-Parr GGA
BLYP	Becke-Lee-Yang-Parr GGA
bcc	body-centered cubic
BO	Born-Oppenheimer
C	Coulomb
CDFT	current density functional theory
CS	Colle-Salvetti
CSDFT	current spin density functional theory
DC	Dirac-Coulomb
DCB	Dirac-Coulomb-Breit
DFT	density functional theory
DIR	direct (matrix element)
EA	electron affinity
ext	external
EXX	exact exchange
fcc	face-centered cubic
FP	full potential
GE	gradient expansion
GGA	generalized gradient approximation
GKS	generalized Kohn-Sham
GK	Gross-Kohn (kernel of TDDFT)
H	Hartree
HDL	high-density limit
HEG	homogeneous (or uniform) electron gas
HF	Hartree-Fock
HK	Hohenberg-Kohn

hom	homogeneous
HOMO	highest occupied molecular orbital
IP	ionization potential
int	interaction
ISI	interaction strength interpolation
KS	Kohn-Sham
KLI	Krieger-Li-Iafrate (approximation)
LAPW	linearized-augmented-plane-wave
LDA	local density approximation
LDA+U	combination of LDA and Hubbard U approach
LH	local hybrid
LHF	localized Hartree-Fock
LM	Langreth-Mehl GGA for correlation energy
LR	linear response
LSDA	local spin-density approximation
LUMO	lowest unoccupied molecular orbital
LYP	Lee-Yang-Parr GGA
MGGA	meta generalized gradient approximation
MP	Møller-Plesset
np	no-pair
NR	nonrelativistic
OPM	optimized (effective) potential method
PP	pseudopotential
P86	Perdew 86 GGA for correlation energy
PBE	Perdew-Burke-Ernzerhof GGA
PBE0	hybrid constructed on basis of Perdew-Burke-Ernzerhof GGA
PKZB	Perdew-Kurth-Zupan-Blaha MGGA
PW86	Perdew-Wang 86 GGA for exchange energy
PW91	Perdew-Wang 91 GGA
PZ	Perdew-Zunger self-interaction correction
QED	quantum electrodynamics
R	retarded
RDFT	relativistic density functional theory
REXX	relativistic exact exchange
RHEG	relativistic homogeneous (or uniform) electron gas
RG	Runge-Gross (theorem)
RGGA	relativistic generalized gradient approximation
RHF	relativistic Hartree-Fock
RLDA	relativistic local density approximation
RLSDA	relativistic local spin density approximation
ROPM	relativistic optimized (effective) potential method
RPA	random phase approximation
RSDFT	relativistic spin density functional theory
SDFT	spin-density functional theory
s	single-particle, noninteracting, Kohn-Sham

SI	self-interaction
SIC	self-interaction correction
SOX	second order exchange
SPP	Stoll-Pavlidou-Preuss self-interaction correction
T	transverse
TDDFT	time-dependent density functional theory
TDKS	time-dependent Kohn-Sham
TDLDA	time-dependent local density approximation
TF	Thomas-Fermi
TPSS	Tao-Perdew-Staroverov-Scuseria MGGA
UV	ultraviolet
vBH	von Barth-Hedin LDA for correlation energy
VWN	Vosko-Wilk-Nusair LDA for correlation energy
WDA	weighted density approximation
xc	exchange-correlation

Chapter 1
Introduction

One of the basic problems in theoretical physics and chemistry is the description of the structure and dynamics of many-electron systems. These systems comprise single atoms, the most elementary building blocks of ordinary matter, all kinds of molecules, ranging from dimers to proteins, as well as mesoscopic systems, for example clusters or quantum dots, and solids, including layered structures, surfaces and quasi-crystals. The following two paragraphs list the properties of such systems which are generally of interest, without, however, aiming at completeness. These properties can roughly be classified as either structural or dynamical.

An important structural property is the electronic shell structure (or band structure in the case of crystals). The shell structure directly determines the stability of a system, but also shows up in a number of other properties—it is, for instance, a key factor in transport properties like the electrical conductivity. Stability manifests itself in various binding energies. These are either of electronic nature, such as the ionization potential and the work function, or they characterize the bonds between atoms, such as the atomization energy of molecules and the cohesive energy of solids. Other structural properties, which are related to shell structure, are electric or magnetic moments. The geometry of poly-atomic systems, that is bond lengths and bond angles as well as the symmetry of the atomic arrangement, constitutes yet another important structural property. Often several geometrical configurations which are almost degenerate (at least compared to typical thermal energies) are observed for such systems. In this case the relative stability of the various configurations is of obvious interest. All these properties have in common that they can be calculated if the relevant electronic ground states are known. Even if more than one ground state is involved, as in the comparison of energies of systems with different electron numbers or with different atomic configurations, there is no need to determine excited electronic states.

The electronic excitation spectrum is the most notable dynamical property. Excitation energies are not only the quantities necessary for an understanding of optical properties, they also feature in all kinds of scattering processes. In addition to the excitation spectrum, a complete description of excitation or ionization requires the evaluation of the associated transition probabilities. In a poly-atomic system the

E. Engel, R.M. Dreizler, Introduction. In: E. Engel, R.M. Dreizler, Density Functional Theory, Theoretical and Mathematical Physics, pp. 1–9 (2011)
DOI 10.1007/978-3-642-14090-7_1

excitation of nuclear motion is also possible, the rotational-vibrational motion in molecules or lattice vibrations (phonons) in solids being the simplest examples. The discussion of these dynamical properties obviously requires knowledge of either excited stationary states of the system, or even of time-dependent states.

For the study of these properties one can distinguish between two basic approaches:

1. The first approach is based on the true, fundamental Hamiltonian of the system. It is therefore called the *ab-initio* or *first principles* approach. As the solution of the resulting many-body Schrödinger equation is an extremely demanding task, some approximations are unavoidable. However, all approximations utilized in this context are of a technical nature, as for instance the determinantal form of the many-body wavefunction in the Hartree-Fock approximation or the representation of its single-particle ingredients in terms of a finite basis set. These technical simplifications can curtail the results of an *ab-initio* calculation: depending on the system and property under investigation, important features can be missed (e.g. a single-determinant wavefunction can not reproduce the fully correlated motion of the electrons). The *ab-initio* approach is thus not characterized by the absence of any approximations, but rather by the fact that the approximations do not introduce adjustable physical(!) parameters. Clearly, a more efficient technical approximation will lead to a computational task that can be handled more easily.
2. In the second approach a suitable *model* Hamiltonian is utilized from the very outset. Relying on available information (often experimental) on the system, the relevant aspects can be isolated and used to construct the model Hamiltonian. For instance, if one is interested in the magnetic ordering of a mono-atomic crystalline solid, one might rely on the *Heisenberg Hamiltonian*. In this model each atomic site α of the crystal is represented by a spin operator $\boldsymbol{s}_\alpha$ with possible eigenvalues corresponding to the available spin per site. The interaction between the spins of two sites α and β is represented by $-J\boldsymbol{s}_\alpha \cdot \boldsymbol{s}_\beta$, so that for positive J energy is gained by the alignment of the two spins. Clearly, the Heisenberg Spin Hamiltonian can not make any statement about the lattice constant of the solid, as it only deals with one particular property, its magnetism. The only remnant of the underlying electronic structure is the size of the coupling constant J. This physical parameter can, however, not be determined within the model, but has to be adjusted to experiment or some prior *ab-initio* results. The advantage of the use of model Hamiltonians is obvious: the resulting many-body problem is much simpler than the corresponding *ab-initio* problem. On the other hand, it is often an art to extract the essential aspects of a system in a way which neither ignores important features nor leads to an overwhelmingly complicated Hamiltonian.

The present text introduces a variant of the first approach, which is at the same time efficient, widely applicable and reasonably accurate.

The starting point of the discussion is the *ab-initio* Hamiltonian of the coupled electron-nucleus system,

$$\hat{H} = \hat{T}_{\rm n} + \hat{V}_{\rm n-n} + \hat{H}_{\rm e} \; (+\hat{V}_{\rm n-field}) \tag{1.1}$$
$$\hat{H}_{\rm e} = \hat{T}_{\rm e} + \hat{V}_{\rm n-e} + \hat{V}_{\rm e-e} \; (+\hat{V}_{\rm e-field}) \,. \tag{1.2}$$

In (1.1) $\hat{T}_{\rm n}$ represents the kinetic energy of the K nuclei of the system,

$$\hat{T}_{\rm n} = \sum_{\alpha=1}^{K} \frac{(-i\hbar \boldsymbol{\nabla}_{\boldsymbol{R}_\alpha})^2}{2M_\alpha} \,, \tag{1.3}$$

where $\boldsymbol{R}_\alpha$ denotes the Cartesian coordinates of nucleus α with mass M_α and $\boldsymbol{\nabla}_{\boldsymbol{R}_\alpha} = \partial/\partial\boldsymbol{R}_\alpha$. The potential $\hat{V}_{\rm n-n}$ represents the repulsion among the nuclei,[1]

$$\hat{V}_{\rm n-n} = \sum_{\alpha,\beta=1;\alpha<\beta}^{K} \frac{Z_\alpha Z_\beta e^2}{|\boldsymbol{R}_\alpha - \boldsymbol{R}_\beta|} \,. \tag{1.4}$$

For the electronic problem $\hat{V}_{\rm n-n}$ just amounts to an additive constant. The third component of $\hat{H}$ is the electronic Hamiltonian $\hat{H}_{\rm e}$, which is decomposed into the kinetic energy of the N electrons,

$$\hat{T}_{\rm e} = \sum_{i=1}^{N} \frac{(-i\hbar \boldsymbol{\nabla}_i)^2}{2m} \tag{1.5}$$

($\boldsymbol{\nabla}_i \equiv \boldsymbol{\nabla}_{\boldsymbol{r}_i}$, with $\boldsymbol{r}_i$ denoting the position of electron i), the interaction between electrons and nuclei (with charges $Z_\alpha e$, $e = |e|$),

$$\hat{V}_{\rm n-e} = -\sum_{\alpha=1}^{K}\sum_{i=1}^{N} \frac{Z_\alpha e^2}{|\boldsymbol{R}_\alpha - \boldsymbol{r}_i|} \,, \tag{1.6}$$

and the interaction among the electrons,

$$\hat{V}_{\rm e-e} = \sum_{i,j=1;i<j}^{N} \frac{e^2}{|\boldsymbol{r}_i - \boldsymbol{r}_j|} \,. \tag{1.7}$$

Finally, depending on the system under consideration, the two types of particles can couple to an external electromagnetic field,

$$\hat{V}_{\rm n-field} = \sum_{\alpha=1}^{K} \left[Z_\alpha e \Phi_{\rm ext}(\boldsymbol{R}_\alpha t) + i\frac{Z_\alpha e\hbar}{M_\alpha c} \boldsymbol{A}_{\rm ext}(\boldsymbol{R}_\alpha t) \cdot \boldsymbol{\nabla}_{\boldsymbol{R}_\alpha} - \boldsymbol{I}_\alpha \cdot \boldsymbol{B}_{\rm ext}(\boldsymbol{R}_\alpha t) \right] \tag{1.8}$$
$$\hat{V}_{\rm e-field} = \sum_{i=1}^{N} \left[-e\Phi_{\rm ext}(\boldsymbol{r}_i t) - 2i\mu_B \boldsymbol{A}_{\rm ext}(\boldsymbol{r}_i t) \cdot \boldsymbol{\nabla}_i + \mu_B \boldsymbol{\sigma}_i \cdot \boldsymbol{B}_{\rm ext}(\boldsymbol{r}_i t) \right] \,. \tag{1.9}$$

Here $\Phi_{\rm ext}(\boldsymbol{r}t)$ and $\boldsymbol{A}_{\rm ext}(\boldsymbol{r}t)$ are the potentials corresponding to the electromagnetic field,

[1] Throughout this text we use Gaussian units for the electric charge.

$$\boldsymbol{E}_{\mathrm{ext}}(\boldsymbol{r}t) = -\boldsymbol{\nabla}\Phi_{\mathrm{ext}}(\boldsymbol{r}t) - \frac{1}{c}\frac{\partial \boldsymbol{A}_{\mathrm{ext}}(\boldsymbol{r}t)}{\partial t} \tag{1.10}$$

$$\boldsymbol{B}_{\mathrm{ext}}(\boldsymbol{r}t) = \boldsymbol{\nabla} \times \boldsymbol{A}_{\mathrm{ext}}(\boldsymbol{r}t) \,. \tag{1.11}$$

$\boldsymbol{I}_\alpha$ characterizes the magnetic moment of nucleus α, μ_B is the Bohr magneton, $\mu_B = e\hbar/(2mc)$, and $\boldsymbol{\sigma}_i$ the spin operator corresponding to electron i, (the vector $\boldsymbol{\sigma}$ denotes the three $2{\times}2$ Pauli matrices—for their explicit forms see Eqs. (2.122)–(2.124)).

A number of comments are appropriate:

- As expressed by (1.3) and (1.5), the motion of both the electrons and the nuclei is treated strictly nonrelativistically. Equations (1.3)–(1.9) also imply that the nuclei are point particles, characterized only by mass, charge and magnetic moment. Of course, the nonrelativistic and elementary treatment of the nuclei is completely legitimate for the range of questions addressed here. On the other hand, the neglect of all relativistic corrections to the kinetic energy of the electrons, most notably of the spin–orbit interaction, is not always legitimate: for heavy atoms relativistic effects modify the structure of the electronic Hamiltonian to an extent which is even noticeable in molecular bonds, so that use of a relativistic $\hat{T}_e$ is required. The issue of relativity will be addressed in Chap. 8. For the time being the interest will, however, be restricted to the nonrelativistic limit.
- Similarly, the interaction between the charged particles in the Hamiltonians (1.4), (1.6) and (1.7) is given by the instantaneous and spin-independent Coulomb interaction, which ignores the transverse nature of light (photons) and its finite speed, manifest e.g. in the Breit interaction. The discussion of relativistic corrections to the Coulomb interaction is also postponed to Chap. 8. However, it seems worthwhile to emphasize already at this point that the neglect of the Breit corrections is in general much less critical than the neglect of the spin–orbit interaction.
- In Eqs. (1.1) and (1.2) the possible presence of an external electromagnetic field is indicated. This field can be either static or time-dependent, as for instance a laser pulse. Again the coupling of the field to the quantized particles is specified on the simplest level available, assuming Coulomb gauge,

$$\boldsymbol{\nabla} \cdot \boldsymbol{A}_{\mathrm{ext}}(\boldsymbol{r}t) = 0 \,. \tag{1.12}$$

  Most of the discussion will, however, be devoted to the field-free situation $\Phi_{\mathrm{ext}} = \boldsymbol{A}_{\mathrm{ext}} = 0$ or to static fields. The Hamiltonian of the coupled system is then stationary. Nevertheless, even in this case a time-dependence can be introduced into the electron problem as soon as the motion of the nuclei is treated classically (as is customary for the discussion of collision problems). An indication of how to deal with time-dependent fields is given in Chap. 7.

The simplest and best-studied systems are single atoms. Here a dynamical treatment is required only for the electrons (at least, as long as the atoms are not subject to electromagnetic fields): after transformation into the center-of-mass frame the Hamiltonian separates into a part governing the trivial translational motion of the complete atom and the motion of the electrons relative to the center-of-mass. As the origin of the center-of-mass frame is (essentially) identical with the position of the nucleus, one ends up with an electronic problem in which the nucleus enters as a (static) external source. This effective factorization of nuclear and electron degrees of freedom is no longer possible for poly-atomic systems. In these systems one has to deal with the coupled dynamics of electrons and nuclei, i.e. with the Schrödinger equation

$$\begin{aligned}&\hat{H}\Psi_a(\boldsymbol{R}_1,\ldots,\boldsymbol{R}_K;\boldsymbol{r}_1\sigma_1,\ldots,\boldsymbol{r}_N\sigma_N)\\&=E_a\Psi_a(\boldsymbol{R}_1,\ldots,\boldsymbol{R}_K;\boldsymbol{r}_1\sigma_1,\ldots,\boldsymbol{r}_N\sigma_N)\,,\end{aligned}\tag{1.13}$$

where σ_i denotes the spin-orientation of electron i with respect to some chosen axis (from now on it is assumed that the external fields are time-independent). As it stands, the partial differential equation (1.13) poses an exceedingly complicated technical problem: the size of the systems of interest ranges from atoms to solids, so that the particle number in (1.13) varies over many orders of magnitudes. At the same time, a quantum mechanical treatment of the electrons is usually unavoidable, even if only the most elementary features of these systems are to be studied.

The standard first step towards a solution of (1.13) is a partial decoupling of the electron from the nuclear motion, which relies on the different time scales of the two types of motion. This is achieved by the *Born-Oppenheimer approximation* [1]. The Born-Oppenheimer approximation amounts to a factorization of the total wavefunction Ψ_a into a nuclear wavefunction $\Psi^{\rm n}_{ik}$ and an electron wavefunction $\Psi^{\rm e}_k$,

$$\begin{aligned}&\Psi_{a\equiv i,k}(\boldsymbol{R}_1,\ldots,\boldsymbol{R}_K;\boldsymbol{r}_1\sigma_1,\ldots,\boldsymbol{r}_N\sigma_N)\\&=\Psi^{\rm n}_{ik}(\boldsymbol{R}_1,\ldots,\boldsymbol{R}_K)\Psi^{\rm e}_k(\boldsymbol{R}_1,\ldots,\boldsymbol{R}_K;\boldsymbol{r}_1\sigma_1,\ldots,\boldsymbol{r}_N\sigma_N)\,.\end{aligned}\tag{1.14}$$

The electron wavefunction depends parametrically on the position of the nuclei. It satisfies the Schrödinger equation

$$\begin{aligned}&\hat{H}_{\rm e}\Psi^{\rm e}_k(\boldsymbol{R}_1,\ldots,\boldsymbol{R}_K;\boldsymbol{r}_1\sigma_1,\ldots,\boldsymbol{r}_N\sigma_N)\\&=E_k(\boldsymbol{R}_1,\ldots,\boldsymbol{R}_K)\Psi^{\rm e}_k(\boldsymbol{R}_1,\ldots,\boldsymbol{R}_K;\boldsymbol{r}_1\sigma_1,\ldots,\boldsymbol{r}_N\sigma_N)\,,\end{aligned}\tag{1.15}$$

which represents a stationary eigenvalue problem for any given set of $\boldsymbol{R}_\alpha$. Together with $\hat{V}_{\rm n-n}$, Eq. (1.4), the eigenvalues $E_k(\boldsymbol{R}_1,\ldots)$ act as potentials in which the nuclei are moving.

The solution of (1.15) is a formidable computational task, even for fixed positions of the nuclei. Not only the large number and the quantum nature of the electrons represent a challenge, but also the complicated geometry of many systems. It is also possible that a number of (meta-)stable arrangements of the nuclei (isomers, conformers) exist, so that the solution of (1.15) for each of these arrangements is required in order to determine the ground state configuration. For these reasons an extremely efficient handling of the electron problem is necessary, even if the possible motion of the nuclei is ignored.

A variety of approaches have been developed to obtain approximate solutions of Eq. (1.15). The oldest and simplest is the *Hartree-Fock* (HF) (or *Self-Consistent-*

Field) approximation, in which the ground state wavefunction $\Psi^{\mathrm{e}}_{k=0}$ is assumed to be a determinant of single-particle states, a so-called *Slater determinant*,

$$\Psi^{\mathrm{e}}_0(\boldsymbol{r}_1\sigma_1,\ldots\boldsymbol{r}_N\sigma_N) \approx \Phi_{1\cdots N}(\boldsymbol{r}_1\sigma_1,\ldots\boldsymbol{r}_N\sigma_N)$$

$$\Phi_{i_1\cdots i_N}(\boldsymbol{r}_1\sigma_1,\ldots\boldsymbol{r}_N\sigma_N) = \frac{1}{\sqrt{N!}}\det\begin{pmatrix} \phi_{i_1}(\boldsymbol{r}_1\sigma_1) & \cdots & \phi_{i_N}(\boldsymbol{r}_1\sigma_1) \\ \vdots & & \vdots \\ \phi_{i_1}(\boldsymbol{r}_N\sigma_N) & \cdots & \phi_{i_N}(\boldsymbol{r}_N\sigma_N) \end{pmatrix} \tag{1.16}$$

(in the following, determinantal wavefunctions will always be denoted by Φ in order to distinguish them from the correlated N-particle wavefunctions Ψ; in addition, the spin variable σ will be displayed explicitly as an argument; the two possible orientations of σ will be either denoted as $\uparrow,\downarrow$ or by $\pm 1/2$). The N single-particle states ϕ_i in the approximate ground state wavefunction $\Phi_{1\ldots N}$ are determined variationally. For this purpose the expectation value of the Hamiltonian (1.2) is evaluated with respect to the wavefunction $\Phi_{1\ldots N}$. Assuming the single-particle states to be orthonormal, the result is

$$\begin{aligned}\langle\Phi_{1\cdots N}|\hat{H}|\Phi_{1\cdots N}\rangle &= \sum_{i=1}^{N}\sum_{\sigma=\uparrow,\downarrow}\int d^3r\,\phi_i^*(\boldsymbol{r}\sigma)\left[\frac{(-i\hbar\boldsymbol{\nabla})^2}{2m}+v_{\mathrm{ext}}(\boldsymbol{r})\right]\phi_i(\boldsymbol{r}\sigma) \\ &+\frac{1}{2}\sum_{i,j=1}^{N}\sum_{\sigma,\sigma'=\uparrow,\downarrow}\int d^3r\int d^3r'\,\phi_i^*(\boldsymbol{r}\sigma)\phi_j^*(\boldsymbol{r}'\sigma')\,w(\boldsymbol{r},\boldsymbol{r}') \\ &\times\left[\phi_i(\boldsymbol{r}\sigma)\phi_j(\boldsymbol{r}'\sigma')-\phi_j(\boldsymbol{r}\sigma)\phi_i(\boldsymbol{r}'\sigma')\right], \end{aligned} \tag{1.17}$$

where $v_{\mathrm{ext}}(\boldsymbol{r}) = -\sum_{\alpha=1}^{K} Z_\alpha e^2/|\boldsymbol{R}_\alpha-\boldsymbol{r}|$ is the total potential generated by the nuclei and $w(\boldsymbol{r},\boldsymbol{r}') = e^2/|\boldsymbol{r}-\boldsymbol{r}'|$ denotes the electron–electron interaction. The contribution of the two-body interaction consists of a direct and an exchange term, as a consequence of the determinantal structure of the state (1.16).

Minimization of the approximate ground state energy (1.17) with respect to the orbitals $\phi_i^*(\boldsymbol{r}\sigma)$, including the constraint of orthonormality,[2] leads to the HF equations ($i = 1,\ldots N$),

---

[2] The constraint is implemented by the introduction of a suitable set of Lagrange multipliers $\{\varepsilon_i\}$ in the functional to be minimized,

$$\langle\Phi_{1\cdots N}|\hat{H}|\Phi_{1\cdots N}\rangle - \sum_{i=1}^{N}\varepsilon_i\left(\sum_\sigma\int d^3r|\phi_i(\boldsymbol{r}\sigma)|^2-1\right).$$

A set of real and diagonal multipliers is sufficient as the resulting single-particle Hamiltonian determining the ϕ_i turns out to be hermitian, which ensures the orthogonality of the ϕ_i. Alternatively, one can explicitly include Lagrange multipliers for the orthogonality constraint and subsequently use the properties of the single-particle equation and the Lagrange multiplier matrix to show that the latter can be diagonalized by a suitable unitary transformation of the solutions of the single-particle equation [2]. The set of single-particle states corresponding to the diagonal Lagrange multiplier matrix is often called canonical.

$$-\frac{\hbar^2\nabla^2}{2m}\phi_i(\boldsymbol{r}\sigma)+\sum_{\sigma'=\uparrow,\downarrow}\int d^3r'\,v_{\rm eff}^{\rm HF}(\boldsymbol{r}\sigma,\boldsymbol{r}'\sigma')\,\phi_i(\boldsymbol{r}'\sigma') = \varepsilon_i\phi_i(\boldsymbol{r}\sigma) \tag{1.18}$$

$$v_{\rm eff}^{\rm HF}(\boldsymbol{r}\sigma,\boldsymbol{r}'\sigma') = \delta_{\sigma\sigma'}\,\delta^{(3)}(\boldsymbol{r}-\boldsymbol{r}')\,[v_{\rm ext}(\boldsymbol{r})+v_{\rm H}(\boldsymbol{r})]+v_{\rm x}^{\rm HF}(\boldsymbol{r}\sigma,\boldsymbol{r}'\sigma') \tag{1.19}$$

(for the variational or functional derivatives required see Appendix A). The HF equations have the structure of the ordinary single-particle Schrödinger equation. However, the total single-particle potential in the HF equations is nonlocal and contains two terms, the Hartree (direct Coulomb) potential

$$v_{\rm H}(\boldsymbol{r}) = \int d^3r'\,w(\boldsymbol{r},\boldsymbol{r}')\sum_{\sigma'=\uparrow,\downarrow}\sum_{j=1}^{N}|\phi_j(\boldsymbol{r}'\sigma')|^2 \tag{1.20}$$

and the exchange potential

$$v_{\rm x}^{\rm HF}(\boldsymbol{r}\sigma,\boldsymbol{r}'\sigma') = -w(\boldsymbol{r},\boldsymbol{r}')\sum_{j=1}^{N}\phi_j(\boldsymbol{r}\sigma)\phi_j^*(\boldsymbol{r}'\sigma'), \tag{1.21}$$

which depend on the solutions to be determined by Eq. (1.18). The standard approach to deal with this nonlinearity is a selfconsistent iteration of Eq. (1.18): starting with an initial guess for the total potential, the states ϕ_i and the potentials (1.20) and (1.21) are improved step by step. In each cycle of the iteration process first the ϕ_i corresponding to the current potentials are determined and then updated potentials are calculated from these ϕ_i. This iterative process is stopped once a suitable accuracy criterion is met (i.e. as soon as selfconsistency is achieved). Since the HF equations (1.18) reflect the Ritz principle, one ends up with the set of ϕ_i which minimizes the total HF energy (1.17). The HF approach yields, in the sense of the Ritz principle, the best determinantal approximation to the ground state.

The determinant (1.16) manifestly satisfies the requirement of antisymmetry. This rather elementary correlation between fermions is often called *Pauli correlation*. The wavefunction (1.16) does, however, not contain the correlation resulting from the interaction between the electrons: while in the HF approach each electron simply experiences the average field of the complete electron cloud, the actual motion of an electron depends, via the Coulomb repulsion, on the individual positions of all other electrons (compare Sect. 3.1). In order to incorporate this *Coulomb correlation* into the many-body wavefunction one has to go beyond the representation by a single determinant. In order to construct such wavefunctions one has to realize that a complete solution of the HF equations not only provides the N *occupied* single-particle states included in the HF ground state (1.16), but also a (finite or infinite) number of further, unoccupied states (the other eigenstates of the differential equation (1.18), obtained for a fixed Hamiltonian on the left-hand side). Together with the occupied ϕ_i, these states form a complete basis of the single-particle Hilbert space. A complete basis in the N-particle Hilbert space is then obtained from the set of all determinants (1.16) which can be built from this complete set of single-particle states (for a more detailed discussion see Appendix B). Using the completeness of the states of type (1.16), correlated wavefunctions can be written as

$$\Psi_k(\boldsymbol{r}_1\sigma_1,\ldots\boldsymbol{r}_N\sigma_N) = \sum_{i_1\cdots i_N} c^k_{i_1\cdots i_N}\Phi_{i_1\cdots i_N}(\boldsymbol{r}_1\sigma_1,\ldots\boldsymbol{r}_N\sigma_N)\,. \tag{1.22}$$

Several strategies are available for the determination of the expansion coefficients $c^k_{i_1\cdots i_N}$ [2, 3]. However, while highly efficient numerical implementations and present day computer power allow the application of these so-called *correlated ab-initio methods* to surprisingly large systems, there is a fundamental drawback, which seriously limits their range of applicability.

In order to understand this point one has to realize that, in one way or another, the single-particle orbitals from which the $\Phi_{i_1\cdots i_N}$ are constructed must be expanded in terms of some finite set of basis functions η_k,

$$\phi_i(\boldsymbol{r}\sigma) = \sum_{k=1}^{M} b_{i,k\sigma}\,\eta_k(\boldsymbol{r})\,. \tag{1.23}$$

Using the matrix elements $\langle\eta_k|\eta_l\rangle$, $\langle\eta_k|\boldsymbol{\nabla}^2|\eta_l\rangle$ etc, Eq. (1.15) can then be recast as an algebraic (eigenvalue) problem, in which the $b_{i,k\sigma}$ and $c^k_{i_1\cdots i_N}$ are determined either sequentially or simultaneously. For instance, in the case of an effective single-particle problem (as the HF scheme) one ends up with

$$\sum_{l=1}^{M}\sum_{\sigma'}\left[\left\langle\eta_k\left|-\frac{\hbar^2\boldsymbol{\nabla}^2}{2m}\delta_{\sigma\sigma'}+\hat{v}_{\mathrm{eff},\sigma\sigma'}\right|\eta_l\right\rangle-\varepsilon_i\langle\eta_k|\eta_l\rangle\right]b_{i,l\sigma'} = 0\,, \tag{1.24}$$

where $\hat{v}_{\mathrm{eff},\sigma\sigma'}$ is the total potential which the electrons experience (the basis functions have not been assumed to be orthonormal nor has the fact been used that the HF effective potential is diagonal with respect to spin). The obvious question is: how do the algebraic formulations of the various many-body methods scale with the size M of the basis set?

The answer to this question is not as straightforward as one might think, as it depends on the technical implementation of the method under investigation. Furthermore, one has to distinguish between the number of operations (i.e. computer time) and the size of arrays (i.e. computer memory) required. The somewhat involved answer is, for the interested reader, displayed in more detail in Appendix C (for an alternative view on the same problem see [4]). The following list gives an overview of the scaling behavior which is obtained for some of the standard many-body methods [5]:

Noninteracting particles	$\sim M^3$	
Density Functional Theory	$\sim M^3$	(in case of the LDA, GGA, MGGA)
Hartree-Fock	$\sim M^4$	(in case of the standard implementation)
Møller-Plesset 2	$\sim M^5$	(2nd order perturbation theory on HF basis)
Møller-Plesset 4	$\sim M^7$	(4th order perturbation theory on HF basis)
CISD(T)	$\sim M^7$	(Configuration Interaction: energy minimization allowing single (S), double (D) and selected triple (T) particle–hole excitations from HF ground state)

It should be mentioned that this table does not take into account reduction techniques, such as the suppression of small matrix elements. An important aspect to be noticed in this respect is the relation between M and the size of the system for multi-center problems: in the case of localized basis functions like Gaussians an increase of M eventually implies an increase of the number of atoms involved, as the number of basis functions required for each atomic center is essentially limited (even if high accuracy is desired). However, the overlap of basis functions centered on atoms which are far apart is very small, as long as the basis functions are well localized. This condition is satisfied most easily, if the basis set only has to represent occupied single-particle orbitals, but no virtual states. As a result, one can devise rather elaborate cut-off schemes for multi-center matrix elements, which reduce the scaling behavior of the HF approximation to M^2. Corresponding speed-ups are found for the correlated approaches. It is nevertheless clear, that the scaling behavior of the correlated *ab-initio* methods is prohibitive, as each additional power of M increases the computational demands by a factor of the order of the electron number N: M scales linearly with N, since for any additional electron at least one additional basis function is required in order to keep up the quality of the basis set representation. While the precise range of applicability of correlated *ab-initio* methods is difficult to assess, there will always be interesting systems whose size is beyond their reach.

For this reason only a method with the optimum scaling behavior is of interest for the treatment of complex systems, i.e. some kind of effective noninteracting approach involving only single-particle operators. This suggests using the HF scheme, as the prototype of an effective single-particle approach. However, as already indicated, correlation plays an important role for many systems of interest, so that one has to go beyond the HF approximation. This immediately leads to the question, whether one can map the fully interacting many-body problem onto an effective single-particle problem in a more complete fashion, including correlation in some way. The answer is yes, this type of mapping is the basic aim of *density functional theory* (DFT).

Chapter 2
Foundations of Density Functional Theory: Existence Theorems

2.1 Hohenberg-Kohn Theorem

The starting point of any discussion of DFT is the Hohenberg-Kohn (HK) theorem [6]. It represents the most basic of a number of existence theorems which ensure that stationary many-particle systems can be characterized (fully) by the ground state density and closely related quantities. As the reasoning leading to the HK theorem is quite instructive, it is worthwhile to study this prototype of an existence theorem in some detail (the discussion of the HK theorem follows [7]).

Let us consider the standard Hamiltonian of a stationary system of N interacting spin-$1/2$ particles (typically electrons),

$$\hat{H} = \hat{T} + \hat{V}_{\text{ext}} + \hat{W}\,, \tag{2.1}$$

which consists of the kinetic energy operator $\hat{T}$,

$$\hat{T} = \sum_{i=1}^{N} \frac{(-i\hbar\boldsymbol{\nabla}_i)^2}{2m} = -\frac{\hbar^2}{2m} \sum_{\sigma=\uparrow,\downarrow} \int d^3r\, \hat{\psi}^\dagger(\boldsymbol{r}\sigma)\boldsymbol{\nabla}^2 \hat{\psi}(\boldsymbol{r}\sigma)\,, \tag{2.2}$$

the interaction of the particles with external sources characterized by a given, time-independent potential $v_{\text{ext}}(\boldsymbol{r})$,

$$\hat{V}_{\text{ext}} = \sum_{i=1}^{N} v_{\text{ext}}(\boldsymbol{r}_i) = \int d^3r\, v_{\text{ext}}(\boldsymbol{r})\,\hat{n}(\boldsymbol{r}) \tag{2.3}$$

$$\hat{n}(\boldsymbol{r}) = \sum_{i=1}^{N} \delta^{(3)}(\boldsymbol{r}-\boldsymbol{r}_i) = \sum_{\sigma=\uparrow,\downarrow} \hat{\psi}^\dagger(\boldsymbol{r}\sigma)\,\hat{\psi}(\boldsymbol{r}\sigma)\,, \tag{2.4}$$

and a particle–particle interaction $\hat{W}$,

E. Engel, R.M. Dreizler, Foundations of Density Functional Theory: Existence Theorems. In:
E. Engel, R.M. Dreizler, Density Functional Theory, Theoretical and Mathematical Physics, pp. 11–56 (2011)
DOI 10.1007/978-3-642-14090-7_2

$$\hat{W} = \sum_{i,j=1\,;\,i<j}^{N} w(\boldsymbol{r}_i,\boldsymbol{r}_j) = \frac{1}{2}\sum_{i,j=1\,;\,i\neq j}^{N} w(\boldsymbol{r}_i,\boldsymbol{r}_j)$$
$$= \frac{1}{2}\sum_{\sigma,\sigma'=\uparrow,\downarrow}\int d^3r \int d^3r'\, \hat{\psi}^\dagger(\boldsymbol{r}\sigma)\hat{\psi}^\dagger(\boldsymbol{r}'\sigma')w(\boldsymbol{r},\boldsymbol{r}')\hat{\psi}(\boldsymbol{r}'\sigma')\hat{\psi}(\boldsymbol{r}\sigma)\,. \quad (2.5)$$

In Eqs. (2.2)–(2.5) $\boldsymbol{r}_i$ represents the position of particle i, σ_i is its spin projection in z-direction (the two possible orientations of σ_i will be either denoted as $\uparrow,\downarrow$ or by $\pm 1/2$). $\hat{\psi}^{(\dagger)}(\boldsymbol{r}\sigma)$ is the field operator which describes the creation/annihilation of one particle with spin σ at point $\boldsymbol{r}$. As we are dealing with spin-1/2 particles, $\hat{\psi}^{(\dagger)}(\boldsymbol{r}\sigma)$ satisfies anticommutation rules,

$$\left\{\hat{\psi}(\boldsymbol{r}\sigma),\hat{\psi}(\boldsymbol{r}'\sigma')\right\} = \left\{\hat{\psi}^\dagger(\boldsymbol{r}\sigma),\hat{\psi}^\dagger(\boldsymbol{r}'\sigma')\right\} = 0 \quad (2.6)$$
$$\left\{\hat{\psi}(\boldsymbol{r}\sigma),\hat{\psi}^\dagger(\boldsymbol{r}'\sigma')\right\} = \delta^{(3)}(\boldsymbol{r}-\boldsymbol{r}')\,\delta_{\sigma\sigma'}\,. \quad (2.7)$$

At this point it is not necessary to specify the interaction $w(\boldsymbol{r},\boldsymbol{r}')$ in detail. The basic DFT formalism is independent of the form of w, as long as w is kept the same throughout the complete discussion. It is only assumed that w is symmetric, $w(\boldsymbol{r},\boldsymbol{r}') = w(\boldsymbol{r}',\boldsymbol{r})$ and independent of spin. In practice, of course, one is primarily interested in electrons with their Coulomb interaction. Similarly, the external potential v_{ext} which is of particular interest in practical applications is given by Eq. (1.6). For the present discussion, however, v_{ext} is considered to be an arbitrary function of $\boldsymbol{r}$.

Some further comments on Eqs. (2.1)–(2.5) seem appropriate.

(a) The Hamiltonian (2.1) is strictly nonrelativistic. The complete formalism presented in this chapter can, however, be extended to the relativistic domain on the basis of quantum electrodynamics. A detailed account of relativistic DFT is given in Chap. 8.
(b) For the time being, the Hamiltonian does not include the presence of external magnetic fields. This extension will be discussed in Sects. 2.5 and 2.6. The absence of magnetic fields does not imply, however, that the eigenstates of (2.1), and in particular the ground state, can not exhibit magnetic properties.
(c) Although it is not obvious from Eqs. (2.1)–(2.5), the following discussion will be restricted to the zero temperature formalism. An extension to systems with $T \neq 0$ may be found in [8, 7].
(d) All energy contributions, which do not depend on the electron degrees of freedom, as the interaction among atomic nuclei (1.4), are irrelevant in the present context. They can be added to the electron part of the Hamiltonian when needed.
(e) An important property of the operator (2.3) is its local (that is multiplicative) character. The potential (1.6) has this character, but the class of legitimate potentials also includes additional electrostatic fields applied to the many-particle system. It does not include, however, the standard type of pseudopotentials utilized in the context of DFT (see e.g. [9, 10]). In order to introduce these

nonlocal pseudopotentials one first establishes DFT on the all-electron level, before making the transition to the pseudopotential description.
It is worthwhile to note at this point that there also exists an extension of the HK-theorem to nonlocal potentials [11, 12]. In this approach the 1-particle density matrix

$$\gamma(\boldsymbol{r}\sigma,\boldsymbol{r}'\sigma') = \sum_{\sigma_2,\ldots,\sigma_N} \int d^3r_2 \ldots d^3r_N \, \Psi_0(\boldsymbol{r}\sigma,\boldsymbol{r}_2\sigma_2,\ldots \boldsymbol{r}_N\sigma_N) \\ \times \Psi_0^*(\boldsymbol{r}'\sigma',\boldsymbol{r}_2\sigma_2,\ldots \boldsymbol{r}_N\sigma_N),$$

where Ψ_0 denotes the non-degenerate ground state wavefunction corresponding to the nonlocal external potential, plays the role that the density plays in DFT. The arguments behind this extension follow closely the lines given below.

(f) As indicated in Eqs. (2.1)–(2.5) we will frequently switch between the first and second quantized formalism, depending on which is more appropriate for a given problem. An introduction to the language of second quantization may be found in Appendix B.

The many-body eigenstates $|\Psi_k\rangle$ corresponding to the Hamiltonian (2.1) are obtained by solution of the stationary Schrödinger equation,

$$\hat{H}|\Psi_k\rangle = E_k|\Psi_k\rangle\,. \tag{2.8}$$

This is true in particular for the ground state $|\Psi_0\rangle$, on which we will focus attention in the following.

Now consider the set of all Hamiltonians of the form (2.1) with non-degenerate ground states, i.e. the set of all local potentials v_{ext} for which (2.8) leads to a non-degenerate $|\Psi_0\rangle$, while the interaction (2.5) is kept fixed. This set does not only contain physically realizable potentials, but also an infinite number of purely mathematical constructs. In addition, it contains an infinite number of trivial copies of any given $v_{\text{ext}}(\boldsymbol{r})$, which are obtained by simply adding an arbitrary constant to $v_{\text{ext}}(\boldsymbol{r})$. These trivial copies, of course, lead to the same ground state, so that they are physically equivalent to $v_{\text{ext}}(\boldsymbol{r})$. On the other hand, there exist physical systems with degenerate ground states which are not included in the present discussion—the restriction to non-degenerate ground states will be removed later.

One can then interpret the solution of the Schrödinger equation as a map between the set $\mathscr{V}$ of external potentials which differ by more than a constant,

$$\mathscr{V} = \Big\{ v_{\text{ext}} \;\Big|\; \text{with: } v_{\text{ext}} \text{ multiplicative, corresponding } |\Psi_0\rangle \text{ exists and} \\ \text{is non-degenerate, } v'_{\text{ext}}(\boldsymbol{r}) \neq v_{\text{ext}}(\boldsymbol{r}) + const \Big\}\,. \tag{2.9}$$

and the set $\mathscr{G}$ of resulting ground states,

$$\mathscr{G} = \Big\{ |\Psi_0\rangle \;\Big|\; \text{with: } |\Psi_0\rangle \text{ ground state corresponding to one element of } \mathscr{V}, \\ |\Psi'_0\rangle \neq e^{i\varphi}|\Psi_0\rangle \text{ with } \varphi \text{ being some global phase} \Big\}\,. \tag{2.10}$$

Let us denote the map from $\mathscr{V}$ to $\mathscr{G}$ as A,

$$A: \mathscr{V} \longrightarrow \mathscr{G}.$$

Introducing the ground state density n_0,

$$\begin{aligned} n_0(\boldsymbol{r}) &= \langle\Psi_0|\hat{n}(\boldsymbol{r})|\Psi_0\rangle \\ &= N \sum_{\sigma_1,\ldots\sigma_N} \int d^3r_2 \ldots d^3r_N \, |(\boldsymbol{r}\sigma_1, \boldsymbol{r}_2\sigma_2 \ldots \boldsymbol{r}_N\sigma_N|\Psi_0\rangle|^2, \end{aligned} \tag{2.11}$$

one can furthermore define a second map B between the set $\mathscr{G}$ and the set $\mathscr{N}$ of all ground state densities obtained from some element of $\mathscr{G}$ via (2.11),

$$\begin{aligned} \mathscr{N} &= \{n_0 \mid n_0(\boldsymbol{r}) = \langle\Psi_0|\hat{n}(\boldsymbol{r})|\Psi_0\rangle, \ |\Psi_0\rangle \in \mathscr{G}\} \\ B: \ \mathscr{G} &\longrightarrow \mathscr{N}. \end{aligned} \tag{2.12}$$

The two maps are illustrated in Fig. 2.1. By construction both maps are surjective.

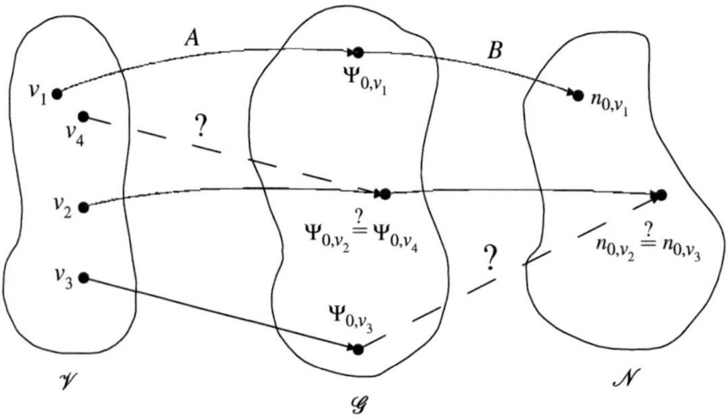

Fig. 2.1 Correspondence between external potentials v_i, associated ground states Ψ_{0,v_i} and ground state densities n_{0,v_i} in the case of non-degenerate ground states.

The question then is: can it happen that the same $|\Psi_0\rangle$ is obtained as ground state for two different potentials or that the same n_0 results from two different ground states? These possibilities are indicated by the dashed lines in Fig. 2.1.

The crucial observation of HK was: both maps are also injective and thus unique. The proof of this statement for map A consists in demonstrating the validity of the following two statements:

(i) For given $v_{\rm ext}$ there exists only one $|\Psi_0\rangle$ in $\mathscr{G}$, i.e. there is no $v_{\rm ext}$ which is mapped onto two elements of $\mathscr{G}$.

(ii) There is no $|\Psi_0\rangle$ which is simultaneously ground state for two different potentials $v_{\rm ext}$ and $v'_{\rm ext}$ which differ by more than a constant.

Statement (i) is trivial due to the restriction to non-degenerate ground states. It remains to prove statement (ii).

The standard proof of (ii) is based on a *reductio ad absurdum*. Let us assume that $|\Psi_0\rangle$ is simultaneously ground state for two different potentials v_{ext} and $v'_{\text{ext}} \neq v_{\text{ext}} + const.$ $|\Psi_0\rangle$ thus satisfies two Schrödinger equations,

$$\hat{H}|\Psi_0\rangle = \left[\hat{T} + \hat{V}_{\text{ext}} + \hat{W}\right]|\Psi_0\rangle = E_0|\Psi_0\rangle \tag{2.13}$$

$$\hat{H}'|\Psi_0\rangle = \left[\hat{T} + \hat{V}'_{\text{ext}} + \hat{W}\right]|\Psi_0\rangle = E'_0|\Psi_0\rangle\,. \tag{2.14}$$

Upon subtraction of (2.13) and (2.14) one obtains

$$\left[\hat{V}_{\text{ext}} - \hat{V}'_{\text{ext}}\right]|\Psi_0\rangle = \left[E_0 - E'_0\right]|\Psi_0\rangle\,. \tag{2.15}$$

Equation (2.15) is best written in first quantized form, in which, due to the multiplicative character of the potentials, one can divide by the ground state wavefunction

$$(\boldsymbol{r}_1\sigma_1,\ldots,\boldsymbol{r}_N\sigma_N|\Psi_0\rangle = \Psi_0(\boldsymbol{r}_1\sigma_1,\ldots,\boldsymbol{r}_N\sigma_N)\,. \tag{2.16}$$

One thus finds

$$\sum_{i=1}^{N}\left[v_{\text{ext}}(\boldsymbol{r}_i) - v'_{\text{ext}}(\boldsymbol{r}_i)\right] = E_0 - E'_0 \tag{2.17}$$

for all points $\boldsymbol{r}_i$ for which the wavefunction $\Psi_0(\boldsymbol{r}_1\sigma_1,\ldots,\boldsymbol{r}_N\sigma_N)$ does not vanish. Keeping $N-1$ of the $\boldsymbol{r}_i$ fixed and letting the remaining position vary, Eq. (2.17) leads to a contradiction (as the right-hand side is constant, while v_{ext} and v'_{ext} are assumed to differ by more than a constant). Consequently, the map A is unique: there is a one-to-one correspondence between the potential v_{ext} and the resulting ground state $|\Psi_0\rangle$ (up to some additive constant in v_{ext}).

Let us now turn to the map B. The definition of B via Eq. (2.11) implies that it is impossible that one $|\Psi_0\rangle$ is mapped onto two different densities n_0 and n'_0. In order to demonstrate the injectivity of B, one has to show that two different $|\Psi_0\rangle \in \mathscr{G}$ can not lead to the same ground state density n_0. The proof again relies on *reductio ad absurdum*. Assume that n_0 is obtained from two different elements $|\Psi_0\rangle$ and $|\Psi'_0\rangle$ of $\mathscr{G}$. From the Ritz variational principle one then obtains an inequality for the ground state energy,

$$E_0 = \langle\Psi_0|\hat{H}|\Psi_0\rangle < \langle\Psi'_0|\hat{H}|\Psi'_0\rangle\,, \tag{2.18}$$

where $\hat{H}$ is the unique Hamiltonian leading to $|\Psi_0\rangle$ (due to the uniqueness of map A) and the strict inequality originates from the non-degeneracy of $|\Psi_0\rangle$ and $|\Psi'_0\rangle$. The right-hand side of (2.18) can be evaluated further by adding and subtracting the unique potential $\hat{V}'_{\text{ext}}$ which corresponds to $|\Psi'_0\rangle$,

$$E_0 < E'_0 + \langle\Psi'_0|\hat{V}_{\text{ext}} - \hat{V}'_{\text{ext}}|\Psi'_0\rangle\,. \tag{2.19}$$

Using the multiplicative form of $\hat{V}_{\text{ext}}$, Eq. (2.3), and the assumption that both states lead to the same density n_0, one obtains

$$E_0 < E_0' + \int d^3r\, n_0(\boldsymbol{r}) \left[v_{\text{ext}}(\boldsymbol{r}) - v_{\text{ext}}'(\boldsymbol{r})\right] . \tag{2.20}$$

However, this argument can equally well be given with primed and unprimed quantities interchanged,

$$E_0' < E_0 + \int d^3r\, n_0(\boldsymbol{r}) \left[v_{\text{ext}}'(\boldsymbol{r}) - v_{\text{ext}}(\boldsymbol{r})\right] . \tag{2.21}$$

Upon addition of Eqs. (2.20) and (2.21), one ends up with a contradiction,

$$E_0 + E_0' < E_0' + E_0 . \tag{2.22}$$

One therefore concludes that the map B is also unique: there is a one-to-one correspondence between $|\Psi_0\rangle$ and n_0.

On the basis of these arguments one can formulate the fundamental statements of the HK theorem.

(a) Taking both maps together, one has a one-to-one correspondence between the external potential v_{ext} in the Hamiltonian, the (non-degenerate) ground state $|\Psi_0\rangle$ resulting from solution of the Schrödinger equation and the associated ground state density n_0,

$$v_{\text{ext}}(\boldsymbol{r}) \underset{\uparrow}{\Longleftrightarrow} |\Psi_0\rangle \underset{\uparrow}{\Longleftrightarrow} n_0(\boldsymbol{r}) = \langle\Psi_0|\hat{n}(\boldsymbol{r})|\Psi_0\rangle . \tag{2.23}$$

unique (up to some constant in v_{ext})

Thus v_{ext}, $|\Psi_0\rangle$ and n_0 determine each other uniquely. In mathematical terms: the ground state is a unique functional of the ground state density, denoted as $|\Psi[n]\rangle$. Upon insertion of one element $n_0 \in \mathcal{N}$, this functional yields the ground state $|\Psi_0\rangle$ associated with this particular n_0,

$$|\Psi_0\rangle = |\Psi[n_0]\rangle . \tag{2.24}$$

The functional $|\Psi[n]\rangle$ is a realization of the map B^{-1}.

Note that no explicit information on v_{ext} is required to construct $|\Psi_0\rangle$ from n_0: $|\Psi[n]\rangle$ has the same functional form for all kinds of many-particle systems with the same interaction (w was kept fixed throughout the complete discussion). The same functional $|\Psi[n]\rangle$ applies to atoms, molecules and solids. The particular geometry of the systems under consideration is mediated by the structure of the density. $|\Psi[n]\rangle$ is therefore called *universal*. Clearly, the functional dependence of $|\Psi[n]\rangle$ on n must be extremely complicated.

(b) The existence of the functional $|\Psi[n]\rangle$ leads to the statement that any ground state observable is a density functional,

$$O[n] := \langle\Psi[n]|\hat{O}|\Psi[n]\rangle . \tag{2.25}$$

This is true, in particular, for the ground state energy, which turns out to be the most important density functional,

$$E[n] := \langle\Psi[n]|\hat{H}|\Psi[n]\rangle = F[n] + \int d^3r\, v_{\rm ext}(\boldsymbol{r})n(\boldsymbol{r}) \tag{2.26}$$

$$F[n] := \langle\Psi[n]|\hat{T} + \hat{W}|\Psi[n]\rangle . \tag{2.27}$$

In (2.26) the universal part $F[n]$ has been extracted from the total energy functional $E[n]$, in order to emphasize the fact that $v_{\rm ext}$ enters $E[n]$ at only one point.

(c) There exists a minimum principle for $E[n]$: if n_0 is the ground state density corresponding to $v_{\rm ext}$, one has for all densities $n_0'(\boldsymbol{r}) \neq n_0(\boldsymbol{r})$

$$E[n_0] < E[n_0'] \qquad \Longleftrightarrow \qquad E_0 = \min_{n \in \mathcal{N}} E[n] . \tag{2.28}$$

This is a direct consequence of the unique relation between n_0 and $|\Psi_0\rangle$ and the Ritz variational principle (the ground state $|\Psi_0'\rangle$ associated with n_0' differs from $|\Psi_0\rangle$). At this point it is worthwhile to emphasize the restricted domain of the functional $E[n]$: only densities in $\mathcal{N}$, i.e. ground state densities which are obtained by solution of the Schrödinger equation, are legitimate, but not arbitrary non-negative functions of $\boldsymbol{r}$ with finite norm.

How can one interpret the last result physically? Obviously the particles react to any (arbitrarily small) change in $v_{\rm ext}$, so that their total energy is minimized, and this response is unique. Due to the locality of the potential there is no modification of $v_{\rm ext}$ which does not require a readjustment of the electron wavefunction and, consequently, of the density distribution.[1] This response is very familiar in a number of cases. If, for instance, the nuclear charge is increased in an atom, the orbitals are contracted in a well-defined way. If the form of the nuclear charge distribution is changed from point-like to a finite shape, the density of the electrons in the vicinity of the nucleus reflects this change of $v_{\rm ext}$. If the atoms in a molecule are taken apart, the wavefunction and the density follow this positional change. The basic result of the HK theorem is therefore intuitively plausible.

Of course, these statements are only correct if the particles actually experience the change in $v_{\rm ext}$. If, for instance, some part of space is separated by a barrier of infinite height and non-vanishing width from the region in which the particles are moving, a modification of $v_{\rm ext}$ in this exterior region does not affect the particles

[1] This implies the uniqueness of the density–density response function of the system.

at all (as there is no communication between the two parts of space). Similarly, if the difference between $v_{\rm ext}$ and $v'_{\rm ext}$ is non-vanishing only at the nodes of $|\Psi_0\rangle$ ($v_{\rm ext} - v'_{\rm ext}$ would have a δ-type form), the proof of the uniqueness of map A breaks down. However, such situations are either trivial (as the case of an infinite barrier) or rather unphysical. Potentials with δ-like structures are not met in first-principles electron structure calculations. For this reason one can safely assume the validity of the HK theorem for all systems of practical interest.

In view of the intuitive physical background of the HK theorem it is not surprising that the first energy functionals were introduced long before HK's existence proof. Thomas [13] and Fermi [14] considered the kinetic energy of a noninteracting electron gas, which turns out to be a simple functional of the gas density—for a derivation see Appendix D. Using this functional with the actual inhomogeneous densities of systems like atoms or molecules (rather than just the uniform gas density) constitutes the so-called *Thomas-Fermi model.* A variety of extensions and properties of the Thomas-Fermi approach have been established with time. A complete review of these aspects of DFT is beyond the scope of this book. The basics are summarized in Appendix D, for further information the interested reader is referred to [15, 7] which provide a rather complete account of the TF approach and extensions.

2.2 Degenerate Ground States

The proof of the HK theorem presented in Sect. 2.1 relies crucially on the restriction to non-degenerate ground states. The argument has to be modified in the case of degenerate states. In this situation two or more ground states $|\Psi_{0,i}\rangle$ originate from the same potential $v_{\rm ext}$. The first step thus is an appropriate redefinition of the set of legitimate potentials $\mathscr{V}$,

$$\mathscr{V} = \left\{ v_{\rm ext} \;\middle|\; \text{with: } v_{\rm ext} \text{ multiplicative, } v'_{\rm ext}(\boldsymbol{r}) \neq v_{\rm ext}(\boldsymbol{r}) + const \right\}, \tag{2.29}$$

the set $\mathscr{G}$ of resulting ground states,

$$\mathscr{G}_{v_{\rm ext}} = \Big\{ |\Psi\rangle \;\Big|\; \text{with: } |\Psi\rangle = \sum_{i=1}^{q} c_i |\Psi_{0,i}\rangle, \quad |\Psi_{0,i}\rangle = \text{degenerate ground states to } v_{\rm ext} \Big\} \tag{2.30}$$

$$\mathscr{G} = \bigcup_{v_{\rm ext}} \mathscr{G}_{v_{\rm ext}}, \tag{2.31}$$

and the set $\mathscr{N}$ of associated ground state densities,

$$\mathscr{N}_{v_{\rm ext}} = \left\{ n(\boldsymbol{r}) \;\middle|\; \text{with: } n(\boldsymbol{r}) = \langle\Psi|\hat{n}(\boldsymbol{r})|\Psi\rangle,\ |\Psi\rangle \in \mathscr{G}_{v_{\rm ext}} \right\} \tag{2.32}$$

$$\mathscr{N} = \bigcup_{v_{\rm ext}} \mathscr{N}_{v_{\rm ext}}. \tag{2.33}$$

The definitions of $\mathscr{G}_{v_{\text{ext}}}$ and $\mathscr{N}_{v_{\text{ext}}}$ include the non-degenerate situation.

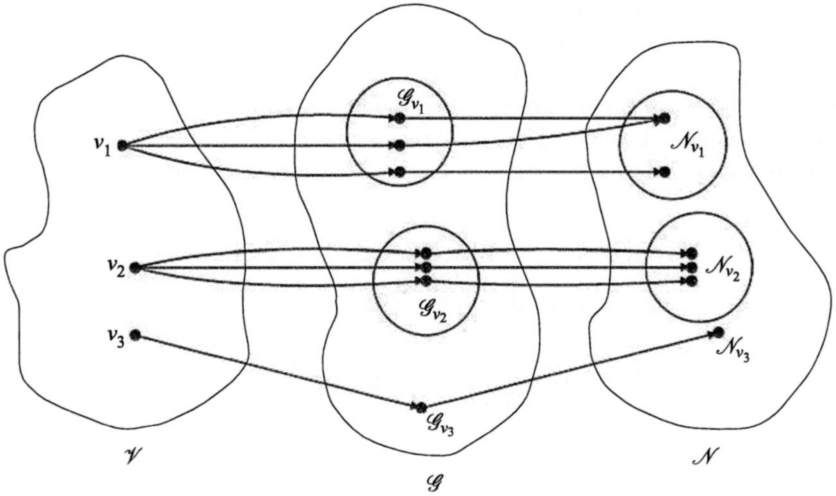

Fig. 2.2 Correspondence between external potentials v_i, associated subsets of ground states $\mathscr{G}_{v_i}$ and subsets of ground state densities $\mathscr{N}_{v_i}$ in the case of degeneracy.

One can then show [16] that there exists a one-to-one correspondence between the potential v_{ext}, the class $\mathscr{G}_{v_{\text{ext}}}$ of all possible ground states resulting from v_{ext} and the class $\mathscr{N}_{v_{\text{ext}}}$ of ground state densities obtained from these ground states and that the subsets $\mathscr{G}_{v_{\text{ext}}}$ and $\mathscr{N}_{v_{\text{ext}}}$ for different v_{ext} are disjoint,

$$v_{\text{ext}}(\boldsymbol{r}) \underset{\uparrow}{\Longleftrightarrow} \mathscr{G}_{v_{\text{ext}}} \underset{\uparrow}{\Longleftrightarrow} \mathscr{N}_{v_{\text{ext}}} \tag{2.34}$$

unique (up to some constant in v_{ext})

$|\Psi_0\rangle \neq |\Psi_0'\rangle$ and $n_0 = \langle\Psi_0|\hat{n}|\Psi_0\rangle \neq n_0' = \langle\Psi_0'|\hat{n}|\Psi_0'\rangle$

for all pairs $|\Psi_0\rangle \in \mathscr{G}_{v_{\text{ext}}}$, $|\Psi_0'\rangle \in \mathscr{G}_{v_{\text{ext}}'}$ with $v_{\text{ext}}' \neq v_{\text{ext}} + const$.

The proof of these statements proceeds as in the non-degenerate case: assuming that two potentials v_{ext} and $v_{\text{ext}}' \neq v_{\text{ext}} + const$ have a common ground state $|\Psi_0\rangle$, one can follow the arguments from (2.13) to (2.17) to end up with a contradiction. Two different v_{ext} always lead to two disjoint sets of ground states $\mathscr{G}_{v_{\text{ext}}}$—a given $|\Psi_0\rangle \in \mathscr{G}$ can only belong to one particular $\mathscr{G}_{v_{\text{ext}}}$ and therefore to one particular v_{ext}. Similarly, one arrives at a contradiction if one assumes that two states $|\Psi_0\rangle \in \mathscr{G}_{v_{\text{ext}}}$ and $|\Psi_0'\rangle \in \mathscr{G}_{v_{\text{ext}}'}$ lead to the same density n_0 and then follows the arguments in Eqs. (2.18) to (2.22). A density $n_0 \in \mathscr{N}$ can only be an element of one single set $\mathscr{N}_{v_{\text{ext}}}$. Two densities from different $\mathscr{N}_{v_{\text{ext}}}$ necessarily correspond to two different

$\mathscr{G}_{v_{\rm ext}}$ and two different $v_{\rm ext}$ and vice versa. A sketch of the situation is given in Fig. 2.2.

In other words: one needs only one element of $\mathscr{N}_{v_{\rm ext}}$ to identify the corresponding $\mathscr{G}_{v_{\rm ext}}$ and $v_{\rm ext}$. On the other hand, it can happen that the same density is obtained for two different elements of one subset $\mathscr{G}_{v_{\rm ext}}$. As a simple example one may consider the boron atom on the noninteracting level, i.e. in terms of a Hamiltonian with only a spherical Coulomb potential. As the $2p$-states are degenerate, a possible classification of the $2p$-states is

$$\begin{aligned} 2p^0: \quad \phi^0_{2p}(\boldsymbol{r}) &= \frac{R_{21}(r)}{r} Y_{1,0}(\Omega) \\ 2p^\pm: \quad \phi^\pm_{2p}(\boldsymbol{r}) &= \frac{R_{21}(r)}{r} Y_{1,\pm1}(\Omega), \end{aligned}$$

where $R_{nl}(r)$ denotes the standard radial orbitals of the hydrogen problem and $Y_{lm}(\Omega)$ the spherical harmonics. However, due to

$$|Y_{1,\pm1}(\Omega)|^2 = \frac{3}{8\pi}\sin^2\Theta\,,$$

one finds that the density

$$n(\boldsymbol{r}) = \sum_{i=1}^{N} |\phi_i(\boldsymbol{r})|^2$$

is identical for the following two 5-particle states (Slater determinants),

$$\begin{aligned} \Psi_+ &= \frac{1}{\sqrt{5!}} \det\Big((1s)^2(2s)^2 2p_+\Big) \\ \Psi_- &= \frac{1}{\sqrt{5!}} \det\Big((1s)^2(2s)^2 2p_-\Big). \end{aligned}$$

A unique functional $|\Psi[n]\rangle$ does not exist in the case of degenerate ground states.

Fortunately, the existence of such a unique functional is not really required in order to extend the most important statement of the HK theorem to degenerate ground states: only the existence of $E[n]$ is of interest [16]. One first notes that, by definition, all the degenerate states $|\Psi_{0,i}\rangle$ lead to the same ground state energy. Even if two degenerate states $|\Psi_1\rangle = \sum_{i=1}^q a_i|\Psi_{0,i}\rangle$ and $|\Psi_2\rangle = \sum_{i=1}^q b_i|\Psi_{0,i}\rangle$ with $|\Psi_1\rangle \neq |\Psi_2\rangle$ lead to the same n, this density nevertheless determines the ground state energy uniquely,

$$E_0 = \langle\Psi_1|\hat{H}|\Psi_1\rangle = \langle\Psi_2|\hat{H}|\Psi_2\rangle\,, \tag{2.35}$$

as, by virtue of (2.34), both $|\Psi_1\rangle$ and $|\Psi_2\rangle$ originate from the same $v_{\rm ext}$, i.e. the same Hamiltonian (up to the constant in $v_{\rm ext}$ which is always undetermined). Similarly, two different densities n_1, n_2 from the same $\mathscr{N}_{v_{\rm ext}}$ give the same ground state energy. As a result, E_0 is a unique functional of n.

Choosing an arbitrary representative $|\Psi\rangle$ among all states which lead to the same given density n, one can define an energy functional $E[n]$ for degenerate states,

$$E[n] = \langle\Psi[n]|\hat{H}|\Psi[n]\rangle\,, \tag{2.36}$$

which is unambiguous for all n in the set (2.33). As in the case of the non-degenerate situation one can then verify the existence of a minimum principle,

$$E_0 = E[n_i] \quad \forall\, n_i \in \mathcal{N}_{v_{\rm ext}} \qquad , \qquad E_0 < E[n] \quad \forall\, n \notin \mathcal{N}_{v_{\rm ext}}\,. \tag{2.37}$$

2.3 Variational Equation, Interacting v-Representability, Functional Differentiability

The minimum principles (2.28) and (2.37) indicate the possibility to determine the ground state density of a many-particle system by a variational equation,

$$\frac{\delta}{\delta n(\boldsymbol{r})}\left\{E[n] - \mu\left(\int d^3r\, n(\boldsymbol{r}) - N\right)\right\}\bigg|_{n(\boldsymbol{r})=n_0(\boldsymbol{r})} = 0\,, \tag{2.38}$$

where the subsidiary condition is required to ensure the proper normalization of the density. Unfortunately, the transition from (2.28), (2.37) to Eq. (2.38) is not completely legitimate from a mathematical point of view. The existence of the variational derivative $\delta E[n]/\delta n$ requires the functional $E[n]$ to be defined on a sufficiently dense set of densities n (just as the differentiation of functions requires more than integer numbers as the domain of definition). So far, however, $E[n]$ is only defined for ground state densities resulting from the solution of the Schrödinger equation (2.8). One thus has to face the question whether there exists a potential $v_{\rm ext}$ for every given normalizable function $n(\boldsymbol{r})$, so that $n(\boldsymbol{r})$ is the ground state density corresponding to $v_{\rm ext}$? If such a potential can be found the density is termed *interacting v-representable.*

Unfortunately, one can give explicit counterexamples, which demonstrate that the issue of v-representability is non-trivial for the functionals (2.26) and (2.36). The discussion of this problem is, however, mathematically somewhat involved. The reader may thus wish to skip the remainder of Sect. 2.3 in a first reading and simply accept the (admittedly vague) statement that the functional derivative $\delta E[n]/\delta n$ exists for all practical purposes, if the definition of $E[n]$ is suitably extended.

1. The first counterexample is based on degeneracy. Let $|\Psi_{0,1}\rangle$ ... $|\Psi_{0,q}\rangle$ be q orthonormal degenerate ground states for some $v_{\rm ext}$. One can then construct a statistical density matrix $\hat{D}$ from these states,

$$\hat{D} = \sum_{i=1}^{q} c_i |\Psi_{0,i}\rangle\langle\Psi_{0,i}| ; \qquad c_i = c_i^* \geq 0; \quad \sum_{i=1}^{q} c_i = 1 . \tag{2.39}$$

The corresponding (ensemble) density is obtained by

$$n_D(\boldsymbol{r}) = \mathrm{tr}\left\{\hat{D}\hat{n}(\boldsymbol{r})\right\} = \sum_{i=1}^{q} c_i \langle\Psi_{0,i}|\hat{n}(\boldsymbol{r})|\Psi_{0,i}\rangle . \tag{2.40}$$

The important observation is: a density of the form (2.40) can not be obtained from a single ground state $|\Psi_0\rangle$ [17, 18].

This statement is worth to be demonstrated explicitly. The most general ground state of the degenerate system has the form

$$|\Psi_0\rangle = \sum_{i=1}^{q} d_i |\Psi_{0,i}\rangle ; \qquad \sum_{i=1}^{q} |d_i|^2 = 1 . \tag{2.41}$$

The corresponding density is

$$n_0(\boldsymbol{r}) = \sum_{i,j=1}^{q} d_i^* d_j \langle\Psi_{0,i}|\hat{n}(\boldsymbol{r})|\Psi_{0,j}\rangle . \tag{2.42}$$

For the densities n_0 and n_D to be identical, the relation

$$\sum_{i,j=1}^{q} \left(c_i\delta_{ij} - d_i^* d_j\right) \langle\Psi_{0,i}|\hat{n}(\boldsymbol{r})|\Psi_{0,j}\rangle = 0$$

has to be satisfied. As the matrix elements $\langle\Psi_{0,i}|\hat{n}(\boldsymbol{r})|\Psi_{0,j}\rangle$ do not vanish and as they all exhibit a different $\boldsymbol{r}$-dependence, all prefactors must vanish individually,

$$c_i\delta_{ij} - d_i^* d_j = 0 . \tag{2.43}$$

The q unknown complex numbers d_i have to satisfy $2q^2$ linearly independent equations. As there is in general no solution to (2.43), one can not find a linear combination of the $|\Psi_{0,i}\rangle$ with the density n_D.

For this reason one distinguishes between *pure-state* and *ensemble* v-representability,

pure-state $\equiv n(\boldsymbol{r})$ results from a single state $|\Psi_0\rangle$
ensemble $\equiv n(\boldsymbol{r})$ results from a density matrix, i.e. a statistical ensemble

Obviously the original HK energy functional is only defined for pure-state representable densities.
In order to cope with ensemble v-representable densities one replaces the set $\mathscr{G}_{v_{\mathrm{ext}}}$ of pure states, Eq. (2.30), by the set of all density matrices which can be built on the basis of the potential v_{ext},

$$\mathscr{D}_{v_{\mathrm{ext}}} = \left\{\hat{D} \,\middle|\, \text{with: } \hat{D} = \sum_{i=1}^{q} c_i |\Psi_{0,i}\rangle\langle\Psi_{0,i}|; \ c_i = c_i^* \geq 0; \ \sum_{i=1}^{q} c_i = 1; \right.$$
$$\left. |\Psi_{0,i}\rangle = \text{degenerate ground states to } v_{\mathrm{ext}} \right\} . \tag{2.44}$$

One can then show that there is a one-to-one correspondence between v_{ext}, the set $\mathscr{D}_{v_{\text{ext}}}$ and the set of ensemble v-representable densities resulting from the elements of $\mathscr{D}_{v_{\text{ext}}}$ via Eq. (2.40) [17, 18]. The sets $\mathscr{D}_{v_{\text{ext}}}$ and v_{ext} are uniquely determined by any given ensemble v-representable n: a given n can not result from more than one $\mathscr{D}_{v_{\text{ext}}}$ and the $\mathscr{D}_{v_{\text{ext}}}$ are disjoint—the underlying argument is the same as in the case of degenerate ground states. This allows a redefinition of $E[n]$, Eq. (2.36), as

$$\begin{aligned} E_{\text{EHK}}[n] &:= \operatorname{tr}\{\hat{D}[n]\hat{H}\} \qquad \text{with} \quad n = \operatorname{tr}\{\hat{D}[n]\hat{n}\} \\ &= \sum_{i=1}^{q} c_i \langle \Psi_{0,i}|\hat{H}|\Psi_{0,i}\rangle \\ &\quad \text{with} \quad n(\boldsymbol{r}) = \sum_{i=1}^{q} c_i \langle \Psi_{0,i}|\hat{n}(\boldsymbol{r})|\Psi_{0,i}\rangle\,; \quad c_i = c_i^* \geq 0\,; \quad \sum_{i=1}^{q} c_i = 1\,; \\ &\qquad |\Psi_{0,i}\rangle = \text{degenerate ground states to } \hat{H}\,. \end{aligned} \tag{2.45}$$

Any $\hat{D}$ which corresponds to the potential v_{ext}, which in turn is determined by a given density n, yields the same energy value, so that $E_{\text{EHK}}[n]$ is a unique density functional. In this way the domain of $E[n]$ is extended to ensemble v-representable densities (for pure-state v-representable densities both functionals coincide, since the minimizing density matrix in (2.45) results from the non-degenerate ground state in this case).

2. The second counterexample [19] is of a more explicit nature: a single particle in one spatial dimension. There is nothing in the HK proof which requires the presence of more than one particle, an interaction or a three-dimensional system. All statements of the HK theorem are also valid in this special limit.
The Schrödinger equation then has the simple form

$$\left\{-\frac{\hbar^2}{2m}\frac{d^2}{dx^2} + v_{\text{ext}}(x)\right\}\Psi(x) = E\Psi(x)\,; \qquad n(x) = |\Psi(x)|^2\,. \tag{2.46}$$

Equation (2.46) represents a direct map between n and v_{ext}, if one chooses $\Psi(x)$ to be real,

$$v_{\text{ext}}(x) = \frac{\hbar^2}{2m}\frac{1}{\sqrt{n(x)}}\frac{d^2}{dx^2}\sqrt{n(x)} + E\,. \tag{2.47}$$

A similar relation can be established for noninteracting particles in three dimensions.
One can now consider the density

$$n(x) = (a + b|x|^{\alpha})^2 f(x)^2\,, \tag{2.48}$$

with

$$a,b>0\,; \qquad \frac{1}{2}<\alpha<1\,; \qquad f(x)=f_0 \quad \forall\, |x|\le x_0\,. \tag{2.49}$$

The function $f(x)$ ensures the normalizability of $n(x)$, so that (2.48) represents a legitimate density for which $E[n]$ should exist (the precise form of $f(x)$ for $|x|>x_0$ is irrelevant at this point).

The derivative of $n(x)$ required for the determination of the corresponding $v_{\rm ext}(x)$ via Eq. (2.47) can be evaluated by use of the general rule[2]

$$\begin{aligned}\frac{d}{dx}|x|^p &= \frac{d}{dx}\Big\{\Theta(x)x^p+\Theta(-x)(-x)^p\Big\}\\ &= \delta(x)\Big\{x^p-(-x)^p\Big\}+px|x|^{p-2} \qquad (2.50)\\ &= px|x|^{p-2} \qquad \text{for } p>0\,. \qquad (2.51)\end{aligned}$$

If one assumes that $df/dx(x_0)=d^2f/dx^2(x_0)=0$ (which is legitimate as a single counterexample is sufficient), one obtains in the interesting region $|x|\le x_0$,

$$\frac{d}{dx}\sqrt{n(x)}=\frac{d}{dx}\Psi(x) = f_0 b\alpha x|x|^{\alpha-2} \tag{2.52}$$

$$\begin{aligned}\frac{d^2}{dx^2}\sqrt{n(x)}=\frac{d^2}{dx^2}\Psi(x) &= f_0 b\alpha(\alpha-1)|x|^{\alpha-2}\\ &\quad +f_0 b\alpha\delta(x)\Big[x^{\alpha-1}+(-x)^{\alpha-1}\Big]\,.\end{aligned} \tag{2.53}$$

The kinetic energy associated with $\Psi(x)$ is finite,

$$\begin{aligned}T &= \frac{\hbar^2}{2m}\int_{-x_0}^{+x_0}dx\left(\frac{d}{dx}\Psi(x)\right)^2+T_R\\ &= \frac{\hbar^2}{2m}f_0^2b^2\alpha^2\int_{-x_0}^{+x_0}dx\,|x|^{2\alpha-2}+T_R \;<\;\infty\,,\end{aligned} \tag{2.54}$$

as $\alpha>1/2$ has been chosen (T_R is the finite contribution from the exterior region $|x|>x_0$). The corresponding potential reads ($|x|\le x_0$) [20]

$$v_{\rm ext}(x)=\frac{\hbar^2}{2m}\frac{b\alpha}{a+b|x|^\alpha}\Big\{(\alpha-1)|x|^{\alpha-2}+\delta(x)\left[x^{\alpha-1}+(-x)^{\alpha-1}\right]\Big\}+E\,, \tag{2.55}$$

As could be expected, $v_{\rm ext}$ has a distributional character. The potential diverges more strongly than $|x|^{-1}$ in the limit $|x|\to 0$ for $\alpha<1$. The corresponding potential energy,

$$V=\int_{-x_0}^{+x_0}dx\,v_{\rm ext}(x)n(x)+V_R\,, \tag{2.56}$$

is not well-defined, the individual contributions being highly divergent. Obviously, potentials of the type (2.55) neither define a proper Hamiltonian, nor are

[2] Note that for $p>0$ one has $\delta'(x)x^pf(x)=-\delta(x)px^{p-1}f(x)$ for any regular $f(x)$.

they particularly realistic. This example calls for a more stringent characterization of the sets of admissible densities and corresponding potentials.

3. The third counterexample [19] is even more simple. Consider a single (noninteracting) particle in a spherical potential. Then any density with a zero is not v-representable, as nodes can only show up for excited states (the ground state is nodeless [21]).

The examples show that the treatment of non-v-representable densities needs closer attention. A rigorous resolution of the problem of interacting v-representability can either be obtained by a more mathematical approach or with a more practically motivated argument.

1. In the practical variant one relies on the fact that any numerical realization of DFT requires the use of grids. On a (finite or infinite) spatial grid, however, any strictly positive ($n(\boldsymbol{r}) > 0$), normalizable density, which is compatible with the Pauli principle, is ensemble v-representable [20]. The crucial point is the representation of the Laplacian by a suitable finite difference formula, as e.g.

$$\nabla^2 f(\boldsymbol{r}) = \frac{1}{h^2}\sum_{i=1}^{3}\left[f(\boldsymbol{r}+h\boldsymbol{e}_i) - 2f(\boldsymbol{r}) + f(\boldsymbol{r}-h\boldsymbol{e}_i)\right], \tag{2.57}$$

where an equidistant mesh and Cartesian coordinates have been used (with mesh spacing h; $\boldsymbol{e}_i$ denotes a Cartesian unit vector). The answer to the question of v-representability is in this line of argumentation associated with the finite resolution of the grid, which does not permit a representation of singular potentials: the finite grid spacing suppresses all singularities.

2. For the mathematical resolution of the v-representability problem one introduces a suitable redefinition of the energy functional [22, 17, 23]. The starting point of this generalization is the *Levy-Lieb functional*

$$E_{\mathrm{LL}}[n] := F_{\mathrm{LL}}[n] + \int d^3r\, v_{\mathrm{ext}}(\boldsymbol{r})\, n(\boldsymbol{r}) \tag{2.58}$$

$$F_{\mathrm{LL}}[n] := \inf_{\Psi\to n} \langle\Psi|\hat{T}+\hat{W}|\Psi\rangle\,. \tag{2.59}$$

The notation $\Psi \longrightarrow n$ indicates that the infimum has to be taken over all N-particle states Ψ which yield the given density n (in fact, the infimum is a minimum, i.e. there always exists some Ψ which minimizes $\langle\Psi|\hat{T}+\hat{W}|\Psi\rangle$ for given n [23]). This restricted minimization procedure is usually referred to as the *Levy-Lieb constrained search.*

The Levy-Lieb functional $E_{\mathrm{LL}}[n]$ represents a consistent extension of the original HK functional: $E_{\mathrm{LL}}[n]$ is identical with $E[n]$ for all pure-state v-representable n and has its minimum for exactly the same density as $E[n]$. This can be shown by use of the Ritz principle,

$$
\begin{aligned}
E_0 &= \inf_{\Psi} \langle \Psi | \hat{H} | \Psi \rangle \\
&= \inf_{n} \left\{ \inf_{\Psi \to n} \langle \Psi | \hat{T} + \hat{W} + \hat{V}_{\text{ext}} | \Psi \rangle \right\} \\
&= \inf_{n} E_{\text{LL}}[n] \,.
\end{aligned} \tag{2.60}
$$

As the infimum is actually reached for the true ground state with the density n_0, $E_{\text{LL}}[n]$ is minimized by this density and one has $E_{\text{LL}}[n_0] = E[n_0]$. Moreover, each state Ψ, which minimizes $\langle \Psi | \hat{H} | \Psi \rangle$ for some potential v_{ext}, is a ground state by construction and therefore leads to a pure-state v-representable density. As a consequence there are no additional densities n which are not pure-state v-representable, but nevertheless give the same energy as the true degenerate ground states. On the complete domain of $E[n]$ one thus finds

$$
E_{\text{LL}}[n] = E[n] \qquad \text{for all pure-state } v\text{-representable } n \,. \tag{2.61}
$$

The same is true for F_{LL}: $F_{\text{LL}}[n] = F[n]$, if n is pure-state v-representable.
So far, the problem of v-representability has, however, only been reformulated by the definition (2.59). The question, still to be answered, is: given some non-negative, normalizable function $n(\boldsymbol{r})$, is there always a suitable antisymmetric, normalizable N-particle function Ψ with $n(\boldsymbol{r}) = \langle \Psi | \hat{n}(\boldsymbol{r}) | \Psi \rangle$? If so, the function $n(\boldsymbol{r})$ is called (pure-state) *N-representable*. Fortunately, the answer is yes [11, 24–26]. In fact, one can explicitly construct such a wavefunction [24].

This construction, ignoring spin for simplicity, starts with the definition of suitable single-particle orbitals,

$$
\phi_{\boldsymbol{k}}(\boldsymbol{r}) := \left(\frac{n(\boldsymbol{r})}{N} \right)^{1/2} e^{i[\boldsymbol{k}\cdot\boldsymbol{f}(\boldsymbol{r}) + \varphi(\boldsymbol{r})]} \,; \qquad \boldsymbol{k} \in \mathbb{Z}^3 \,, \tag{2.62}
$$

with the Cartesian components of the vector field $\boldsymbol{f}$ given by

$$
f_1(\boldsymbol{r}) = 2\pi \frac{\int_{-\infty}^{x} dx'\, n(x',y,z)}{\int_{-\infty}^{\infty} dx'\, n(x',y,z)} \tag{2.63}
$$

$$
f_2(\boldsymbol{r}) = 2\pi \frac{\int_{-\infty}^{\infty} dx' \int_{-\infty}^{y} dy'\, n(x',y',z)}{\int_{-\infty}^{\infty} dx' \int_{-\infty}^{\infty} dy'\, n(x',y',z)} \tag{2.64}
$$

$$
f_3(\boldsymbol{r}) = 2\pi \frac{\int_{-\infty}^{\infty} dx' \int_{-\infty}^{\infty} dy' \int_{-\infty}^{z} dz'\, n(x',y',z')}{\int_{-\infty}^{\infty} dx' \int_{-\infty}^{\infty} dy' \int_{-\infty}^{\infty} dz'\, n(x',y',z')} \,, \tag{2.65}
$$

and a real, but otherwise arbitrary scalar field $\varphi(\boldsymbol{r})$. Different choices of $\phi_{\boldsymbol{k}}$ are possible, uniqueness is not required at this point.
The functions $\phi_{\boldsymbol{k}}$ define an orthonormal and complete basis. Orthonormality is verified as follows. One starts with

$$
\int d^3r\, \phi_{\boldsymbol{k}}^{\dagger}(\boldsymbol{r}) \phi_{\boldsymbol{q}}(\boldsymbol{r}) = \frac{1}{N} \int d^3r\, n(\boldsymbol{r})\, e^{i(\boldsymbol{q}-\boldsymbol{k})\cdot\boldsymbol{f}(\boldsymbol{r})} \,. \tag{2.66}
$$

For the evaluation of (2.66) the integration over $\boldsymbol{r}$ has to be replaced by one over $\boldsymbol{f}$. Consider first the x-coordinate. As long as $n(\boldsymbol{r})$ does not vanish identically in some finite region of space, f_1 is a monotonically increasing function of x (for fixed y and z). One can therefore substitute

$$
\begin{aligned}
&\int_{-\infty}^{\infty} dz \int_{-\infty}^{\infty} dy \int_{-\infty}^{\infty} dx\, n(x,y,z) \\
&= \int_{-\infty}^{\infty} dz \int_{-\infty}^{\infty} dy \int_{f_1(-\infty)}^{f_1(+\infty)} df_1 \left(\frac{\partial f_1(x,y,z)}{\partial x} \right)^{-1} n(x,y,z) \\
&= \frac{1}{2\pi} \int_{-\infty}^{\infty} dz \int_{-\infty}^{\infty} dy \int_0^{2\pi} df_1 \int_{-\infty}^{\infty} dx'\, n(x',y,z)\,.
\end{aligned}
$$

This procedure can be repeated with y and f_2. For fixed z, the function (2.64) is a monotonically increasing function of y,

$$
\begin{aligned}
&\int_{-\infty}^{\infty} dz \int_{-\infty}^{\infty} dy \int_{-\infty}^{\infty} dx\, n(x,y,z) \\
&= \frac{1}{2\pi} \int_0^{2\pi} df_1 \int_{-\infty}^{\infty} dz \int_{f_2(-\infty)}^{f_2(+\infty)} df_2 \left(\frac{\partial f_2(y,z)}{\partial y} \right)^{-1} \int_{-\infty}^{\infty} dx'\, n(x',y,z) \\
&= \frac{1}{(2\pi)^2} \int_0^{2\pi} df_1 \int_{-\infty}^{\infty} dz \int_0^{2\pi} df_2 \int_{-\infty}^{\infty} dx' \int_{-\infty}^{\infty} dy'\, n(x',y',z)\,.
\end{aligned}
$$

Finally, z is replaced by f_3,

$$
\int_{-\infty}^{\infty} dz \int_{-\infty}^{\infty} dy \int_{-\infty}^{\infty} dx\, n(x,y,z) = \frac{N}{(2\pi)^3} \int_0^{2\pi} df_1 \int_0^{2\pi} df_2 \int_0^{2\pi} df_3\,. \tag{2.67}
$$

Equation (2.67) allows a direct evaluation of (2.66),

$$
\int d^3r\, \phi_{\boldsymbol{k}}^{\dagger}(\boldsymbol{r}) \phi_{\boldsymbol{q}}(\boldsymbol{r}) = \frac{1}{(2\pi)^3} \prod_{j=1}^{3} \int_0^{2\pi} df_j\, e^{i(q_j - k_j) f_j} = \delta_{\boldsymbol{k},\boldsymbol{q}}\,. \tag{2.68}
$$

Equation (2.68) is the desired orthonormality relation. In a similar fashion one can establish the completeness of the set of functions $\phi_{\boldsymbol{k}}$,

$$
\begin{aligned}
\sum_{\boldsymbol{k} \in \mathbb{Z}^3} \phi_{\boldsymbol{k}}(\boldsymbol{r}) \phi_{\boldsymbol{k}}^{\dagger}(\boldsymbol{r}') &= \frac{\sqrt{n(\boldsymbol{r}) n(\boldsymbol{r}')}}{N} e^{i\varphi(\boldsymbol{r}) - i\varphi(\boldsymbol{r}')} \sum_{\boldsymbol{k} \in \mathbb{Z}^3} e^{i\boldsymbol{k} \cdot [\boldsymbol{f}(\boldsymbol{r}) - \boldsymbol{f}(\boldsymbol{r}')]} \\
&= \frac{\sqrt{n(\boldsymbol{r}) n(\boldsymbol{r}')}}{N} e^{i\varphi(\boldsymbol{r}) - i\varphi(\boldsymbol{r}')} (2\pi)^3 \delta^{(3)}(\boldsymbol{f}(\boldsymbol{r}) - \boldsymbol{f}(\boldsymbol{r}')) \\
&= \frac{\sqrt{n(\boldsymbol{r}) n(\boldsymbol{r}')}}{N} e^{i\varphi(\boldsymbol{r}) - i\varphi(\boldsymbol{r}')} (2\pi)^3 \delta^{(3)}(\boldsymbol{r} - \boldsymbol{r}') \left| \frac{\partial(f_1, f_2, f_3)}{\partial(x,y,z)} \right|^{-1} \\
&= \delta^{(3)}(\boldsymbol{r} - \boldsymbol{r}')\,,
\end{aligned} \tag{2.69}
$$

using the Jacobi determinant of (2.67).

With the $\phi_{\boldsymbol{k}}$ one can construct an antisymmetric N-particle state. The Slater determinant

$$
\Phi_{\boldsymbol{k}_1 \dots \boldsymbol{k}_N} = \frac{1}{\sqrt{N}} \det\Big(\phi_{\boldsymbol{k}_1} \dots \phi_{\boldsymbol{k}_N} \Big) \tag{2.70}
$$

has all the properties required for the Levy-Lieb construction (2.59). In particular, one obtains as the density corresponding to $\Phi_{\boldsymbol{k}_1 \dots \boldsymbol{k}_N}$ the desired result

$$
n(\boldsymbol{r}) = \sum_{i=1}^{N} |\phi_{\boldsymbol{k}_i}(\boldsymbol{r})|^2 = \frac{n(\boldsymbol{r})}{N} \cdot N\,. \tag{2.71}
$$

On the basis of this explicit construction of suitable N-particle states one concludes that the Levy-Lieb functional (2.58) is mathematically well-defined for arbitrary non-negative functions $n(\boldsymbol{r})$.

The Levy-Lieb construction solves the question of v-representability: $E_{\rm LL}[n]$ is well-defined for any density n in the vicinity of some ground state density n_0. Unfortunately, this does not automatically imply that the functional derivative of $E_{\rm LL}[n]$ at n_0 exists. This is analogous to the situation for ordinary functions, for which differentiability at some point x_0 requires more than the existence of the function in a neighborhood of x_0.

In order to settle the question of functional differentiability an even more general extension of the HK functional $E[n]$ than $E_{\rm LL}[n]$ has to be introduced. A complete and mathematically rigorous discussion of this extension, the *Lieb functional* [23], requires a substantial background in functional analysis. Therefore only the basic concepts and the main results will be outlined here (for an extended and coherent review of Lieb's work see [27]; a complementary approach, emphasizing the aspect of Legendre transforms and generalizing the Lieb functional to non-integer particle number, is presented in [28]). After a characterization of the admissible densities and potentials in section (a) the Lieb functional is defined in section (b). Its functional differentiability is discussed in section (c). Finally, in section (d) an alternative form of the Lieb functional is introduced, which turns out to be most useful for establishing the Kohn-Sham equations in a rigorous way. The hasty reader may proceed directly to the summarizing statement at the end of section (c).

(a) Admissible densities and potentials

In the first step of this extension the sets of admissible densities and potentials are specified more precisely, as differentiability can only be demonstrated for a mathematically well-defined domain. In fact, not every non-negative, normalizable function $n(\boldsymbol{r})$ is a reasonable candidate for which a ground state energy functional should be defined. Rather one requires the components of the ground state energy to be finite separately,

$$\langle\Psi|\hat{T}|\Psi\rangle < \infty \tag{2.72}$$

$$|\langle\Psi|\hat{V}_{\rm ext}|\Psi\rangle| < \infty \tag{2.73}$$

$$\langle\Psi|\hat{W}|\Psi\rangle < \infty\,. \tag{2.74}$$

According to condition (2.72) the gradients of the admissible N-particle wavefunctions $\Psi(\boldsymbol{r}_1\sigma_1,\ldots)$ have to be square-integrable,

$$T(\Psi) := \sum_{i=1}^{N}\sum_{\sigma_1,\ldots\sigma_N}\int d^3r_1\ldots d^3r_N\,|\boldsymbol{\nabla}_i\Psi(\boldsymbol{r}_1\sigma_1,\ldots\boldsymbol{r}_N\sigma_N)|^2 \tag{2.75}$$

$$= N\sum_{\sigma_1,\ldots\sigma_N}\int d^3r_1\ldots d^3r_N\,|\boldsymbol{\nabla}_1\Psi(\boldsymbol{r}_1\sigma_1,\ldots\boldsymbol{r}_N\sigma_N)|^2 \;<\infty\,. \tag{2.76}$$

This condition leads to a constraint for the corresponding density [23],

$$\int d^3r \left[\boldsymbol{\nabla} n^{1/2}(\boldsymbol{r})\right]^2 < \infty . \tag{2.77}$$

In mathematical language, each component of $\boldsymbol{\nabla}_i\Psi$ is in the set $\mathscr{L}^2(\mathbb{R}^{3N})$ of square-integrable functions over $\mathbb{R}^{3N}$, each component of $\boldsymbol{\nabla} n(\boldsymbol{r})^{1/2}$ is in the set $\mathscr{L}^2(\mathbb{R}^3)$. Quite generally, $\mathscr{L}^p(\mathbb{R}^q)$ denotes the set of all functions of q real variables for which the norm

$$\|f\|_p := \left[\int dx_1 \dots dx_q \, |f(x_1,\dots x_q)|^p\right]^{1/p} \tag{2.78}$$

is finite,

$$\mathscr{L}^p(\mathbb{R}^q) := \left\{f(x_1,\dots x_q) \mid \|f\|_p < \infty\right\} . \tag{2.79}$$

If one combines (2.76) and (2.77) with the square-integrability of Ψ,

$$\|\Psi\| := (\|\Psi\|_2)^2 = \sum_{\sigma_1,\dots\sigma_N} \int d^3r_1 \dots d^3r_N \, |\Psi(\boldsymbol{r}_1\sigma_1,\dots \boldsymbol{r}_N\sigma_N)|^2 = 1 \tag{2.80}$$

$$\Longrightarrow \quad \Psi \in \mathscr{L}^2(\mathbb{R}^{3N}) , \tag{2.81}$$

and of $n^{1/2}$,

$$\left(\|n^{1/2}\|_2\right)^2 = \|n\|_1 = \int d^3r n(\boldsymbol{r}) = N \tag{2.82}$$

$$\Longrightarrow \quad n \in \mathscr{L}^1(\mathbb{R}^3); \; n^{1/2} \in \mathscr{L}^2(\mathbb{R}^3) , \tag{2.83}$$

one arrives at the statement that both Ψ and $n^{1/2}$ belong to the Sobolev space $\mathscr{H}^1$, $\Psi \in \mathscr{H}^1(\mathbb{R}^{3N})$ and $n^{1/2} \in \mathscr{H}^1(\mathbb{R}^3)$. This space,

$$\mathscr{H}^1 = \left\{f \mid f \in \mathscr{L}^2, \boldsymbol{\nabla} f \in \mathscr{L}^2\right\} ,$$

is a Hilbert space with the inner product

$$\langle f|g\rangle = \int f^* g + \int \boldsymbol{\nabla} f^* \cdot \boldsymbol{\nabla} g .$$

As soon as $\Psi \in \mathscr{H}^1(\mathbb{R}^{3N})$ the constraint (2.74) is automatically satisfied [23]. The sets of relevant wavefunctions $\mathscr{W}$ and densities $\mathscr{S}$ are thus given by

$$\mathscr{W} = \left\{\Psi \mid \|\psi\| = 1, \; T(\Psi) < \infty\right\} \tag{2.84}$$

$$\mathscr{S} = \left\{n \mid n(\boldsymbol{r}) \geq 0, \; \int d^3r n(\boldsymbol{r}) = N, \; n^{1/2} \in \mathscr{H}^1(\mathbb{R}^3)\right\} . \tag{2.85}$$

$\mathscr{S}$ has the important property that it is *convex*: with the densities n_1 and n_2 all densities on the "straight connection line" between the two densities are in $\mathscr{S}$,

$$n_1, n_2 \in \mathscr{S};\ 0 \le \lambda \le 1 \quad \Longrightarrow \quad n = [\lambda n_1 + (1-\lambda) n_2] \in \mathscr{S}\,. \tag{2.86}$$

Since $\mathscr{S} \subset \mathscr{H}^1(\mathbb{R}^3)$, this result follows directly from the definition of the norm $\|f\| = \langle f|f\rangle^{1/2}$ in $\mathscr{H}^1(\mathbb{R}^3)$, if Schwartz's inequality,

$$\begin{aligned}
\left(\nabla n^{1/2}\right)^2 &= \frac{(\nabla n)^2}{4n} \\
&= \left[\lambda \frac{n_1^{1/2}}{n^{1/2}} \nabla n_1^{1/2} + (1-\lambda) \frac{n_2^{1/2}}{n^{1/2}} \nabla n_2^{1/2}\right]^2 \\
&\le \lambda \left(\frac{\lambda n_1}{n}\right) \left(\nabla n_1^{1/2}\right)^2 + (1-\lambda) \left(\frac{(1-\lambda) n_2}{n}\right) \left(\nabla n_2^{1/2}\right)^2 \\
&\le \lambda \left(\nabla n_1^{1/2}\right)^2 + (1-\lambda) \left(\nabla n_2^{1/2}\right)^2 ,
\end{aligned}$$

is used.

On this basis one can now consider the constraint (2.73), which leads to a characterization of the set of admissible external potentials. One first has to realize that any density in $\mathscr{S}$ also belongs to the larger space $\mathscr{L}^1(\mathbb{R}^3) \cap \mathscr{L}^3(\mathbb{R}^3)$, which is a Banach space (i.e. a complete normed vector space). This is a result of Sobolev's inequality, which (in 3 dimensions) states

$$\|n\|_3 = \left[\int d^3r\, |n(\boldsymbol{r})|^3\right]^{1/3} \le \frac{1}{3}\left(\frac{2}{\pi}\right)^{4/3} \int d^3r \left[\nabla n^{1/2}(\boldsymbol{r})\right]^2 .$$

However, a function which belongs to both $\mathscr{L}^1(\mathbb{R}^3)$ and $\mathscr{L}^3(\mathbb{R}^3)$, also belongs to $\mathscr{L}^2(\mathbb{R}^3)$. The integral (2.73),

$$\langle \Psi | \hat{V}_{\mathrm{ext}} | \Psi \rangle = \int d^3r\, n(\boldsymbol{r}) v_{\mathrm{ext}}(\boldsymbol{r}) = \|n v_{\mathrm{ext}}\|_1 , \tag{2.87}$$

is thus finite, if

$$v_{\mathrm{ext}} \in \mathscr{L}^{3/2}(\mathbb{R}^3) + \mathscr{L}^{\infty}(\mathbb{R}^3) , \tag{2.88}$$

where $\mathscr{L}^{\infty}(\mathbb{R}^3)$ denotes the space of all bounded functions,

$$\|f\|_\infty := \sup_{(x_1,\dots x_q)\in\mathbb{R}^q} |f(x_1,\dots x_q)| \tag{2.89}$$

$$\mathscr{L}^{\infty}(\mathbb{R}^q) := \left\{ f(x_1,\dots x_q) \mid \|f\|_\infty < M \right\} . \tag{2.90}$$

It is immediately clear that potentials from the set $\mathscr{L}^{\infty}$ (we will now drop the indication of the coordinate space over which the functions are defined, as it should be obvious at this point) yield a finite $\|n v_{\mathrm{ext}}\|_1$. For the potentials from the set $\mathscr{L}^{3/2}$

one relies on the Hölder inequality [29],

$$\|fg\|_1 \leq \|f\|_p \, \|f\|_q \qquad \text{with} \qquad \frac{1}{p} + \frac{1}{q} = 1 \tag{2.91}$$

$$\Longrightarrow \quad \|nv_{\text{ext}}\|_1 \leq \|n\|_3 \, \|v_{\text{ext}}\|_{3/2} < \infty \, .$$

The set $\mathscr{L}^{3/2} + \mathscr{L}^{\infty}$ includes in particular the Coulomb potential, which may be decomposed as[3]

$$\frac{1}{|\boldsymbol{r}|} = \frac{\Theta(R - |\boldsymbol{r}|)}{|\boldsymbol{r}|} + \frac{1 - \Theta(R - |\boldsymbol{r}|)}{|\boldsymbol{r}|}$$

$$\frac{\Theta(R - |\boldsymbol{r}|)}{|\boldsymbol{r}|} \in \mathscr{L}^{3/2}$$

$$\frac{1 - \Theta(R - |\boldsymbol{r}|)}{|\boldsymbol{r}|} \in \mathscr{L}^{\infty} \, .$$

$\mathscr{L}^{3/2} + \mathscr{L}^{\infty}$ is a Banach space with the norm

$$\|v_{\text{ext}}\| = \inf_{f \in \mathscr{L}^{3/2}, g \in \mathscr{L}^{\infty}} \left\{ \|f\|_{3/2} + \|g\|_{\infty} \; \middle| \; v_{\text{ext}} = f + g \right\} . \tag{2.92}$$

The characterization of the spaces involved is completed by the observation that the dual space of $\mathscr{L}^1 \cap \mathscr{L}^3$, i.e. the space of all continuous linear functionals on the space of functions $\mathscr{L}^1 \cap \mathscr{L}^3$, is exactly $\mathscr{L}^{3/2} + \mathscr{L}^{\infty}$—all continuous linear functionals on $\mathscr{L}^1 \cap \mathscr{L}^3$ have the form $\int d^3r \, v(\boldsymbol{r}) n(\boldsymbol{r})$ with $v \in \mathscr{L}^{3/2} + \mathscr{L}^{\infty}$ [29].

(b) Definition of Lieb functional and basic properties

At this point one can start the discussion of energy functionals. One first defines the energy $E[v]$ of an N-particle system subject to the external potential v in the most natural way,

$$E[v] := \inf_{\Psi} \left\{ \langle \Psi | \hat{H}_v | \Psi \rangle \; \middle| \; \Psi \in \mathscr{W} \right\} . \tag{2.93}$$

Here the index v at $\hat{H}_v$ characterizes the external potential in $\hat{V}_{\text{ext}}$. This definition is obviously legitimate for all $v \in \mathscr{L}^{3/2} + \mathscr{L}^{\infty}$. Of course, the existence of $E[v]$ for some v does not imply that that there is a minimizing state Ψ which satisfies a Schrödinger equation with the operator $\hat{H}_v$. This is only the case if v belongs to the set of potentials for which a ground state exists: the ground state then minimizes the right-hand side of (2.93). The set of all densities, for which one finds such a ground state is the domain of the original HK-functional,

---

[3] On the other hand, unbounded potentials such as the harmonic oscillator require an additional restriction on the set of densities.

$$\mathscr{A} := \left\{ n = \langle\Psi|\hat{n}|\Psi\rangle \mid \exists\Psi \in \mathscr{W} \text{ with } E[v] = \langle\Psi|\hat{H}_v|\Psi\rangle \right\} . \tag{2.94}$$

The second domain of interest is that of $E_{\rm EHK}[n]$, i.e. the set of all ensemble v-representable densities,

$$\mathscr{B} := \left\{ n = \sum_{i=1}^{q} c_i \langle\Psi_{0,i}|\hat{n}|\Psi_{0,i}\rangle \mid c_i = c_i^* \geq 0; \ \sum_{i=1}^{q} c_i = 1; \right.$$
$$\left. |\Psi_{0,i}\rangle = \text{degenerate ground states for some } v \in \mathscr{L}^{3/2} + \mathscr{L}^{\infty} \right\} . \tag{2.95}$$

With the energy functional (2.93) the Lieb functional is then defined as [23]

$$F_{\rm L}[n] := \sup_v \left\{ E[v] - \int d^3r\, v(\boldsymbol{r}) n(\boldsymbol{r}) \,\Big|\, v \in \mathscr{L}^{3/2} + \mathscr{L}^{\infty} \right\} \quad \text{with } n \in \mathscr{S} . \tag{2.96}$$

The domain of this functional can be extended to the complete space $\mathscr{L}^1 \cap \mathscr{L}^3$, if one allows $F_{\rm L}[n]$ to assume the value $+\infty$ (which is no problem in functional analysis). However, only the densities in $\mathscr{S}$ are really relevant in the following. The energy (2.93) for any external potential $v \in \mathscr{L}^{3/2} + \mathscr{L}^{\infty}$ is then obtained by

$$E[v] = \inf_n \left\{ F_{\rm L}[n] + \int d^3r\, v(\boldsymbol{r}) n(\boldsymbol{r}) \,\Big|\, n \in \mathscr{S} \right\} , \tag{2.97}$$

which is a consequence of the definitions of $E[v]$ and $F_{\rm L}[n]$ as mutual Legendre transforms.[4]

This identity can be verified by establishing two suitable inequalities. On the one hand, one has by definition of $F_{\rm L}[n]$ for any given $v_0 \in \mathscr{L}^{3/2} + \mathscr{L}^{\infty}$

$$\begin{aligned} & E[v_0] - \int d^3r\, v_0\, n \ \leq\ F_{\rm L}[n] && \forall\ n \in \mathscr{S},\ v_0 \in \mathscr{L}^{3/2} + \mathscr{L}^{\infty} \\ \Longrightarrow\quad & \inf_n \left\{ \left[E[v_0] - \int d^3r\, v_0\, n \right] + \int d^3r\, v\, n \,\Big|\, n \in \mathscr{S} \right\} && \\ & \leq\ \inf_n \left\{ F_{\rm L}[n] + \int d^3r\, v\, n \,\Big|\, n \in \mathscr{S} \right\} && \forall\ v, v_0 \in \mathscr{L}^{3/2} + \mathscr{L}^{\infty} \end{aligned}$$

---

[4] An extended version of $F_{\rm L}[n]$ in which the pure-state energy (2.93) is replaced by an ensemble energy,

$$F[n,N] := \sup_v \left\{ E[v,N] - \int d^3r\, v(\boldsymbol{r}) n(\boldsymbol{r}) \,\Big|\, v \in \mathscr{L}^{3/2} + \mathscr{L}^{\infty} \right\}$$
$$E[v,N] := \inf_{\hat{D}} \left\{ \mathrm{tr}\left\{\hat{D}\hat{H}\right\} \,\Big|\, \mathrm{tr}\left\{\hat{D}\hat{N}\right\} = N \right\} ,$$

has been introduced by Eschrig [28]. It shares many properties with $F_{\rm L}[n]$ and allows a consistent handling of non-integer particle numbers. In accordance with the more general definition of $E[v,N]$, one finds

$$F[n,N] \leq F_{\rm L}[n] = F_{\rm HK}[n] \qquad \text{for} \quad n \in \mathscr{A} .$$

This inequality also holds for $v_0 = v$,

$$E[v] \;\le\; \inf_n \left\{ F_{\rm L}[n] + \int d^3r\, v\, n \,\middle|\, n \in \mathscr{S} \right\} \qquad \forall\; v \in \mathscr{L}^{3/2} + \mathscr{L}^{\infty}. \tag{2.98}$$

On the other hand, one has by definition of $E[v]$

$$E[v] \;\le\; \inf_\Psi \{ \langle\Psi|\hat{H}_v|\Psi\rangle \mid \Psi \in \mathscr{W},\ \langle\Psi|\hat{n}|\Psi\rangle = n \} \quad \forall\; n \in \mathscr{S},\ v \in \mathscr{L}^{3/2} + \mathscr{L}^{\infty},$$

so that

$$\begin{aligned}
& E[v] - \int d^3r\, v\, n \\
& \le \inf_\Psi \{ \langle\Psi|\hat{T}+\hat{W}|\Psi\rangle \mid \Psi \in \mathscr{W},\ \langle\Psi|\hat{n}|\Psi\rangle = n \} \quad \forall\; n \in \mathscr{S},\ v \in \mathscr{L}^{3/2} + \mathscr{L}^{\infty} \\
\Longrightarrow\quad & F_{\rm L}[n] = \sup_v \left\{ E[v] - \int d^3r\, v\, n \,\middle|\, v \in \mathscr{L}^{3/2} + \mathscr{L}^{\infty} \right\} \\
& \le \inf_\Psi \{ \langle\Psi|\hat{T}+\hat{W}|\Psi\rangle \mid \Psi \in \mathscr{W},\ \langle\Psi|\hat{n}|\Psi\rangle = n \} \qquad \forall\; n \in \mathscr{S} \\
\Longrightarrow\quad & F_{\rm L}[n] + \int d^3r\, v\, n \\
& \le \inf_\Psi \{ \langle\Psi|\hat{H}_v|\Psi\rangle \mid \Psi \in \mathscr{W},\ \langle\Psi|\hat{n}|\Psi\rangle = n \} \qquad \forall\; n \in \mathscr{S},\ v \in \mathscr{L}^{3/2} + \mathscr{L}^{\infty}.
\end{aligned}$$

However, for any $\Psi \in \mathscr{W}$ the resulting expectation value $\langle\Psi|\hat{n}|\Psi\rangle$ is in $\mathscr{S}$, so that the infimum of the right-hand side of this inequality is exactly $E[v]$,

$$\begin{aligned}
& \inf_n \left\{ F_{\rm L}[n] + \int d^3r\, v\, n \,\middle|\, n \in \mathscr{S} \right\} \\
& \le \inf_\Psi \{ \langle\Psi|\hat{H}_v|\Psi\rangle \mid \Psi \in \mathscr{W} \} \;=\; E[v] \qquad \forall\; v \in \mathscr{L}^{3/2} + \mathscr{L}^{\infty}.
\end{aligned} \tag{2.99}$$

Combination of this inequality with (2.98) proves (2.97).

As a direct consequence of (2.97) one obtains

$$F_{\rm L}[n] \;=\; F_{\rm LL}[n] \;=\; F_{\rm HK}[n] \qquad \forall\; n \in \mathscr{A}, \tag{2.100}$$

so that $F_{\rm L}[n]$ is a consistent extension of the initial HK functional.

(c) Functional differentiability of Lieb functional

For the functional $F_{\rm L}[n]$ one can prove the following properties [23, 30, 31]:

1. $F_{\rm L}[n]$ is *convex*: for $n_0, n_1 \in \mathscr{S}$ and $0 \le \lambda \le 1$ one has

$$F_{\rm L}[\lambda n_1 + (1-\lambda) n_0] \;\le\; \lambda F_{\rm L}[n_1] + (1-\lambda) F_{\rm L}[n_0]. \tag{2.101}$$

   Note that the convexity of $\mathscr{S}$ ensures that $F_{\rm L}[n]$ is defined (i.e. finite) for all $n = \lambda n_1 + (1-\lambda) n_0$. The property (2.101) results from the linearity of $F_{\rm L}[n]$ in n in combination with the definition of $F_{\rm L}[n]$ as a supremum.

2. $F_{\rm L}[n]$ is *weakly lower semicontinuous*: for any sequence n_k which converges weakly against a limit n, i.e. for which

$$\|(n_k - n)v\|_1 \xrightarrow{k\to\infty} 0 \qquad \forall\ v \in \mathscr{L}^{3/2} + \mathscr{L}^{\infty}\,,$$

   one has

$$F_{\rm L}[n] \le \lim_{k\to\infty}\ \inf_{l\ge k} F_{\rm L}[n_l] = \liminf_{k\to\infty} F_{\rm L}[n_k]\,.$$

   This relation is a "weak" version of the standard concept of continuity which requires that $|F_{\rm L}[n] - F_{\rm L}[n_k]| \xrightarrow{k\to\infty} 0$ if n_k is norm convergent against n (i.e. if $\|n_k - n\|_1 \xrightarrow{k\to\infty} 0$ and simultaneously $\|n_k - n\|_3 \xrightarrow{k\to\infty} 0$).
3. $F_{\rm L}[n]$ has a *unique continuous tangent functional* on $\mathscr{B}$: for all $n_0 \in \mathscr{B}$ there exists a continuous linear functional $\delta F_{n_0}[n]$ with domain $\mathscr{L}^1 \cap \mathscr{L}^3$, i.e. a functional of the form

$$\delta F_{n_0}[n] = -\int d^3r\, v(\boldsymbol{r})\, n(\boldsymbol{r}) \qquad \text{with} \quad v \in \mathscr{L}^{3/2} + \mathscr{L}^{\infty}\,, \tag{2.102}$$

   so that

$$F_{\rm L}[n] \ \ge\ F_{\rm L}[n_0] + \delta F_{n_0}[n - n_0] \qquad \forall\ n \in \mathscr{L}^1 \cap \mathscr{L}^3\,, \tag{2.103}$$

   and the kernel v in $\delta F_{n_0}[n - n_0]$ is unique up to a constant (the uniqueness is a consequence of the HK theorem). The tangent functional vanishes for $n = n_0$, so that (2.103) becomes an equality at this density, which explains the name.
4. The *set of densities in $\mathscr{B}$ is norm dense in $\mathscr{S}$*: for arbitrary $n_0 \in \mathscr{S}$ there exists a sequence $n_k \in \mathscr{B}$ such that $\|n_k - n_0\|_1 \xrightarrow{k\to\infty} 0$ and simultaneously $\|n_k - n_0\|_3 \xrightarrow{k\to\infty} 0$.
   This statement ensures that for any $n_0 \in \mathscr{S}$ one can find a unique continuous tangent functional for a density n_1 which is infinitesimally close to n_0, so that, from a practical point of view, continuous tangent functionals exist for all $n_0 \in \mathscr{S}$.

These properties finally allow a resolution of the initial question concerning functional differentiability. The point is: for any convex, finite, lower semicontinuous functional with a unique continuous tangent functional the existence of the functional derivative is guaranteed, the functional derivative being identical with the kernel $-v(\boldsymbol{r})$ of the tangent functional [30, 31, 29] (for a more precise formulation see in particular Corollary 2.5 and Proposition 5.3 of [32] or [27]). Since the first step of the proof of this statement is both simple and instructive, it is worthwhile to present it here. Consider the densities

$$n = n_0 + \lambda(n_1 - n_0) \qquad \text{with} \quad n_0 \in \mathscr{B},\ n_1 \in \mathscr{S},\ 0 \le \lambda \le 1\,.$$

Due to the convexity of $\mathscr{S}$ the density n is also in $\mathscr{S}$. Now combine the convexity of $F_{\rm L}[n]$ (property 1.) with the existence of the tangent functional (property 3.),

$$\lambda F_{\rm L}[n_1]+(1-\lambda)F_{\rm L}[n_0] \geq F_{\rm L}[n_0+\lambda(n_1-n_0)] \geq F_{\rm L}[n_0]+\delta F_{n_0}[\lambda(n_1-n_0)] .$$

If one subtracts $F_{\rm L}[n_0]$, uses $\delta F_{n_0}[\lambda(n_1-n_0)] = \lambda\delta F_{n_0}[n_1-n_0]$ and divides by λ, one arrives at

$$F_{\rm L}[n_1]-F_{\rm L}[n_0] \geq \frac{F_{\rm L}[n_0+\lambda(n_1-n_0)]-F_{\rm L}[n_0]}{\lambda} \geq \delta F_{n_0}[n_1-n_0] . \qquad (2.104)$$

Both $F_{\rm L}[n_1]-F_{\rm L}[n_0]$ and $\delta F_{n_0}[n_1-n_0]$ are well defined and finite, so that the so-called *Gâteaux differential*

$$F'[n_0,n_1] := \lim_{\lambda\to 0}\frac{F_{\rm L}[n_0+\lambda(n_1-n_0)]-F_{\rm L}[n_0]}{\lambda}$$

of $F_{\rm L}[n]$ at $n=n_0$ exists.

In order to complete the proof of the existence of the functional derivative one has to demonstrate that the Gâteaux differential is linear and continuous in (n_1-n_0), i.e. that it is identical with $\delta F_{n_0}[n_1-n_0]$. This second step of the proof is more involved, so that the reader is referred to the literature [32, 27] for a mathematically rigorous discussion. The linearity of the Gâteaux differential is, however, intuitively clear on geometrical grounds, if one interprets the functionals involved as simple functions as in Fig. 2.3: the right-hand inequality in (2.104) shows that the Gâteaux differential approaches the limit $n_1=n_0$ at least as slowly as the linear functional $\delta F_{n_0}[n_1-n_0]$ when n_1 approaches n_0. So, $F'[n_0,n_1]$ can not vanish faster than n_1-n_0

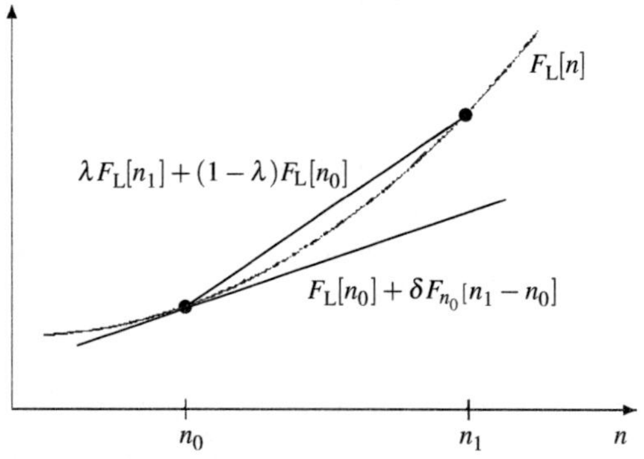

Fig. 2.3 Illustration of convex functional $F_{\rm L}[n]$ with unique Gâteaux differential $\delta F_{n_0}[n_1-n_0]$.

for $n_1\to n_0$. On the other hand, the convexity of $F_{\rm L}[n]$ does not allow $F'[n_0,n_1]$ to approach its limit zero more slowly than n_1-n_0 (as the straight line between $F_{\rm L}[n_1]$ and $F_{\rm L}[n_0]$ must be above $F_{\rm L}[n_0+\lambda(n_1-n_0)]$ for arbitrary n_1—this excludes that $F_{\rm L}[n_0+\lambda(n_1-n_0)]$ jumps above this line for some sequence of λ, so that semicontinuity implies continuity). Consequently, $F'[n_0,n_1]$ must be linear in n_1-n_0. Moreover, as the direction of $n_1 - n_0$ is arbitrary and the tangent functional is unique, $F'[n_0, n_1]$ must coincide with $\delta F_{n_0}[n_1 - n_0]$.

In summary: The functional derivative of $F_L[n]$ exists for all ensemble v-representable densities and is identical with a potential v_{ext} from the dual space $\mathscr{L}^{3/2} + \mathscr{L}^{\infty}$,

$$\left.\frac{F_L[n]}{\delta n(\boldsymbol{r})}\right|_{n=n_0} = -v_{ext}(\boldsymbol{r}) \qquad \text{with} \quad v_{ext} \in \mathscr{L}^{3/2} + \mathscr{L}^{\infty} . \tag{2.105}$$

Moreover, for any other "reasonable" density n (i.e. for any $n \in \mathscr{S}$) one can find an ensemble v-representable density which is arbitrarily close to n, so that the functional derivative of $F_L[n]$ again exists.

In order to complete the picture, it is worthwhile to reconsider the Levy-Lieb functional (2.59) at this point. The existence of a unique continuous tangent functional for all $n \in \mathscr{A}$ has also been demonstrated for $F_{LL}[n]$ [23]. However, this functional is not convex on $\mathscr{S}$ [23]. Since convexity is crucial for establishing the existence of the functional derivative of $F_L[n]$ via Eq. (2.104), functional differentiability has not been rigorously proven for $F_{LL}[n]$ so far.

(d) Representation of Lieb functional in terms of density matrices

So, as a matter of principle, the subsequent development of the DFT formalism should therefore be based explicitly on the Lieb functional. We will nevertheless often ignore the issue of functional differentiability in the following and will not distinguish between the various flavors of the energy functional. A prominent exception is the derivation of the Kohn-Sham equations. For this purpose, an alternative form of $F_L[n]$ is most useful. In fact, for all $n \in \mathscr{S}$ the Lieb functional $F_L[n]$ can be recast in the form [23]

$$F_L[n] = \inf_{\hat{D} \to n} \operatorname{tr}\{\hat{D}(\hat{T} + \hat{W})\} , \tag{2.106}$$

with the density matrices restricted to the form

$$\hat{D} = \sum_k d_k |\Psi_k\rangle\langle\Psi_k|, \; d_k^* = d_k \geq 0, \; \sum_k d_k = 1, \; \langle\Psi_k|\Psi_l\rangle = \delta_{kl}, \; \Psi_k \in \mathscr{H}^1$$

$$n(\boldsymbol{r}) = \sum_k d_k \langle\Psi_k|\hat{n}(\boldsymbol{r})|\Psi_k\rangle$$

(the $|\Psi_k\rangle$ denote a set of orthonormal wavefunctions in the N-particle Hilbert space). One can also show that there always exists a minimizing $\hat{D}$ for all $n \in \mathscr{S}$ [23]. The form (2.106) will be used in particular to establish the Kohn-Sham equations for degenerate systems (see Sect. 3.3).

2.4 Fractional Particle Numbers, Derivative Discontinuity

The variational equation (2.38) raises one further question: in this equation the particle number is determined via a subsidiary condition, which implies the existence of $E[n]$ for non-integer particle numbers. However, so far all energy functionals are only defined for integer N. Therefore the question has to be addressed, how to extend the energy functional to fractional particle numbers.

Assume that a density integrates up to $N+\eta$,

$$\int d^3r n(\boldsymbol{r}) = N+\eta \; ; \qquad N=1,2,\ldots \; ; \qquad 0 \le \eta < 1 \,. \tag{2.107}$$

The simplest definition of an energy functional for such a density is a statistical superposition of the lowest possible energies of two states $|\Psi_N\rangle$ and $|\Psi_{N+1}\rangle$ with the neighboring integer particle numbers N and $N+1$ [33],

$$E_f[n] := F_f[n] + \int d^3r v_{\text{ext}}(\boldsymbol{r}) n(\boldsymbol{r}) \tag{2.108}$$

$$F_f[n] := \min_{\Psi_N,\Psi_{N+1}} \left\{(1-\eta)\langle\Psi_N|\hat{T}+\hat{W}|\Psi_N\rangle + \eta\langle\Psi_{N+1}|\hat{T}+\hat{W}|\Psi_{N+1}\rangle\right\} \tag{2.109}$$

$$\text{with } n(\boldsymbol{r}) = (1-\eta)\langle\Psi_N|\hat{n}(\boldsymbol{r})|\Psi_N\rangle + \eta\langle\Psi_{N+1}|\hat{n}(\boldsymbol{r})|\Psi_{N+1}\rangle \tag{2.110}$$

$$\langle\Psi_N|\Psi_N\rangle = \langle\Psi_{N+1}|\Psi_{N+1}\rangle = 1 \,.$$

The condition (2.110) restricts the variational search for the minimum in (2.109) to those combinations of normalizable states $|\Psi_N\rangle$ and $|\Psi_{N+1}\rangle$, which yield the prescribed density n (constrained search). On the other hand, $|\Psi_N\rangle$ and $|\Psi_{N+1}\rangle$ need not be related to the potential v_{ext} in any way. Equation (2.110) automatically guarantees the desired particle number (2.107). The definition (2.109) is an obvious extension of the Levy-Lieb functional (2.59) to fractional particle numbers. Correspondingly, $F_f[n]$ becomes identical with the Levy-Lieb functional for $\eta = 0$. On the basis of $E_f[n]$ the variational equation (2.38) is also well-defined for fractional particle numbers,[5]

$$\frac{\delta E_f[n]}{\delta n(\boldsymbol{r})} = \mu_{\text{L}} \,, \tag{2.111}$$

as $F_f[n]$ exists for any non-negative $n(\boldsymbol{r})$ which integrates up to $N+\eta$. For any given particle number $N+\eta$ one finds a corresponding Lagrange multiplier $\mu_{\text{L}}(N+\eta)$.

For integer particle number this Lagrange multiplier is identical with the chemical potential

$$\mu(N) := \frac{\partial E}{\partial N}(N) \,, \tag{2.112}$$

---

[5] Precisely speaking, the extension of the Lieb functional to non-integer particle number [28] is required at this point. We will, however, not distinguish between the two functionals in the following.

where $E(N+\eta)$ denotes the minimum of the functional (2.108) for given, fractional particle number,

$$E(N+\eta) = \min_n E_f[n]. \tag{2.113}$$

Equation (2.112) can be verified by functional Taylor expansion of the total energy of a system with fractional particle number around the ground state density n_N of the N-particle system,

$$\begin{aligned}\frac{\partial E}{\partial N}(N) &\\ = \lim_{\eta\to 0}\frac{1}{\eta}\big[E(N+\eta)-E(N)\big] &\\ = \lim_{\eta\to 0}\frac{1}{\eta}\left[E_f[n_N] + \int d^3r\, \frac{\delta E_f[n]}{\delta n(\boldsymbol{r})}\bigg|_{n=n_N}[n_{N+\eta}(\boldsymbol{r})-n_N(\boldsymbol{r})]+\ldots-E(N)\right],\end{aligned}$$

where $n_{N+\eta}$ denotes the ground state density for particle number $N+\eta$ and $E_f[n_N]=E(N)$. One can now use (2.111) to identify the Lagrange multiplier with $\partial E/\partial N$,

$$\mu(N) = \frac{\partial E}{\partial N}(N) = \mu_{\rm L}(N)\lim_{\eta\to 0}\frac{1}{\eta}\int d^3r\,[n_{N+\eta}(\boldsymbol{r})-n_N(\boldsymbol{r})] = \mu_{\rm L}(N). \tag{2.114}$$

It is worthwhile to examine $\mu(N)$ more closely. Consider the ground state energy of a system with $N+\eta$ particles,

$$\begin{aligned}E(N+\eta) &= \min_n E_f[n]\\ &= \min_n \min_{\Psi_N,\Psi_{N+1}} \left\{(1-\eta)\langle\Psi_N|\hat{H}|\Psi_N\rangle + \eta\langle\Psi_{N+1}|\hat{H}|\Psi_{N+1}\rangle\right\}\\ &\quad\text{with } n(\boldsymbol{r}) = (1-\eta)\langle\Psi_N|\hat{n}(\boldsymbol{r})|\Psi_N\rangle + \eta\langle\Psi_{N+1}|\hat{n}(\boldsymbol{r})|\Psi_{N+1}\rangle,\end{aligned} \tag{2.115}$$

where the search for the minimum over n is restricted to densities with the fractional particle number $N+\eta$. However, the expression

$$(1-\eta)\langle\Psi_N|\hat{H}|\Psi_N\rangle + \eta\langle\Psi_{N+1}|\hat{H}|\Psi_{N+1}\rangle$$

becomes minimal if $|\Psi_N\rangle$ is the ground state of the N particle system and $|\Psi_{N+1}\rangle$ is the ground state of the $N+1$ particle system (for the same $v_{\rm ext}$—one now assumes that the normalizable ground state $|\Psi_{N+1}\rangle$ exists, i.e. that $v_{\rm ext}$ is capable of binding $N+1$ particles). The minimum of (2.115) is therefore obtained if n is given by a superposition of the ground state density n_N of the N-particle system and the ground state density n_{N+1} of the $(N+1)$-particle system,

$$n(\boldsymbol{r}) = (1-\eta)\,n_N(\boldsymbol{r}) + \eta\,n_{N+1}(\boldsymbol{r}). \tag{2.116}$$

A similar superposition is found for the total energy of the system with $N+\eta$ particles,

$$E(N+\eta) = (1-\eta)E(N) + \eta E(N+1)\,. \tag{2.117}$$

This dependence of the energy on the particle number is illustrated in Fig. 2.4. The

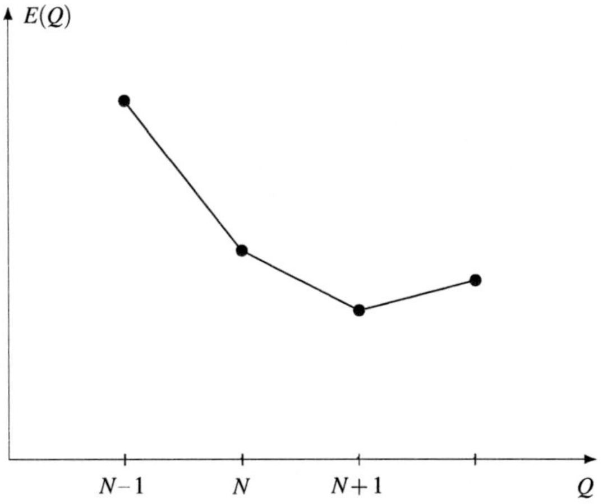

Fig. 2.4 Particle number dependence of the total energy $E(Q)$ for fractional particle number Q (the external potential v_{ext} is the same for all Q).

total energy is a linear function between two neighboring integer particle numbers. According to Eq. (2.114), $\mu(N)$ is the slope of this piecewise linear curve. One thus obtains

$$\mu(N-\eta) = E(N) - E(N-1) = -\text{IP} \quad \text{(ionization potential)} \tag{2.118}$$

$$\mu(N+\eta) = E(N+1) - E(N) = -\text{EA} \quad \text{(electron affinity)}\,, \tag{2.119}$$

if N is the number of electrons required for charge neutrality (and $\eta > 0$). As a consequence, $\mu(N)$ is discontinuous at all integer particle numbers. In view of Eqs. (2.114) and (2.111) this implies that the functional derivative of the total energy functional $E_f[n]$ itself must have discontinuities at these particle numbers, the so-called *derivative discontinuities* (for an extended discussion of this and related aspects see [15]).

2.5 Spin-Polarized Systems

Up to now the discussion focused on systems of interacting particles moving in an electrostatic external potential. Density functional theory can, however, also be extended to systems in which the particles are subject to a static magnetic field [34, 35]. In the simplest version the corresponding Hamiltonian is given by

$$\hat{H} = \hat{T} + \hat{W} + \int d^3r \left\{ v_{\text{ext}}(\boldsymbol{r})\hat{n}(\boldsymbol{r}) + \boldsymbol{B}_{\text{ext}}(\boldsymbol{r}) \cdot \hat{\boldsymbol{m}}(\boldsymbol{r}) \right\} . \tag{2.120}$$

Here $\hat{\boldsymbol{m}}$ is the operator of the magnetization density,[6]

$$\hat{\boldsymbol{m}}(\boldsymbol{r}) = \mu_B \sum_{i=1}^{N} \boldsymbol{\sigma}_i \delta^{(3)}(\boldsymbol{r} - \boldsymbol{r}_i) = \mu_B \sum_{\sigma,\sigma'=\uparrow,\downarrow} \hat{\psi}^\dagger(\boldsymbol{r}\sigma) \boldsymbol{\sigma}_{\sigma\sigma'} \hat{\psi}(\boldsymbol{r}\sigma') , \tag{2.121}$$

where $\boldsymbol{\sigma}$ denotes the 2×2 Pauli matrices,

$$\sigma_x = \begin{pmatrix} 0 & 1 \\ 1 & 0 \end{pmatrix} \iff \sigma_{x,\sigma\sigma'} = \delta_{\sigma',-\sigma} \tag{2.122}$$

$$\sigma_y = \begin{pmatrix} 0 & -i \\ i & 0 \end{pmatrix} \iff \sigma_{y,\sigma\sigma'} = i\,\text{sign}(\sigma')\,\delta_{\sigma',-\sigma} \tag{2.123}$$

$$\sigma_z = \begin{pmatrix} 1 & 0 \\ 0 & -1 \end{pmatrix} \iff \sigma_{z,\sigma\sigma'} = \text{sign}(\sigma)\,\delta_{\sigma',\sigma} \quad , \tag{2.124}$$

and μ_B is the Bohr magneton, $\mu_B = e\hbar/(2mc)$ $(e = |e|)$. In addition to the conventional single-particle density n_0, now a second density variable, the ground state magnetization density

$$\boldsymbol{m}_0(\boldsymbol{r}) = \langle \Psi_0 | \hat{\boldsymbol{m}}(\boldsymbol{r}) | \Psi_0 \rangle , \tag{2.125}$$

offers itself for a characterization of the ground state $|\Psi_0\rangle$ of the Hamiltonian (2.120). And indeed, one can verify the following two statements [34, 35]:

(a) Two different non-degenerate ground states $|\Psi_0\rangle$ and $|\Psi_0'\rangle$ from the set of all ground states resulting from Hamiltonians of type (2.120) always lead to two different sets of ground state densities $(n_0, \boldsymbol{m}_0)$ and $(n_0', \boldsymbol{m}_0')$, i.e. at least one of the four density components differs. As a conse-

[6] Both the components of bispinors and spin quantum numbers will be characterized by either the numerical values $\pm 1/2$ or, alternatively, the symbolic equivalents $\uparrow, \downarrow$, depending on which notation is more convenient or clear.

In Eqs. (2.120) and (2.121) the sign of the magnetization density has been chosen so that $\hat{\boldsymbol{m}}$ differs from the conventional magnetization density of electrons introduced in classical electrodynamics (see e.g. [36]) by a minus sign.

quence, $|\Psi_0\rangle$ is uniquely determined by the set $(n_0, \boldsymbol{m}_0)$, i.e. is a unique functional of $(n, \boldsymbol{m})$,

$$|\Psi_0\rangle \underset{\substack{\uparrow \\ \text{one-to-one}}}{\Longleftrightarrow} (n_0, \boldsymbol{m}_0) \qquad \Longrightarrow \qquad |\Psi_0\rangle = |\Psi[n_0, \boldsymbol{m}_0]\rangle . \tag{2.126}$$

The proof via *reductio ad absurdum* proceeds as for the purely electrostatic potential. Assume that the ground states $|\Psi_0\rangle$ and $|\Psi_0'\rangle$ corresponding to the Hamiltonians $\hat{H}$ and $\hat{H}'$ yield the same set $(n_0, \boldsymbol{m}_0)$. Then consider the ground state energy and use both the Ritz variational principle and the fact that the two states are non-degenerate,

$$\begin{aligned}\langle\Psi_0|\hat{H}|\Psi_0\rangle &< \langle\Psi_0'|\hat{H}|\Psi_0'\rangle \\ &= \langle\Psi_0'|\hat{H}'|\Psi_0'\rangle + \int d^3r\left\{\left[v_{\rm ext} - v_{\rm ext}'\right]n_0 + \left[\boldsymbol{B}_{\rm ext} - \boldsymbol{B}_{\rm ext}'\right]\cdot\boldsymbol{m}_0\right\} \end{aligned} \tag{2.127}$$

(as $|\Psi_0\rangle \neq |\Psi_0'\rangle$ and both states are non-degenerate one necessarily has $\hat{H} \neq \hat{H}'$, i.e. $(v_{\rm ext}, \boldsymbol{B}_{\rm ext}) \neq (v_{\rm ext}', \boldsymbol{B}_{\rm ext}')$). Interchanging primed and unprimed quantities one finds

$$\langle\Psi_0'|\hat{H}'|\Psi_0'\rangle < \langle\Psi_0|\hat{H}|\Psi_0\rangle + \int d^3r\left\{\left[v_{\rm ext}' - v_{\rm ext}\right]n_0 + \left[\boldsymbol{B}_{\rm ext}' - \boldsymbol{B}_{\rm ext}\right]\cdot\boldsymbol{m}_0\right\} . \tag{2.128}$$

Addition of (2.127) and (2.128) leads to the desired contradiction.

(b) The functional $|\Psi[n, \boldsymbol{m}]\rangle$ allows the definition of a ground state energy functional,

$$E[n, \boldsymbol{m}] = F[n, \boldsymbol{m}] + \int d^3r\{v_{\rm ext}(\boldsymbol{r})n(\boldsymbol{r}) + \boldsymbol{B}_{\rm ext}(\boldsymbol{r})\cdot\boldsymbol{m}(\boldsymbol{r})\} \tag{2.129}$$

$$F[n, \boldsymbol{m}] = \langle\Psi[n, \boldsymbol{m}]|\hat{T} + \hat{W}|\Psi[n, \boldsymbol{m}]\rangle , \tag{2.130}$$

which satisfies a minimum principle,

$$E[n_0, \boldsymbol{m}_0] < E[n, \boldsymbol{m}] \qquad \forall\ (n, \boldsymbol{m}) \neq (n_0, \boldsymbol{m}_0) , \tag{2.131}$$

with $(n_0, \boldsymbol{m}_0)$ being the ground state densities corresponding to $(v_{\rm ext}, \boldsymbol{B}_{\rm ext})$.

The proof follows the same pattern as in the original argument by HK. Note that it is usual to call $E[n, \boldsymbol{m}]$ a density functional, in spite of the fact that this functional not only depends on n, but also on the magnetization density. Ignoring the issue of v-representability, one can furthermore reformulate (2.131) as a set of four coupled variational equations,[7]

$$\left.\frac{\delta E[n,\boldsymbol{m}]}{\delta n(\boldsymbol{r})}\right|_{n=n_0,\boldsymbol{m}=\boldsymbol{m}_0} = \mu\,; \qquad \left.\frac{\delta E[n,\boldsymbol{m}]}{\delta \boldsymbol{m}(\boldsymbol{r})}\right|_{n=n_0,\boldsymbol{m}=\boldsymbol{m}_0} = \boldsymbol{0}\,, \tag{2.132}$$

with the Lagrange parameter μ ensuring the correct particle number as in (2.38).

A number of points should be noted:

1. The reader will have noticed that the existence theorem formulated for systems subject to magnetic fields is much more restrictive than the original HK theorem. In particular, nothing has been said about the relation between the ground state $|\Psi_0\rangle$ and the set of external potentials $(v_{\rm ext}, \boldsymbol{B}_{\rm ext})$. Obviously, the corresponding proof used for the original HK theorem does not apply in the present situation, as the operator $\boldsymbol{B}_{\rm ext}\cdot\hat{\boldsymbol{m}}$ does not allow a factorization of the ground state wavefunction (which was utilized in Eq. (2.17)).
   In fact, one can give a rather simple argument that two different sets of $(v_{\rm ext}, \boldsymbol{B}_{\rm ext})$ can lead to the same ground state [34, 37, 38]: whenever one has a magnetic field of the *collinear* form
   $$\boldsymbol{B}_{\rm ext}(\boldsymbol{r}) = (0,0,B)$$
   with constant B, any ground state $|\Psi_0\rangle$ of some Hamiltonian $\hat{H}$,
   $$\hat{H}|\Psi_0\rangle = E_0|\Psi_0\rangle\,,$$
   which is simultaneously an eigenstate of
   $$\hat{S}_z = \sum_{i=1}^{N}\sigma_{i,z}\,; \qquad \hat{S}_z|\Psi_0\rangle = S_z|\Psi_0\rangle\,,$$
   is also an eigenstate of the extended Hamiltonian $\hat{H}+\mu_B B\hat{S}_z$,
   $$\left(\hat{H}+\mu_B B\hat{S}_z\right)|\Psi_0\rangle = \left(E_0+\mu_B BS_z\right)|\Psi_0\rangle\,,$$
   irrespective of the magnitude of B. Moreover, as long as B is sufficiently small, $|\Psi_0\rangle$ remains the ground state of the system characterized by $\hat{H}+\mu_B B\hat{S}_z$, demonstrating the non-uniqueness of the map between all ground states and the set of external potentials $(v_{\rm ext}, \boldsymbol{B}_{\rm ext})$.
   However, the above example obviously relies on the fact that B is constant. Correspondingly, one can show that two pairs of spin-dependent potentials,
   $$v_\pm(\boldsymbol{r}) = v_{\rm ext}(\boldsymbol{r}) \pm \mu_B B_{\rm ext}(\boldsymbol{r}) \qquad \text{and} \qquad v'_\pm(\boldsymbol{r}) = v'_{\rm ext}(\boldsymbol{r}) \pm \mu_B B'_{\rm ext}(\boldsymbol{r})\,,$$

---

[7] Here and in the following the derivative with respect to a vector is to be understood as the vector which results from differentiation with respect to the components,

$$\frac{\delta E[n,\boldsymbol{m}]}{\delta \boldsymbol{m}(\boldsymbol{r})} = \left(\frac{\delta E[n,\boldsymbol{m}]}{\delta m_1(\boldsymbol{r})}, \frac{\delta E[n,\boldsymbol{m}]}{\delta m_2(\boldsymbol{r})}, \frac{\delta E[n,\boldsymbol{m}]}{\delta m_3(\boldsymbol{r})}\right)\,.$$

for which v_σ differs from v'_σ by more than a σ-dependent constant (for both $\sigma = \pm$) always have different ground states [38–40].
Returning to the general case of a magnetic field $\boldsymbol{B}_{\rm ext}$ with more than one non-vanishing component, the question of uniqueness then reduces to the question whether one can find a local rotation in spin space which transforms $\boldsymbol{B}_{\rm ext}$ to the collinear form $(0,0,B_{\rm ext})$? The answer is no [40] (with the exception of some rather special cases [41]), so that the map between the ground state $|\Psi_0\rangle$ and the set of external potentials $(v_{\rm ext},\boldsymbol{B}_{\rm ext})$ is invertible in the general situation.

2. Often the actual magnetic field has only one non-vanishing component,

$$\boldsymbol{B}_{\rm ext}(\boldsymbol{r}) = (0,0,B_{\rm ext}(\boldsymbol{r}))\,. \tag{2.133}$$

Of course, one can go through the existence theorem for a field of the form (2.133) as in the general situation. One ends up with the statement that there is a one-to-one correspondence between $|\Psi_0\rangle$ and the set $(n_0,m_{z,0})$,

$$|\Psi_0\rangle \iff (n_0,m_{z,0}) \qquad \Longrightarrow \qquad |\Psi_0\rangle = |\Psi[n_0,m_{z,0}]\rangle\,. \tag{2.134}$$

Correspondingly, the energy becomes a functional of n and the z-component of the magnetization density, m_z.
It is standard to reformulate this functional in terms of the spin-densities,

$$\hat{n}_\sigma(\boldsymbol{r}) = \frac{1}{2}\sum_{i=1}^{N}[1+{\rm sign}(\sigma)\sigma_{z,i}]\,\delta^{(3)}(\boldsymbol{r}-\boldsymbol{r}_i) = \hat{\psi}^\dagger(\boldsymbol{r}\sigma)\hat{\psi}(\boldsymbol{r}\sigma) \tag{2.135}$$

$$\hat{n}(\boldsymbol{r}) = \sum_{\sigma=\uparrow,\downarrow}\hat{n}_\sigma(\boldsymbol{r})\,. \tag{2.136}$$

In fact, for the z-component of Eq. (2.121) one immediately obtains

$$\hat{m}_z(\boldsymbol{r}) = \mu_B\left[\hat{\psi}^\dagger(\boldsymbol{r}\uparrow)\hat{\psi}(\boldsymbol{r}\uparrow) - \hat{\psi}^\dagger(\boldsymbol{r}\downarrow)\hat{\psi}(\boldsymbol{r}\downarrow)\right] = \mu_B\left[\hat{n}_\uparrow(\boldsymbol{r}) - \hat{n}_\downarrow(\boldsymbol{r})\right]. \tag{2.137}$$

Thus the set $(n_\uparrow,n_\downarrow)$ is completely equivalent to (n,m_z). All statements can equally well be formulated in terms of $(n_\uparrow,n_\downarrow)$. The most frequently used form of the spin-density functional then reads

$$E[n_\uparrow,n_\downarrow] = F[n_\uparrow,n_\downarrow] + \int d^3r\left\{v_{\rm ext}[n_\uparrow+n_\downarrow] + \mu_B B_{\rm ext}[n_\uparrow-n_\downarrow]\right\} \tag{2.138}$$

$$F[n_\uparrow,n_\downarrow] = \langle\Psi[n_\uparrow,n_\downarrow]|\hat{T}+\hat{W}|\Psi[n_\uparrow,n_\downarrow]\rangle\,. \tag{2.139}$$

The associated variational equations rely on the fact that in the case of the reduced magnetic field (2.133) the Hamiltonian (2.120) commutes with the total particle number for given spin,

$$\hat{N}_\sigma = \int d^3r\,\hat{n}_\sigma(\boldsymbol{r}) \qquad \Longrightarrow \qquad \left[\hat{H},\hat{N}_\sigma\right] = 0 \tag{2.140}$$

(for the basic commutators involved see Appendix L). It is thus possible to fix the individual numbers N_σ of particles with spin σ, as long as the total particle number is N,

$$N_\sigma = \langle \Psi_0 | \hat{N}_\sigma | \Psi_0 \rangle \, ; \qquad N_\uparrow + N_\downarrow = N \, . \tag{2.141}$$

Each pair $N_\uparrow, N_\downarrow$ defines one sector of Fock space for which the Ritz variational principle is separately valid, so that the DFT minimum principle also applies to each sector separately. Consequently, the variational equations have the form

$$\left. \frac{\delta E[n_\uparrow, n_\downarrow]}{\delta n_\sigma(\boldsymbol{r})} \right|_{n_\sigma = n_{\sigma,0}} = \mu_\sigma \, , \tag{2.142}$$

with the Lagrange parameters μ_σ controlling the individual particle numbers N_σ. The true ground state then corresponds to the pair of N_σ which gives the minimum energy.

3. One can finally reconsider the many-particle problem without any magnetic field. For this problem it is nevertheless legitimate, though not formally necessary, to use the spin-dependent functional

$$E[n_\uparrow, n_\downarrow] = F[n_\uparrow, n_\downarrow] + \int d^3r \, v_{\rm ext}[n_\uparrow + n_\downarrow] \, ,$$

rather than the original form (2.26): the class of systems discussed in the present section also contains the more restricted class considered in the original HK theorem. In practice, Eqs. (2.138)–(2.142) with $B_{\rm ext} = 0$ represent the standard DFT approach to spin-polarized systems, i.e. systems with non-vanishing magnetic moment.

Why is this extended description of spin-polarized systems superior to the more direct approach in terms of only the density? In principle, $n_\uparrow$ and $n_\downarrow$ are functionals of the complete n as long as $B_{\rm ext} = 0$. However, their functional dependence on n is definitely complicated. Consequently, it is advantageous to resolve this unknown functional dependence by explicit use of the spin-densities via the functional (2.138). In this way, one effectively introduces an exact symmetry of the system into the energy functional.

One can go even further and apply the full magnetization density formalism (2.129)–(2.132) for the description of systems without any external magnetic field. As a matter of fact, many systems, ranging from open subshell atoms to large classes of solids, show a local variation of the direction of $\boldsymbol{m}$, i.e. a *non-collinear* magnetization density.

The most prominent example for a system with non-collinear $\boldsymbol{m}$ is γ-Fe, which crystallizes in the fcc structure.[8] When γ-Fe is synthesized as precipitates in a Cu matrix, the local magnetic moments $\boldsymbol{M}_i$ of the iron atoms, i.e. the integrals over $\boldsymbol{m}$ around single sites,

$$\boldsymbol{M}_i = \int_{\Omega_i} d^3r \, \boldsymbol{m}(\boldsymbol{r}) \, ,$$

[8] γ-Fe is the standard form of iron in the temperature range between 910°C and 1390°C.

do not align ferromagnetically, as one would expect from the bcc phase of iron and other 3*d* elements. Rather the local moments form a spiral wave [42], referred to as *spin-spiral* or spiral *spin-density wave* (SDW),

$$\boldsymbol{M}_i = M\big(\cos(\boldsymbol{Q}\cdot\boldsymbol{R}_i+\phi)\sin(\theta), \sin(\boldsymbol{Q}\cdot\boldsymbol{R}_i+\phi)\sin(\theta), \cos(\theta)\big)\,,$$

in which the direction of the $\boldsymbol{M}_i$ precesses around one of the cubic axes, if one proceeds from site to site along this axis ($\boldsymbol{R}_i$ denotes the position of site i). Moreover, the spin-spiral is incommensurate with the crystal lattice, i.e. its wavelength can not be expressed as a rational number times the lattice constant a of the fcc lattice. Experimentally one observes a wave vector of $\boldsymbol{Q} \approx (0.1,0,1)(2\pi/a)$ and $\theta = \pi/2$ [42], corresponding to a planar spiral wave[9] (helical SDW).

Early DFT calculations for γ-Fe (see e.g. [44]) relied on the inclusion of non-collinearity on an intermediate level: assuming $\boldsymbol{m}$ to be collinear around the individual sites, only the spin-quantization axis was allowed to vary from site to site. A spin-density functional approach allowing for this inter-atomic non-collinearity has been formulated quite early [45, 46] (for a review see [47]). More recently, however, it became clear that the variation of $\boldsymbol{m}$ on the intra-atomic scale (intra-atomic non-collinearity) plays an important role[10] for γ-Fe [48, 43, 49, 50]. Non-collinearity has also been observed for the ground states of α-Mn (bulk Mn at room temperature and ambient pressure—see [51] and references therein), a number of Mn and Fe compounds (see e.g. [52]), several Uranium compounds (see [47] and references therein) as well as thin Cr and Fe films (see [53–55] and references therein). Mono-layers of Mn on a tungsten surface develop a cycloidal spin-spiral, resulting from the spin–orbit interaction in a system lacking inversion symmetry [56].

Finite systems exhibit non-collinear magnetism as well: for instance, non-collinear calculations for several low-lying states of free small iron clusters show a variation of the direction and size of $\boldsymbol{m}$ on the constituent atoms [57]. A simple example is the ground state of Fe_5. In this trigonal bipyramid structure the magnetic moments in the basis of the pyramids are aligned, the moments of the two tips, however, are tilted with respect to the majority spin direction defined by the atoms of the basis. When deposited on a Cu surface, on the other hand, iron clusters seem to prefer ferromagnetic ordering [58]. However, even for clusters on a surface geometric frustration can lead to non-collinear magnetic moments in the case of Mn and Cr clusters [58], consistent with the vanishing total magnetic moments observed in experiment [59, 60].

It is obvious that use of the non-collinear formalism (2.129)–(2.132) is mandatory for systems for which the non-collinearity of $\boldsymbol{m}$ is an indispensable feature of the electronic structure, as, for instance, γ-Fe. For most problems, however, the corrections resulting from non-collinearity are rather small, in particular for open-subshell atoms [61] (compare also [62, 63]). For these systems the application of the more complicated calculational scheme resulting from Eqs. (2.129)–(2.132) is usually too high a price to be paid for the more accurate representation of $\boldsymbol{m}$. Spin-density functional theory in the form (2.138)–(2.142) represents the standard approach to magnetic systems for this reason. It seems worthwhile to emphasize that this approach rigorously covers antiferromagnetic ordering of

[9] The value $\theta = \pi/2$ follows rigorously from symmetry constraints as long as spin–orbit coupling is neglected [43].

[10] At the same time even the early calculations revealed a very high sensitivity of the magnetic structure of the ground state to the equilibrium volume.

magnetic moments, including situations as the linearly polarized spin-density wave observed for the ground state[11] of bulk Cr.

2.6 Current Density Functional Theory

The Hamiltonian (2.120) accounts for the dominant coupling mechanism between magnetic fields and electrons, at least for low electron velocities. It neglects, however, the Lorentz force exerted on the electrons by the magnetic field. This effect is included in *current density functional theory* (CDFT) and its extension, *current spin density functional theory* (CSDFT) [68–71]. The starting point for the discussion of CSDFT is the Pauli Hamiltonian,

$$\hat{H} = \frac{1}{2m}\sum_{\sigma=\uparrow,\downarrow}\int d^3r\,\hat{\psi}^\dagger(\boldsymbol{r}\sigma)\left[-i\hbar\nabla+\frac{e}{c}\boldsymbol{A}_{\rm ext}(\boldsymbol{r})\right]^2\hat{\psi}(\boldsymbol{r}\sigma)$$
$$+\int d^3r\left[v_{\rm ext}(\boldsymbol{r})\hat{n}(\boldsymbol{r})+\boldsymbol{B}_{\rm ext}(\boldsymbol{r})\cdot\hat{\boldsymbol{m}}(\boldsymbol{r})\right]+\hat{W}\,, \qquad (2.143)$$

where the magnetization density $\hat{\boldsymbol{m}}$ is defined as in Eq. (2.121) and $\boldsymbol{A}_{\rm ext}$ denotes the vector potential which generates the magnetic field,

$$\boldsymbol{B}_{\rm ext}(\boldsymbol{r}) = \nabla\times\boldsymbol{A}_{\rm ext}(\boldsymbol{r})\,. \qquad (2.144)$$

The Hamiltonian (2.143) can be systematically derived from the fully relativistic Hamiltonian of Dirac theory,[12] either by an expansion in powers of $1/c$ or, alternatively, by a low order Foldy-Wouthuysen transformation.

The formulation of a density functional approach for the Hamiltonian (2.143) has to be preceded by some remarks on the issue of gauge transformations. The magnetic field (2.144) does not change under the gauge transformation[13]

---

[11] The ground state of bulk Cr (which crystallizes in the bcc structure) exhibits a static SDW along the (100) direction of the conventional cubic unit cell (for an overview see [64]): while the directions of the local magnetic moments on nearest neighbor sites (corner and body-center of the bcc structure) are antiparallel (corresponding to an anti-ferromagnetic ordering), the amplitudes μ_i of these local moments are modulated in an almost sinusoidal form, $\mu_i = M_1\sin(\boldsymbol{Q}\cdot\boldsymbol{R}_i)+\ldots$ (with $M_1 = 0.62\,\mu_B$). Moreover, the wavelength of the SDW is incommensurate with the crystal lattice: the dominant wave vector in the SDW is found to be $|\boldsymbol{Q}| = 0.952\,\frac{2\pi}{a}$, the period of the complete spin-density pattern is as long as $20.83\,a$. A longitudinal SDW is observed below a spin-flip temperature of 123 K, a transverse SDW between this and the Neél temperature of 311 K. However, in both cases linear polarization is energetically favored over some helical SDW, so that $\boldsymbol{m}$ remains collinear (for corresponding SDFT calculations see [65–67]).

[12] More precisely speaking, the Hamiltonian (2.143) is obtained from quantum electrodynamics, i.e. from the QED Hamiltonian (8.39). In the present context, however, only an expansion of the relativistic kinetic energy operator is relevant.

[13] Quite generally, the electromagnetic fields (1.10), (1.11) do not change under the combined gauge transformation

$$A'_{\mathrm{ext}}(\boldsymbol{r}) = A_{\mathrm{ext}}(\boldsymbol{r}) - \nabla\lambda(\boldsymbol{r}) . \tag{2.148}$$

On the other hand, the Hamiltonian (2.143), which depends on the vector potential itself, is not invariant under the transformation (2.148) (unlike the reduced Hamiltonian (2.120)). The same statement necessarily also applies to the ground state $|\Psi_0\rangle$ corresponding to (2.143),

$$\hat{H}|\Psi_0\rangle = E_0|\Psi_0\rangle . \tag{2.149}$$

However, it is easy to show that the transformation (2.148) simply leads to a phase transformation of the corresponding ground state: if the wavefunction

$$\Psi_0(\boldsymbol{r}_1\sigma_1,\ldots\boldsymbol{r}_N\sigma_N) \equiv (\boldsymbol{r}_1\sigma_1,\ldots\boldsymbol{r}_N\sigma_N|\Psi_0\rangle$$

is a solution of (2.149) for the potential A_{ext}, the gauge transformed wavefunction

$$\Psi'_0(\boldsymbol{r}_1\sigma_1,\ldots\boldsymbol{r}_N\sigma_N) = \exp\left[\frac{ie}{\hbar c}\sum_{k=1}^{N}\lambda(\boldsymbol{r}_k)\right]\Psi_0(\boldsymbol{r}_1\sigma_1,\ldots\boldsymbol{r}_N\sigma_N) \tag{2.150}$$

is a solution of (2.149) for the potential A'_{ext} obtained by the gauge transformation (2.148)—this can be verified by insertion into (2.149) in first quantized form. All pairs $(A_{\mathrm{ext}},\Psi_0)$ which differ by no more than the combined transformation (2.148), (2.150) are physically equivalent.[14] As a result they lead to the same ground state energy E_0, the same ground state density n_0 and the same ground state magnetization density $\boldsymbol{m}_0$,

$$E'_0 = E_0 \tag{2.152}$$

$$v'_{\mathrm{ext}}(\boldsymbol{r}t) = v_{\mathrm{ext}}(\boldsymbol{r}t) - \frac{e}{c}\frac{\partial}{\partial t}\Lambda(\boldsymbol{r}t) \tag{2.145}$$

$$A'_{\mathrm{ext}}(\boldsymbol{r}t) = A_{\mathrm{ext}}(\boldsymbol{r}t) - \nabla\Lambda(\boldsymbol{r}t) \tag{2.146}$$

of the set of potentials. As the present discussion is restricted to static external fields, the class of admissible gauge transformations is given by

$$\Lambda(\boldsymbol{r}t) = \lambda_0 t + \lambda(\boldsymbol{r}) . \tag{2.147}$$

The first term corresponds to the addition of the constant λ_0 to the external potential v_{ext}. If only this potential is present, $\Lambda(\boldsymbol{r}t) = \lambda_0 t$ is the only legitimate gauge transformation. One can therefore identify the class of v_{ext} which differ by more than a constant with the class of potentials which differ by more than a (static) gauge transformation.

[14] Note that this equivalence also manifests itself as the invariance of the Hamiltonian (2.143) under the simultaneous transformation of the vector potential by (2.148) and of the field operator by

$$\hat{\psi}'(\boldsymbol{r}\sigma) = e^{ie\lambda(\boldsymbol{r})/(\hbar c)}\,\hat{\psi}(\boldsymbol{r}\sigma) . \tag{2.151}$$

In fact, this invariance of $\hat{H}$ is sometimes even identified with the actual gauge invariance of the corresponding expectation value $\langle\Psi_0|\hat{H}|\Psi_0\rangle$. The same statements apply to the complete ground state current (2.156). Similarly, the gauge dependence of the paramagnetic current is reflected by the lack of invariance of $\hat{j}_{\mathrm{p}}$ under the combined transformations (2.148) and (2.151).

$$n_0'(\mathbf{r}) = n_0(\mathbf{r}) \tag{2.153}$$
$$\mathbf{m}_0'(\mathbf{r}) = \mathbf{m}_0(\mathbf{r}) . \tag{2.154}$$

As required, gauge invariance is also observed for the physical ground state current $\mathbf{j}_0$,

$$\mathbf{j}_0'(\mathbf{r}) \;=\; \mathbf{j}_0(\mathbf{r}) \;=\; \langle\Psi_0|\hat{\mathbf{j}}(\mathbf{r})|\Psi_0\rangle\,, \tag{2.155}$$

which, in the case of the Hamiltonian (2.143), is obtained from the operator[15]

$$\hat{\mathbf{j}}(\mathbf{r}) = \hat{\mathbf{j}}_{\mathrm{p}}(\mathbf{r}) + \frac{c}{e}\nabla\times\mathbf{m}(\mathbf{r}) + \frac{e}{mc}\mathbf{A}_{\mathrm{ext}}(\mathbf{r})\hat{n}(\mathbf{r})\,. \tag{2.156}$$

Here $\hat{\mathbf{j}}_{\mathrm{p}}$ denotes the (canonical) paramagnetic current density,[16]

$$\hat{\mathbf{j}}_{\mathrm{p}}(\mathbf{r}) = \frac{-i\hbar}{2m}\sum_{i=1}^{N}\left[\nabla_i\delta^{(3)}(\mathbf{r}-\mathbf{r}_i) + \delta^{(3)}(\mathbf{r}-\mathbf{r}_i)\nabla_i\right] \tag{2.157}$$
$$= -\frac{i\hbar}{2m}\sum_{\sigma=\uparrow,\downarrow}\left[\hat{\psi}^\dagger(\mathbf{r}\sigma)\left(\nabla\hat{\psi}(\mathbf{r}\sigma)\right) - \left(\nabla\hat{\psi}^\dagger(\mathbf{r}\sigma)\right)\hat{\psi}(\mathbf{r}\sigma)\right]. \tag{2.158}$$

If one wants to set up a DFT scheme for the Hamiltonian (2.143), the first issue to be addressed is an appropriate choice of the basic variables. As usual, the coupling between densities and external potentials in the Hamiltonian allows the identification of the basic DFT variables. However, using (2.156), the Hamiltonian (2.143) can be rewritten in two alternative ways,

$$\hat{H} = \hat{T} + \hat{W} + \int d^3r\left\{\frac{e}{c}\mathbf{A}_{\mathrm{ext}}\cdot\hat{\mathbf{j}} + \left[v_{\mathrm{ext}} - \frac{e^2}{2mc^2}\mathbf{A}_{\mathrm{ext}}^2\right]\hat{n}\right\} \tag{2.159}$$
$$= \hat{T} + \hat{W} + \int d^3r\left\{\frac{e}{c}\mathbf{A}_{\mathrm{ext}}\cdot\left[\hat{\mathbf{j}}_{\mathrm{p}} + \frac{c}{e}\nabla\times\hat{\mathbf{m}}\right] + \left[v_{\mathrm{ext}} + \frac{e^2}{2mc^2}\mathbf{A}_{\mathrm{ext}}^2\right]\hat{n}\right\}, \tag{2.160}$$

where $\hat{T}$ is the standard kinetic energy operator (2.2). The form (2.159) suggests to employ the density plus the complete physical current as basic variables of a DFT scheme. However, the discussion of gauge transformations has shown that the set $n_0, \mathbf{j}_0$ does not determine the ground state uniquely. One is thus bound to base CSDFT on the combination of the density operator $\hat{n}$ with the current operator [72, 73]

[15] The operator (2.156) can either be derived by a Gordon decomposition of the fully relativistic current (compare Sect. 8.7), followed by the limit $c\to\infty$. Alternatively, $\hat{\mathbf{j}}$ can be identified by an analysis of the time-dependent Schrödinger equation obtained from the Hamiltonian (2.143) as the current for which a continuity equation holds.

[16] For the transition from first to second quantized form note that

$$\langle\mathbf{r}'\sigma'|\hat{\mathbf{j}}_{\mathrm{p}}(\mathbf{r})|\mathbf{r}''\sigma''\rangle = \frac{-i\hbar}{2m}\delta_{\sigma'\sigma''}\,\delta^{(3)}(\mathbf{r}'-\mathbf{r}'')\left[\nabla''\delta^{(3)}(\mathbf{r}-\mathbf{r}'') + \delta^{(3)}(\mathbf{r}-\mathbf{r}'')\nabla''\right].$$

$$\hat{\boldsymbol{j}}_{\rm g}(\boldsymbol{r}) = \hat{\boldsymbol{j}}_{\rm p}(\boldsymbol{r}) + \frac{c}{e}\boldsymbol{\nabla}\times\hat{\boldsymbol{m}}(\boldsymbol{r}), \tag{2.161}$$

whose ground state expectation value

$$\boldsymbol{j}_{\rm g,0}(\boldsymbol{r}) = \langle\Psi_0|\hat{\boldsymbol{j}}_{\rm g}(\boldsymbol{r})|\Psi_0\rangle \tag{2.162}$$

is as gauge-dependent as the ground state itself: under the gauge transformation (2.150) the ground state current $\boldsymbol{j}_{\rm g,0}$ transforms just as the paramagnetic current,

$$\boldsymbol{j}'_{\rm p,0}(\boldsymbol{r}) = \boldsymbol{j}_{\rm p,0}(\boldsymbol{r}) + \frac{e}{mc}n(\boldsymbol{r})\boldsymbol{\nabla}\lambda(\boldsymbol{r}), \tag{2.163}$$

since the magnetization density is gauge invariant, Eq. (2.154).

In the second step one has to prove an existence theorem for these variables. The core of the proof of the HK-theorem is the strict inequality (2.20). In order to derive an equivalent inequality for the Hamiltonian (2.160), one considers two different non-degenerate ground states $|\Psi_0\rangle$ and $|\Psi'_0\rangle$ with associated sets of potentials $v_{\rm ext},\boldsymbol{A}_{\rm ext}$ and $v'_{\rm ext},\boldsymbol{A}'_{\rm ext}$ (the sets $v_{\rm ext},\boldsymbol{A}_{\rm ext}$ and $v'_{\rm ext},\boldsymbol{A}'_{\rm ext}$ also differ, since $|\Psi_0\rangle$ and $|\Psi'_0\rangle$ are non-degenerate). Now assume that $|\Psi_0\rangle$ and $|\Psi'_0\rangle$ lead to the same ground state density n_0 and current

$$\boldsymbol{j}_{\rm g,0}(\boldsymbol{r}) = \langle\Psi_0|\hat{\boldsymbol{j}}_{\rm g}(\boldsymbol{r})|\Psi_0\rangle = \langle\Psi'_0|\hat{\boldsymbol{j}}_{\rm g}(\boldsymbol{r})|\Psi'_0\rangle\,. \tag{2.164}$$

The expectation value of the Hamiltonian (2.160) with respect to the ground state $|\Psi'_0\rangle$ corresponding to $v'_{\rm ext},\boldsymbol{A}'_{\rm ext}$ can then be expressed as

$$\begin{aligned}\langle\Psi'_0|\hat{H}|\Psi'_0\rangle = \langle\Psi'_0|\hat{H}'|\Psi'_0\rangle + \int d^3r\left[v_{\rm ext} - v'_{\rm ext} + \frac{e^2}{2mc^2}\left(\boldsymbol{A}^2_{\rm ext} - \boldsymbol{A}'^2_{\rm ext}\right)\right]n_0 \\ + \frac{e}{c}\int d^3r\left[\boldsymbol{A}_{\rm ext} - \boldsymbol{A}'_{\rm ext}\right]\cdot\boldsymbol{j}_{\rm g,0}, \end{aligned}\tag{2.165}$$

where $\hat{H}'$ is the Hamiltonian (2.160) with primed potentials. If one could now state that the unprimed ground state energy $\langle\Psi_0|\hat{H}|\Psi_0\rangle$ is strictly lower than $\langle\Psi'_0|\hat{H}|\Psi'_0\rangle$,

$$E_0 = \langle\Psi_0|\hat{H}|\Psi_0\rangle < \langle\Psi'_0|\hat{H}|\Psi'_0\rangle \quad \forall \quad |\Psi'_0\rangle \neq |\Psi_0\rangle, \tag{2.166}$$

the desired inequality of type (2.20) would have been found,

$$\begin{aligned}E_0 < E'_0 + \int d^3r\left[v_{\rm ext} - v'_{\rm ext} + \frac{e^2}{2mc^2}\left(\boldsymbol{A}^2_{\rm ext} - \boldsymbol{A}'^2_{\rm ext}\right)\right]n_0 \\ + \frac{e}{c}\int d^3r\left[\boldsymbol{A}_{\rm ext} - \boldsymbol{A}'_{\rm ext}\right]\cdot\boldsymbol{j}_{\rm g,0}\,. \end{aligned}\tag{2.167}$$

However, $|\Psi_0\rangle$ and $|\Psi'_0\rangle$ can be related by a gauge transformation and the actual ground state energies associated with these states are identical, Eq. (2.152). This raises the question whether the strict inequality (2.166), i.e. the Ritz principle, still holds if $|\Psi'_0\rangle$ differs from $|\Psi_0\rangle$ only by the phase transformation (2.150). Let us convince ourselves that this is indeed the case. If $|\Psi_0'\rangle$ and $|\Psi_0\rangle$ are related by (2.150), the difference between $\langle\Psi_0'|\hat{H}|\Psi_0'\rangle$ and the ground state energy is given by

$$\begin{aligned}
&\langle\Psi_0'|\hat{H}|\Psi_0'\rangle - \langle\Psi_0|\hat{H}|\Psi_0\rangle \\
&= \frac{1}{2m}\sum_{\sigma_1,\cdots\sigma_N}\int d^3r_1\cdots d^3r_N\,\Psi_0^*(\boldsymbol{r}_1\sigma_1,\ldots\boldsymbol{r}_N\sigma_N) \\
&\times\sum_{k=1}^{N}\left[e^{-\frac{ie}{\hbar c}\lambda(\boldsymbol{r}_k)}\left(-i\hbar\boldsymbol{\nabla}_k+\frac{e}{c}\boldsymbol{A}_{\rm ext}(\boldsymbol{r}_k)\right)^2 e^{\frac{ie}{\hbar c}\lambda(\boldsymbol{r}_k)}-\left(-i\hbar\boldsymbol{\nabla}_k+\frac{e}{c}\boldsymbol{A}_{\rm ext}(\boldsymbol{r}_k)\right)^2\right] \\
&\times\Psi_0(\boldsymbol{r}_1\sigma_1,\ldots\boldsymbol{r}_N\sigma_N)\,,
\end{aligned}$$

which is easily evaluated to

$$\begin{aligned}
&\langle\Psi_0'|\hat{H}|\Psi_0'\rangle - \langle\Psi_0|\hat{H}|\Psi_0\rangle \\
&= \frac{1}{2m}\sum_{\sigma_1,\cdots\sigma_N}\int d^3r_1\cdots d^3r_N\,\Psi_0^*(\boldsymbol{r}_1\sigma_1,\ldots\boldsymbol{r}_N\sigma_N) \\
&\times\sum_{k=1}^{N}\left[2\frac{e}{c}[\boldsymbol{\nabla}_k\lambda(\boldsymbol{r}_k)]\cdot\left(-i\hbar\boldsymbol{\nabla}_k+\frac{e}{c}\boldsymbol{A}_{\rm ext}(\boldsymbol{r}_k)\right)-i\hbar\frac{e}{c}\left[\boldsymbol{\nabla}_k^2\lambda(\boldsymbol{r}_k)\right]+\left(\frac{e}{c}\boldsymbol{\nabla}_k\lambda(\boldsymbol{r}_k)\right)^2\right] \\
&\times\Psi_0(\boldsymbol{r}_1\sigma_1,\ldots\boldsymbol{r}_N\sigma_N)\,.
\end{aligned}$$

After partial integration of half of the first term in the square brackets one can rewrite this expression in terms of the paramagnetic current (2.158),

$$\begin{aligned}
&\langle\Psi_0'|\hat{H}|\Psi_0'\rangle - \langle\Psi_0|\hat{H}|\Psi_0\rangle \\
&= \int d^3r\,\langle\Psi_0|\left[\frac{e}{c}[\boldsymbol{\nabla}\lambda(\boldsymbol{r})]\cdot\left(\hat{\boldsymbol{j}}_{\rm p}(\boldsymbol{r})+\frac{e}{mc}\boldsymbol{A}_{\rm ext}(\boldsymbol{r})\hat{n}(\boldsymbol{r})\right)+\left(\frac{e}{c}\boldsymbol{\nabla}\lambda(\boldsymbol{r})\right)^2\hat{n}(\boldsymbol{r})\right]|\Psi_0\rangle\,.
\end{aligned}$$

One can now use Eq. (2.156) to replace $\boldsymbol{j}_{\rm p}$ by the physical current,

$$\langle\Psi_0'|\hat{H}|\Psi_0'\rangle - \langle\Psi_0|\hat{H}|\Psi_0\rangle = \int d^3r\left[\frac{e}{c}(\boldsymbol{\nabla}\lambda)\cdot\left(\boldsymbol{j}_0-\frac{c}{e}\boldsymbol{\nabla}\times\boldsymbol{m}_0\right)+\left(\frac{e}{c}\boldsymbol{\nabla}\lambda\right)^2 n_0\right].$$

Finally, use of current conservation and Gauss' theorem leads to

$$\langle\Psi_0'|\hat{H}|\Psi_0'\rangle - \langle\Psi_0|\hat{H}|\Psi_0\rangle = \int d^3r\left(\frac{e}{c}\boldsymbol{\nabla}\lambda(\boldsymbol{r})\right)^2 n_0(\boldsymbol{r}) > 0\,. \qquad (2.168)$$

This confirms the strict inequality (2.166) and thus Eq. (2.167) (provided that λ is not a simple constant and does not vanish wherever n_0 is non-zero).

The inequality (2.167) can then be used in the standard fashion (combination with the same relation with primed and unprimed quantities interchanged) to derive a contradiction of the type (2.22).

(a) One concludes that there exists a one-to-one correspondence between the set of ground states and the set of densities $(n_0, \boldsymbol{j}_{g,0})$ [68],

$$|\Psi_0\rangle \iff \{n_0(\boldsymbol{r}), \boldsymbol{j}_{g,0}(\boldsymbol{r})\}\,. \tag{2.169}$$

This correspondence establishes the existence of a unique functional $|\Psi[n, \boldsymbol{j}_g]\rangle$, which, by insertion of the actual $(n_0, \boldsymbol{j}_{g,0})$, reproduces the ground state of the system, $|\Psi_0\rangle = |\Psi[n_0, \boldsymbol{j}_{g,0}]\rangle$.

An extended variant of CSDFT is obtained, if one gives up the relation (2.144) between $\boldsymbol{B}_{\text{ext}}$ and $\boldsymbol{A}_{\text{ext}}$ and considers these two fields as independent (which is legitimate from a mathematical point of view—for an even more general form of CSDFT see [68]). The Hamiltonian then contains three independent coupling terms,

$$\hat{H} = \hat{T} + \hat{W} + \frac{e}{c}\int d^3r\, \hat{\boldsymbol{j}}_p(\boldsymbol{r})\cdot \boldsymbol{A}_{\text{ext}}(\boldsymbol{r}) + \int d^3r\, \boldsymbol{B}_{\text{ext}}(\boldsymbol{r})\cdot \hat{\boldsymbol{m}}(\boldsymbol{r}) \\ + \int d^3r \left[v_{\text{ext}}(\boldsymbol{r}) + \frac{e^2}{2mc^2}\boldsymbol{A}^2_{\text{ext}}(\boldsymbol{r}) \right] \hat{n}(\boldsymbol{r})\,. \tag{2.170}$$

In this case one finds (by the standard argument) an extended one-to-one correspondence [68],

$$|\Psi_0\rangle \iff \{n_0(\boldsymbol{r}), \boldsymbol{j}_{p,0}(\boldsymbol{r}), \boldsymbol{m}_0(\boldsymbol{r})\}\,, \tag{2.171}$$

i.e. a ground state functional of the form $|\Psi[n_0, \boldsymbol{j}_{p,0}, \boldsymbol{m}_0]\rangle$. This approach is particularly legitimate if the external magnetic field vanishes anyway, $\boldsymbol{B}_{\text{ext}} = \boldsymbol{A}_{\text{ext}} = \boldsymbol{0}$. The use of $\boldsymbol{j}_{p,0}$ and $\boldsymbol{m}_0$ as independent variables could potentially introduce additional flexibility into the representation of the ground state, compared to the reduced form (2.169). This point is further investigated in the context of the KS equations of CSDFT in Sect. 3.7.

If one neglects the coupling between $\boldsymbol{m}$ and $\boldsymbol{B}_{\text{ext}}$ completely, one arrives at the existence theorem of current density functional theory (CDFT) [74]: the ground state $|\Psi_0\rangle$ of a system governed by the Hamiltonian (2.170) with $\boldsymbol{B}_{\text{ext}} = \boldsymbol{0}$ is uniquely determined by the ground state density and the paramagnetic current density,

$$|\Psi_0\rangle \iff \{n_0(\boldsymbol{r}), \boldsymbol{j}_{p,0}(\boldsymbol{r})\}\,, \tag{2.172}$$

so that it can be written as a functional of these quantities, $|\Psi_0\rangle = |\Psi[n_0, \boldsymbol{j}_{p,0}]\rangle$.

Either of the correspondences (2.169), (2.171) and (2.172) allows a representation of the ground state expectation value of any operator $\hat{O}$ as a functional of the corresponding independent combination of variables.[17]

[17] For a constrained search formulation of CDFT see [75].

(b) Restricting the explicit discussion to CSDFT in the form (2.169), one has

$$O[n,\boldsymbol{j}_{\mathrm{g}}] = \langle\Psi[n,\boldsymbol{j}_{\mathrm{g}}]|\hat{O}|\Psi[n,\boldsymbol{j}_{\mathrm{g}}]\rangle \,. \tag{2.173}$$

In particular, one obtains for the ground state energy

$$E[n,\boldsymbol{j}_{\mathrm{g}}] = F[n,\boldsymbol{j}_{\mathrm{g}}] + \frac{e}{c}\int d^3r\, \boldsymbol{j}_{\mathrm{g}}(\boldsymbol{r})\cdot\boldsymbol{A}_{\mathrm{ext}}(\boldsymbol{r}) + \int d^3r \left[v_{\mathrm{ext}}(\boldsymbol{r}) + \frac{e^2}{2mc^2}\boldsymbol{A}^2_{\mathrm{ext}}(\boldsymbol{r})\right] n(\boldsymbol{r}) \tag{2.174}$$

$$F[n,\boldsymbol{j}_{\mathrm{g}}] = \langle\Psi[n,\boldsymbol{j}_{\mathrm{g}}]|\hat{T}+\hat{W}|\Psi[n,\boldsymbol{j}_{\mathrm{g}}]\rangle \,. \tag{2.175}$$

The basic variational principle of CSDFT then follows, as usual, from Ritz's principle. It states that the functional (2.174) has a minimum for the true current and density distributions $n_0, \boldsymbol{j}_{\mathrm{g},0}$ corresponding to the given external fields,

$$E[n_0,\boldsymbol{j}_{\mathrm{g},0}] < E[n,\boldsymbol{j}_{\mathrm{g}}] \quad \forall\ (n,\boldsymbol{j}_{\mathrm{g}}) \neq (n_0,\boldsymbol{j}_{\mathrm{g},0}) \,. \tag{2.176}$$

The resulting variational equations,

$$\left.\frac{\delta E[n,\boldsymbol{j}_{\mathrm{g}}]}{\delta n(\boldsymbol{r})}\right|_{n_0,\boldsymbol{j}_{\mathrm{g},0}} = 0 \tag{2.177}$$

$$\left.\frac{\delta E[n,\boldsymbol{j}_{\mathrm{g}}]}{\delta \boldsymbol{j}_{\mathrm{g}}(\boldsymbol{r})}\right|_{n_0,\boldsymbol{j}_{\mathrm{g},0}} = \boldsymbol{0} \,, \tag{2.178}$$

have to be solved under the constraint of particle number conservation,

$$\int d^3r\, n(\boldsymbol{r}) = N \,,$$

and a constraint expressing the (static) continuity equation for the physical current,

$$\boldsymbol{\nabla}\cdot\boldsymbol{j}(\boldsymbol{r}) = 0 \,, \tag{2.179}$$

which may be resolved as

$$\boldsymbol{\nabla}\cdot\boldsymbol{j}_{\mathrm{p}}(\boldsymbol{r}) = -\frac{e}{mc}\boldsymbol{\nabla}\cdot[\boldsymbol{A}_{\mathrm{ext}}(\boldsymbol{r})n(\boldsymbol{r})] \,. \tag{2.180}$$

As in the case of SDFT, the existence theorems of C(S)DFT only involve the relation between the current and density variables and the ground state. No statement is made concerning a possible unique correspondence between the external potentials and the ground state. The reason for this restriction is the same as for SDFT: one can give explicit counterexamples which demonstrate that one can find more than one set of potentials which yield a given ground state [76]. In fact, two different types of counterexamples are available, an extension of the symmetry argument of Sect. 2.5 (which may be characterized as *systematic non-uniqueness*) and a second class (referred to as *accidental non-uniqueness*).

Let us first consider the symmetry-related counterexample for the case of CDFT, i.e. ignoring the spin degree of freedom for simplicity. Assume that $|\Psi_0\rangle$ is the ground state corresponding to the CDFT Hamiltonian

$$\hat{H} = \hat{T} + \hat{W} + \int d^3r \left\{ \hat{n}(\boldsymbol{r}) \left[v_{\text{ext}}(\boldsymbol{r}) + \frac{e^2}{2mc^2}\boldsymbol{A}^2_{\text{ext}}(\boldsymbol{r}) \right] + \frac{e}{c}\hat{\boldsymbol{j}}_{\text{p}}(\boldsymbol{r}) \cdot \boldsymbol{A}_{\text{ext}}(\boldsymbol{r}) \right\} \tag{2.181}$$

$$\hat{H}\,|\Psi_0\rangle = E_0\,|\Psi_0\rangle\,. \tag{2.182}$$

The question then is: can $|\Psi_0\rangle$ also be the ground state of a CDFT Hamiltonian in which v_{ext} and $\boldsymbol{A}_{\text{ext}}$ are replaced by two different potentials

$$v'_{\text{ext}} = v_{\text{ext}} + \Delta v_{\text{ext}} \tag{2.183}$$

$$\boldsymbol{A}'_{\text{ext}} = \boldsymbol{A}_{\text{ext}} + \Delta\boldsymbol{A}_{\text{ext}}\,. \tag{2.184}$$

In other words: can one find Δv_{ext} and $\Delta\boldsymbol{A}_{\text{ext}}$, so that

$$\Delta\hat{H}\,|\Psi_0\rangle = \Delta E_0\,|\Psi_0\rangle\,, \tag{2.185}$$

with $\Delta\hat{H}$ given by

$$\Delta\hat{H} = \int d^3r \left\{ \hat{n} \left[\Delta v_{\text{ext}} + \frac{e^2}{2mc^2}\left(\Delta\boldsymbol{A}^2_{\text{ext}} + 2\Delta\boldsymbol{A}_{\text{ext}} \cdot \boldsymbol{A}_{\text{ext}}\right) \right] + \frac{e}{c}\hat{\boldsymbol{j}}_{\text{p}} \cdot \Delta\boldsymbol{A}_{\text{ext}} \right\} ? \tag{2.186}$$

For Eq. (2.185) to hold, the operator $\Delta\hat{H}$ has to be a constant of motion, i.e. has to commute with the Hamiltonian (2.181). An explicit example for such a constant of motion is the angular momentum[18] [76]

$$\hat{L}_z = m \int d^3r\, (\boldsymbol{e}_z \times \boldsymbol{r}) \cdot \hat{\boldsymbol{j}}_{\text{p}}(\boldsymbol{r}) \tag{2.187}$$

in the case of systems which are invariant under rotations about the z-axis. However, choosing

---

[18] In second quantization the angular momentum operator is given by

$$\hat{\boldsymbol{L}} = -i\hbar \sum_{\sigma=\uparrow,\downarrow} \int d^3r\, \left[\hat{\psi}^\dagger(\boldsymbol{r}\sigma)(\boldsymbol{r} \times \boldsymbol{\nabla})\hat{\psi}(\boldsymbol{r}\sigma)\right]\,,$$

which may be expressed in terms of the paramagnetic current after suitable partial integration,

$$\hat{\boldsymbol{L}} = m \int d^3r\, \left[\boldsymbol{r} \times \hat{\boldsymbol{j}}_{\text{p}}(\boldsymbol{r})\right]\,.$$

$$\Delta \boldsymbol{A}_{\text{ext}}(\boldsymbol{r}) = \frac{\Delta B}{2}(\boldsymbol{e}_z \times \boldsymbol{r}) \tag{2.188}$$

$$\Delta v_{\text{ext}}(\boldsymbol{r}) = -\frac{e^2}{2mc^2}\left(\Delta \boldsymbol{A}_{\text{ext}}^2(\boldsymbol{r}) + 2\Delta \boldsymbol{A}_{\text{ext}}(\boldsymbol{r}) \cdot \boldsymbol{A}_{\text{ext}}(\boldsymbol{r})\right), \tag{2.189}$$

with constant ΔB, the operator (2.186) becomes a simple multiple of $\hat{L}_z$, so that Eq. (2.185) is satisfied for all Hamiltonians for which $[\hat{H}, \hat{L}_z] = 0$. Thus, as long as ΔE_0 is smaller than the energy gap between the ground and first excited state, $|\Psi_0\rangle$ remains the ground state corresponding to the primed potentials. The size of ΔE_0, however, is easily controlled by the size of ΔB. Consequently, one finds a complete set of potentials which lead to the same ground state.

As an example for an accidental non-uniqueness consider a single particle within CDFT, again ignoring spin. Assume that the non-degenerate ground state orbital $\phi_0(\boldsymbol{r})$ satisfies the Schrödinger equation

$$\left\{\frac{1}{2m}\left[-i\hbar\nabla + \frac{e}{c}\boldsymbol{A}_{\text{ext}}(\boldsymbol{r})\right]^2 + v_{\text{ext}}(\boldsymbol{r})\right\}\phi_0(\boldsymbol{r}) = \varepsilon_0\phi_0(\boldsymbol{r}) \tag{2.190}$$

for

$$\boldsymbol{A}_{\text{ext}} = \boldsymbol{0}$$

(this example includes all vector potentials which differ from $\boldsymbol{0}$ by a gauge transformation, $\boldsymbol{A}_{\text{ext}} = -\nabla\lambda$). The (nodeless) orbital $\phi_0(\boldsymbol{r})$ can then be chosen real,

$$\phi_0^*(\boldsymbol{r}) = \phi_0(\boldsymbol{r}). \tag{2.191}$$

Now one again asks the question whether one can find some further set of potentials v'_{ext}, $\boldsymbol{A}'_{\text{ext}}$ for which $\phi_0(\boldsymbol{r})$ remains the ground state? For this to be the case, Eq. (2.190) must also be valid with the unprimed potentials replaced by the primed ones. It is straightforward to show, however, that the Schrödinger equation with the potentials

$$\boldsymbol{A}'_{\text{ext}}(\boldsymbol{r}) = \frac{\nabla \times \boldsymbol{C}(\boldsymbol{r})}{\phi_0(\boldsymbol{r})^2} \tag{2.192}$$

$$v'_{\text{ext}}(\boldsymbol{r}) = v_{\text{ext}}(\boldsymbol{r}) - \frac{e^2}{2mc^2}\left(\boldsymbol{A}'_{\text{ext}}(\boldsymbol{r})\right)^2 \tag{2.193}$$

is satisfied by $\phi_0(\boldsymbol{r})$ for arbitrary $\boldsymbol{C}(\boldsymbol{r})$, provided that Eq. (2.190) holds for the unprimed potentials [76]. In fact, insertion of (2.193) into the Schrödinger equation (with primed potentials) and use of (2.190) for the unprimed potentials yields the condition

$$\left[\left(\nabla \cdot \boldsymbol{A}'_{\text{ext}}(\boldsymbol{r})\right) + 2\boldsymbol{A}'_{\text{ext}}(\boldsymbol{r}) \cdot \nabla\right]\phi_0(\boldsymbol{r}) = 0, \tag{2.194}$$

which is trivially satisfied by the form (2.192). For given v_{ext} and thus ϕ_0 one has found an infinite number of potentials v'_{ext}, $\boldsymbol{A}'_{\text{ext}}$, for which ϕ_0 is an eigenstate. Again, if the magnitude of $\boldsymbol{C}(\boldsymbol{r})$ is chosen sufficiently small, ϕ_0 remains the ground state.

In order to provide some physical background of this accidental non-uniqueness let us fix the gauge of the potential (2.192): among all physically admissible vector potentials giving a particular magnetic field $\boldsymbol{B}(\boldsymbol{r}) = \boldsymbol{\nabla} \times \boldsymbol{A}'_{\text{ext}}(\boldsymbol{r})$ one can always choose to work with the one which satisfies Coulomb gauge,

$$\boldsymbol{\nabla} \cdot \boldsymbol{A}'_{\text{ext}}(\boldsymbol{r}) = 0 \,. \tag{2.195}$$

All other potentials which give the same $\boldsymbol{B}(\boldsymbol{r})$ can only differ from this $\boldsymbol{A}'_{\text{ext}}$ by gauge transformations. Restricting the set of potentials (2.192) to those which satisfy Eq. (2.195) requires

$$[\boldsymbol{\nabla}\phi_0(\boldsymbol{r})] \cdot [\boldsymbol{\nabla} \times \boldsymbol{C}(\boldsymbol{r})] = 0 \tag{2.196}$$

for all $\boldsymbol{r}$, so that the potentials of the type (2.192) ultimately also correspond to a specific symmetry of the ground state.

A final remark on C(S)DFT addresses the limit $\boldsymbol{A}_{\text{ext}} = \boldsymbol{0}$. In this limit all variants of C(S)DFT reduce to standard DFT, as all spin- and spin-current-densities are now unique functionals of n. For instance, one has for the approach (2.171),

$$\boldsymbol{j}_{\text{p}}(\boldsymbol{r}) = \boldsymbol{j}_{\text{p}}[n](\boldsymbol{r}) \,; \;\; \boldsymbol{m}(\boldsymbol{r}) = \boldsymbol{m}[n](\boldsymbol{r}) \quad \Longrightarrow \quad E[n, \boldsymbol{j}_{\text{p}}, \boldsymbol{m}] = \tilde{E}[n] \,.$$

In analogy to the improved description of spin-dependent systems with $\boldsymbol{B}_{\text{ext}} = \boldsymbol{0}$ by $E[n_\uparrow, n_\downarrow]$, there might be some advantage in retaining the more flexible form $E[n, \boldsymbol{j}_{\text{p}}, \boldsymbol{m}]$ also for $\boldsymbol{A}_{\text{ext}} = \boldsymbol{0}$. Similar statements apply to all other variants of C(S)DFT.

2.7 Excited States: Part 1

Let us return to the Hamiltonian (2.1) for which the original HK-theorem is valid. The ground state density n_0 does not only determine the ground state $|\Psi_0\rangle$ uniquely, but also all excited states $|\Psi_i\rangle$ $(i > 0)$. The reason for this, at first glance surprising, statement is the one-to-one correspondence of n_0 and the external potential v_{ext}. Once v_{ext} is known, all excited states are unambiguously determined by solution of the Schrödinger equation,

$$\underset{\substack{\uparrow \\ \text{HK theorem}}}{n_0 \Longrightarrow} v_{\text{ext}} \underset{\substack{\uparrow \\ \text{solution of the Schrödinger equation}}}{\Longrightarrow} |\Psi_i\rangle; \quad i = 0, 1, \ldots \infty \,. \tag{2.197}$$

All matrix elements of excited states $|\Psi_i\rangle$ can, in view of this correspondence, be regarded as functionals of the ground state(!) density. This conclusion applies in particular to all excitation energies.

The practical exploitation of (2.197) is complicated by two aspects: it is more difficult to derive acceptable approximations for the energy functional of excited states and there exists no suitable variational principle for these states. An exception to the latter statement is found if the complete Fock space can be decomposed into subspaces with different symmetry, i.e. if the Hamiltonian commutes with some symmetry operator. In this case there exists a minimum principle for each of the subspaces with a given symmetry (an argument, which has already been used in the case of the spin). It is possible to formulate a HK theorem for the energetically lowest state of each subspace, in analogy to the original version [77]: there is a one-to-one correspondence between the energetically lowest state $|\Psi_i\rangle$ of the subspace and the associated density $n_i = \langle\Psi_i|\hat{n}|\Psi_i\rangle$, defining the functional $|\Psi_i[n_i]\rangle$ (the spaces $\mathcal{G}$ and $\mathcal{N}$ of the original HK proof have to be restricted to the symmetry under consideration). The functional dependence of $|\Psi_i\rangle$ on n_i, however, in general differs from that of the ground state functional $|\Psi[n]\rangle$. Lacking any alternative, ground state density functionals are nonetheless often used for the discussion of excited states as, for example, atomic states with different angular momenta. However, this approach runs into difficulties with the resolution of the individual states of multiplets, as long as the density functional does not reflect the specific symmetry under consideration (for a scheme to deal with multiplet states see [78, 79]).

Chapter 3
Effective Single-Particle Equations

3.1 Kohn-Sham Equations

Let us return to the simplest situation considered so far, i.e. to a Hamiltonian of type (2.1) with a non-degenerate ground state. For this system the Hohenberg-Kohn theorem states that knowledge of the ground state density is sufficient to determine all ground state observables. In addition, the ground state energy functional $E[n]$ allows the determination of the ground state density itself via the variational equation (2.38) Unfortunately, the HK-theorem does not give any hint concerning the explicit form of $E[n]$ (or $F[n]$). Moreover, the initial motivation for the discussion of DFT was to introduce an exact mapping of the interacting N-particle problem onto a suitable effective noninteracting system. The connection of the existence theorems established so far with this mapping is provided by the Kohn-Sham (KS) scheme.

In order to introduce the KS equations in a systematic fashion we first consider a system of noninteracting electrons with a multiplicative external potential v_s,

$$\hat{H}_s = \hat{T} + \hat{V}_s \,; \qquad \hat{V}_s = \int d^3r\, \hat{n}(\boldsymbol{r}) v_s(\boldsymbol{r}) \,. \tag{3.1}$$

The corresponding N-particle ground state $|\Phi_0\rangle$, assumed to be non-degenerate at this point, is a Slater determinant,

$$\hat{H}_s|\Phi_0\rangle = E_{s,0}|\Phi_0\rangle \tag{3.2}$$

$$\begin{aligned}(\boldsymbol{r}_1\sigma_1,\ldots \boldsymbol{r}_N\sigma_N|\Phi_0\rangle &\equiv \Phi_0(\boldsymbol{r}_1\sigma_1,\ldots \boldsymbol{r}_N\sigma_N) \\ &= \frac{1}{\sqrt{N!}} \det \begin{pmatrix} \phi_1(\boldsymbol{r}_1\sigma_1) & \cdots & \phi_N(\boldsymbol{r}_1\sigma_1) \\ \vdots & & \vdots \\ \phi_1(\boldsymbol{r}_N\sigma_N) & \cdots & \phi_N(\boldsymbol{r}_N\sigma_N) \end{pmatrix} ,\end{aligned} \tag{3.3}$$

constructed from the energetically lowest solutions ϕ_i of the single-particle Schrödinger equation,

E. Engel, R.M. Dreizler, Effective Single-Particle Equations. In: E. Engel, R.M. Dreizler, Density Functional Theory, Theoretical and Mathematical Physics, pp. 57–108 (2011)
DOI 10.1007/978-3-642-14090-7_3

$$\left\{-\frac{\hbar^2\nabla^2}{2m}+v_{\rm s}(\boldsymbol{r})\right\}\phi_i(\boldsymbol{r}\sigma)=\varepsilon_i\phi_i(\boldsymbol{r}\sigma). \tag{3.4}$$

The eigenvalues ε_i are assumed to be ordered as

$$\varepsilon_1 \le \varepsilon_2 \le \ldots \le \varepsilon_N = \varepsilon_{\rm F} < \varepsilon_{N+1} \le \ldots \quad , \tag{3.5}$$

where, as usual, the Fermi energy $\varepsilon_{\rm F}$ is identified with the eigenvalue ε_N of the highest occupied single-particle level.

In (3.4) the spin degree of freedom has been expressed in terms of the variable σ, rather than by understanding ϕ_i as a two-component spinor. Correspondingly, the quantum number i represents both spatial and spin quantum numbers. The single-particle states ϕ_i can be characterized further by taking into account the fact that the Hamiltonian (3.1) commutes with the spin-projection on the z-axis,

$$\hat{S}_z=\frac{1}{2}\sum_{i=1}^{N}\sigma_{z,i}=\frac{1}{2}\sum_{\sigma,\sigma'=\uparrow,\downarrow}\int d^3r\,\hat{\psi}^\dagger(\boldsymbol{r}\sigma)\,\sigma_{z,\sigma\sigma'}\,\hat{\psi}(\boldsymbol{r}\sigma'), \tag{3.6}$$

where σ_z is the z-component of the vector of Pauli matrices, Eq. (2.124) (for the basic commutators required see Appendix L). The spin-projection can be rewritten in terms of the particle numbers (2.140) of the two spin orientations,

$$\hat{S}_z=\frac{1}{2}\left\{\hat{N}_\uparrow-\hat{N}_\downarrow\right\}, \tag{3.7}$$

so that the commutation relation

$$\left[\hat{H}_{\rm s},\hat{S}_z\right]=0 \tag{3.8}$$

follows directly from $[\hat{H}_{\rm s},\hat{N}_\sigma]=0$. The orbitals ϕ_i can be chosen as eigenstates of σ_z, i.e. they can be factorized into a spatial wavefunction and the Pauli spinor $\chi_\pm$. As the potential $v_{\rm s}$ does not distinguish between the two spin orientations, the spatial wavefunction is independent of spin,

$$\phi_i(\boldsymbol{r}\sigma)=\phi_\alpha(\boldsymbol{r})\,\chi_s(\sigma) \quad\Longleftrightarrow\quad i\equiv(\alpha,s) \tag{3.9}$$

$$\chi_\uparrow\equiv\chi_+=\begin{pmatrix}1\\0\end{pmatrix}; \quad \chi_\downarrow\equiv\chi_-=\begin{pmatrix}0\\1\end{pmatrix} \quad\Longleftrightarrow\quad \chi_s(\sigma)=\delta_{s\sigma} \tag{3.10}$$

$$\sigma_z\,\chi_\pm=\pm\chi_\pm \quad\Longleftrightarrow\quad \sum_{\sigma'}\sigma_{z,\sigma\sigma'}\,\chi_s(\sigma')={\rm sign}(s)\,\delta_{s\sigma}\,. \tag{3.11}$$

Any single-particle level is at least twofold degenerate due to spin. As long as $|\Phi_0\rangle$ is assumed to be non-degenerate one has $\varepsilon_N<\varepsilon_{N+1}$. For a non-degenerate ground state one thus necessarily finds $N_\uparrow=N_\downarrow$, so that N must be even. This implies that the ground state must have a vanishing magnetic moment, as $\langle\Phi_0|\hat{S}_z|\Phi_0\rangle=0$. The issues of degeneracy at the Fermi level and of spin-polarized systems will be discussed in Sects. 3.3 and 3.5.

The ground state density corresponding to (3.3) is given by

$$n_{0s}(\boldsymbol{r}) = \sum_{\sigma=\uparrow,\downarrow} \sum_{i=1}^{N} |\phi_i(\boldsymbol{r}\sigma)|^2 = 2\sum_{\alpha=1}^{N/2} |\phi_\alpha(\boldsymbol{r})|^2 . \tag{3.12}$$

In the following the sum over the N energetically lowest single-particle states will nevertheless be written in the more general form

$$n_{0s}(\boldsymbol{r}) = \sum_{\sigma=\uparrow,\downarrow} \sum_{i} \Theta_i |\phi_i(\boldsymbol{r}\sigma)|^2 . \tag{3.13}$$

In the present context the occupation function Θ_i denotes the standard step function,

$$\Theta_i \equiv \Theta(\varepsilon_F - \varepsilon_i) = \begin{cases} 1 & \text{for } \varepsilon_F \geq \varepsilon_i \\ 0 & \text{elsewhere} \end{cases} . \tag{3.14}$$

The formulation (3.13) has the advantage that it can equally well be used for a system at finite temperature $T > 0$. In this case the step function simply has to be replaced by a Fermi distribution,

$$\Theta_i = \left[1 + \exp\left(\frac{\varepsilon_i - \mu}{k_B T}\right)\right]^{-1} , \tag{3.15}$$

with μ being the chemical potential, which is chosen so that $\sum_i \Theta_i = N$. The ground state energy corresponding to (3.3) is

$$E_{s,0} = \sum_i \Theta_i \sum_{\sigma=\uparrow,\downarrow} \int d^3r\, \phi_i^*(\boldsymbol{r}\sigma) \frac{(-i\hbar\boldsymbol{\nabla})^2}{2m} \phi_i(\boldsymbol{r}\sigma) + \int d^3r\, v_s(\boldsymbol{r}) n_{0s}(\boldsymbol{r}) . \tag{3.16}$$

At this point one recollects the fact that the HK theorem is valid for arbitrary many-particle systems, irrespective of their particle–particle interaction w. It applies in particular to noninteracting systems, for which $w = 0$. The non-degenerate ground state (3.3) is thus uniquely determined by the ground state density (3.13). In other words: The ground state of a noninteracting system is a unique functional of the ground state density, $|\Phi[n]\rangle$. The particular ground state (3.3) is obtained if the associated ground state density n_{0s} is inserted into this functional,

$$|\Phi[n]\rangle : \quad \text{with} \quad |\Phi_0\rangle = |\Phi[n_{0s}]\rangle . \tag{3.17}$$

The functional $|\Phi[n]\rangle$ is not identical with the functional (2.24) obtained for nonvanishing interaction w. Using $|\Phi[n]\rangle$, one can define the HK ground state energy functional of noninteracting systems,

$$E_s[n] = \langle\Phi[n]|\hat{T} + \hat{V}_s|\Phi[n]\rangle = \langle\Phi[n]|\hat{T}|\Phi[n]\rangle + \int d^3r\, v_s(\boldsymbol{r}) n(\boldsymbol{r}) . \tag{3.18}$$

It has the same properties as the functional (2.26). In particular, the minimum principle (2.28) applies,

$$E_s[n_{0s}] < E_s[n] \quad \forall\ n \neq n_{0s}\,; \qquad E_{s,0} = E_s[n_{0s}]\,. \tag{3.19}$$

Equation (3.18) defines the kinetic energy functional $T_s[n]$ of noninteracting particles for any ground state density n resulting from a Hamiltonian of type (3.1), i.e. for any n which is *noninteracting v-representable*,

$$T_s[n] := \langle \Phi[n]|\hat{T}|\Phi[n]\rangle\,. \tag{3.20}$$

As $|\Phi[n]\rangle$ is universal (i.e. independent of v_s) the same is true for $T_s[n]$. The kinetic energy has, however, already been expressed in terms of the single-particle orbitals ϕ_i in Eq. (3.16),

$$T_s[n] = \sum_i \Theta_i \sum_{\sigma=\uparrow,\downarrow} \int d^3r\, \phi_i^*(\boldsymbol{r}\sigma)\frac{(-i\hbar\boldsymbol{\nabla})^2}{2m}\phi_i(\boldsymbol{r}\sigma)\,, \tag{3.21}$$

which implies that the ϕ_i themselves are functionals of the density.

How can the orbitals be density functionals? Clearly, a change of n can only be induced by a corresponding change of the external potential v_s. Any modification of v_s leads to a unique change of all orbitals via Eq. (3.4). This statement is immediately obvious if one reconsiders the HK argument for the case of a single-particle problem. If two potentials v_s and $v_s' \neq v_s + const$ would have a common eigenstate ϕ_i, the combination of the corresponding single-particle equations (3.4) gives

$$\left[v_s(\boldsymbol{r}) - v_s'(\boldsymbol{r})\right]\phi_i(\boldsymbol{r}\sigma) = \left[\varepsilon_i - \varepsilon_i'\right]\phi_i(\boldsymbol{r}\sigma)\,,$$

so that one ends up with a contradiction. Not only the many-particle ground state Φ_0, but also its ingredients $\phi_i(\boldsymbol{r}\sigma)$ are unique functionals of n,

$$\phi_i(\boldsymbol{r}\sigma) = \phi_i[n](\boldsymbol{r}\sigma)\,. \tag{3.22}$$

Equation (3.21) identifies the first non-trivial density functional, although in an implicit form.

Let us return to the discussion of the interacting system (2.1). Assume that for any admissible potential $v_{\rm ext}$ the ground state density $n_0(\boldsymbol{r})$ of the interacting system is simultaneously the ground state density of some yet to be determined noninteracting system with a single-particle potential v_s (different from $v_{\rm ext}$). This auxiliary noninteracting system with a Hamiltonian of the form (3.1) is called the *KS system*. The question whether such a KS system actually exists for arbitrary n_0, the question of the *noninteracting v-representability*, will be addressed in Sect. 3.2. If the assumption stated is correct, one can represent the ground state density of the interacting system in terms of the single-particle orbitals ϕ_i of the KS system,

$$n_0(\boldsymbol{r}) \equiv n_{0s}(\boldsymbol{r}) = \sum_{\sigma=\uparrow,\downarrow}\sum_i \Theta_i|\phi_i(\boldsymbol{r}\sigma)|^2\,. \tag{3.23}$$

It is important to realize the difference between (3.13) and (3.23): Equation (3.13) is a straightforward result for the density of a noninteracting system, while (3.23) is a highly non-trivial representation of the interacting density!

Of course, the structure of v_s must reflect the nature of the interacting system. In order to determine v_s one decomposes the total energy functional $E[n]$ in a suitable fashion. One first realizes that the kinetic energy functional $T_s[n]$, Eq. (3.20), is well-defined for the ground state densities of all interacting systems for which a v_s exists. $T_s[n]$ can be explicitly expressed in terms of the auxiliary orbitals ϕ_i, which satisfy Eq. (3.4), as in Eq. (3.21). One can therefore rewrite $E[n]$ as

$$E[n] = T_s[n] + E_H[n] + E_{ext}[n] + E_{xc}[n] . \tag{3.24}$$

The functional $E_H[n]$ is the classical (Hartree) interaction energy between the N particles with density n,

$$E_H[n] = \frac{1}{2}\int d^3r \int d^3r'\, n(\boldsymbol{r})w(\boldsymbol{r},\boldsymbol{r}')n(\boldsymbol{r}') , \tag{3.25}$$

including their self-interaction energy. $E_H[n]$ is usually referred to as *Hartree term.* $E_{ext}[n]$ characterizes the coupling between the particles and the external potential,

$$E_{ext}[n] = \int d^3r\, v_{ext}(\boldsymbol{r})n(\boldsymbol{r}) . \tag{3.26}$$

Finally, the *exchange-correlation* (xc) *energy functional* $E_{xc}[n]$ is *defined* by (3.24). It absorbs all the complicated many-body effects not contained in T_s, E_H and E_{ext}. $E_{xc}[n]$ is a density functional as, on the one hand, $E[n]$ is a density functional by virtue of the HK theorem for interacting particles and, on the other hand, T_s is a density functional by virtue of the HK theorem for noninteracting particles, while E_H and E_{ext} are explicit density functionals. However, one should realize that, from a mathematical point of view, the decomposition (3.24) is only legitimate for densities which are simultaneously interacting and noninteracting v-representable.

Before proceeding with the derivation of v_s, some comments seem appropriate:

- The decomposition (3.24) isolates exactly those contributions to the total energy functional, which can be treated rigorously: E_H and E_{ext} are simple functionals of the density whose evaluation is straightforward. The exact handling of T_s will be discussed below. At the same time, these components of $E[n]$ usually dominate over the exchange-correlation energy. Even the complete neglect of E_{xc} often leads to an electronic structure which is qualitatively correct. One could thus hope that an approximate account of E_{xc} allows a sufficiently accurate description of most many-particle systems.
- As $F[n]$, $T_s[n]$, $E_H[n]$ and $E_{ext}[n]$ are universal, the same is true for $E_{xc}[n]$. Therefore the same xc-functional applies to all systems in which the particle–particle interaction is given by the Coulomb force. An approximation for E_{xc}, which has been obtained for some model system with the Coulomb interaction can, from

a mathematical point of view, legitimately be utilized for all kinds of Coulomb systems.

- The DFT xc-energy E_{xc} is not identical with the conventional xc-energy E_{xc}^{QC}, which is usually employed in standard many-body theory and quantum chemistry. A more detailed discussion of this point requires the separation of E_{xc}^{QC} into an exchange and a correlation component and is therefore postponed until Sect. 4.1. At this point, we only emphasize that E_{xc} is not identical with the difference between the complete interaction energy of the interacting system and its classical counterpart,

$$E_{xc}^{|\Psi_0\rangle} = \langle\Psi_0|\hat{W}|\Psi_0\rangle - E_H[n_0]. \tag{3.27}$$

In fact, Eq. (3.24) may be rewritten as

$$E_{xc} = \langle\Psi_0|\hat{T}+\hat{W}|\Psi_0\rangle - T_s[n_0] - E_H[n_0], \tag{3.28}$$

so that the difference between E_{xc} and $E_{xc}^{|\Psi_0\rangle}$ originates from the difference between the kinetic energy of an interacting system with density n_0 and that of a noninteracting system with the same density,

$$E_{xc} - E_{xc}^{|\Psi_0\rangle} = \langle\Psi_0|\hat{T}|\Psi_0\rangle - T_s[n_0] = \langle\Psi_0|\hat{T}|\Psi_0\rangle - \langle\Phi_0|\hat{T}|\Phi_0\rangle. \tag{3.29}$$

As the interacting ground state $|\Psi_0\rangle$ differs from the Slater determinant $|\Phi_0\rangle$ which gives the same density, one has $\langle\Psi_0|\hat{T}|\Psi_0\rangle \neq \langle\Phi_0|\hat{T}|\Phi_0\rangle$.
In fact, the interacting ground state $|\Psi_0\rangle$ minimizes the total energy of the interacting system, $\langle\Psi_0|\hat{T}+\hat{W}+\hat{V}_{ext}|\Psi_0\rangle$, while the KS ground state $|\Phi_0\rangle = |\Phi[n_0]\rangle$ corresponding to the same n_0 minimizes the expectation value of the kinetic energy operator,

$$\langle\Phi_0|\hat{T}|\Phi_0\rangle = \inf_{\Psi\to n_0} \langle\Psi|\hat{T}|\Psi\rangle$$

(since for fixed n_0 the external potential term $\langle\Phi[n_0]|\hat{V}_{ext}|\Phi[n_0]\rangle = \int d^3r\, n_0 v_s$ is also fixed, the minimization of the total energy reduces to a minimization of $\langle\Phi[n_0]|\hat{T}|\Phi[n_0]\rangle$). As a results one finds

$$T_s[n_0] = \langle\Phi_0|\hat{T}|\Phi_0\rangle \leq \langle\Psi_0|\hat{T}|\Psi_0\rangle, \tag{3.30}$$

and thus

$$E_{xc} \geq E_{xc}^{|\Psi_0\rangle}. \tag{3.31}$$

This result is illustrated by a look at the difference between E_{xc} and $E_{xc}^{|\Psi_0\rangle}$ in terms of numbers. These quantities have been evaluated rigorously only for rather simple systems such as closed-subshell atoms. For instance, for helium one obtains $T = 2.903724$ Hartree [80, 81] (nonrelativistically) and $T_s = 2.867082$ Hartree [82, 83], so that $T - T_s = E_{xc} - E_{xc}^{|\Psi_0\rangle} = 0.036642$ Hartree. Similarly, for beryllium the difference amounts to $E_{xc} - E_{xc}^{|\Psi_0\rangle} = 0.0737$ Hartree [84, 82]. The difference $E_{\text{xc}} - E_{\text{xc}}^{|\Psi_0\rangle}$ is thus found to be of the same order of magnitude as the complete correlation energy (see below). Note, however, that $E_{\text{xc}}^{|\Psi_0\rangle}$ is not identical with the conventional xc-energy $E_{\text{xc}}^{\text{QC}}$.

The real task after these remarks is the specification of v_s and thus of the orbitals ϕ_i which reproduce the interacting density n_0. For this purpose one considers the ground state energy functional $E[n]$ of an interacting system for a density n close to the true ground state density n_0,

$$n(\boldsymbol{r}) = n_0(\boldsymbol{r}) + \delta n(\boldsymbol{r}). \tag{3.32}$$

Using a functional Taylor expansion of $E[n]$ about $n = n_0$ together with the variational equation (2.38),

$$\begin{aligned} E[n] &= E[n_0] + \int d^3r \left.\frac{\delta E[n]}{\delta n(\boldsymbol{r})}\right|_{n=n_0} [n(\boldsymbol{r}) - n_0(\boldsymbol{r})] + \mathcal{O}(\delta n^2) \\ &= E[n_0] + \int d^3r\, \mu\, \delta n(\boldsymbol{r}) + \mathcal{O}(\delta n^2), \end{aligned} \tag{3.33}$$

and restricting the variation of $E[n]$ to fixed particle number,

$$\int d^3r\, n(\boldsymbol{r}) = \int d^3r\, n_0(\boldsymbol{r}) \qquad \Longleftrightarrow \qquad \int d^3r\, \delta n(\boldsymbol{r}) = 0, \tag{3.34}$$

one finds that the deviation of $E[n_0 + \delta n]$ from the true ground state energy $E[n_0]$ is of second order in the density deviation δn,

$$E[n_0 + \delta n] - E[n_0] = \mathcal{O}(\delta n^2). \tag{3.35}$$

This variational property is the reason why the total energy functional is rather insensitive to errors in the density which is inserted, a fact that is often utilized in applications.

The expansion (3.33) requires the existence of the functional derivative of $E[n]$ at n_0. As discussed earlier, however, many densities are not v-representable, so that $\delta E[n]/\delta n|_{n=n_0}$ can not be expected to exist in general. The issue of functional differentiability of $E[n]$ (and its components) is nevertheless set aside for a moment—we will return to this point in Sect. 3.2.

Now the individual components of $E[n_0 + \delta n] - E[n_0]$ have to be examined. In the case of T_s one resorts to the implicit density functional representation (3.21) via the ϕ_i. Since the ϕ_i are unique functionals of the density, any variation $\delta n(\boldsymbol{r})$ corresponds to a unique variation $\delta\phi_i$ of ϕ_i, i.e. any density $n_0 + \delta n$ corresponds to a unique set of orbitals $\phi_i + \delta\phi_i$—provided that $n_0 + \delta n$ is noninteracting v-representable. At this point one thus needs the assumption that all relevant interacting densities are also noninteracting v-representable (for a discussion of this point see below). To first order in the deviation $\delta\phi_i$ and thus $\delta n(\boldsymbol{r})$ one obtains

$$
\begin{aligned}
&T_s[n_0+\delta n]-T_s[n_0]\\
&=\frac{(-i\hbar)^2}{2m}\sum_{i,\sigma}\Theta_i\int d^3r\left\{\delta\phi_i^*(\boldsymbol{r}\sigma)\nabla^2\phi_i(\boldsymbol{r}\sigma)+\phi_i^*(\boldsymbol{r}\sigma)\nabla^2\delta\phi_i(\boldsymbol{r}\sigma)\right\}+O(\delta\phi^2)\\
&=\frac{(-i\hbar)^2}{2m}\sum_{i,\sigma}\Theta_i\int d^3r\left\{\delta\phi_i^*(\boldsymbol{r}\sigma)\nabla^2\phi_i(\boldsymbol{r}\sigma)+\delta\phi_i(\boldsymbol{r}\sigma)\nabla^2\phi_i^*(\boldsymbol{r}\sigma)\right\}+O(\delta\phi^2),
\end{aligned}
$$

where partial integration has been utilized for the last line, assuming all surface contributions to vanish (or to cancel). One can now use the single-particle equations (3.4) in order to eliminate the derivatives

$$
\begin{aligned}
&T_s[n_0+\delta n]-T_s[n_0]\\
&=\sum_{i,\sigma}\Theta_i\int d^3r\left[\varepsilon_i-v_s(\boldsymbol{r})\right]\left\{\delta\phi_i^*(\boldsymbol{r}\sigma)\phi_i(\boldsymbol{r}\sigma)+\phi_i^*(\boldsymbol{r}\sigma)\delta\phi_i(\boldsymbol{r}\sigma)\right\}+O(\delta\phi^2).
\end{aligned}
$$

Applying Eq. (3.23) to $n_0+\delta n$,

$$
\delta n(\boldsymbol{r})=\sum_i\Theta_i\sum_{\sigma=\uparrow,\downarrow}\left\{\delta\phi_i^*(\boldsymbol{r}\sigma)\phi_i(\boldsymbol{r}\sigma)+\phi_i^*(\boldsymbol{r}\sigma)\delta\phi_i(\boldsymbol{r}\sigma)\right\}+O(\delta\phi^2),
$$

and utilizing proper normalization,

$$
\begin{aligned}
0&=\int d^3r\,|\phi_i(\boldsymbol{r}\sigma)+\delta\phi_i(\boldsymbol{r}\sigma)|^2-\int d^3r\,|\phi_i(\boldsymbol{r}\sigma)|^2\\
&=\int d^3r\left\{\delta\phi_i^*(\boldsymbol{r}\sigma)\phi_i(\boldsymbol{r}\sigma)+\phi_i^*(\boldsymbol{r}\sigma)\delta\phi_i(\boldsymbol{r}\sigma)\right\}+O(\delta\phi^2),
\end{aligned}
$$

one can express $T_s[n_0+\delta n]-T_s[n_0]$ in terms of δn,

$$
\begin{aligned}
T_s[n_0+\delta n]-T_s[n_0]&=\int d^3r\left[\varepsilon_i-v_s(\boldsymbol{r})\right]\delta n(\boldsymbol{r})+O(\delta\phi^2)\\
&=-\int d^3r\,v_s(\boldsymbol{r})\delta n(\boldsymbol{r})+O(\delta n^2).
\end{aligned}
\tag{3.36}
$$

As one might have expected, the individual components of $E[n]$ do not have the variational property (3.35), but rather contain terms of first order in δn.

The next component to be considered is the Hartree term (3.25),

$$
E_H[n_0+\delta n]-E_H[n_0]=\int d^3r\int d^3r'\,\delta n(\boldsymbol{r})w(\boldsymbol{r},\boldsymbol{r}')n_0(\boldsymbol{r}')+O(\delta n^2).\tag{3.37}
$$

The variation of the external potential energy is even simpler,

$$
E_{ext}[n_0+\delta n]-E_{ext}[n_0]=\int d^3r\,v_{ext}(\boldsymbol{r})\delta n(\boldsymbol{r}).\tag{3.38}
$$

On the other hand, for the unknown xc-functional one can only write down the formal functional Taylor expansion,

$$E_{\rm xc}[n_0+\delta n]-E_{\rm xc}[n_0]=\int d^3r\,\frac{\delta E_{\rm xc}[n]}{\delta n(\boldsymbol{r})}\bigg|_{n=n_0}\delta n(\boldsymbol{r})+O(\delta n^2)\,. \tag{3.39}$$

Collecting the results (3.35)–(3.39), one arrives at

$$0=\int d^3r\left\{v_{\rm s}(\boldsymbol{r})-\int d^3r'\,w(\boldsymbol{r},\boldsymbol{r}')n_0(\boldsymbol{r}')-v_{\rm ext}(\boldsymbol{r})-\frac{\delta E_{\rm xc}[n]}{\delta n(\boldsymbol{r})}\bigg|_{n=n_0}\right\}\delta n(\boldsymbol{r})\,. \tag{3.40}$$

Equation (3.40) is valid for arbitrary variations $\delta n(\boldsymbol{r})$ which satisfy (3.34), so that one ends up with the identity

$$v_{\rm s}(\boldsymbol{r})=v_{\rm ext}(\boldsymbol{r})+v_{\rm H}[n_0](\boldsymbol{r})+v_{\rm xc}[n_0](\boldsymbol{r})\,, \tag{3.41}$$

where $v_{\rm H}$ is the Hartree (direct Coulomb) potential,

$$v_{\rm H}[n](\boldsymbol{r})=\int d^3r'\,w(\boldsymbol{r},\boldsymbol{r}')n(\boldsymbol{r}')\,, \tag{3.42}$$

and the xc-potential $v_{\rm xc}$ is defined as

$$v_{\rm xc}[n](\boldsymbol{r})=\frac{\delta E_{\rm xc}[n]}{\delta n(\boldsymbol{r})}\,. \tag{3.43}$$

It is apparent that the unknown potential $v_{\rm s}$ is a density functional itself, so that the single-particle equations (3.4) are nonlinear: The solutions ϕ_i determine the density via Eq. (3.23), which then determines the effective potential $v_{\rm s}$ via Eq. (3.41). The celebrated *Kohn-Sham equations* [85] (see also [86]) resulting from insertion of (3.41) into (3.4),

$$\left\{-\frac{\hbar^2\boldsymbol{\nabla}^2}{2m}+v_{\rm ext}(\boldsymbol{r})+v_{\rm H}[n](\boldsymbol{r})+v_{\rm xc}[n](\boldsymbol{r})\right\}\phi_i(\boldsymbol{r}\sigma)\;=\;\varepsilon_i\phi_i(\boldsymbol{r}\sigma)\,, \tag{3.44}$$

have to be solved in a self-consistent fashion, similar to the HF equations.

One starts with some trial density $n^{(1)}$, which allows the construction of the starting potential

$$v_{\rm s}^{(1)}(\boldsymbol{r})=v_{\rm ext}(\boldsymbol{r})+v_{\rm H}[n^{(1)}](\boldsymbol{r})+v_{\rm xc}[n^{(1)}](\boldsymbol{r})$$

(assuming the functional derivative of $E_{\rm xc}[n]$ to be given). Only a rough estimate of the actual density is required. For instance, (suitably screened) hydrogenic orbitals could be used to generate $n^{(1)}$ in calculations for atoms. For molecules or solids the trial density is often set up as a superposition of atomic densities. Alternatively, one can start with a direct guess for the total potential $v_{\rm s}^{(1)}(\boldsymbol{r})$. Solution of (3.44) with $v_{\rm s}^{(1)}(\boldsymbol{r})$ then provides some orbitals $\phi_i^{(2)}$ which lead to an improved density $n^{(2)}$ via Eq. (3.23). The density $n^{(2)}$ can in turn be used to obtain improved potentials $v_{\rm H}[n^{(2)}](\boldsymbol{r})$ and $v_{\rm xc}[n^{(2)}](\boldsymbol{r})$. This iterative procedure is repeated until the difference between the densities obtained in two successive iterations falls below some predefined accuracy criterion. In practice, the control of this self-consistency procedure is non-trivial.

A number of comments are appropriate.

1. The KS equations are equivalent to a minimization of the total ground state energy, which is the reason why the self-consistent iteration in the spirit of the HF equations converges. In fact, the KS equations can be derived in a somewhat more direct way by minimization of the total energy functional (3.24) with respect to N single-particle orbitals which are assumed to form the density via Eq. (3.23), while proper normalization is ensured by a subsidiary condition for the orbitals. This approach relies explicitly on the one-to-one correspondence between the ground state density of a noninteracting system and the corresponding set of single-particle orbitals (as well as on the noninteracting v-representability of the interacting density).
2. The Slater determinant constructed from the KS orbitals, Eq. (3.3), does not represent the exact ground state Ψ_0 of the interacting system. Rather, the KS orbitals ϕ_i should be understood as purely mathematical constructs which only serve as a basis for the representation of the density. The mapping between the original interacting system and the effective noninteracting system, which is established by the KS approach, only ensures that the ground state density and energy are reproduced, but not the interacting ground state itself. The price for the reduction of the complexity of the many-body problem is the limited information that can be extracted via the KS scheme (at least from a formal point of view). However, for many structural questions in condensed matter physics and quantum chemistry this restricted information on the energetics and the electronic charge distribution is completely sufficient to give an answer.
   Obviously, Φ_0 approaches the true Ψ_0 for vanishing particle–particle interaction. This implies that the agreement between Φ_0 and Ψ_0 depends on the relative importance of the external and Hartree potentials compared to xc-effects. For that reason Φ_0 is often used as an approximation to the true ground state in the case of systems whose electronic structure is dominated by single-particle aspects.
   One should nevertheless keep in mind that the difference between Φ_0 and Ψ_0 becomes immediately apparent even for a weak interaction w, if one looks at the 2-particle density

$$n^{(2)}(\boldsymbol{r},\boldsymbol{r}') = \frac{N(N-1)}{2}\sum_{\sigma_1\ldots\sigma_N}\int d^3r_3\ldots d^3r_N\,|\langle\boldsymbol{r}\sigma_1,\boldsymbol{r}'\sigma_2,\boldsymbol{r}_3\sigma_3,\ldots\boldsymbol{r}_N\sigma_N|\Psi_0\rangle|^2. \tag{3.45}$$

$n^{(2)}(\boldsymbol{r},\boldsymbol{r}')$ is the probability to find one particle with arbitrary spin at position $\boldsymbol{r}$ and simultaneously a second particle with arbitrary spin at position $\boldsymbol{r}'$. For the Slater determinant Φ_0 one obtains

$$2n_s^{(2)}(\boldsymbol{r},\boldsymbol{r}') = n(\boldsymbol{r})n(\boldsymbol{r}') - \sum_{\sigma,\sigma'}\left|\sum_{i=1}^{N}\phi_i(\boldsymbol{r}\sigma)\phi_i^*(\boldsymbol{r}'\sigma')\right|^2. \tag{3.46}$$

The first term in Eq. (3.46) is the direct product of the probability of finding one particle at point $\boldsymbol{r}$ with the probability to find a second particle at $\boldsymbol{r}'$. This product of single-particle probabilities would be the exact result if the motion of the particles was completely uncorrelated. However, in contrast to classical particles spin-1/2 fermions are correlated by the Pauli principle, even if no particle–particle interaction is present. This so-called *Pauli correlation* is expressed by the second term in Eq. (3.46). The role of this term is quite easily seen if one considers the probability to find two noninteracting particles at the same point $\boldsymbol{r} = \boldsymbol{r}'$,

$$
\begin{aligned}
2n_s^{(2)}(\boldsymbol{r},\boldsymbol{r}) &= n(\boldsymbol{r})^2 - \sum_{\sigma} \left| \sum_{i=1}^{N} |\phi_i(\boldsymbol{r}\sigma)|^2 \right|^2 \\
&= \sum_{\sigma,\sigma'=\uparrow,\downarrow} n_\sigma(\boldsymbol{r}) n_{\sigma'}(\boldsymbol{r}) - \sum_{\sigma=\uparrow,\downarrow} n_\sigma(\boldsymbol{r})^2 \\
&= 2n_\uparrow(\boldsymbol{r}) n_\downarrow(\boldsymbol{r}) . \qquad (3.47)
\end{aligned}
$$

In (3.47) the total density has been decomposed into the contributions of the two spin orientations as in (2.136). In addition, one has used the fact that a given orbital either contains a spin-up or a spin-down electron and that $\phi_i^*(\boldsymbol{r}\uparrow)\phi_i(\boldsymbol{r}\downarrow) = 0$ (spin-projection is assumed to be a good quantum number). As expected, the probability to find two particles at the same point is given by the probability to find one particle with spin up and one particle with spin down.

As soon as a particle–particle interaction is present, the motion of two particles with different spins is no longer independent. This is reflected by the so-called *electron–electron cusp* in $n^{(2)}(\boldsymbol{r},\boldsymbol{r}')$. This cusp is illustrated in Figs. 3.1 and 3.2 for helium. In the case of helium the exact ground state corresponding to the Hamiltonian (2.1) is a product of a symmetric spatial wavefunction ψ and an antisymmetric spin-wavefunction χ_{00} in which the two spins are coupled to a total spin of zero,

$$
(\boldsymbol{r}_1\sigma_1\boldsymbol{r}_2\sigma_2|\Psi_0\rangle = \psi(\boldsymbol{r}_1,\boldsymbol{r}_2)\,\chi_{00}(\sigma_1,\sigma_2) \qquad (3.48)
$$

$$
\chi_{00}(\sigma_1,\sigma_2) = \frac{1}{\sqrt{2}}\left\{\chi_+(\sigma_1)\chi_-(\sigma_2) - \chi_-(\sigma_1)\chi_+(\sigma_2)\right\} . \qquad (3.49)
$$

The spatial function ψ can be characterized further by use of the inherent symmetries of the helium ground state. Given the triangle defined by the positions of the nucleus and the two electrons, any rotation of this triangle about the corner at which the nucleus is located can not lead to a change in the wavefunction. As a consequence, there are only three independent coordinates on which ψ depends. One usually chooses the distances r_1 and r_2 between the electrons and the nucleus (located at the origin) as well as the distance $|\boldsymbol{r}_1 - \boldsymbol{r}_2|$ between the two electrons,

$$
\psi(\boldsymbol{r}_1,\boldsymbol{r}_2) = \psi(r_1,r_2,|\boldsymbol{r}_1 - \boldsymbol{r}_2|) .
$$

Extremely accurate forms for $\psi(r_1,r_2,|\boldsymbol{r}_1 - \boldsymbol{r}_2|)$ have been obtained by use of a variational approach [80].

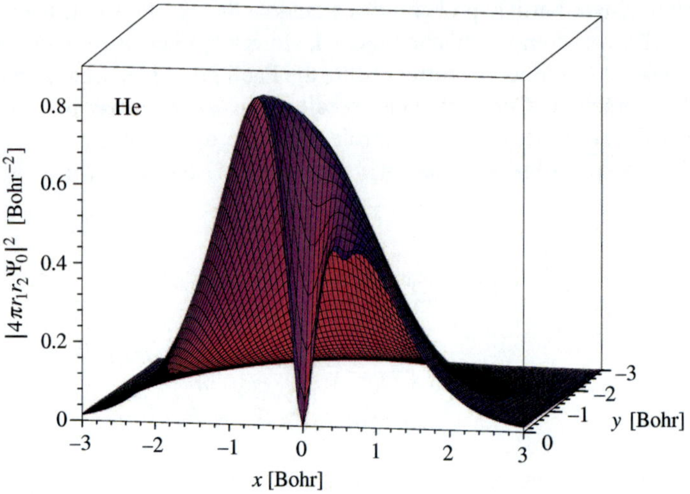

Fig. 3.1 Exact 2-particle density of helium as function of $\boldsymbol{r}_2 = (x,y,0)$: The nucleus is located at the origin, while the second electron is kept fixed at $\boldsymbol{r}_1 = (0.559\,\text{Bohr}, 0, 0)$.

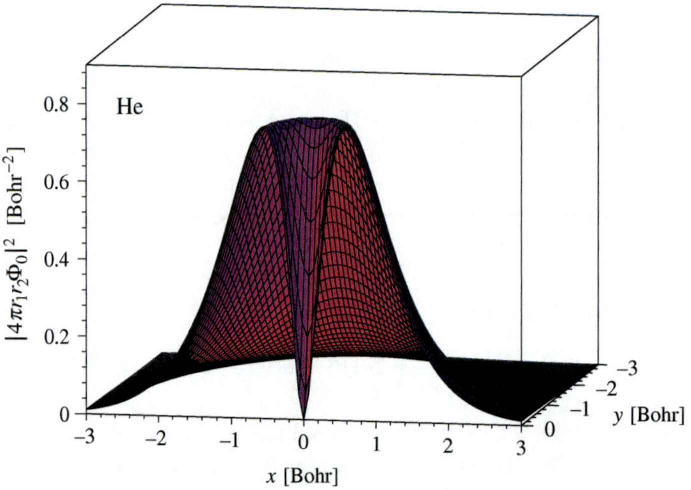

Fig. 3.2 Kohn-Sham 2-particle density of helium as function of $\boldsymbol{r}_2 = (x,y,0)$: The nucleus is located at the origin, while the second electron is kept fixed at $\boldsymbol{r}_1 = (0.559\,\text{Bohr}, 0, 0)$.

For the ground state (3.48) the 2-particle density is simply given by

$$n^{(2)}(r_1,r_2,|\boldsymbol{r}_1-\boldsymbol{r}_2|) = |\psi(r_1,r_2,|\boldsymbol{r}_1-\boldsymbol{r}_2|)|^2 . \qquad (3.50)$$

This function is shown in Fig. 3.1, normalized so that integration over $\boldsymbol{r}_2$ yields the standard radial density,

$$4\pi r_1^2 n(r_1)$$
$$= (4\pi r_1^2)\, 4\pi \int_0^\infty r_2^2 dr_2 \int_{-1}^{+1} d(\cos\Theta)\, n^{(2)}(r_1,r_2,[r_1^2+r_2^2-2r_1r_2\cos\Theta]^{1/2}),$$

with Θ denoting the angle between $\boldsymbol{r}_1$ and $\boldsymbol{r}_2$. In Fig. 3.1 one electron is fixed at a distance of $r_1 = 0.559$ Bohr from the nucleus. The position of the second electron is then varied in a plane containing the nucleus and the first electron. For $r_2 = 0.559$ Bohr and $\Theta = 0$ the two electrons sit on top of each other.
One can see that the two electrons preferably move on opposite sides of the nucleus: the likelihood to come close to each other is only half as large as that of remaining on opposite sides. The 2-particle density clearly shows the electron–electron cusp at $\boldsymbol{r}_1 = \boldsymbol{r}_2$, reflecting the short-range Coulomb repulsion. Figure 3.2 provides the corresponding result obtained with the KS determinant Φ_0. As this determinant only contains Pauli, but no Coulomb correlations, the electrons move independently,

$$n_s^{(2)}(r_1,r_2,[r_1^2+r_2^2-2r_1r_2\cos\Theta]^{1/2}) = |\phi_1(r_1)|^2|\phi_1(r_2)|^2 ,$$

which is consistent with (3.47). The probability for the electrons to sit on top of each other is as high as that for sitting at the same radial distance on opposite sides of the nucleus. One finds that the KS determinant can not reproduce the interacting 2-particle density even qualitatively, although the KS orbitals reproduce the interacting ground state density exactly. The KS equations yield only two quantities correctly, the ground state density and, by insertion into $E[n]$, the ground state energy.

3. On this basis a brief comparison of the KS approach with the HF method seems worthwhile. The core of the HF scheme is the approximation of the many-body ground state wavefunction in terms of a Slater determinant of effective single-particle orbitals. The equations for the determination of these orbitals are obtained from minimization of the total energy. They are nonlinear (they have to be solved by iteration) and nonlocal (due to the form of the HF exchange potential). The HF energy and the ground state wavefunction do not contain any correlation effects. On the other hand, all ground state expectation values for observables can be evaluated directly in this approximation, as the many-body wavefunction is available.
The core of the KS approach is the representation of the ground state density and thus, via the HK energy functional, of the ground state energy in terms of effective single-particle orbitals. The equations which determine the orbitals are again obtained from minimization of the total energy. They are nonlinear, but local (as

the KS xc-potential has this property). The quality of the results depends on the choice for the energy functional. However, both the ground state density and, in particular, the energy include correlation effects even in the simplest available approximations for the energy functional. On the other hand, no approximation for the ground state wavefunction emerges from the KS scheme, at least in principle. Other ground state observables can therefore only be calculated if corresponding density functionals are available.
A more detailed comparison of the HF and KS approaches can be found in subsequent chapters of this book. A formal analysis of the underlying variational procedures is given in Sects. 4.1 and 6.2.3. The consequences of the differences observed are investigated quantitatively in Sect. 6.3.

4. For the ground state energy an alternative form to (3.24) is obtained by rewriting T_s with (3.4),

$$T_s = \sum_i \Theta_i \sum_{\sigma=\uparrow,\downarrow} \int d^3r\, \phi_i^*(\boldsymbol{r}\sigma)\{\varepsilon_i - v_s(\boldsymbol{r})\}\phi_i(\boldsymbol{r}\sigma) \tag{3.51}$$

$$E_0 = \sum_i \Theta_i \varepsilon_i - \int d^3r\, v_{xc}(\boldsymbol{r}) n_0(\boldsymbol{r}) - E_H[n_0] + E_{xc}[n_0]\,. \tag{3.52}$$

In comparison with the original form (3.24) the expression (3.52) has the advantage that one does not have to evaluate gradients of the KS orbitals. For this reason it is almost exclusively used in all numerical implementations. The relation is valid for arbitrary xc-functionals, as long as self-consistency has been reached.

3.2 Noninteracting *v*-Representability

The question whether one can always find a noninteracting system with potential v_s, for which the density n_0 of the interacting system is the ground state density, has already been raised. In order to answer this question of *noninteracting v-representability*, one begins with a reformulation. Consider the energy functional of noninteracting particles,

$$E_s[n] = T_s[n] + \int d^3r\, v_s(\boldsymbol{r}) n(\boldsymbol{r})\,. \tag{3.53}$$

Use of the variational equation (2.38) for $E_s[n]$ yields

$$v_s(\boldsymbol{r}) = -\left.\frac{\delta T_s[n]}{\delta n(\boldsymbol{r})}\right|_{n(\boldsymbol{r})=n_0(\boldsymbol{r})} + \mu\,. \tag{3.54}$$

One thus finds that the existence of v_s for some given n_0 is intimately related to the existence of the functional derivative $\delta T_s[n]/\delta n$ for this density. The existence of $\delta T_s[n]/\delta n$ for the functional (3.21), which originates from a single Slater determinant ground state has not been demonstrated. The same statement applies to the Levy-Lieb extension of $T_s[n]$, $T_{s,LL}[n] = \inf_{\Psi\to n}\langle\Psi|\hat{T}|\Psi\rangle$.

On the other hand, the existence of the functional derivative has been established [30, 31, 87] for the noninteracting limit of the Lieb functional (2.106),

$$T_{s,L}[n] := \inf_{\hat{D}\to n} \operatorname{tr}\{\hat{D}\hat{T}\}\,, \tag{3.55}$$

with

$$\hat{D} = \sum_k d_k |\Psi_k\rangle\langle\Psi_k|,\ d_k^* = d_k \geq 0,\ \sum_k d_k = 1,\ \langle\Psi_k|\Psi_l\rangle = \delta_{kl},\ \Psi_k \in \mathscr{H}^1 \tag{3.56}$$

so that

$$n(\boldsymbol{r}) = \sum_k d_k \langle\Psi_k|\hat{n}(\boldsymbol{r})|\Psi_k\rangle\,.$$

Since the proofs given for $F_L[n]$ do not depend on the presence of the electron–electron interaction, all statements on $F_L[n]$ carry over to $T_{s,L}[n]$, with the appropriate redefinitions of the sets involved (see Sect. 2.3). In particular, the functional derivative of $T_{s,L}[n]$ exists for all densities which are noninteracting ensemble v-representable, i.e. all densities of the set

$$\mathscr{B}_0 := \left\{ n = \sum_{i=1}^q c_i\langle\Psi_{0,i}|\hat{n}|\Psi_{0,i}\rangle \,\middle|\, \text{with: } c_i = c_i^* \geq 0;\ \sum_{i=1}^q c_i = 1; \right.$$
$$\left. |\Psi_{0,i}\rangle = \text{degenerate ground states of } \hat{T} + \textstyle\int v\hat{n},\ v \in \mathscr{L}^{3/2} + \mathscr{L}^\infty \right\}.$$

For each density $n_0 \in \mathscr{B}_0$ there exists a unique potential v_s from $\mathscr{L}^{3/2} + \mathscr{L}^\infty$, so that

$$\left.\frac{\delta T_{s,L}[n]}{\delta n(\boldsymbol{r})}\right|_{n(\boldsymbol{r})=n_0(\boldsymbol{r})} = -v_s(\boldsymbol{r}) \qquad \text{with} \quad v_s \in \mathscr{L}^{3/2} + \mathscr{L}^\infty \tag{3.57}$$

(of course, v_s is only unique up to an irrelevant additive constant).

Unfortunately, it is not clear whether $\mathscr{B}_0$ is identical with the set of interacting ensemble v-representable densities $\mathscr{B}$, Eq. (2.95). While there are some hints indicating that these sets might coincide [27], their identity remains questionable in view of the fact that the set of noninteracting pure-state v-representable densities is not identical [23] with the set of interacting pure-state v-representable densities $\mathscr{A}$, Eq. (2.94). However, as for the interacting situation, the set of densities in $\mathscr{B}_0$ is dense in the larger set $\mathscr{S}$, Eq. (2.85). For any density $n \in \mathscr{S}$ with $n \notin \mathscr{B}_0$ one can thus find a noninteracting ensemble v-representable density which is arbitrarily close to n. So, for all practical purposes the functional derivative of $T_{s,L}[n]$ exists for all interacting densities of interest.

Even though it is clear from Eq. (2.100), it seems worthwhile to convince oneself explicitly that $T_{s,L}[n]$ is a consistent extension of the HK functional $T_s[n]$. Consider a noninteracting v-representable density n_0 resulting from a non-degenerate KS ground state $|\Phi_0\rangle$ (the set of all such densities is the domain of $T_s[n]$). The question then is: what is the density matrix $\hat{D}_s$ for which $\inf_{\hat{D}\to n_0} \mathrm{tr}\{\hat{D}\hat{T}\}$ is obtained?

The answer to this question can be given in a rather general form. Let us for the moment consider an arbitrary fixed density n. We want to find the density matrix $\hat{D}_s$ with the properties

$$\begin{aligned}&(a)\ \ n = \mathrm{tr}\{\hat{D}_s\hat{n}\}\\ &(b)\ \ \mathrm{tr}\{\hat{D}\hat{T}\}\ \text{minimal}.\end{aligned}$$

However, for fixed n the minimum of $\mathrm{tr}\{\hat{D}\hat{T}\}$ is obtained for exactly the same density matrix as the minimum of

$$\mathrm{tr}\{\hat{D}\hat{H}_s\} = \mathrm{tr}\{\hat{D}\hat{T}\} + \int d^3r\, v_s(\boldsymbol{r})\,\mathrm{tr}\{\hat{D}\hat{n}(\boldsymbol{r})\} = \mathrm{tr}\{\hat{D}\hat{T}\} + \int d^3r\, v_s(\boldsymbol{r})\, n(\boldsymbol{r}), \quad (3.58)$$

since for fixed n the contribution of the second term on the right-hand side is also fixed. Thus, quite generally, the density matrix $\hat{D}_s$ which minimizes the total KS energy $\mathrm{tr}\{\hat{D}\hat{H}_s\}$ simultaneously minimizes its kinetic energy component.

Let us now return to our special case, in which the non-degenerate $|\Phi_0\rangle$ minimizes the energy of the KS system. This means that $\hat{D}_s = |\Phi_0\rangle\langle\Phi_0|$ is the density matrix which minimizes $\mathrm{tr}\{\hat{D}\hat{T}\}$ for given

$$n_0(\boldsymbol{r}) = \mathrm{tr}\left\{\hat{D}_s\hat{n}\right\} = \langle\Phi_0|\hat{n}|\Phi_0\rangle = \sum_i \Theta_i \sum_{\sigma=\uparrow,\downarrow} |\phi_i(\boldsymbol{r}\sigma)|^2 . \quad (3.59)$$

With this $\hat{D}_s$ one obtains

$$T_{s,L}[n_0] = \mathrm{tr}\left\{\hat{D}_s\hat{T}\right\} = \sum_i \Theta_i \sum_{\sigma=\uparrow,\downarrow} \int d^3r\, \phi_i^*(\boldsymbol{r}\sigma)\frac{(-i\hbar\boldsymbol{\nabla})^2}{2m}\phi_i(\boldsymbol{r}\sigma), \quad (3.60)$$

for the Lieb kinetic energy, in agreement with the original KS kinetic energy functional for non-degenerate KS systems, Eq. (3.21).

The kinetic energy functional $T_{s,L}[n]$ induces a corresponding definition for the xc-functional,

$$E_{xc,L}[n] := F_L[n] - T_{s,L}[n] - E_H[n] . \quad (3.61)$$

Strictly speaking, functional differentiability of $E_{xc,L}[n]$ is restricted to the set of densities, for which both $F_L[n]$ and $T_{s,L}[n]$ are differentiable, i.e. the densities which are simultaneously interacting and noninteracting ensemble v-representable, $n \in (\mathscr{B}_0 \cap \mathscr{B})$. However, as in the case of $F_L[n]$ and $T_{s,L}[n]$ one can rely on the fact that for any $n \in \mathscr{S}$ there is some $n_1 \in \mathscr{B}$ and some $n_2 \in \mathscr{B}_0$ which are both infinitesimally close to n, so that the existence of the functional derivative is ensured for all practical purposes.

3.3 Degenerate Kohn-Sham Ground States

The discussion in Sect. 3.1 is based on the assumption that the ground state of the auxiliary KS system is non-degenerate. The question remains: How can one handle the situation, if there are more degenerate KS states at the Fermi level than there are electrons to be distributed among them? The answer to this question is provided by an analysis of $T_{s,L}[n]$, as only this form of the kinetic energy functional guarantees the existence of the functional derivative $\delta T_s[n]/\delta n$ for arbitrary n.

Let us assume that the ground state density n_0 of the interacting system, when inserted into $\delta T_{s,L}[n]/\delta n$, leads to a v_s (via (3.57)) which yields q degenerate N-particle KS Slater determinants $|\Phi_{0,k}\rangle$,

$$\hat{H}_s|\Phi_{0,k}\rangle = E_s|\Phi_{0,k}\rangle \; ; \qquad k = 1,\ldots q. \tag{3.62}$$

The single-particle spectrum of such a degenerate KS system is shown in Fig. 3.3. All KS states with eigenvalues below the highest occupied KS level (with energy ε_F)

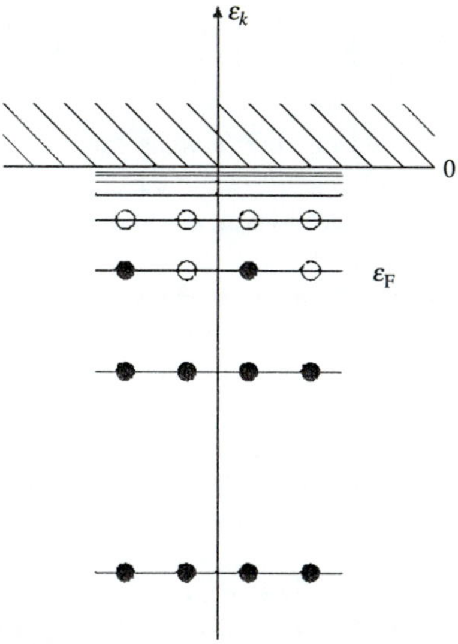

Fig. 3.3 Sketch of the single-particle spectrum of a degenerate KS system. *Filled circles* indicate occupied KS states, open circles refer to unoccupied states.

are occupied in all the determinants $|\Phi_{0,k}\rangle$. However, the occupation of the highest KS level differs for each of the states $|\Phi_{0,k}\rangle$. The presence of ϕ_i in $|\Phi_{0,k}\rangle$ can be characterized by an occupation factor Θ_{ik},

$$\Theta_{ik} = \begin{cases} 1 & \text{if } \phi_i \text{ present in } |\Phi_{0,k}\rangle \\ 0 & \text{otherwise} \end{cases} . \tag{3.63}$$

In total there are q possibilities to distribute the most weakly bound KS particles among the degenerate orbitals ϕ_i at the Fermi level (in the example of Fig. 3.3 one has $q = 6$, ignoring spin).

As long as one does not fix the density, the total energy functional of noninteracting particles $\mathrm{tr}\{\hat{D}\hat{H}_\mathrm{s}\}$, Eq. (3.58), is then minimized by an arbitrary linear superposition of all degenerate states $|\Phi_{0,k}\rangle$, i.e. the minimizing density matrix has the form

$$\hat{D}_\mathrm{s} = \sum_{k=1}^{q} d_k \, |\Phi_{0,k}\rangle\langle\Phi_{0,k}| \, ; \qquad d_k^* = d_k \geq 0 \, ; \qquad \sum_{k=1}^{q} d_k = 1 \, . \tag{3.64}$$

$\hat{D}_\mathrm{s}$ is constructed from the degenerate ground state determinants only—any admixture of a higher lying state gives an energy $\mathrm{tr}\{\hat{D}\hat{H}_\mathrm{s}\}$ above E_s. The density corresponding to the ensemble (3.64) reads

$$\begin{aligned} n_\mathrm{s}(\boldsymbol{r}) &= \sum_{k=1}^{q} d_k \, \langle\Phi_{0,k}|\hat{n}(\boldsymbol{r})|\Phi_{0,k}\rangle \\ &= \sum_{\sigma=\uparrow,\downarrow} \left\{ \sum_{\varepsilon_i<\varepsilon_\mathrm{F}} |\phi_i(\boldsymbol{r}\sigma)|^2 + \sum_{\varepsilon_i=\varepsilon_\mathrm{F}} \left(\sum_{k=1}^{q} d_k \, \Theta_{ik} \right) |\phi_i(\boldsymbol{r}\sigma)|^2 \right\} . \end{aligned} \tag{3.65}$$

For the derivation of the second line the fact has been used that all single-particle levels with an eigenvalue below ε_F are occupied in all degenerate Slater determinants $|\Phi_{0,k}\rangle$, while those at the Fermi level, i.e. with $\varepsilon_i = \varepsilon_\mathrm{F}$, are distributed among the states $|\Phi_{0,k}\rangle$ according to (3.63).

Any set of d_k with proper normalization leads to the same ground state energy E_s, so that the d_k are not determined by the requirement of energy minimization for the KS system. Even if some of the corresponding densities (3.65) are identical, one generally finds a number of different ground state densities for the degenerate KS system. Only one of them agrees with the actual density n_0 of the interacting system. It is the identity of the ensemble density (3.65) with n_0 which fixes the weights d_k.

Equation (3.65) suggests a redefinition of the KS occupation factor (3.14) as

$$\Theta_i = \begin{cases} 1 & \text{for } \varepsilon_i < \varepsilon_\mathrm{F} \\ \sum_{k=1}^{q} d_k \, \Theta_{ik} & \text{for } \varepsilon_i = \varepsilon_\mathrm{F} \\ 0 & \text{for } \varepsilon_i > \varepsilon_\mathrm{F} \end{cases} ; \qquad \sum_i \Theta_i = N \, . \tag{3.66}$$

With this redefinition the ensemble density (3.65) and the associated kinetic energy (3.55) have the same form as the corresponding non-degenerate quantities (3.13) and (3.21). This is immediately obvious for the density (3.65),

$$n_\mathrm{s}(\boldsymbol{r}) = \mathrm{tr}\left\{\hat{D}_\mathrm{s}\hat{n}\right\} = \sum_i \Theta_i \sum_{\sigma=\uparrow,\downarrow} |\phi_i(\boldsymbol{r}\sigma)|^2 \, . \tag{3.67}$$

In the case of $T_\mathrm{s,L}$ one again has to use the fact that, among all density matrices of the general form (3.56) which yield the density (3.67), the expression $\mathrm{tr}\{\hat{D}\hat{T}\}$ is minimized by exactly the same $\hat{D}_s$ as the total KS energy (3.58). Consequently, one obtains for any admissible ground state density n_s,

$$T_{s,L}[n_s] = \inf_{\hat{D}\to n_s} \operatorname{tr}\{\hat{D}\hat{T}\} = \operatorname{tr}\{\hat{D}_s\hat{T}\}$$
$$= \sum_i \Theta_i \sum_{\sigma=\uparrow,\downarrow} \int d^3r\, \phi_i^*(\boldsymbol{r}\sigma) \frac{(-i\hbar\boldsymbol{\nabla})^2}{2m} \phi_i(\boldsymbol{r}\sigma). \tag{3.68}$$

On the basis of (3.67) and (3.68) as well as the standard decomposition (3.24) of the total energy (with T_s and E_{xc} replaced by (3.55) and (3.61)) one can now go through the explicit derivation of the KS potential as in Sect. 3.1. One ends up with the KS equations for degenerate KS systems,

$$\left\{ -\frac{\hbar^2\boldsymbol{\nabla}^2}{2m} + v_{\rm ext}(\boldsymbol{r}) + v_{\rm H}[n](\boldsymbol{r}) + v_{\rm xc}[n](\boldsymbol{r}) \right\} \phi_i(\boldsymbol{r}\sigma) = \varepsilon_i \phi_i(\boldsymbol{r}\sigma), \tag{3.69}$$

where the eigenvalues at the Fermi level are now allowed to be degenerate,

$$\varepsilon_i \le \varepsilon_{i+1} \qquad \text{(for arbitrary } i\text{)}. \tag{3.70}$$

The density is given by (3.67) with occupation factors of the form

$$\left.\begin{array}{rl} \Theta_i = 1 & \text{for} \quad \varepsilon_i < \varepsilon_{\rm F} \\ 0 \le \Theta_i \le 1 & \text{for} \quad \varepsilon_i = \varepsilon_{\rm F} \\ \Theta_i = 0 & \text{for} \quad \varepsilon_i > \varepsilon_{\rm F} \end{array}\right\} \qquad \sum_i \Theta_i = N, \tag{3.71}$$

$v_{\rm H}$ is defined as in Eq. (3.42) and $v_{\rm xc}$ by

$$v_{\rm xc}[n](\boldsymbol{r}) = \frac{\delta E_{\rm xc,L}[n]}{\delta n(\boldsymbol{r})}. \tag{3.72}$$

The alternative form of the total energy, Eq. (3.52), also remains valid with Θ_i now given by (3.71).

It is worthwhile pointing out the difference between Eqs. (3.66) and (3.71): in order to set up a KS scheme for the case of degenerate KS ground states the mere existence of weights d_k which reproduce any interacting ground state density via (3.65) is sufficient. On the other hand, in a self-consistent scheme the occupation factors at the Fermi level can no longer be simply evaluated from a given set of d_k and $|\Phi_{0,k}\rangle$ as these quantities are the outcome of the calculation. One needs a constructive scheme for the determination of the d_k, i.e. for choosing the Θ_i for the KS levels with $\varepsilon_i = \varepsilon_{\rm F}$ during the self-consistent iteration. As the correct ground state density of the KS system also minimizes the energy of the interacting system, the Θ_i have to be chosen in accordance with this requirement. If the interacting system is degenerate, several degenerate KS densities and several sets of minimizing Θ_i exist.

On the other hand, fractional values of Θ_i at the Fermi level are only legitimate if the resulting self-consistent KS potential does not break the degeneracy, which has led to the non-integer Θ_i. For instance, in the boron atom with electron–electron interaction only the occupation

$$\Theta_{2p^0} = \Theta_{2p^\pm} = \frac{1}{3}$$

leads to a spherically symmetric n and thus v_s, which is the basis for the degeneracy of the KS $2p^0$ and $2p^\pm$ levels. Any other choice for Θ_i gives a non-spherical potential which lifts the degeneracy of the three p-states. A fractional occupation always has to be chosen so that the degeneracy remains unbroken, i.e. in an internally consistent fashion. This requirement usually determines the Θ_i completely (together with the correct particle number). Nevertheless, often several internally consistent sets of Θ_i must be compared (usually both fractional and integer sets). The ground state is then identified with the set which yields the lowest total energy.

In practice, the Fermi distribution (3.15) is the most elegant way to handle degenerate or nearly degenerate systems. As long as the temperature T is chosen sufficiently small, one essentially obtains the same result as for zero temperature.

3.4 Janak's Theorem, Fractional Particle Numbers

In the previous sections the occupation numbers Θ_i of the individual KS states emerged from the total energy minimization of DFT. Let us now step back for a moment and consider the situation in which the occupation of the single-particle states generating the density is specified in an arbitrary fashion [88],

$$n(\boldsymbol{r}) := \sum_i \Theta_i \sum_{\sigma=\uparrow,\downarrow} |\phi_i(\boldsymbol{r}\sigma)|^2 \qquad \text{with} \quad 0 \le \Theta_i \le 1 \quad \forall\, i. \tag{3.73}$$

At this point, the occupation numbers Θ_i may be fractional and may not sum up to an integer particle number N. There may be gaps in the occupation of the underlying single-particle spectrum, i.e. there may be states with vanishing Θ_i scattered among states with non-zero occupation numbers. The number of Θ_i which are non-zero may be finite or infinite. The corresponding total energy is then defined as

$$E_{\mathrm{J}}(\Theta_i) := \sum_i \Theta_i t_i + \int d^3r\, v_{\mathrm{ext}}(\boldsymbol{r}) n(\boldsymbol{r}) + E_{\mathrm{H}}[n] + E_{\mathrm{xc,L}}[n] \tag{3.74}$$

$$t_i := \sum_{\sigma=\uparrow,\downarrow} \int d^3r\, \phi_i^*(\boldsymbol{r}\sigma) \frac{(-i\hbar\boldsymbol{\nabla})^2}{2m} \phi_i(\boldsymbol{r}\sigma), \tag{3.75}$$

where $E_{\mathrm{H}}[n]$ and $E_{\mathrm{xc,L}}[n]$ are the standard DFT functionals evaluated with the density (3.73). Minimization of E_{J} with respect to the yet undetermined orbitals ϕ_i for fixed Θ_i and under the constraint of normalization[1] leads to KS-type single-particle equations,

$$\left\{-\frac{\hbar^2\nabla^2}{2m}+v_{\mathrm{ext}}(\boldsymbol{r})+v_{\mathrm{H}}[n](\boldsymbol{r})+v_{\mathrm{xc}}[n](\boldsymbol{r})\right\}\phi_i(\boldsymbol{r}\sigma)=\varepsilon_i\phi_i(\boldsymbol{r}\sigma)\,. \tag{3.76}$$

In the present situation these equations have to be solved at least for all states ϕ_i with non-vanishing Θ_i, not just for the lowest N orbitals.

Clearly, the energy (3.74) coincides with the ground state energy $E_0(N)$ of the N-particle system subject to v_{ext} if (i) $\sum_i\Theta_i$ yields the correct particle number and (ii) the energy $E_{\mathrm{J}}(\Theta_i)$ is simultaneously minimized with respect to the Θ_i,

$$E_{\mathrm{J}}(\Theta_{i,0}) = E_0(N) \quad \text{if} \quad \sum_i\Theta_{i,0} = N \quad \text{and} \quad \left.\frac{\partial E_{\mathrm{J}}}{\partial\Theta_i}\right|_{\Theta_i=\Theta_{i,0}} = 0\,. \tag{3.77}$$

On the other hand, for non-integer particle number $N+\eta$ (with $0\leq\eta\leq 1$) $E_{\mathrm{J}}(\Theta_i)$ provides a continuous interpolation between the ground state energies of the neighboring integer particle numbers N and $N+1$, as long as $E_{\mathrm{J}}(\Theta_i)$ is minimized with respect to the Θ_i for all η.

Let us evaluate the partial derivative $\partial E_{\mathrm{J}}/\partial\Theta_i$, which controls the occupation, explicitly. Given the definition of E_{J}, it is obvious that the orbitals ϕ_i which minimize E_{J} vary, if the Θ_i are varied. The desired partial derivative therefore consists of two contributions, the explicit dependence of E_{J} on Θ_i and the implicit dependence via the orbitals,

$$\frac{\partial E_{\mathrm{J}}}{\partial\Theta_i} = t_i+\sum_j\Theta_j\frac{\partial t_j}{\partial\Theta_i}+\int d^3r\,[v_{\mathrm{ext}}+v_{\mathrm{H}}+v_{\mathrm{xc}}]\sum_{\sigma=\uparrow,\downarrow}\left[|\phi_i|^2+\sum_j\Theta_j\frac{\partial|\phi_j|^2}{\partial\Theta_i}\right]. \tag{3.78}$$

The first term, the kinetic energy of orbital i, can be rewritten by use of the single-particle equation (3.76), in analogy to Eq. (3.51). Multiplication of (3.76) with ϕ_i^* and subsequent integration yields

$$t_i+\sum_{\sigma=\uparrow,\downarrow}\int d^3r\,[v_{\mathrm{ext}}(\boldsymbol{r})+v_{\mathrm{H}}(\boldsymbol{r})+v_{\mathrm{xc}}(\boldsymbol{r})]\,|\phi_i(\boldsymbol{r}\sigma)|^2=\varepsilon_i\,. \tag{3.79}$$

Insertion of this relation (reflecting the minimization of E_{J} with respect to ϕ_i) and of the definition (3.75) into Eq. (3.78) then gives

$$\frac{\partial E_{\mathrm{J}}}{\partial\Theta_i}=\varepsilon_i+\sum_j\Theta_j\sum_{\sigma=\uparrow,\downarrow}\int d^3r\,\frac{\partial\phi_j^*}{\partial\Theta_i}\left[\frac{(-i\hbar\nabla)^2}{2m}+v_{\mathrm{ext}}+v_{\mathrm{H}}+v_{\mathrm{xc}}\right]\phi_j+c.c..$$

If the single-particle equations (3.76) are used once more, one finally arrives at Janak's theorem [88],

[1] There is no need to constrain different orbitals to be orthogonal as the single-particle operator on the left-hand side of Eq. (3.76) is hermitian anyway.

$$\frac{\partial E_J}{\partial \Theta_i} = \varepsilon_i + \sum_j \Theta_j \varepsilon_j \frac{\partial}{\partial \Theta_i} \sum_{\sigma=\uparrow,\downarrow} \int d^3r |\phi_j(\boldsymbol{r}\sigma)|^2 = \varepsilon_i \,. \tag{3.80}$$

Equation (3.80) is a generalization of Slater's analogous result obtained for the $X\alpha$-method [89, 90].

It remains to analyze which distribution of occupation numbers emerges from the minimization of E_J with respect to the Θ_i for fixed particle number $N+\eta$. Implementing the particle number and normalization constraints by Lagrange parameters (as usual) and the constraint $0 \leq \Theta_i \leq 1$ by the transformation

$$\Theta_i = \cos^2(\alpha_i) \qquad \text{with} \quad 0 \leq \alpha_i \leq \frac{\pi}{2} \,,$$

the functional to be minimized is given by

$$E_J(\cos^2(\alpha_i)) - \sum_i \cos^2(\alpha_i)\varepsilon_i \left[\sum_{\sigma=\uparrow,\downarrow} \int d^3r |\phi_i(\boldsymbol{r}\sigma)|^2 - 1 \right]$$
$$-\mu \left[\sum_i \cos^2(\alpha_i) - (N+\eta) \right] .$$

Variation with respect to ϕ_i leads to Eq. (3.76), as before. Minimization with respect to α_i, on the other hand, gives the condition

$$\sin(2\alpha_i) \left[\frac{\partial E_J}{\partial \Theta_i} - \mu \right] = \sin(2\alpha_i) \left[\varepsilon_i - \mu \right] = 0 \,.$$

This relation can only be satisfied in three ways:

$$\begin{array}{lll} \alpha_i = 0 \iff \Theta_i = 1 & \varepsilon_i \text{ arbitrary} \\ \alpha_i = \frac{\pi}{2} \iff \Theta_i = 0 & \varepsilon_i \text{ arbitrary} \\ \alpha_i \text{ arbitrary} \iff \Theta_i \text{ arbitrary} & \varepsilon_i = \mu \end{array}$$

Thus there is only one particular energy for which a fractional occupancy can occur. All other states are either completely filled or empty. From the KS approach one would expect that the levels below μ are filled, the ones above μ are empty.

This expectation can be verified by use of Janak's theorem. As discussed, E_J recovers the true ground state energy in the case of integer particle numbers, if minimized with respect to the Θ_i. For this reason the present Θ_i must be identical with the KS occupation numbers (3.71) for integer particle numbers. Let us convince ourselves that Janak's theorem (3.80) actually yields the KS occupation in this limit. For fixed particle number any increase of one of the Θ_i has to be compensated by a reduction of the others. So, starting from the KS occupation, an infinitesimal increase of the occupation Θ_u of an unoccupied state ϕ_u from its initial value of zero leads to the total energy change

$$\delta E = \frac{\partial E_{\mathrm{J}}}{\partial \Theta_u}\Theta_u - \sum_{\mathrm{occ}.i} \frac{\partial E_{\mathrm{J}}}{\partial \Theta_i}\eta_i \qquad \text{with} \quad \sum_{\mathrm{occ}.i} \eta_i = \Theta_u ,$$

where $\sum_{\mathrm{occ}.i}$ only extends over the initially occupied states (It is irrelevant at this point whether the highest occupied state is fractionally occupied or not). The shift δE includes the infinitesimal rearrangement of the ϕ_i which results from the modified occupation numbers. However, as the occupation is assumed to change only infinitesimally, there is no energetic reordering of eigenvalues as compared to the initial configuration—the eigenvalues themselves also change only infinitesimally.[2] Use of Eq. (3.80) and of the fact that the eigenvalues of the occupied states are all lower than that of the highest occupied state ϕ_{ho} therefore allows to determine the sign of δE,

$$\delta E = \Theta_u \varepsilon_u - \sum_{\mathrm{occ}.i} \eta_i \varepsilon_i \geq \Theta_u(\varepsilon_u - \varepsilon_{ho}) \geq 0 .$$

Any increase of Θ_u from zero raises the energy.

In the same way one can analyze an initial occupation with a gap at state j, $\Theta_j = 0$, and occupied states with energies higher than ε_j. In this case an infinitesimal increase of the occupation Θ_j yields the energy change

$$\delta E = \frac{\partial E_{\mathrm{J}}}{\partial \Theta_j}\Theta_j - \sum_{i \neq j} \frac{\partial E_{\mathrm{J}}}{\partial \Theta_i}\eta_i \qquad \text{with} \quad \sum_{i \neq j} \eta_i = \Theta_j .$$

As long as all levels with eigenvalues lower than ε_j are completely filled, one finds

$$\delta E = \Theta_j \varepsilon_j - \sum_{i \neq j; \varepsilon_i > \varepsilon_j} \eta_i \varepsilon_i \leq \Theta_j(\varepsilon_j - \varepsilon_k) \leq 0 ,$$

where k is the energetically lowest state above the initially empty state j. Thus the state j begins to be filled due to energy minimization. This process continues at least as long as the ordering of ε_j and ε_k is preserved. If the point $\varepsilon_j = \varepsilon_k$ is reached without both states being fully occupied, one either starts to fill up both of them simultaneously from non-empty states higher than k or one is facing the fractional occupation of a degenerate KS Fermi level. In both cases one finally ends up with the KS occupation.

The same basic mechanism applies for fractional particle number, so that only the highest occupied state (or the set of degenerate highest occupied states) can be partially filled. Integrating up Eq. (3.80) between two neighboring integer particle numbers thus yields a relation for the ionization energy,

$$E(N+1) - E(N) = \int_0^1 d\Theta_l \, \varepsilon_l , \qquad (3.81)$$

[2] Of course, if $\varepsilon_i = \varepsilon_j$ holds before the modification of the occupation numbers, one might find some ordering $\varepsilon_i < \varepsilon_j$ after the transfer of norm. However, this splitting of previously degenerate states does not affect the argument.

where l denotes the unoccupied state of the N-particle system which is filled in the $N+1$-particle system.[3]

Let us finally go beyond the actual realm of DFT for a moment and try to utilize Eq. (3.80) for the calculation of excitation energies. In fact, the functional (3.74) can be used to set up a first approximation for the energy of excited states. For instance, a single-electron excitation from the many-particle ground state is obtained, if the occupation numbers are chosen to differ from the ground state occupation for exactly two single-particle states i and j: If Θ_i and Θ_j correspond to the ground state, the occupation $\Theta_i - 1$ and $\Theta_j + 1$ (with all other occupation numbers kept fixed) represents an excited many-particle state in which an electron is transfered from the occupied state i to the unoccupied state j. If the symmetries of the two states differ, this representation is in the spirit of the DFT for excited states discussed in Sect. 2.7. Repeating the argument behind Eq. (3.81), one can express the excitation energy of this transition as

$$E_1(N) - E_0(N) = \int_0^1 d\Theta_j\, \varepsilon_j - \int_0^1 d\Theta_i\, \varepsilon_i\,. \tag{3.82}$$

In practice, the evaluation of the right-hand side of Eq. (3.82) relies on a discretization of the $\Theta_{i/j}$-integrals. The simplest approximation is to use the value of the integrand at the midpoint $\Theta_i = \Theta_j = \frac{1}{2}$ for the complete range of the integration. This choice minimizes the error introduced by the discretization, if only a single mesh point is to be utilized for the discretization. The eigenvalues ε_i and ε_j then have to be obtained from a selfconsistent calculation for the occupation $\Theta_i = \Theta_j = \frac{1}{2}$,

$$E_1(N) - E_0(N) \approx \varepsilon_j(\Theta_i = \Theta_j = \frac{1}{2}) - \varepsilon_i(\Theta_i = \Theta_j = \frac{1}{2})\,.$$

In this approximation Eq. (3.82) simply represents Slater's transition state approach [90], applied to the KS scheme.

3.5 Kohn-Sham Equations for Spin-Polarized Systems

Similar to the HK theorem, one can also extend the KS scheme to systems in external magnetic fields [34, 91]. One starts with the assumption of the existence of a noninteracting system,

$$\hat{H}_s = \hat{T} + \int d^3r \left\{ v_s(\boldsymbol{r})\hat{n}(\boldsymbol{r}) + \boldsymbol{B}_s(\boldsymbol{r}) \cdot \hat{\boldsymbol{m}}(\boldsymbol{r}) \right\}, \tag{3.83}$$

with the same ground state density n and ground state magnetization density $\boldsymbol{m}$,

[3] This argument assumes, of course, that there is no complete rearrangement of states when going from the N- to the $N+1$-particle system.

$$n(\boldsymbol{r}) = \sum_{\sigma=\uparrow,\downarrow} \sum_i \Theta_i |\phi_i(\boldsymbol{r}\sigma)|^2 \tag{3.84}$$

$$\boldsymbol{m}(\boldsymbol{r}) = \mu_B \sum_{\sigma,\sigma'=\uparrow,\downarrow} \sum_i \Theta_i \phi_i^*(\boldsymbol{r}\sigma)\boldsymbol{\sigma}_{\sigma\sigma'}\phi_i(\boldsymbol{r}\sigma'), \tag{3.85}$$

as the interacting system (2.120). The kinetic energy of this KS system, which is a functional of n and $\boldsymbol{m}$, is given by

$$T_{\mathrm{s,L}}[n,\boldsymbol{m}] = \sum_i \Theta_i \sum_{\sigma=\uparrow,\downarrow} \int d^3r\, \phi_i^*(\boldsymbol{r}\sigma) \frac{(-i\hbar\nabla)^2}{2m} \phi_i(\boldsymbol{r}\sigma), \tag{3.86}$$

with the Θ_i chosen in accordance with (3.71). $T_{\mathrm{s,L}}$ then induces a decomposition of the total energy of the interacting system,

$$E[n,\boldsymbol{m}] = T_{\mathrm{s,L}}[n,\boldsymbol{m}] + \int d^3r\, \{v_{\mathrm{ext}} n + \boldsymbol{B}_{\mathrm{ext}} \cdot \boldsymbol{m}\} + E_{\mathrm{H}}[n] + E_{\mathrm{xc,L}}[n,\boldsymbol{m}]. \tag{3.87}$$

Minimization of $E[n,\boldsymbol{m}]$ with respect to the orbitals ϕ_i yields the associated KS equations,

$$\sum_{\sigma'} \left\{ \left[-\frac{\hbar^2\nabla^2}{2m} + v_{\mathrm{s}}[n,\boldsymbol{m}](\boldsymbol{r}) \right] \delta_{\sigma\sigma'} + \mu_B \boldsymbol{\sigma}_{\sigma\sigma'} \cdot \boldsymbol{B}_{\mathrm{s}}[n,\boldsymbol{m}](\boldsymbol{r}) \right\} \phi_i(\boldsymbol{r}\sigma')$$
$$= \varepsilon_i \phi_i(\boldsymbol{r}\sigma) \tag{3.88}$$

$$v_{\mathrm{s}}[n,\boldsymbol{m}](\boldsymbol{r}) = v_{\mathrm{ext}}(\boldsymbol{r}) + v_{\mathrm{H}}[n](\boldsymbol{r}) + \frac{\delta E_{\mathrm{xc,L}}[n,\boldsymbol{m}]}{\delta n(\boldsymbol{r})} \tag{3.89}$$

$$\boldsymbol{B}_{\mathrm{s}}[n,\boldsymbol{m}](\boldsymbol{r}) = \boldsymbol{B}_{\mathrm{ext}}(\boldsymbol{r}) + \frac{\delta E_{\mathrm{xc,L}}[n,\boldsymbol{m}]}{\delta \boldsymbol{m}(\boldsymbol{r})}. \tag{3.90}$$

In addition to the external magnetic field the KS orbitals experience some internally generated magnetic field, originating from the xc-functional.

The standard form of spin-density functional theory is obtained if the magnetic field is restricted to a pure z-component, as in Eq. (2.133). In this limit the spin-densities are the fundamental variables of DFT. The Hamiltonian of the noninteracting KS system can thus be written as

$$\hat{H}_{\mathrm{s}} = \hat{T} + \sum_{\sigma=\uparrow,\downarrow} \int d^3r\, v_{\mathrm{s}}^{\sigma}(\boldsymbol{r}) \hat{n}_{\sigma}(\boldsymbol{r}). \tag{3.91}$$

In contrast to the general form (3.83) the Hamiltonian (3.91) obtained for $\boldsymbol{B}_{\mathrm{ext}} = (0,0,B_{\mathrm{ext}})$ commutes with the operator $\hat{S}_z$ given in Eq. (3.6), as $\hat{H}_{\mathrm{s}}$ commutes with the total particle number for given spin,

$$\left[\hat{H}_{\mathrm{s}}, \hat{N}_{\sigma}\right] = 0. \tag{3.92}$$

However, the effective KS potentials for the two spin orientations differ, so that the corresponding spatial single-particle orbitals depend on the spin quantum number,

$$\phi_i(\boldsymbol{r}\sigma) = \phi_{\alpha s}(\boldsymbol{r})\chi_s(\sigma) \qquad i \equiv (\alpha, s)\,. \tag{3.93}$$

The spin-densities are obtained as

$$n_\sigma(\boldsymbol{r}) = \sum_i \Theta_i |\phi_i(\boldsymbol{r}\sigma)|^2 = \sum_\alpha \Theta_{\alpha\sigma} |\phi_{\alpha\sigma}(\boldsymbol{r})|^2\,, \tag{3.94}$$

where (3.10) has been utilized to evaluate the sum over s. The spin-dependence of the occupation numbers reflects the commutation relation (3.92). Each pair of particle numbers N_σ, with $N_\uparrow + N_\downarrow = N$, represents a different sector of Fock space, for which the existence theorem (2.134) applies separately. The absolute ground state is then determined by minimization of the total energy of the interacting system with respect to N_σ. In the most general situation the occupation numbers are thus given by

$$\Theta_{\alpha\sigma} = \begin{cases} 1 & \text{for } \varepsilon_{\alpha\sigma} < \varepsilon_{F\sigma} \\ 0 \le \Theta_{\alpha\sigma} \le 1 & \text{for } \varepsilon_{\alpha\sigma} = \varepsilon_{F\sigma}\,; \\ 0 & \text{for } \varepsilon_{\alpha\sigma} > \varepsilon_{F\sigma} \end{cases} \qquad \text{with } \sum_\alpha \Theta_{\alpha\sigma} = N_\sigma\,. \tag{3.95}$$

The particle number N_σ for spin σ determines the corresponding Fermi energy $\varepsilon_{F\sigma}$. The kinetic energy of the KS system is

$$\begin{aligned} T_{\mathrm{s,L}}[n_\uparrow, n_\downarrow] &= \sum_{\sigma=\uparrow,\downarrow} \sum_i \Theta_i \int d^3r\, \phi_i^*(\boldsymbol{r}\sigma) \frac{(-i\hbar\boldsymbol{\nabla})^2}{2m} \phi_i(\boldsymbol{r}\sigma) \\ &= \sum_{\sigma=\uparrow,\downarrow} \sum_\alpha \Theta_{\alpha\sigma} \int d^3r\, \phi_{\alpha\sigma}^*(\boldsymbol{r}) \frac{(-i\hbar\boldsymbol{\nabla})^2}{2m} \phi_{\alpha\sigma}(\boldsymbol{r})\,. \end{aligned} \tag{3.96}$$

With the decomposition

$$\begin{aligned} E[n_\uparrow, n_\downarrow] = T_{\mathrm{s,L}}[n_\uparrow, n_\downarrow] + \int d^3r \Big\{ \big[v_{\mathrm{ext}} + \mu_B B_{\mathrm{ext}}\big] n_\uparrow + \big[v_{\mathrm{ext}} - \mu_B B_{\mathrm{ext}}\big] n_\downarrow \Big\} \\ + E_{\mathrm{H}}[n] + E_{\mathrm{xc,L}}[n_\uparrow, n_\downarrow] \end{aligned} \tag{3.97}$$

one finds the KS equations of spin-density functional theory,

$$\left\{ -\frac{\hbar^2\boldsymbol{\nabla}^2}{2m} + v_{\mathrm{s}}^\sigma[n_\uparrow, n_\downarrow](\boldsymbol{r}) \right\} \phi_{\alpha\sigma}(\boldsymbol{r}) = \varepsilon_{\alpha\sigma}\phi_{\alpha\sigma}(\boldsymbol{r}) \tag{3.98}$$

$$v_{\mathrm{s}}^\sigma[n_\uparrow, n_\downarrow](\boldsymbol{r}) = v_{\mathrm{ext}}(\boldsymbol{r}) + \mathrm{sign}(\sigma)\mu_B B_{\mathrm{ext}}(\boldsymbol{r}) + v_{\mathrm{H}}[n](\boldsymbol{r}) + v_{\mathrm{xc}}^\sigma[n_\uparrow, n_\downarrow](\boldsymbol{r}) \tag{3.99}$$

$$v_{\mathrm{xc}}^\sigma[n_\uparrow, n_\downarrow](\boldsymbol{r}) = \frac{\delta E_{\mathrm{xc,L}}[n_\uparrow, n_\downarrow]}{\delta n_\sigma(\boldsymbol{r})}\,. \tag{3.100}$$

Both spins experience different effective potentials v_{s}^σ. In addition to the external magnetic field the internal spin-effects included in $E_{\mathrm{xc,L}}[n_\uparrow, n_\downarrow]$ contribute to the difference between the spin up and spin down potentials. For completeness we also note the spin-dependent analog of Eq. (3.52),

$$E_0 = \sum_{\sigma=\uparrow,\downarrow} \left\{ \sum_{\alpha} \Theta_{\alpha\sigma} \varepsilon_{\alpha\sigma} - \int d^3r\, v_{\rm xc}^{\sigma}[n_\uparrow, n_\downarrow](\boldsymbol{r}) n_\sigma(\boldsymbol{r}) \right\} - E_{\rm H}[n] + E_{\rm xc,L}[n_\uparrow, n_\downarrow]. \tag{3.101}$$

Of course, for unpolarized systems Eqs. (3.94)–(3.101) are identical with the KS scheme of Sect. 3.3.

It is worthwhile to emphasize that the spin-dependent scheme (3.94)–(3.101) is most often applied in the limit $B_{\rm ext} = 0$. In fact, the explicit inclusion of the spin degree of freedom via the spin-dependent KS equations (3.98) (or even (3.88)) is mandatory, whenever the ground state of interest exhibits some nonzero magnetic moment. Given the claim that the original HK theorem guarantees a purely density-dependent description of spin-polarized systems, this statement requires some explanation. The critical point is the degree to which the physical properties of the true ground state are transferred to the ground state of the auxiliary noninteracting system. The KS system needed is necessarily a system of spin-$1/2$ particles—otherwise one would have to introduce *ad hoc* rules for the occupation of the KS levels. A non-degenerate KS system therefore automatically implies an even particle number N (as discussed in Sect. 3.1). A description of spin-polarized states on the basis of a scheme that depends only on the density (rather than the spin-densities) must thus employ the formalism for degenerate situations outlined in Sect. 3.3, even if no spatial symmetry is present.

Let us analyze the consequences of this ensemble approach for a noninteracting physical system with $N_\uparrow = N_\downarrow + 1$ and without spatial symmetries. The KS potential is identical with $v_{\rm ext}$ in this case, so that all KS levels are exactly twofold degenerate. All KS levels up to $N - 1 = 2N_\downarrow$ are doubly occupied, while the degenerate levels N and $N+1$ contain only one KS particle. Since a spin-independent potential can never lift the spin-degeneracy, the occupation numbers Θ_N and Θ_{N+1} can have arbitrary fractional values consistent with $\Theta_N + \Theta_{N+1} = 1$. Moreover, the density does not depend on the actual values of Θ_N and Θ_{N+1}, as ϕ_N and ϕ_{N+1} only differ in their spin-dependence. There is thus no intrinsic requirement which determines these occupation numbers. On the other hand, there is no need to fix them individually, as the complete formalism only depends on $\Theta_N + \Theta_{N+1}$ and the magnetic moment need not be reproduced by the KS ground state in the unpolarized scheme. As a consequence, the density and the energy of a noninteracting system with $N_\uparrow = N_\downarrow + 1$ are correctly obtained from Eqs. (3.67)–(3.72), while the magnetic moment remains undetermined. In this sense $T_{\rm s,L}[n]$ can describe ground states with a total spin-projection 1/2.

Let us next consider a noninteracting physical system with $N_\uparrow = N_\downarrow + 2$ and no spatial symmetries. The corresponding physical state is not the absolute ground state of the underlying Hamiltonian, but the energetically lowest state in one particular sector of Fock space, for which the Ritz principle is valid. It is immediately obvious that the density of this state can not be obtained from the ensemble KS scheme (3.67)–(3.72), as the highest two relevant single-particle states are only singly occupied in the physical system. Clearly, $T_{\rm s,L}[n]$ is not applicable to individual spin sectors of Fock space, only the absolute ground state can be described.

Consider finally an interacting physical system and assume that the internal spin-effects lead to an absolute ground state with $N_\uparrow = N_\downarrow + 2$ for full interaction strength w. Imagine that this strength is obtained by switching on the interaction from the value $w = 0$. There is one critical interaction strength $\lambda_c w$ $(0 < \lambda_c < 1)$ during this process, for which the energy of the lowest state with $N_\uparrow = N_\downarrow + 2$ falls below that of the lowest state with $N_\uparrow = N_\downarrow = N/2$. This critical strength λ_c depends on the external potential $v_{\rm ext}$—the same basic effect could be observed by varying $v_{\rm ext}$ for fixed interaction strength. In order to be generally applicable, the ensemble KS scheme has to reproduce both types of ground state densities, the one with $N_\uparrow = N_\downarrow = N/2$ below λ_c and the one with $N_\uparrow = N_\downarrow + 2$ for $\lambda \geq \lambda_c$. However, the absolute ground state density is discontinuous when crossing λ_c. This discontinuity must be generated by the xc-functional $E_{\rm xc,L}[n]$, i.e. the corresponding $v_{\rm xc}$ must favor spatially different densities for $\lambda < \lambda_c$ and $\lambda > \lambda_c$. And it must do so for a large class of potentials $v_{\rm ext}$ without depending on $v_{\rm ext}$ itself (as $E_{\rm xc,L}[n]$ is universal). While this is not impossible in principle, no mathematical realization of a functional with these properties is conceivable at present (not to speak of the approximations for $E_{\rm xc,L}[n]$ available to date).

In summary: The description of spin-polarized systems within the unpolarized KS scheme is possible in principle, as the magnetic moment of the KS system need not agree with the magnetic moment of the interacting system. However, the irrelevance of the magnetic moment of the KS system is also the major drawback of the unpolarized framework. The burden of distinguishing between different spin states must solely be carried by the purely density-dependent xc-functional. No suitable xc-functional for this task is known so far.

As a final remark, one should take note of the fact that the scheme (3.94)–(3.100) only applies to the energetically lowest state for given $N_\uparrow, N_\downarrow$. For instance, in the case of helium the choice $N_\uparrow = N_\downarrow = 1$ and thus $S_z = 0$ necessarily corresponds to the Singlet ground state. The lowest Triplet state with $S_z = 0$ is an excited state of this sector of Fock space. On the other hand, the lowest state with $N_\uparrow = 2$ and $N_\downarrow = 0$ is the KS representation of the lowest $S_z = 1$ Triplet state. In this way, a certain amount of information on excited states is accessible via the KS approach.

3.6 Interpretation of Kohn-Sham Eigenvalues: Relation to Ionization Potential, Fermi Surface and Band Gap

3.6.1 Ionization Potential

It has been emphasized in Sect. 3.1 that the KS ground state Φ_0 is not identical with the true ground state Ψ_0 of the interacting system. The KS formalism is a mathematical tool to generate the exact ground state density and energy in the most efficient way. This implies that the individual KS orbitals and eigenvalues have, in general, no physical meaning. There is, however, one exception to this statement:

The eigenvalue of the highest occupied KS level of a finite system, i.e. a system for which the density decays exponentially outside a finite region of space, is identical with the ionization potential (IP) of the interacting system [92].

The proof of this identity relies on the concept of the *quasi-particle amplitudes*. In order to introduce these quantities one starts with the full many-body Hamiltonian,

$$\hat{H} = \sum_{\sigma=\uparrow,\downarrow} \int d^3r\, \hat{\psi}^\dagger(\boldsymbol{r}\sigma) \left\{ -\frac{\hbar^2\boldsymbol{\nabla}^2}{2m} + v_{\rm ext}(\boldsymbol{r}) \right\} \hat{\psi}(\boldsymbol{r}\sigma) + \frac{1}{2} \sum_{\sigma,\sigma'=\uparrow,\downarrow} \int d^3r \int d^3r'\, \hat{\psi}^\dagger(\boldsymbol{r}\sigma)\hat{\psi}^\dagger(\boldsymbol{r}'\sigma') w(\boldsymbol{r},\boldsymbol{r}') \hat{\psi}(\boldsymbol{r}'\sigma')\hat{\psi}(\boldsymbol{r}\sigma). \quad (3.102)$$

The N-particle eigenstates of this Hamiltonian are denoted as $|\Psi_k^N\rangle$,

$$\hat{H}\,|\Psi_k^N\rangle = E_k^N\,|\Psi_k^N\rangle\,; \qquad E_k^N \le E_{k+1}^N\,. \quad (3.103)$$

As indicated in (3.103), the $|\Psi_k^N\rangle$ are assumed to be energetically ordered, so that $|\Psi_0^N\rangle$ represents the ground state in the N-particle sector of Fock space (assumed to be non-degenerate). The quasi-particle amplitudes are then defined as

$$f_k(\boldsymbol{r}\sigma) := \langle\Psi_k^{N-1}|\hat{\psi}(\boldsymbol{r}\sigma)|\Psi_0^N\rangle \quad (3.104)$$

$$g_k^*(\boldsymbol{r}\sigma) := \langle\Psi_k^{N+1}|\hat{\psi}^\dagger(\boldsymbol{r}\sigma)|\Psi_0^N\rangle\,. \quad (3.105)$$

f_k is the overlap of the $N-1$-particle state obtained by taking out an electron with spin σ at point $\boldsymbol{r}$ from the N-particle ground state with the k-th eigenstate in the $N-1$-particle sector. Similarly, g_k is the overlap of the $N+1$-particle state obtained by adding an electron to $|\Psi_0^N\rangle$ with the k-th eigenstate in the $N+1$-particle sector.

The relevance of these coefficients for the present discussion becomes more clear, if one compares the definition (3.104) with that of the ground state density of the interacting system, Eq. (2.11): the ground state density can be directly expressed in terms of the quasi-particle amplitudes f_k by use of the completeness relation in the $(N-1)$-particle Hilbert space, $\sum_k |\Psi_k^{N-1}\rangle\langle\Psi_k^{N-1}| = \hat{1}$,

$$n_0(\boldsymbol{r}) = \sum_{\sigma=\uparrow,\downarrow} \langle\Psi_0^N|\hat{\psi}^\dagger(\boldsymbol{r}\sigma)\hat{\psi}(\boldsymbol{r}\sigma)|\Psi_0^N\rangle = \sum_{\sigma=\uparrow,\downarrow}\sum_k |f_k(\boldsymbol{r}\sigma)|^2. \quad (3.106)$$

Therefore the properties of n_0 are intrinsically related to the properties of the f_k.

The quasi-particle amplitudes f_k satisfy differential equations which have some similarity with the Hartree-Fock (self-consistent field) equations. They are most easily derived by evaluation of the commutator $[\hat{\psi},\hat{H}]$,

$$\left[\hat{\psi}(\boldsymbol{r}\sigma),\hat{H}\right] = \left\{ -\frac{\hbar^2\boldsymbol{\nabla}^2}{2m} + v_{\rm ext}(\boldsymbol{r}) \right\} \hat{\psi}(\boldsymbol{r}\sigma) + \int d^3r'\, w(\boldsymbol{r},\boldsymbol{r}') \sum_{\sigma'=\uparrow,\downarrow} \hat{\psi}^\dagger(\boldsymbol{r}'\sigma')\hat{\psi}(\boldsymbol{r}'\sigma')\hat{\psi}(\boldsymbol{r}\sigma)\,. \quad (3.107)$$

After identification of the density operator (2.4) one arrives at

$$\begin{aligned}
&\langle\Psi_k^{N-1}|\left[\hat{\psi}(\boldsymbol{r}\sigma),\hat{H}\right]|\Psi_0^N\rangle \\
&= \langle\Psi_k^{N-1}|\left\{-\frac{\hbar^2\nabla^2}{2m}+v_{\text{ext}}(\boldsymbol{r})\right\}\hat{\psi}(\boldsymbol{r}\sigma)+\int d^3r'\,w(\boldsymbol{r},\boldsymbol{r}')\hat{n}(\boldsymbol{r}')\hat{\psi}(\boldsymbol{r}\sigma)|\Psi_0^N\rangle \\
&= \left[E_0^N-E_k^{N-1}\right]\langle\Psi_k^{N-1}|\hat{\psi}(\boldsymbol{r}\sigma)|\Psi_0^N\rangle\,, \qquad (3.108)
\end{aligned}$$

where the last line has been obtained with Eq. (3.103). Use of the definition (3.104) then yields

$$\begin{aligned}
&\left\{-\frac{\hbar^2\nabla^2}{2m}+v_{\text{ext}}(\boldsymbol{r})\right\}f_k(\boldsymbol{r}\sigma)+\int d^3r'\,w(\boldsymbol{r},\boldsymbol{r}')\langle\Psi_k^{N-1}|\hat{n}(\boldsymbol{r}')\hat{\psi}(\boldsymbol{r}\sigma)|\Psi_0^N\rangle \\
&= \hbar\omega_k\,f_k(\boldsymbol{r}\sigma)\,, \qquad (3.109)
\end{aligned}$$

where the excitation energy

$$\hbar\omega_k := E_0^N-E_k^{N-1} \qquad \Longrightarrow \qquad \omega_k \geq \omega_{k+1} \qquad (3.110)$$

has been introduced. The ordering of the ω_k results directly from the energetic ordering of the states, Eq. (3.103). Moreover, if $|\Psi_0^N\rangle$ is a normalizable bound state, its energy is usually lower than that of $|\Psi_0^{N-1}\rangle$. This is true in particular if $|\Psi_0^N\rangle$ is the ground state of a neutral atom or molecule. In this case one immediately identifies the IP as first excitation energy,

$$\text{IP} = -\hbar\omega_0\,; \qquad \omega_0 < 0\,. \qquad (3.111)$$

One can finally insert the completeness relation in Fock space, $\sum_N\sum_l|\Psi_l^N\rangle\langle\Psi_l^N| = \hat{1}$, between $\hat{n}(\boldsymbol{r}')$ and $\hat{\psi}(\boldsymbol{r}\sigma)$ and utilize the fact that $\hat{\psi}|\Psi_k^N\rangle$ is a state in the $N-1$-particle sector,

$$\begin{aligned}
0 &= \left\{-\frac{\hbar^2\nabla^2}{2m}+v_{\text{ext}}(\boldsymbol{r})-\hbar\omega_k\right\}f_k(\boldsymbol{r}\sigma) \\
&\quad +\sum_l\int d^3r'\,w(\boldsymbol{r},\boldsymbol{r}')\,\langle\Psi_k^{N-1}|\hat{n}(\boldsymbol{r}')|\Psi_l^{N-1}\rangle\,f_l(\boldsymbol{r}\sigma)\,. \qquad (3.112)
\end{aligned}$$

One has thus found an infinite set of coupled single-particle equations for the amplitudes f_k.

In the limit of vanishing interaction w all states $|\Psi_k^N\rangle$ are Slater determinants built from the single-particle orbitals corresponding to the potential v_{ext}. As a consequence, the overlap integral $\langle\Psi_k^{N-1}|\hat{\psi}(\boldsymbol{r}\sigma)|\Psi_0^N\rangle$ vanishes if $|\Psi_k^{N-1}\rangle$ contains orbitals which are not present in $|\Psi_0^N\rangle$. Only the N amplitudes for which $|\Psi_k^{N-1}\rangle$ is obtained by taking out one of the single-particle orbitals from $|\Psi_0^N\rangle$ are nonzero. The N non-vanishing f_k are identical with the N energetically lowest single-particle orbitals, i.e. the orbitals which are occupied in the N-particle ground state. The associated $\hbar\omega_k$ agree with the corresponding single-particle eigenvalues. This result can also be obtained directly from (3.112), as for $w = 0$ Eq. (3.112) is nothing but the single-particle Schrödinger equation for particles experiencing the potential $v_{\rm ext}$.

Starting from $[\hat{\psi}^\dagger, \hat{H}]$ one can also derive a differential equation for the g_k. However, these amplitudes are irrelevant for the present discussion, so that no details are given here. The only point worth noting is the fact that the g_k approach the unoccupied single-particle orbitals in the limit of vanishing interaction.

The asymptotic decay of the quasi-particle amplitudes f_k for the case of finite systems is the key for the identification of the eigenvalue of the highest occupied KS state with the IP of the interacting system. The asymptotic behavior of the f_k can be determined by an analysis of the differential equation (3.112) for large $|\boldsymbol{r}|$—an explicit discussion is given in Appendix E. One finds that the asymptotically leading amplitude is obtained for $k = 0$,

$$f_0(\boldsymbol{r}\sigma) \xrightarrow[|\boldsymbol{r}|\to\infty]{} \left[\sum_{lm} c_\sigma^{lm} Y_{lm}(\Omega)\right] r^{b_\sigma} e^{-\sqrt{-2m\omega_0/\hbar}\,r}$$

(the values of the coefficients c_σ^{lm} and the exponent b_σ are not relevant at this point). In the asymptotic region the density (3.106) is dominated by this leading amplitude,

$$n_0(\boldsymbol{r}) \xrightarrow[|\boldsymbol{r}|\to\infty]{} \sum_{\sigma=\uparrow,\downarrow} \left|\sum_{lm} c_\sigma^{lm} Y_{lm}(\Omega)\right|^2 r^{2b_\sigma} e^{-2\sqrt{-2m\omega_0/\hbar}\,r}. \tag{3.113}$$

This result may be compared with the asymptotic form of the KS density (3.23), which is dominated by the most weakly decaying KS orbital. The asymptotic form of the latter can be directly extracted from the KS equations (3.44). Defining the origin of the energy scale by

$$v_s(\boldsymbol{r}) \xrightarrow[|\boldsymbol{r}|\to\infty]{} 0, \tag{3.114}$$

one obtains by the same analysis as for the quasi-particle amplitude,

$$\phi_i(\boldsymbol{r}\sigma) \xrightarrow[|\boldsymbol{r}|\to\infty]{} \left[\sum_{lm} d_{i\sigma}^{lm} Y_{lm}(\Omega)\right] r^{\gamma_{i\sigma}} e^{-\sqrt{-2m\varepsilon_i}\,r/\hbar} \tag{3.115}$$

$$n_0(\boldsymbol{r}) \xrightarrow[|\boldsymbol{r}|\to\infty]{} \sum_{\sigma=\uparrow,\downarrow} \left|\sum_{lm} d_{N\sigma}^{lm} Y_{lm}(\Omega)\right|^2 r^{2\gamma_{N\sigma}} e^{-2\sqrt{-2m\varepsilon_N}\,r/\hbar} \tag{3.116}$$

(the KS levels are assumed to be energetically ordered). As the exact density is identical with the KS density one has

$$\varepsilon_N = \hbar\omega_0 = -\mathrm{IP}. \tag{3.117}$$

The relation (3.117) not only applies to the first IP, i.e. to a neutral system with $N = Z$, but also to all higher IPs (and the EA, if a negative ion exists).

3.6.2 Fermi Surface

This result automatically raises the question whether an equivalent statement can be made for extended systems. In the case of metals one could hope that the KS Fermi surface, as the analog of the highest occupied KS eigenvalue, is identical with the exact Fermi surface. In fact, this hope is fostered by two limiting cases, for which the agreement is obvious.

(a) For noninteracting systems the KS Fermi surface is trivially identical with the exact Fermi surface (for magnetic systems the spin-dependent version of the KS approach has to be used).
(b) In the case of the uniform electron gas both Fermi surfaces are spherical and contain the same volume of k-space, according to Luttinger's theorem [93]. Consequently the Fermi surfaces coincide.

The question to be answered is, whether this result also holds for real systems, which are simultaneously interacting and inhomogeneous?

In order to answer this question one first has to define the exact Fermi surface in a precise manner. For this purpose one needs an equation which allows a more direct determination of the excitation energies $\hbar\omega_k$ than (3.112). The derivation of this alternative equation starts from the *single-particle Green's function* of the interacting system,

$$G(\boldsymbol{r}\sigma t, \boldsymbol{r}'\sigma' t') := -i \langle \Psi_0 | T \hat{\psi}(\boldsymbol{r}\sigma t) \hat{\psi}^\dagger(\boldsymbol{r}'\sigma' t') | \Psi_0 \rangle , \tag{3.118}$$

where $\hat{\psi}(\boldsymbol{r}\sigma t)$ is the field operator in the Heisenberg representation ($\hat{H}$ is time-independent),

$$\hat{\psi}(\boldsymbol{r}\sigma t) = e^{i\hat{H}t/\hbar} \, \hat{\psi}(\boldsymbol{r}\sigma) \, e^{-i\hat{H}t/\hbar} , \tag{3.119}$$

and T denotes time-ordering of the two field operators. Quite generally, the time-ordering symbol T reorders arbitrary sets of elementary creation and annihilation operators $\hat{O}_i(t_i)$ chronologically, with the latest time to the left: if P is the permutation of the operators required to achieve this chronological ordering, T is defined as

$$T \, \hat{O}_1(t_1) \ldots \hat{O}_n(t_n) := S^{N_P} \, \hat{O}_{P(1)}(t_{P(1)}) \, \hat{O}_{P(2)}(t_{P(2)}) \, \ldots \, \hat{O}_{P(n)}(t_{P(n)}) \tag{3.120}$$

$$\text{with} \quad t_{P(1)} > t_{P(2)} > \ldots > t_{P(n)}$$

$$S = \begin{cases} -1 & \text{for fermions} \\ +1 & \text{for bosons} \end{cases}$$

$$N_P = \text{number of pairwise permutations required to set up } P .$$

The prefactor S^{N_P} accounts for the fermionic or bosonic character of the $\hat{O}_i$: the overall sign is always positive for bosons and equal to the sign of the permutation P for fermions. Using Eqs. (3.119) and (3.120), the definition (3.118) can be written more explicitly as

$$G(\boldsymbol{r}\sigma t,\boldsymbol{r}'\sigma' t') = -i\Theta(t-t')\langle\Psi_0|\hat{\psi}(\boldsymbol{r}\sigma)\,e^{-i\hat{H}(t-t')/\hbar}\,\hat{\psi}^\dagger(\boldsymbol{r}'\sigma')|\Psi_0\rangle\,e^{iE_0^N(t-t')/\hbar}$$
$$+i\Theta(t'-t)\langle\Psi_0|\hat{\psi}^\dagger(\boldsymbol{r}'\sigma')\,e^{-i\hat{H}(t'-t)/\hbar}\,\hat{\psi}(\boldsymbol{r}\sigma)|\Psi_0\rangle\,e^{iE_0^N(t'-t)/\hbar},$$

which leads to the following interpretation of G: for $t > t'$ a particle with spin σ' is added at position $\boldsymbol{r}'$ and time t' to the N-particle ground state $|\Psi_0\rangle$. The resulting $(N+1)$-particle state then evolves in time according to the Hamiltonian $\hat{H}$ until the time t, at which one particle with spin σ is taken out of the system at position $\boldsymbol{r}$. Similarly, for the case $t < t'$ a particle with spin σ is annihilated from the ground state at position $\boldsymbol{r}$ and time t, leaving a $(N-1)$-particle state. After propagation until t' the missing N-th particle is added at position $\boldsymbol{r}'$ with spin σ'. Thus, the term with $t > t'$ describes the propagation of an additional particle in the presence of the N particles of the ground state, the second term describes the propagation of a missing particle, called a *hole*. Taking both transition amplitudes together, the single-particle Green's function reflects the complete structure and dynamics of the system. As a consequence, many physical observables can be extracted from G.

G can be expressed in terms of the quasi-particle amplitudes (3.104), (3.105) by use of the completeness relation in the $(N\pm 1)$-particle Hilbert space,

$$G(\boldsymbol{r}\sigma t,\boldsymbol{r}'\sigma' t') = \int\frac{d\omega}{2\pi}\,e^{-i\omega(t-t')}\,G(\boldsymbol{r}\sigma,\boldsymbol{r}'\sigma',\omega) \tag{3.121}$$

$$G(\boldsymbol{r}\sigma,\boldsymbol{r}'\sigma',\omega) = \hbar\sum_l\left\{\frac{g_l(\boldsymbol{r}\sigma)g_l^*(\boldsymbol{r}'\sigma')}{\hbar\omega - E_l^{N+1}+E_0^N+i\eta} + \frac{f_l(\boldsymbol{r}\sigma)f_l^*(\boldsymbol{r}'\sigma')}{\hbar\omega - E_0^N+E_l^{N-1}-i\eta}\right\}, \tag{3.122}$$

where the standard representation of the step function as a contour integral, Eq. (L.5), has been used to obtain the Fourier decomposition. The pole shifts $\pm i\eta$ in the denominator of Eq. (3.122), which is referred to as *Lehmann representation* of G, have to be understood as infinitesimal, with $\eta > 0$ in all cases. These shifts only serve to define the contour in the complex ω-plane which implements the correct time-ordering for the individual contributions to G. Once the integration over ω is performed, the limit $\eta \to 0$ is taken. This limit is, however, never noted explicitly in equations as (3.121), but automatically implied whenever pole shifts show up. The same notation is subsequently used for all other quantities for which some time-ordering has to be incorporated in their Fourier representation, in particular for the response functions introduced in Sect. 4.2.1.

The Green's function satisfies the *Dyson equation* [94, 95],

$$G(\boldsymbol{r}\sigma t, \boldsymbol{r}'\sigma' t') = G_0(\boldsymbol{r}\sigma t, \boldsymbol{r}'\sigma' t') + \sum_{\sigma_3,\sigma_4} \int d^3r_3 d^3r_4 \int dt_3 dt_4$$
$$\times G_0(\boldsymbol{r}\sigma t, \boldsymbol{r}_3\sigma_3 t_3)\Sigma(\boldsymbol{r}_3\sigma_3 t_3, \boldsymbol{r}_4\sigma_4 t_4) G(\boldsymbol{r}_4\sigma_4 t_4, \boldsymbol{r}'\sigma' t'). \quad (3.123)$$

Here G_0 represents the noninteracting Green's function which is obtained from G in the limit $w \to 0$. An explicit form of G_0 is obtained by realizing that for vanishing w the amplitudes f_k and g_k approach the single-particle orbitals ϕ_k of the noninteracting system with external potential $v_{\rm ext}$ (compare Eq. (3.112), which, for $w = 0$, is nothing but the single-particle equation corresponding to the noninteracting system). Correspondingly, the energy differences $\hbar\omega_l = E_0^N - E_l^{N-1}$ go over into the associated eigenvalues ε_l (which, for the moment, should not be understood as KS eigenvalues). Denoting the occupation of the single-particle states by Θ_l, one therefore obtains for the Lehmann representation of G_0

$$G_0(\boldsymbol{r}\sigma, \boldsymbol{r}'\sigma', \omega) = \sum_l \left\{ (1-\Theta_l) \frac{\phi_l(\boldsymbol{r}\sigma)\phi_l^*(\boldsymbol{r}'\sigma')}{\omega - \varepsilon_l/\hbar + i\eta} + \Theta_l \frac{\phi_l(\boldsymbol{r}\sigma)\phi_l^*(\boldsymbol{r}'\sigma')}{\omega - \varepsilon_l/\hbar - i\eta} \right\}. \quad (3.124)$$

Σ denotes the *proper* (or *irreducible*) *self-energy*, for which Eq. (3.123) represents one possible definition. Σ contains all non-trivial interaction contributions to G, i.e. all terms which are not simply obtained by repetition of more elementary contributions to Σ. This statement becomes clear if Eq. (3.123) is iterated, i.e. if the full Green's function on the right-hand side of (3.123) is repeatedly replaced by use of Eq. (3.123) itself,

$$\begin{aligned}
&G(\boldsymbol{r}\sigma t, \boldsymbol{r}'\sigma' t') \\
&= G_0(\boldsymbol{r}\sigma t, \boldsymbol{r}'\sigma' t') \\
&\quad + \sum_{\sigma_3\sigma_4} \int d^3r_3 d^3r_4 \int dt_3 dt_4\, G_0(\boldsymbol{r}\sigma t, \boldsymbol{r}_3\sigma_3 t_3)\Sigma(\boldsymbol{r}_3\sigma_3 t_3, \boldsymbol{r}_4\sigma_4 t_4) \\
&\qquad \times G_0(\boldsymbol{r}_4\sigma_4 t_4, \boldsymbol{r}'\sigma' t') \\
&\quad + \sum_{\sigma_3\sigma_4} \int d^3r_3 d^3r_4 \int dt_3 dt_4\, G_0(\boldsymbol{r}\sigma t, \boldsymbol{r}_3\sigma_3 t_3)\Sigma(\boldsymbol{r}_3\sigma_3 t_3, \boldsymbol{r}_4\sigma_4 t_4) \\
&\quad \times \sum_{\sigma_5\sigma_6} \int d^3r_5 d^3r_6 \int dt_5 dt_6\, G_0(\boldsymbol{r}_4\sigma_4 t_4, \boldsymbol{r}_5\sigma_5 t_5)\Sigma(\boldsymbol{r}_5\sigma_5 t_5, \boldsymbol{r}_6\sigma_6 t_6) \\
&\qquad \times G_0(\boldsymbol{r}_6\sigma_6 t_6, \boldsymbol{r}'\sigma' t') + \ldots .
\end{aligned}$$

This equation demonstrates how the basic building blocks of G, absorbed in Σ, are repeated infinitely often in the form of a geometric series. The structure of this series immediately suggests the use of perturbation theory with respect to w: Any approximation to Σ is resummed to infinite order by the geometric series, so that a perturbative treatment of Σ corresponds to a non-perturbative treatment of G. In fact, the perturbative evaluation of Σ via the Feynman diagram technique is the standard approach in Green's function based many-body theory. To lowest order (first order in w) one finds (see e.g. [94])

$$
\begin{aligned}
&\hbar\Sigma^{(1)}(\boldsymbol{r}\sigma t,\boldsymbol{r}'\sigma' t') \\
&= -i\delta_{\sigma,\sigma'}\delta^{(3)}(\boldsymbol{r}-\boldsymbol{r}')\delta(t-t')\sum_{\sigma''}\int d^3r''\, w(\boldsymbol{r},\boldsymbol{r}'')\lim_{\eta\to 0^+} G_0(\boldsymbol{r}''\sigma'' t,\boldsymbol{r}''\sigma'' t+\eta) \\
&\quad +i\delta(t-t')w(\boldsymbol{r},\boldsymbol{r}')\lim_{\eta\to 0^+} G_0(\boldsymbol{r}\sigma t,\boldsymbol{r}'\sigma' t+\eta) \\
&= \delta_{\sigma,\sigma'}\delta^{(3)}(\boldsymbol{r}-\boldsymbol{r}')\delta(t-t')v_{\mathrm{H}}[n_0](\boldsymbol{r})+\hbar\Sigma_{\mathrm{x}}(\boldsymbol{r}\sigma t,\boldsymbol{r}'\sigma' t'),
\end{aligned}
\tag{3.125}
$$

where, after identification of the density (2.11) as a contraction of the Green's function,

$$
-i\sum_{\sigma}\lim_{\eta\to 0^+} G_0(\boldsymbol{r}\sigma t,\boldsymbol{r}\sigma t+\eta) = n_0(\boldsymbol{r}),
$$

Eq. (3.42) has been utilized and an abbreviation has been introduced for the second term (n_0 is the ground state density of the noninteracting system with potential v_{ext} in the present discussion). To lowest order Σ thus consists of a direct Coulomb term and an exchange contribution Σ_{x}.

On the basis of the Dyson equation (3.123) one can define the exact Fermi surface, provided one adapts (3.124) to a periodic system, i.e. a system with

$$
v_{\mathrm{ext}}(\boldsymbol{r}+\boldsymbol{R_n}) = v_{\mathrm{ext}}(\boldsymbol{r}) \tag{3.126}
$$

for all translations $\boldsymbol{R_n}$ which leave the periodic arrangement of the nuclei in the crystal unchanged (readers familiar with elementary quantum mechanics of periodic systems may proceed directly to Eq. (3.141)). The *Bravais vectors* $\boldsymbol{R_n}$ can be expressed in terms of the three *primitive vectors* $\boldsymbol{a}_i$ of the lattice which reflect three characteristic axes of the lattice with respect to which periodicity is observed,

$$
\boldsymbol{R_n} = n_1\boldsymbol{a}_1+n_2\boldsymbol{a}_2+n_3\boldsymbol{a}_3\,; \qquad n_i\in\mathbb{Z}. \tag{3.127}
$$

The parallelepiped spanned by the primitive vectors $\boldsymbol{a}_i$ is the *unit cell* of the lattice: repetition of this elementary building block of the lattice with volume

$$
\Omega = |\boldsymbol{a}_1\cdot(\boldsymbol{a}_2\times\boldsymbol{a}_3)| \tag{3.128}
$$

fills all space without leaving any voids.

As soon as the external potential satisfies the periodicity condition (3.126), the complete Hamiltonian commutes with all translations by $\boldsymbol{R_n}$, since the kinetic energy operator and the Coulomb interaction remain unaffected if all electronic coordinates are shifted by the same $\boldsymbol{R_n}$. As a result all eigenstates of the Hamiltonian may be classified by quantum numbers according to the translational symmetry. The consequences of this translational symmetry are most easily analyzed for noninteracting particles: the single-particle states ϕ_l corresponding to the potential (3.126),

$$
\left[-\frac{\hbar^2\nabla^2}{2m}+v_{\mathrm{ext}}(\boldsymbol{r})\right]\phi_{\boldsymbol{k}\alpha}(\boldsymbol{r}\sigma) = \varepsilon_{\boldsymbol{k}\alpha}\phi_{\boldsymbol{k}\alpha}(\boldsymbol{r}\sigma), \tag{3.129}
$$

satisfy Bloch's theorem [96],

$$\phi_{\boldsymbol{k}\alpha}(\boldsymbol{r}\sigma) = e^{i\boldsymbol{k}\cdot\boldsymbol{r}}\, u_{\boldsymbol{k}\alpha}(\boldsymbol{r}\sigma)\,. \tag{3.130}$$

The quantum number $\boldsymbol{k}$, the *crystal momentum*, reflects the periodicity of the system, the quantum number α, the *band index*, originates from the atomic states which form the *Bloch states* $\phi_{\boldsymbol{k}\alpha}$: for any value of $\boldsymbol{k}$ one finds a complete set of bands α. If the eigenvalues $\varepsilon_{\boldsymbol{k}\alpha}$ are plotted along a specific direction in $\boldsymbol{k}$-space, one obtains the *band structure* of the solid (for an example see Fig. 6.3). Finally, $u_{\boldsymbol{k}\alpha}(\boldsymbol{r}\sigma)$ is a function which is strictly periodic with the lattice,

$$u_{\boldsymbol{k}\alpha}(\boldsymbol{r}+\boldsymbol{R}_{\boldsymbol{n}},\sigma) = u_{\boldsymbol{k}\alpha}(\boldsymbol{r}\sigma)\,. \tag{3.131}$$

As a direct consequence of (3.126) and (3.131) $v_{\rm ext}$ and $u_{\boldsymbol{k}\alpha}$ (and, in fact, all observable (real) fields of the system) can be Fourier expanded in terms of the *reciprocal lattice vectors* $\boldsymbol{G}_{\boldsymbol{m}}$,

$$\boldsymbol{G}_{\boldsymbol{m}} = m_1\boldsymbol{b}_1 + m_2\boldsymbol{b}_2 + m_3\boldsymbol{b}_3\,; \qquad m_i \in \mathbb{Z}\,, \tag{3.132}$$

where the vectors $\boldsymbol{b}_i$ denote the primitive vectors of the reciprocal lattice,

$$\boldsymbol{b}_i = 2\pi\,\frac{\boldsymbol{a}_j\times\boldsymbol{a}_k}{\boldsymbol{a}_1\cdot(\boldsymbol{a}_2\times\boldsymbol{a}_3)} \qquad \text{with} \quad i,j,k = \text{cyclic permutations of 1,2,3.} \tag{3.133}$$

By construction one has

$$\boldsymbol{b}_i\cdot\boldsymbol{a}_j = 2\pi\,\delta_{ij}\,, \tag{3.134}$$

so that Eqs. (3.132), (3.133) implement the periodicity requirement

$$\exp[i\boldsymbol{G}_{\boldsymbol{m}}\cdot(\boldsymbol{r}+\boldsymbol{R}_{\boldsymbol{n}})] = \exp[i\boldsymbol{G}_{\boldsymbol{m}}\cdot\boldsymbol{r}] \quad\Longleftrightarrow\quad \exp[i\boldsymbol{G}_{\boldsymbol{m}}\cdot\boldsymbol{R}_{\boldsymbol{n}}] = 1\,; \qquad \forall\boldsymbol{R}_{\boldsymbol{n}}\,, \tag{3.135}$$

which results from the Fourier expansion of Eqs. (3.126) and (3.131). One thus has

$$u_{\boldsymbol{k}\alpha}(\boldsymbol{r}\sigma) = \sum_{\boldsymbol{m}} e^{i\boldsymbol{G}_{\boldsymbol{m}}\cdot\boldsymbol{r}}\, u_{\boldsymbol{k}\alpha}(\boldsymbol{G}_{\boldsymbol{m}}\sigma) \tag{3.136}$$

and analogous expansions for $v_{\rm ext}$ and all other strictly periodic functions.

It remains to discuss the quantum number $\boldsymbol{k}$ in more detail. Its character is determined by the requirement of normalization. Obviously, proper normalization of the Bloch states (3.130), (3.136) is not possible, if arbitrary values of $\boldsymbol{k}$ are allowed. This observation expresses the fact that one can not have more than countably many electrons in a periodic system. Boundary conditions are needed to ensure proper normalization and the correct particle number. In order to specify these boundary conditions consider a large block of unit cells, consisting of N_1 repetitions of the unit cell in the direction of $\boldsymbol{a}_1$, N_2 repetitions in the direction of $\boldsymbol{a}_2$ and N_3 repetitions in the direction of $\boldsymbol{a}_3$. The full crystal is then obtained by periodic repetition of this large block of unit cells, referred to as Born-von Karman block in the following. On the other hand, each individual block approaches the complete crystal in the limit $N_1, N_2, N_3 \to \infty$. Proper normalization is then ensured if one requires strict periodicity of the $\phi_{\boldsymbol{k}\alpha}$ with respect to repetitions of the Born-von Karman block of unit cells,

$$\phi_{\boldsymbol{k}\alpha}(\boldsymbol{r}+N_i\boldsymbol{a}_i,\sigma) = \phi_{\boldsymbol{k}\alpha}(\boldsymbol{r}\sigma) \qquad \text{for } i=1,2,3. \tag{3.137}$$

From this *Born-von Karman boundary condition* one obtains as admissible $\boldsymbol{k}$-values

$$\boldsymbol{k}_l = \frac{l_1}{N_1}\boldsymbol{b}_1 + \frac{l_2}{N_2}\boldsymbol{b}_2 + \frac{l_3}{N_3}\boldsymbol{b}_3 \qquad \text{with} \quad -\frac{N_i}{2} \le l_i < \frac{N_i}{2}, \quad l_i \in \mathbb{Z} \tag{3.138}$$

(alternatively, one may use the range $0 \le l_i \le N_i - 1$). In total, a multiplicity of $N_1N_2N_3$ is associated with each atomic state contributed by the atoms in the actual unit cell, allowing exactly as many electrons to fill these states as provided by the atoms in the complete Born-von Karman block. In the limit $N_1, N_2, N_3 \to \infty$ the ratio l_i/N_i can be any rational number between $-\frac{1}{2}$ and $+\frac{1}{2}$. The set of $\boldsymbol{k}$-values characterized by (3.138) is called the *first Brillouin zone*.

On this basis one can then establish the orthonormality and completeness relations for the Bloch states,

$$\sum_\sigma \int_{N_1N_2N_3\Omega} d^3r\, \phi^*_{\boldsymbol{k}\alpha}(\boldsymbol{r}\sigma)\, \phi_{\boldsymbol{k}'\alpha'}(\boldsymbol{r}\sigma) = \delta_{\boldsymbol{k}\boldsymbol{k}'}\, \delta_{\alpha\alpha'} \tag{3.139}$$

$$\sum_{\boldsymbol{k}\alpha} \phi_{\boldsymbol{k}\alpha}(\boldsymbol{r}\sigma)\phi^*_{\boldsymbol{k}\alpha}(\boldsymbol{r}'\sigma') = \delta_{\sigma\sigma'}\, \delta^{(3)}(\boldsymbol{r}-\boldsymbol{r}')\,. \tag{3.140}$$

Here the short-hand notation $\boldsymbol{k} \equiv \boldsymbol{k}_l$ has been introduced, $\sum_{\boldsymbol{k}} \equiv \sum_{\boldsymbol{l}}$, and $\delta_{\boldsymbol{k}\boldsymbol{k}'}$ implies $l_i = l'_i$ $(i = 1,2,3)$ in the representation (3.138). The Bloch states are normalized to 1 in the Born-von Karman block with volume $N_1N_2N_3\Omega$.

After this preparation, the discussion of the single-particle Green's function can be resumed. In view of the translational symmetry of the system a Fourier representation is most suitable. The basis for this Fourier representation is an expansion of the field operator in terms of creation/annihilation operators $\hat{c}^{(\dagger)}_{\boldsymbol{k}\alpha}$ for the noninteracting Bloch states $\phi_{\boldsymbol{k}\alpha}$,

$$\hat{\psi}(\boldsymbol{r}\sigma) = \sum_{\boldsymbol{k}\alpha} \phi_{\boldsymbol{k}\alpha}(\boldsymbol{r}\sigma)\, \hat{c}_{\boldsymbol{k}\alpha} \qquad \hat{\psi}^\dagger(\boldsymbol{r}\sigma) = \sum_{\boldsymbol{k}\alpha} \phi^*_{\boldsymbol{k}\alpha}(\boldsymbol{r}\sigma)\, \hat{c}^\dagger_{\boldsymbol{k}\alpha} \tag{3.141}$$

$$\left\{\hat{c}_{\boldsymbol{k}\alpha}, \hat{c}_{\boldsymbol{k}'\alpha'}\right\} = \left\{\hat{c}^\dagger_{\boldsymbol{k}\alpha}, \hat{c}^\dagger_{\boldsymbol{k}'\alpha'}\right\} = 0 \qquad \left\{\hat{c}_{\boldsymbol{k}\alpha}, \hat{c}^\dagger_{\boldsymbol{k}'\alpha'}\right\} = \delta_{\boldsymbol{k}\boldsymbol{k}'}\, \delta_{\alpha\alpha'} \tag{3.142}$$

(the commutation relations follow from Eqs. (2.6), (2.7)). Insertion into (3.118) defines the single-particle Green's function in momentum representation,

$$G(\boldsymbol{r}\sigma t, \boldsymbol{r}'\sigma' t') = \sum_{\boldsymbol{k}\alpha} \sum_{\boldsymbol{k}'\alpha'} \phi_{\boldsymbol{k}\alpha}(\boldsymbol{r}\sigma)\, \phi^*_{\boldsymbol{k}'\alpha'}(\boldsymbol{r}'\sigma')\, G(\boldsymbol{k}\alpha t, \boldsymbol{k}'\alpha' t') \tag{3.143}$$

$$G(\boldsymbol{k}\alpha t, \boldsymbol{k}'\alpha' t') = -i\,\langle\Psi_0|T\,\hat{c}_{\boldsymbol{k}\alpha}(t)\hat{c}^\dagger_{\boldsymbol{k}'\alpha'}(t')|\Psi_0\rangle \tag{3.144}$$

$$\hat{c}_{\boldsymbol{k}\alpha}(t) = e^{i\hat{H}t/\hbar}\, \hat{c}_{\boldsymbol{k}\alpha}\, e^{-i\hat{H}t/\hbar}\,. \tag{3.145}$$

In the state $\hat{c}^\dagger_{\boldsymbol{k}\alpha}|\Psi_0\rangle$ a bare (i.e. noninteracting) Bloch electron is added to the ground state of the N-particle system. Of course, this state is not an eigenstate of the Hamiltonian in the $(N+1)$-particle sector, but rather a superposition of eigenstates,

$$\hat{c}^\dagger_{\boldsymbol{k}\alpha}|\Psi_0\rangle = \sum_l g_l^*(\boldsymbol{k}\alpha)\,|\Psi_l^{N+1}\rangle \tag{3.146}$$

$$g_l^*(\boldsymbol{k}\alpha) := \langle\Psi_l^{N+1}|\hat{c}^\dagger_{\boldsymbol{k}\alpha}|\Psi_0^N\rangle\,. \tag{3.147}$$

Similarly, the state $\hat{c}_{\boldsymbol{k}\alpha}|\Psi_0\rangle$ is a superposition of $(N-1)$-particle eigenstates,

$$\hat{c}_{\boldsymbol{k}\alpha}|\Psi_0\rangle = \sum_l f_l(\boldsymbol{k}\alpha)\,|\Psi_l^{N-1}\rangle \tag{3.148}$$

$$f_l(\boldsymbol{k}\alpha) := \langle\Psi_l^{N-1}|\hat{c}_{\boldsymbol{k}\alpha}|\Psi_0^N\rangle\,. \tag{3.149}$$

$G(\boldsymbol{k}\alpha t,\boldsymbol{k}'\alpha' t')$ describes the propagation of these wave packets in time, with $\hat{c}^\dagger_{\boldsymbol{k}\alpha}|\Psi_0\rangle$ being propagated for $t > t'$ and $\hat{c}_{\boldsymbol{k}\alpha}|\Psi_0\rangle$ for $t < t'$. The quasi-particle amplitudes $g_l^*(\boldsymbol{k}\alpha)$ and $f_l(\boldsymbol{k}\alpha)$ represent the probabilities to find the wave packets in particular eigenstates of the $(N\pm 1)$-particle systems.

As in the case of its real space equivalent, the quasi-particle amplitudes can be used to establish the Lehmann representation of $G(\boldsymbol{k}\alpha t,\boldsymbol{k}'\alpha' t')$ by implementation of time-ordering via suitable contours in frequency space,

$$G(\boldsymbol{k}\alpha t,\boldsymbol{k}'\alpha' t') = \int \frac{d\omega}{2\pi}\, e^{-i\omega(t-t')}\, G(\boldsymbol{k}\alpha,\boldsymbol{k}'\alpha',\omega) \tag{3.150}$$

$$G(\boldsymbol{k}\alpha,\boldsymbol{k}'\alpha',\omega) = \hbar\sum_l \Bigg\{ \frac{g_l(\boldsymbol{k}\alpha)g_l^*(\boldsymbol{k}'\alpha')}{\hbar\omega - E_l^{N+1} + E_0^N + i\eta} + \frac{f_l(\boldsymbol{k}\alpha)f_l^*(\boldsymbol{k}'\alpha')}{\hbar\omega - E_0^N + E_l^{N-1} - i\eta} \Bigg\}. \tag{3.151}$$

For extended and, in particular, for periodic systems, it is convenient to recast the representation (3.151) by use of the chemical potentials and excitation energies of the $(N-1)$- and $(N+1)$-particle systems,

$$\mu(N) = E_0^N - E_0^{N-1} \tag{3.152}$$

$$\mu(N+1) = E_0^{N+1} - E_0^N \tag{3.153}$$

$$\Omega_l^{N-1} := E_l^{N-1} - E_0^{N-1} \quad\Longrightarrow\quad \Omega_l^{N-1} \geq 0 \tag{3.154}$$

$$\Omega_l^{N+1} := E_l^{N+1} - E_0^{N+1} \quad\Longrightarrow\quad \Omega_l^{N+1} \geq 0\,. \tag{3.155}$$

In fact, for large N (for solids $N \sim 10^{23}$) one has

$$\mu(N) = \mu(N+1) + \mathcal{O}\left(\frac{1}{N}\right) \equiv \mu\,, \tag{3.156}$$

so that the Lehmann representation (3.151) may be written as

$$G(\boldsymbol{k}\alpha,\boldsymbol{k}'\alpha',\omega) = \hbar\sum_l \left\{ \frac{g_l(\boldsymbol{k}\alpha)g_l^*(\boldsymbol{k}'\alpha')}{\hbar\omega-\mu-\Omega_l^{N+1}+i\eta} + \frac{f_l(\boldsymbol{k}\alpha)f_l^*(\boldsymbol{k}'\alpha')}{\hbar\omega-\mu+\Omega_l^{N-1}-i\eta} \right\}. \quad (3.157)$$

The form (3.157) exhibits most clearly that the threshold energy $\hbar\omega = \mu$ distinguishes between the particle propagation described by the first term and the hole propagation in the second.

The important observation now is: $G(\boldsymbol{k}\alpha,\boldsymbol{k}'\alpha',\omega)$ is diagonal with respect to $\boldsymbol{k},\boldsymbol{k}'$. This property is easily verified for the noninteracting Green's function, whose momentum space representation is defined in analogy to Eq. (3.143). Insertion of the Bloch states (3.130) into (3.124) and subsequent comparison with (3.143)–(3.151) directly yields

$$G_0(\boldsymbol{k}\alpha,\boldsymbol{k}'\alpha',\omega) = \delta_{\boldsymbol{k}\boldsymbol{k}'}\,\delta_{\alpha\alpha'}\,G_0(\boldsymbol{k},\alpha,\omega) \quad (3.158)$$

$$G_0(\boldsymbol{k},\alpha,\omega) = \frac{1-\Theta(\varepsilon_{\mathrm{F}}-\varepsilon_{\boldsymbol{k}\alpha})}{\omega-\varepsilon_{\boldsymbol{k}\alpha}/\hbar+i\eta} + \frac{\Theta(\varepsilon_{\mathrm{F}}-\varepsilon_{\boldsymbol{k}\alpha})}{\omega-\varepsilon_{\boldsymbol{k}\alpha}/\hbar-i\eta}, \quad (3.159)$$

where the occupation number has now been expressed in terms of the Fermi energy ε_{F} of the noninteracting system.

However, diagonality with respect to $\boldsymbol{k},\boldsymbol{k}'$ can also be demonstrated for the interacting Green's function (compare [97, 98]). For this purpose one first notes that all interacting states have to be periodic with respect to the Born-von Karman block: any translation by $N_i\boldsymbol{a}_i$ (or multiples of it) must leave the states invariant,

$$\begin{aligned}&\Psi_l^N(\boldsymbol{r}_1+N_i\boldsymbol{a}_i,\sigma_1,\ldots\boldsymbol{r}_N+N_i\boldsymbol{a}_i,\sigma_N)\\ &= \Psi_l^N(\boldsymbol{r}_1\sigma_1,\ldots\boldsymbol{r}_N\sigma_N) \qquad \text{for } i=1,2,3\,. \end{aligned} \quad (3.160)$$

The interacting states can thus be expanded in terms of the N-particle basis provided by the Slater determinants $|\Phi_{\boldsymbol{k}_1\ldots\boldsymbol{k}_N,\alpha_1\ldots\alpha_N}\rangle$ constructed from the Bloch states (3.130),

$$|\Psi_l^N\rangle = \sum_{\boldsymbol{k}_1\ldots\boldsymbol{k}_N,\alpha_1\ldots\alpha_N} a_l(\boldsymbol{k}_1\ldots\boldsymbol{k}_N,\alpha_1\ldots\alpha_N)\,|\Phi_{\boldsymbol{k}_1\ldots\boldsymbol{k}_N,\alpha_1\ldots\alpha_N}\rangle \quad (3.161)$$

$$|\Phi_{\boldsymbol{k}_1\ldots\boldsymbol{k}_N,\alpha_1\ldots\alpha_N}\rangle = \hat{c}^\dagger_{\boldsymbol{k}_1\alpha_1}\cdots\hat{c}^\dagger_{\boldsymbol{k}_N\alpha_N}|0\rangle\,. \quad (3.162)$$

In addition, Ψ_l^N can be chosen to be an eigenstate of all translations by Bravais vectors $\boldsymbol{R}_{\boldsymbol{n}}$, as the Hamiltonian commutes with these translations,

$$\Psi_l^N(\boldsymbol{r}_1+\boldsymbol{R}_{\boldsymbol{n}},\sigma_1,\ldots\boldsymbol{r}_N+\boldsymbol{R}_{\boldsymbol{n}},\sigma_N) = e^{i\boldsymbol{R}_{\boldsymbol{n}}\cdot\boldsymbol{\kappa}_l^N}\,\Psi_l^N(\boldsymbol{r}_1,\sigma_1,\ldots\boldsymbol{r}_N,\sigma_N)\,. \quad (3.163)$$

If one applies this relation to the right-hand side of the expansion (3.161), one finds a condition for the total crystal momentum of the basis functions, which is valid for each of the terms in the expansion,

$$\boldsymbol{\kappa}_l^N = \sum_{i=1}^N \boldsymbol{k}_i\,. \quad (3.164)$$

Now consider the amplitude which defines $G(\boldsymbol{k}\alpha t, \boldsymbol{k}'\alpha' t')$,

$$\langle\Psi_0|\hat{c}^\dagger_{\boldsymbol{k}'\alpha'}(t')\hat{c}_{\boldsymbol{k}\alpha}(t)|\Psi_0\rangle = \sum_l \langle\Psi_0^N|\hat{c}^\dagger_{\boldsymbol{k}'\alpha'}|\Psi_l^{N-1}\rangle\langle\Psi_l^{N-1}|\hat{c}_{\boldsymbol{k}\alpha}|\Psi_0^N\rangle \\ \times e^{-i(E_0^N-E_l^{N-1})(t-t')} . \tag{3.165}$$

If one inserts the expansion (3.161) into the amplitude $\langle\Psi_l^{N-1}|\hat{c}_{\boldsymbol{k}\alpha}|\Psi_0^N\rangle$,

$$\langle\Psi_l^{N-1}|\hat{c}_{\boldsymbol{k}\alpha}|\Psi_0^N\rangle = \sum_{\boldsymbol{k}'_1\ldots\boldsymbol{k}'_{N-1},\alpha'_1\ldots\alpha'_{N-1}} \sum_{\boldsymbol{k}_1\ldots\boldsymbol{k}_N,\alpha_1\ldots\alpha_N} a_l^*(\boldsymbol{k}'_1\ldots\boldsymbol{k}'_{N-1},\alpha'_1\ldots\alpha'_{N-1}) \\ \times a_0(\boldsymbol{k}_1\ldots\boldsymbol{k}_N,\alpha_1\ldots\alpha_N) \\ \times\langle 0|\hat{c}_{\boldsymbol{k}'_{N-1}\alpha'_{N-1}}\cdots\hat{c}_{\boldsymbol{k}'_1\alpha'_1}\hat{c}_{\boldsymbol{k}\alpha}\hat{c}^\dagger_{\boldsymbol{k}_1\alpha_1}\cdots\hat{c}^\dagger_{\boldsymbol{k}_N\alpha_N}|0\rangle ,$$

one notices that only those terms in the expansion are non-vanishing, for which

$$\boldsymbol{\kappa}_l^{N-1} = \sum_{i=1}^{N-1}\boldsymbol{k}'_i = \sum_{i=1}^{N}\boldsymbol{k}_i - \boldsymbol{k} = \boldsymbol{\kappa}_0^N - \boldsymbol{k} .$$

This argument can be repeated for the second amplitude on the right-hand side of (3.165), $\langle\Psi_0^N|\hat{c}^\dagger_{\boldsymbol{k}'\alpha'}|\Psi_l^{N-1}\rangle$, with the result

$$\boldsymbol{\kappa}_l^{N-1} = \sum_{i=1}^{N-1}\boldsymbol{k}'_i = \sum_{i=1}^{N}\boldsymbol{k}_i - \boldsymbol{k}' = \boldsymbol{\kappa}_0^N - \boldsymbol{k}' .$$

Since both conditions have to be satisfied simultaneously in the amplitude (3.165), one concludes that $\langle\Psi_0|\hat{c}_{\boldsymbol{k}\alpha}(t)\hat{c}^\dagger_{\boldsymbol{k}'\alpha'}(t')|\Psi_0\rangle$ can only be non-zero, if $\boldsymbol{k} = \boldsymbol{k}'$: the interacting Green's function is also diagonal,

$$G(\boldsymbol{k}\alpha, \boldsymbol{k}'\alpha', \omega) = \delta_{\boldsymbol{k}\boldsymbol{k}'}\, G(\boldsymbol{k}, \alpha\alpha', \omega) . \tag{3.166}$$

The connection between $G(\boldsymbol{k}, \alpha\alpha', \omega)$ and its noninteracting limit $G_0(\boldsymbol{k}, \alpha, \omega)$ is again given by the Dyson equation. Taking into account that the full G is not diagonal with respect to the band index, one obtains

$$G(\boldsymbol{k}, \alpha\alpha', \omega) = \delta_{\alpha\alpha'}\, G_0(\boldsymbol{k}, \alpha, \omega) \\ + G_0(\boldsymbol{k}, \alpha, \omega)\sum_{\alpha''}\Sigma(\boldsymbol{k}, \alpha\alpha'', \omega)G(\boldsymbol{k}, \alpha''\alpha', \omega) . \tag{3.167}$$

This relation may either be considered as definition of $\Sigma(\boldsymbol{k}, \alpha\alpha'', \omega)$ or it may be derived from (3.123) by expansion of the self-energy in terms of Bloch states, similar to Eq. (3.143), and use of the orthonormality relation (3.139) as well as of the diagonality of both G and G_0 with respect to $\boldsymbol{k}, \boldsymbol{k}'$. If a matrix notation for the band index is introduced,

$$\underline{\underline{G}}(\boldsymbol{k}\omega) = \big(G(\boldsymbol{k}, \alpha\alpha', \omega)\big) \qquad \underline{\underline{\Sigma}}(\boldsymbol{k}\omega) = \big(\Sigma(\boldsymbol{k}, \alpha\alpha', \omega)\big) , \tag{3.168}$$

the Dyson equation may be rewritten as

$$\underline{\underline{G}}^{-1}(\boldsymbol{k}\omega) = \underline{\underline{G}}_0^{-1}(\boldsymbol{k}\omega) - \underline{\underline{\Sigma}}(\boldsymbol{k}\omega) \qquad (3.169)$$

($\underline{\underline{G}}_0$ is a diagonal matrix, so that $\underline{\underline{G}}_0^{-1}$ is directly available).

However, the diagonal elements of $\Sigma(\boldsymbol{k},\alpha\alpha'',\omega)$ are usually much larger than its off-diagonal elements. If the off-diagonal terms of $\Sigma(\boldsymbol{k},\alpha\alpha'',\omega)$ are neglected, Eq. (3.167) reduces to

$$G(\boldsymbol{k},\alpha\alpha,\omega) = G_0(\boldsymbol{k},\alpha,\omega) + G_0(\boldsymbol{k},\alpha,\omega)\Sigma(\boldsymbol{k},\alpha\alpha,\omega)G(\boldsymbol{k},\alpha\alpha,\omega). \qquad (3.170)$$

As in the general situation, this Dyson equation may be resummed by use of G_0^{-1},

$$G(\boldsymbol{k},\alpha\alpha,\omega) = \frac{1}{\omega - \varepsilon_{\boldsymbol{k}\alpha}/\hbar - \Sigma(\boldsymbol{k},\alpha\alpha,\omega)}, \qquad (3.171)$$

where the imaginary part of $\Sigma(\boldsymbol{k},\alpha\alpha,\omega)$ now plays the role of the pole shift η. Moreover, even if the off-diagonal elements of $\underline{\underline{\Sigma}}$ are kept, an equation of this type can be obtained by diagonalization of (3.169), which implies the transition to a new set of quasi-particles/holes.

Now compare the result (3.171) with the diagonal element of the Lehmann representation (3.157),

$$G(\boldsymbol{k},\alpha\alpha,\omega) = \sum_l \left\{ \frac{|g_l(\boldsymbol{k}\alpha)|^2}{\omega - (\mu + \Omega_l^{N+1})/\hbar + i\eta} + \frac{|f_l(\boldsymbol{k}\alpha)|^2}{\omega - (\mu - \Omega_l^{N-1})/\hbar - i\eta} \right\}. \qquad (3.172)$$

As indicated before, $G(\boldsymbol{k},\alpha\alpha,\omega)$ represents the propagation of a superposition of infinitely many eigenstates of the $(N \pm 1)$-particle systems,

$$\begin{aligned} G(\boldsymbol{k},\alpha\alpha,t-t') = &-i\Theta(t-t')\sum_l |g_l(\boldsymbol{k}\alpha)|^2 e^{-i(\mu+\Omega_l^{N+1})(t-t')/\hbar} \\ &+i\Theta(t'-t)\sum_l |f_l(\boldsymbol{k}\alpha)|^2 e^{-i(\mu-\Omega_l^{N-1})(t-t')/\hbar}. \end{aligned} \qquad (3.173)$$

In general, this superposition is not an eigenstate itself, but rather a wave packet which is damped out in time, a so-called *quasi-particle* (for $t > t'$) or *quasi-hole* (for $t < t'$)—which finally explains the names of the corresponding amplitudes. The longer the lifetime of this wave packet is, the closer is the wave packet to an actual eigenstate. The energies and lifetimes of the quasi-particles/holes are determined by the complex poles

$$\omega_{\mathrm{p}}(\boldsymbol{k}\alpha) = \tilde{\varepsilon}(\boldsymbol{k}\alpha)/\hbar - i\Gamma(\boldsymbol{k}\alpha); \qquad \tilde{\varepsilon}(\boldsymbol{k}\alpha), \Gamma(\boldsymbol{k}\alpha) \in \mathbb{R} \qquad (3.174)$$

of the Dyson denominator (3.171),

$$\omega_{\rm p}(\boldsymbol{k}\alpha) - \varepsilon_{\boldsymbol{k}\alpha}/\hbar - \Sigma(\boldsymbol{k},\alpha\alpha,\omega_{\rm p}(\boldsymbol{k}\alpha)) = 0\,. \tag{3.175}$$

For frequencies close to the pole an expansion of Σ around this pole can be used,

$$\Sigma(\boldsymbol{k},\alpha\alpha,\omega) = \Sigma(\boldsymbol{k},\alpha\alpha,\omega_{\rm p}(\boldsymbol{k}\alpha)) + \Sigma'(\boldsymbol{k},\alpha\alpha,\omega_{\rm p}(\boldsymbol{k}\alpha))(\omega-\omega_{\rm p}(\boldsymbol{k}\alpha)) + \ldots\,.$$

This allows to rewrite the Green's function close to the pole as

$$G(\boldsymbol{k},\alpha\alpha,\omega) = \frac{z(\boldsymbol{k}\alpha)}{\omega-\tilde{\varepsilon}(\boldsymbol{k}\alpha)/\hbar+i\Gamma(\boldsymbol{k}\alpha)} + \Delta G(\omega) \tag{3.176}$$

$$z(\boldsymbol{k}\alpha) = \frac{1}{1-\Sigma'(\boldsymbol{k},\alpha\alpha,\omega_{\rm p}(\boldsymbol{k}\alpha))}\,, \tag{3.177}$$

where ΔG denotes the remaining terms of the Laurent expansion, which are finite at $\omega = \omega_{\rm p}(\boldsymbol{k}\alpha)$. Transformation of this result to the time domain,

$$\begin{aligned} G(\boldsymbol{k},\alpha\alpha,t-t') = {} & -iz(\boldsymbol{k}\alpha)\Theta(t-t')\Theta(\Gamma(\boldsymbol{k}\alpha))e^{-i\tilde{\varepsilon}(\boldsymbol{k}\alpha)(t-t')/\hbar-\Gamma(\boldsymbol{k}\alpha)(t-t')} \\ & +iz(\boldsymbol{k}\alpha)\Theta(t'-t)\Theta(-\Gamma(\boldsymbol{k}\alpha))e^{i\tilde{\varepsilon}(\boldsymbol{k}\alpha)(t'-t)/\hbar+\Gamma(\boldsymbol{k}\alpha)(t'-t)} \\ & +\Delta G(t-t')\,, \end{aligned} \tag{3.178}$$

provides a direct interpretation of $\tilde{\varepsilon}$ and Γ: $\tilde{\varepsilon}(\boldsymbol{k}\alpha)$ represents the quasi-particle/hole energy, while $1/|\Gamma(\boldsymbol{k}\alpha)|$ is the lifetime of the resonance.

As is clear from (3.173), the threshold between quasi-particle and quasi-hole energies is exactly the chemical potential μ (remember that $\Omega_l^{N\pm1} \geq 0$). The Fermi surface of the interacting system is therefore determined by the solutions of (3.175) for which

$$\Gamma(\boldsymbol{k}\alpha) = \Im\left\{\Sigma(\boldsymbol{k},\alpha\alpha,\omega_{\rm p}(\boldsymbol{k}\alpha))\right\} = 0\,.$$

The associated real part of $\hbar\omega_{\rm p}(\boldsymbol{k}\alpha)$,

$$\tilde{\varepsilon}(\boldsymbol{k}\alpha) = \varepsilon_{\boldsymbol{k}\alpha} + \hbar\Re\left\{\Sigma(\boldsymbol{k},\alpha\alpha,\omega_{\rm p}(\boldsymbol{k}\alpha))\right\}\,,$$

is the chemical potential μ. For this particular energy the quasi-particle/hole in (3.178) is not damped out, but represents an eigenstate of the interacting system, i.e. has an infinite lifetime.[4] The Fermi surface of the interacting system is thus defined as the set of $\boldsymbol{k}$-values for which[5]

---

[4] Quite generally, the lifetime of the quasi-particles/holes increases if $\tilde{\varepsilon}(\boldsymbol{k}\alpha)$ approaches μ. For energies sufficiently close to the Fermi surface the quasi-particles/holes are therefore accurate representations of the true excitations of the system (which provides the basis for Landau's Fermi-liquid theory).

[5] The value of μ may alternatively be fixed by the particle number constraint. If there are N_Ω particles in the volume Ω of the unit cell, Eq. (3.106) yields

$$N_\Omega = \int_\Omega d^3r\, n(\boldsymbol{r}) = -i\sum_\sigma\int_\Omega d^3r\, G(\boldsymbol{r}\sigma t,\boldsymbol{r}\sigma t^+) = \sum_\sigma\int_\Omega d^3r\sum_l |f_l(\boldsymbol{r}\sigma)|^2\,.$$

$$\mu - \varepsilon_{k\alpha} - \hbar\Re\{\Sigma(\boldsymbol{k},\alpha\alpha,\mu/\hbar)\} = 0 \quad \Longleftrightarrow \quad \tilde{\varepsilon}_{\mathrm{F}}(\boldsymbol{k}\alpha) = \mu\,. \tag{3.179}$$

Equation (3.179) implies that different bands α may be cut by the Fermi surface for different momenta $\boldsymbol{k}$. Different bands α define different branches of the Fermi surface, provided that (3.179) has a solution for a given band α.

The KS Fermi surface is, in analogy to Eq. (3.179), defined by the set of all KS eigenvalues $\varepsilon_{k\alpha}$ (from now on ε_l again denotes a KS eigenvalue) which agree with the KS Fermi level ε_{F}, i.e. the KS chemical potential,

$$\varepsilon_{k\alpha} = \varepsilon_{\mathrm{F}}\,. \tag{3.180}$$

For given E_{xc} the KS equations thus allow the determination of the KS Fermi surface.

This concludes the discussion of the definition of the two Fermi surfaces. The question concerning their identity can now be answered. It is sufficient to examine the simplest periodic system, the weakly inhomogeneous electron gas, since this system represents an explicit counterexample. The weakly inhomogeneous gas is an extension of the homogeneous gas (for details see Chap. 4), which is obtained by introducing a weak perturbation

$$v_{\mathrm{ext}}(\boldsymbol{r}) = v_0 \cos(\boldsymbol{Q}\cdot\boldsymbol{r})\,.$$

This perturbation induces a one-dimensional Bravais lattice in the direction of $\boldsymbol{Q}$. The Fermi surfaces (3.179) and (3.180) of this system can be evaluated explicitly to first order in the electron–electron interaction (utilizing (3.125)) and to lowest order in v_0 (via the linear response formalism). One finds that the two surfaces differ in the order $v_0^2 w$ [99, 100] (there is no contribution linear in v_0 to the Fermi surface). Notwithstanding this exact result, one finds in practice that the KS Fermi surface is often rather close to the exact Fermi surface.

3.6.3 Band Gap

Turning from metals with their Fermi surface to insulators and semiconductors, the obvious question is whether the energy gap in the spectrum of these systems can be described by the KS scheme? The so-called *fundamental band gap* E_{g} is precisely defined as the difference between the binding energy of the most weakly

However, using the strict periodicity of the product $\phi^*_{k\alpha'}(\boldsymbol{r}\sigma)\phi_{k\alpha}(\boldsymbol{r}\sigma)$ as well as the orthonormality relation (3.139), one can express N_Ω also in terms of the $f_l(\boldsymbol{k}\alpha)$,

$$\begin{aligned} N_\Omega &= -i\,\frac{1}{N_1N_2N_3}\sum_{k\alpha\alpha'} G(\boldsymbol{k},\alpha\alpha',t-t^+)\sum_\sigma \int_{N_1N_2N_3\Omega} d^3r\,\phi^*_{k\alpha'}(\boldsymbol{r}\sigma)\phi_{k\alpha}(\boldsymbol{r}\sigma) \\ &= \frac{1}{N_1N_2N_3}\sum_{k\alpha}\sum_l |f_l(\boldsymbol{k}\alpha)|^2\,. \end{aligned}$$

bound electron in the ground state of the insulator or semiconductor and that of the most weakly bound electron in the ground state of the system obtained by adding one electron. Identifying the neutral solid with the N-particle system, one can write [101, 102]

$$E_g = -\left\{\left[E_0^N - E_0^{N-1}\right] - \left[E_0^{N+1} - E_0^N\right]\right\}, \tag{3.181}$$

where the sign convention ensures that E_g is positive. The name band gap originates from the fact that the most weakly bound electron of the insulator or semiconductor belongs to the highest occupied band, the valence band, while the $(N+1)$-th electron goes into the lowest empty band, the conduction band. The band gap is easily evaluated for noninteracting systems as the KS system,

$$\Delta_s = \varepsilon_{N+1} - \varepsilon_N , \tag{3.182}$$

as the ionization energies are identical with the orbital energies, $E_0^N - E_0^{N-1} = \varepsilon_N$ (we now return to the short-hand notation for the KS quantum numbers).

Using the earlier results (2.118) and (2.119), the exact band gap can be rewritten as

$$E_g = \mathrm{IP} - \mathrm{EA} = \lim_{\eta \to 0^+} \left\{\mu(N+\eta) - \mu(N-\eta)\right\}. \tag{3.183}$$

On the basis of the variational equation (2.111) one then finds

$$E_g = \lim_{\eta \to 0^+} \left\{ \left.\frac{\delta E[n]}{\delta n(\boldsymbol{r})}\right|_{N+\eta} - \left.\frac{\delta E[n]}{\delta n(\boldsymbol{r})}\right|_{N-\eta} \right\}, \tag{3.184}$$

with the understanding that $\delta E[n]/\delta n$ is evaluated with the ground state density corresponding to the particle number $N \pm \eta$. Finally, the total energy functional can be decomposed in the standard fashion (3.24). Utilizing the continuous dependence of $E_{\rm ext}$ and $E_{\rm H}$ on the particle number, one arrives at

$$\begin{aligned} E_g &= \lim_{\eta \to 0^+} \left\{ \left[\left.\frac{\delta T_s[n]}{\delta n(\boldsymbol{r})}\right|_{N+\eta} - \left.\frac{\delta T_s[n]}{\delta n(\boldsymbol{r})}\right|_{N-\eta} \right] + \left[\left.\frac{\delta E_{\rm xc}[n]}{\delta n(\boldsymbol{r})}\right|_{N+\eta} - \left.\frac{\delta E_{\rm xc}[n]}{\delta n(\boldsymbol{r})}\right|_{N-\eta} \right] \right\} \\ &= \Delta_s + \Delta_{\rm xc} . \end{aligned} \tag{3.185}$$

In the second line the contribution of T_s to the band gap has already been identified with Δ_s, Eq. (3.182), as T_s is the only part of the total energy of the KS system which shows a derivative discontinuity. The total band gap thus consists of the difference between the highest occupied and the lowest unoccupied KS eigenvalue and an xc-contribution [101, 102],

$$E_g = \varepsilon_{N+1} - \varepsilon_N + \Delta_{\rm xc} . \tag{3.186}$$

In practice, $\Delta_{\rm xc}$ is often ignored. This is completely legitimate for approximate xc-functionals which do not show a derivative discontinuity, such as the LDA (for details on this xc-functional, see Sect. 4.3). On the other hand, the exact $E_{\rm xc}$ leads to a non-vanishing $\Delta_{\rm xc}$. This can be explicitly demonstrated to first order in the electron–electron coupling constant e^2, i.e. for the exchange contribution $\Delta_{\rm x}$ (for the precise definition of exchange within DFT, see Sect. 4.1). Table 3.1 lists $\Delta_{\rm x}$ together with the band gaps resulting from LDA calculations and the experimental data for two prototype systems. These numbers show, that the eigenvalue difference

Table 3.1 Band gap of insulators and semiconductors: Exact $\Delta_{\rm x}$ [103] versus $\Delta_{\rm s}$ obtained within the LDA and experimental gap.

System	$\Delta_{\rm s}$(LDA) [eV]	$\Delta_{\rm x}$ [eV]	Expt. [eV]
C	4.16	8.70	5.48
Si	0.49	5.62	1.17

$\Delta_{\rm s}$ obtained with the standard approximations to $E_{\rm xc}[n]$ underestimates the true band gaps of semiconductors considerably. On the other hand, $\Delta_{\rm x}$ is surprisingly large, indicating that the correlation contribution to $\Delta_{\rm xc}$ must also be rather large (and negative) in order to compensate $\Delta_{\rm x}$ (for a more detailed discussion see at the end of Sect. 6.3).

3.7 Kohn-Sham Equations of Current Density Functional Theory

The extension of the KS scheme to the case of C(S)DFT (Sect. 2.6) proceeds along the same lines as the extension to spin-density functional theory [74, 68]. One starts with introducing auxiliary orbitals to represent all relevant ground state densities of the interacting system (assuming noninteracting v-representability, as usual). In the case of CDFT, which is considered first, these are the density,

$$n(\boldsymbol{r}) = \sum_{k\sigma} \Theta_k \phi_k^*(\boldsymbol{r}\sigma)\phi_k(\boldsymbol{r}\sigma) , \tag{3.187}$$

and the paramagnetic current density,

$$\boldsymbol{j}_{\rm p}(\boldsymbol{r}) = -\frac{i\hbar}{2m}\sum_{k\sigma} \Theta_k \{\phi_k^*(\boldsymbol{r}\sigma)\,[\boldsymbol{\nabla}\phi_k(\boldsymbol{r}\sigma)] - [\boldsymbol{\nabla}\phi_k^*(\boldsymbol{r}\sigma)]\,\phi_k(\boldsymbol{r}\sigma)\} . \tag{3.188}$$

In the next step, the standard decomposition (3.24) is used for the total energy functional (2.174),

$$E[n, \boldsymbol{j}_\mathrm{p}] = T_\mathrm{s}[n, \boldsymbol{j}_\mathrm{p}] + E_\mathrm{ext}[n, \boldsymbol{j}_\mathrm{p}] + E_\mathrm{H}[n] + E_\mathrm{xc}[n, \boldsymbol{j}_\mathrm{p}] , \quad (3.189)$$

with T_s and E_H given by Eqs. (3.21) and (3.25), respectively, and E_ext defined as

$$\begin{aligned} E_\mathrm{ext}[n, \boldsymbol{j}_\mathrm{p}] &= \int d^3r \left[v_\mathrm{ext}(\boldsymbol{r}) + \frac{e^2}{2mc^2} \boldsymbol{A}^2_\mathrm{ext}(\boldsymbol{r}) \right] n(\boldsymbol{r}) \\ &\quad + \frac{e}{c} \int d^3r\, \boldsymbol{j}_\mathrm{p}(\boldsymbol{r}) \cdot \boldsymbol{A}_\mathrm{ext}(\boldsymbol{r}) . \end{aligned} \quad (3.190)$$

As usual, the decomposition (3.189) serves as definition of the xc-functional, which is $\boldsymbol{j}_\mathrm{p}$-dependent in the present situation, $E_\mathrm{xc}[n, \boldsymbol{j}_\mathrm{p}]$.

One arrives at the KS equations of CDFT [74], if one follows the argument in Sect. 3.1 or, alternatively, minimizes $E[n, \boldsymbol{j}_\mathrm{p}]$ with respect to the auxiliary orbitals under the constraint of proper normalization,

$$\begin{aligned} &\left\{ \frac{1}{2m} \left[-i\hbar\boldsymbol{\nabla} + \frac{e}{c} (\boldsymbol{A}_\mathrm{ext}(\boldsymbol{r}) + \boldsymbol{A}_\mathrm{xc}(\boldsymbol{r})) \right]^2 - \frac{e^2}{mc^2} \left[\frac{1}{2} \boldsymbol{A}_\mathrm{xc}(\boldsymbol{r})^2 + \boldsymbol{A}_\mathrm{ext}(\boldsymbol{r}) \cdot \boldsymbol{A}_\mathrm{xc}(\boldsymbol{r}) \right] \right. \\ &\left. + v_\mathrm{ext}(\boldsymbol{r}) + v_\mathrm{H}(\boldsymbol{r}) + v_\mathrm{xc}(\boldsymbol{r}) \right\} \phi_k(\boldsymbol{r}\sigma) = \varepsilon_k \phi_k(\boldsymbol{r}\sigma) . \end{aligned} \quad (3.191)$$

Here the xc-potentials are defined as

$$v_\mathrm{xc}(\boldsymbol{r}) = \left. \frac{\delta E_\mathrm{xc}[n, \boldsymbol{j}_\mathrm{p}]}{\delta n(\boldsymbol{r})} \right|_{\boldsymbol{j}_\mathrm{p}} \quad (3.192)$$

$$\frac{e}{c} \boldsymbol{A}_\mathrm{xc}(\boldsymbol{r}) = \left. \frac{\delta E_\mathrm{xc}[n, \boldsymbol{j}_\mathrm{p}]}{\delta \boldsymbol{j}_\mathrm{p}(\boldsymbol{r})} \right|_n . \quad (3.193)$$

The notation indicates that the second density variable has to be kept constant in the variation of E_xc with respect to the other (v_H is given by Eq. (3.42)).

In view of the gauge freedom of $\boldsymbol{A}_\mathrm{ext}$ the gauge properties of the KS formalism for CDFT are of obvious interest. The crucial statement is: the KS equations (3.191) are gauge covariant [68], as will be shown in the following. Let us start with a closer look at the ground state Φ_0 of noninteracting particles, expressed as a current-density functional, $\Phi_0[n, \boldsymbol{j}_\mathrm{p}]$. The gauge transformation (2.148) corresponds to the phase transformation (2.150) of the interacting ground state wavefunction. An analogous transformation has to hold for the corresponding ground state current-density functional of interacting particles,

$$\Psi_0[n', \boldsymbol{j}'_\mathrm{p}](\boldsymbol{r}_1, \ldots) = \exp\left[\frac{ie}{\hbar c} \sum_{k=1}^N \lambda(\boldsymbol{r}_k) \right] \Psi_0[n, \boldsymbol{j}_\mathrm{p}](\boldsymbol{r}_1, \ldots) . \quad (3.194)$$

The same transformation is also valid for $\Phi_0[n, \boldsymbol{j}_\mathrm{p}]$, which is a simple limit of $\Psi_0[n, \boldsymbol{j}_\mathrm{p}]$,

$$\Phi_0'[n',\boldsymbol{j}_p'](\boldsymbol{r}_1,\ldots) = \exp\left[\frac{ie}{\hbar c}\sum_{k=1}^{N}\lambda(\boldsymbol{r}_k)\right]\Phi_0[n,\boldsymbol{j}_p](\boldsymbol{r}_1,\ldots)\,. \tag{3.195}$$

This result can, alternatively, be derived explicitly for the KS ground state determinant Φ_0 (consisting of the single-particle orbitals of Eq. (3.191)) by the following argument: Since Φ_0 yields the exact density, the invariance of $n(\boldsymbol{r})$ under the transformation (2.148) only allows a local phase transformation of Φ_0. Due to the single-particle nature of both Φ_0 and the transformed function Φ_0' the local phase has to be a simple sum over single-particle phase functions depending on one coordinate only. The phase function is therefore uniquely determined by the transformation behavior (2.163) of $\boldsymbol{j}_p$, which must be reproduced by Φ_0, Φ_0'.

As a consequence, the noninteracting kinetic energy functional $T_s[n,\boldsymbol{j}_p]$ transforms as

$$\begin{aligned} T_s[n',\boldsymbol{j}_p'] &= T_s[n,\boldsymbol{j}_p] + \frac{e}{c}\int d^3r\,\boldsymbol{j}_p(\boldsymbol{r})\cdot\nabla\lambda(\boldsymbol{r}) \\ &\quad + \int d^3r\,n(\boldsymbol{r})\frac{e^2}{2mc^2}[\nabla\lambda(\boldsymbol{r})]^2\,. \end{aligned} \tag{3.196}$$

However, both the total energy (3.189) and the combination $T_s + E_{ext}$ are invariant under the gauge transformation (2.148) plus the associated phase transformations of the ground state wavefunctions. This gauge invariance is reflected by the corresponding functionals,

$$\begin{aligned} E_{\boldsymbol{A}_{ext}'}[n',\boldsymbol{j}_p'] &= E_{\boldsymbol{A}_{ext}}[n,\boldsymbol{j}_p] \\ T_s[n',\boldsymbol{j}_p'] + E_{ext,\boldsymbol{A}_{ext}'}[n',\boldsymbol{j}_p'] &= T_s[n,\boldsymbol{j}_p] + E_{ext,\boldsymbol{A}_{ext}}[n,\boldsymbol{j}_p]\,, \end{aligned}$$

where the additional index indicates the vector potential present in the external energy. As a result the xc-energy functional is gauge invariant as well,

$$E_{xc}[n',\boldsymbol{j}_p'] = E_{xc}[n,\boldsymbol{j}_p]\,. \tag{3.197}$$

As n is gauge invariant by itself, one arrives at

$$E_{xc}[n,\boldsymbol{j}_p'] = E_{xc}[n,\boldsymbol{j}_p]\,. \tag{3.198}$$

Moreover, as $E_{xc}[n,\boldsymbol{j}_p]$ is universal in the sense that the functional dependence of E_{xc} on n and $\boldsymbol{j}_p$ is identical for all external potentials, the functional dependence on $\boldsymbol{j}_p$ itself must have a form which ensures Eq. (3.198). The only local, gauge invariant quantity which can be constructed from $\boldsymbol{j}_p$ with its transformation behavior (2.163) is the vorticity

$$\boldsymbol{v}(\boldsymbol{r}) = \nabla\times\left[\frac{\boldsymbol{j}_p(\boldsymbol{r})}{n(\boldsymbol{r})}\right]\,. \tag{3.199}$$

E_{xc} can thus be expressed as a functional of $\boldsymbol{v}$,

$$E_{xc}[n,\boldsymbol{j}_p] \equiv \tilde{E}_{xc}[n,\boldsymbol{v}]\,. \tag{3.200}$$

The functional form (3.200) has important consequences. Let us first consider the functional derivatives (3.192) and (3.193). The derivative (3.193) can be directly evaluated with the chain rule, keeping track of the quantities which have to be kept constant in the variational process,

$$\frac{e}{c}\boldsymbol{A}_{\mathrm{xc}}(\boldsymbol{r}) = \int d^3r' \left.\frac{\delta\tilde{E}_{\mathrm{xc}}[n,\boldsymbol{v}]}{\delta\boldsymbol{v}(\boldsymbol{r}')}\right|_n \cdot \left.\frac{\delta\boldsymbol{v}(\boldsymbol{r}')}{\delta\boldsymbol{j}_{\mathrm{p}}(\boldsymbol{r})}\right|_n$$
$$= \frac{1}{n(\boldsymbol{r})}\boldsymbol{\nabla}\times\left.\frac{\delta\tilde{E}_{\mathrm{xc}}[n,\boldsymbol{v}]}{\delta\boldsymbol{v}(\boldsymbol{r})}\right|_n . \tag{3.201}$$

With this result one obtains for the derivative (3.192)

$$v_{\mathrm{xc}}(\boldsymbol{r}) = \left.\frac{\delta\tilde{E}_{\mathrm{xc}}[n,\boldsymbol{v}]}{\delta n(\boldsymbol{r})}\right|_{\boldsymbol{v}} + \int d^3r' \left.\frac{\delta\tilde{E}_{\mathrm{xc}}[n,\boldsymbol{v}]}{\delta\boldsymbol{v}(\boldsymbol{r}')}\right|_n \cdot \left.\frac{\delta\boldsymbol{v}(\boldsymbol{r}')}{\delta n(\boldsymbol{r})}\right|_{\boldsymbol{j}_{\mathrm{p}}}$$
$$= \left.\frac{\delta\tilde{E}_{\mathrm{xc}}[n,\boldsymbol{v}]}{\delta n(\boldsymbol{r})}\right|_{\boldsymbol{v}} - \frac{\boldsymbol{j}_{\mathrm{p}}(\boldsymbol{r})}{n(\boldsymbol{r})^2}\cdot\boldsymbol{\nabla}\times\left.\frac{\delta\tilde{E}_{\mathrm{xc}}[n,\boldsymbol{v}]}{\delta\boldsymbol{v}(\boldsymbol{r})}\right|_n$$
$$= \left.\frac{\delta\tilde{E}_{\mathrm{xc}}[n,\boldsymbol{v}]}{\delta n(\boldsymbol{r})}\right|_{\boldsymbol{v}} - \frac{\boldsymbol{j}_{\mathrm{p}}(\boldsymbol{r})}{n(\boldsymbol{r})}\cdot\frac{e}{c}\boldsymbol{A}_{\mathrm{xc}}(\boldsymbol{r}) . \tag{3.202}$$

Equations (3.201) and (3.202) allow the determination of the gauge properties of the xc-potentials. In Eq. (3.201) $\boldsymbol{A}_{\mathrm{xc}}$ is expressed completely in terms of gauge invariant quantities, so that it is gauge invariant itself,

$$\boldsymbol{A}_{\mathrm{xc}}[n',\boldsymbol{j}'_{\mathrm{p}}] = \boldsymbol{A}_{\mathrm{xc}}[n,\boldsymbol{j}_{\mathrm{p}}] . \tag{3.203}$$

Similarly, Eq. (3.202) shows that v_{xc} transforms as

$$v_{\mathrm{xc}}[n',\boldsymbol{j}'_{\mathrm{p}}] = v_{\mathrm{xc}}[n,\boldsymbol{j}_{\mathrm{p}}] - \frac{e^2}{mc^2}\boldsymbol{A}_{\mathrm{xc}}[n,\boldsymbol{j}_{\mathrm{p}}]\cdot\boldsymbol{\nabla}\lambda , \tag{3.204}$$

where the transformation behavior of the paramagnetic current, Eq. (2.163), and (3.203) have been utilized.

With this information one can finally analyze the gauge properties of the KS equations (3.191), in order to confirm Eq. (3.195) explicitly. After insertion of Eq. (3.202) one can introduce the physical current (2.156) (for the case of an unpolarized system) and rewrite (3.191) as

$$\left\{\frac{1}{2m}\left[-i\hbar\boldsymbol{\nabla} + \frac{e}{c}\left(\boldsymbol{A}_{\mathrm{ext}}(\boldsymbol{r}) + \boldsymbol{A}_{\mathrm{xc}}(\boldsymbol{r})\right)\right]^2 + v_{\mathrm{ext}}(\boldsymbol{r}) + v_{\mathrm{H}}(\boldsymbol{r}) + \left.\frac{\delta\tilde{E}_{\mathrm{xc}}[n,\boldsymbol{v}]}{\delta n(\boldsymbol{r})}\right|_{\boldsymbol{v}}\right.$$
$$\left. - \frac{e}{c}\boldsymbol{A}_{\mathrm{xc}}(\boldsymbol{r})\left[\frac{\boldsymbol{j}(\boldsymbol{r})}{n(\boldsymbol{r})} + \frac{e}{2mc}\boldsymbol{A}_{\mathrm{xc}}(\boldsymbol{r})\right]\right\}\phi_k(\boldsymbol{r}\sigma) = \varepsilon_k\phi_k(\boldsymbol{r}\sigma) . \tag{3.205}$$

Gauge covariance of the KS equations then follows directly from the *gauge invariance* of $\boldsymbol{A}_{\rm xc}$, $\boldsymbol{v}$, n and $\boldsymbol{j}$—under a gauge transformation of the external vector potential the KS orbitals transform in the same way as the orbitals of a truly noninteracting system, consistent with the starting point of the analysis, Eq. (3.195).[6]

Combination of Eqs. (3.201) and (3.202) with the KS equations also allows a verification of the conservation of the physical current (2.156). From Eq. (3.201) one immediately finds

$$\nabla \cdot [n(\boldsymbol{r})\boldsymbol{A}_{\rm xc}(\boldsymbol{r})] = 0 \,. \tag{3.206}$$

On the other hand, the standard combination of (3.191) with its hermitian conjugate yields

$$\nabla \cdot \left\{ \boldsymbol{j}_{\rm p}(\boldsymbol{r}) + \frac{e}{mc} n(\boldsymbol{r}) \left[\boldsymbol{A}_{\rm ext}(\boldsymbol{r}) + \boldsymbol{A}_{\rm xc}(\boldsymbol{r}) \right] \right\} = 0 \,.$$

Use of (3.206) and (2.156) confirms the validity of current conservation, Eq. (2.179), in the KS approach to CDFT.

The KS equations of CSDFT, which are listed below, complement the results for CDFT. We first consider the extended variant (2.171), in which n, $\boldsymbol{m}$ and $\boldsymbol{j}_{\rm p}$ are independent variables. For their derivation one combines Eqs. (3.187) and (3.188) with the representation (3.85) of the magnetization density, assuming that all three densities n, $\boldsymbol{m}$ and $\boldsymbol{j}_{\rm p}$ can be reproduced by an effective noninteracting system. In the standard decomposition of the total energy, Eq. (3.189), $T_{\rm s}$, $E_{\rm xc}$ and $E_{\rm ext}$ are now

---

[6] The gauge invariance of $\boldsymbol{v}$, $\boldsymbol{A}_{\rm xc}$ and the effective scalar potential in the KS equations,

$$v_{\rm eff}(\boldsymbol{r}) = v_{\rm xc}(\boldsymbol{r}) - \frac{e^2}{mc^2}\boldsymbol{A}_{\rm ext}(\boldsymbol{r}) \cdot \boldsymbol{A}_{\rm xc}(\boldsymbol{r}) = \left. \frac{\delta \tilde{E}_{\rm xc}[n,\boldsymbol{v}]}{\delta n(\boldsymbol{r})} \right|_{\boldsymbol{v}} - \frac{e}{c}\boldsymbol{A}_{\rm xc}(\boldsymbol{r}) \left[\frac{\boldsymbol{j}(\boldsymbol{r})}{n(\boldsymbol{r})} + \frac{e}{2mc}\boldsymbol{A}_{\rm xc}(\boldsymbol{r}) \right] ,$$

has an interesting consequence for periodic systems [68]. Consider a periodic system with primitive vectors $\boldsymbol{a}_i$, $i = 1,2,3$. In this system all observable densities must be invariant against a translation by some arbitrary lattice vector $\boldsymbol{R}_{\boldsymbol{n}} = n_1\boldsymbol{a}_1 + n_2\boldsymbol{a}_2 + n_3\boldsymbol{a}_3$, $n_i \in \mathbb{Z}$. If the magnetic field vanishes, $\boldsymbol{A}_{\rm ext} = \boldsymbol{0}$, both n and $\boldsymbol{j}_{\rm p} = \boldsymbol{j}$ are periodic, so that also $v_{\rm xc}$ and $\boldsymbol{A}_{\rm xc}$ are periodic by construction. As a consequence, Bloch's theorem directly applies to the KS states.

The situation is no longer as simple, if a homogeneous magnetic field $\boldsymbol{B}_{\rm ext}$ is present. The vector potential corresponding to $\boldsymbol{B}_{\rm ext}$,

$$\boldsymbol{A}_{\rm ext}(\boldsymbol{r}) = \frac{1}{2}\boldsymbol{B}_{\rm ext} \times \boldsymbol{r} \,,$$

is obviously not periodic, so that the same applies to the total Hamiltonian. Fortunately, the term violating periodicity in case of a translation by $\boldsymbol{R}_{\boldsymbol{n}}$,

$$\boldsymbol{A}_{\rm ext}(\boldsymbol{r} + \boldsymbol{R}_{\boldsymbol{n}}) = \boldsymbol{A}_{\rm ext}(\boldsymbol{r}) + \frac{1}{2}\boldsymbol{B}_{\rm ext} \times \boldsymbol{R}_{\boldsymbol{n}} \,,$$

can be viewed as a gauge transformation of $\boldsymbol{A}_{\rm ext}$ by the gauge function

$$\lambda(\boldsymbol{r}) = -\frac{1}{2}(\boldsymbol{B}_{\rm ext} \times \boldsymbol{R}_{\boldsymbol{n}}) \cdot \boldsymbol{r} \,.$$

The ground state wavefunction then acquires an additional phase factor according to (2.150) upon translation by $\boldsymbol{R}_{\boldsymbol{n}}$. However, due to its equivalence to a gauge transformation, this translation does not affect $\boldsymbol{v}$, $\boldsymbol{A}_{\rm xc}$ and $v_{\rm eff}$. It then follows that the conditions for the applicability of Bloch's theorem, which are quite restrictive for non-vanishing $\boldsymbol{B}_{\rm ext}$ [104], are not changed by the presence of the effective potentials.

functionals of n, $\boldsymbol{m}$ and $\boldsymbol{j}_{\rm p}$,

$$E[n,\boldsymbol{j}_{\rm p},\boldsymbol{m}] = T_{\rm s}[n,\boldsymbol{j}_{\rm p},\boldsymbol{m}]+E_{\rm ext}[n,\boldsymbol{j}_{\rm p},\boldsymbol{m}]+E_{\rm H}[n]+E_{\rm xc}[n,\boldsymbol{j}_{\rm p},\boldsymbol{m}]\,, \tag{3.207}$$

and $E_{\rm ext}$ is given by

$$\begin{aligned} E_{\rm ext}[n,\boldsymbol{j}_{\rm p},\boldsymbol{m}] = &\int d^3r\left[v_{\rm ext}(\boldsymbol{r})+\frac{e^2}{2mc^2}\boldsymbol{A}^2_{\rm ext}(\boldsymbol{r})\right]n(\boldsymbol{r}) \\ &+\frac{e}{c}\int d^3r\,\boldsymbol{A}_{\rm ext}(\boldsymbol{r})\cdot\boldsymbol{j}_{\rm p}(\boldsymbol{r})+\int d^3r\,\boldsymbol{B}_{\rm ext}(\boldsymbol{r})\cdot\boldsymbol{m}(\boldsymbol{r})\,. \end{aligned} \tag{3.208}$$

With the definition of the xc-magnetic field,

$$\boldsymbol{B}_{\rm xc}(\boldsymbol{r}) = \left.\frac{\delta E_{\rm xc}[n,\boldsymbol{j}_{\rm p},\boldsymbol{m}]}{\delta\boldsymbol{m}(\boldsymbol{r})}\right|_{n,\boldsymbol{j}_{\rm p}}\,, \tag{3.209}$$

and the xc-potentials,

$$v_{\rm xc}(\boldsymbol{r}) = \left.\frac{\delta E_{\rm xc}[n,\boldsymbol{j}_{\rm p},\boldsymbol{m}]}{\delta n(\boldsymbol{r})}\right|_{\boldsymbol{j}_{\rm p},\boldsymbol{m}} \tag{3.210}$$

$$\frac{e}{c}\boldsymbol{A}_{\rm xc}(\boldsymbol{r}) = \left.\frac{\delta E_{\rm xc}[n,\boldsymbol{j}_{\rm p},\boldsymbol{m}]}{\delta\boldsymbol{j}_{\rm p}(\boldsymbol{r})}\right|_{n,\boldsymbol{m}}\,, \tag{3.211}$$

one arrives (via minimization of the total energy) at the KS equations of CSDFT,

$$\begin{aligned} \sum_{\sigma'}\Bigg\{&\left[\frac{1}{2m}\left[-i\hbar\boldsymbol{\nabla}+\frac{e}{c}\boldsymbol{A}_{\rm ext}\right]^2-i\mu_B\left[(\boldsymbol{\nabla}\cdot\boldsymbol{A}_{\rm xc})+2\boldsymbol{A}_{\rm xc}\cdot\boldsymbol{\nabla}\right]+v_{\rm ext}+v_{\rm H}+v_{\rm xc}\right]\delta_{\sigma\sigma'} \\ &+\mu_B\boldsymbol{\sigma}_{\sigma\sigma'}\cdot(\boldsymbol{B}_{\rm ext}+\boldsymbol{B}_{\rm xc})\Bigg\}\phi_k(\boldsymbol{r}\sigma') = \varepsilon_k\phi_k(\boldsymbol{r}\sigma)\,. \end{aligned} \tag{3.212}$$

Exactly the same form of KS equations is obtained for the CSDFT variant (2.169), using n and $\boldsymbol{j}_{\rm g}$ as basic variables,

$$E[n,\boldsymbol{j}_{\rm g}] = T_{\rm s}[n,\boldsymbol{j}_{\rm g}]+E_{\rm ext}[n,\boldsymbol{j}_{\rm g}]+E_{\rm H}[n]+E_{\rm xc}[n,\boldsymbol{j}_{\rm g}] \tag{3.213}$$

$$\begin{aligned} E_{\rm ext}[n,\boldsymbol{j}_{\rm g}] = &\int d^3r\left[v_{\rm ext}(\boldsymbol{r})+\frac{e^2}{2mc^2}\boldsymbol{A}^2_{\rm ext}(\boldsymbol{r})\right]n(\boldsymbol{r}) \\ &+\frac{e}{c}\int d^3r\,\boldsymbol{A}_{\rm ext}(\boldsymbol{r})\cdot\boldsymbol{j}_{\rm g}(\boldsymbol{r})\,, \end{aligned} \tag{3.214}$$

provided that $v_{\rm xc}$, $\boldsymbol{A}_{\rm xc}$ and $\boldsymbol{B}_{\rm xc}$ are understood as

$$v_{\rm xc}(\boldsymbol{r}) = \left.\frac{\delta E_{\rm xc}[n,\boldsymbol{j}_{\rm g}]}{\delta n(\boldsymbol{r})}\right|_{\boldsymbol{j}_{\rm g}} \tag{3.215}$$

$$\frac{e}{c}\boldsymbol{A}_{\mathrm{xc}}(\boldsymbol{r}) = \left.\frac{\delta E_{\mathrm{xc}}[n, \boldsymbol{j}_{\mathrm{g}}]}{\delta \boldsymbol{j}_{\mathrm{g}}(\boldsymbol{r})}\right|_n \tag{3.216}$$

$$\boldsymbol{B}_{\mathrm{xc}}(\boldsymbol{r}) = \boldsymbol{\nabla} \times \boldsymbol{A}_{\mathrm{xc}}(\boldsymbol{r}) \,. \tag{3.217}$$

The similarity of the KS equations raises the question about the relation between the two CSDFT schemes. It can be answered by an analysis of the corresponding xc-functionals [73]. The functional E_{xc} is gauge invariant in both CSDFT variants, due to the gauge invariance of the total energy and of $T_{\mathrm{s}} + E_{\mathrm{ext}}$,

$$E_{\mathrm{xc}}[n, \boldsymbol{j}'_{\mathrm{p}}, \boldsymbol{m}] = E_{\mathrm{xc}}[n, \boldsymbol{j}_{\mathrm{p}}, \boldsymbol{m}] \tag{3.218}$$

$$E_{\mathrm{xc}}[n, \boldsymbol{j}'_{\mathrm{g}}] = E_{\mathrm{xc}}[n, \boldsymbol{j}_{\mathrm{g}}] \,. \tag{3.219}$$

As both functionals are universal, their gauge invariance requires that they depend on suitable vorticities. In the case of the functional (3.218) this vorticity is given by Eq. (3.199),

$$E_{\mathrm{xc}}[n, \boldsymbol{j}_{\mathrm{p}}, \boldsymbol{m}] \equiv \tilde{E}_{\mathrm{xc}}[n, \boldsymbol{\nu}, \boldsymbol{m}] \,. \tag{3.220}$$

On the other hand, for $E_{\mathrm{xc}}[n, \boldsymbol{j}_{\mathrm{g}}]$ it is the extended form

$$\boldsymbol{\nu}_{\mathrm{g}}(\boldsymbol{r}) = \boldsymbol{\nabla} \times \left[\frac{\boldsymbol{j}_{\mathrm{g}}(\boldsymbol{r})}{n(\boldsymbol{r})}\right] \tag{3.221}$$

$$E_{\mathrm{xc}}[n, \boldsymbol{j}_{\mathrm{g}}] \equiv \bar{E}_{\mathrm{xc}}[n, \boldsymbol{\nu}_{\mathrm{g}}] \,. \tag{3.222}$$

The relation between the two functionals can then be established by use of (2.161),

$$\boldsymbol{\nu}_{\mathrm{g}}(\boldsymbol{r}) = \boldsymbol{\nu}(\boldsymbol{r}) + \frac{c}{e}\boldsymbol{\nabla} \times \left[\frac{\boldsymbol{\nabla} \times \boldsymbol{m}(\boldsymbol{r})}{n(\boldsymbol{r})}\right] , \tag{3.223}$$

and the fact that the total energies of both approaches have to be identical. In fact, for arbitrary, given $\boldsymbol{A}_{\mathrm{ext}}$ the external energies (3.208) and (3.214) of both schemes coincide as Eqs. (2.161) and (2.144) must hold,

$$E_{\mathrm{ext}}[n, \boldsymbol{j}_{\mathrm{p}}, \boldsymbol{m}] = E_{\mathrm{ext}}[n, \boldsymbol{j}_{\mathrm{g}}] \,. \tag{3.224}$$

Moreover, the total energies have to be identical even for vanishing interaction, so that

$$T_{\mathrm{s}}[n, \boldsymbol{j}_{\mathrm{p}}, \boldsymbol{m}] = T_{\mathrm{s}}[n, \boldsymbol{j}_{\mathrm{g}}] \,. \tag{3.225}$$

As a result one finds [73]

$$\tilde{E}_{\mathrm{xc}}[n, \boldsymbol{\nu}, \boldsymbol{m}] = \bar{E}_{\mathrm{xc}}\left[n, \boldsymbol{\nu} + \frac{c}{e}\boldsymbol{\nabla} \times \left(\frac{\boldsymbol{\nabla} \times \boldsymbol{m}}{n}\right)\right] . \tag{3.226}$$

Combining this relation with Eq. (3.220) and using the fact that due to Eq. (3.199) fixed n and $\boldsymbol{j}_{\mathrm{p}}$ are equivalent to fixed $\boldsymbol{\nu}$, one then obtains

$$\left.\frac{\delta E_{\rm xc}[n,\boldsymbol{j}_{\rm p},\boldsymbol{m}]}{\delta\boldsymbol{m}(\boldsymbol{r})}\right|_{n,\boldsymbol{j}_{\rm p}} = \left.\frac{\delta\tilde{E}_{\rm xc}[n,\boldsymbol{v},\boldsymbol{m}]}{\delta\boldsymbol{m}(\boldsymbol{r})}\right|_{n,\boldsymbol{v}}$$
$$= \left.\frac{\delta}{\delta\boldsymbol{m}(\boldsymbol{r})}\bar{E}_{\rm xc}\left[n,\boldsymbol{v}+\frac{c}{e}\boldsymbol{\nabla}\times\left(\frac{\boldsymbol{\nabla}\times\boldsymbol{m}}{n}\right)\right]\right|_{n,\boldsymbol{v}}.$$

At this point one can go back to the original form of $\bar{E}_{\rm xc}[n,\boldsymbol{v}_{\rm g}]$ via Eq. (3.222), again using the fact that fixed n and $\boldsymbol{v}$ imply fixed $\boldsymbol{j}_{\rm p}$,

$$\left.\frac{\delta E_{\rm xc}[n,\boldsymbol{j}_{\rm p},\boldsymbol{m}]}{\delta\boldsymbol{m}(\boldsymbol{r})}\right|_{n,\boldsymbol{j}_{\rm p}} = \left.\frac{\delta E_{\rm xc}[n,\boldsymbol{j}_{\rm g}]}{\delta\boldsymbol{m}(\boldsymbol{r})}\right|_{n,\boldsymbol{j}_{\rm p}}.$$

Finally, one can use the unique correspondence of $\boldsymbol{j}_{\rm g}$ and $\boldsymbol{m}$ established by (2.161) for fixed $\boldsymbol{j}_{\rm p}$,

$$\left.\frac{\delta E_{\rm xc}[n,\boldsymbol{j}_{\rm p},\boldsymbol{m}]}{\delta\boldsymbol{m}(\boldsymbol{r})}\right|_{n,\boldsymbol{j}_{\rm p}} = \int d^3r'\left.\frac{\delta E_{\rm xc}[n,\boldsymbol{j}_{\rm g}]}{\delta\boldsymbol{j}_{\rm g}(\boldsymbol{r}')}\right|_{n}\left.\frac{\delta\boldsymbol{j}_{\rm g}(\boldsymbol{r}')}{\delta\boldsymbol{m}(\boldsymbol{r})}\right|_{\boldsymbol{j}_{\rm p}},$$

and evaluate $\delta\boldsymbol{j}_{\rm g}/\delta\boldsymbol{m}$ via Eq. (2.161)

$$\left.\frac{\delta E_{\rm xc}[n,\boldsymbol{j}_{\rm p},\boldsymbol{m}]}{\delta\boldsymbol{m}(\boldsymbol{r})}\right|_{n,\boldsymbol{j}_{\rm p}} = \frac{c}{e}\boldsymbol{\nabla}\times\left.\frac{\delta E_{\rm xc}[n,\boldsymbol{j}_{\rm g}]}{\delta\boldsymbol{j}_{\rm g}(\boldsymbol{r})}\right|_{n}. \tag{3.227}$$

The magnetic field $\boldsymbol{B}_{\rm xc}$ of the formalism based on n, $\boldsymbol{j}_{\rm p}$ and $\boldsymbol{m}$, Eq. (3.209), is therefore identical with the curl of the xc-potential $\boldsymbol{A}_{\rm xc}$ of the scheme based on n and $\boldsymbol{j}_{\rm g}$, Eq. (3.216), exactly as required for the identity of the corresponding KS equations by Eq. (3.217). Since the xc-potentials of both schemes agree,

$$\left.\frac{\delta E_{\rm xc}[n,\boldsymbol{j}_{\rm p},\boldsymbol{m}]}{\delta\boldsymbol{j}_{\rm p}(\boldsymbol{r})}\right|_{n,\boldsymbol{m}} = \left.\frac{\delta E_{\rm xc}[n,\boldsymbol{j}_{\rm g}]}{\delta\boldsymbol{j}_{\rm p}(\boldsymbol{r})}\right|_{n,\boldsymbol{m}} = \left.\frac{\delta E_{\rm xc}[n,\boldsymbol{j}_{\rm g}]}{\delta\boldsymbol{j}_{\rm g}(\boldsymbol{r})}\right|_{n} \tag{3.228}$$
$$\left.\frac{\delta E_{\rm xc}[n,\boldsymbol{j}_{\rm p},\boldsymbol{m}]}{\delta n(\boldsymbol{r})}\right|_{\boldsymbol{j}_{\rm p},\boldsymbol{m}} = \left.\frac{\delta E_{\rm xc}[n,\boldsymbol{j}_{\rm g}]}{\delta n(\boldsymbol{r})}\right|_{\boldsymbol{j}_{\rm g}}, \tag{3.229}$$

one has verified the equivalence of both CSDFT variants.

With Eq. (3.212) our collection of nonrelativistic, stationary KS equations is complete. Depending on the phenomenon under consideration, one can apply either (3.44), (3.69), (3.88), (3.98), (3.191) or (3.212). We have finally reached the point where the xc-functional has to be studied more closely.

Chapter 4
Exchange-Correlation Energy Functional

In this chapter we introduce the most frequently used approximations for the xc-energy functional on the basis of a number of rigorous results for $E_{xc}[n]$. The presentation focuses on the derivation of the various functionals, their performance will only be discussed in so far as it serves as motivation for improvements. However, some prototype results will be given for the most important functionals, the local density and generalized gradient approximation. An excellent overview of the performance of many of the approximations in the field of quantum chemistry is given in [105] (see also [106]); for solids corresponding information can be found in [107–111] and references therein.

4.1 Definition of Exact Exchange within DFT

It is usual to decompose the total xc-energy functional $E_{xc}[n]$ into an exchange part $E_x[n]$ and a correlation functional $E_c[n]$, in analogy to conventional many-body theory. In view of the difference between $E_{xc}[n]$ and the standard xc-energy discussed in Sect. 3.1, the precise definition of $E_x[n]$ is somewhat arbitrary. It is nevertheless the natural first choice to define the exchange functional in such a way that the total energy E^{HF} and density n_{HF} of the Hartree-Fock (HF) approximation are reproduced if the correlation functional is completely neglected. The corresponding HF-only ground state energy functional $\tilde{E}[n]$,

$$\tilde{E}[n] = T_s[n] + E_{ext}[n] + E_H[n] + \tilde{E}_x[n] , \tag{4.1}$$

is hence to be minimized by n_{HF},

$$E^{HF} = \tilde{E}[n_{HF}] , \tag{4.2}$$

while for any other density one must have

$$E^{HF} < \tilde{E}[n] \quad \forall \ n \neq n_{HF} . \tag{4.3}$$

E. Engel, R.M. Dreizler, Exchange-Correlation Energy Functional. In: E. Engel, R.M. Dreizler, Density Functional Theory, Theoretical and Mathematical Physics, pp. 109–217 (2011)
DOI 10.1007/978-3-642-14090-7_4

The existence of such an exchange functional is guaranteed by the HK theorem, which remains valid in the exchange-only limit [112, 22]. One can then set up a KS scheme on the basis of $\tilde{E}_{\rm x}[n]$. However, contrary to naive expectation, the resulting KS orbitals and eigenvalues do not agree with the original HF orbitals and eigenvalues, as the KS potential is always multiplicative: the functional derivative of $\tilde{E}_{\rm x}[n]$ with respect to $n(\boldsymbol{r})$ yields a multiplicative exchange potential which has nothing to do with the nonlocal HF exchange operator, only the resulting densities coincide by definition.

Unfortunately, the definition (4.1)–(4.3) leads to a number of difficulties:

- No explicit expression for the exact functional $\tilde{E}_{\rm x}[n]$ is available, neither in terms of the density, nor in terms of some suitable N-particle wavefunction.
- With this definition of the exchange functional no virial relation can be formulated (compare Sect. 5.3).
- A gradient expansion does not exist for $\tilde{E}_{\rm x}[n]$ (compare Sect. 4.4.3).

For these reasons an alternative definition [113, 114] of the exchange energy functional has become the standard in DFT,

$$E_{\rm x}[n] := \langle\Phi_0|\hat{W}|\Phi_0\rangle - E_{\rm H}[n]\,, \tag{4.4}$$

where $|\Phi_0\rangle$ is the KS Slater determinant (3.3) (a non-degenerate KS system is assumed). Equation (4.4) can easily be evaluated explicitly. One obtains the standard Fock expression, written, however, in terms of the KS orbitals ϕ_k,

$$E_{\rm x}[n] = -\frac{1}{2}\sum_{kl}\Theta_k\Theta_l\sum_{\sigma\sigma'}\int d^3r\int d^3r'\,\phi_k^*(\boldsymbol{r}\sigma)\phi_l(\boldsymbol{r}\sigma)w(\boldsymbol{r},\boldsymbol{r}')\phi_l^*(\boldsymbol{r}'\sigma')\phi_k(\boldsymbol{r}'\sigma')\,. \tag{4.5}$$

The right-hand side of Eq. (4.5) is a density functional in the same sense as the kinetic energy functional $T_{\rm s}[n]$. As indicated in Eq. (3.22), the ϕ_k are uniquely determined by the density n, since n uniquely determines $v_{\rm s}$ (according to the Hohenberg-Kohn theorem for noninteracting systems), which in turn allows the unambiguous calculation of the ϕ_k. $E_{\rm x}[n]$ represents an implicit density functional, in contrast to the explicit density functionals to be discussed below.

In spite of the agreement of the functional (4.5) with the Fock expression, the density which minimizes the total exchange-only (x-only) energy functional of DFT,

$$E^{\rm x-only}[n] := T_{\rm s}[n] + E_{\rm ext}[n] + E_{\rm H}[n] + E_{\rm x}[n]\,, \tag{4.6}$$

is not identical with $n_{\rm HF}$. This statement becomes immediately clear if one analyzes x-only DFT from the viewpoint of an energy minimization with respect to the orbitals ϕ_k. Taking Eqs. (4.6), (3.21), (3.25), (3.26) and (4.5) together, the energy expression to be minimized in x-only DFT is identical with its HF counterpart (1.17). However, in DFT the single-particle orbitals have to satisfy the KS equations (3.44), with $v_{\rm x}$ given by the multiplicative potential $\delta E_{\rm x}[n]/\delta n(\boldsymbol{r})$. The multiplicative nature of the total KS potential $v_{\rm s}$ represents a subsidiary condition in the minimization procedure, which is not present in the HF-type minimization with its nonlocal exchange potential (1.21). In the HF approach one thus has some additional variational freedom, which, in general, leads to a lower energy minimum,

$$E^{\mathrm{HF}} \leq \min_n E^{\mathrm{x-only}}[n]. \tag{4.7}$$

The HF orbitals are not included in the variational space available to x-only DFT. This implies that the insertion of the HF ground state density into $E^{\mathrm{x-only}}[n]$ does not yield the HF ground state energy,

$$E^{\mathrm{HF}} \neq E^{\mathrm{x-only}}[n_{\mathrm{HF}}] \geq \min_n E^{\mathrm{x-only}}[n].$$

The x-only DFT minimization coincides with the HF scheme only in special situations in which the HF potential (1.19) can be recast as a local potential for the occupied state, i.e. for spin-saturated 2-particle systems. The use of Eq. (4.5) within DFT—either in the x-only limit or in the general situation—leads to exchange energies which differ from their HF counterparts, due to the difference between KS and HF orbitals.

This point is illustrated quantitatively in Table 4.1. Three values obtained by

Table 4.1 Exchange and correlation energies resulting from different definitions: Eq. (4.5) evaluated with HF, exact KS and x-only KS orbitals as well as DFT correlation energy versus E_c^{QC}, Eq. (4.9) [115, 83, 116] (all energies in mHartree).

Atom	$-E_x$			$-E_c$	$-E_c^{\mathrm{QC}}$
	HF	KS: exact	KS: x-only		
H^-	395.49	380.90	395.49	41.99	39.82
He	1025.77	1024.57	1025.77	42.11	42.04
Be	2666.91	2673.98	2665.77	96.2	94.34
Ne	12108.35	12083.93	12105.01	394.	390.47

evaluation of (4.5) with three different types of orbitals are listed: the standard HF exchange energy is compared with the DFT exchange resulting from insertion of the KS orbitals which reproduce the exact density of the interacting system[1] (referred to as exact KS orbitals in the following) and the DFT exchange energy obtained by insertion of the KS orbitals which minimize (4.6) (called exact x-only orbitals in the following). Table 4.1 indicates that HF and x-only DFT exchange energies are always rather close to each other (they are identical for spin-saturated 2-electron systems). However, the same is not true for the E_x calculated from the exact KS

[1] The exact densities of interacting systems with only few electrons can be obtained with high accuracy by Quantum Monte Carlo or Configuration Interaction calculations. Once the density is available, one can determine the total KS potential which gives this density by a suitably constrained solution of the single-particle equations [117–122].

orbitals. Even the sign of the deviation from E_x^{HF} and E_x^{x-only} changes from atom to atom.

The definition (4.5) induces a corresponding definition of the correlation energy functional $E_c[n]$,

$$E_c[n] := E_{xc}[n] - E_x[n]\,. \tag{4.8}$$

The difference between (4.5) and the HF exchange energy is absorbed into E_c. The DFT correlation energy is thus not identical with the conventional correlation energy E_c^{QC}, which is employed in standard many-body theory and quantum chemistry. The latter energy is defined as the difference between the total ground state energy of the interacting system and its HF counterpart,

$$E_c^{QC} = \langle\Psi_0|\hat{H}|\Psi_0\rangle - \langle\Phi_0^{HF}|\hat{H}|\Phi_0^{HF}\rangle\,, \tag{4.9}$$

with $|\Phi_0^{HF}\rangle$ denoting the HF ground state Slater determinant. On the other hand, Eq. (4.8) may be rewritten as

$$E_c = \langle\Psi_0|\hat{H}|\Psi_0\rangle - \langle\Phi_0|\hat{H}|\Phi_0\rangle\,. \tag{4.10}$$

The difference between E_c and E_c^{QC} hence originates from the difference between the HF ground state energy and $E^{x-only}[n]$, Eq. (4.6), evaluated with the density n_0 of the fully interacting system,

$$\begin{aligned} E_c - E_c^{QC} &= \langle\Phi_0^{HF}|\hat{H}|\Phi_0^{HF}\rangle - \langle\Phi_0|\hat{H}|\Phi_0\rangle \\ &= E^{HF} - E^{x-only}[n_0]\,. \end{aligned} \tag{4.11}$$

The size of $E_c - E_c^{QC}$ can be characterized further by introducing the x-only ground state energy [116],

$$E_c - E_c^{QC} = \left(E^{HF} - E^{x-only}[n_0^{x-only}]\right) + \left(E^{x-only}[n_0^{x-only}] - E^{x-only}[n_0]\right)\,, \tag{4.12}$$

where n_0^{x-only} represents the exact x-only ground state density. The first term on the right-hand side of Eq. (4.12) is known for a variety of systems and turns out to be small for all of them (for explicit numbers see Sect. 6.3). The second expression allows a functional Taylor series expansion, similar to the expansion of the total ground state energy in Eq. (3.33). As n_0^{x-only} minimizes $E^{x-only}[n]$, the difference between $E^{x-only}[n_0^{x-only}]$ and $E^{x-only}[n_0]$ is of second order in $n_0^{x-only} - n_0$,

$$E^{x-only}[n_0^{x-only}] - E^{x-only}[n_0] = \mathcal{O}\left((n_0^{x-only} - n_0)^2\right)\,. \tag{4.13}$$

As a consequence, the difference $E_c - E_c^{QC}$ is generally quite small. This point is demonstrated explicitly for a few light atoms in Table 4.1 [116].

The most important property of the functional (4.5) is the exact cancellation of the self-interaction contained in E_H. The Coulomb integral (3.25) with the total density contains the interaction of the orbital density $|\phi_k|^2$ with itself,

$$E_{\rm H} = \frac{1}{2}\sum_{kl}\Theta_k\Theta_l \sum_{\sigma,\sigma'}\int d^3r \int d^3r' \, |\phi_k(\boldsymbol{r}\sigma)|^2 w(\boldsymbol{r},\boldsymbol{r}')|\phi_l(\boldsymbol{r}'\sigma')|^2$$

$$\Longrightarrow \quad E_{\rm H}^{\rm SI} = \frac{1}{2}\sum_{k}\Theta_k^2 \sum_{\sigma,\sigma'}\int d^3r \int d^3r' \, |\phi_k(\boldsymbol{r}\sigma)|^2 w(\boldsymbol{r},\boldsymbol{r}')\,|\phi_k(\boldsymbol{r}'\sigma')|^2 . \tag{4.14}$$

The self-interaction energy $E_{\rm H}^{\rm SI}$ is exactly compensated by the terms with $l = k$ of (4.5),

$$E_{\rm x}^{\rm SI} = -E_{\rm H}^{\rm SI} . \tag{4.15}$$

A second important feature of (4.5) is its additivity with respect to the two spin orientations [123]. Utilizing the definition (4.5) within (collinear) SDFT, one can insert the form (3.93) of the corresponding KS orbitals (and $w(\boldsymbol{r},\boldsymbol{r}') = e^2/|\boldsymbol{r}-\boldsymbol{r}'|$) to obtain

$$E_{\rm x}[n_\uparrow,n_\downarrow] \;=\; -\frac{e^2}{2}\sum_{\sigma}\sum_{\alpha\beta}\Theta_{\alpha\sigma}\Theta_{\beta\sigma}\int d^3r \int d^3r' \, \frac{\phi_{\alpha\sigma}^*(\boldsymbol{r})\phi_{\beta\sigma}(\boldsymbol{r})\phi_{\beta\sigma}^*(\boldsymbol{r}')\phi_{\alpha\sigma}(\boldsymbol{r}')}{|\boldsymbol{r}-\boldsymbol{r}'|} . \tag{4.16}$$

One thus finds a simple separation of spins in $E_{\rm x}[n_\uparrow,n_\downarrow]$ on the level of the orbital representation (4.16),

$$E_{\rm x}[n_\uparrow,n_\downarrow] = E_{x\uparrow} + E_{x\downarrow} . \tag{4.17}$$

For an unpolarized system with $n_\uparrow = n_\downarrow$ one has

$$E_{\rm x}[n] = E_{\rm x}[2n_\uparrow] = E_{\rm x}[n_\uparrow,n_\uparrow] = 2E_{x\uparrow} , \tag{4.18}$$

where $E_{\rm x}[n]$ denotes the functional (4.5) and the factor of 2 in the right-most equation follows directly from (4.17). Now consider a spin-polarized system: for a given n_σ of this polarized system one can imagine an auxiliary unpolarized system with $\tilde{n}_\uparrow = \tilde{n}_\downarrow = n_\sigma$. In this auxiliary system both spin-up and spin-down KS orbitals are identical with the orbitals $\phi_{\alpha\sigma}$ of the actual polarized system. According to Eq. (4.18), the exchange energy of this unpolarized system satisfies $E_{\rm x}[2n_\sigma] = E_{\rm x}[2\tilde{n}_\sigma] = 2E_{x\sigma}$. Upon insertion of this relation into (4.17) one arrives at

$$E_{\rm x}[n_\uparrow,n_\downarrow] \;=\; \frac{1}{2}\Big\{E_{\rm x}[2n_\uparrow] + E_{\rm x}[2n_\downarrow]\Big\} . \tag{4.19}$$

The third important property of (4.5) and (4.16) is the behavior of the corresponding potentials in the asymptotic region of finite systems (for which the density decays exponentially for large $\boldsymbol{r}$). Applying the standard normalization $v_{\rm x}^{\sigma}(\boldsymbol{r}) \xrightarrow{|\boldsymbol{r}|\to\infty} 0$, one finds

$$v_{\rm x}^{\sigma}(\boldsymbol{r}) \;\xrightarrow[|\boldsymbol{r}|\to\infty]{}\; -\frac{e^2}{|\boldsymbol{r}|} . \tag{4.20}$$

For the verification of (4.20) one needs a scheme which allows the explicit evaluation of the functional derivative $\delta/\delta n$ of an implicit functional of type (4.5). Such a scheme is provided by the *optimized potential method* (OPM) [124, 125]. An explicit account of the OPM is given in Chap. 6. It will be shown in this chapter that v_{x}^{σ} is obtained as solution of a linear integral equation. Examination of this integral equation for large r establishes the validity of (4.20) (see Sect. 6.2.5).

The physical reason behind this result is very simple: if one electron of a system of N electrons and a corresponding assembly of nuclei is sufficiently far away from the other electrons bound to the nuclei, it must experience the net charge of the remaining system. The total electronic potential is given by the sum of v_{xc}^{σ} and the Hartree potential (3.42). However, v_{H} contains the Coulomb self-repulsion of the isolated electron far out. This self-interaction has to be eliminated by v_{x}^{σ}. The asymptotic limit (4.20) just reflects the exact cancellation of the self-interaction in E_{H} by E_{x}. No other contribution to v_{xc}^{σ} can be proportional to $1/r$, as this behavior can only result from a monopole term.

This physical origin of (4.20) can be illustrated directly for a spin-saturated two-electron system like the helium atom. In this case there is only one occupied KS level for each spin,

$$n_{\sigma}(\boldsymbol{r}) = |\phi_1(\boldsymbol{r})|^2\,; \qquad\qquad n(\boldsymbol{r}) = 2n_{\sigma}(\boldsymbol{r})\,,$$

and the exchange energy only has to cancel the self-interaction. Equation (4.16) thus reduces to

$$E_{\mathrm{x}}^{\mathrm{He}}[n_{\uparrow},n_{\downarrow}] = -\frac{e^2}{2}\sum_{\sigma}\int d^3r\int d^3r'\,\frac{n_{\sigma}(\boldsymbol{r})n_{\sigma}(\boldsymbol{r}')}{|\boldsymbol{r}-\boldsymbol{r}'|}\,. \tag{4.21}$$

The functional derivative of $E_{\mathrm{x}}^{\mathrm{He}}$ is then trivially given by

$$v_{\mathrm{x}}^{\sigma,\mathrm{He}}(\boldsymbol{r}) = -e^2\int d^3r'\,\frac{n_{\sigma}(\boldsymbol{r}')}{|\boldsymbol{r}-\boldsymbol{r}'|}\,, \tag{4.22}$$

so that (4.20) is obtained in the asymptotic limit.

Closely related to (4.20) is the asymptotic behavior of the exchange energy density e_{x} (we will always denote the energy per volume corresponding to some energy E_{a} by e_{a}). As any other energy density, e_{x} has the disadvantage that it can not be defined uniquely: one can always add terms which integrate up to zero without affecting E_{x}. Nevertheless, the Fock expression (4.5) suggests as a natural definition for e_{x},

$$e_{\mathrm{x}}(\boldsymbol{r}) := -\frac{e^2}{2}\sum_{\sigma}\sum_{\alpha\beta}\Theta_{\alpha\sigma}\Theta_{\beta\sigma}\int d^3r'\,\frac{\phi_{\alpha\sigma}^*(\boldsymbol{r})\phi_{\beta\sigma}(\boldsymbol{r})\phi_{\beta\sigma}^*(\boldsymbol{r}')\phi_{\alpha\sigma}(\boldsymbol{r}')}{|\boldsymbol{r}-\boldsymbol{r}'|}\,, \tag{4.23}$$

which has become standard within DFT. If one now assumes the orbitals to decay exponentially for large $\boldsymbol{r}$, one finds for the asymptotic behavior of the exchange energy density

$$e_x(\mathbf{r}) \xrightarrow[|\mathbf{r}|\to\infty]{} -\frac{e^2 n(\mathbf{r})}{2|\mathbf{r}|}. \tag{4.24}$$

The pair of Eqs. (4.20) and (4.24) reflects the quadratic structure of the self-interaction energy with respect to the orbital density of the most weakly bound electron.

With Eqs. (4.15), (4.19), (4.20) and (4.24) the list of important rigorous properties of the exact exchange (4.5) is complete. Before discussing approximations for $E_x[n]$ we first establish some exact results for the complete xc-functional.

4.2 Exact Representations of $E_{xc}[n]$

Given the exact representation of $E_x[n]$ in terms of the KS orbitals, Eq. (4.5), the question concerning an equivalent exact expression for the correlation functional comes up quite naturally. This is also in line with the initial motivation for DFT: the aim was to establish an exact mapping of the interacting many-body problem onto an effective single-particle system. Of course, one should not expect an exact result for E_c to be directly applicable in practice, as this would imply the exact solution of the many-body Schrödinger equation. However, a formally exact representation provides an ideal starting point for the derivation of systematic approximations. In this context the term "systematic" means that, at least in principle, one can improve a given approximation by successive inclusion of well-defined corrections until the exact result is approached. This strategy reflects the basic understanding of DFT as a first-principles approach.

In this section two different exact expressions for the xc-functional will be discussed, which are usually referred to as

(a) Kohn-Sham perturbation theory [126, 127], and
(b) adiabatic connection [77, 128].

Both of them will be derived explicitly, as they serve as starting points for two quite different classes of approximate functionals.

4.2.1 *Variant (a): Kohn-Sham Perturbation Theory*

Let us for a moment assume that v_s, the total KS potential, is known. In this case the KS Hamiltonian $\hat{H}_s$, Eq. (3.1), can be utilized as any other given noninteracting N-particle Hamiltonian—the fact that in reality v_s emerges from a self-consistent calculation will have to be addressed subsequently. The ground state $|\Phi_0\rangle$ (assumed to be nondegenerate) corresponding to $\hat{H}_s$ is then obtained by solution of Eq. (3.2), the ground state energy E_s and density n_0 are given by Eqs. (3.16) and (3.13), respectively. The density n_0 is—by construction—identical with the density of the interacting system, in which we are interested.

One can then use $\hat{H}_s$ as noninteracting reference Hamiltonian in order to derive an exact relation for E_{xc}. First the total Hamiltonian of the interacting system is decomposed into $\hat{H}_s$ and a remainder $\hat{H}_1$, for which the main component is the electron–electron interaction $\hat{W}$. In addition, $\hat{H}_1$ has to compensate the parts of $\hat{H}_s$ which are not contained in $\hat{H}$,

$$\hat{H}_1 = \hat{W} - \int d^3r\,\hat{n}(\boldsymbol{r})v_{\rm Hxc}(\boldsymbol{r}), \tag{4.25}$$

where $v_{\rm Hxc}$ represents the electron–electron interaction components in v_s,

$$v_{\rm Hxc}(\boldsymbol{r}) = v_s(\boldsymbol{r}) - v_{\rm ext}(\boldsymbol{r}) = v_{\rm H}(\boldsymbol{r}) + v_{\rm xc}(\boldsymbol{r}). \tag{4.26}$$

In the second step a running coupling constant λ is introduced into the total Hamiltonian,

$$\hat{H}(\lambda) = \hat{H}_s + \lambda \hat{H}_1, \tag{4.27}$$

which allows the use of the coupling constant integration technique. The ground state $|\Psi_0(\lambda)\rangle$ corresponding to $\hat{H}(\lambda)$ (also assumed to be nondegenerate) is obtained from the interacting Schrödinger equation,

$$\hat{H}(\lambda)|\Psi_0(\lambda)\rangle = E_0(\lambda)|\Psi_0(\lambda)\rangle. \tag{4.28}$$

One can now apply the method of coupling constant integration to the λ-dependent ground state energy,

$$E_0(\lambda) = \langle\Psi_0(\lambda)|\hat{H}(\lambda)|\Psi_0(\lambda)\rangle. \tag{4.29}$$

One starts by differentiating $E_0(\lambda)$ with respect to λ,

$$\begin{aligned}\frac{d}{d\lambda}E_0(\lambda) &= \langle\frac{d\Psi_0(\lambda)}{d\lambda}|\hat{H}(\lambda)|\Psi_0(\lambda)\rangle + \langle\Psi_0(\lambda)|\hat{H}(\lambda)|\frac{d\Psi_0(\lambda)}{d\lambda}\rangle \\ &\quad +\langle\Psi_0(\lambda)|\hat{H}_1|\Psi_0(\lambda)\rangle \\ &= E_0(\lambda)\left[\langle\frac{d\Psi_0(\lambda)}{d\lambda}|\Psi_0(\lambda)\rangle + \langle\Psi_0(\lambda)|\frac{d\Psi_0(\lambda)}{d\lambda}\rangle\right] \\ &\quad +\langle\Psi_0(\lambda)|\hat{H}_1|\Psi_0(\lambda)\rangle.\end{aligned}$$

If one uses the fact that $|\Psi_0(\lambda)\rangle$ is normalized for all λ,

$$\langle\Psi_0(\lambda)|\Psi_0(\lambda)\rangle = 1 \quad \Longrightarrow \quad \frac{d}{d\lambda}\langle\Psi_0(\lambda)|\Psi_0(\lambda)\rangle = 0, \tag{4.30}$$

one arrives at

$$\frac{d}{d\lambda}E_0(\lambda) = \langle\Psi_0(\lambda)|\hat{H}_1|\Psi_0(\lambda)\rangle. \tag{4.31}$$

Next, one integrates (4.31) with respect to λ. The integration starts at $\lambda = 0$, for which $\hat{H}(\lambda)$ agrees with the KS Hamiltonian, and ends at $\lambda = 1$, where $\hat{H}(\lambda)$ is identical with the true interacting Hamiltonian,

$$\hat{H}(\lambda) = \begin{cases} \hat{H}_s \text{ for } \lambda = 0 \\ \hat{H} \text{ for } \lambda = 1 \end{cases} .$$

On the left-hand side of (4.31), the integration leads to the difference between the energy $E_0(1) = E_0$ of the interacting system (which is the energy one is interested in) and the energy $E_0(0) = E_s$ of the KS reference system,

$$E_0(1) - E_0(0) = E_0 - E_s = \int_0^1 d\lambda \, \langle \Psi_0(\lambda) | \hat{H}_1 | \Psi_0(\lambda) \rangle . \tag{4.32}$$

The expectation value on the right-hand side of Eq. (4.32) can be rewritten in several ways. It is instructive to consider first a less successful option. For this one uses the equal-time commutation relations for the field operator, Eqs. (2.6) and (2.7), to show that

$$\begin{aligned} \hat{\psi}^\dagger(\boldsymbol{r}\sigma)\hat{\psi}^\dagger(\boldsymbol{r}'\sigma')\hat{\psi}(\boldsymbol{r}'\sigma')\hat{\psi}(\boldsymbol{r}\sigma) &= \hat{\psi}^\dagger(\boldsymbol{r}\sigma)\hat{\psi}(\boldsymbol{r}\sigma)\hat{\psi}^\dagger(\boldsymbol{r}'\sigma')\hat{\psi}(\boldsymbol{r}'\sigma') \\ &\quad -\delta^{(3)}(\boldsymbol{r}-\boldsymbol{r}')\,\delta_{\sigma\sigma'}\hat{\psi}^\dagger(\boldsymbol{r}\sigma)\hat{\psi}(\boldsymbol{r}\sigma) . \end{aligned} \tag{4.33}$$

Combination of (4.33) with (4.25) and (2.5) leads to

$$\begin{aligned} \langle \Psi_0(\lambda) | \hat{H}_1 | \Psi_0(\lambda) \rangle &= \frac{1}{2} \int d^3r \int d^3r' \, w(\boldsymbol{r},\boldsymbol{r}') \\ &\quad \times \langle \Psi_0(\lambda) | \hat{n}(\boldsymbol{r})\hat{n}(\boldsymbol{r}') - \delta^{(3)}(\boldsymbol{r}-\boldsymbol{r}')\hat{n}(\boldsymbol{r}) | \Psi_0(\lambda) \rangle \\ &\quad - \int d^3r \, v_{\mathrm{Hxc}}(\boldsymbol{r}) \langle \Psi_0(\lambda) | \hat{n}(\boldsymbol{r}) | \Psi_0(\lambda) \rangle . \end{aligned} \tag{4.34}$$

The kernel of (4.34) can be expressed in terms of the so-called *density–density response function* of the interacting system.

The concept of response functions is so fundamental to many-body theory and is utilized so often in this text that it is worth a little detour. Several kinds of response functions (alternatively called *correlation functions* or *susceptibilities*) can be found in the literature: depending on the context, different forms turn out to be particularly useful. Of course, all of them are closely related. Two variants of response functions are particularly relevant in the context of DFT. For the present discussion the *time-ordered* response function is the most appropriate. For a stationary system it is defined as

$$\begin{aligned} \chi(\boldsymbol{r}t,\boldsymbol{r}'t') &:= -\frac{i}{\hbar} \langle \Psi_0 | T \, \hat{\tilde{n}}(\boldsymbol{r}t) \hat{\tilde{n}}(\boldsymbol{r}'t') | \Psi_0 \rangle \\ &= -\frac{i}{\hbar} \Big[\langle \Psi_0 | T \, \hat{n}(\boldsymbol{r}t) \hat{n}(\boldsymbol{r}'t') | \Psi_0 \rangle - n_0(\boldsymbol{r}) n_0(\boldsymbol{r}') \Big] \end{aligned} \tag{4.35}$$

(for the definition of the time-ordering symbol T see Eq. (3.120)). The basic ingredient of χ is the density operator in the Heisenberg representation,

$$\hat{\tilde{n}}(\boldsymbol{r}t) = \hat{n}(\boldsymbol{r}t) - n_0(\boldsymbol{r}) \tag{4.36}$$

$$\hat{n}(\boldsymbol{r}t) = e^{i\hat{H}t/\hbar}\, \hat{n}(\boldsymbol{r})\, e^{-i\hat{H}t/\hbar} = \sum_\sigma \hat{\psi}^\dagger(\boldsymbol{r}\sigma t)\hat{\psi}(\boldsymbol{r}\sigma t) \tag{4.37}$$

$$n_0(\boldsymbol{r}) = \langle\Psi_0|\hat{n}(\boldsymbol{r}t)|\Psi_0\rangle = \langle\Psi_0|\hat{n}(\boldsymbol{r})|\Psi_0\rangle\,. \tag{4.38}$$

The time-ordered response has to be distinguished from the *retarded* response function,

$$\begin{aligned}\chi_{\mathrm{R}}(\boldsymbol{r}t,\boldsymbol{r}'t') &:= -\frac{i}{\hbar}\,\Theta(t-t')\,\langle\Psi_0|[\hat{\tilde{n}}(\boldsymbol{r}t),\hat{\tilde{n}}(\boldsymbol{r}'t')]|\Psi_0\rangle \\ &= -\frac{i}{\hbar}\,\Theta(t-t')\,\langle\Psi_0|[\hat{n}(\boldsymbol{r}t),\hat{n}(\boldsymbol{r}'t')]|\Psi_0\rangle\,,\end{aligned} \tag{4.39}$$

on which the discussion of Sect. 4.4 and, in particular, of Sects. 7.4–7.6 relies. The retarded response function is somewhat closer to physical reality, in that it connects the response of an observable directly with the perturbation which causes the response (compare the discussion in Sect. 4.4). The time-ordered response function, on the other hand, is more easily dealt with in mathematical terms. In particular, it allows the application of the complete machinery of Green's function theory. For this reason the relation between both functions is of obvious interest, as one often would like to eliminate χ_{R} in favor of χ.

Let us therefore collect the basic properties of these functions. Due to the stationarity of the Hamiltonian both response functions only depend on the time difference $t-t'$. This is immediately clear by insertion of $\hat{n}(\boldsymbol{r}t)$ into the building block of both functions,

$$\begin{aligned}\langle\Psi_0|\hat{n}(\boldsymbol{r}t)\hat{n}(\boldsymbol{r}'t')|\Psi_0\rangle &= \langle\Psi_0|e^{i\hat{H}t/\hbar}\,\hat{n}(\boldsymbol{r})\,e^{-i\hat{H}t/\hbar}e^{i\hat{H}t'/\hbar}\,\hat{n}(\boldsymbol{r}')\,e^{-i\hat{H}t'/\hbar}|\Psi_0\rangle \\ &= \langle\Psi_0|\hat{n}(\boldsymbol{r})\,e^{-i\hat{H}(t-t')/\hbar}\,\hat{n}(\boldsymbol{r}')|\Psi_0\rangle e^{iE_0(t-t')/\hbar} \\ \Longrightarrow \qquad \chi_{(\mathrm{R})}(\boldsymbol{r}t,\boldsymbol{r}'t') &= \chi_{(\mathrm{R})}(\boldsymbol{r},\boldsymbol{r}',t-t')\,.\end{aligned} \tag{4.40}$$

It is therefore convenient to analyze these functions in frequency space,

$$\chi_{(\mathrm{R})}(\boldsymbol{r},\boldsymbol{r}',\omega) = \int_{-\infty}^{+\infty} d(t-t')\,e^{i\omega(t-t')}\,\chi_{(\mathrm{R})}(\boldsymbol{r}t,\boldsymbol{r}'t')\,. \tag{4.41}$$

The intimate relation between χ and χ_{R} can be established on the basis of the *Lehmann representation*, in analogy to Eq. (3.122). Let us denote the N-particle eigenstates of the interacting Hamiltonian by $|\Psi_n\rangle$, with the quantum number n chosen according to their energy ordering,

$$\hat{H}\,|\Psi_n\rangle = E_n\,|\Psi_n\rangle\,; \qquad E_n \le E_{n+1}\,. \tag{4.42}$$

Let us furthermore assume that the ground state is non-degenerate. The Lehmann representations of χ and χ_R are then obtained from the definitions (4.35) and (4.39) by insertion of the completeness relation in the N-particle Hilbert space,

$$\sum_n |\Psi_n\rangle\langle\Psi_n| = \hat{1}\,, \tag{4.43}$$

use of (4.42) and subsequent Fourier transformation,[2]

$$\left\{ \begin{array}{l} \chi(\boldsymbol{r},\boldsymbol{r}',\omega) \\ \chi_R(\boldsymbol{r},\boldsymbol{r}',\omega) \end{array} \right\} = \sum_{n\neq 0} \frac{\langle\Psi_0|\hat{n}(\boldsymbol{r})|\Psi_n\rangle\langle\Psi_n|\hat{n}(\boldsymbol{r}')|\Psi_0\rangle}{\hbar\omega - (E_n - E_0) + i\eta} - \sum_{n\neq 0} \frac{\langle\Psi_0|\hat{n}(\boldsymbol{r}')|\Psi_n\rangle\langle\Psi_n|\hat{n}(\boldsymbol{r})|\Psi_0\rangle}{\hbar\omega + (E_n - E_0) \mp i\eta}\,. \tag{4.44}$$

The only difference between the Lehmann representations of χ and χ_R is found for the pole shift of the second contribution, for which the (upper) minus sign refers to χ, while the (lower) plus sign is correct for χ_R (concerning the handling of the pole shifts see the remark following Eq. (3.122)). As $E_n - E_0 > 0$ for all states with $n \neq 0$, the pole structures of χ and χ_R have the forms given in Fig. 4.1. In this

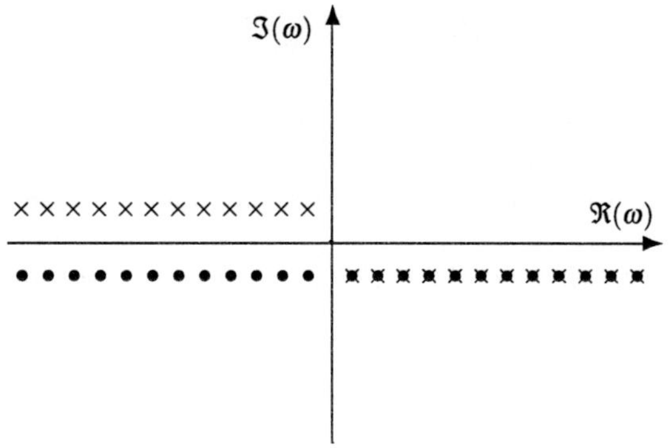

Fig. 4.1 Analytic structure of χ ($\times$) and χ_R ($\bullet$), Eq. (4.44), in the complex ω-plane.

figure crosses ($\times$) denote the poles of χ, filled circles ($\bullet$) those of χ_R. Figure 4.1 demonstrates that neither of the functions has poles on the imaginary axis and in the upper-right quadrant of the complex ω-plane. One can directly verify that

[2] Note that the term with $n = 0$, which is present in the completeness relation (4.43), is missing in the final Lehmann representation (4.44). In the case of the time-ordered response function this contribution drops out due to the definition of χ in terms of the density deviation operator $\hat{\tilde{n}}$, i.e. due to the subtraction of $n_0(\boldsymbol{r})n_0(\boldsymbol{r}')$ from the actual time-ordered product of density operators in Eq. (4.35). For the retarded response function the definition in terms of the commutator of two density operators, Eq. (4.39), is responsible for the elimination of the term with $n = 0$.

$$\chi_{\rm R}(\boldsymbol{r},\boldsymbol{r}',\omega) = \begin{cases} \chi(\boldsymbol{r},\boldsymbol{r}',\omega) & \text{for } \Re(\omega) \geq 0 \\ \chi^*(\boldsymbol{r}',\boldsymbol{r},\omega) & \text{for } \Re(\omega) \leq 0 \end{cases}, \tag{4.45}$$

as the pole shift in the second term of (4.44) becomes irrelevant for $\omega \geq 0$, while for $\omega \leq 0$ the same is true for the pole shift in the first denominator ($E_n - E_0 > 0$). Direct inspection of (4.44) leads to the symmetry relations

$$\chi_{\rm R}(\boldsymbol{r},\boldsymbol{r}',-\omega) = \chi^*_{\rm R}(\boldsymbol{r},\boldsymbol{r}',\omega) \tag{4.46}$$

$$\chi(\boldsymbol{r},\boldsymbol{r}',-\omega) = \chi(\boldsymbol{r}',\boldsymbol{r},\omega)\,. \tag{4.47}$$

Equations (4.45)–(4.47) imply that both functions are identical and real for $\omega = 0$,

$$\chi_{\rm R}(\boldsymbol{r},\boldsymbol{r}',\omega=0) = \chi(\boldsymbol{r},\boldsymbol{r}',\omega=0) = \chi^*(\boldsymbol{r},\boldsymbol{r}',\omega=0)\,. \tag{4.48}$$

Moreover, for systems with a finite number of particles integration over space leads to

$$\int d^3r\,\chi_{\rm (R)}(\boldsymbol{r},\boldsymbol{r}',\omega=0) = \int d^3r'\,\chi_{\rm (R)}(\boldsymbol{r},\boldsymbol{r}',\omega=0) = 0\,. \tag{4.49}$$

This identity, however, relies on

$$\int d^3r\,\langle\Psi_0|\hat{n}(\boldsymbol{r})|\Psi_n\rangle = \langle\Psi_0|\hat{N}|\Psi_n\rangle = N\delta_{n0}\,.$$

As a consequence, Eq. (4.49) does not apply to infinite systems, for which the particle number diverges, so that the integral over space is not well-defined as it stands.

Additional information on both types of response functions will be given in subsequent sections, in particular in Sects. 4.3 and 4.4. For the present discussion the limit $t,t' \to 0$ is of interest. One immediately recognizes, from either the definition (4.39) or the Lehmann representation (4.44), that the retarded response function vanishes in this limit,

$$\begin{aligned} \lim_{t'\to 0}\left[\lim_{t\to 0}\chi_{\rm R}(\boldsymbol{r}t,\boldsymbol{r}'t')\right] &= \lim_{t\to 0}\left[\lim_{t'\to 0}\chi_{\rm R}(\boldsymbol{r}t,\boldsymbol{r}'t')\right] \\ &= \int_{-\infty}^{+\infty}\frac{d\omega}{2\pi}\,\chi_{\rm R}(\boldsymbol{r},\boldsymbol{r}',\omega) \\ &= 0\,. \end{aligned} \tag{4.50}$$

All poles of $\chi_{\rm R}(\boldsymbol{r},\boldsymbol{r}',\omega)$ lie in the lower half of the complex plane. On the other hand, the limit $t,t' \to 0$ is well-defined and non-zero in the case of the time-ordered response function,

$$\begin{aligned} \lim_{t'\to 0}\left[\lim_{t\to 0}\chi(\boldsymbol{r}t,\boldsymbol{r}'t')\right] &= \lim_{t\to 0}\left[\lim_{t'\to 0}\chi(\boldsymbol{r}t,\boldsymbol{r}'t')\right] \\ &= \int_{-\infty}^{+\infty}\frac{d\omega}{2\pi}\,\chi(\boldsymbol{r},\boldsymbol{r}',\omega) \end{aligned}$$

$$= -\frac{i}{\hbar}\sum_{n\neq 0}\langle\Psi_0|\hat{n}(\boldsymbol{r})|\Psi_n\rangle\langle\Psi_n|\hat{n}(\boldsymbol{r}')|\Psi_0\rangle$$

$$= -\frac{i}{\hbar}[\langle\Psi_0|\hat{n}(\boldsymbol{r})\hat{n}(\boldsymbol{r}')|\Psi_0\rangle - n(\boldsymbol{r})n(\boldsymbol{r}')]\,. \tag{4.51}$$

With this result let us resume the discussion of the energy shift (4.34). The limit (4.51) can be used to reformulate (4.34) in terms of the time-ordered response function. Combining Eqs. (4.32), (4.34) and (4.51), one arrives at

$$\begin{aligned}E_0(1) = E_0(0) &- \int_0^1 d\lambda \int d^3r\, n_\lambda(\boldsymbol{r})\left\{v_{\mathrm{Hxc}}(\boldsymbol{r}) - \frac{1}{2}\int d^3r'\, w(\boldsymbol{r},\boldsymbol{r}')n_\lambda(\boldsymbol{r}')\right\}\\ &+\frac{1}{2}\int d^3r\int d^3r'\, w(\boldsymbol{r},\boldsymbol{r}')\\ &\times\int_0^1 d\lambda\left\{i\hbar\lim_{t,t'\to 0}\chi_\lambda(\boldsymbol{r}t,\boldsymbol{r}'t') - \delta^{(3)}(\boldsymbol{r}-\boldsymbol{r}')n_\lambda(\boldsymbol{r})\right\},\end{aligned} \tag{4.52}$$

where n_λ and χ_λ denote the density and response function resulting from the ground state $|\Psi_0(\lambda)\rangle$ for given λ. One can now use Eq. (3.18) for $E_0(0)$,

$$E_0(0) = E_{\mathrm{s}} = \langle\Phi_0|\hat{H}_{\mathrm{s}}|\Phi_0\rangle = T_{\mathrm{s}} + \int d^3r\, v_{\mathrm{s}}(\boldsymbol{r})n(\boldsymbol{r})\,, \tag{4.53}$$

and (3.24) to end up with

$$\begin{aligned}E_{\mathrm{xc}} = &\frac{1}{2}\int d^3r\int d^3r'\, w(\boldsymbol{r},\boldsymbol{r}')\int_0^1 d\lambda\left\{i\hbar\chi_\lambda(\boldsymbol{r}0,\boldsymbol{r}'0) - \delta^{(3)}(\boldsymbol{r}-\boldsymbol{r}')n_\lambda(\boldsymbol{r})\right\}\\ &+\int_0^1 d\lambda\left\{E_{\mathrm{H}}[n_\lambda] - E_{\mathrm{H}}[n] + \int d^3r\big[v_{\mathrm{s}}(\boldsymbol{r}) - v_{\mathrm{ext}}(\boldsymbol{r})\big]\big[n(\boldsymbol{r}) - n_\lambda(\boldsymbol{r})\big]\right\}.\end{aligned} \tag{4.54}$$

The expression (4.54) is a density functional in the sense that the ground state $|\Psi_0(\lambda)\rangle$ is a unique functional of the density n_λ for any λ. However, as the strength of the particle–particle interaction varies with λ, the functional dependence of $|\Psi_0\rangle$ on the density also varies, $|\Psi_0(\lambda)\rangle = |\Psi_\lambda[n_\lambda]\rangle$. One has to perform an integration over λ, in order to obtain the desired functional of the actually interesting density n. While (4.54) is an exact representation of E_{xc}, it is difficult to imagine how this functional could be used in practice.

Fortunately, Eq. (4.54) can be reformulated in a more useful form. With this aim in mind, it is most convenient to go back to the coupling constant integration formula (4.32). For its evaluation the concept of adiabatic switching [94] is applied to $\hat{H}_1$. This means that $\hat{H}_1$ is switched off for large positive and negative times, using some exponential switching factor,

$$\hat{H}_1 \longrightarrow e^{-\varepsilon|t|}\hat{H}_1\,, \tag{4.55}$$

so that asymptotically one has $\lim_{t\to\pm\infty}\hat{H}(\lambda) = \hat{H}_{\mathrm{s}}$ (while for $t=0$ the original Hamiltonian is retained). The limit $\varepsilon\to 0$ is taken at the very end of the discussion, in order to return to the original Hamiltonian for all times. On this basis the machinery of standard many-body theory, i.e. the Gell-Mann-Low theorem [129] applied to the present decomposition of the total Hamiltonian, leads to an expression which connects the interacting ground state with the KS ground state,

$$|\Psi_0(\lambda)\rangle = A \lim_{\varepsilon\to 0} \frac{\hat{U}_{I,\varepsilon}(0,\mp\infty)|\Phi_0\rangle}{\langle\Phi_0|\hat{U}_{I,\varepsilon}(0,\mp\infty)|\Phi_0\rangle} \tag{4.56}$$

$$A = \lim_{\varepsilon_1,\varepsilon_2\to 0} \left[\frac{\langle\Phi_0|\hat{U}_{I,\varepsilon_1}(+\infty,0)|\Phi_0\rangle\,\langle\Phi_0|\hat{U}_{I,\varepsilon_2}(0,-\infty)|\Phi_0\rangle}{\langle\Phi_0|\hat{U}_{I,\varepsilon_1}(+\infty,0)\,\hat{U}_{I,\varepsilon_2}(0,-\infty)|\Phi_0\rangle} \right]^{1/2}, \tag{4.57}$$

where A ensures the normalization (4.30). The main ingredient of (4.56) is the time-evolution operator in the interaction picture (defined by $\hat{H}_{\mathrm{s}}$), which is given by a power series in the perturbation,

$$\begin{aligned} \hat{U}_{I,\varepsilon}(t,t') = \sum_{n=0}^{\infty} \frac{(-i\lambda)^n}{\hbar^n n!} \int_{t'}^{t} dt_1 \cdots \int_{t'}^{t} dt_n\, e^{-\varepsilon(|t_1|+\cdots+|t_n|)} \\ \times T\left[\hat{H}_{1,I}(t_1)\cdots\hat{H}_{1,I}(t_n)\right] \end{aligned} \tag{4.58}$$

$$\begin{aligned} \hat{H}_{1,I}(t) &= e^{i\hat{H}_{\mathrm{s}}t}\,\hat{H}_1\,e^{-i\hat{H}_{\mathrm{s}}t} \\ &= \frac{1}{2}\sum_{\sigma,\sigma'} \int d^3r \int d^3r'\, w(\boldsymbol{r},\boldsymbol{r}')\,\hat{\psi}_0^\dagger(\boldsymbol{r}\sigma t)\,\hat{\psi}_0^\dagger(\boldsymbol{r}'\sigma' t)\,\hat{\psi}_0(\boldsymbol{r}'\sigma' t)\,\hat{\psi}_0(\boldsymbol{r}\sigma t) \\ &\quad -\sum_{\sigma} \int d^3r\, v_{\mathrm{Hxc}}(\boldsymbol{r})\,\hat{\psi}_0^\dagger(\boldsymbol{r}\sigma t)\,\hat{\psi}_0(\boldsymbol{r}\sigma t)\,. \end{aligned} \tag{4.59}$$

Here $\hat{\psi}_0$ represents the field operator in the interaction picture, which is identical with the field operator in the Heisenberg representation defined by the KS Hamiltonian,

$$\hat{\psi}_0(\boldsymbol{r}\sigma t) = e^{i\hat{H}_{\mathrm{s}}t/\hbar}\,\hat{\psi}(\boldsymbol{r}\sigma)\,e^{-i\hat{H}_{\mathrm{s}}t/\hbar} = \sum_i \hat{b}_i\phi_i(\boldsymbol{r}\sigma)e^{-i\varepsilon_i t/\hbar}\,. \tag{4.60}$$

The operator $\hat{b}_i^{(\dagger)}$ denotes the annihilation (creation) operator for the single-particle KS state ϕ_i

$$\phi_i(\boldsymbol{r}\sigma) = \langle \boldsymbol{r}\sigma|\hat{b}_i^\dagger|0\rangle \tag{4.61}$$

$$0 = \hat{b}_i|0\rangle \tag{4.62}$$

$$|\Phi_0\rangle = \prod_{\varepsilon_i\le\varepsilon_{\mathrm{F}}} \hat{b}_i^\dagger\,|0\rangle\,. \tag{4.63}$$

The field operator and the KS Hamiltonian may thus be expanded as

$$\hat{\psi}(\boldsymbol{r}\sigma) = \sum_i \hat{b}_i\,\phi_i(\boldsymbol{r}\sigma)\,; \qquad \hat{H}_{\mathrm{s}} = \sum_i \varepsilon_i\,\hat{b}_i^\dagger\hat{b}_i\,,$$

which leads to the simple commutation relations

$$\left[\hat{H}_{\rm s},\hat{b}_i\right] = -\varepsilon_i\hat{b}_i\,; \qquad \left[\hat{H}_{\rm s},\hat{b}_i^\dagger\right] = \varepsilon_i\hat{b}_i^\dagger\,.$$

Use of the identity (L.12) then explains the right-hand equality in (4.60). The reader is referred to a text on many-body theory (for instance [94]) for the derivation of Eq. (4.56), which is completely independent of the form of $\hat{H}_1$.

Insertion of (4.56) into the coupling constant integral (4.32) leads to the energy correction which results from switching on the perturbation. In order to express this energy in standard form one first uses the fact that $\hat{H}_1 = \hat{H}_{1,I}(t=0)$ can be absorbed into the time-ordered product in $\hat{U}_{I,\varepsilon}$,

$$\begin{aligned}
&\hat{U}_{I,\varepsilon}(+\infty,0)\hat{H}_1\hat{U}_{I,\varepsilon}(0,-\infty)\\
&= \sum_{n=0}^{\infty}\frac{(-i\lambda)^n}{\hbar^n n!}\int_0^{+\infty}dt_1\cdots\int_0^{+\infty}dt_n\, e^{-\varepsilon(|t_1|+\cdots+|t_n|)}\\
&\quad\times\sum_{k=0}^{\infty}\frac{(-i\lambda)^k}{\hbar^k k!}\int_{-\infty}^{0}dt_1'\cdots\int_{-\infty}^{0}dt_k'\, e^{-\varepsilon(|t_1'|+\cdots+|t_k'|)}\\
&\quad\times T\left[\hat{H}_{1,I}(t_1)\cdots\hat{H}_{1,I}(t_n)\right]\hat{H}_{1,I}(0)T\left[\hat{H}_{1,I}(t_1')\cdots\hat{H}_{1,I}(t_k')\right]\\
&= \sum_{n=0}^{\infty}\frac{(-i\lambda)^n}{\hbar^n n!}\int_0^{+\infty}dt_1\cdots\int_0^{+\infty}dt_n\, e^{-\varepsilon(|t_1|+\cdots+|t_n|)}\\
&\quad\times\sum_{k=0}^{\infty}\frac{(-i\lambda)^k}{\hbar^k k!}\int_{-\infty}^{0}dt_1'\cdots\int_{-\infty}^{0}dt_k'\, e^{-\varepsilon(|t_1'|+\cdots+|t_k'|)}\\
&\quad\times T\left[\hat{H}_{1,I}(t_1)\cdots\hat{H}_{1,I}(t_n)\hat{H}_{1,I}(0)\hat{H}_{1,I}(t_1')\cdots\hat{H}_{1,I}(t_k')\right]\\
&= T\left[\hat{H}_{1,I}(0)\hat{U}_{I,\varepsilon}(+\infty,0)\hat{U}_{I,\varepsilon}(0,-\infty)\right].
\end{aligned}$$

Together with the additivity of the time-evolution operator,

$$\hat{U}_{I,\varepsilon}(+\infty,0)\,\hat{U}_{I,\varepsilon}(0,-\infty) = \hat{U}_{I,\varepsilon}(+\infty,-\infty)\,,$$

one arrives at

$$\begin{aligned}
E_0 - E_{\rm s} = \lim_{\varepsilon\to 0}\int_0^1 d\lambda\sum_{n=0}^{\infty}\frac{(-i\lambda)^n}{\hbar^n n!}\int_{-\infty}^{\infty}dt_1\cdots\int_{-\infty}^{\infty}dt_n\, e^{-\varepsilon(|t_1|+\cdots+|t_n|)}\\
\times\frac{\langle\Phi_0|T\hat{H}_{1,I}(0)\hat{H}_{1,I}(t_1)\cdots\hat{H}_{1,I}(t_n)|\Phi_0\rangle}{\langle\Phi_0|\hat{U}_{I,\varepsilon}(+\infty,-\infty)|\Phi_0\rangle}\,. \qquad (4.64)
\end{aligned}$$

Using Eqs. (3.18) and (3.24) one can finally extract $E_{\rm xc}$ and perform the coupling constant integration,

$$
\begin{aligned}
E_{\rm xc} = \frac{1}{2}\int d^3r \int d^3r'\, w(\boldsymbol{r},\boldsymbol{r}') \\
\times \Big\{ \sum_{\sigma,\sigma'} \langle \Phi_0 | \hat{\psi}_0^\dagger(\boldsymbol{r}\sigma)\hat{\psi}_0^\dagger(\boldsymbol{r}'\sigma')\hat{\psi}_0(\boldsymbol{r}'\sigma')\hat{\psi}_0(\boldsymbol{r}\sigma)|\Phi_0\rangle - n(\boldsymbol{r})n(\boldsymbol{r}') \Big\} \\
+ \lim_{\varepsilon\to 0} \sum_{n=1}^{\infty} \frac{(-i)^n}{\hbar^n (n+1)!} \int_{-\infty}^{\infty} dt_1 \cdots \int_{-\infty}^{\infty} dt_n\, e^{-\varepsilon(|t_1|\cdots+|t_n|)} \\
\times \langle \Phi_0 | T \hat{H}_{1,I}(0)\hat{H}_{1,I}(t_1)\cdots \hat{H}_{1,I}(t_n)|\Phi_0\rangle_l \,. \qquad (4.65)
\end{aligned}
$$

The index l indicates that in the order-by-order evaluation of (4.65) only those contributions have to be included which can not be factorized into one term involving $\hat{H}_{1,I}(0)$ and a remainder. Expressed in the language of Feynman diagrams, only linked diagrams are to be included in the evaluation of (4.65) via Wick's theorem (this restriction corresponds to the cancellation of the denominator of (4.64)).

The first term on the right-hand side of (4.65) emerges from the perturbative contribution linear in $\hat{H}_1$ and is easily identified as the exchange energy (4.5). In order to prove this assertion, one can either insert the noninteracting field operator (4.60) and the KS ground state (4.63) to evaluate the expectation value directly. Alternatively, one can commute the operators $\hat{\psi}_0$ as in (4.33) and subsequently use (4.51),

$$
\begin{aligned}
\frac{1}{2}\int d^3r \int d^3r'\, w(\boldsymbol{r},\boldsymbol{r}') \\
\times \Big[\sum_{\sigma,\sigma'} \langle \Phi_0 | \hat{\psi}_0^\dagger(\boldsymbol{r}\sigma)\hat{\psi}_0^\dagger(\boldsymbol{r}'\sigma')\hat{\psi}_0(\boldsymbol{r}'\sigma')\hat{\psi}_0(\boldsymbol{r}\sigma)|\Phi_0\rangle - n(\boldsymbol{r})n(\boldsymbol{r}') \Big] \\
= \frac{1}{2}\int d^3r \int d^3r'\, w(\boldsymbol{r},\boldsymbol{r}') \Big\{ i\hbar \int_{-\infty}^{+\infty} \frac{d\omega}{2\pi} \chi_{\rm s}(\boldsymbol{r},\boldsymbol{r}',\omega) - \delta^{(3)}(\boldsymbol{r}-\boldsymbol{r}')\,n(\boldsymbol{r}) \Big\} \,. \qquad (4.66)
\end{aligned}
$$

Here $\chi_{\rm s}$ denotes the time-ordered response function of the KS system, which is defined in complete analogy to the full response function (4.35), with the ground state and density operator being replaced by their KS counterparts,

$$
\begin{aligned}
\chi_{\rm s}(\boldsymbol{r}t,\boldsymbol{r}'t') \\
= -\frac{i}{\hbar} \Big[\sum_{\sigma,\sigma'} \langle \Phi_0 | T\, \hat{\psi}_0^\dagger(\boldsymbol{r}\sigma t)\hat{\psi}_0(\boldsymbol{r}\sigma t)\hat{\psi}_0^\dagger(\boldsymbol{r}'\sigma' t')\hat{\psi}_0(\boldsymbol{r}'\sigma' t')|\Phi_0\rangle - n(\boldsymbol{r})n(\boldsymbol{r}') \Big] \,. \qquad (4.67)
\end{aligned}
$$

The function $\chi_{\rm s}$ and its retarded version $\chi_{\rm s,R}$ (defined in analogy to Eq. (4.39)) satisfy the same relations as the interacting response functions, Eqs. (4.40)–(4.49). In particular, there also exists a Lehmann representation for $\chi_{\rm s}$. In fact, this Lehmann representation becomes particularly simple, due to the determinantal structure of the KS N-particle states. As $\hat{\psi}_0^\dagger \hat{\psi}_0$ can only excite a single KS particle from the ground state Slater determinant $|\Phi_0\rangle$, one finds

$$
\begin{aligned}
\chi_s(\boldsymbol{r},\boldsymbol{r}',\omega) = \; & \sum_{kl}\Theta_k(1-\Theta_l)\sum_{\sigma,\sigma'}\frac{\phi_k^*(\boldsymbol{r}\sigma)\phi_l(\boldsymbol{r}\sigma)\phi_l^*(\boldsymbol{r}'\sigma')\phi_k(\boldsymbol{r}'\sigma')}{\hbar\omega-(\varepsilon_l-\varepsilon_k)+i\eta} \\
& -\sum_{kl}\Theta_k(1-\Theta_l)\sum_{\sigma,\sigma'}\frac{\phi_k^*(\boldsymbol{r}'\sigma')\phi_l(\boldsymbol{r}'\sigma')\phi_l^*(\boldsymbol{r}\sigma)\phi_k(\boldsymbol{r}\sigma)}{\hbar\omega+(\varepsilon_l-\varepsilon_k)-i\eta}\,.
\end{aligned} \tag{4.68}
$$

Insertion of (4.68) into (4.66) then allows a direct evaluation of the frequency integral by contour integration techniques. One ends up with the exchange (4.5), establishing two alternative representations of E_x,

$$
\begin{aligned}
E_x &= \frac{1}{2}\int d^3r\int d^3r'\, w(\boldsymbol{r},\boldsymbol{r}') \\
&\quad\times\left\{\sum_{\sigma,\sigma'}\langle\Phi_0|\hat\psi_0^\dagger(\boldsymbol{r}\sigma)\hat\psi_0^\dagger(\boldsymbol{r}'\sigma')\hat\psi_0(\boldsymbol{r}'\sigma')\hat\psi_0(\boldsymbol{r}\sigma)|\Phi_0\rangle - n(\boldsymbol{r})n(\boldsymbol{r}')\right\} \\
&= \frac{1}{2}\int d^3r\int d^3r'\, w(\boldsymbol{r},\boldsymbol{r}')\left[i\hbar\int_0^\infty\frac{d\omega}{\pi}\chi_s(\boldsymbol{r},\boldsymbol{r}',\omega)-n(\boldsymbol{r})\,\delta^{(3)}(\boldsymbol{r}-\boldsymbol{r}')\right],
\end{aligned} \tag{4.69}
$$

where (4.47) has been used to restrict the frequency integration to the positive range.

The second term in (4.65), which incorporates all higher orders of $\hat{H}_1$, provides an exact expression for the correlation energy E_c,

$$
\begin{aligned}
E_c = \lim_{\varepsilon\to 0}\sum_{n=1}^{\infty}\frac{(-i)^n}{\hbar^n(n+1)!}\int_{-\infty}^{\infty}dt_1\cdots\int_{-\infty}^{\infty}dt_n\, e^{-\varepsilon(|t_1|\cdots+|t_n|)} \\
\times\langle\Phi_0|T\hat{H}_{1,I}(0)\hat{H}_{1,I}(t_1)\cdots\hat{H}_{1,I}(t_n)|\Phi_0\rangle_l\,.
\end{aligned} \tag{4.70}
$$

The quantities required for the evaluation of this expression via Wick's theorem, i.e. the elementary building blocks of many-body perturbation theory based on $\hat{H}_s$, are the KS Green's function,

$$
\begin{aligned}
& G_s(\boldsymbol{r}\sigma t,\boldsymbol{r}'\sigma' t') \\
&:= -i\langle\Phi_0|T\,\hat\psi_0(\boldsymbol{r}\sigma t)\,\hat\psi_0^\dagger(\boldsymbol{r}'\sigma' t')|\Phi_0\rangle \\
&= -i\sum_i\left\{\Theta(t-t')(1-\Theta_i)-\Theta(t'-t)\Theta_i\right\}\phi_i(\boldsymbol{r}\sigma)\,\phi_i^\dagger(\boldsymbol{r}'\sigma')\,e^{-i\varepsilon_i(t-t')/\hbar},
\end{aligned} \tag{4.71}
$$

the Coulomb interaction w and v_{Hxc} (as $\hat{H}_1$ depends on this potential). The Hartree component of $v_{Hxc} = v_H + v_{xc}$ is readily calculated from the KS orbitals, so that (4.65) depends on three basic quantities, ϕ_k, ε_k and v_{xc}.

The result obtained is hence an exact representation of E_{xc} in terms of the KS orbitals and eigenvalues as well as in terms of its own functional derivative. Consequently, Eq. (4.65) does not assign a well-defined expression to E_{xc}, but rather represents a highly nonlinear functional equation. This result is, however, still consistent with the basic statements of DFT: as v_{xc} is a density functional itself, the right-hand side of Eq. (4.65) represents an implicit density functional.

While the expression (4.65) is more transparent than (4.54), its usefulness for practical applications is not obvious at all. Both its dependence on the KS states as well as its nonlinearity prevent a direct exploitation. Nevertheless, the representation (4.65) serves as basis for a complete class of approximations for $E_{\rm xc}$. As these approximations rely on the explicit representation of xc-effects in terms of KS states, further discussion is postponed until Chap. 6, in which the handling of orbital-dependent functionals is explained.

4.2.2 Variant (b): Adiabatic Connection

The starting point of the discussion is once more the representation of $E_{\rm xc}$ in terms of the response function χ. Obviously, the result (4.54), obtained by the decomposition (4.27) of the total Hamiltonian, is rather difficult to handle, because the ground state density changes with the switching factor λ. One might thus ask whether one can modify this procedure in a way which ensures that the ground state density remains the same along the path from the KS system to the fully interacting system? This invariance is the crucial feature behind the adiabatic connection [77, 128].

The basic assumption, that one uses here, is: the ground state density n of the interacting system is v-representable for any strength of the interaction λw with $\lambda \in [0,1]$—until now, only the two end points have been utilized. In other words: one assumes that one can find an external potential, so that the same ground state density is obtained for any interaction strength $\lambda \in [0,1]$,

$$n_\lambda(\boldsymbol{r}) = \langle \Psi_0(\lambda)|\hat{n}(\boldsymbol{r})|\Psi_0(\lambda)\rangle \equiv n(\boldsymbol{r}) \qquad \text{for all } 0 \le \lambda \le 1 . \tag{4.72}$$

This external potential will be denoted by u_λ,

$$u_\lambda(\boldsymbol{r}) = \begin{cases} v_{\rm s}(\boldsymbol{r}) & \text{for } \lambda = 0 \\ \text{unknown} & \text{for } 0 < \lambda < 1 \\ v_{\rm ext}(\boldsymbol{r}) & \text{for } \lambda = 1 \end{cases} . \tag{4.73}$$

The total Hamiltonian for given λ then reads

$$\hat{H}(\lambda) = \hat{T} + \int d^3r\, u_\lambda(\boldsymbol{r})\hat{n}(\boldsymbol{r}) + \lambda \hat{W} , \tag{4.74}$$

so that one obtains again

$$\hat{H}(\lambda) = \begin{cases} \hat{H}_{\rm s} & \text{for } \lambda = 0 \\ \hat{H} & \text{for } \lambda = 1 \end{cases} .$$

However, in contrast to the Hamiltonian (4.27) the present $\hat{H}(\lambda)$ is only known for the two limiting cases $\lambda = 0$ and $\lambda = 1$. This complicates the evaluation of the energy for intermediate λ.

For the Hamiltonian (4.74) one can use coupling constant integration in the same fashion as for (4.27). If one denotes the ground state, which results from (4.74), by $|\Psi_0(\lambda)\rangle$, one obtains for the ground state energy,

$$
\begin{aligned}
E_0(1)-E_0(0) &= \int_0^1 d\lambda \left\langle \Psi_0(\lambda) \middle| \int d^3r\, \hat{n}(\boldsymbol{r}) \frac{du_\lambda(\boldsymbol{r})}{d\lambda} + \hat{W} \middle| \Psi_0(\lambda) \right\rangle \\
&= \int d^3r \left[u_{\lambda=1}(\boldsymbol{r}) - u_{\lambda=0}(\boldsymbol{r}) \right] n(\boldsymbol{r}) \\
&\quad + \frac{1}{2} \sum_{\sigma,\sigma'} \int d^3r \int d^3r'\, w(\boldsymbol{r},\boldsymbol{r}') \int_0^1 d\lambda \\
&\quad \times \langle \Psi_0(\lambda) | \hat{\psi}^\dagger(\boldsymbol{r}\sigma) \hat{\psi}^\dagger(\boldsymbol{r}'\sigma') \hat{\psi}(\boldsymbol{r}'\sigma') \hat{\psi}(\boldsymbol{r}\sigma) | \Psi_0(\lambda) \rangle \,.
\end{aligned}
$$

In the next step, the time-ordered response function corresponding to the Hamiltonian (4.74) can be introduced and $E_{\rm xc}$ can be extracted,

$$
E_{\rm xc}[n] = \frac{1}{2} \int d^3r \int d^3r'\, w(\boldsymbol{r},\boldsymbol{r}') \int_0^1 d\lambda \left[i\hbar \chi_\lambda(\boldsymbol{r}0,\boldsymbol{r}'0) - n(\boldsymbol{r})\, \delta^{(3)}(\boldsymbol{r}-\boldsymbol{r}') \right] . \tag{4.75}
$$

Of course, the response function in (4.75) is not identical with the χ_λ in (4.54), as the underlying Hamiltonians differ. Nevertheless, use of the HK theorem for all $\lambda \in [0,1]$ leads again to the statement that the ground state $|\Psi_0(\lambda)\rangle$ is a unique functional of the density n_λ. The important advantage of (4.75), compared to the expression (4.54), results from the fact that n_λ is now identical for all λ, so that $|\Psi_0(\lambda)\rangle$ and hence χ_λ as well as (4.75) are functionals of the actual interacting ground state density n only ($|\Psi_0(\lambda)\rangle = |\Psi_\lambda[n]\rangle$). Equation (4.75) represents the *adiabatic connection formula* for $E_{\rm xc}$.

Often an alternative form of (4.75) is found in the literature, which relies on a variant of the response function, the so-called *pair-correlation function*,

$$
g(\boldsymbol{r},\boldsymbol{r}') := \frac{\langle \Psi_0 | \hat{n}(\boldsymbol{r}) \hat{n}(\boldsymbol{r}') | \Psi_0 \rangle - n(\boldsymbol{r})\, \delta^{(3)}(\boldsymbol{r}-\boldsymbol{r}')}{n(\boldsymbol{r})\, n(\boldsymbol{r}')} \,. \tag{4.76}
$$

$g(\boldsymbol{r},\boldsymbol{r}')-1$ is the percentage deviation of the probability to find one particle at point $\boldsymbol{r}$ and simultaneously a second particle at $\boldsymbol{r}'$ from the uncorrelated product of single-particle probabilities,

$$
n^{(2)}(\boldsymbol{r},\boldsymbol{r}') = \frac{1}{2} n(\boldsymbol{r}) n(\boldsymbol{r}') g(\boldsymbol{r},\boldsymbol{r}') \,, \tag{4.77}
$$

where $n^{(2)}$ is given by (3.45). The connection between $n^{(2)}(\boldsymbol{r},\boldsymbol{r}')$ and $g(\boldsymbol{r},\boldsymbol{r}')$ can be easily established via the 2-particle density operator

$$\hat{n}^{(2)}(\boldsymbol{r},\boldsymbol{r}') = \frac{1}{2}\sum_{i,j=1;i\neq j}^{N} \delta^{(3)}(\boldsymbol{r}-\hat{\boldsymbol{r}}_i)\,\delta^{(3)}(\boldsymbol{r}'-\hat{\boldsymbol{r}}_j) \tag{4.78}$$

$$= \frac{1}{2}\sum_{\sigma,\sigma'} \hat{\psi}^\dagger(\boldsymbol{r}\sigma)\hat{\psi}^\dagger(\boldsymbol{r}'\sigma')\hat{\psi}(\boldsymbol{r}'\sigma')\hat{\psi}(\boldsymbol{r}\sigma) \tag{4.79}$$

(the equivalence of (4.78) and (4.79) can be verified on the basis of the general relation between the first and second quantized representation of 2-particle operators—see Appendix B). Insertion of $g(\boldsymbol{r},\boldsymbol{r}')$ into (4.75) yields

$$E_{\mathrm{xc}}[n] = \frac{1}{2}\int d^3r \int d^3r'\, n(\boldsymbol{r})\, w(\boldsymbol{r},\boldsymbol{r}')\, n(\boldsymbol{r}') \int_0^1 d\lambda\, [g_\lambda(\boldsymbol{r},\boldsymbol{r}')-1]\,. \tag{4.80}$$

$g_\lambda(\boldsymbol{r},\boldsymbol{r}')$ is a unique functional of n, in the same way as χ_λ.

There exists yet another reformulation of (4.75) which is frequently met within DFT. The definition of the *exchange-correlation hole* h_{xc},

$$h_{\mathrm{xc}}(\boldsymbol{r},\boldsymbol{r}') = n(\boldsymbol{r}') \int_0^1 d\lambda\, [g_\lambda(\boldsymbol{r},\boldsymbol{r}')-1]\,, \tag{4.81}$$

allows to rewrite (4.75) as

$$E_{\mathrm{xc}}[n] = \frac{1}{2}\int d^3r\, n(\boldsymbol{r}) \int d^3r'\, w(\boldsymbol{r},\boldsymbol{r}')\, h_{\mathrm{xc}}(\boldsymbol{r},\boldsymbol{r}')\,. \tag{4.82}$$

In the form (4.82) the xc-energy can be interpreted as the interaction energy of the charge density n with its associated xc-hole. Any approximation for h_{xc}, g_λ or χ_λ then defines an approximation for $E_{\mathrm{xc}}[n]$.

For practical purposes it is convenient to recast (4.75) using the fact that the response function $\chi_\lambda(\boldsymbol{r}t,\boldsymbol{r}'t')$ only depends on $t-t'$, which allows the insertion of the Fourier representation (4.41),

$$E_{\mathrm{xc}} = \frac{1}{2}\int d^3r \int d^3r'\, \frac{e^2}{|\boldsymbol{r}-\boldsymbol{r}'|} \times \int_0^1 d\lambda \left[i\hbar \int_{-\infty}^{\infty} \frac{d\omega}{2\pi}\, \chi_\lambda(\boldsymbol{r},\boldsymbol{r}',\omega) - n(\boldsymbol{r})\,\delta^{(3)}(\boldsymbol{r}-\boldsymbol{r}')\right]. \tag{4.83}$$

Using Eq. (4.47) to restrict the frequency integration to the positive regime and subtracting the exact exchange (4.69), one ends up with an exact representation of the correlation energy functional,

$$E_{\mathrm{c}} = \frac{i\hbar}{2}\int d^3r \int d^3r'\, \frac{e^2}{|\boldsymbol{r}-\boldsymbol{r}'|} \int_0^1 d\lambda \int_0^{\infty} \frac{d\omega}{\pi}\, [\chi_\lambda(\boldsymbol{r},\boldsymbol{r}',\omega) - \chi_{\mathrm{s}}(\boldsymbol{r},\boldsymbol{r}',\omega)]\,. \tag{4.84}$$

For $\Re(\omega) \geq 0$, however, $\chi_\lambda(\boldsymbol{r},\boldsymbol{r}',\omega)$ does not have any poles in the upper half of the complex ω-plane, as can be explicitly verified via the Lehmann representation (4.44). The same statement is true for $\chi_{\mathrm{s}}(\boldsymbol{r},\boldsymbol{r}',\omega)$, see Eq. (4.68). As a consequence, integrals along arbitrary closed contours in the upper-right quadrant of the ω-plane vanish. This finally allows us to rewrite (4.84) as an integral over imaginary frequencies, i.e. an integral along the imaginary axis,

$$E_c = -\frac{\hbar}{2}\int d^3r \int d^3r' \frac{e^2}{|\boldsymbol{r}-\boldsymbol{r}'|}\int_0^1 d\lambda \int_0^\infty \frac{d\omega}{\pi}\left[\chi_\lambda(\boldsymbol{r},\boldsymbol{r}',i\omega)-\chi_s(\boldsymbol{r},\boldsymbol{r}',i\omega)\right]. \tag{4.85}$$

The form (4.85) has the advantage that one does not come close to any pole of $\chi_\lambda(\boldsymbol{r},\boldsymbol{r}',\omega)$ and $\chi_s(\boldsymbol{r},\boldsymbol{r}',\omega)$ along the imaginary axis, so that the pole shifts in their Lehmann representations become irrelevant. On this basis one can directly demonstrate that $\chi_\lambda(\boldsymbol{r},\boldsymbol{r}',i\omega)$ and $\chi_s(\boldsymbol{r},\boldsymbol{r}',i\omega)$ are manifestly real, so that also the integrand of (4.85) is real.

4.3 Local Density Approximation (LDA)

In view of the complex structure of the time-ordered response function one might ask whether the adiabatic connection (4.75) can be of any use? Clearly, it does not directly define some explicit density functional: only the mere existence of the functional $\chi_\lambda[n]$ is ensured, while the actual density dependence of χ_λ is as unknown as that of $|\Psi_\lambda[n]\rangle$. However, there exists a prominent interacting many-body system for which the density dependence of χ_λ is rather well-known, the *homogeneous* (or *uniform*) *electron gas* (HEG). The HEG is a system of infinitely many interacting electrons which do not experience a spatially varying external potential. Of course, the total energy of infinitely many particles is infinite, only the energy density is a meaningful quantity, or, alternatively, the energy per particle. However, even the energy density of infinitely many electrons diverges, due to the long range of the Coulomb interaction. In order to cure this problem, the charge of the electrons has to be neutralized by a homogeneous, positive background charge density n_+, consistent with the requirement that the external potential has to be a constant. As a result, the net charge in any (suitably chosen) volume of space is zero, so that the long-range Coulomb forces vanish and a finite energy density is obtained. If the number of electrons per volume exceeds a certain threshold, the electron ground state density n_0 of the HEG is constant throughout all space, i.e. homogeneous, with $n_0 = n_+$.

As a direct consequence, the spatial xc-energy density e_{xc} of the HEG (or the xc-energy per particle $\varepsilon_{xc} = e_{xc}/n_0$) is an ordinary function of n_0, rather than an actual functional. In addition, the response function of the HEG only depends on the distance between the points $\boldsymbol{r}$ and $\boldsymbol{r}'$,

$$\chi_\lambda(\boldsymbol{r}t,\boldsymbol{r}'t') \equiv \chi_\lambda(|\boldsymbol{r}-\boldsymbol{r}'|,t-t'), \tag{4.86}$$

as a result of the translational and rotational invariance of the HEG (in the following the response function of the HEG will always be distinguished from the general χ

by its arguments). The adiabatic connection thus takes the form[3]

$$e_{\rm xc}^{\rm HEG}(n_0) = \frac{1}{2}\int d^3r'\, w(\boldsymbol{r}-\boldsymbol{r}') \int_0^1 d\lambda \left\{ i\hbar\chi_\lambda(\boldsymbol{r}-\boldsymbol{r}',0;n_0) - n_0\,\delta^{(3)}(\boldsymbol{r}-\boldsymbol{r}')\right\}, \tag{4.87}$$

where the density dependence of χ_λ has been explicitly indicated. In addition, one has used the fact that the Coulomb interaction only depends on the distance between the two particles. A simpler form is obtained by Fourier transformation,

$$e_{\rm xc}^{\rm HEG}(n_0) = \frac{1}{2}\int \frac{d^3q}{(2\pi)^3}\, w(\boldsymbol{q}) \left\{ i\hbar \int_0^1 d\lambda \int \frac{d\omega}{2\pi}\, \chi_\lambda(\boldsymbol{q},\omega;n_0) - n_0 \right\}, \tag{4.88}$$

with

$$w(\boldsymbol{q}) = \int d^3r\, e^{-i\boldsymbol{q}\cdot\boldsymbol{r}}\, w(\boldsymbol{r}-\boldsymbol{r}') \tag{4.89}$$

$$\chi_\lambda(\boldsymbol{q},\omega) = \int dt \int d^3r\, e^{i\omega t}\, e^{-i\boldsymbol{q}\cdot\boldsymbol{r}}\, \chi_\lambda(\boldsymbol{r}-\boldsymbol{r}',t-t'). \tag{4.90}$$

The explicit form of $w(\boldsymbol{q})$ for the case of the Coulomb interaction is[4]

$$w(\boldsymbol{q}) = \frac{4\pi e^2}{\boldsymbol{q}^2}. \tag{4.91}$$

Equation (4.88) can be evaluated for a number of approximations for χ_λ. However, a detailed account of the state-of-the-art knowledge on the HEG, and in particular on χ_λ and $e_{\rm xc}^{\rm HEG}$, is beyond the scope of this text. In the following we will only make use of the *Dyson equation* for the response function,

$$\begin{aligned}
\chi(\boldsymbol{q},\omega) &= \Pi(\boldsymbol{q},\omega) + \Pi(\boldsymbol{q},\omega)\, w(\boldsymbol{q})\, \chi(\boldsymbol{q},\omega) && (4.92)\\
&= \frac{\Pi(\boldsymbol{q},\omega)}{1-w(\boldsymbol{q})\,\Pi(\boldsymbol{q},\omega)} && (4.93)\\
&= \Pi(\boldsymbol{q},\omega) + \Pi(\boldsymbol{q},\omega)\, w(\boldsymbol{q})\, \Pi(\boldsymbol{q},\omega)\\
&\quad + \Pi(\boldsymbol{q},\omega)\, w(\boldsymbol{q})\, \Pi(\boldsymbol{q},\omega)\, w(\boldsymbol{q})\, \Pi(\boldsymbol{q},\omega) + \ldots,
\end{aligned}$$

in which χ is expanded in powers of its basic buildings blocks, the *irreducible* (or *proper*) polarization insertion Π. The concept behind Eq. (4.92) is closely related to that of the Dyson equation for the single-particle Green's function, introduced in Sect. 3.6: Equation (4.92) separates the core contributions to χ, as the propagation and annihilation of a single particle–hole pair created by the incoming momentum

[3] Note that $e_{\rm xc}$ is the xc-energy per volume, not the xc-energy per particle, which is often used to characterize the HEG. The latter quantity will be denoted by $\epsilon_{\rm xc}$.

[4] In actual calculations the long range of the Coulomb interaction often requires use of the screened form (L.10) in intermediate steps of the calculation. The limit of vanishing screening is then usually taken at the end of the calculation.

$\boldsymbol{q}$, from simple repetitions of the same process. Many further details on χ can be found in standard textbooks as [94, 95].

4.3.1 Exchange

We first consider the lowest order contribution to χ_λ, i.e. the response function χ_0 of a noninteracting HEG with density n_0. This noninteracting response function is independent of λ. When inserted into the general expression (4.88), χ_0 yields a contribution to $e_{\rm xc}^{\rm HEG}$ which is linear in e^2. By definition, this is the exchange energy of the HEG,

$$e_{\rm x}^{\rm HEG}(n_0) = \frac{1}{2}\int \frac{d^3q}{(2\pi)^3}\, w(\boldsymbol{q}) \left\{ i\hbar \int \frac{d\omega}{2\pi}\, \chi_0(\boldsymbol{q},\omega) - n_0 \right\}. \tag{4.94}$$

χ_0, the so-called *Lindhard function* [130, 94], can be written in terms of the single-particle Green's function G_0 of the noninteracting HEG,

$$\begin{aligned}
\chi_0(\boldsymbol{r}-\boldsymbol{r}',t-t') &\equiv \Pi^{(0)}(\boldsymbol{r}-\boldsymbol{r}',t-t') \\
&= -\frac{i}{\hbar}\sum_{\sigma,\sigma'} G_0(\boldsymbol{r}-\boldsymbol{r}',t-t',\sigma\sigma')\, G_0(\boldsymbol{r}'-\boldsymbol{r},t'-t,\sigma'\sigma) \\
\Pi^{(0)}(\boldsymbol{q},\omega) &= -\frac{i}{\hbar}\sum_{\sigma,\sigma'}\int \frac{d^3k}{(2\pi)^3}\frac{dk^0}{2\pi}\, G_0(\boldsymbol{k}+\boldsymbol{q},k^0+\omega,\sigma\sigma')\, G_0(\boldsymbol{k},\omega,\sigma'\sigma)\,.
\end{aligned} \tag{4.95}$$

The explicit form of G_0, Eq. (L.29), is obtained from the general result (3.124) by insertion of the eigenstates of the noninteracting HEG (and use of a suitable limiting procedure in order to keep the normalization of the eigenstates and the density under control—see Appendix D). These eigenstates are simple plane waves,

$$\phi_l(\boldsymbol{r}\sigma) \equiv \phi_{\boldsymbol{k}s}(\boldsymbol{r}\sigma) = C e^{i\boldsymbol{k}\cdot\boldsymbol{r}} \chi_s(\sigma)\,, \qquad \varepsilon_l \equiv \varepsilon_{\boldsymbol{k}} = \frac{\hbar^2\boldsymbol{k}^2}{2m}\,,$$

which are occupied for all $|\boldsymbol{k}|$ below the *Fermi momentum*

$$k_{\rm F} = \frac{\sqrt{2m\varepsilon_{\rm F}}}{\hbar}\,. \tag{4.96}$$

The Fermi momentum, in turn, is determined by the density of the gas,

$$n_0 = \frac{k_{\rm F}^3}{3\pi^2}\,. \tag{4.97}$$

The expression (4.95) indicates why it is often preferable to represent Green's and response functions in terms of Feynman diagrams, rather than by explicit specification of Feynman integrals. The diagrammatic equivalent of (4.95),

$$i\hbar\Pi^{(0)} =$$  , (4.98)

in which the solid line represents G_0 (either in real or in momentum space), is much more compact. In the following the diagrammatic form will therefore be utilized frequently. The technical details of the Feynman rules, required to translate diagrams into integrals such as (4.95) and vice versa are given in Appendix L (the rules used are identical with those of [94], Chaps. 9–12, to which the reader is referred for background information).

Insertion of (4.95) and (L.29) into (4.94) and subsequent integration over $\boldsymbol{k}, k^0$ leads to [131]

$$e_{\mathrm{x}}^{\mathrm{HEG}}(n_0) = -\frac{3(3\pi^2)^{1/3}}{4\pi} e^2 n_0^{4/3} \,. \tag{4.99}$$

An alternative, very elementary derivation of this result is given in Appendix D.

4.3.2 *Correlation: High-Density Limit*

The leading term of the high-density limit of the correlation energy $e_{\mathrm{c}}^{\mathrm{HEG}}$ is contained in a set of contributions termed *random phase approximation* (RPA). The RPA corresponds to the approximation

$$\chi_\lambda^{\mathrm{RPA}}(\boldsymbol{q},\omega) = \frac{\Pi^{(0)}(\boldsymbol{q},\omega)}{1-\lambda w(\boldsymbol{q})\Pi^{(0)}(\boldsymbol{q},\omega)} \tag{4.100}$$

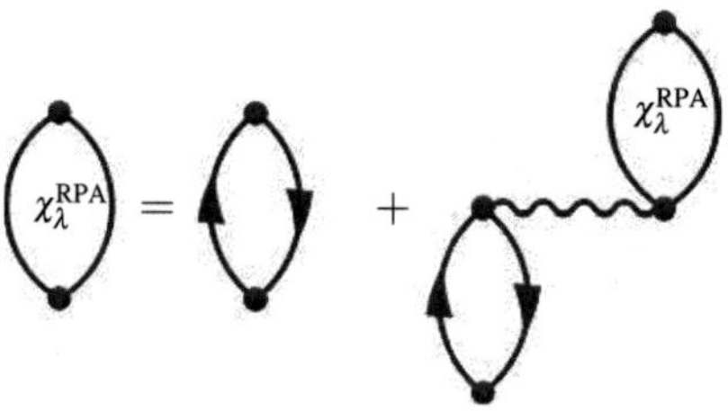

This form allows one to perform the λ-integration in (4.88),

$$e_{\mathrm{c}}^{\mathrm{HEG,RPA}} = -\frac{i\hbar}{2}\int \frac{d^3q}{(2\pi)^3}\frac{d\omega}{2\pi}\left\{ \ln\left|1-w(\boldsymbol{q})\Pi^{(0)}(\boldsymbol{q},\omega)\right| + w(\boldsymbol{q})\Pi^{(0)}(\boldsymbol{q},\omega)\right\} \tag{4.101}$$

(the exchange term (4.99) has been subtracted from the total xc energy (4.88)). An exact analytical evaluation of the remaining integrals is not possible. However, in the limit of high density one obtains [132, 133],

$$e_{\mathrm{c}}^{\mathrm{HEG,RPA}}(n_0) \xrightarrow{n_0\to\infty} \frac{e^2}{a_0}n_0\left\{\frac{1-\ln 2}{\pi^2}\ln(r_s)-0.071+\ldots\right\}, \qquad (4.102)$$

with the dimensionless *Wigner-Seitz radius*

$$r_{\mathrm{s}} = \left(\frac{3}{4\pi n_0}\right)^{1/3}\frac{me^2}{\hbar^2}. \qquad (4.103)$$

The Wigner-Seitz radius is the ratio between the radius of the sphere which is on the average occupied by a single electron of the gas and the Bohr radius

$$a_0 = \frac{\hbar^2}{me^2}. \qquad (4.104)$$

The leading term of the high-density expansion of $e_{\mathrm{c}}^{\mathrm{HEG}}$ is seen to increase faster than linear with n_0, due to the $\ln(r_{\mathrm{s}})$-factor. In terms of the electron–electron coupling constant e^2 the leading term scales as $e^4\ln(e^2)$, which indicates that the high-density limit emerges from a resummation of the geometric series inherent in (4.100). This result reflects the fact that there is no gap between the highest occupied and the lowest unoccupied state in the case of the HEG. The high-density limit is therefore not identical with the second order perturbative correction in w (which diverges for the HEG). A complete numerical evaluation of (4.101) has been carried out by von Barth and Hedin [34] as well as by Vosko, Wilk and Nusair [134].

The result (4.102) raises two questions:

- For which values of r_{s} does (4.102) provide reliable results?
- Which range of r_{s} is relevant for realistic systems?

Even though the HEG only serves as a model system for the construction of approximate xc-functionals, it is obvious that any functional relying on expressions like (4.102) can only be accurate for systems which have densities in the range of r_{s} covered by (4.102). The average density obtained from the valence bands of a number of solids are listed in Table 4.2. These examples show that the range $1 < r_{\mathrm{s}} < 6$ is probed by the valence densities of solids. An impression of inner shell densities of atoms is given in Table 4.3. The data given in this table are local density values, in contrast to the average valence densities in Table 4.2. As expected, one finds rather high densities in the inner shell regions, so that r_{s} can be as small as 0.01 for very heavy atoms. However, even for atoms the density in the valence regime is more relevant for all kinds of physical processes, so that an approximation should primarily address the much lower valence densities.

From the values in Tables 4.2 and 4.3 it is immediately clear that, even for the highest valence densities, the leading, logarithmic term in (4.102) does not dominate over the next to leading term,

$$(1-\ln 2)\frac{\ln(r_{\mathrm{s}}=2)}{\pi^2} = 0.021 < 0.071\,.$$

One thus has to go beyond the leading term.

Table 4.2 Average valence densities n_0 of prototype solids. The number of valence electrons is given for the conventional cubic unit cell.

Solid	Lattice constant [Bohr]	n_0 (valence) [Bohr^{-3}]	r_s (valence)	Lattice type	Number of valence electrons per atom	
					atom	unit cell
Li	6.60	0.00696	3.25	bcc	1	2
Na	7.99	0.00392	3.93	bcc	1	2
Cs	11.43	0.00134	5.62	bcc	1	2
Fe	5.42	0.0252	2.12	bcc	2	4
Al	7.65	0.0268	2.07	fcc	3	12
Au	7.71	0.00873	3.01	fcc	1	4
C	6.75	0.1040	1.32	dia	4	32
Si	10.26	0.0296	2.00	dia	4	32

Table 4.3 Densities of prototype atoms at the r-expectation values $\langle r \rangle$ of various subshells.

Atom	Subshell		$\langle r \rangle$	n_0	r_s
	n	l	[Bohr]	[Bohr^{-3}]	
Li	1	s	0.59	0.50	0.78
	2	s	3.93	0.0017	5.90
Cs	1	s	0.026	7841.	0.031
	2	s	0.11	496.	0.078
	3	s	0.30	56.4	0.16
	4	s	0.70	4.99	0.36
	5	s	1.74	0.15	1.16
	6	s	5.55	0.00055	7.59
Fe	1	s	0.059	639.	0.072
	2	s	0.27	28.5	0.20
	3	s	0.82	1.77	0.51
	3	p	0.86	1.51	0.54
	3	d	1.13	0.54	0.76
	4	s	3.00	0.0075	3.17
Si	1	s	0.11	86.7	0.14
	2	s	0.57	2.75	0.44
	3	s	2.16	0.029	2.02
	3	p	2.79	0.012	2.72

The next important contribution is obtained from the so-called *second order exchange term*. It results from the first order contributions to the response function which are not contained in the RPA, Eq. (4.100). These terms can be depicted diagrammatically as

$i\hbar\Pi^{(1)} =$ 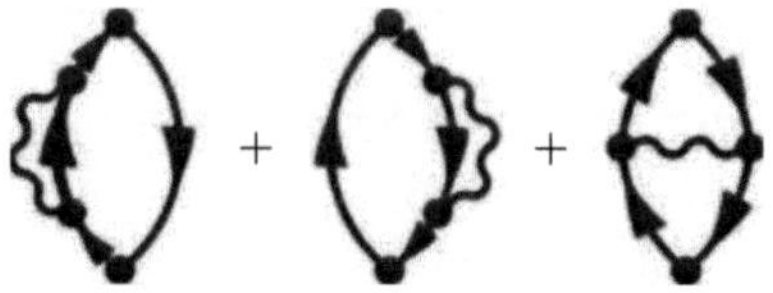 (4.105)

and contribute to the complete χ according to (4.93),

$$\chi = \Pi^{(0)} + \Pi^{(1)} + \Pi^{(0)} w \Pi^{(0)} + \mathcal{O}(e^4) .$$

When inserted into (4.88), $\Pi^{(1)}$ leads to an additional contribution to e_c^{HEG} which is linear in n_0 [133, 135], the second order exchange (SOX) term. If one adds the SOX term to the RPA energy, one obtains the complete high-density limit,

$$e_c^{\mathrm{HEG}}(n_0) \xrightarrow{n_0 \to \infty} \frac{e^2}{a_0} n_0 \left\{ \frac{1-\ln 2}{\pi^2} \ln(r_s) - 0.047 + \ldots \right\} . \tag{4.106}$$

Unfortunately, the next order in this perturbation series scales as $r_s \ln(r_s)$ [136, 137], so that (4.106) is still not sufficient to cover the relevant range of densities. In one way or another one needs more information on e_c^{HEG} than provided by the high-density limit.

4.3.3 Correlation: Low-Density Limit

For very low densities the HEG crystallizes in the form of a bcc lattice, the so-called *Wigner crystal* [138–140]. In this limit the kinetic energy of the electrons is much smaller than the Coulomb repulsion between them. It is thus energetically favorable that the electrons permanently remain at the largest possible separation from each other, which the given density allows. Their motion is restricted to small vibrations around their equilibrium positions, similar to the motion of the nuclei in a standard lattice. The resulting correlation energy density has the form

$$e_c^{\mathrm{HEG}}(n_0) \xrightarrow{n_0 \to 0} \frac{e^2}{a_0} n_0 \left\{ -\frac{0.438}{r_s} + \frac{1.33}{r_s^{3/2}} - \frac{1.47}{r_s^2} + \ldots \right\} , \tag{4.107}$$

where the last two terms originate from the zero-point energy corresponding to the vibrational motion. The transition from the regular (unpolarized) HEG to the Wigner solid is expected to take place around $r_s \approx 80$ [141]. This limiting behavior of the electron gas is obviously not relevant for ordinary matter. Unfortunately, this implies that neither the limit (4.106) nor the limit (4.107) is particularly useful in practice.

4.3.4 Correlation: Interpolation Between High- and Low-Density Regime

One is thus led to consider the intermediate regime of densities. The simplest approach to the correlation energy at these densities is an interpolation between the high- and the low-density limit. The first interpolation of this type has been suggested by Wigner [138],

$$e_c^{\rm Wigner}(n_0) = -\frac{e^2}{a_0} n_0 \frac{0.44}{r_s + 7.8} .$$

This formula neither includes the leading, logarithmic high-density term, nor does it take into account the fact that the high- and the low-density limits correspond to different physical phases, i.e. the liquid high-density phase and Wigner crystal. It is nevertheless surprisingly accurate in the relevant density range of $1 \le r_s \le 6$.

A more sophisticated form for $e_c^{\rm HEG}$ can only be obtained by an explicit evaluation for intermediate densities. In practice, it turns out to be simpler to perform a Monte Carlo simulation of the HEG and extract $e_c^{\rm HEG}$ from the resulting total energy, than to evaluate (4.88). Rather accurate results have been obtained in this way [141, 142]. In order to facilitate their practical use, they are usually combined with the analytical results for the high- and low-density limits. For this purpose one chooses an analytical ansatz which reproduces the forms of the high- and low-density limits (4.106) and (4.107), but simultaneously shows sufficient flexibility at intermediate densities. For instance, Vosko, Wilk and Nusair (VWN) [134] rely on the ansatz

$$\begin{aligned} e_c^{\rm VWN}(n_0) = \frac{e^2}{a_0} n_0 \frac{1-\ln 2}{\pi^2} \Bigg\{ & \ln\frac{x^2}{X(x)} + \frac{2b}{(4c-b^2)^{1/2}} \tan^{-1}\frac{(4c-b^2)^{1/2}}{2x+b} \\ & - \frac{b x_0}{X(x_0)} \left[\ln\frac{(x-x_0)^2}{X(x)} + \frac{2(2x_0+b)}{(4c-b^2)^{1/2}} \tan^{-1}\frac{(4c-b^2)^{1/2}}{2x+b} \right] \Bigg\} \\ X(x) = x^2 + bx + c \qquad ; \qquad & x = \sqrt{r_s} . \end{aligned} \tag{4.108}$$

By construction, this formula reproduces the leading logarithmic term of the high-density limit (4.106) exactly. In addition, the analytical form of all other known terms of the high-density expansion is preserved,

$$e_c^{\rm VWN}(n_0) \xrightarrow{n_0\to\infty} \frac{e^2}{a_0} n_0 \left\{ \frac{1-\ln 2}{\pi^2} \ln(r_s) + A + r_s \Big[B\ln(r_s) + C \Big] + \ldots \right\} .$$

However, VWN do not require that the exact numerical values of the coefficients A, B and C (as e.g. $A = -0.047$) are reproduced. The same is true for the low-density expansion (4.107), whose analytical form is recovered after an expansion of the VWN formula for large r_s. The numerical values given in (4.107), on the other hand, correspond to the Wigner crystal, so that they are not very useful for the liquid phase anyway. The parameters x_0, b and c are then optimized to reproduce the exact coefficient $A = -0.047$ of (4.106) and the Monte Carlo data (the result is $x_0 = -0.10498$, $b = 3.72744$ and $c = 12.9352$—see Table V of [134]). The concepts behind most other parameterizations are quite similar.

Some of the more accurate parameterizations of this type are plotted in Fig. 4.2. This figure confirms the earlier statement that the high-density limit (4.106) can

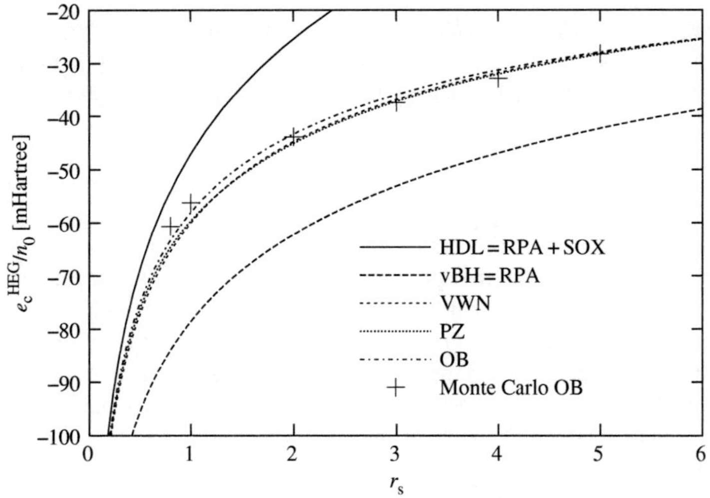

Fig. 4.2 Correlation energy per electron ($\varepsilon_c = e_c^{HEG}/n_0$) of the spin-saturated homogeneous electron gas as a function of the Wigner-Seitz radius r_s, Eq. (4.103), for several parameterizations: vBH—[34], VWN—[134], PZ—[143], OB—[142]. Also plotted is the high-density limit (4.106) (HDL $\equiv$ RPA plus second order exchange) and the Monte Carlo data [142].

not be used for realistic systems: one finds a 25% difference between (4.106) and the complete e_c^{HEG} for densities as large as $r_s = 1$. Moreover, the difference between the RPA and (4.106) observed for small r_s emphasizes the importance of the SOX contribution. The three interpolation formulae shown differ by the use of different Monte Carlo reference energies and/or by the analytical ansatz chosen. The VWN form [134] and the Perdew-Zunger (PZ) [143] interpolation both employ the Ceperley-Alder data [141], but differ in the analytical structure of the analytical ansatz. On the other hand, the Ortiz-Ballone (OB) [142] interpolation is based on their own Monte Carlo data, but uses the same ansatz as VWN. One recognizes that (i) the form of the ansatz is irrelevant (as long as it reproduces the limits and is otherwise sufficiently flexible) and, (ii) the Monte Carlo results have converged within a few percent. It turns out that the remaining uncertainty in the Monte Carlo data and the resulting interpolation formulae is irrelevant in practice (see below).

4.3.5 Density Functional: Local Density Approximation (LDA)

So far, we have considered the xc-energy of the HEG. The important question to be addressed at this stage is: how can one utilize these results within the framework of DFT, given the fact that all systems of interest are inhomogeneous? The answer to this question is provided by the *local density approximation* (LDA) [6]. In the LDA the xc-energy density of the inhomogeneous system with density $n(\boldsymbol{r})$ is locally approximated by the xc-energy density of an electron gas with density $n_0 = n(\boldsymbol{r})$,

$$E_{\rm xc}^{\rm LDA}[n] = \int d^3r\, e_{\rm xc}^{\rm HEG}(n_0 = n(\boldsymbol{r}))\,. \tag{4.109}$$

A number of comments are appropriate:

(a) The LDA is a universal first-principles functional of n in the sense that it does not depend on any free parameters which introduce some physical scale. It would be a misinterpretation to regard the coefficients in the interpolation formulae as free parameters: they are completely fixed by the properties of the HEG. One could, as an alternative, directly use numerical Monte Carlo results in the LDA (4.109). The LDA is thus consistent with the *ab-initio* concept of DFT.

(b) In the LDA any system is locally treated as an electron gas. The LDA can be applied to arbitrarily inhomogeneous systems without encountering any technical difficulties. However, its formal universality and technical applicability do not imply validity for just any system. Clearly, one would expect the LDA to be particularly appropriate for systems which share some properties with the HEG, as for instance simple metals. A more precise characterization of its regime of validity will be given below.

(c) The LDA is easily utilized in the KS equations, as the corresponding xc-potential is a simple function of the local density,

$$\begin{aligned} v_{\rm xc}^{\rm LDA}(\boldsymbol{r}) &= \frac{\delta E_{\rm xc}^{\rm LDA}[n]}{\delta n(\boldsymbol{r})} = \int d^3r' \left.\frac{de_{\rm xc}^{\rm HEG}(n_0)}{dn_0}\right|_{n_0=n(\boldsymbol{r}')} \frac{\delta n(\boldsymbol{r}')}{\delta n(\boldsymbol{r})} \\ &= \left.\frac{de_{\rm xc}^{\rm HEG}(n_0)}{dn_0}\right|_{n_0=n(\boldsymbol{r})}\,. \end{aligned} \tag{4.110}$$

One observes, however, that $v_{\rm xc}^{\rm LDA}$ has an extremely short range, as it only depends on the local density. As an important consequence, the LDA potential vanishes exponentially in the asymptotic region of finite systems. This property can easily be demonstrated for its exchange component, which is obtained by differentiation of (4.99),

$$v_{\rm x}^{\rm LDA}(\boldsymbol{r}) = -\frac{(3\pi^3)^{1/3}}{\pi}\, e^2 n^{1/3}(\boldsymbol{r})\,. \tag{4.111}$$

As soon as the density decays exponentially, the same is true for the LDA exchange potential,

$$n(\boldsymbol{r}) \sim e^{-\alpha r} \qquad \Longrightarrow \qquad v_{\rm x}^{\rm LDA}(\boldsymbol{r}) \sim e^{-\alpha r/3}\,. \tag{4.112}$$

This behavior is in obvious contradiction to the exact result (4.20).

A direct consequence is the fact that the KS spectrum obtained with the LDA for finite system like atoms does not contain Rydberg states, which are characteristic of the Coulombic $-1/r$-behavior. This implies that the LDA necessarily predicts atomic negative ions to be unbound, which is one of its most important deficiencies.

The asymptotic behavior of the LDA correlation potential for finite systems is determined by the low-density limit of the correlation energy density $e_c^{\rm HEG}(n_0)$. In view of (4.110) any functional with the same density-dependence as (4.107) shows an exponential decay,

$$n(\boldsymbol{r}) \sim e^{-\alpha r} \qquad \Longrightarrow \qquad v_c^{\rm LDA}(\boldsymbol{r}) \sim e^{-\alpha r/3}\,, \tag{4.113}$$

rather than the correct power law behavior [92].

(d) The LDA has been the workhorse of DFT applications for decades, so that an overwhelming number of LDA results can be found in the literature. Any attempt to give an overview of this vast body of material is bound to fail. At this point the discussion is restricted to the most basic numbers which can be quoted, i.e. the xc-energies of atoms. Atoms are not only the elementary building blocks of matter, they are also quite critical test systems due to the piecewise exponential behavior of atomic densities (which reflects the orbital structure). In addition, a number of exact results for atomic xc-energies are available, allowing an unambiguous comparison.

Atomic exchange energies from various sources are compared in Table 4.4. The corresponding reference values are obtained by an exact treatment of the DFT exchange (4.5) and a complete neglect of correlation. These exact *exchange-*

Table 4.4 Exchange energies of spin-saturated, closed-subshell atoms: LDA, second order GE (GE2), PW91-GGA, PBE-GGA and B88-GGA energies (the results including gradient corrections are discussed in Sect. 4.5.5) obtained by insertion of exact exchange-only densities into the functionals (4.109), (4.178), (4.255), (4.256) and (4.286) in comparison with exact values (all energies are in Hartree). Also given is the corresponding percentage error Δ.

Atom	Exact	LDA		GE2		PW91		PBE		B88	
	$-E_x$	$-E_x$	Δ	$-E_x$	Δ	$-E_x$	Δ	$-E_x$	Δ	$-E_x$	Δ
He	1.026	0.884	−13.82	1.007	−1.86	1.017	−0.88	1.014	−1.19	1.025	−0.03
Be	2.666	2.312	−13.26	2.581	−3.19	2.645	−0.77	2.636	−1.13	2.658	−0.30
Ne	12.105	11.033	−8.85	11.775	−2.73	12.115	0.08	12.067	−0.32	12.138	0.27
Mg	15.988	14.612	−8.61	15.510	−2.99	15.980	−0.05	15.915	−0.46	16.000	0.08
Ar	30.175	27.863	−7.66	29.293	−2.92	30.123	−0.17	29.996	−0.59	30.153	−0.07
Ca	35.199	32.591	−7.41	34.183	−2.89	35.165	−0.10	35.016	−0.52	35.192	−0.02
Zn	69.619	65.645	−5.71	68.109	−2.17	69.834	0.31	69.531	−0.13	69.867	0.36
Kr	93.833	88.624	−5.55	91.651	−2.33	93.831	0.00	93.426	−0.43	93.872	0.04
Sr	101.926	96.362	−5.46	99.560	−2.32	101.918	−0.01	101.479	−0.44	101.956	0.03
Cd	148.880	141.543	−4.93	145.702	−2.13	148.885	0.00	148.260	−0.42	148.930	0.03
Xe	179.064	170.566	−4.75	175.304	−2.10	178.991	−0.04	178.245	−0.46	179.043	−0.01

only (x-only) calculations require a fully numerical evaluation of the potential corresponding to the expression (4.5) by the OPM (see Chap. 6). Self-consistent KS calculations with this potential provide a large amount of information on the exact x-only ground state. In particular, one obtains the exact x-only density which can then be inserted into any approximate exchange functional. The resulting LDA values (as well as energies obtained with gradient corrected functionals—see Sect. 4.4.3), are also given in Table 4.4, together with their percentage deviation from the exact E_x. The LDA turns out to be moderately accurate, with a consistent underestimation of the exact E_x.

The picture which emerges from Table 4.4 should not be generalized without caution to other systems or to more subtle atomic quantities. It has already been pointed out that the LDA does not predict any negative atomic ion to be stable. This emphasizes the fact that there is a crucial difference between integrated quantities like the energy and local quantities like the potential. Integrated quantities can be more accurate due to a cancellation of local errors. In addition, the accuracy of the LDA clearly depends on the specific system under consideration. The less the density varies spatially, the higher is the accuracy of the LDA. This point is investigated in more detail later.

Table 4.5 offers a similar comparison for the correlation energy. In this case it is much more difficult to generate exact reference data. As the exact correlation functional is not known, exact values for E_c can only be calculated indirectly by subtraction of the x-only ground state energy obtained by insertion of the exact KS orbitals into the functional (4.6) from the exact total ground state energy obtained with a non-DFT method. Although a few numbers of this type are available, the comparison in Table 4.5 and all subsequent comparisons of correlation energies rely on the conventional quantum chemical correlation energy E_c^{QC}, defined by (4.9). This procedure is legitimate, since the difference between E_c and E_c^{QC} is found to be extremely small (compare the discussion of Table 4.1 in Sect. 4.1). This statement is true in particular for the present analysis: the LDA correlation energies obtained by insertion of the exact x-only densities into Eq. (4.109) differ from E_c^{QC} by a factor of two. The small differences between E_c and E_c^{QC} as well as the small differences between the LDA values calculated from the x-only densities and the LDA energies resulting from the exact densities are thus irrelevant. The same is true for the differences between the various parameterizations of e_c^{HEG}.

While an error of 100% seems to exclude the application of the LDA to atomic systems, one should be aware of the fact that the absolute size of E_c is usually much smaller than that of E_x, so that the accuracy of the exchange component dominates. In addition, the LDA xc-energies often profit from error cancellation between the underestimated E_x and the overestimated E_c.

Overall, one can state that the LDA is surprisingly accurate for a good number of quite inhomogeneous systems. This accuracy can be understood on the basis of (i) the size of the inhomogeneity corrections (see Sect. 4.4), and (ii) the sum rules which the LDA satisfies (see Sect. 4.5). Further discussion of the merits

Table 4.5 Correlation energies ($-E_c$) of unpolarized, closed-subshell atoms: LDA, second order GE (GE2), PW91-GGA, LYP-GGA, PKZB-MGGA, TPSS-MGGA, B3LYP- and PBE0-hybrid results obtained by insertion of exact exchange-only densities into (4.109) (using the VWN parameterization [134]), (4.183), (4.274), (4.289), (4.316), (4.326), (6.128) and (6.135) in comparison with second order perturbation theory on HF basis (MP2) [144, 145] and the exact E_c^{QC} [115] (all energies are given in mHartree). The correlation component of the PBE0-hybrid is identical with that of the PBE-GGA.

Atom	MP2	Exact	LDA	GE2	PW91	LYP	PKZB	TPSS	B3LYP	PBE(0)
He	37	42	113	−82	46	44	42	43	64	42
Be	76	94	225	−213	94	94	86	87	133	86
Ne	388	391	746	−487	382	384	351	354	491	351
Mg	428	438	892	−607	450	459	411	415	587	411
Ar	709	722	1431	−962	771	751	707	711	950	707
Ca	798		1581	−1087	847	830	774	779	1049	774
Zn	1678		2668	−1471	1526	1431	1406	1410	1785	1406
Kr	1891		3284	−1801	1914	1749	1767	1771	2184	1767
Cd	2618		4571	−2425	2739	2423	2536	2540	3025	2536
Xe	3088		5199	−2768	3149	2744	2918	2920	3428	2918

and failures of the LDA, as well as additional numerical results, will be given in connection with the extensions of the LDA.

(e) In the literature one sometimes meets some variants of the LDA for the exchange potential. The first of these is the *Slater exchange* [146],

$$v_x^{\mathrm{Slater}}(\boldsymbol{r}) = \frac{3}{2} v_x^{\mathrm{LDA}}(\boldsymbol{r}) .$$

This form is obtained if one identifies the exchange potential with $2e_x^{\mathrm{HEG}}/n$, rather than with the functional derivative of E_x^{LDA}. This identification follows the pattern set by the Hartree term, for which $v_{\mathrm{H}} = 2e_{\mathrm{H}}/n$. In applications, v_x^{Slater} is usually multiplied by an adjustable prefactor α, which defines the so-called $X\alpha$-method [147],

$$v_x^{X\alpha}(\boldsymbol{r}) = \frac{3}{2} \alpha\, v_x^{\mathrm{LDA}}(\boldsymbol{r}) . \tag{4.114}$$

The standard value for α is 0.7, which essentially leads back to the LDA exchange.

The second modification of v_x^{LDA} represents an attempt to resolve the problem expressed by (4.112): the *Latter-correction* [148] reestablishes the correct asymptotic behavior (4.20) by hand,

$$v_x^{\mathrm{Latter}}(\boldsymbol{r}) = \begin{cases} v_x^{\mathrm{LDA}}(\boldsymbol{r}) & \text{for } |\boldsymbol{r}| \le R(\hat{\boldsymbol{r}}) \\ -e^2/|\boldsymbol{r}| & \text{for } |\boldsymbol{r}| > R(\hat{\boldsymbol{r}}) \end{cases} , \tag{4.115}$$

where $R(\hat{\boldsymbol{r}})$ ($\hat{\boldsymbol{r}} = \boldsymbol{r}/|\boldsymbol{r}|$) denotes the distance of the largest closed surface around the finite system for which

$$v_{\rm x}^{\rm LDA}(R\hat{\boldsymbol{r}}) = -\frac{e^2}{R(\hat{\boldsymbol{r}})}$$

from the center of the system.

(f) Already the first density functional established in the literature [13, 14], an approximation for the kinetic energy $T_s[n]$, was based on the same concept as the LDA for E_{xc} (for details see Appendix D). For the discussion of $T_s[n]$ the kinetic energy density of the noninteracting HEG has to be considered,

$$t_{\rm s}^{\rm HEG}(n_0) = \frac{3(3\pi^2)^{2/3}}{10}\frac{\hbar^2}{m}n_0^{5/3}. \tag{4.116}$$

The functional (4.116) can be utilized for inhomogeneous systems, if $t_{\rm s}^{\rm HEG}$ is evaluated locally with the actual $\boldsymbol{r}$-dependent densities of these systems,

$$T_{\rm s}^{\rm TF}[n] = \int d^3r\, t_{\rm s}^{\rm HEG}(n(\boldsymbol{r})). \tag{4.117}$$

$T_{\rm s}^{\rm TF}[n]$ constitutes the basis for the *Thomas-Fermi approach*, in which the many-body problem is tackled by an explicit solution of the variational equation (2.38).

4.3.6 Spin-Polarized Electron Gas: Local Spin-Density Approximation (LSDA)

The discussion of the previous section is based on the assumption that one has as many electrons with spin-up as with spin-down in the electron gas, $n_\uparrow = n_\downarrow = n_0/2$. On the other hand, for actual applications one often needs an approximation for the spin-density functional $E_{xc}[n_\uparrow, n_\downarrow]$. A spin-density-dependent version of the LDA (*local spin-density approximation*—LSDA) is easily constructed for the exchange functional. In this case one can combine the exact relation (4.19) with (4.109),

$$E_{\rm x}^{\rm LSDA}[n_\uparrow, n_\downarrow] = \frac{1}{2}\left\{E_{\rm x}^{\rm LDA}[2n_\uparrow] + E_{\rm x}^{\rm LDA}[2n_\downarrow]\right\}. \tag{4.118}$$

The correlation energy of a spin-polarized HEG is much more difficult to extract. It can, however, be evaluated numerically within the RPA [34, 134]. The result is usually expressed in terms of the total gas density n and the fractional polarization

$$\zeta = \frac{n_\uparrow - n_\downarrow}{n_\uparrow + n_\downarrow}, \tag{4.119}$$

rather than in terms of $n_\uparrow$ and $n_\downarrow$. One finds that the spin-dependence of the correlation energy has some similarity with that of the exchange energy (4.118), which can be written as

$$e_{\mathrm{x}}^{\mathrm{HEG}}(n,\zeta) = e_{\mathrm{x}}^{\mathrm{HEG}}(n,\zeta=0) + \left[e_{\mathrm{x}}^{\mathrm{HEG}}(n,\zeta=1) - e_{\mathrm{x}}^{\mathrm{HEG}}(n,\zeta=0)\right] f(\zeta) \quad (4.120)$$

$$f(\zeta) = \frac{(1+\zeta)^{4/3} + (1-\zeta)^{4/3} - 2}{2(2^{1/3}-1)} \,. \quad (4.121)$$

This similarity has first been noticed by von Barth and Hedin in their numerical RPA data [34], which suggests an analytical parameterization of the form

$$e_{\mathrm{c}}^{\mathrm{HEG,RPA}}(n,\zeta) \approx e_{\mathrm{c}}^{\mathrm{HEG,RPA}}(n,0) + \left[e_{\mathrm{c}}^{\mathrm{HEG,RPA}}(n,1) - e_{\mathrm{c}}^{\mathrm{HEG,RPA}}(n,0)\right] f(\zeta) \,. \quad (4.122)$$

However, VWN found that the form (4.122) is rather inaccurate for small ζ, even in the case of the RPA [134]. They suggested the improved parameterization

$$\begin{aligned} e_{\mathrm{c}}^{\mathrm{HEG}}(n,\zeta) &= e_{\mathrm{c}}^{\mathrm{HEG}}(n,0) + \alpha_{\mathrm{c}}(n)\frac{f(\zeta)}{f''(0)}[1-\zeta^4] \\ &\quad + \left[e_{\mathrm{c}}^{\mathrm{HEG}}(n,1) - e_{\mathrm{c}}^{\mathrm{HEG}}(n,0)\right] f(\zeta)\zeta^4 \,, \end{aligned} \quad (4.123)$$

with the spin-stiffness

$$\alpha_{\mathrm{c}}(n) = \left.\frac{d^2 e_{\mathrm{c}}^{\mathrm{HEG}}(n,\zeta)}{d\zeta^2}\right|_{\zeta=0} \,. \quad (4.124)$$

The form (4.123) reproduces the exact ζ-dependence of $e_{\mathrm{c}}^{\mathrm{HEG}}(n,\zeta)$ to lowest order. This can be verified by a Taylor expansion for small ζ, using the fact that $f(\zeta)/f''(0) = \zeta^2/2$ for $\zeta \to 0$,

$$e_{\mathrm{c}}^{\mathrm{HEG}}(n,\zeta) = e_{\mathrm{c}}^{\mathrm{HEG}}(n,\zeta=0) + \frac{1}{2}\alpha_{\mathrm{c}}(n)\zeta^2 + \mathscr{O}(\zeta^4) \,. \quad (4.125)$$

There is no term linear in ζ as there is no preferred spin direction in the HEG. For $\zeta \to 1$, on the other hand, the term in the second line of (4.123) is dominant.

The ingredients of (4.123) are the correlation energy of the unpolarized gas, $e_{\mathrm{c}}^{\mathrm{HEG}}(n,\zeta=0)$, the corresponding energy of a fully spin-polarized gas with the same total density, $e_{\mathrm{c}}^{\mathrm{HEG}}(n,\zeta=1)$, and the spin-stiffness. For $e_{\mathrm{c}}^{\mathrm{HEG}}(n,\zeta=0)$ one can use the Monte Carlo data already presented. Analogous results are available for $e_{\mathrm{c}}^{\mathrm{HEG}}(n,\zeta=1)$ [141, 142]. They can be represented in terms of some analytical interpolation formula, just as $e_{\mathrm{c}}^{\mathrm{HEG}}(n,\zeta=0)$. Figure 4.3 provides some explicit numbers. While the accuracy of these results is slightly lower than that for the unpolarized situation, it is nevertheless sufficient for all practical purposes. It is worthwhile to mention in this context that the Monte Carlo data suggest the fully spin-polarized HEG to be energetically lower than both the unpolarized gas and the Wigner crystal in the density range $75 < r_{\mathrm{s}} < 100$.

The third quantity which enters (4.123), the spin-stiffness, is known analytically only in the high-density limit [150–152, 134],

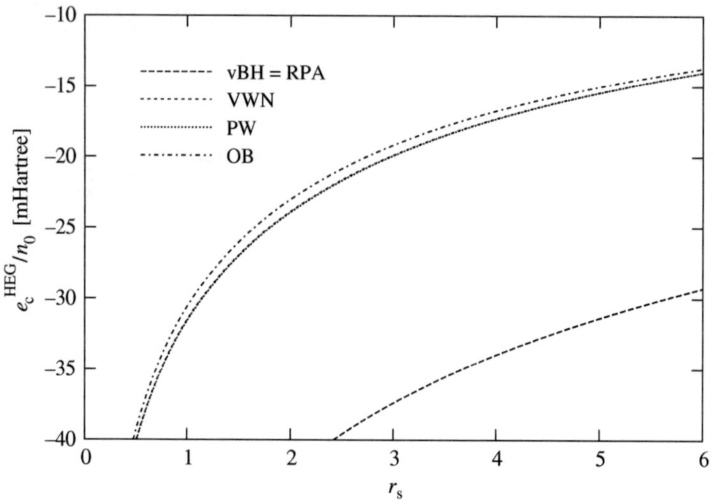

Fig. 4.3 Correlation energy per electron ($\varepsilon_c = e_c^{\rm HEG}/n_0$) of the fully spin-polarized homogeneous electron gas as a function of the Wigner-Seitz radius r_s, Eq. (4.103), for several parameterizations: vBH—[34], VWN—[134], PW—[149], OB—[142] (VWN and PW data essentially coincide).

$$\alpha_c(n) \xrightarrow{n_0 \to \infty} \frac{e^2}{a_0}\left(-\frac{1}{6\pi^2}\ln(r_s) + 0.03547 + \ldots\right) . \tag{4.126}$$

In addition, rather accurate numerical information is available within the RPA. These RPA data are reproduced extremely well by an analytical ansatz for $\alpha_c(n)$, which obeys the limit (4.126) and has the same n-dependence as the interpolation formula used for $\varepsilon_c^{\rm HEG}(n,0)$: for all densities the error is below 0.2%. This leads VWN to the expectation that the same analytical form is suitable for the spin-stiffness corresponding to the complete $\varepsilon_c^{\rm HEG}(n,\zeta)$. In order to utilize this ansatz for intermediate densities, they fit its parameters to the numerical data for

$$\frac{2}{\zeta^2}\,\frac{[\varepsilon_c^{\rm HEG}(n,1) - \varepsilon_c^{\rm HEG}(n,0)]}{[\varepsilon_c^{\rm HEG,RPA}(n,1) - \varepsilon_c^{\rm HEG,RPA}(n,0)]}\,[\varepsilon_c^{\rm HEG,RPA}(n,\zeta) - \varepsilon_c^{\rm HEG,RPA}(n,0)]\,,$$

for a number of low densities. This means that they use the ζ-dependence of the RPA for all densities, but rescale the RPA spin-stiffness by the overall sensitivity of $\varepsilon_c^{\rm HEG}(n,\zeta)$ to ζ. It turns out that the resulting $\alpha_c(n)$ also reproduces the few available Monte Carlo data for the partially spin-polarized HEG fairly accurately [142] (when combined with the optimum parameterizations of $\varepsilon_c^{\rm HEG}(n,1)$ and $\varepsilon_c^{\rm HEG}(n,0)$).

The VWN ansatz for the spin-stiffness is also utilized in other modern parameterizations of $\varepsilon_c^{\rm HEG}(n,\zeta)$, as e.g. the Perdew-Wang form [149], the only exception being the Perdew-Zunger functional [143]. The overall accuracy of the resulting $\varepsilon_c^{\text{HEG}}(n,\zeta)$ is of the order of 1–3%.

The LSDA for correlation is then defined by use of (4.123) with the inhomogeneous spin-densities of the actual system,

$$E_c^{\text{LSDA}}[n_\uparrow,n_\downarrow] = \int d^3r\, e_c^{\text{HEG}}(n(\boldsymbol{r}),\zeta(\boldsymbol{r})) \ . \tag{4.127}$$

It shares the properties (a)–(d) with the unpolarized form of the LDA. In the following, we no longer distinguish between LDA and LSDA: the term LDA should automatically be understood as LSDA whenever spin-polarization is present.

4.4 Nonlocal Corrections to the LDA

It has already been emphasized that the universal applicability of the LDA does not imply that this functional covers the physics of all kinds of systems. Real systems are more or less inhomogeneous, so that the adequacy of the LDA is not clear *a priori*. Quite naturally the question arises whether one can derive corrections to the LDA in a systematic fashion? The answer is yes, the underlying model system being the weakly inhomogeneous electron gas [6]. The resulting contributions to $E_{\text{xc}}[n]$ and the potential v_{xc} are often referred to as nonlocal corrections, as the corresponding energy density $e_{\text{xc}}(\boldsymbol{r})$ and potential $v_{\text{xc}}(\boldsymbol{r})$ are not simple functions of the local density $n(\boldsymbol{r})$. This terminology should, however, not be understood in the sense that v_{xc} would constitute a nonlocal potential, only its density dependence is nonlocal.

4.4.1 Weakly Inhomogeneous Electron Gas

The inhomogeneous electron gas is characterized by the Hamiltonian

$$\hat{H} = \hat{T} + \hat{W} + \int d^3r\, \hat{n}(\boldsymbol{r})\, \delta v(\boldsymbol{r}) \ , \tag{4.128}$$

with the external potential $\delta v(\boldsymbol{r})$ assumed to be a weak perturbation (in some yet to be specified sense). In order to keep control of the charge balance in the gas, one has to think of the perturbation as being generated by a small positive charge density $\delta n_+(\boldsymbol{r})$,

$$\delta v(\boldsymbol{r}) = -\int d^3r'\, w(\boldsymbol{r}-\boldsymbol{r}')\, \delta n_+(\boldsymbol{r}') \ , \tag{4.129}$$

where $w(\boldsymbol{r}-\boldsymbol{r}')$ denotes, as usual, the Coulomb interaction. Equation (4.129) covers both the case of a localized perturbation as well as some periodic structure. At this point there is no need to restrict δn_+ any further, so that the net charge δN_+

associated with δn_+,

$$\delta N_+ = \int d^3r\, \delta n_+(\mathbf{r}) \,, \tag{4.130}$$

could be nonzero in the case of a localized perturbing charge. For a periodic perturbation, on the other hand, only the choice $\delta N_+ = 0$ ensures that the energy scale remains unchanged, which requires that the average potential vanishes,

$$\int d^3r\, \delta v(\mathbf{r}) = 0 \,. \tag{4.131}$$

The presence of δv leads to a redistribution of the electronic density, so that the total $n(\mathbf{r})$ deviates from the unperturbed value n_0 by an induced shift $\delta n(\mathbf{r})$,

$$n(\mathbf{r}) = n_0 + \delta n(\mathbf{r}) \,. \tag{4.132}$$

The gas thus consists of electrons with their two density components n_0 and δn, the constant positive background charge $n_+ = n_0$ and the given perturbing charge δn_+. A weakly inhomogeneous electron gas is characterized by

$$\delta n(\mathbf{r}) \ll n_0 \,. \tag{4.133}$$

As soon as δv is assumed to be sufficiently small to ensure (4.133), a perturbative treatment is legitimate. All pertinent quantities can be expanded in powers of δv, most notably the induced density,

$$\delta n(\mathbf{r}) = \delta n^{(1)}(\mathbf{r}) + \delta n^{(2)}(\mathbf{r}) + \ldots \,, \tag{4.134}$$

where $\delta n^{(i)}$ is understood to be of the order $(\delta v)^i$. For extended many-body systems the perturbation expansion for δn is usually formulated in terms of the retarded response functions of the system. The first order shift $\delta n^{(1)}$ is obtained via the linear response function (4.39),

$$\delta n^{(1)}(\mathbf{r}) = \int d^3r'\, \chi_{\mathrm{R}}(\mathbf{r}-\mathbf{r}', \omega = 0)\, \delta v(\mathbf{r}') \,. \tag{4.135}$$

The ground state of the unperturbed system defining χ_{R} in Eq. (4.39) represents, in the present situation, the ground state of the HEG. All higher order contributions to δn can, in complete analogy to (4.135), be expressed in terms of the associated higher order retarded response functions. However, even for the second order (quadratic) response function only very limited information is available. In addition, the response formalism becomes more and more involved with increasing order. The present discussion is therefore restricted to linear response, which is expected to dominate anyway in the case of a truly weak perturbation. As long as one keeps an eye on the consistency of all subsequent results in δv, one can then simply identify the complete induced density shift with $\delta n^{(1)}$,

$$\delta n(\boldsymbol{r}) \cong \delta n^{(1)}(\boldsymbol{r}) . \tag{4.136}$$

At this point it is convenient to go over to the momentum space representation of all relevant quantities, as for instance

$$\delta n(\boldsymbol{r}) = \int \frac{d^3q}{(2\pi)^3} e^{i\boldsymbol{q}\cdot\boldsymbol{r}} \, \delta n(\boldsymbol{q}) \, ; \qquad \delta n(\boldsymbol{q})^* = \delta n(-\boldsymbol{q}) . \tag{4.137}$$

The relation between $\delta n(\boldsymbol{q})^*$ and $\delta n(-\boldsymbol{q})$ follows from the fact that $\delta n(\boldsymbol{r})$ is real (the same holds for the Fourier transforms of δv and δn_+). To first order the induced density deviation can then be written as

$$\delta n(\boldsymbol{q}) = \chi(\boldsymbol{q}, \omega = 0) \, \delta v(\boldsymbol{q}) , \tag{4.138}$$

where Eq. (4.48) has been utilized to replace χ_{R} by the time-ordered response function χ (see Eq. (4.90) for the Fourier transformation of χ). Ultimately, the response function will have to be evaluated in some detail, for which the standard machinery of Green's function theory (which can only be exploited for χ) provides an adequate framework.

In any real, closed physical system the number of electrons remains unchanged after switching on some perturbation. However, the HEG with its uncountable reservoir of electrons, can add a finite number of localized electrons without a change of the average density n_0. As a consequence, the net induced charge,

$$\delta N = \int d^3r \, \delta n(\boldsymbol{r}) = \delta n(\boldsymbol{q} = \boldsymbol{0}) , \tag{4.139}$$

does not necessarily vanish. In fact, insertion of (4.129) and of the Dyson equation (4.93) into (4.138) demonstrates that the induced density concentrates as much charge around a localized perturbing charge δn_+ as required for complete screening,

$$\begin{aligned} \delta n(\boldsymbol{q}) &= -\chi(\boldsymbol{q}, \omega = 0) \, w(\boldsymbol{q}) \, \delta n_+(\boldsymbol{q}) & (4.140) \\ &= \frac{\Pi(\boldsymbol{q}, \omega = 0)}{\Pi(\boldsymbol{q}, \omega = 0) - w^{-1}(\boldsymbol{q})} \delta n_+(\boldsymbol{q}) & \\ \Longrightarrow \qquad \delta n(\boldsymbol{q} = \boldsymbol{0}) &= \delta n_+(\boldsymbol{q} = \boldsymbol{0}) = \delta N_+ . & (4.141) \end{aligned}$$

Here $w(\boldsymbol{q}) = 4\pi e^2/\boldsymbol{q}^2$ has been used for the last line ($\lim_{\boldsymbol{q}\to\boldsymbol{0}} \Pi(\boldsymbol{q}, \omega = 0) \neq 0$). In the case of the noninteracting HEG with $\chi = \Pi^{(0)}$ the complete Rydberg series of states generated by δv would be filled, as the bound electrons do not repel each other (which corresponds to $\delta N = \infty$). As a consequence, a rigorous conservation of the electronic norm,

$$\int d^3r \, \delta n(\boldsymbol{r}) = \delta n(\boldsymbol{q} = \boldsymbol{0}) = 0 , \tag{4.142}$$

can only be ensured by a vanishing perturbing charge,

$$\int d^3r\,\delta n_+(\boldsymbol{r}) \;=\; \delta n_+(\boldsymbol{q}=\boldsymbol{0}) \;=\; 0\,. \tag{4.143}$$

For the present purpose, however, a strict conservation of the electron number is essential. The ground states of the unperturbed and of the perturbed system must correspond to exactly the same, mathematically well-defined particle number, in order to allow the derivation of a first-principles density functional. For this reason the class of legitimate external potentials has to be restricted to those satisfying Eqs. (4.142), (4.143). This neither excludes localized perturbations δv nor does it require $\int d^3r\delta v(\boldsymbol{r}) = \delta v(\boldsymbol{q}=\boldsymbol{0})$ to vanish as $\chi(\boldsymbol{q}=\boldsymbol{0},\omega=0)=0$. Of course, $\int d^3r\delta v(\boldsymbol{r})=0$ is completely sufficient to ensure (4.142).

After these preliminaries one can consider the shift in the total energy which is induced by δv. The xc-component of this energy shift serves as basis for the construction of inhomogeneity corrections to the LDA. In order to extract the xc-component all other contributions to the total energy must be known. The most critical component is the electrostatic energy, which requires some care due to the long range of the Coulomb force. In order to discuss the electrostatic energy of the inhomogeneous electron gas, one therefore works with a screened Coulomb interaction,

$$w(\boldsymbol{r},\boldsymbol{r}') \longrightarrow w_{\mathrm{s}}(\boldsymbol{r},\boldsymbol{r}') = e^2\,\frac{e^{-\mu|\boldsymbol{r}-\boldsymbol{r}'|}}{|\boldsymbol{r}-\boldsymbol{r}'|}\,, \tag{4.144}$$

which suppresses all long-range singularities at intermediate steps. The limit $\mu\to 0$ leads back to the system of interest at the end of the calculation. With the interaction (4.144) the total electrostatic energy is given by

$$\begin{aligned} E_{\mathrm{es}} = e^2\int d^3r\int d^3r'\,\frac{e^{-\mu|\boldsymbol{r}-\boldsymbol{r}'|}}{|\boldsymbol{r}-\boldsymbol{r}'|}\bigg\{&\frac{1}{2}[n_+ + \delta n_+(\boldsymbol{r})][n_+ + \delta n_+(\boldsymbol{r}')] \\ &+\frac{1}{2}[n_0+\delta n(\boldsymbol{r})][n_0+\delta n(\boldsymbol{r}')] \\ &-[n_0+\delta n(\boldsymbol{r})][n_+ + \delta n_+(\boldsymbol{r}')]\bigg\}\,. \end{aligned} \tag{4.145}$$

In addition to the electronic charge density $n = n_0 + \delta n$, the complete positive background charge density $n_+ + \delta n_+$ has been included in (4.145). The electrostatic energy will not be finite in the limit $\mu\to 0$ without taking into account n_+. The inclusion of δn_+, on the other hand, expresses the fact that δv is, as indicated by Eq. (4.129), generated by some charge. Without this specification, the external potential term $\int d^3r\hat{n}\delta v$ would have an origin different than electrostatics. In this case E_{es} would be given by the above expression with δn_+ set to zero. The two approaches (inclusion of δn_+ or not) just differ by the treatment of the external source. They are completely equivalent in the electronic sector, as long as they are applied consistently and mere additive constants in the Hamiltonian are ignored. The approach chosen here appears to be somewhat more transparent in a physical sense.

Using $n_+ = n_0$ one obtains

$$E_{\text{es}} = \frac{e^2}{2} \int d^3r \int d^3r' \, \frac{e^{-\mu|\boldsymbol{r}-\boldsymbol{r}'|}}{|\boldsymbol{r}-\boldsymbol{r}'|} \\ \times \left\{ \delta n_+(\boldsymbol{r}) \delta n_+(\boldsymbol{r}') + \delta n(\boldsymbol{r}) \delta n(\boldsymbol{r}') - 2\delta n(\boldsymbol{r}) \delta n_+(\boldsymbol{r}') \right\} .$$

As expected, the electron density n_0 and the background charge n_+ completely neutralize each other. The limit $\mu \to 0$ can now be taken without any problem,

$$E_{\text{es}} = E_{\text{H}}[\delta n_+] + E_{\text{H}}[\delta n] + \int d^3r \, \delta n(\boldsymbol{r}) \, \delta v(\boldsymbol{r}) \, . \tag{4.146}$$

The first contribution in (4.146) is the self-interaction of the fixed background charge, which represents an irrelevant additive constant for the electronic problem and can simply be ignored. The remaining two terms have exactly the form of the electronic Hartree energy and the external energy functional. In the case of a localized perturbation they are finite as long as $\delta N_+ < \infty$ (for periodic perturbations the energy density is the only relevant quantity).

The total energy of the weakly inhomogeneous electron gas is now easily obtained via the coupling constant integration technique. The appropriate Hamiltonian $\hat{H}(\lambda)$, which reproduces the HEG in the limit $\lambda = 0$ and the inhomogeneous gas for $\lambda = 1$, is given by

$$\hat{H}(\lambda) \; = \; \hat{T} + \hat{W}^{\text{HEG}} + \lambda \left[\int d^3r \, [\hat{n}(\boldsymbol{r}) - n_+] \, \delta v(\boldsymbol{r}) + E_{\text{H}}[\delta n_+] \right] . \tag{4.147}$$

In addition to the coupling between the electrons and the external potential, it contains an additive constant which reproduces the total electrostatic energy (4.145)—the interaction of the positive background charge n_+ with itself and the HEG density n_0 is understood to be included in $\hat{W}^{\text{HEG}}$. Coupling constant integration then yields (compare Eq. (4.32))

$$\begin{aligned} & E_0(1) - E_0(0) \\ &= \int_0^1 d\lambda \left[\int d^3r \, [n_\lambda(\boldsymbol{r}) - n_+] \, \delta v(\boldsymbol{r}) + E_{\text{H}}[\delta n_+] \right] \\ &= \int_0^1 d\lambda \int d^3r \left[n_0 + \lambda \, \delta n^{(1)}(\boldsymbol{r}) + \mathcal{O}(\delta v^2) - n_+ \right] \delta v(\boldsymbol{r}) + E_{\text{H}}[\delta n_+] \\ &= \frac{1}{2} \int d^3r \, \delta n^{(1)}(\boldsymbol{r}) \, \delta v(\boldsymbol{r}) + E_{\text{H}}[\delta n_+] + \mathcal{O}(\delta v^3) \, . \end{aligned} \tag{4.148}$$

One should note that Eq. (4.148) includes all second order contributions in δv consistently, although the density is only treated consistently to first order. This result reflects the fact that knowledge of the first order wave function is sufficient to determine the second order corrections to the energy within Rayleigh-Schrödinger perturbation theory. On the other hand, all contributions beyond the order δv^2 are ignored in Eq. (4.148). Insertion of the HEG energy $E_0(0)$ and use of (4.135) leads to

$$E_0(1) = T_s^{\rm LDA}[n_0] + E_{\rm xc}^{\rm LDA}[n_0] + \int d^3r\, \delta n(\boldsymbol{r})\delta v(\boldsymbol{r}) + E_{\rm H}[\delta n_+]$$
$$-\frac{1}{2}\int d^3r \int d^3r'\, \chi(\boldsymbol{r}-\boldsymbol{r}',\omega=0)\,\delta v(\boldsymbol{r})\,\delta v(\boldsymbol{r}') + \mathcal{O}(\delta v^3)\,. \qquad (4.149)$$

The correction $\delta n^{(1)}$ has now been identified with the full δn, as the associated error in $E_0(1)$ is of the order $(\delta v)^3$.

In the next step one has to extract the xc-contribution to Eq. (4.149). One starts with the exact energy functional for the weakly inhomogeneous electron gas, obtained by inclusion of the total electrostatic energy (4.145),

$$E_0(1) \;=\; T_s[n_0+\delta n] + \int d^3r\, \delta n(\boldsymbol{r})\,\delta v(\boldsymbol{r}) + E_{\rm H}[\delta n_+] + E_{\rm H}[\delta n] + E_{\rm xc}[n_0+\delta n]\,. \qquad (4.150)$$

This exact expression has to be compared with the result (4.149). One directly identifies the external potential energy and the self-interaction energy of the background charge. In order to deal with the Hartree energy it is advantageous to go to momentum space and to introduce the temporary abbreviation

$$E_{\rm sxc}^{\rm (LDA)}[n] = T_s^{\rm (LDA)}[n] + E_{\rm xc}^{\rm (LDA)}[n]\,. \qquad (4.151)$$

Combination of Eqs. (4.149) and (4.150), together with use of (4.137) and (4.138), then leads to

$$E_{\rm sxc}[n_0+\delta n] + \frac{1}{2}\int \frac{d^3q}{(2\pi)^3}\,|\delta n(\boldsymbol{q})|^2 w(\boldsymbol{q})$$
$$= E_{\rm sxc}^{\rm LDA}[n_0] - \frac{1}{2}\int \frac{d^3q}{(2\pi)^3}\,|\delta v(\boldsymbol{q})|^2 \chi(\boldsymbol{q},0) + \mathcal{O}(\delta v^3)$$
$$= E_{\rm sxc}^{\rm LDA}[n_0] - \frac{1}{2}\int \frac{d^3q}{(2\pi)^3}\,\frac{|\delta n(\boldsymbol{q})|^2}{\chi(\boldsymbol{q},0)} + \mathcal{O}(\delta n^3)\,, \qquad (4.152)$$

where $w(\boldsymbol{q})$ is given by (4.91). The last line indicates that, in view of (4.138), the order in δv is equivalent to the order in δn. Use of the Dyson equation (4.93) in the form

$$\frac{1}{\chi(\boldsymbol{q},\omega)} = \frac{1}{\Pi(\boldsymbol{q},\omega)} - w(\boldsymbol{q})$$

allows the elimination of $E_{\rm H}[\delta n]$ from (4.152),

$$E_{\rm sxc}[n_0+\delta n] = E_{\rm sxc}^{\rm LDA}[n_0] - \frac{1}{2}\int \frac{d^3q}{(2\pi)^3}\,\frac{|\delta n(\boldsymbol{q})|^2}{\Pi(\boldsymbol{q},0)} + \mathcal{O}(\delta n^3)\,. \qquad (4.153)$$

The right-hand side of (4.153) is a functional of n_0 and δn, but not yet a functional of $n = n_0 + \delta n$, as desired. This form can, however, easily be obtained for the LDA components in (4.153): one can use the functional Taylor expansion of $E_{\rm sxc}^{\rm LDA}[n]$ about the unperturbed density to rewrite (4.153) as

$$
\begin{aligned}
&E_{\rm sxc}[n_0+\delta n] \\
&= E_{\rm sxc}^{\rm LDA}[n_0+\delta n] - \int d^3r \left.\frac{\delta E_{\rm sxc}^{\rm LDA}[n]}{\delta n(\boldsymbol{r})}\right|_{n=n_0} \delta n(\boldsymbol{r}) \\
&\quad -\frac{1}{2}\int d^3r d^3r' \left.\frac{\delta^2 E_{\rm sxc}^{\rm LDA}[n]}{\delta n(\boldsymbol{r})\delta n(\boldsymbol{r}')}\right|_{n=n_0} \delta n(\boldsymbol{r})\delta n(\boldsymbol{r}') - \ldots - \frac{1}{2}\int \frac{d^3q}{(2\pi)^3}\frac{|\delta n(\boldsymbol{q})|^2}{\Pi(\boldsymbol{q},0)} \\
&= E_{\rm sxc}^{\rm LDA}[n_0+\delta n] - \int d^3r \frac{de_{\rm sxc}^{\rm HEG}(n_0)}{dn_0}\delta n(\boldsymbol{r}) \\
&\quad -\frac{1}{2}\int d^3r d^3r' \frac{d^2 e_{\rm sxc}^{\rm HEG}(n_0)}{dn_0^2}\delta^{(3)}(\boldsymbol{r}-\boldsymbol{r}')\delta n(\boldsymbol{r})\delta n(\boldsymbol{r}') - \frac{1}{2}\int \frac{d^3q}{(2\pi)^3}\frac{|\delta n(\boldsymbol{q})|^2}{\Pi(\boldsymbol{q},0)}\,.
\end{aligned}
$$

One now uses norm conservation, Eq. (4.142), to eliminate the term linear in δn. In addition, one can apply the compressibility sum rule [153], which relates the second derivative of the energy per volume to the chemical potential $\mu(n_0)$ and the long-wavelength limit of the static irreducible polarization insertion of the interacting HEG,

$$
\begin{aligned}
\frac{d^2 e_{\rm sxc}^{\rm HEG}(n_0)}{dn_0^2} &= \frac{d^2}{dn_0^2}\left[t_{\rm s}^{\rm HEG}(n_0) + e_{\rm xc}^{\rm HEG}(n_0)\right] \\
&= \frac{d}{dn_0}\mu(n_0) = \frac{-1}{\Pi(\boldsymbol{q}=\boldsymbol{0},\omega=0)}\,. \qquad (4.154)
\end{aligned}
$$

In this fashion one is led to

$$
\begin{aligned}
E_{\rm sxc}[n_0+\delta n] &= E_{\rm sxc}^{\rm LDA}[n_0+\delta n] \\
&\quad -\frac{1}{2}\int \frac{d^3q}{(2\pi)^3}|\delta n(\boldsymbol{q})|^2\left\{\frac{1}{\Pi(\boldsymbol{q},0)} - \frac{1}{\Pi(\boldsymbol{0},0)}\right\}. \qquad (4.155)
\end{aligned}
$$

Equation (4.155) can be split into a noninteracting (kinetic) component, obtained in the limit $e^2 \to 0$,

$$
\begin{aligned}
T_{\rm s}[n_0+\delta n] &= T_{\rm s}^{\rm LDA}[n_0+\delta n] \\
&\quad -\frac{1}{2}\int \frac{d^3q}{(2\pi)^3}|\delta n(\boldsymbol{q})|^2\left\{\frac{1}{\Pi^{(0)}(\boldsymbol{q},0)} - \frac{1}{\Pi^{(0)}(\boldsymbol{0},0)}\right\}, \qquad (4.156)
\end{aligned}
$$

and a remainder, which contributes to the xc-functional,

$$
E_{\rm xc}[n_0+\delta n] = E_{\rm xc}^{\rm LDA}[n_0+\delta n] - \frac{1}{2}\int \frac{d^3q}{(2\pi)^3}|\delta n(\boldsymbol{q})|^2 K(\boldsymbol{q}^2,n_0) \qquad (4.157)
$$

$$
K(\boldsymbol{q}^2,n_0) = \frac{1}{\Pi(\boldsymbol{q},0)} - \frac{1}{\Pi^{(0)}(\boldsymbol{q},0)} - \frac{1}{\Pi(\boldsymbol{0},0)} + \frac{1}{\Pi^{(0)}(\boldsymbol{0},0)}\,. \qquad (4.158)
$$

Equations (4.156) and (4.157) represent systematic inhomogeneity corrections to the LDA. All ingredients are completely determined by the ground state of the HEG, no free parameters are involved.

The only problem is that these expressions still depend on n_0 and δn separately, rather than on $n_0 + \delta n$. The elimination of n_0 and δn in favor of $n_0 + \delta n$ turns out to be a nontrivial problem, if one wants to avoid any further approximation. For this reason the linear response kernel in (4.157) is usually reduced to its long-wavelength limit, which greatly simplifies the problem of elimination. Nevertheless, we first take a closer look at the complete linear response functional, restricting the discussion to the xc-component. Of course, the same steps can be taken for T_s (for a brief discussion see Appendix D).

4.4.2 Complete Linear Response

The first step of the transition from (4.157) to a density functional is a back-transformation to real space. The result for the nonlocal (i.e. non-LDA) contribution to (4.157), denoted by ΔE_{xc}, reads

$$\Delta E_{xc} = -\frac{1}{2}\int d^3r d^3r' \, \delta n(\boldsymbol{r}) K(|\boldsymbol{r}-\boldsymbol{r}'|, n_0)\, \delta n(\boldsymbol{r}') \tag{4.159}$$

$$K(|\boldsymbol{r}|, n_0) = \int \frac{d^3q}{(2\pi)^3} e^{i\boldsymbol{q}\cdot\boldsymbol{r}} K(\boldsymbol{q}^2, n_0). \tag{4.160}$$

One now has two options for the elimination of δn which lead to slightly different results. The first option [6, 154–157] relies on the fact that

$$\int d^3r K(|\boldsymbol{r}-\boldsymbol{r}'|, n_0) = K(\boldsymbol{q}^2 = 0, n_0) = 0. \tag{4.161}$$

This relation allows adding two vanishing contributions proportional to $\delta n(\boldsymbol{r})^2$ and $\delta n(\boldsymbol{r}')^2$ to (4.159),

$$\begin{aligned}\Delta E_{xc}^{(a1)} &= \frac{1}{4}\int d^3r d^3r' \left[\delta n(\boldsymbol{r}) - \delta n(\boldsymbol{r}')\right]^2 K(|\boldsymbol{r}-\boldsymbol{r}'|, n_0) \\ &= \frac{1}{4}\int d^3r d^3r' \left[n(\boldsymbol{r}) - n(\boldsymbol{r}')\right]^2 K(|\boldsymbol{r}-\boldsymbol{r}'|, n_0).\end{aligned} \tag{4.162}$$

In the second approach [158] one uses

$$\boldsymbol{\nabla}\delta n(\boldsymbol{r}) = \boldsymbol{\nabla}\Big(n_0 + \delta n(\boldsymbol{r})\Big) = \boldsymbol{\nabla} n(\boldsymbol{r}) \tag{4.163}$$

to rewrite (4.157) as

$$\Delta E_{\rm xc}^{(a2)} = -\frac{1}{2}\int d^3r d^3r' \Big(\boldsymbol{\nabla} n(\boldsymbol{r})\cdot\boldsymbol{\nabla}' n(\boldsymbol{r}')\Big) L(|\boldsymbol{r}-\boldsymbol{r}'|,n_0) \qquad (4.164)$$

$$L(|\boldsymbol{r}|,n_0) = \int \frac{d^3q}{(2\pi)^3}\, e^{i\boldsymbol{q}\cdot\boldsymbol{r}}\, \frac{K(\boldsymbol{q}^2,n_0)}{\boldsymbol{q}^2}\,. \qquad (4.165)$$

Up to this point the forms (4.162) and (4.164) are not really different, rather they can be transformed into each other by partial integration (for finite systems). In both cases it remains to eliminate n_0. Any substitution of n_0, which reduces to n_0 in the limit of vanishing δn, is admissible for this step. On the other hand, both forms immediately reveal the difficulty of the task: the kernel of the fully nonlocal linear response functional depends on two positions $\boldsymbol{r}$ and $\boldsymbol{r}'$, so that the replacement of n_0 is ambiguous. In fact, an infinite number of possibilities exists, as one is not limited to use some kind of local substitution. Nevertheless, two local substitutions offer themselves quite naturally,

$$\text{(i)} \qquad n_0 \longrightarrow n_0 + \delta n\left(\frac{\boldsymbol{r}+\boldsymbol{r}'}{2}\right) = n\left(\frac{\boldsymbol{r}+\boldsymbol{r}'}{2}\right)$$

$$\text{(ii)} \qquad n_0 \longrightarrow \frac{1}{2}\Big[n_0 + \delta n(\boldsymbol{r}) + n_0 + \delta n(\boldsymbol{r}')\Big] = \frac{1}{2}\Big[n(\boldsymbol{r}) + n(\boldsymbol{r}')\Big]$$

They correspond to two different partial resummations of the complete response expansion. Both forms have been tested only for a very limited number of systems. It turns out that (i) leads to divergences [155], while (ii) does not. Neither the few results obtained with (4.162) for solids [154, 159, 155, 156] nor those obtained with (4.164) for atoms [158] are completely satisfactory. A more suitable replacement for n_0 remains to be found. For this reason this type of functional is not used in practice.[5]

4.4.3 Gradient Expansion

The situation is completely different for the second strategy to generate a density functional from (4.157). In this approach [6] one assumes δn to vary only very slowly with $\boldsymbol{r}$. This implies that the Fourier transform $\delta n(\boldsymbol{q})$ is strongly peaked at small $\boldsymbol{q}$ and decays very rapidly with increasing $|\boldsymbol{q}|$—note, however, that norm conservation requires $\delta n(\boldsymbol{q}=\boldsymbol{0}) = 0$. One can then replace the static response function in the kernel of (4.157) by its long-wavelength expansion,

$$\Pi(\boldsymbol{q},0) = a(n_0) + b(n_0)\boldsymbol{q}^2 + \ldots\,, \qquad (4.166)$$

and carry through the inversion,

[5] The expression (4.159), provides, however, the basis for a density functional approach to dispersion forces [160–170] (compare Sect. 6.1.2).

$$K(\boldsymbol{q}^2,n_0) = -\left[\frac{b(n_0)}{a(n_0)^2} - \frac{b_0(n_0)}{a_0(n_0)^2}\right]\boldsymbol{q}^2 + \mathcal{O}(\boldsymbol{q}^4), \tag{4.167}$$

as $\delta n(\boldsymbol{q})$ is essentially zero for all $\boldsymbol{q}$ for which (4.166) and (4.167) are not valid ($a_0(n_0)$ and $b_0(n_0)$ denote the expansion coefficients of the noninteracting response function).

A more precise characterization of this approximation is the following: use of the representation (4.167) of $1/\Pi$, i.e. of $\Pi(\boldsymbol{q})^{-1} \approx \Pi(0)^{-1}[1-(\Pi(\boldsymbol{q})-\Pi(0))/\Pi(0)]$, in (4.157) requires

$$\int \frac{d^3q}{(2\pi)^3}\,|\delta n(\boldsymbol{q})|^2 \gg \frac{b(n_0)}{a(n_0)}\int \frac{d^3q}{(2\pi)^3}\,\boldsymbol{q}^2\,|\delta n(\boldsymbol{q})|^2, \tag{4.168}$$

while the neglect of all higher order terms in (4.166) relies on

$$\frac{b(n_0)}{a(n_0)}\int \frac{d^3q}{(2\pi)^3}\,\boldsymbol{q}^2\,|\delta n(\boldsymbol{q})|^2 \gg \frac{c(n_0)}{a(n_0)}\int \frac{d^3q}{(2\pi)^3}\,\boldsymbol{q}^4\,|\delta n(\boldsymbol{q})|^2 \ldots, \tag{4.169}$$

where $c(n_0)$ denotes the coefficient of the $\boldsymbol{q}^4$-term of the expansion (4.166). In order to analyze the conditions (4.168) and (4.169) further, one has to take a closer look at the static response function. Quite generally, Π can be written as

$$\Pi(\boldsymbol{q},0) = -\frac{mk_\mathrm{F}}{\hbar^2\pi^2}J(Q,r_\mathrm{s}). \tag{4.170}$$

J is a dimensionless function of the dimensionless momentum

$$Q = \frac{|\boldsymbol{q}|}{2k_\mathrm{F}} \tag{4.171}$$

and the dimensionless Wigner-Seitz radius (4.103). J has a rather simple form in the noninteracting limit [94] (it is the static Lindhard function),

$$\Pi^{(0)}(\boldsymbol{q},0) = -\frac{mk_\mathrm{F}}{\hbar^2\pi^2}\left\{\frac{1}{2}+\frac{1-Q^2}{4Q}\ln\left|\frac{1+Q}{1-Q}\right|\right\}, \tag{4.172}$$

and is also known [171] analytically for the first order diagrams (4.105),

$$\Pi^{(1)}(\boldsymbol{q},0) = -\frac{mk_\mathrm{F}}{\hbar^2\pi^2}\left[\left(\frac{4}{9\pi}\right)^{1/3}\frac{r_\mathrm{s}}{\pi}I(Q)\right] = -\frac{m^2e^2}{\hbar^4\pi^3}I(Q), \tag{4.173}$$

with

$$I(Q) = \frac{1-Q^4}{48Q^3}\left(\ln\left|\frac{1+Q}{1-Q}\right|\right)^3 - \frac{1-Q^2}{24Q^2}\int_0^Q dx\,\frac{1-x^2}{x^2}\left(\ln\left|\frac{1+x}{1-x}\right|\right)^3$$
$$+\frac{1}{8}\left[\frac{1}{Q}+\frac{1-Q^2}{2Q^2}\ln\left|\frac{1+Q}{1-Q}\right|\right]\int_0^Q dx\,\frac{1-x^2}{x^2}\left(\ln\left|\frac{1+x}{1-x}\right|\right)^2. \tag{4.174}$$

The kernel (4.157) resulting from the approximation $\Pi \approx \Pi^{(0)} + \Pi^{(1)}$ is plotted in Fig. 4.4. One recognizes the infinite slope at $Q = 1$, which originates from the

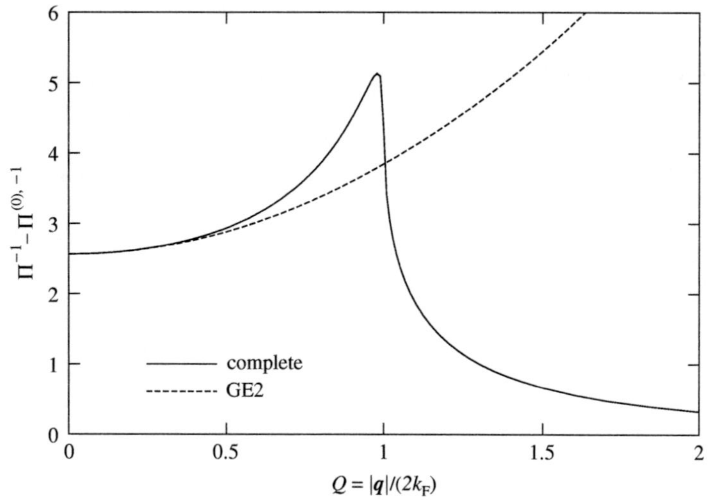

Fig. 4.4 Kernel of nonlocal xc-functional (4.157) as a function of $\boldsymbol{q}$ (for $r_s = 2$): Complete inverse polarization insertion (*solid line*) versus long-wavelength expansion $\Pi^{-1} = 1/a - b\boldsymbol{q}^2/a^2$ (*dashed line*) corresponding to second order gradient expansion, Eqs. (4.166), (4.167). The polarization insertion has been approximated by its x-only limit, $\Pi \approx \Pi^{(0)} + \Pi^{(1)}$.

derivative discontinuity of both $\Pi^{(0)}$ and $\Pi^{(1)}$ at $Q = 1$. While the derivative discontinuity of the first order diagrams (4.105) is damped by inclusion of screening effects (see e.g. [172]), the quantity $1/(2k_F)$ nevertheless represents the characteristic length scale of the linear response functional (and that of its long-wavelength expansion). It is obvious that the second order expansion (4.167) completely misrepresents the correct kernel for $|\boldsymbol{q}| \gtrsim 2k_F$ (In fact, the Taylor expansion in powers of $\boldsymbol{q}$ does not even converge for $|\boldsymbol{q}| > 2k_F$). Consequently, the long-wavelength expansion of (4.157) is only legitimate, i.e. the conditions (4.168) and (4.169) can only be satisfied, if $\delta n(\boldsymbol{q})$ decays rapidly for $|\boldsymbol{q}| \gtrsim 2k_F$.

Insert now (4.167) into (4.157), keeping the conditions (4.168) and (4.169) in mind. After Fourier transformation to real space and use of (4.163) the inhomogeneity correction to the LDA reads

$$\Delta E_{xc}^{(b)} = \frac{1}{2}\int d^3r \left[\frac{b(n_0)}{a(n_0)^2} - \frac{b_0(n_0)}{a_0(n_0)^2}\right] \Big[\boldsymbol{\nabla} n(\boldsymbol{r})\Big]^2 + \mathcal{O}(\boldsymbol{\nabla}^4). \qquad (4.175)$$

As indicated, the terms neglected involve at least four gradient operators (order $\boldsymbol{q}^4$). The long-wavelength expansion of Π induces a *gradient expansion* (GE) for E_{xc}.

The elimination of n_0 is now trivial. As the complete energy (4.157) is only consistent to second order in δn, one can simply use a local density substitution,

$$\Delta E_{\rm xc}^{(b)} = \frac{1}{2}\int d^3r \left[\frac{b(n(\boldsymbol{r}))}{a(n(\boldsymbol{r}))^2} - \frac{b_0(n(\boldsymbol{r}))}{a_0(n(\boldsymbol{r}))^2}\right] [\boldsymbol{\nabla} n(\boldsymbol{r})]^2 + \mathcal{O}(\boldsymbol{\nabla}^4) + \mathcal{O}(\delta n^3) . \quad (4.176)$$

This local density substitution represents a partial resummation of the expansion in powers of δn, just like the LDA. The resummation has the potential to extend the range of validity of the (second order) gradient correction (4.176).

This point is illustrated in Fig. 4.5. The linear response expression (4.157) has

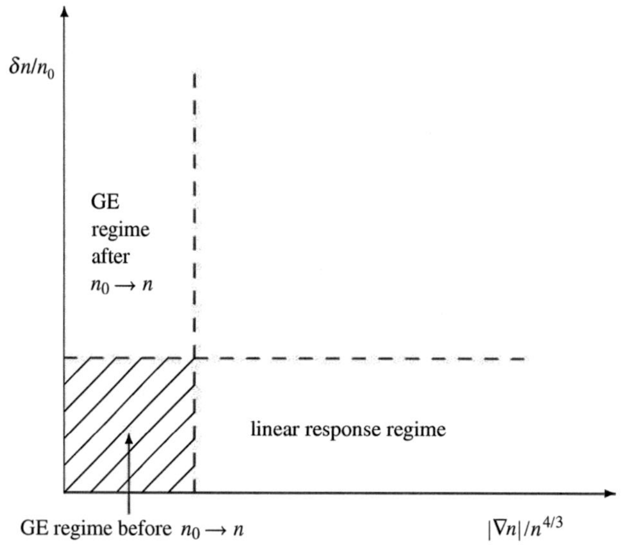

Fig. 4.5 Regimes of validity of linear response and gradient expansion.

been derived under the assumption $\delta n/n_0 \ll 1$. This requirement has then been combined with (4.168) and (4.169) to arrive at (4.175). The validity of the functional (4.175) is therefore restricted to the shaded region in Fig. 4.5. However, one can also approach this functional in a different way. If one first assumes $(\boldsymbol{\nabla}^i n)/n^{1+i/3} \ll 1$, one can establish the GE expression (4.176) without any recourse to the properties of the HEG. The density dependence of the unknown function b/a^2 is completely determined by dimensional arguments, only numerical prefactors remain open. However, for $n = n_0 + \delta n$ and $\delta n/n_0 \ll 1$ the resulting expression must agree with (4.175), which allows the determination of b/a^2. In other words: the linear response regime is only exploited to evaluate the numerical parameters in an otherwise fixed, universal functional. This argument reflects the fact that alternative methods for the derivation of gradient corrections, as the real-space GE [173], are available. If viewed from this perspective the GE is valid as long as $(\boldsymbol{\nabla}^i n)/n^{1+i/3} \ll 1$, in spite of its relation to the weakly inhomogeneous electron gas.

The crucial step in the explicit construction of the GE is the evaluation of the long-wavelength expansion of $\Pi(\boldsymbol{q}^2,0)$. This is trivial for the Lindhard-function, which yields the second order GE for T_s,

$$\Delta T_s^{[2]} = \frac{1}{72m}\int d^3r\,\frac{(\nabla n)^2}{n}. \tag{4.177}$$

This functional has originally been suggested by von Weizsäcker [174] with a somewhat different prefactor, resulting from the kinetic energy of a system with a single particle (like the hydrogen atom—see Appendix D). The correct prefactor is also obtained with the commutator expansion of Kirzhnits [175], which is one variant of the real-space GE. Both results illustrate the argument that the form of the gradient correction is fixed by dimensional requirements, so that a model system is only needed to determine the numerical parameters in this general form. The functional (4.177) provides the basis for all kinds of extended Thomas-Fermi-type approximations, in which the variational equation (2.38) is solved in its original form, rather than via the KS scheme (for a detailed discussion of this approach see [7, 15]).

Unfortunately, the evaluation of the long-wavelength expansion of $\Pi(\boldsymbol{q}^2,0)$ is much more involved in the case of E_{xc}. We first consider the exchange contribution which is linear in e^2. It is thus obtained from an expansion of the kernel (4.158) to first order in e^2,

$$\frac{1}{\Pi(\boldsymbol{q},0)} = \frac{1}{\Pi^{(0)}(\boldsymbol{q},0)}\left\{1-\frac{\Pi^{(1)}(\boldsymbol{q},0)}{\Pi^{(0)}(\boldsymbol{q},0)}+\dots\right\},$$

The relevant first order diagrams of Π have already been displayed in Eq. (4.105). The loop integrations occurring in these diagrams are rather intricate, so that initially the small-$\boldsymbol{q}$ expansion was employed for the integrands of the Feynman integrals (4.105) and the loop integrations were carried out afterwards [176]. As the resulting integrals exhibit an infrared divergence for $\boldsymbol{q}^2 \to 0$, Yukawa-screening of the Coulomb integration, Eq. (4.144), was utilized for all intermediate steps. After all loop integrations were performed the screening was removed. In this way one finds

$$\Delta E_x^{[2]} = \frac{7}{81}\int d^3r\, e_x^{\mathrm{HEG}}(n)\,\xi, \tag{4.178}$$

with the dimensionless characteristic density gradient

$$\xi = \left(\frac{\nabla n(\boldsymbol{r})}{2[3\pi^2 n(\boldsymbol{r})]^{1/3} n(\boldsymbol{r})}\right)^2. \tag{4.179}$$

The same result was obtained by the real-space GE, for which intermediate Yukawa-screening was also used [177].

Later, it was observed that the small-$\boldsymbol{q}$ expansion does not commute with the loop integration [178], due to the singular structure of the Coulomb interaction for small momenta. A complete calculation of $\Pi^{(1)}$ without Yukawa-screening, followed by a small-$\boldsymbol{q}$ expansion of the final $\Pi^{(1)}(\boldsymbol{q},0)$, leads to a different exchange coefficient [178–180, 171],

$$\Delta E_{\mathrm{x}}^{[2]} = C_{\mathrm{x}} \int d^3r\, e_{\mathrm{x}}^{\mathrm{HEG}}(n)\, \xi\,; \qquad C_{\mathrm{x}} = \frac{10}{81}\,. \tag{4.180}$$

As the form of the functional is completely determined from the very outset, only its prefactors can be a matter of debate. Moreover, Eqs. (4.179), (4.180) demonstrate explicitly that after the local density replacement the appropriate length scale for measuring gradients is given by $2[3\pi^2 n(\boldsymbol{r})]^{-1/3}$. The form of ξ is a direct consequence of the form of the response function (4.173).

It should be pointed out that the functional (4.180) is only determined up to some partial integration. For instance, partial integration of one of the density gradients in (4.180) yields

$$\Delta E_{\mathrm{x}}^{[2]} = -C_{\mathrm{x}} \int d^3r\, e_{\mathrm{x}}^{\mathrm{HEG}}(n) \left[\eta - \frac{4}{3}\xi\right] \tag{4.181}$$

$$\eta = \frac{\boldsymbol{\nabla}^2 n(\boldsymbol{r})}{4[3\pi^2 n(\boldsymbol{r})]^{2/3} n(\boldsymbol{r})}\,, \tag{4.182}$$

where the surface contribution is assumed to vanish (which requires a suitable decay of $n(\boldsymbol{r})$ for large $\boldsymbol{r}$). The expressions (4.180) and (4.181) are completely equivalent—they lead to the same energy and the same potential. While the energy density is only defined up to contributions which integrate up to zero, the total energy and the xc-potential are unambiguous quantities.

The second order gradient contribution has also been studied for the density functional $\tilde{E}_{\mathrm{x}}[n]$ for the HF exchange, defined by Eqs. (4.1)–(4.3). The corresponding approximation for Π consists of $\Pi^{(1)}$ plus additional exchange-type diagrams of higher order in e^2. For the evaluation of the corresponding Feynman integrals again Yukawa-screening has been introduced, in order to carry through the small-$\boldsymbol{q}$ expansion before performing any loop integration. One finds that the resulting gradient coefficient diverges in the limit of vanishing screening [181–184], the divergence originating from the higher order diagrams. Given the fact that an erroneous gradient coefficient is obtained by application of the same procedure to $\Pi^{(1)}$, one might wonder whether this negative result is a mathematical artefact. In order to give a definitive answer to this question, one has to evaluate the relevant Feynman diagrams for Π without utilizing the small-$\boldsymbol{q}$ expansion before loop integration. The result of a corresponding fully numerical study [185] shows a completely smooth behavior in the small-$\boldsymbol{q}$ regime. Unfortunately, the analytical form in this regime has not been extracted from the numerical data. It is thus not clear whether the leading term of the small-$\boldsymbol{q}$ expansion is proportional to $\boldsymbol{q}^2$ (as implied by Eq. (4.166)) or approaches zero in a different fashion (e.g. as $\boldsymbol{q}^2 \ln|\boldsymbol{q}|$).

The evaluation of the correlation contributions to Π is even more involved than that of $\Pi^{(1)}$. In this case only the first strategy, in which the small-$\boldsymbol{q}$ expansion is applied inside the Feynman integrals, has been pursued [186, 187]. The net result reads [188],

$$\Delta E_{\rm xc}^{[2]} = \int d^3r\, e_{\rm x}^{\rm HEG}(n)\, \xi\, C_{\rm xc}(n) \tag{4.183}$$

$$C_{\rm xc}(n) = -\underbrace{\frac{16\pi(3\pi^2)^{1/3}}{3000}}_{=0.0518349} \frac{2.568 + 23.266\, r_{\rm s} + 0.007389\, r_{\rm s}^2}{1 + 8.723\, r_{\rm s} + 0.472\, r_{\rm s}^2 + 0.07389\, r_{\rm s}^3}, \tag{4.184}$$

where (4.184) represents a parameterization of the original numerical data by Geldart and Rasolt [187]. Equation (4.183) still contains the exchange term (4.180), which, however, is easily subtracted,

$$C_{\rm c}(n) = C_{\rm xc}(n) - C_{\rm x}\,. \tag{4.185}$$

$C_{\rm c}$ is plotted in Fig. 4.6 as a function of $r_{\rm s}$, rather than of the density. $C_{\rm c}$ turns out to

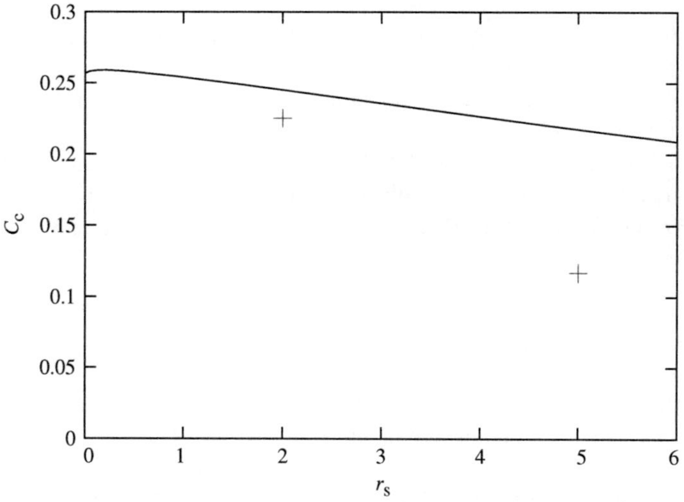

Fig. 4.6 Second order gradient coefficient $C_{\rm c}$, Eq. (4.185) as a function of $r_{\rm s}$. Also shown are two data points for $C_{\rm c}$ based on Monte Carlo simulations for the local field factor [172].

be only weakly dependent on $r_{\rm s}$.

A number of comments are required at this stage:

1. The derivation of the functional (4.183) starts with the gradient corrections resulting from the RPA energy (4.101) by the replacement of G_0 by the propagator of the weakly inhomogeneous electron gas and subsequent expansion to second order in δv. Expressed in terms of Π, this procedure amounts to the inclusion of the screened exchange as well as of the so-called fluctuation diagrams. Taken together, these diagrams constitute the RPA-approximation for Π,

$i\hbar\Pi^{\mathrm{RPA}} =$ 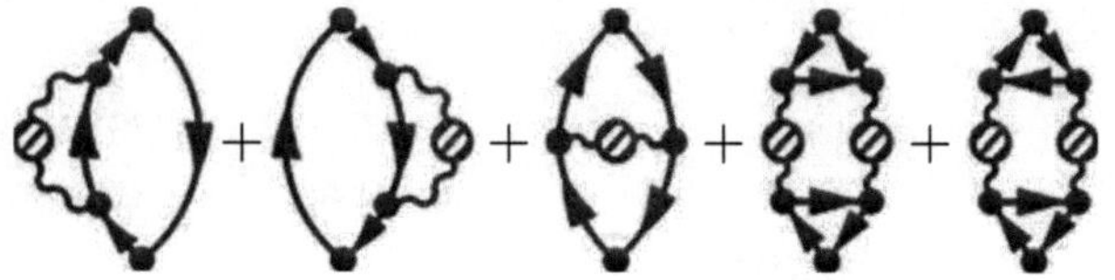 , (4.186)

in which the Coulomb interaction only enters in RPA-screened form,

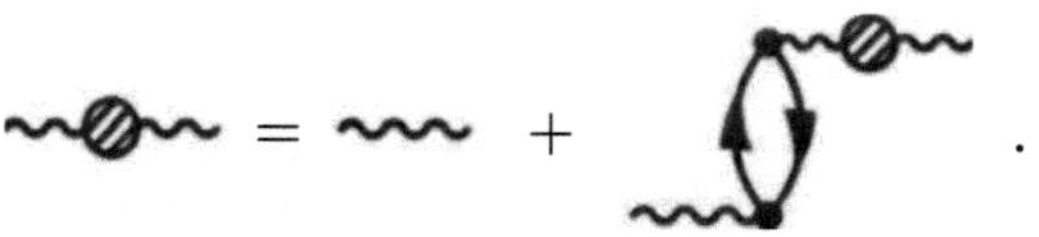 (4.187)

In this way the high-density limit of $\Delta E_{\mathrm{c}}^{[2]}$, which gives the dominant contribution to (4.183), is treated rigorously via standard Green's function techniques. The result is afterwards extended to metallic densities, relying on a Hubbard-type approximation [187].

The final gradient coefficient (4.184) can, however, be compared with results for C_{c} obtained from fixed-node diffusion Monte Carlo calculations for the *local field correction* G [172]. This quantity is equivalent to the irreducible polarization insertion,

$$G(\boldsymbol{q},\omega) = \frac{1}{w(\boldsymbol{q})}\left[\frac{1}{\Pi(\boldsymbol{q},\omega)} - \frac{1}{\Pi^{(0)}(\boldsymbol{q},\omega)}\right], \tag{4.188}$$

so that its behavior for small $\boldsymbol{q}$ allows the determination of the gradient coefficient. Two such data points are included in Fig. 4.6. They have been extracted under the assumption that the numerical value for G at $|\boldsymbol{q}| = 1.01\,k_{\mathrm{F}}$ is completely due to the leading term of the long-wavelength expansion which contributes to C_{c}. Given the uncertainty in this procedure, the resulting data points are consistent with the parameterization (4.184), at least for moderately high metallic densities as $r_{\mathrm{s}} = 2$.

2. According to present knowledge, the interchange of the small-$\boldsymbol{q}$ expansion with the loop integration is legitimate for the diagrams contributing to (4.186). The RPA-screening suppresses the infrared singularity of the Coulomb interaction, which is responsible for the problem with the exchange diagrams (4.105). In other words: the screened exchange diagrams are regularized in a physical fashion, so that a mathematical regularization via Yukawa-screening is no longer necessary.
3. The result (4.183) represents an inhomogeneity correction to the LDA, which has been derived systematically and is not restricted to high densities. There are

no adjustable parameters in $\Delta E_c^{[2]}$. The expression (4.183) reappears as the hard core of all *first-principles* gradient functionals.
4. In the high-density limit the functional (4.183) leads to a term linear in e^2, just as the exchange. This is a result of the resummation of the RPA-contributions to Π, similar to the $\ln(r_s)$-term in the correlation energy of the HEG.
5. The gradient contribution to the correlation energy dominates over the exchange term. This becomes more obvious if the high-density limit of $C_{xc}(n)$ is compared with $C_x = 0.123$,

$$C_{xc}(n \to \infty) = -0.133 \qquad \Longrightarrow \qquad C_c(n \to \infty) = -0.257\,. \qquad (4.189)$$

As a consequence $\Delta E_c^{[2]}$ has the opposite sign as $\Delta E_x^{[2]}$ and E_c^{LDA}.

Before examining the validity of the GE we extend the results (4.180) and (4.183) to spin-density functional theory. This extension is straightforward for the exchange, if one uses the general relation (4.19),

$$\Delta E_x^{[2]}[n_\uparrow, n_\downarrow] = \frac{1}{2}\left\{\Delta E_x^{[2]}[2n_\uparrow] + \Delta E_x^{[2]}[2n_\downarrow]\right\}. \qquad (4.190)$$

On the other hand, an evaluation of the response function of the spin-polarized electron gas would be required in the case of correlation. This is an extremely involved task, so that only the high-density limit has been examined [189]. Fortunately, the general form of $\Delta E_c^{[2]}[n_\uparrow, n_\downarrow]$ is already fixed by the symmetry of the system. Including the exchange, one has

$$\begin{aligned}\Delta E_{xc}^{[2]}[n_\uparrow, n_\downarrow] = \int d^3r \Big\{ & e_x^{\mathrm{HEG}}(2n_\uparrow)\, \xi_{\uparrow\uparrow}\, C_{xc}^{\uparrow\uparrow}(n_\uparrow, n_\downarrow) \\ & + e_x^{\mathrm{HEG}}(2\sqrt{n_\uparrow n_\downarrow})\, \xi_{\uparrow\downarrow}\, C_{xc}^{\uparrow\downarrow}(n_\uparrow, n_\downarrow) \\ & + e_x^{\mathrm{HEG}}(2n_\downarrow)\, \xi_{\downarrow\downarrow}\, C_{xc}^{\downarrow\downarrow}(n_\uparrow, n_\downarrow)\Big\}, \end{aligned} \qquad (4.191)$$

with the dimensionless gradient given by

$$\xi_{\sigma\sigma'} = \left(\frac{\nabla n_\sigma}{2[6\pi^2 n_\sigma]^{1/3} n_\sigma}\right) \cdot \left(\frac{\nabla n_{\sigma'}}{2[6\pi^2 n_{\sigma'}]^{1/3} n_{\sigma'}}\right). \qquad (4.192)$$

For $n_\uparrow = n_\downarrow$ the gradient $\xi_{\sigma\sigma'}$ reduces to (4.179), the complete energy (4.191) approaches (4.183). Moreover, the spin-symmetry requires

$$C_{xc}^{\uparrow\uparrow}(n_\downarrow, n_\uparrow) = C_{xc}^{\downarrow\downarrow}(n_\uparrow, n_\downarrow) \qquad \Longleftrightarrow \qquad C_{xc}^{\uparrow\uparrow}(n, -\zeta) = C_{xc}^{\downarrow\downarrow}(n, \zeta) \qquad (4.193)$$

$$C_{xc}^{\uparrow\downarrow}(n_\downarrow, n_\uparrow) = C_{xc}^{\uparrow\downarrow}(n_\uparrow, n_\downarrow) \qquad \Longleftrightarrow \qquad C_{xc}^{\uparrow\downarrow}(n, -\zeta) = C_{xc}^{\uparrow\downarrow}(n, \zeta)\,. \qquad (4.194)$$

The spin-density-dependent functions $C_{xc}^{\sigma\sigma'}$ which result from the numerical calculation of the high-density limit can be parameterized as [190],

$$\lim_{n\to\infty} C_{\rm xc}^{\sigma\sigma'}(n_\uparrow,n_\downarrow) = -\frac{8\pi(6\pi^2)^{1/3}}{3000}\sum_{k=0}^{7} a_k^{\sigma\sigma'}\left(\frac{n_\uparrow}{n_\downarrow}\right)^k , \tag{4.195}$$

where $\downarrow$ is assumed to be the majority spin and the coefficients $a_k^{\sigma\sigma'}$ are given in Table 1 of [190]. If $\uparrow$ is the majority spin, the expressions for $C_{\rm xc}^{\uparrow\uparrow}$ and $C_{\rm xc}^{\downarrow\downarrow}$ have to be interchanged according to (4.193), (4.194). It is generally believed that the spin-dependence observed in the high-density limit is also quite accurate for the more relevant intermediate densities, which leads to the functional

$$\begin{aligned}\Delta E_{\rm xc}^{[2]}[n_\uparrow,n_\downarrow] = \int d^3r \Bigg\{ & e_{\rm x}^{\rm HEG}(2n_\uparrow)\,\xi_{\uparrow\uparrow}\left[\lim_{n\to\infty}\frac{C_{\rm xc}^{\uparrow\uparrow}(n_\uparrow,n_\downarrow)}{C_{\rm xc}(n_\uparrow+n_\downarrow)}\right] \\ & + e_{\rm x}^{\rm HEG}(2\sqrt{n_\uparrow n_\downarrow})\,\xi_{\uparrow\downarrow}\left[\lim_{n\to\infty}\frac{C_{\rm xc}^{\uparrow\downarrow}(n_\uparrow,n_\downarrow)}{C_{\rm xc}(n_\uparrow+n_\downarrow)}\right] \\ & + e_{\rm x}^{\rm HEG}(2n_\downarrow)\,\xi_{\downarrow\downarrow}\left[\lim_{n\to\infty}\frac{C_{\rm xc}^{\downarrow\downarrow}(n_\uparrow,n_\downarrow)}{C_{\rm xc}(n_\uparrow+n_\downarrow)}\right]\Bigg\} \\ & \times C_{\rm xc}(n_\uparrow+n_\downarrow), \end{aligned} \tag{4.196}$$

with $C_{\rm xc}(n)$ being the coefficient obtained for the unpolarized gas, Eq. (4.184).

A reformulation of (4.196) paves the way to an even simpler, but still quite accurate approximation [191],

$$\begin{aligned}\Delta E_{\rm xc}^{[2]}[n_\uparrow,n_\downarrow] = \int d^3r\, e_{\rm x}^{\rm HEG}(n) \Bigg\{ & \left(\frac{\nabla n}{2(3\pi^2 n)^{1/3} n}\right)^2 C_{\rm xc}(n,\zeta) \\ & + \left(\frac{\nabla n\cdot\nabla\zeta}{4(3\pi^2 n)^{2/3} n}\right)\tilde{C}_{\rm xc}(n,\zeta) \\ & + \left(\frac{\nabla\zeta}{2(3\pi^2 n)^{1/3}}\right)^2 \tilde{\tilde{C}}_{\rm xc}(n,\zeta)\Bigg\}. \end{aligned} \tag{4.197}$$

The coefficients $C_{\rm xc}$, $\tilde{C}_{\rm xc}$ and $\tilde{\tilde{C}}_{\rm xc}$ can be derived directly from the original functions $C_{\rm xc}^{\sigma\sigma'}$. One finds that in the high-density limit $C_{\rm xc}$ is much larger than the other two coefficients,

$$|C_{\rm xc}(n,\zeta)| \gg |\tilde{C}_{\rm xc}(n,\zeta)| \gg |\tilde{\tilde{C}}_{\rm xc}(n,\zeta)| .$$

This suggests that $\tilde{C}_{\rm xc}$ and $\tilde{\tilde{C}}_{\rm xc}$ may be neglected (see also [192]). In addition, one finds that the spin-dependence of the remaining coefficient $C_{\rm xc}(n,\zeta)$ is very well represented by

$$\frac{C_{\rm xc}(n,\zeta)}{C_{\rm xc}(n,0)} \approx g(\zeta) = \frac{1}{2}\left[(1+\zeta)^{2/3}+(1-\zeta)^{2/3}\right], \tag{4.198}$$

which is the exact spin-dependence of the x-only limit,

$$\frac{C_{\mathrm{x}}(n,\zeta)}{C_{\mathrm{x}}(n,0)} = g(\zeta)\,. \tag{4.199}$$

The ζ-dependence of C_{xc} deviates by less than 1% from that of C_{x} [191]. With these approximations the spin-dependent second order gradient correction finally reads

$$\Delta E_{\mathrm{xc}}^{[2]}[n_\uparrow,n_\downarrow] = \int d^3r\, e_{\mathrm{x}}^{\mathrm{HEG}}(n)\,\xi\, C_{\mathrm{xc}}(n)\, g(\zeta)\,, \tag{4.200}$$

with $C_{\mathrm{xc}}(n)$ given by (4.184).

The applicability of the GE is limited by the fact that it requires all density gradients to be small. This point is illustrated by consideration of the lowest order (fourth order) gradient corrections which have been neglected in (4.180) and (4.183). Fourth order corrections have been derived in various ways in the case of T_{s} [193–195], but are known only partially even for E_{x} [171, 196]. Of course, the general form of the fourth order gradient corrections for E_{x} is again clear on dimensional grounds [197],

$$\Delta E_{\mathrm{x}}^{[4]} = C_{\mathrm{x}}^{[4]} \int d^3r\, e_{\mathrm{x}}^{\mathrm{HEG}}(n)\, \left[\eta^2 + a\eta\xi + b\xi^2\right]. \tag{4.201}$$

The coefficient $C_{\mathrm{x}}^{[4]} = 146/2025$ can be extracted from $\Pi^{(1)}$ [171], $a = -2.5 \pm 0.5$ from its quadratic response counterpart [196]. If one combines (4.109), (4.180) and (4.201), the GE for $E_{\mathrm{x}}[n]$ reads

$$E_{\mathrm{x}}^{\mathrm{GE}}[n] = \int d^3r\, e_{\mathrm{x}}^{\mathrm{HEG}}(n) \left\{1 + \frac{10}{81}\xi + \frac{146}{2025}\left[\eta^2 - \frac{5}{2}\eta\xi + b\xi^2\right] + \ldots\right\}. \tag{4.202}$$

Equation (4.202) reveals some sufficient criteria for the validity of the LDA and the second order GE (GE2) in a quantitative form: the LDA is an accurate approximation if the local gradient corrections $10\xi/81$, $146\eta^2/2025$ etc are all small compared to 1. The GE2 is accurate if $146\eta^2/2025 \ll 10\xi/81$ etc. This argument can be easily extended to higher gradients. Of course, these criteria are nothing but the real-space equivalents of (4.168) and (4.169), interpreted locally in accordance with the local density replacement.

The two lowest gradients can be explicitly examined for all interesting systems. As only an estimate of the form and magnitude of ξ and η is required, it is not necessary to use the exact density. Any approximate n, resulting e.g. from an exchange-only or an LDA calculation, is sufficient. Figures 4.7–4.10 provide some illustrative examples. Figure 4.7 shows both gradients for the calcium atom. Three spatial regions can be distinguished:

- For very small r, i.e. in the vicinity of the nucleus, ξ is finite, in accordance with the cusp condition for point nuclei (Kato's theorem [198]),

$$\frac{n'(0)}{n(0)} = -2\frac{Z}{a_0} \quad\Longrightarrow\quad \xi(0) = \frac{Z^2}{a_0^2[3\pi^2 n(0)]^{2/3}}\,. \tag{4.203}$$

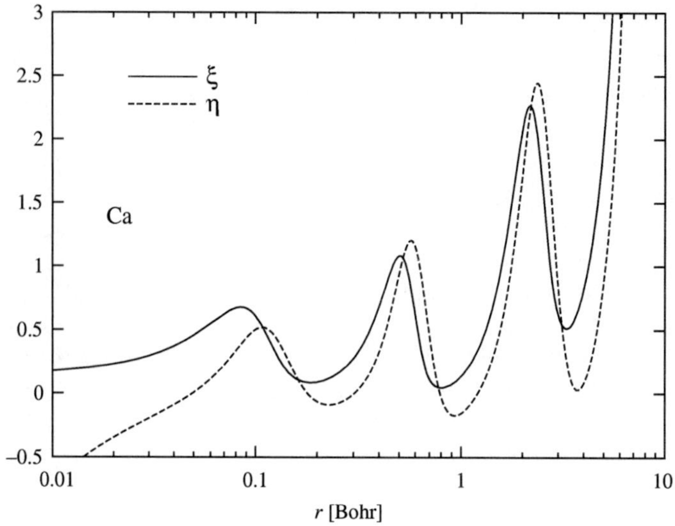

Fig. 4.7 Characteristic gradients of the density for ground state of Ca: $\xi = (\nabla n)^2/[4(3\pi^2 n)^{2/3} n^2]$, $\eta = \nabla^2 n/[4(3\pi^2 n)^{2/3} n]$. The density was obtained by a KS calculation with the exact exchange.

Here n' denotes the derivative of the radially symmetric $n(r)$ with respect to r and a_0 is the Bohr radius (4.104). On the other hand, η diverges at the origin,

$$\eta(r \to 0) \sim \frac{1}{4[3\pi^2 n(0)]^{2/3}} \left[\frac{n''(0)}{n(0)} + \frac{2}{r} \frac{n'(0)}{n(0)} \right] . \tag{4.204}$$

- For intermediate r-values both ξ and η are of the order of 1. Both gradients clearly exhibit the atomic shell structure.
- Finally, for large r, for which the density decays exponentially, the gradients diverge exponentially,

$$n(\mathbf{r}) \sim e^{-\alpha r} \qquad \Longrightarrow \qquad \xi(r), \eta(r) \sim e^{+2\alpha r/3} . \tag{4.205}$$

The same behavior is observed for all other atoms. In the case of spin-polarized atoms the spin-densities follow this pattern individually, as is shown for chromium in Fig. 4.8. ξ and η also remain of the order of 1–3 in the bonding region of molecules and the interstitial region of solids, as can be seen in Figs. 4.9 and 4.10. In fact, ξ necessarily vanishes at some point in the bonding region. On the other hand, in the inner shell regime the atomic behavior is reproduced by the poly-atomic density gradients.

These figures lead to the following conclusions:

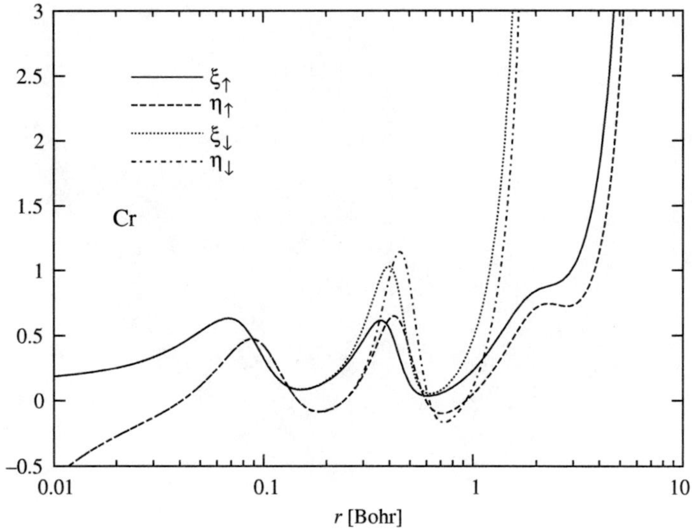

Fig. 4.8 Characteristic gradients of the spin-densities for ground state of Cr: $\xi_\sigma = (\nabla n_\sigma)^2/[4(6\pi^2 n_\sigma)^{2/3} n_\sigma^2]$, $\eta_\sigma = \nabla^2 n_\sigma/[4(6\pi^2 n_\sigma)^{2/3} n_\sigma]$. The spin-densities were obtained by a KS calculation with the exact exchange.

(a) The GE cannot converge locally in the vicinity of the nucleus.
Fortunately, the contribution of this region to the energy is suppressed by the volume element $4\pi r^2$, so that the energy (4.180), but also (4.201), does not diverge in the small-r region. On the other hand, the potential obtained from (4.180) by functional differentiation has the form

$$\Delta v_{\mathrm{x}}^{[2]} = v_{\mathrm{x}}^{\mathrm{LDA}}(n)\, C_{\mathrm{x}} \left[\xi - \frac{3}{2}\eta \right], \qquad (4.206)$$

so that the second order potential is proportional to $1/r$ for small r. This divergence is somewhat unpleasant, but does not cause any serious problems, as the $1/r$-term in the total potential is dominated by the nuclear charge. In other words: the "effective charge"

$$Z_{\mathrm{GE}} = \frac{5}{27\pi} \frac{Z}{[3\pi^2 n(0)]^{1/3}} \frac{m}{\hbar^2} \qquad (Z_{\mathrm{total}} = Z + Z_{\mathrm{GE}}) \qquad (4.207)$$

introduced by the GE is much smaller than the nuclear charge Z. For this reason a regularization of $\Delta v_{\mathrm{x}}^{[2]}$ for small r is neither required nor usual.
This is no longer true for the fourth order potential obtained from $\Delta E_{\mathrm{x}}^{[4]}$, Eq. (4.201),

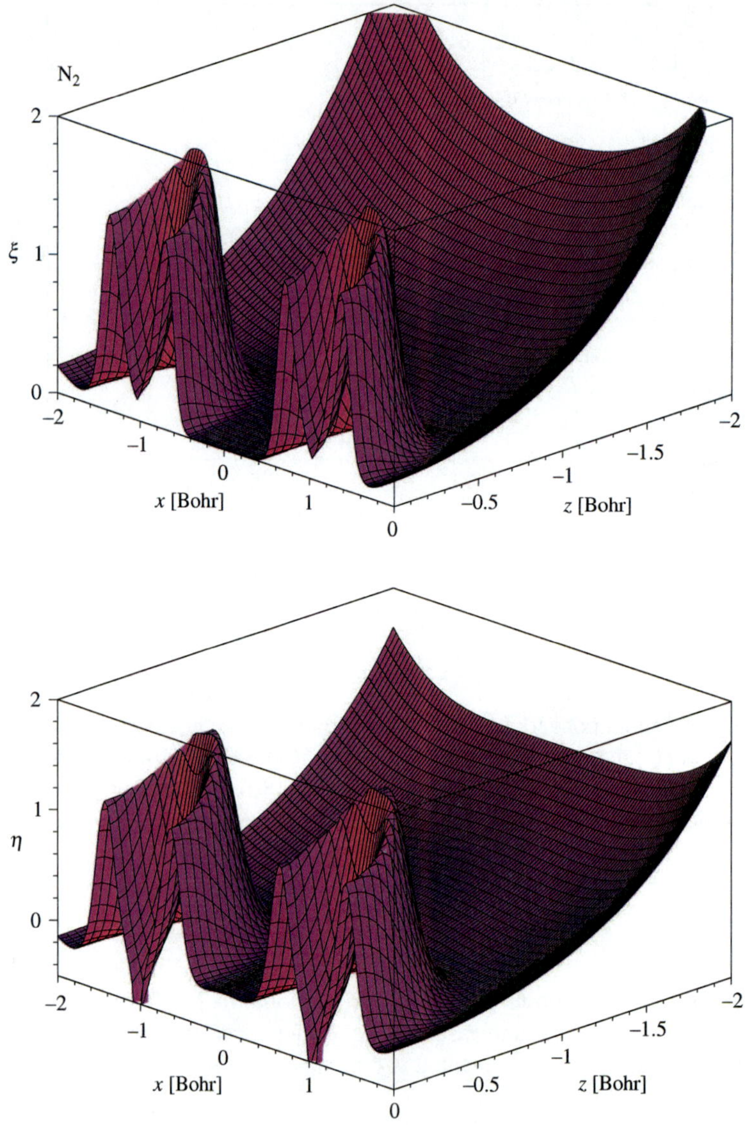

Fig. 4.9 Characteristic gradients of the density for ground state of N_2: $\xi = (\boldsymbol{\nabla} n)^2/[4(3\pi^2 n)^{2/3} n^2]$, $\eta = \boldsymbol{\nabla}^2 n/[4(3\pi^2 n)^{2/3} n]$. The density was obtained by a KS calculation with the exact exchange. The two nuclei are located at $x = \pm 1.035$ Bohr, $y = z = 0$.

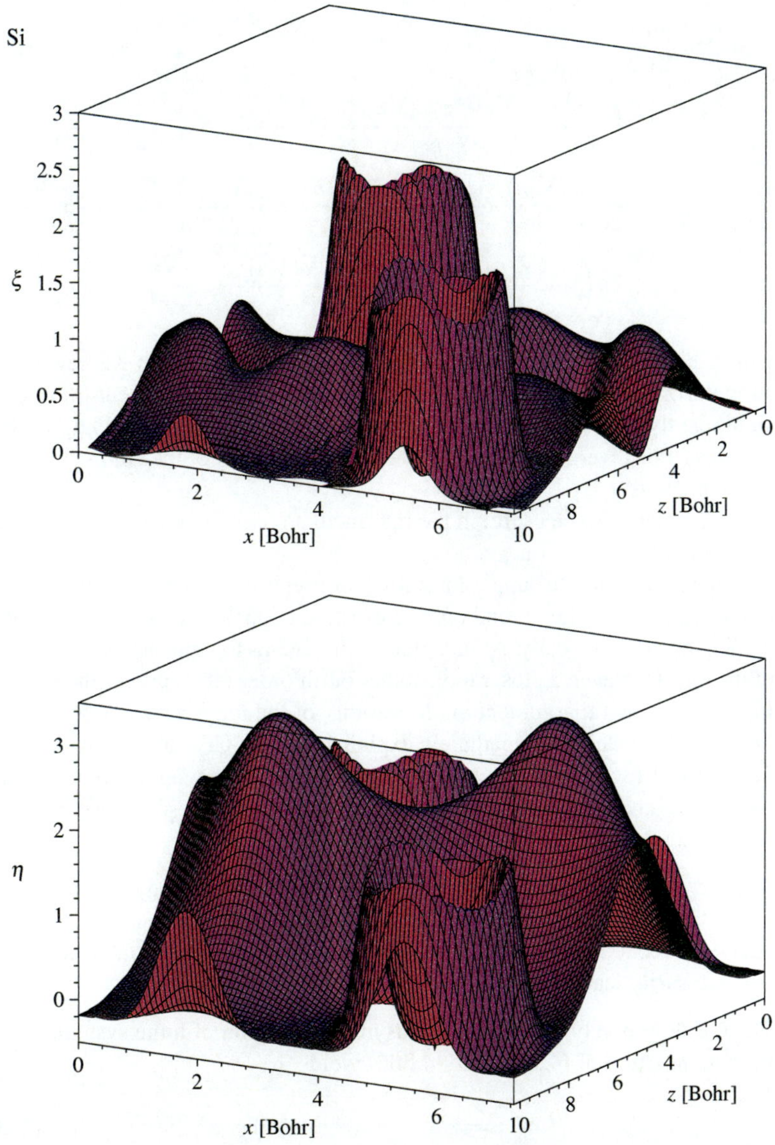

Fig. 4.10 Characteristic gradients of density for ground state of bulk silicon (*diamond structure*): $\xi = (\nabla n)^2/[4(3\pi^2 n)^{2/3} n^2]$, $\eta = \nabla^2 n/[4(3\pi^2 n)^{2/3} n]$, The density was obtained by a KS pseudopotential calculation with the LDA.

$$\Delta v_{\rm x}^{[4]} = \frac{146}{2025}\,\frac{1}{16(3\pi^2 n)^{4/3}}$$
$$\times\left[\frac{3}{2}\left(\frac{(\nabla^4 n)}{n} - 4\frac{(\nabla^3 n)\cdot(\nabla n)}{n^2} - 3\frac{(\nabla^2 n)^2}{n^2} + 6\frac{(\nabla^2 n)(\nabla n)^2}{n^3}\right)\right.$$
$$-\frac{3}{2}a\left(\frac{(\nabla^2 n)^2}{n^2} - \frac{(\nabla\nabla n)\cdot(\nabla\nabla n)}{n^2} + 3\frac{(\nabla n)\cdot\nabla(\nabla n)^2}{n^3} - 6\frac{(\nabla n)^4}{n^4}\right)$$
$$\left.+3b\left(-\frac{(\nabla^2 n)(\nabla n)^2}{n^3} - \frac{(\nabla n)\cdot\nabla(\nabla n)^2}{n^3} + 3\frac{(\nabla n)^4}{n^4}\right)\right], \quad (4.208)$$

which diverges like $1/r^2$. In order to utilize this expression in KS calculations a regularization is required for small r [199]. Fortunately, this cut-off does not significantly affect the potential and orbitals for intermediate and large r. Nevertheless, this divergence clearly shows the inadequacy of the GE in the vicinity of the nucleus.

(b) The GE appears to be useful in the intermediate regime, at least in the sense of an asymptotic expansion.
Taking into account the upper limit for ξ of roughly 3 and its prefactor of 10/81 in the GE (4.202), the second order correction contributes less than 40% of the leading term even locally. At first glance, this seems to indicate the convergence of the GE. However, a closer look at the fourth order terms proves the opposite. In fact, the general argument on the validity of the long-wavelength expansion (4.167), given earlier, immediately explains why the GE can not converge even for $\xi, \eta \approx 1$ (corresponding to $|\boldsymbol{q}| \approx 2k_{\rm F}$). The GE is not applicable as soon as the density changes substantially over a length scale of $1/[2(3\pi^2 n)^{1/3}]$.
One should keep in mind, however, that

(i) $\xi \gtrapprox 1$ does not automatically imply that the GE energy density and potential diverge, but rather that they become inaccurate, and
(ii) the local divergence of the GE energy density or potential does not necessarily lead to a divergence of the integrated GE energy.

(c) The GE definitively diverges in the asymptotic region of finite systems.
In fact, insertion of (4.205) into (4.206) yields

$$n(\boldsymbol{r}) \sim e^{-\alpha r} \qquad \Longrightarrow \qquad \Delta v_{\rm x}^{[2]}(r\to\infty) \sim e^{+\alpha r/3}\,. \quad (4.209)$$

While the exponential divergence is suppressed by the vanishing $C_{\rm xc}(n)$ in the total xc-functional, $\Delta v_{\rm xc}^{[2]}$ nevertheless does not approach zero for large r,

$$\Delta v_{\rm xc}^{[2]} = v_{\rm x}^{\rm LDA}(n)\left\{\left(C_{\rm xc}(n) - \frac{3}{4}n\frac{dC_{\rm xc}}{dn}(n)\right)\xi - \frac{3}{2}C_{\rm xc}(n)\eta\right\} \quad (4.210)$$

$$\Longrightarrow \qquad \Delta v_{\rm xc}^{[2]}(r\to\infty) \sim const\,. \quad (4.211)$$

An indication of the net accuracy of the second order GE (GE2) for atoms is given in Tables 4.4 and 4.5. In the case of the exchange the inclusion of $\Delta E_{\rm x}^{[2]}$ leads to an improvement, reducing the error of the LDA by roughly 50%. Consistent with $C_{\rm x}\xi \ll 1$, the integrated $\Delta E_{\rm x}^{[2]}$ represents a small correction to the LDA. Unfortunately, the same is not true for $\Delta E_{\rm c}^{[2]}$: here one finds that the second order energy is larger than the LDA energy itself. As $\Delta E_{\rm c}^{[2]}$ is positive, the resulting atomic correlation energies even have the wrong sign. This indicates a complete failure of the GE for correlation. Obviously, the criteria for the legitimate size of ξ and higher gradients derived from the exchange contribution to Π are not applicable to correlation.

In summary, one can state that the GE can not be applied to atoms, molecules or surfaces without suitable modification. On the other hand, there are no technical reasons which prevent GE calculations for solids. The adequacy of this approach can only be established by explicit applications. Only few such calculations with the original GE have been reported in the literature, the reason being the failure of the GE2 for metallic iron. The GE2 predicts, as the LDA, an incorrect paramagnetic fcc ground state [200]. For this reason one is forced to consider extensions of the GE2. In fact, already very early the failure of the GE2 has prompted suggestions for the regularization of $\Delta E_{\rm xc}^{[2]}$ [186, 197] in the form

$$\Delta E_{\rm xc}[n] = \int d^3r\, e_{\rm x}^{\rm HEG}(n)\,\xi\, g_{\rm xc}(n,\xi) \tag{4.212}$$

$$g_{\rm xc}(n,\xi) = \begin{cases} C_{\rm xc}(n) & \text{for } \xi \to 0 \\ 0 & \text{for } \xi \to \infty \end{cases} . \tag{4.213}$$

Functionals of this form are called *Generalized Gradient Approximations* (GGAs). They represent a *de facto* standard in DFT applications. GGAs will therefore be discussed in detail in the next section.

4.5 Generalized Gradient Approximation (GGA)

Let us summarize the main problems with the second order GE:

(i) The GE2 exchange potential diverges for exponentially decaying densities.
(ii) The GE2 gives only moderately accurate exchange energies and unacceptable atomic correlation energies.
(iii) The GE2 does not lead to an improvement over the LDA even in the case of solids.

On the other hand, the semi-local structure of gradient-dependent functionals is very attractive for applications. They are computationally much less demanding than calculations with fully nonlocal functionals like (4.162). For this reason much effort has been spent on the systematic construction of improved gradient-based functionals, which do not suffer from the shortcomings of the GE2. Two complementary variants have been investigated, in order to understand the failure of the GE on a formal level and, with this information, to formulate extensions of the GE. While the first of these is based on a momentum space analysis of the GE [201, 202, 114, 203, 192], the second relies on real space arguments [204–207]. Both variants are quite instructive, so that both will be discussed explicitly in the following.

4.5.1 Momentum Space Variant

Let us first go through the momentum space analysis [201, 202, 114, 203, 192]. The starting point for the derivation of gradient corrections is the linear response expression (4.157) with its kernel depending on the inverse polarization insertion. In the high-density limit, which dominates the final $\Delta E_{xc}^{[2]}$, Π^{-1} can be approximated by

$$\frac{1}{\Pi(\boldsymbol{q}^2,0)} \approx \frac{1}{\Pi^{(0)}(\boldsymbol{q}^2,0)}\left\{1-\frac{\Pi^{\mathrm{RPA}}(\boldsymbol{q}^2,0)}{\Pi^{(0)}(\boldsymbol{q}^2,0)}+\ldots\right\}, \tag{4.214}$$

with Π^{RPA} given by (4.186). Equation (4.214) contains all contributions of the order e^2, which is the only order which will be treated consistently in the following. Insertion into (4.157) leads to

$$\Delta E_{xc}^{[2+]} = -\int \frac{d^3q}{(2\pi)^3}\, e_x^{\mathrm{HEG}}(n_0)\, \frac{\boldsymbol{q}^2\,\delta n(\boldsymbol{q})\,\delta n(-\boldsymbol{q})}{(2(3\pi^2 n_0)^{1/3} n_0)^2}\, Z_{xc}(\boldsymbol{q}^2,n_0), \tag{4.215}$$

with the dimensionless kernel

$$Z_{xc}(\boldsymbol{q}^2,n_0) = \frac{2k_F^2}{9\pi e^2}\frac{4k_F^2}{\boldsymbol{q}^2}\left[\frac{\Pi^{\mathrm{RPA}}(\boldsymbol{q}^2,0,n_0)}{\Pi^{(0)}(\boldsymbol{q}^2,0,n_0)^2}-\frac{\Pi^{\mathrm{RPA}}(0,0,n_0)}{\Pi^{(0)}(0,0,n_0)^2}\right]. \tag{4.216}$$

The value of Z_{xc} at $\boldsymbol{q}^2=0$ determines the GE2 (in the high-density limit). One can now consider a *wavevector decomposition* of Z_{xc} with respect to the momentum $\boldsymbol{k}$ which runs through the interaction line in the Feynman diagrams (4.186) which constitute Π^{RPA}:

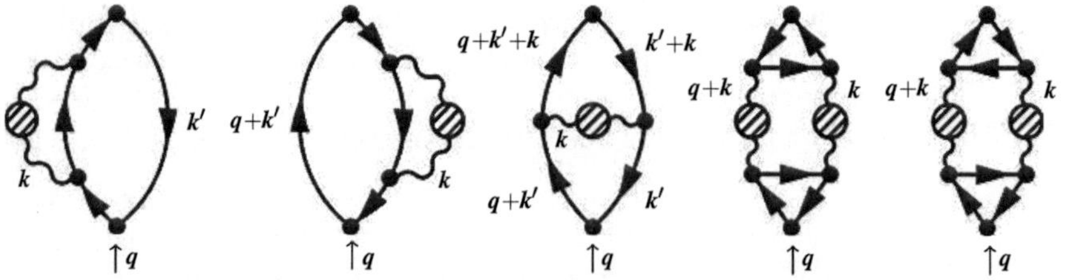

Separation of the integration over $|\boldsymbol{k}|$ from all other loop integrations leads to the wavevector decomposition z_{xc},

$$Z_{xc}(\boldsymbol{q}^2,n_0) = \int_0^\infty dk\, z_{xc}(\boldsymbol{q}^2,k,n_0). \tag{4.217}$$

A sketch of the correlation contribution to the quantity z_{xc} as a function of k is shown in Fig. 4.11 for several values of

$$Q = \sqrt{3}\frac{|\boldsymbol{q}|}{k_{TF}}, \tag{4.218}$$

where k_{TF} denotes the (inverse) *Thomas-Fermi screening length*,

$$k_{TF} = \left(\frac{4}{\pi}\left(\frac{4}{9\pi}\right)^{1/3} r_s\right)^{1/2} k_F = \left(\frac{4k_F}{\pi a_0}\right)^{1/2} = 2\left(\frac{3}{\pi}\right)^{1/6}\left(\frac{n_0}{a_0^3}\right)^{1/6}. \tag{4.219}$$

One immediately notices the extreme sensitivity of z_c on $\boldsymbol{q}$ for small $\boldsymbol{q}$-values. In

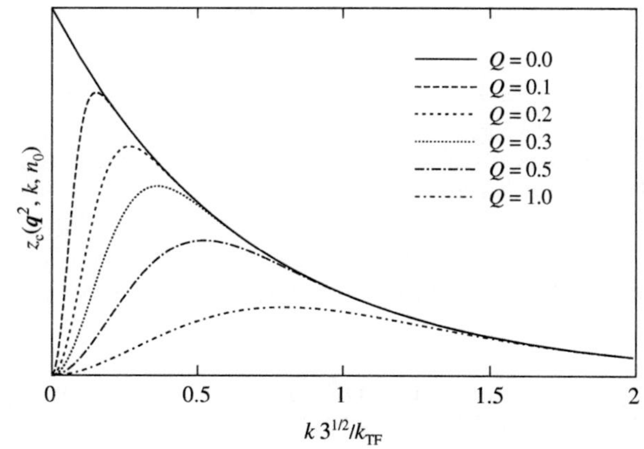

Fig. 4.11 Schematic plot of the wavevector-decomposition of the nonlocal corrections to the xc-energy in the high-density limit, Eq. (4.217), for different external momenta $Q = \sqrt{3}|\boldsymbol{q}|/k_{TF}$ (adopted from [203]).

particular, one finds

$$\lim_{k\to 0}\lim_{q\to 0} z_{xc}(\boldsymbol{q}^2, k, n_0) \neq \lim_{q\to 0}\lim_{k\to 0} z_{xc}(\boldsymbol{q}^2, k, n_0) = 0. \tag{4.220}$$

The second order gradient correction results from the curve with $|\boldsymbol{q}| = 0$. One thus realizes that the wavevector decomposition of the GE2 is quite different from that of the complete linear response kernel. This sensitive $\boldsymbol{q}$-dependence is transfered into the integrated quantity Z_c, which is plotted in Fig. 4.12. For $|\boldsymbol{q}| < k_{TF}/\sqrt{3}$ the total Z_{xc} is well approximated by

$$Z_{xc}(\boldsymbol{q}^2, n_0) = \frac{1}{9}\left[1.1978 + 0.77Q\ln(Q) - 1.25Q\right]. \tag{4.221}$$

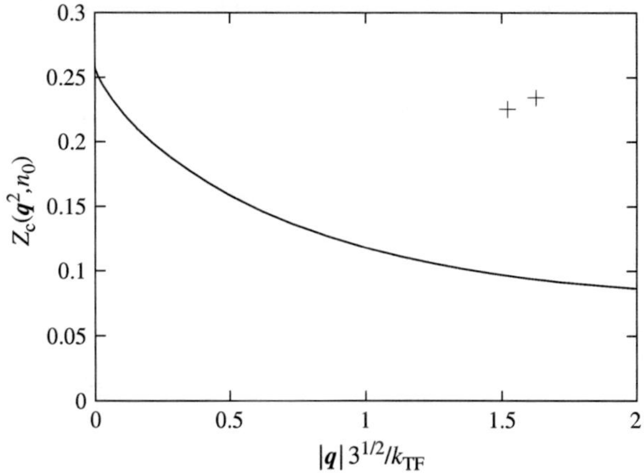

Fig. 4.12 Correlation contribution to the linear response kernel in the high-density limit, Eq. (4.216), according to the parameterization (4.221) [203]. Also given are two data points for Z_c extracted from fixed-node diffusion Monte Carlo simulations for the local field factor [172]. They correspond to the (metallic) density $r_s = 2$.

with Q given by (4.218) [208] (note that $9C_{xc}(n=\infty) = -1.1978$). Figure 4.12 is then obtained by subtraction of Z_x,

$$Z_x(\boldsymbol{q}^2, n_0) = -\frac{1}{2}\frac{(2k_F n_0)^2}{e_x^{HEG}(n_0)}\frac{1}{\boldsymbol{q}^2}\left[\frac{\Pi^{(1)}(\boldsymbol{q}^2,0,n_0)}{\Pi^{(0)}(\boldsymbol{q}^2,0,n_0)^2} - \frac{\Pi^{(1)}(0,0,n_0)}{\Pi^{(0)}(0,0,n_0)^2}\right]$$
$$= -\frac{10}{81} - \frac{146}{2025}\frac{\boldsymbol{q}^2}{4k_F^2} - \ldots . \tag{4.222}$$

In order to appreciate the strong $\boldsymbol{q}$-dependence of Z_c, one has to relate it to the $\boldsymbol{q}$-dependence of Z_x. The appropriate scale for the $\boldsymbol{q}$-dependence of Z_c is k_{TF}, while the Fermi momentum occurs in Z_x. In view of Eq. (4.219) an explicit comparison of Fig. 4.12 with (4.222) is only possible for a given density. For instance, for $r_s = 2$ one has $k_{TF} = 1.15k_F$. The corresponding Z_x is plotted in Fig. 4.13. Z_x is almost constant over the range of Q for which Z_c is given in Fig. 4.12. While $Z_x(\boldsymbol{q}^2)$ is comparatively well approximated by its long-wavelength limit $Z_x(0)$, this is not the case for Z_c. In other words: while $1/(2k_F)$ is the appropriate length scale for exchange, the much larger $\sqrt{3}/k_{TF}$ is the inherent length scale for correlation. One thus concludes that higher order gradient corrections are much more important for correlation than for exchange.

In fact, the GE2-limit $z_c(\boldsymbol{q}^2 = 0, k)$ is much less characteristic for the inhomogeneity corrections in real systems than the behavior of $z_c(\boldsymbol{q}^2, k)$ for $\boldsymbol{q}^2 > 0$, most notably the fact that $z_c(\boldsymbol{q}^2, k=0) = 0$. On the other hand, the $z_c(\boldsymbol{q}^2, k)$ obtained for dif-

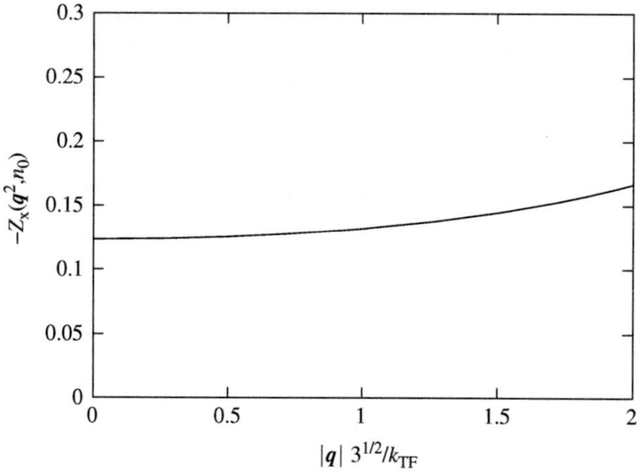

Fig. 4.13 Exchange contribution to (4.216) for $r_s = 2$.

ferent $\boldsymbol{q}^2$ soon approach each other with increasing k. This observation has prompted Langreth and Mehl to introduce a cut-off into the wavevector-decomposition of Z_c in such a way, that the behavior of $z_c(\boldsymbol{q}^2,k,n_0)$ for vanishing k is enforced even for $\boldsymbol{q}^2 = 0$,

$$z_c^{LM}(\boldsymbol{q}^2,k,n_0) = z_c(\boldsymbol{q}^2 = 0,k,n_0)\,\Theta\left(k - f\frac{|\boldsymbol{\nabla} n|}{n}\right). \tag{4.223}$$

The step function cuts off the GE2-result $z_c(\boldsymbol{q}^2 = 0,k)$ for all k smaller than $f|\boldsymbol{\nabla} n|/n$. The quantity $|\boldsymbol{\nabla} n|/n$ serves as a measure of the inhomogeneity of the real system under consideration and thus as a measure of the momentum $|\boldsymbol{q}|$ below which $z_c(\boldsymbol{q}^2,k)$ should drop to zero. The smaller $|\boldsymbol{\nabla} n|/n$ is, the closer is $z_c^{LM}(\boldsymbol{q}^2,k,n_0)$ to the GE2. f is a fit parameter which absorbs all numerical prefactors involved. Langreth and Mehl then use an analytical parameterization of their numerical RPA data for $z_c(\boldsymbol{q}^2 = 0,k)$,

$$z_c(\boldsymbol{q}^2 = 0,k,n_0) \approx -C_c(n=\infty)\,\frac{2\sqrt{3}}{k_{TF}}\,\exp\left[-2\sqrt{3}\frac{k}{k_{TF}}\right], \tag{4.224}$$

to carry through the k-integration in (4.217). The result reads

$$E_c^{LM}[n] = C_c(n=\infty)\int d^3r\, e_x^{HEG}(n)\,\xi\,\exp\left[-(9\pi a_0^3)^{1/6} f\frac{|\boldsymbol{\nabla} n|}{n^{7/6}}\right]. \tag{4.225}$$

For f a value of approximately 0.15 turned out to be a good compromise between the optimum f found for different systems like surfaces and atoms. As it stands,

Eq. (4.225) covers only the high-density limit of the RPA (order e^2) consistently. It has later been extended by inclusion of the contributions of the second order exchange graphs [192]. In addition, a spin-dependent version has been put forward by Hu and Langreth [209],

$$E_{\mathrm{c}}^{\mathrm{HLM}}[n_\uparrow,n_\downarrow] = C_{\mathrm{c}}(n=\infty)\int d^3r\,\frac{e_{\mathrm{x}}^{\mathrm{HEG}}(n)}{d(\zeta)}\,\xi\,\exp\left[-(9\pi a_0^3)^{1/6} f\frac{|\nabla n|}{n^{7/6}}\right], \quad (4.226)$$

with

$$d(\zeta) = 2^{1/3}\left[\left(\frac{1+\zeta}{2}\right)^{5/3}+\left(\frac{1-\zeta}{2}\right)^{5/3}\right] \quad (4.227)$$

and ζ given by Eq. (4.119).

While LM have established the GGA approach, $E_{\mathrm{c}}^{\mathrm{LM}}[n]$ is no longer used in applications today. One reason is the fact that $E_{\mathrm{c}}^{\mathrm{LM}}[n]$ does not cover the metallic density range (in analogy to the observation made for RPA for $e_{\mathrm{c}}^{\mathrm{HEG}}$). This statement is corroborated by Fig. 4.12, in which two data points extracted from fixed-node diffusion Monte Carlo simulations for the local field factor G, Eq. (4.188), at $r_{\mathrm{s}}=2$ are given [172]. The relation between the xc-kernel and G is established after extension of Z_{xc} to arbitrary densities. Using the compressibility sum rule (4.154), one finds

$$\begin{aligned} Z_{\mathrm{xc}}(\boldsymbol{q}^2,n_0) &= -\frac{2k_{\mathrm{F}}^2}{9\pi e^2}\frac{4k_{\mathrm{F}}^2}{\boldsymbol{q}^2}\left[\frac{1}{\Pi(\boldsymbol{q}^2,0)}-\frac{1}{\Pi^{(0)}(\boldsymbol{q}^2,0)}-\frac{1}{\Pi(0,0)}+\frac{1}{\Pi^{(0)}(0,0)}\right] \\ &= -\frac{2k_{\mathrm{F}}^2}{9\pi e^2}\frac{4k_{\mathrm{F}}^2}{\boldsymbol{q}^2}\left[\frac{4\pi e^2}{\boldsymbol{q}^2}G(\boldsymbol{q})+\frac{d^2e_{\mathrm{xc}}^{\mathrm{HEG}}}{dn_0^2}(n_0)\right]. \end{aligned} \quad (4.228)$$

The derivative of the xc-energy density can be rewritten in terms of the correlation energy per particle ε_{c},

$$\frac{d^2e_{\mathrm{xc}}^{\mathrm{HEG}}}{dn_0^2}(n_0) = -\frac{\pi e^2}{k_{\mathrm{F}}^2}-\frac{r_{\mathrm{s}}}{3n_0}\frac{d}{dr_{\mathrm{s}}}\left[\varepsilon_{\mathrm{c}}(r_{\mathrm{s}})-\frac{r_{\mathrm{s}}}{3}\frac{d}{dr_{\mathrm{s}}}\varepsilon_{\mathrm{c}}(r_{\mathrm{s}})\right],$$

which allows an evaluation of $d^2e_{\mathrm{xc}}^{\mathrm{HEG}}/dn_0^2$ for instance via the VWN parameterization for ε_{c}, Eq. (4.108). It is obvious that the Monte Carlo results for $r_{\mathrm{s}}=2$ are quite different from the high-density limit.

For this reason the first important step to be taken is a generalization to arbitrary densities. Such a generalization has been suggested by Perdew [210], relying on the density dependence of $\Delta E_{\mathrm{c}}^{[2]}$, Eq. (4.183),

$$E_{\mathrm{c}}^{\mathrm{P86}}[n] = \int d^3r\,C_{\mathrm{c}}(n)\,e_{\mathrm{x}}^{\mathrm{HEG}}(n)\,\xi\,\exp\left[-(9\pi a_0^3)^{1/6}\tilde{f}\,\frac{C_{\mathrm{c}}(\infty)}{C_{\mathrm{c}}(n)}\frac{|\nabla n|}{n^{7/6}}\right]. \quad (4.229)$$

The factor $C_{\mathrm{c}}(n)$ is a very natural extension of the LM prefactor $C_{\mathrm{c}}(\infty)$ to intermediate densities, as it reintroduces the complete second order gradient correction. The underlying assumption is that the small-k behavior of the wavevector-decomposition observed for the RPA is also roughly correct for the complete $z_c(\boldsymbol{q}^2,k,n_0)$. The additional density dependence in the exponential function leads to a faster cut-off for low densities. The fit parameter $\tilde{f}$ was adjusted to the correlation energy of the neon atom, which led to $\tilde{f}=0.11$. For spin-dependent systems Perdew suggests the utilization of the same overall scaling as in Eqs. (4.226), (4.227). The P86-GGA for correlation is still used today in applications.

4.5.2 Real Space Variant

The starting point of the complementary real-space analysis of the GE [204–207] is the adiabatic connection formula, expressed in terms of the xc-hole h_{xc}. After explicit insertion of the Coulomb interaction and a suitable shift of the integration variable $\boldsymbol{r}'$, the corresponding relation (4.82) can be written as

$$E_{xc}[n] = \frac{e^2}{2}\int d^3r\, n(\boldsymbol{r}) \int d^3u \frac{1}{|\boldsymbol{u}|} h_{xc}(\boldsymbol{r},\boldsymbol{r}+\boldsymbol{u}), \tag{4.230}$$

with h_{xc} given by Eq. (4.81). h_{xc} satisfies a number of exact relations [77]. In view of the definition (4.76) of the pair-correlation function, integration over $\boldsymbol{u}$ yields

$$\begin{aligned}
&\int d^3u\, h_{xc}(\boldsymbol{r},\boldsymbol{r}+\boldsymbol{u}) \\
&= \int d^3r' \left\{ \frac{1}{n(\boldsymbol{r})}\int_0^1 d\lambda\, \langle\Psi_0(\lambda)|[\hat{n}(\boldsymbol{r})-n_0(\boldsymbol{r})][\hat{n}(\boldsymbol{r}')-n_0(\boldsymbol{r}')]|\Psi_0(\lambda)\rangle \right. \\
&\qquad \left. - \delta^{(3)}(\boldsymbol{r}-\boldsymbol{r}') \right\} \\
&= \frac{1}{n(\boldsymbol{r})}\int_0^1 d\lambda\, \langle\Psi_0(\lambda)|[\hat{n}(\boldsymbol{r})-n_0(\boldsymbol{r})][\hat{N}-N]|\Psi_0(\lambda)\rangle - 1 \\
&= -1\,.
\end{aligned} \tag{4.231}$$

This sum rule can then be decomposed into an exchange and a correlation component, $h_{xc}=h_x+h_c$. The exchange component (linear in e^2) is obtained from the KS pair-correlation function (i.e. the noninteracting limit—compare Eq. (3.46)),

$$g_x(\boldsymbol{r},\boldsymbol{r}') = 1 - \sum_{\sigma\sigma'} \frac{|\sum_i \Theta_i \phi_i(\boldsymbol{r}\sigma)\phi_i^*(\boldsymbol{r}'\sigma')|^2}{n(\boldsymbol{r})n(\boldsymbol{r}')} \tag{4.232}$$

$$\Rightarrow \qquad h_x(\boldsymbol{r},\boldsymbol{r}') = -\sum_{\sigma} \frac{|\sum_\alpha \Theta_{\alpha\sigma}\phi_{\alpha\sigma}(\boldsymbol{r})\phi_{\alpha\sigma}^*(\boldsymbol{r}')|^2}{n(\boldsymbol{r})}, \tag{4.233}$$

where the ϕ_i represent KS orbitals. Integration immediately leads to

$$\int d^3u\, h_x(\boldsymbol{r},\boldsymbol{r}+\boldsymbol{u}) = -1\,, \tag{4.234}$$

so that (4.231) requires

$$\int d^3u\, h_c(\boldsymbol{r},\boldsymbol{r}+\boldsymbol{u}) = 0\,. \tag{4.235}$$

Two further exact relations directly emerge from (4.233),

$$h_x(\boldsymbol{r},\boldsymbol{r}+\boldsymbol{u}) \le 0 \qquad \forall\, \boldsymbol{r},\boldsymbol{u} \tag{4.236}$$

$$h_x(\boldsymbol{r},\boldsymbol{r}) = -\frac{n(\boldsymbol{r})}{2} \qquad \text{(for unpolarized systems)}\,. \tag{4.237}$$

Both the sum rules (4.234) and (4.235) as well as the local conditions (4.236) and (4.237) are satisfied by the LDA xc-hole. The compliance with these rigorous properties is one of the major advantages of the LDA. In particular, the compliance with the sum rule (4.231) has been identified as the main reason for the surprising performance of the LDA in the case of atoms [155]: due to the validity of Eq. (4.231) the spherically averaged xc-hole,

$$\int \frac{d\Omega_{\boldsymbol{u}}}{4\pi} h_{xc}(\boldsymbol{r},\boldsymbol{r}+\boldsymbol{u})\,,$$

which determines E_{xc} via Eq. (4.230), is well reproduced by the LDA.

The situation is quite different for the exchange hole resulting from the gradient expansion. Within the real-space method of Kirzhnits one finds [177],

$$h_x(\boldsymbol{r},\boldsymbol{r}+\boldsymbol{u}) = -\frac{n(\boldsymbol{r})}{2}\left\{f_x^{[0]}(\boldsymbol{r},\boldsymbol{u}) + f_x^{[1]}(\boldsymbol{r},\boldsymbol{u}) + f_x^{[2]}(\boldsymbol{r},\boldsymbol{u}) + \ldots\right\} \tag{4.238}$$

$$f_x^{[0]}(\boldsymbol{r},\boldsymbol{u}) = J(z) \tag{4.239}$$

$$f_x^{[1]}(\boldsymbol{r},\boldsymbol{u}) = 12L(z)\frac{\hat{\boldsymbol{u}}\cdot\boldsymbol{\nabla}n}{2k_F n} \tag{4.240}$$

$$\begin{aligned} f_x^{[2]}(\boldsymbol{r},\boldsymbol{u}) &= \left[M(z) - \frac{4}{3}zL(z)\right]\left(\frac{\hat{\boldsymbol{u}}\cdot\boldsymbol{\nabla}n}{2k_F n}\right)^2 + 4zL(z)\frac{(\hat{\boldsymbol{u}}\cdot\boldsymbol{\nabla})^2 n}{4k_F^2 n} \\ &\quad - \left[N(z) + \frac{1}{3}zL(z)\right]\left(\frac{\boldsymbol{\nabla}n}{2k_F n}\right)^2 + 2N(z)\frac{\boldsymbol{\nabla}^2 n}{4k_F^2 n}\,, \end{aligned} \tag{4.241}$$

with the abbreviations

$$z = 2k_F|\boldsymbol{u}| \qquad k_F = (3\pi^2 n)^{1/3} \qquad \hat{\boldsymbol{u}} = \frac{\boldsymbol{u}}{|\boldsymbol{u}|} \tag{4.242}$$

$$J(z) = \frac{72}{z^6}\left[4 + z^2 + (z^2-4)\cos(z) - 4z\sin(z)\right] \tag{4.243}$$

$$L(z) = \frac{1}{z^3}\left[2 - 2\cos(z) - z\sin(z)\right] \tag{4.244}$$

$$M(z) = \frac{1}{z}\left[-z\cos(z) + \sin(z)\right] \tag{4.245}$$

$$N(z) = \frac{2}{z^4}\left[-4-z^2+(4-z^2)\cos(z)+4z\sin(z)\right]. \tag{4.246}$$

The real-space GE of $h_\mathrm{x}(\boldsymbol{r},\boldsymbol{r}+\boldsymbol{u})$ essentially represents a Taylor series expansion with respect to powers of the operator $\boldsymbol{u}\cdot\boldsymbol{\nabla}$. For this reason Yukawa-screening has to be utilized to regularize the $\boldsymbol{u}$-integration in (4.230), if the GE2 for E_x is evaluated by insertion of (4.238) [177]—without screening the convergence of the integral for large $|\boldsymbol{u}|$ is not ensured (note, however, that an incorrect gradient coefficient is obtained in this way).

The important observation [204] is that the second order gradient correction to h_x neither satisfies the sum rule (4.234) nor the negativity condition (4.236). In fact, the individual terms in (4.238) even lead to divergent $\boldsymbol{u}$-integrals, as the functions (4.244)–(4.246) do not decay sufficiently fast for large $|\boldsymbol{u}|$. These integrals can no longer be kept under control by some regularization: as soon as the screening is removed after the $\boldsymbol{u}$-integration, the divergences show up again. While the $\boldsymbol{u}$-integral obtained by insertion of (4.238) into (4.230) is just not well-defined without Yukawa-screening, the $\boldsymbol{u}$-integral in (4.234) simply diverges.

In view of their importance Perdew therefore suggests to reinforce the conditions (4.234) and (4.236) *a posteriori* [204]. In order to introduce a corresponding real-space cut-off first an integration by parts is performed in (4.230) (with respect to the uncritical variable $\boldsymbol{r}$). This leads to a modified exchange hole,

$$\tilde{h}_\mathrm{x}(\boldsymbol{r},\boldsymbol{r}+\boldsymbol{u}) = -\frac{n(\boldsymbol{r})}{2}\tilde{f}(\boldsymbol{r},\boldsymbol{u}) \tag{4.247}$$

$$\tilde{f}(\boldsymbol{r},\boldsymbol{u}) = J + 12L\frac{\hat{\boldsymbol{u}}\cdot\boldsymbol{\nabla}n}{2k_\mathrm{F}n} - \frac{M}{3}\left(\frac{\hat{\boldsymbol{u}}\cdot\boldsymbol{\nabla}n}{2k_\mathrm{F}n}\right)^2 + [zL+N]\left(\frac{\boldsymbol{\nabla}n}{2k_\mathrm{F}n}\right)^2 + \ldots. \tag{4.248}$$

The form (4.247) avoids the presence of second density gradients, which simplifies the subsequent analysis. Of course, $\tilde{h}_\mathrm{x}$ violates the conditions (4.234) and (4.236) just as the original h_x does. For this reason two cut-offs are applied to (4.247),

$$h_\mathrm{x}^\mathrm{GGA}(\boldsymbol{r},\boldsymbol{r}+\boldsymbol{u}) := \tilde{h}_\mathrm{x}^{[2]}(\boldsymbol{r},\boldsymbol{r}+\boldsymbol{u})\,\Theta\left(\tilde{f}(\boldsymbol{r},\boldsymbol{u})\right)\,\Theta\left(R_\mathrm{c}(\boldsymbol{r})-|\boldsymbol{u}|\right). \tag{4.249}$$

While (4.236) is reinstalled by $\Theta(\tilde{f}(\boldsymbol{r},\boldsymbol{u}))$, the sum rule (4.234) is implemented via the function $R_\mathrm{c}(\boldsymbol{r})$. R_c is that value of $|\boldsymbol{u}|$ for which

$$\int d^3u\, h_\mathrm{x}^\mathrm{GGA}(\boldsymbol{r},\boldsymbol{r}+\boldsymbol{u}) = \frac{-1}{48\pi^2}\int_0^{2k_\mathrm{F}R_\mathrm{c}} z^2dz \int d\Omega_u\, \tilde{f}(\boldsymbol{r},\boldsymbol{u})\,\Theta\left(\tilde{f}(\boldsymbol{r},\boldsymbol{u})\right) = -1 \tag{4.250}$$

holds for a given $\boldsymbol{r}$ ($z = 2k_\mathrm{F}|\boldsymbol{u}|$). Insertion of (4.248) demonstrates that the violation of (4.234) becomes more and more serious with increasing gradient of n: the larger the prefactor $\xi = (\boldsymbol{\nabla}n)^2/(2k_\mathrm{F}n)^2$ is, the earlier does the $|\boldsymbol{u}|$-integral in (4.250) exhaust the sum rule. On the other hand, there is no additional explicit $\boldsymbol{r}$-dependence found in the exchange hole (4.247). Consequently, the cut-off is a function of ξ only, which can be determined by solution of (4.250) for arbitrary ξ. Insertion of (4.249) into (4.230) then defines a GGA,

$$E_{\mathrm{x}}^{\mathrm{GGA}}[n] = \frac{e^2}{2}\int d^3r\,n(\boldsymbol{r})\int d^3u\,\frac{1}{|\boldsymbol{u}|}\,h_{\mathrm{x}}^{\mathrm{GGA}}(\boldsymbol{r},\boldsymbol{r}+\boldsymbol{u})$$
$$= \int d^3r\,e_{\mathrm{x}}^{\mathrm{HEG}}(n)\,f(R_{\mathrm{c}}(\xi)) \tag{4.251}$$
$$f(R_{\mathrm{c}}(\xi)) = \frac{1}{36\pi}\int_0^{2k_{\mathrm{F}}R_{\mathrm{c}}} z dz \int d\Omega_{\boldsymbol{u}}\,\tilde{f}(\boldsymbol{r},\boldsymbol{u})\,\Theta\left(\tilde{f}(\boldsymbol{r},\boldsymbol{u})\right)\,. \tag{4.252}$$

The numerical results for $f(R_{\mathrm{c}}(\xi))$ can finally be parameterized in the form

$$f^{\mathrm{PW86}}(\xi) = \left[1+15C_{\mathrm{x}}\xi+14\xi^2+0.2\xi^3\right]^{1/15}\,. \tag{4.253}$$

Initially, the Sham coefficient $C_{\mathrm{x}}^{\mathrm{Sham}} = 7/81$ has been used in (4.253). However, the sum rule (4.250) is not very sensitive to ξ for very small gradients, so that $f(R_{\mathrm{c}}(\xi))$ can not be determined with high accuracy in the limit $\xi \to 0$. In fact, the numerical results for small ξ even suggest a dependence on $\sqrt{\xi}$ [207]. For consistency the small-ξ behavior has therefore later been adjusted to the correct gradient coefficient $C_{\mathrm{x}} = 10/81$. The GGA (4.251), (4.253) is the exchange complement to the GGA (4.229) for $E_{\mathrm{c}}[n]$.

The concept of the real-space cut-off of $h_{\mathrm{x}}^{[2]}$ has subsequently been refined by Perdew and Wang [206, 211] with the aim to incorporate the Lieb-Oxford lower bound for the exchange energy [212],

$$E_{\mathrm{x}} \geq E_{\mathrm{xc}} \geq E_{\mathrm{xc}}^{|\Psi_0\rangle} \geq 2.273\int d^3r\,e_{\mathrm{x}}^{\mathrm{HEG}}(n) \tag{4.254}$$

(here the inequality (3.31) between E_{xc} and the non-classical part of the interaction energy of the interacting system, $E_{\mathrm{xc}}^{|\Psi_0\rangle} = \langle\Psi_0|\hat{W}|\Psi_0\rangle - E_{\mathrm{H}}[n_0]$, Eq. (3.27), has been used). Interpreted locally, this exact relation implies $f(\xi) \leq 2.273$. Combining this requirement with an improved analytical parameterization of the numerical $f(R_{\mathrm{c}}(\xi))$ and the correct gradient coefficient, one ends up with the PW91-GGA for exchange [206],

$$f^{\mathrm{PW91}}(\xi) = 1+\frac{[a_1+(C_{\mathrm{x}}-a_1)e^{-100\xi}]\xi - a_2\xi^2}{1+b_1\sqrt{\xi}\,\mathrm{arsh}(b_2\sqrt{\xi})+a_2\xi^2}\,, \tag{4.255}$$

with $a_1 = 0.2743 \approx 2.2C_{\mathrm{x}}$, $a_2 = 0.004$, $b_1 = 9a_1/(4\pi)$ and $b_2 = 2(6\pi^2)^{1/3}b_1$. While the kernel (4.255) reproduces the exact second order gradient term for $\xi = 0$, f^{PW91} shows a very sensitive ξ-dependence for small ξ (which is nevertheless consistent with the results of the real-space cut-off procedure (4.250) in the small-ξ regime). However, the strong variation of f^{PW91} for small ξ leads to undesired (though uncritical) peak structures in the corresponding exchange potential [213].

In a later contribution, a simplified form of f^{PW91} has been suggested [207],

$$f^{\text{PBE}}(\xi) = 1 + \frac{a\xi}{1+b\xi}\,, \tag{4.256}$$

with

$$a = -\left[C_c(\infty) + \frac{3}{81}\right] = 0.21951 \approx 1.8C_x \qquad \text{and} \qquad b = 0.2730\,.$$

In this GGA the gradient coefficient has been chosen so that exchange and correlation cancel each other in the limit of small ξ, i.e. that the total second order gradient correction vanishes. This construction essentially relies on the argument that the LDA is more accurate for weakly inhomogeneous systems than the GE2.

It is obvious from this discussion that, in spite of the first-principles background of GGAs, there remains some uncertainty in the determination of their precise form. The final form depends on the set of criteria which are used for its construction and the priorities between the criteria. On the other hand, practical results show only a moderate dependence on the precise form of the GGA, if one keeps a realistic perspective of the overall accuracy of this type of functional (compare the discussion below).

The transition to spin-density functional theory is again achieved by (4.19).

4.5.3 Combination of Momentum and Real Space Variants

In view of the successful real-space cut-off procedure for h_x the question arises whether a similar scheme can be applied to the correlation hole? As a direct real-space GE of h_c is not available, information on the correlation hole can only be obtained indirectly. In order to extract this information one first has to remember that the xc-energy only depends on the spherically averaged xc-hole

$$E_{xc}[n] = \frac{1}{2}\int d^3r\, n(\boldsymbol{r}) \int_0^\infty 4\pi u^2 du\, \frac{e^2}{u}\, \overline{h}_{xc}(\boldsymbol{r},u) \tag{4.257}$$

$$\overline{h}_{xc}(\boldsymbol{r},u) = \int \frac{d\Omega_{\boldsymbol{u}}}{4\pi}\, h_{xc}(\boldsymbol{r},\boldsymbol{r}+\boldsymbol{u})\,. \tag{4.258}$$

One can now use the Fourier representation of the Coulomb interaction,

$$\begin{aligned} \frac{e^2}{|\boldsymbol{u}|} &= \int \frac{d^3k}{(2\pi)^3}\, e^{i\boldsymbol{k}\,\boldsymbol{u}}\, \frac{4\pi e^2}{k^2} \\ &= \int_0^\infty \frac{4\pi k^2 dk}{(2\pi)^3}\, \frac{\sin(ku)}{ku}\, \frac{4\pi e^2}{k^2}\,, \end{aligned} \tag{4.259}$$

to rewrite E_{xc} as

$$E_{xc}[n] = \frac{1}{2}\int d^3r\, n(\boldsymbol{r}) \int_0^\infty \frac{4\pi k^2 dk}{(2\pi)^3}\, \frac{4\pi e^2}{k^2}\, \overline{h}_{xc}(\boldsymbol{r},k) \tag{4.260}$$

$$\overline{h}_{\rm xc}(\boldsymbol{r},k) = \int_0^\infty 4\pi u^2 du \, \frac{\sin(ku)}{ku} \, \overline{h}_{\rm xc}(\boldsymbol{r},u) \, . \tag{4.261}$$

This result can be compared with the wavevector decomposition (4.215)–(4.217),

$$\begin{aligned} \Delta E_{\rm c}^{[2+]} &= -\int \frac{d^3q}{(2\pi)^3} \, e_{\rm x}^{\rm HEG}(n_0) \, \frac{\boldsymbol{q}^2 \, \delta n(\boldsymbol{q}) \, \delta n(-\boldsymbol{q})}{4(3\pi^2 n_0)^{2/3} n_0^2} \int_0^\infty dk \, z_{\rm c}(\boldsymbol{q}^2,k,n_0) \\ &= \int d^3r \int d^3r' \, \delta n(\boldsymbol{r}) \, \delta n(\boldsymbol{r}') \int_0^\infty \frac{4\pi k^2 dk}{(2\pi)^3} \frac{4\pi e^2}{k^2} \\ &\quad \times \int \frac{d^3q}{(2\pi)^3} \, e^{-i\boldsymbol{q}\cdot(\boldsymbol{r}-\boldsymbol{r}')} \, \boldsymbol{q}^2 \, \frac{-e_{\rm x}^{\rm HEG}(n_0)}{4(3\pi^2 n_0)^{2/3} n_0^2} \frac{(2\pi)^3}{(4\pi e)^2} \, z_{\rm c}(\boldsymbol{q}^2,k,n_0) \, . \end{aligned} \tag{4.262}$$

The $\boldsymbol{q}$-integral can easily be carried through, if one restricts (4.262) to the lowest order gradient contribution. The usual local density substitution then leads to

$$\begin{aligned} \Delta E_{\rm c}^{[2]} &= \frac{1}{2} \int d^3r \, n(\boldsymbol{r}) \int_0^\infty \frac{4\pi k^2 dk}{(2\pi)^3} \frac{4\pi e^2}{k^2} \\ &\quad \times [\boldsymbol{\nabla} n(\boldsymbol{r})]^2 \frac{-e_{\rm x}^{\rm HEG}(n(\boldsymbol{r}))}{4(3\pi^2 n(\boldsymbol{r}))^{2/3} n(\boldsymbol{r})^3} \frac{\pi}{e^2} \, z_{\rm c}(0,k,n(\boldsymbol{r})) \, . \end{aligned} \tag{4.263}$$

Comparison with (4.261) allows the direct identification of the high-density limit of the spherically averaged $h_{\rm c}$,

$$\overline{h}_{\rm c}^{\rm HDL}(\boldsymbol{r},u) = (\boldsymbol{\nabla} n)^2 \frac{-e_{\rm x}^{\rm HEG}(n)}{4(3\pi^2 n)^{2/3} n^3} \frac{1}{2\pi e^2} \int_0^\infty k^2 dk \, \frac{\sin(ku)}{ku} \, z_{\rm c}(0,k,n) \, . \tag{4.264}$$

With the analytical parameterization of $z_{\rm c}(0,k,n)$, Eq. (4.224), one can perform the k-integration (see [214], Eq. 2.667.5),

$$\overline{h}_{\rm c}^{\rm HDL}(\boldsymbol{r},u) = -C_{\rm c}(n=\infty) \left(\frac{\boldsymbol{\nabla} n}{2k_{\rm TF} n}\right)^2 \frac{36}{\pi^3 a_0} \frac{k_{\rm TF}^2}{[12+(k_{\rm TF} u)^2]^2} \, . \tag{4.265}$$

Insertion into (4.257) and subsequent evaluation of the $\boldsymbol{u}$-integration leads back to the high-density limit of the GE2, Eq. (4.183). In order to recover the complete $E_{\rm c}^{[2]}$, one has to extend (4.265) to arbitrary densities. To this aim one notes that $\overline{h}_{\rm c}$ essentially depends on the product $k_{\rm TF} u$, so that high densities are equivalent to large particle–particle separations u. Similarly, low densities are equivalent to small u, so that the low-density limit of $\overline{h}_{\rm c}$ corresponds to its short-range behavior. Perdew and Wang [206] suggest the following ansatz for the short-range component, which is missing in $\overline{h}_{\rm c}^{\rm HDL}$,

$$\overline{h}_{\rm c}^{[2]}(\boldsymbol{r},u) = \overline{h}_{\rm c}^{\rm HDL}(\boldsymbol{r},u) - \frac{1}{18\pi^3 a_0} \left(\frac{\boldsymbol{\nabla} n}{2k_{\rm TF} n}\right)^2 k_{\rm TF}^2 \, e^{-b(k_{\rm TF} u)^2} \tag{4.266}$$

$$b = \frac{1}{54[C_{\rm c}(n) - C_{\rm c}(\infty)]} \, . \tag{4.267}$$

It is worthwhile to verify explicitly that (4.266) reproduces the second order gradient correction (4.183). In fact, insertion into (4.257) gives

$$\begin{aligned}
\Delta E_{\mathrm{c}}^{[2]} &= \frac{1}{2}\int d^3r\, n(\boldsymbol{r}) \int_0^\infty 4\pi u^2 du \frac{e^2}{u} \overline{h}_{\mathrm{c}}^{[2]}(\boldsymbol{r},u) \\
&= -\frac{e^2}{2}\int d^3r\, n(\boldsymbol{r}) \int_0^\infty u du \left(\frac{\boldsymbol{\nabla} n}{n}\right)^2 \frac{1}{\pi^2 a_0}\left[\frac{36 C_{\mathrm{c}}(\infty)}{[12+(k_{\mathrm{TF}}u)^2]^2} + \frac{e^{-b(k_{\mathrm{TF}}u)^2}}{18}\right] \\
&= -e^2\int d^3r\, n(\boldsymbol{r}) \left(\frac{\boldsymbol{\nabla} n}{2k_{\mathrm{F}} n}\right)^2 \frac{k_{\mathrm{F}}}{4\pi}\int_0^\infty dy \left[\frac{36 C_{\mathrm{c}}(\infty)}{[12+y]^2} + \frac{e^{-by}}{18}\right] \\
&= \int d^3r\, e_{\mathrm{x}}^{\mathrm{HEG}}(n)\, C_{\mathrm{c}}(n)\, \xi\,,
\end{aligned} \tag{4.268}$$

which is the desired result.

Once a reasonably accurate representation of $\overline{h}_{\mathrm{c}}^{[2]}(\boldsymbol{r},u)$ has been established, one can check whether it satisfies the sum rule (4.235). It is no surprise that (4.235) is violated,

$$\int_0^\infty 4\pi u^2 du\, \overline{h}_{\mathrm{c}}^{[2]}(\boldsymbol{r},u) \neq 0\,. \tag{4.269}$$

The sum rule is therefore reinforced by introduction of a real-space cut-off R_{c} in complete analogy to (4.249),

$$\overline{h}_{\mathrm{c}}^{\mathrm{GGA}}(\boldsymbol{r},u) := \overline{h}_{\mathrm{c}}^{[2]}(\boldsymbol{r},u)\,\Theta(R_{\mathrm{c}}-u)\,, \tag{4.270}$$

with R_{c} determined by

$$\int_0^{R_{\mathrm{c}}} 4\pi u^2 du \left[\overline{h}_{\mathrm{c}}^{[0]}(\boldsymbol{r},u) + \overline{h}_{\mathrm{c}}^{[2]}(\boldsymbol{r},u)\right] = 0\,. \tag{4.271}$$

If more than one value of R_{c} solves (4.271), the solution with the largest value is chosen. Insertion of (4.266) into (4.271) immediately exhibits the quantities on which R_{c} depends,

$$\int_0^{R_{\mathrm{c}}} 4\pi u^2 du\, \overline{h}_{\mathrm{c}}^{[2]}(\boldsymbol{r},u) = -\left(\frac{\boldsymbol{\nabla} n}{2k_{\mathrm{TF}} n}\right)^2 \frac{4}{\pi^2 k_{\mathrm{TF}}}\int_0^{k_{\mathrm{TF}}R_{\mathrm{c}}} x^2 dx \left[\frac{36 C_{\mathrm{c}}(\infty)}{[12+x^2]^2} + \frac{e^{-bx^2}}{18}\right].$$

One identifies

$$t := \frac{|\boldsymbol{\nabla} n|}{2k_{\mathrm{TF}} n} \tag{4.272}$$

as characteristic dimensionless density gradient of the correlation functional. In addition, R_{c} depends on the density. The GGA for $E_{\mathrm{c}}[n]$ is defined by

$$\Delta E_{\rm c}^{\rm GGA} = \frac{1}{2}\int d^3r n(\boldsymbol{r}) \int_0^{R_{\rm c}} 4\pi u^2 du \frac{e^2}{u} \overline{h}_{\rm c}^{[2]}(\boldsymbol{r},u) . \tag{4.273}$$

The construction of the GGA is completed by an analytical parameterization of the t- and n-dependent kernel resulting from the condition (4.271),

$$\Delta E_{\rm c}^{\rm PW91} = \int d^3r n(\boldsymbol{r}) \left[f_1(t,n) + f_2(t,n) \right] \tag{4.274}$$

$$f_1(t,n) = \frac{\beta^2}{2\alpha} \ln\left[1 + \frac{2\alpha}{\beta} \frac{t^2 + At^4}{1 + At^2 + A^2t^4}\right] \tag{4.275}$$

$$f_2(t,n) = -\frac{3}{\pi^2}\frac{e^2}{a_0}\left[C_{\rm c}(n) - C_{\rm c}(\infty) - \frac{3}{81}\right] t^2 e^{-100\xi} \tag{4.276}$$

$$\beta = -\frac{3}{\pi^2}\frac{e^2}{a_0}\left[C_{\rm c}(\infty) + \frac{3}{81}\right] \quad ; \quad \alpha = 0.09 \frac{e^2}{a_0} \tag{4.277}$$

$$A = \frac{2\alpha}{\beta}\left[\exp\left(-\frac{2\alpha}{\beta^2}\frac{e_{\rm c}^{\rm HEG}(n)}{n}\right) - 1\right]^{-1} . \tag{4.278}$$

Equation (4.274) is to be combined with the PW91-GGA for exchange, Eq. (4.255).

It remains to make the transition to spin-dependent systems. Relying on the neglect of $\nabla\zeta$ and the approximation (4.198) (which led to the gradient term (4.200)) and expressing the f_i in terms of the actually occurring variables, the transition can be made by

$$f_i\left(t, \xi, C_{\rm c}(n), \frac{e_{\rm c}^{\rm HEG}(n)}{n}\right) \longrightarrow g^3 f_i\left(\frac{t}{g}, g^2\xi, C_{\rm c}(n), \frac{e_{\rm c}^{\rm HEG}(n)}{g^3 n}\right) , \tag{4.279}$$

with $g(\zeta)$ given by Eq. (4.198).

A simplified variant of (4.274) has been put forward by Perdew, Burke and Ernzerhof [207],

$$\Delta E_{\rm c}^{\rm PBE} = \int d^3r n(\boldsymbol{r}) f_1(t,n) . \tag{4.280}$$

It is based on the concept of a complete cancellation of the total second order gradient term, which originates from the observation that the LDA is more accurate for weakly inhomogeneous systems than the GE2. Together with the PBE-GGA for exchange, Eq. (4.256), the functional (4.280) leads to a vanishing second order gradient coefficient for the complete $E_{\rm xc}$.

4.5.4 Semi-Empirical Construction of GGAs

The fact that the basic form of GGA-type functionals is already determined by dimensional arguments has prompted quite a number of semi-empirical schemes for their construction. This route has been pursued in particular in quantum chemistry, often aiming at specific classes of systems. Alternatively, GGA-type expressions are combined with a fraction of the exact exchange (4.5), which constitutes the class of *hybrid functionals* [215–217] discussed in Sect. 6.5.4. It is beyond the scope of this work to review the wealth of semi-empirical functionals which can be found in the literature. Only the original and best-known GGAs of this type will be discussed.

The structure of the GGA is most obvious for the exchange part, as $E_{\rm x}$ is linear in e^2,

$$E_{\rm x}^{\rm GGA}[n] = \int d^3r\, e_{\rm x}^{\rm HEG}(n)\, f_{\rm x}(\xi)\,. \tag{4.281}$$

The dimensionless function $f_{\rm x}$ can only depend on the dimensionless gradient ξ, as any further density dependence requires the presence of e^2—otherwise it is not possible to generate a dimensionless quantity (as $r_{\rm s}$). Moreover, in analogy to the first-principles GGAs the dependence on higher gradients of the density has been neglected. Of course, $f_{\rm x}(\xi)$ has to satisfy a number of criteria. Some of them are of fundamental nature, others are motivated by practical considerations. Given the universality of $E_{\rm x}$ and the rigorous information available for the HEG, $f_{\rm x}$ has to approach the LDA for vanishing ξ,

$$(a)\quad f_{\rm x}(\xi = 0) = 1\,. \tag{4.282}$$

In addition, the structure of the second order gradient term has to be reproduced,

$$(b)\quad f_{\rm x}(\xi \ll 1) = 1 + a\xi + \ldots\,. \tag{4.283}$$

While the weakly inhomogeneous electron gas requires $a = C_{\rm x}$, this parameter is often treated as adjustable, only the functional form of the second order gradient correction is kept. In addition to these more fundamental requirements one wants to obtain a non-diverging exchange potential for exponentially decaying densities, which restricts the behavior of $f_{\rm x}$ for large ξ,

$$(c)\quad f_{\rm x}(\xi \to \infty) \le c\sqrt{\xi}\,. \tag{4.284}$$

The simplest realization of a kernel which obeys the conditions (a)–(c) is a [1,1]-Padé function

$$f_{\rm x}^{\rm B86}(\xi) = 1 + \frac{a\xi}{1 + b\xi}\,, \tag{4.285}$$

as suggested by Becke in his seminal contribution [218]. The gradient coefficient a of the B86-GGA was not restricted to the exact coefficient $C_{\rm x}$. Rather both parameters a and b were fitted to a set of atomic exchange energies, with the result $a = 0.2351 \approx 1.9C_{\rm x}$ and $b = 0.24308$.

An extension of this concept leads to the most frequently used semi-empirical GGA (B88) [219]. With the alternative ansatz

$$f_{\mathrm{x}}^{\mathrm{B88}}(\xi) = 1 + \frac{a\xi}{1 + b\sqrt{\xi}\,\mathrm{Arsh}[2(6\pi^2)^{1/3}\sqrt{\xi}]} \tag{4.286}$$

one can not only reproduce atomic E_{x} even more accurately than with (4.285) (with $a = 0.2743 \approx 2.2C_{\mathrm{x}}$), the choice $b = 9a/(4\pi)$ also allows the incorporation of the correct asymptotic behavior of the exchange energy density, Eq. (4.24). In fact, insertion of the asymptotic forms of n and ξ, Eq. (4.205), into (4.286) leads to

$$e_{\mathrm{x}}^{\mathrm{HEG}}(n) f_{\mathrm{x}}^{\mathrm{B88}}(\xi) \xrightarrow[|\boldsymbol{r}|\to\infty]{} -\frac{e^2 n(\boldsymbol{r})}{2|\boldsymbol{r}|}. \tag{4.287}$$

However, as emphasized earlier, the energy density is not uniquely defined, so that the physical relevance of (4.287) remains open at this point. In contrast to e_{xc} the xc-potential is an unambiguous quantity. Asymptotically the exact v_{x} behaves like $1/r$, Eq. (4.20). On the other hand, one finds for the B88-GGA (see e.g. [220]),

$$v_{\mathrm{x}}^{\mathrm{B88}}(\boldsymbol{r}) \xrightarrow[|\boldsymbol{r}|\to\infty]{} -\frac{C}{|\boldsymbol{r}|^2}. \tag{4.288}$$

As the asymptotic relations (4.20) and (4.24) both originate from the nonlocal character of the self-interaction contribution to the exact E_{x}, the mismatch between (4.287) and (4.288) indicates that (4.287) does not have the desired physical background.

The discussion of semi-empirical GGAs is closed by a remark on the functional suggested by Lee, Yang and Parr (LYP) [221], which often serves as the counterpart to the B88-GGA (4.286) (the combined functional is termed BLYP). The LYP-GGA is based on the orbital-dependent correlation functional of Colle and Salvetti [222] (see Sect. 6.5.2), whose crucial ingredient is the Laplacian of the noninteracting 2-particle density (3.46). The Laplacian of $n_{\mathrm{s}}^{(2)}$ can be rewritten as the sum of simple gradients of the density and the KS kinetic energy density (relying on the fact that normalizable KS states can always be chosen to be real). LYP then approximate the KS kinetic energy density by its second order GE, so that the Laplacian of $n_{\mathrm{s}}^{(2)}$ is expressed as a density functional. The final functional reads (in atomic units)

$$\begin{aligned} E_{\mathrm{c}}^{\mathrm{LYP}}[n_\uparrow, n_\downarrow] = &-a \int d^3r\, \gamma \frac{n}{\alpha} - \frac{3ab}{5} \int d^3r\, \gamma\beta \sum_\sigma (6\pi^2 n_\sigma)^{2/3} n_\sigma^2 \\ &- \frac{ab}{4} \int d^3r\, \gamma\beta \left\{ \frac{1}{9} \sum_\sigma \left[(\nabla n_\sigma)^2 + 3n_\sigma \nabla^2 n_\sigma \right] - (\nabla n)^2 + n\nabla^2 n \right\}, \end{aligned} \tag{4.289}$$

with the abbreviations

$$\alpha(\boldsymbol{r}) = 1 + dn(\boldsymbol{r})^{-1/3}$$

$$\beta(\boldsymbol{r}) = \frac{n(\boldsymbol{r})^{-5/3} \exp[-cn(\boldsymbol{r})^{-1/3}]}{\alpha(\boldsymbol{r})}$$

$$\gamma(\boldsymbol{r}) = 4\frac{n_\uparrow(\boldsymbol{r})n_\downarrow(\boldsymbol{r})}{n(\boldsymbol{r})^2} .$$

The parameters used by LYP are the original values of Colle and Salvetti [222] (see Sect. 6.5.2), whose $a = 0.04918\, me^4/\hbar^2$, $b = 0.132$, $c = 0.2533/a_0$, $d = 0.349/a_0$, which are optimized to reproduce the correlation energy of the helium atom.

4.5.5 Merits and Limitations of GGAs

Some illustrative GGA results for atoms are included in Tables 4.4 and 4.5. For the comparison with the LDA and GE2 the most frequently used GGAs have been chosen. An impressive improvement is observed both for exchange and for correlation. Compared to the LDA, the net error is reduced by more than an order of magnitude in both cases. Moreover, while the differences originating from the different kernels of the GGAs are not completely negligible, they are much smaller than the differences between any of the GGAs and the LDA.

The accuracy of the xc-energies is directly transfered to the total ground state energies obtained by selfconsistent calculations with these functionals. This is demonstrated in Table 4.6 for the case of closed-subshell atoms. Exact total energies for

Table 4.6 Exchange-only ground state energies of closed-subshell atoms: Selfconsistent LDA, PW91-GGA, PBE-GGA and B88-GGA energies versus exact results [223] (all energies in mHartree).

Atom	E_{tot}	$E_{\mathrm{tot}} - E_{\mathrm{tot}}^{\mathrm{exact}}$			
	Exact	LDA	PW91	PBE	B88
He	−2861.7	138.0	6.5	9.6	−1.7
Be	−14572.4	349.1	18.2	27.4	6.1
Ne	−128545.4	1054.7	−23.5	25.3	−44.7
Mg	−199611.6	1362.8	−0.5	64.7	−20.4
Ar	−526812.2	2294.8	41.2	169.5	12.4
Ca	−676751.9	2591.8	25.7	175.9	−1.0
Zn	−1777834.4	3924.5	−252.6	51.5	−285.2
Kr	−2752042.9	5176.8	−18.4	389.3	−57.7
Sr	−3131533.4	5535.4	−8.8	432.2	−45.7
Pd	−4937906.0	6896.0	−65.2	524.8	−109.6
Cd	−5465114.4	7292.6	−31.9	595.6	−76.3
Xc	−7232121.1	8463.8	54.9	803.1	4.5

these systems are available in the x-only limit, while highly accurate numbers for the complete energy are only known for a few, very light atoms. In order to allow an unambiguous comparison Table 4.6 is therefore restricted to the x-only limit. The improvement provided by GGAs is obvious.

In fact, quite convincing numbers are found for a wide range of systems and properties (see e.g. [211, 224, 225]). Some illustrative results for diatomic molecules are given in Table 4.7. This table lists all first and second row diatomic molecules

Table 4.7 Bond lengths R_e, dissociation energies D_e (including the zero-point energies [226]) and harmonic frequencies ω_e of first and second row diatomic molecules: Selfconsistent LDA [134], PBE-GGA [207] and BLYP-GGA [219, 221] results versus experimental data (taken from [227–229]—see also http://cccbdb.nist.gov/). For the atomic ground state non-spherical densities have been allowed.
A) Hydrogen and first row homonuclear diatomics.

Molecule	Method	R_e [Bohr]	D_e [eV]	ω_e [cm^{-1}]
H_2	Exact	1.401	4.747	4401
$^1\Sigma$	LDA	1.446	4.913	4203
	PBE	1.418	4.538	4318
	BLYP	1.410	4.749	4347
Li_2	Expt.	5.051	1.058	351
$^1\Sigma$	LDA	5.118	1.027	332
	PBE	5.153	0.868	325
	BLYP	5.119	0.895	327
B_2	Expt.	3.005	3.071	1051
$^3\Sigma$	LDA	3.033	3.855	1032
	PBE	3.057	3.340	1007
	BLYP	3.057	2.911	987
C_2	Expt.	2.348	6.333	1855
$^1\Sigma$	LDA	2.353	7.249	1878
	PBE	2.369	6.258	1824
	BLYP	2.371	5.866	1799
N_2	Expt.	2.074	9.905	2359
$^1\Sigma$	LDA	2.068	11.601	2393
	PBE	2.082	10.583	2344
	BLYP	2.082	10.440	2327
O_2	Expt.	2.282	5.213	1580
$^3\Sigma$	LDA	2.274	7.590	1621
	PBE	2.301	6.248	1552
	BLYP	2.321	5.902	1488
F_2	Expt.	2.668	1.658	917
$^1\Sigma$	LDA	2.614	3.395	1043
	PBE	2.672	2.295	948
	BLYP	2.710	2.127	894
Mean abs.	LDA	0.030	1.101	66
deviation	PBE	0.032	0.442	37
	BLYP	0.034	0.355	49

contained in the *G2 test set* [225]. The G2 set is a collection of experimentally well-studied molecules, for which the spectroscopic constants can serve as reference data for benchmarking theoretical approaches. There exists a number of further test sets,

Table 4.7 continued: B) Hydrogen and first row heteronuclear diatomics.

Molecule	Method	R_e [Bohr]	D_e [eV]	ω_e [cm^{-1}]
LiH	Expt.	3.015	2.501	1406
$^1\Sigma$	LDA	3.031	2.640	1377
	PBE	3.031	2.322	1375
	BLYP	3.018	2.525	1387
FH	Expt.	1.732	6.126	4138
$^1\Sigma$	LDA	1.761	7.039	4045
	PBE	1.758	6.167	4012
	BLYP	1.762	6.129	3972
CO	Expt.	2.132	11.243	2170
$^1\Sigma$	LDA	2.128	12.968	2179
	PBE	2.145	11.674	2127
	BLYP	2.145	11.372	2113
NO	Expt.	2.175	6.625	1904
$^2\Pi$	LDA	2.165	8.633	1944
	PBE	2.185	7.413	1883
	BLYP	2.193	7.150	1840
OH	Expt.	1.832	4.642	3738
$^2\Pi$	LDA	1.861	5.350	3628
	PBE	1.857	4.552	3605
	BLYP	1.860	4.563	3568
NH	Expt.	1.958	3.605	3282
$^3\Sigma$	LDA	1.992	4.141	3150
	PBE	1.983	3.844	3177
	BLYP	1.984	3.889	3140
CH	Expt.	2.116	3.629	2858
$^2\Pi$	LDA	2.154	3.984	2720
	PBE	2.147	3.544	2735
	BLYP	2.138	3.558	2732
CN	Expt.	2.214	7.849	2069
$^2\Sigma$	LDA	2.202	9.519	2135
	PBE	2.217	8.576	2087
	BLYP	2.218	8.308	2067
LiF	Expt.	2.955	6.022	910
$^1\Sigma$	LDA	2.930	6.785	942
	PBE	2.975	6.034	918
	BLYP	2.979	6.097	918
BeH	Expt.	2.537	2.373	2061
$^2\Sigma$	LDA	2.568	2.620	1981
	PBE	2.557	2.409	1994
	BLYP	2.541	2.474	2022
Mean abs.	LDA	0.023	0.906	73
deviation	PBE	0.019	0.263	68
	BLYP	0.017	0.175	79

Table 4.7 continued: C) Second row homonuclear diatomics.

Molecule	Method	R_e [Bohr]	D_e [eV]	ω_e [cm^{-1}]
Na_2	Expt.	5.818	0.745	159
$^1\Sigma$	LDA	5.668	0.879	161
	PBE	5.834	0.770	151
	BLYP	5.788	0.768	150
Al_2	Expt.	4.660	1.628	350
$^3\Sigma$	LDA	4.648	1.978	349
	PBE	4.702	1.674	340
	BLYP	4.761	1.309	312
Si_2	Expt.	4.244	3.258	511
$^3\Sigma$	LDA	4.283	4.040	492
	PBE	4.315	3.531	482
	BLYP	4.341	3.316	464
P_2	Expt.	3.578	5.080	781
$^1\Sigma$	LDA	3.570	6.224	796
	PBE	3.594	5.283	767
	BLYP	3.609	5.290	762
S_2	Expt.	3.570	4.414	726
$^3\Sigma$	LDA	3.577	5.876	719
	PBE	3.606	5.013	703
	BLYP	3.639	4.684	668
Cl_2	Expt.	3.757	2.514	560
$^1\Sigma$	LDA	3.738	3.626	567
	PBE	3.783	2.863	539
	BLYP	3.849	2.540	500
Mean abs.	LDA	0.039	0.831	8
deviation	PBE	0.035	0.249	18
	BLYP	0.070	0.151	38

which either cover larger collections of molecules or specialize in particular types of bonds. However, for the present purpose of illustrating the overall accuracy of xc-functionals the G2 set appears to be best suited.

Before discussing its content, a few remarks on the technical details behind Table 4.7 seem to be appropriate. The theoretical data in Table 4.7 have been obtained by solution of Eq. (3.98) with a basis set expansion for the KS orbitals. Large two-center basis sets [230] have been employed in order to make sure that the basis set limit is reached. For each diatomic molecule the ground state energy has been calculated for a number of internuclear distances. The resulting discrete values for the energy surface have finally been fitted to a Morse-type potential, from which the spectroscopic constants can be directly extracted. It is this last step which limits the accuracy of the data in Table 4.7 most severely. Their uncertainty is roughly given by 0.002 Bohr in the case of R_e, 0.002 eV for D_e and 10–20 cm^{-1} for ω_e, respectively.

Turning to the actual results in Table 4.7, one immediately notices that the molecular binding energies are substantially improved by use of the GGA, at least for the

Table 4.7 continued: D) Second row and mixed hydrogen/first row—second row heteronuclear diatomics.

Molecule	Method	R_e [Bohr]	D_e [eV]	ω_e [cm^{-1}]
HCl	Expt.	2.409	4.618	2991
$^1\Sigma$	LDA	2.440	5.228	2888
	PBE	2.434	4.621	2890
	BLYP	2.438	4.538	2846
ClF	Expt.	3.077	2.610	786
$^1\Sigma$	LDA	3.063	4.144	807
	PBE	3.115	3.149	754
	BLYP	3.156	2.934	715
ClO	Expt.	2.966	2.801	854
$^2\Pi$	LDA	2.927	4.564	927
	PBE	2.974	3.470	873
	BLYP	3.012	3.181	821
SiO	Expt.	2.853	8.306	1242
$^1\Sigma$	LDA	2.856	9.731	1240
	PBE	2.882	8.542	1204
	BLYP	2.888	8.498	1195
SO	Expt.	2.799	5.426	1149
$^3\Sigma$	LDA	2.807	7.284	1148
	PBE	2.835	6.148	1109
	BLYP	2.852	5.882	1072
CS	Expt.	2.901	7.468	1285
$^1\Sigma$	LDA	2.894	8.763	1288
	PBE	2.919	7.800	1259
	BLYP	2.923	7.483	1240
PN	Expt.	2.817	6.446	1337
$^1\Sigma$	LDA	2.805	7.784	1372
	PBE	2.828	6.800	1335
	BLYP	2.831	6.840	1321
NaCl	Expt.	4.461	4.245	366
$^1\Sigma$	LDA	4.413	4.516	378
	PBE	4.491	4.104	354
	BLYP	4.512	3.984	352
Mean abs.	LDA	0.020	1.262	31
deviation	PBE	0.024	0.375	34
	BLYP	0.041	0.263	56

vast majority of molecules. In fact, the binding energy is always reduced by inclusion of the gradient corrections. While this reduction sometimes overshoots the desired correction, it definitively represents progress. This statement is true for both GGAs included in Table 4.7, the first-principles PBE form, Eqs. (4.256), (4.280), and the semi-empirical BLYP-GGA, Eqs. (4.286), (4.289). Moreover, the improvement of D_e is roughly the same for all subsets of molecules grouped together in Table 4.7, which reflects the universal character of the GGAs (for completeness the corresponding average absolute deviations for the complete set of molecules are listed in Table 4.8).

Table 4.8 Bond lengths R_e, dissociation energies D_e (including zero-point energies [226]) and harmonic frequencies ω_e of first and second row diatomic molecules: Mean absolute deviation of selfconsistent LDA [134], PBE-GGA [207] and BLYP-GGA [219, 221] results from experimental data (taken from [227–229]—see also http://cccbdb.nist.gov/) for the set of molecules listed in Table 4.7. The corresponding mean absolute errors obtained by *a posteriori* application of the PKZB [231] and TPSS [232] MGGAs on the basis of selfconsistent PBE-GGA solutions are also given (see Sect. 4.8).

Method	R_e [Bohr]	D_e [eV]	ω_e [cm^{-1}]
LDA	0.027	1.027	48
PBE	0.026	0.329	43
BLYP	0.037	0.234	59
PKZB	0.051	0.172	52
TPSS	0.025	0.168	31

On the other hand, the GGA bond lengths are not really superior to the corresponding numbers obtained with the LDA (at least for this, somewhat arbitrarily chosen, set of molecules). The same is true for the form of the potential well, for which the vibrational frequencies provide some measure. Looking at the individual numbers, one finds that the gradient corrections stretch the bonds almost consistently. While this improves the agreement with experiment for a number of molecules, the GGA also increases the bond lengths in many cases in which the LDA already overestimates the experimental values. The only exception from this general tendency are the hydrogen bonds, which are significantly shortened by the inclusion of gradient terms. As this bond length reduction is accompanied by a reduction of the dissociation energy, GGAs account for hydrogen bonds much better than the LDA (for additional examples see [233–236]).

An analogous comparison for a number of metals is given in Table 4.9. Again the GGA results compare quite favorably with the LDA values. As a general trend the GGA leads to larger lattice constants, which goes hand in hand with a reduction of the bulk moduli. GGAs most notably predict the correct ferromagnetic bcc ground state of metallic iron [200], in contrast to the LDA for which the paramagnetic fcc phase has a lower energy.

The brief overview of GGA results given so far makes it quite clear why GGAs represent the *de facto* standard approximation in present-day applications of DFT. One should, however, keep a realistic perspective of the predictive power of GGAs. The point to be noted in this context is the fact that physical and chemical processes are controlled by energy differences, rather than absolute energies (as those given in Tables 4.4–4.6). The best-known examples are atomic ionization potentials (IP) and electron affinities (EA), molecular atomization energies and the cohesive energies of solids. In these energy differences the major contribution to the total energies

Table 4.9 Lattice constants a and bulk moduli B of metals: Selfconsistent LDA and PW91-GGA results versus experimental data. All calculations rely on the full potential linearized-augmented-plane-wave scheme (a: [237], b: [238], c: [239]).

Solid	a [Bohr]			B [GPa]		
	LDA	GGA	Expt.	LDA	GGA	Expt.
Li^a (bcc)	6.36	6.49	6.57	15	12	13
Al^a (fcc)	7.54	7.65	7.65	84	74	72
V^c (bcc)	5.56	5.69	5.71	187	183	159
Fe^a (bcc)		5.36	5.41		215	172
Cu^b (fcc)	6.65	6.84	6.81	192	151	138
Nb^a (bcc)	6.17	6.27	6.24	199	177	170
Pd^c (fcc)	7.30	7.49	7.35	222	175	195

due to the core electrons usually cancels out (as the core electrons do not really participate in the process of interest). The percentage error in energy differences is therefore primarily determined by the description of the valence states, for which xc-effects play a much more prominent role than for the core electrons. One finds that the improvement of GGAs over the LDA is often less impressive for energy differences. As a first illustration of this statement one can compare the accuracy of the total energies in Table 4.6 with that of the dissociation energies in Table 4.7. The most prominent example, however, are atomic IPs, for which some numbers are listed in Table 4.10. This table (relying again on the x-only limit) demonstrates that GGAs can yield worse results for IPs than the LDA.

Table 4.10 also lists the eigenvalues of the highest occupied KS states obtained with the three functionals. These eigenvalues are generally referred to as HOMO (highest occupied molecular orbital) eigenvalues, in spite of the fact that they are of atomic nature. We will follow this terminology. According to Sect. 3.6.1, the exact KS HOMO eigenvalues are identical with the corresponding IPs. Table 4.10 demonstrates that the LDA, but also the GGA underestimate the exact results by roughly a factor of 2, with a marginal improvement by the GGA. As the HOMO eigenvalue is particularly sensitive to the form of the KS potential for large r, this drastic underestimation indicates a rather poor behavior of the LDA and GGA xc-potential in the asymptotic region. We will come back to this point below.

Closely related to the quality of HOMO eigenvalues is that of the band gap of semiconductors. As discussed in Sect. 3.6.3, the band gap reduces to the difference between the highest occupied and the lowest unoccupied KS eigenvalue, if the xc-functional applied does not show a derivative discontinuity. Neither the LDA nor the GGA do that, so that the value for the gap directly reflects the KS eigenvalues. Numbers for some prototype semiconductors are given in Table 4.11. Similar to the situation for the HOMO eigenvalue of finite systems, the LDA and GGA band gaps underestimate the experimental values drastically. This fact has obvious implications for all observables related to virtual or real excitation processes, such as the optical conductivity.

Table 4.10 Exchange-only ionization potentials of atoms: Selfconsistent LDA, PW91-GGA and B88-GGA data versus exact results (all energies in mHartree). All atoms have been treated nonrelativistically.

Atom	IP	$IP-IP^{exact}$			$-\varepsilon_{HOMO}$		
	Exact	LDA	PW91	B88	LDA	PW91	B88
He	862	−51	4	7	517	555	554
Be	295	−15	5	1	170	182	181
Mg	242	−4	12	6	142	150	149
Ca	188	1	12	7	111	117	116
Sr	171	3	13	8	103	107	106
Ba	152	4	12	8	91	95	94
Cu	231	47	54	52	159	164	163
Ag	215	36	41	38	142	145	143
Au	216	38	43	39	145	148	146
Li	196	−11	3	1	100	110	109
Na	181	−2	10	6	97	103	102
K	147	2	10	7	80	85	84
Rb	137	4	11	8	76	80	79
Cs	123	4	11	8	69	72	72
Zn	276	34	44	39	185	191	190
Cd	252	30	37	32	168	172	170
Hg	248	33	39	34	169	172	170
Yb	174	8	18	12	107	112	111

Table 4.11 Band gap of semiconductors: LDA and PBE-GGA results obtained with self-consistent LAPW [240] and plane-wave pseudopotential [103, 241] calculations versus experimental data (taken from [242]) (all energies in eV).

E_{xc}	Method	C	Si	GaAs
LDA	PW-PP	4.16	0.49	0.32
LDA	LAPW	4.11	0.47	0.30
PBE	LAPW	4.15	0.57	0.53
Expt.		5.45	1.17	1.52

As in the case of the LDA, the quality of GGA results is to some extent based on error cancellation between exchange and correlation. This statement is illustrated in Table 4.12, in which a subset of the molecules in Table 4.7 is considered in the x-only limit. In this limit the spectroscopic constants can be evaluated exactly via the OPM (Chap. 6). It turns out that the x-only bond lengths and vibrational frequencies obtained with the GGA are consistently further away from the exact numbers than their LDA counterparts, with the (only) exception of hydrogen bonds. Even the error in the binding energies reduces by only 30–50%, when going from the LDA to the GGA. The mean absolute deviations for the set of molecules chosen amount to 2.76 eV in the case of the LDA and to 1.87 eV for the GGA, which has to be compared to 1.03 eV (LDA) and 0.33 eV (GGA) for the average deviations including

Table 4.12 Exchange-only spectroscopic constants of diatomic first row molecules: Selfconsistent LDA, PW91-GGA and TPSS-MGGA results versus exact x-only data [243].

Molecule	Method	R_e [Bohr]	D_e [eV]	ω_e [cm^{-1}]
H_2	Exact	1.386	3.637	4585
	LDA	1.476	3.552	4059
	PW91	1.424	3.686	4254
	TPSS	1.409	3.758	4359
Li_2	Exact	5.264	0.169	337
	LDA	5.300	0.292	305
	PW91	5.267	0.343	310
	TPSS	5.298	0.262	310
B_2	Exact	3.069	0.608	971
	LDA	3.085	3.450	987
	PW91	3.103	2.516	945
	TPSS	3.107	2.157	939
C_2	Exact	2.332	0.282	1932
	LDA	2.384	5.743	1839
	PW91	2.398	4.328	1746
	TPSS	2.399	3.779	1733
N_2	Exact	2.012	4.970	2737
	LDA	2.089	9.191	2338
	PW91	2.099	7.995	2262
	TPSS	2.096	7.511	2278
O_2	Exact	2.184	1.440	1980
	LDA	2.304	6.384	1554
	PW91	2.352	4.584	1422
	TPSS	2.355	3.910	1416
F_2	Exact	2.497	−1.607	1283
	LDA	2.653	2.752	1015
	PW91	2.755	1.430	905
	TPSS	2.763	1.138	875
LiH	Exact	3.037	1.483	1427
	LDA	3.114	1.477	1290
	PW91	3.086	1.639	1312
	TPSS	3.057	1.610	1356
FH	Exact	1.693	4.202	4499
	LDA	1.782	5.902	3846
	PW91	1.777	5.118	3857
	TPSS	1.777	4.817	3873
CO	Exact	2.080	7.521	2445
	LDA	2.149	11.348	2088
	PW91	2.163	9.740	2055
	TPSS	2.162	9.180	2063
Mean abs. deviation	LDA	0.078	2.757	291
	PW91	0.087	1.867	313
	TPSS	0.087	1.542	299

correlation (see Table 4.8). Error cancellation between exchange and correlation improves the net accuracy by a factor of 3–4. A similar behavior is observed for the mean deviations of bond lengths and vibrational frequencies.

In order to understand these observations one has to analyze the accuracy of xc-functionals on a local level, for which the corresponding potential provides an unambiguous measure. The exact exchange potential is known for a number of systems ranging from atoms to solids (see Chap. 6). Even the exact correlation potential is available for a few atoms [83]. Unfortunately, the comparison of LDA and GGA potentials with these reference results turns out to be disappointing. While GGAs come closer to the exact potential than the LDA in the case of exchange, no improvement at all is observed for v_c [83]. The local error of the GGA exchange potential,[6]

$$v_x^{GGA}(\boldsymbol{r}) = v_x^{LDA}(n)\left\{f_x(\xi) - \frac{3}{2}\frac{df_x}{d\xi}(\xi)\eta - \frac{3}{2}\frac{d^2 f_x}{d\xi^2}(\xi)\frac{\boldsymbol{\nabla}n\cdot\boldsymbol{\nabla}\xi}{4(3\pi^2 n)^{2/3}n}\right\}, \quad (4.290)$$

is much larger than deviation of the GGA exchange energy from the corresponding exact E_x [223, 244]. Examples for these statements are given in Figs. 4.14–4.17. Figure 4.14 shows the exchange potential of the calcium atom, for which the closed-

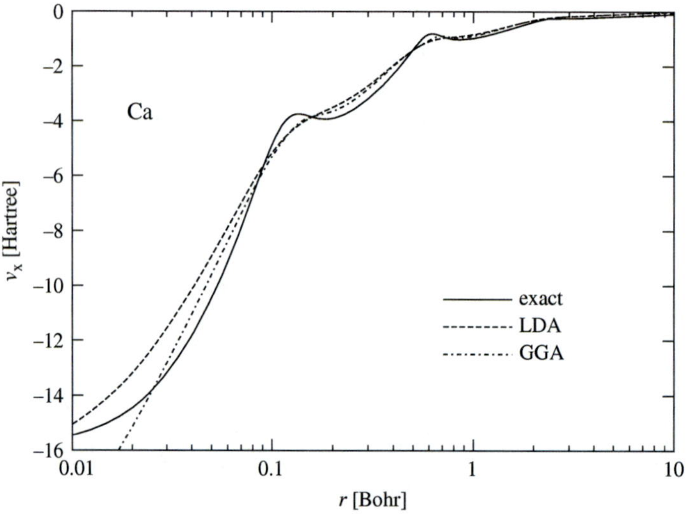

Fig. 4.14 Exchange potential for atomic calcium. The LDA and GGA potentials have been generated by insertion of the exact exchange-only density corresponding to the exact v_x. For the GGA the PW91-form has been used.

subshell structure leads to a spherically symmetric density and potential. The exact exchange potential has been generated by the OPM, as explained in Chap. 6. The

[6] The result (4.290) is obtained by direct functional differentiation of the general form (4.281)).

resulting density, i.e. the exact x-only density, has then been inserted into the LDA and the PW91-GGA in order to evaluate the corresponding potentials.

One first of all observes a clear oscillatory structure in the exact v_x, which reflects the individual atomic shells. For small r the exact potential approaches a finite value, for large r one rediscovers the $-1/r$-decay, Eq. (4.20). The LDA potential, on the other hand, averages over the shell oscillations and vanishes exponentially for large r. This behavior is more clearly seen in Fig. 4.15, which shows the outermost shell and the asymptotic region on an enlarged scale. The GGA clearly improves the

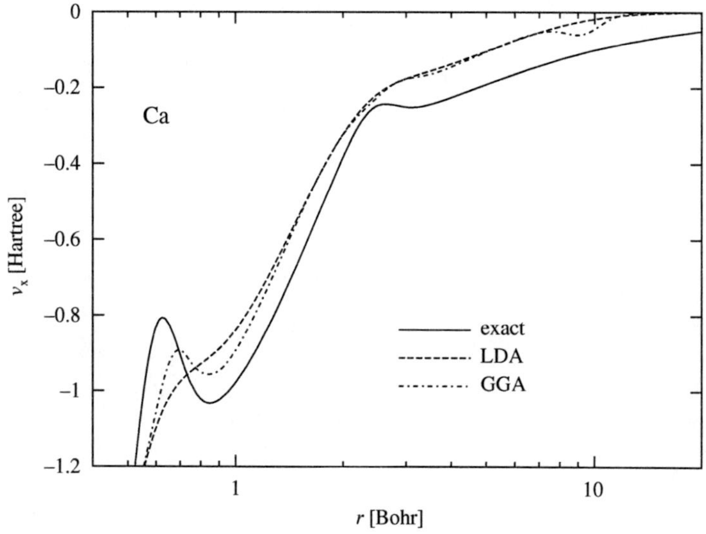

Fig. 4.15 Valence and asymptotic region of Fig. 4.14 on enlarged scale.

situation in comparison with the LDA. While the GGA does not reproduce the exact shell structure very accurately, there is at least some resemblance of this feature. The GGA potential goes like $1/r$ for $r \to 0$, in accordance with the behavior of the second order gradient term, Eq. (4.206). Finally, for large r the GGA potential decays as fast as its LDA counterpart. In fact, one can show quite generally that GGAs cannot simultaneously reproduce the asymptotic behavior of the exact e_x and the exact v_x for finite systems [220]. This result simply reflects their semi-local functional form. As a consequence, GGAs cannot describe atomic negative ions.

There is one further, related consequence of the semi-locality of the LDA and GGA worth to be noted. As soon as one considers open-shell systems the difference between the highest occupied eigenvalues of the majority spin (spin-up) and the minority spin (spin-down) channels comes into play. This difference determines the relative stability of the various spin states and therefore the magnetization. As the highest occupied eigenvalues strongly depend on v_x, the balance between the spin-up and the spin-down exchange potential has a major impact on the local magnetic moments. In Fig. 4.16 the difference between the spin-up and the spin-down exchange potentials of Cr is shown. The deviations of the LDA and the GGA from

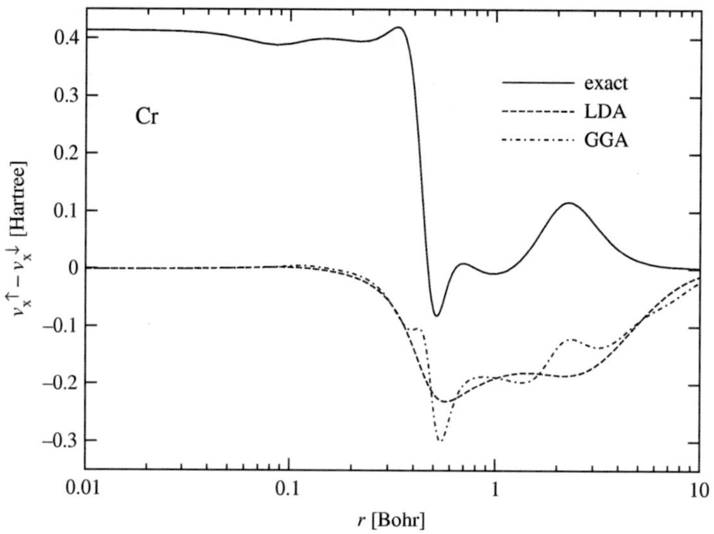

Fig. 4.16 Exchange potential of Cr: Spin balance obtained with LDA and GGA versus exact result. The LDA and GGA potentials have been generated by insertion of the exact exchange-only density corresponding to the exact v_x. For the GGA the PBE-form has been used.

the exact result are obvious [223]. For large r the exact $v_{x,\sigma}$ is dominated by the $-1/r$-tail in case of both spins. In the valence regime the exact difference $v_{x,\uparrow} - v_{x,\downarrow}$ is thus less affected by the actual positions of the spin-aligned 4s- and 3d-electrons than $v_{x,\uparrow}^{\mathrm{LDA/GGA}} - v_{x,\downarrow}^{\mathrm{LDA/GGA}}$, which directly reflect the structures of the valence spin-densities: the exact $v_{x,\uparrow} - v_{x,\downarrow}$ is close to zero and repulsive, while its LDA and GGA counterparts necessarily have to be attractive. The difference between the spin-up and spin-down densities essentially vanishes in the L- and K-shell regime, so that $v_{x,\uparrow} - v_{x,\downarrow}$ approaches zero in the LDA and GGA. The exact result, on the other hand, shows an almost constant shift between $v_{x,\uparrow}$ and $v_{x,\downarrow}$. The nonlocality of the exact E_x propagates the differences between the spin-up and spin-down densities in the valence regime into the inner shell region.

An idea of the behavior of the GGA exchange potential in the bonding region of molecules is given in Fig. 4.17, in which results for the hydrogen dimer are plotted. The individual potentials have been generated in the same way as for the atomic case. It is obvious that the GGA-potential is not particularly close to the exact v_x in the bonding region. As H_2 is a spin-saturated two-electron system, the exact v_x only provides the required self-interaction correction (as in the case of helium, Eq. (4.22)). This Coulomb integral of the molecular density has a broad minimum in the center between the two protons. The LDA exchange potential (4.111), on

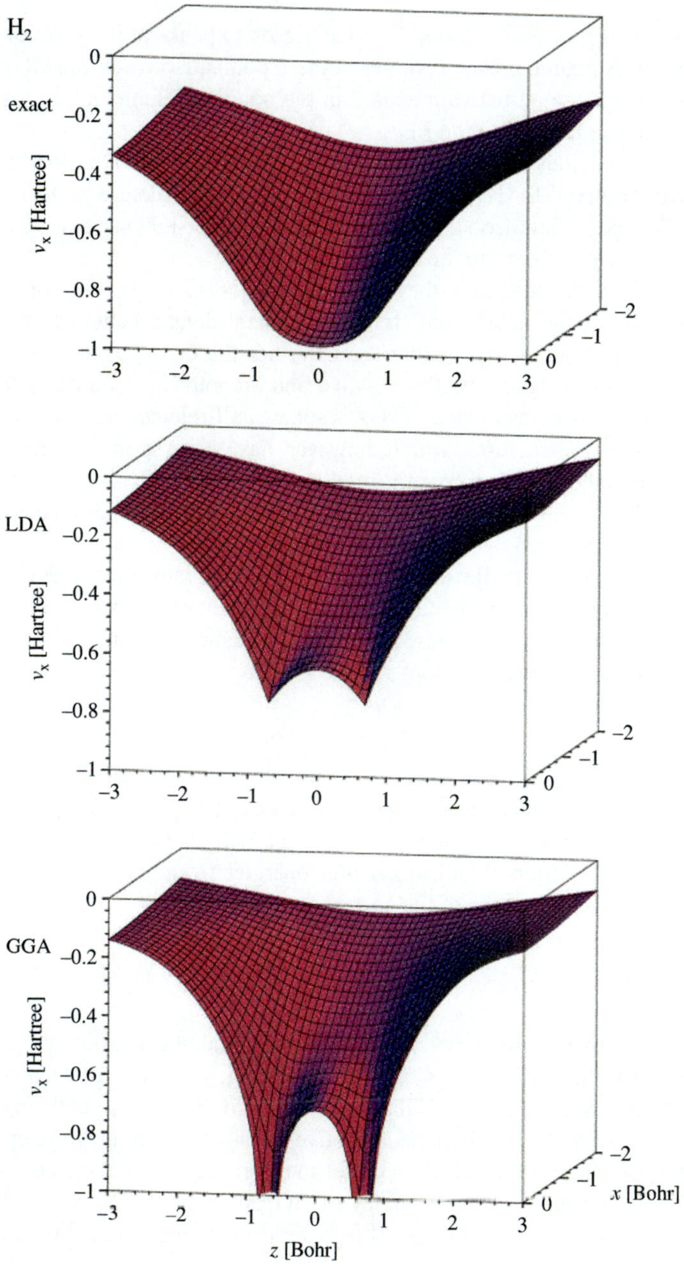

Fig. 4.17 Exchange potential of H_2. LDA and PBE-GGA versus exact v_x. The LDA and GGA potentials have been generated by insertion of the exact exchange-only density. The two protons are located at $z = \pm 0.7$ Bohr, $x = y = 0$.

the other hand, is proportional to $n^{1/3}$, which leads to peaks at the positions of the nuclei. The GGA potential even diverges at these points. However, the GGA generates a somewhat more attractive potential in the bonding region and, in that sense, improves results compared to the LDA.

Figure 4.18 provides a corresponding comparison for N_2. In addition to the features already observed for H_2, one can now see the shell structure in the molecular v_x.[7] The GGA potential also shows an indication of the shell structure, while this feature is completely absent in the LDA.

Finally, Fig. 4.19 illustrates the statement on the GGA correlation potential for the example of the neon atom (for the technical details behind Fig. 4.19 see Sect. 6.6.3). It is obvious that neither the LDA nor the GGA agree with the exact v_c [83]. The LDA is again strictly negative and smooth, in contrast to the exact potential, which can be positive and shows some shell-related structure. The GGA also generates some structures, which, however, have nothing in common with the exact data. The large local errors in v_x and v_c explain why GGAs improve energy differences often much less than total energies, as these differences usually imply the removal of a localized part of the electronic density.

How can one understand the discrepancy between the rather accurate GGA energies and the partial failure of the GGA potentials? The answer is seen most clearly on the basis of the virial relation (5.31), which E_x satisfies. For finite systems partial integration allows the reformulation

$$E_x = \int d^3r \, [3n(\boldsymbol{r}) + \boldsymbol{r} \cdot \boldsymbol{\nabla} n(\boldsymbol{r})] v_x(\boldsymbol{r}) \, . \tag{4.291}$$

The virial relation originates from a scaling analysis of $E_x[n]$, i.e. from dimensional arguments. It is thus not only satisfied by the exact $E_x[n]$, but also by approximations like GGAs. The deviation of GGA exchange energies from the exact E_x can, using (4.291), be expressed in terms of the corresponding potentials,

$$E_x - E_x^{\mathrm{GGA}} = \int d^3r \, [3n(\boldsymbol{r}) + \boldsymbol{r} \cdot \boldsymbol{\nabla} n(\boldsymbol{r})][v_x(\boldsymbol{r}) - v_x^{\mathrm{GGA}}(\boldsymbol{r})] \approx 0 \, .$$

While the local error $v_x(\boldsymbol{r}) - v_x^{\mathrm{GGA}}(\boldsymbol{r})$ is quite substantial, the integrated error is very small [244]. An illustration of this fact is given in Fig. 4.20, in which the zinc atom is shown. The local error in v_x oscillates according to the atomic shell structure. In the virial relation (4.291) positive and negative deviations then integrate up to zero. As the GGA potential is only slightly closer to the exact v_x than the LDA potential, one concludes that the extreme accuracy of GGAs for atomic E_x results from the cancellation of local errors and does not fully reflect the inherent quality of this type of functional.

[7] In the case of the exact exchange the KLI approximation (6.63) is used for the evaluation of v_x. The result is, however, very close to the exact v_x (see Chap. 6).

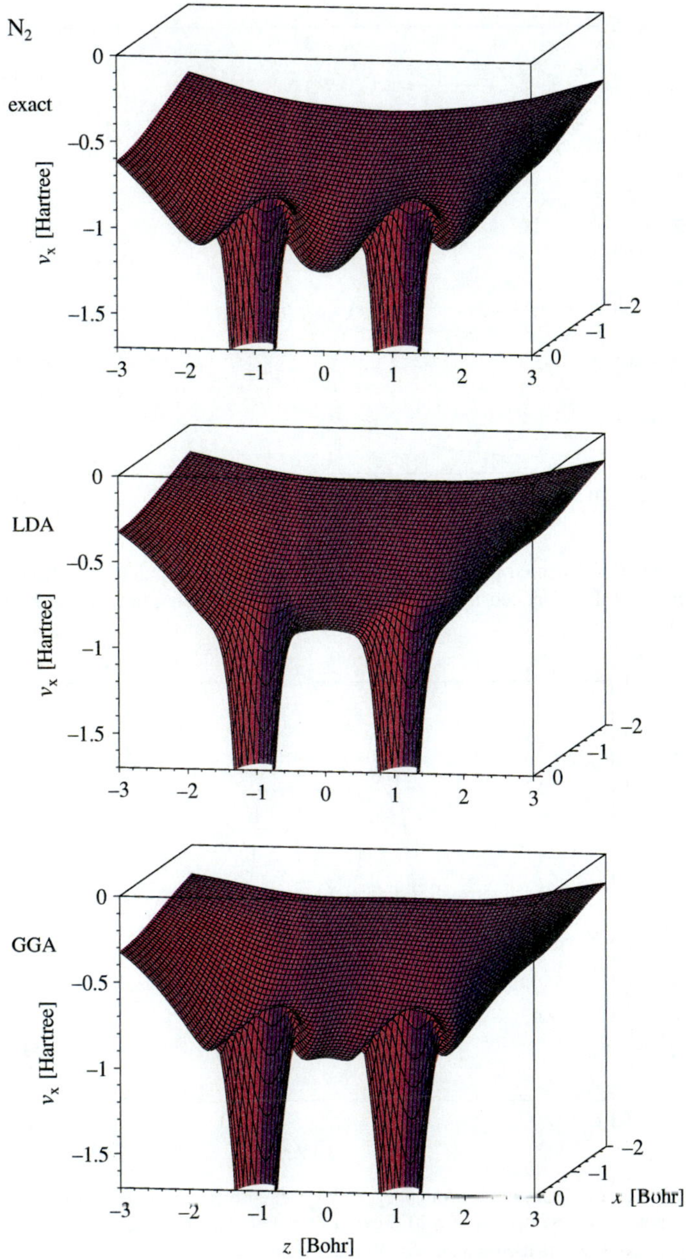

Fig. 4.18 Exchange potential of N_2. LDA and PBE-GGA versus exact v_x. The LDA and GGA potentials have been generated by insertion of the density corresponding to the exact v_x. The two nitrogen nuclei are located at $z = \pm 1.05\,\mathrm{Bohr}$, $x = y = 0$.

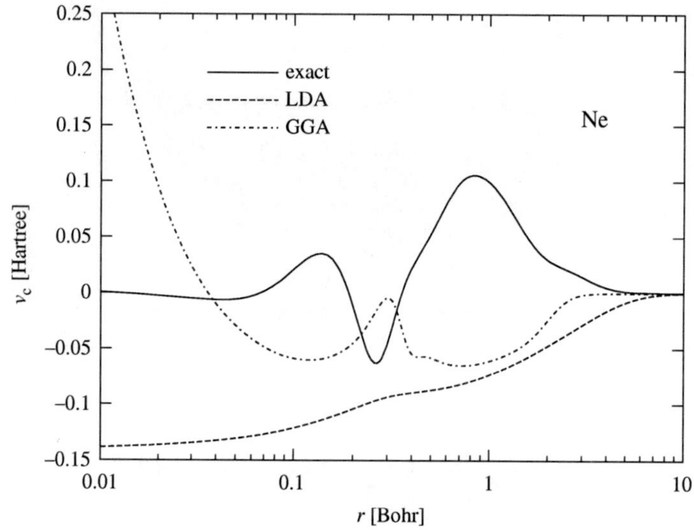

Fig. 4.19 Correlation potential for atomic neon. The LDA and GGA potentials have been generated by insertion of the exact density. For the GGA the PW91-form has been used (the exact v_c is taken from [83]).

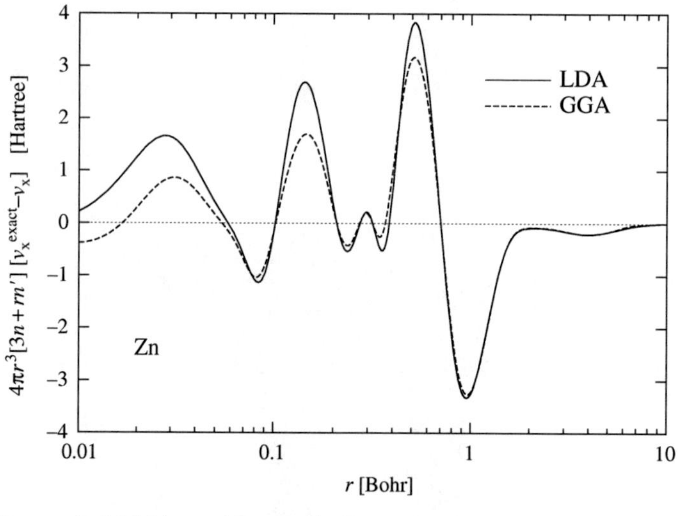

Fig. 4.20 Integrand of virial integral for spherically symmetric ground state density and potential of zinc. n' denotes the first derivative with respect to r. For the GGA the PW91-form has been utilized. Very similar results are obtained for all other standard GGAs.

4.6 Weighted Density Approximation (WDA)

There exists yet another class of nonlocal xc-functionals which is based on the adiabatic connection (4.80), the *weighted density approximation* (WDA) [245, 246, 155] and its precursor, the *average density approximation* (ADA) [247]. Starting from (4.80), the WDA is based on an approximate expression for the coupling constant integrated pair-correlation function

$$\tilde{g}(\boldsymbol{r},\boldsymbol{r}') = \int_0^1 d\lambda\, g_\lambda(\boldsymbol{r},\boldsymbol{r}') \,. \tag{4.292}$$

The function $\tilde{g}$ is then subject to the sum rule (4.231), which can be rewritten as

$$\int d^3r'\, n(\boldsymbol{r}')\,[\tilde{g}(\boldsymbol{r},\boldsymbol{r}') - 1] = -1 \,. \tag{4.293}$$

This procedure represents a systematic extension of the LDA if g is approximated by the pair-correlation function of the HEG.

This concept is most easily demonstrated in the case of the exchange. The pair-correlation function of the noninteracting HEG is obtained from Eq. (4.232), with the ϕ_i given by plane-wave states. It already featured as the lowest order contribution to the GE of the exchange hole in Eqs. (4.238), (4.239). The function $J(2k_F|\boldsymbol{r}-\boldsymbol{r}'|)$ can be written in a more suitable form, which leads to

$$g_{\mathrm{x}}^{\mathrm{HEG}}(\boldsymbol{r}-\boldsymbol{r}',n_0) - 1 = -\frac{9}{2}\left[\frac{\sin(k_F|\boldsymbol{r}-\boldsymbol{r}'|)}{(k_F|\boldsymbol{r}-\boldsymbol{r}'|)^2}\right]^2 . \tag{4.294}$$

The coupling constant integration in (4.292) is trivial in the case of exchange, $\tilde{g}_{\mathrm{x}} = g_{\mathrm{x}}$. The remaining task is to find a suitable local density replacement for the gas density n_0, which is reminiscent of the analogous problem for the full linear response functional (4.157). Within the WDA this task is performed via the sum rule (4.293). Choosing the local replacement to be a function of $\boldsymbol{r}$ only,

$$n_0 \longrightarrow \overline{n}(\boldsymbol{r}) \,,$$

one can determine $\overline{n}$ by the requirement

$$\int d^3r'\, n(\boldsymbol{r}')\left[g_{\mathrm{x}}^{\mathrm{HEG}}(\boldsymbol{r}-\boldsymbol{r}',\overline{n}(\boldsymbol{r})) - 1\right] = -1 \,, \tag{4.295}$$

which has to be satisfied for all $\boldsymbol{r}$. The resulting local $\overline{n}(\boldsymbol{r})$ is then inserted into the functional (4.80),

$$E_{\mathrm{x}}^{\mathrm{WDA}}[n] = \frac{1}{2}\int d^3r \int d^3r'\, n(\boldsymbol{r})\, w(\boldsymbol{r},\boldsymbol{r}')\, n(\boldsymbol{r}')\left[g_{\mathrm{x}}^{\mathrm{HEG}}(\boldsymbol{r}-\boldsymbol{r}',\overline{n}(\boldsymbol{r})) - 1\right] . \tag{4.296}$$

An older variant of this approach is the ADA in which the sum rule (4.293) is utilized in the form

$$\overline{n}(\boldsymbol{r}) \int d^3r' \left[g_{\mathrm{x}}^{\mathrm{HEG}}(\boldsymbol{r}-\boldsymbol{r}',\overline{n}(\boldsymbol{r})) - 1 \right] = -1 \,. \tag{4.297}$$

Of course, the screening density $\overline{n}(\boldsymbol{r})$ depends on the system under consideration. The WDA and ADA thus require the solution of (4.293) in each step of a KS calculation.

The results obtained with the functional (4.296) for atoms are rather disappointing. In addition, little is known about the correlation contribution to the pair-correlation function. For this reason often empirical forms are used for $\tilde{g}$.

An additional problem with the WDA is its inherent violation of the symmetry of $\tilde{g}(\boldsymbol{r},\boldsymbol{r}')$ with respect to interchange of $\boldsymbol{r}$ and $\boldsymbol{r}'$. As a result the corresponding exchange potential does not satisfy the exact relation (4.20),

$$v_{\mathrm{x}}^{\mathrm{WDA}}(\boldsymbol{r}) \quad \xrightarrow[|\boldsymbol{r}|\to\infty]{} \quad -\frac{e^2}{2|\boldsymbol{r}|} \,. \tag{4.298}$$

in spite of the fully nonlocal character of the WDA functional. Due to the facts that the WDA is much more difficult to handle numerically than gradient-based functionals and that its performance is not very satisfactory, the WDA has rarely been used in applications so far (for some results see [248–253]). Extensions of the WDA have, however, been successfully applied to the problem of freezing [254–256], in particular in the theory of nonuniform *classical* liquids [257, 258].

4.7 Self-Interaction Corrections (SIC)

There is one important deficiency which is shared by all functionals based on the HEG: none of these approximations can describe a 1-particle system such as the hydrogen atom properly. In order to demonstrate this fact, let us look at the spin-density dependent version of the LDA, $E_{\mathrm{xc}}^{\mathrm{LDA}}[n_\uparrow,n_\downarrow]$. In contrast to the exact exchange (4.16), the LSDA exchange does not reduce to a pure Coulomb self-interaction integral, if only a single particle with spin up and density $n_\uparrow = |\phi_{1,\uparrow}|^2$ is present,

$$E_{\mathrm{x}}^{\mathrm{LDA}}[|\phi_{1,\uparrow}|^2,0] \neq E_{\mathrm{x}}^{\mathrm{exact}}[|\phi_{1,\uparrow}|^2,0] = -\frac{e^2}{2}\int d^3r d^3r' \frac{|\phi_{1,\uparrow}(\boldsymbol{r})|^2|\phi_{1,\uparrow}(\boldsymbol{r}')|^2}{|\boldsymbol{r}-\boldsymbol{r}'|} \,. \tag{4.299}$$

In addition, the LDA for correlation does not vanish in this limit,

$$E_{\mathrm{c}}^{\mathrm{LDA}}[|\phi_{1,\uparrow}|^2,0] \neq E_{\mathrm{c}}^{\mathrm{exact}}[|\phi_{1,\uparrow}|^2,0] = 0 \,. \tag{4.300}$$

In a many-particle system this self-interaction error is present for each of the particles in the system. The same observation is made for GGA-type functionals.

Given the fact that the LDA and the GGA are exact in the opposite limit of an infinite HEG, it is tempting to try to eliminate the self-interaction in an *ad hoc* fashion [259, 260, 143, 261]. In the *self-interaction corrected LDA* (SIC-LDA) [259, 143] the erroneous terms are subtracted *a posteriori* for the individual KS states of both spins,

$$\begin{aligned} E_{\mathrm{xc}}^{\mathrm{SIC-LDA}} &= E_{\mathrm{xc}}^{\mathrm{LDA}}[n_\uparrow, n_\downarrow] - \sum_{k,\sigma} \Theta_{k,\sigma} \Big\{ E_{\mathrm{H}}\big[|\phi_{k,\sigma}|^2\big] + E_{\mathrm{xc}}^{\mathrm{LDA}}\big[|\phi_{k,\sigma}|^2, 0\big] \Big\} \\ &= E_{\mathrm{xc}}^{\mathrm{LDA}}[n_\uparrow, n_\downarrow] - \sum_{k,\sigma} \Theta_{k,\sigma} E_{\mathrm{xc}}^{\mathrm{LDA}}\big[|\phi_{k,\sigma}|^2, 0\big] \\ &\quad - \frac{e^2}{2} \sum_{k,\sigma} \Theta_{k,\sigma} \int d^3r\, d^3r' \, \frac{|\phi_{k,\sigma}(\boldsymbol{r})|^2 |\phi_{k,\sigma}(\boldsymbol{r}')|^2}{|\boldsymbol{r}-\boldsymbol{r}'|} \, . \end{aligned} \tag{4.301}$$

By construction, $E_{\mathrm{xc}}^{\mathrm{SIC-LDA}}$ reproduces the exact xc-energy of a single particle. The same scheme can be applied to any other density functional, in particular to GGAs.

A closer inspection of (4.301), however, reveals that the SIC-LDA is not a density functional, but rather an orbital-dependent functional, just as the exact exchange. It thus belongs to the class of implicit functionals, for which, as a matter of principle, the OPM (for details see Chap. 6) has to be utilized for the calculation of the corresponding xc-potential [262]. On the other hand, the standard procedure for the application of the SIC-LDA relies on the use of orbital-dependent KS potentials, derived by minimization of (4.301) with respect to the $\phi_{k,\sigma}$: a separate KS equation is solved for each individual KS state. This procedure leads to non-orthogonal KS orbitals, so that an *a posteriori* orthogonalization is required [259]. In practice, the orthogonality of the KS orbitals turns out to be only weakly violated by the orbital-dependent SIC-LDA potential, so that the explicit orthogonalization is often neglected. In any case, the problem of non-orthogonality is automatically avoided by use of the OPM, which produces the multiplicative potential corresponding to a given orbital-dependent expression.

On the other hand, the OPM does not resolve the unitarity problem which is inherent in the SIC functional [263–265]: if one performs a unitary transformation among the KS orbitals, the individual orbital densities will change, even if the transformation only couples degenerate KS states. Consequently, the value of $E_{\mathrm{xc}}^{\mathrm{SIC-LDA}}$ also changes. An additional prescription which defines a suitable representation of the KS orbitals (which usually implies a localization) is necessary for practical calculations [266]. While the results for atoms are not very sensitive to this problem of unitarity, it becomes more important in extended systems (for some numbers see Sect. 6.6.2). For this reason the SIC-LDA has rarely been applied to molecules [267–269]. It has mainly been used to handle the localized d-and f-electrons in transition metal elements (see, for instance, [270–272]).

One way to overcome the unitarity problem is an approximate elimination of the orbital densities in favor of the spin-densities. The first suggestion of this type goes back to Stoll, Pavlidou and Preuss [260], who modify the correlation component of (4.301) as

$$E_{\mathrm{c}}^{\mathrm{SIC-SPP}}[n_\uparrow,n_\downarrow] = E_{\mathrm{c}}^{\mathrm{LDA}}[n_\uparrow,n_\downarrow] - \sum_\sigma E_{\mathrm{c}}^{\mathrm{LDA}}[n_\sigma,0]\,. \tag{4.302}$$

It is obvious that (4.302) vanishes in the case of a single particle. However, the SIC-functional (4.302) is no longer exact for a truly homogeneous system. This problem is resolved by the most straightforward replacement of the orbital density [261],

$$|\phi_{i,\sigma}(\boldsymbol{r})|^2 \longrightarrow \frac{n_\sigma(\boldsymbol{r})}{N_\sigma}\,. \tag{4.303}$$

By construction, this replacement is exact for a single particle. For correlation one obtains in this way

$$E_{\mathrm{c}}^{\mathrm{SIC-VW}}[n_\uparrow,n_\downarrow] = E_{\mathrm{c}}^{\mathrm{LDA}}[n_\uparrow,n_\downarrow] - \sum_\sigma N_\sigma E_{\mathrm{c}}^{\mathrm{LDA}}\left[\frac{n_\sigma}{N_\sigma},0\right]\,. \tag{4.304}$$

Since

$$\lim_{n_0\to 0} \frac{e_{\mathrm{c}}^{\mathrm{HEG}}(n_0)}{n_0} = 0$$

(see Eq. (4.107)), the SIC-correction in (4.304) vanishes for the HEG. On the other hand, application of (4.303) to the exchange part of (4.301) yields

$$E_{\mathrm{x}}^{\mathrm{SIC-FA}}[n_\uparrow,n_\downarrow] = \frac{1}{2}\sum_\sigma \left\{ E_{\mathrm{x}}^{\mathrm{LDA}}[2n_\sigma] - N_\sigma E_{\mathrm{x}}^{\mathrm{LDA}}\left[2\frac{n_\sigma}{N_\sigma}\right] - \frac{2}{N_\sigma} E_{\mathrm{H}}[n_\sigma] \right\}. \tag{4.305}$$

The main ingredient of (4.305) is the *Fermi-Amaldi term* $\sum_\sigma E_{\mathrm{H}}[n_\sigma]/N_\sigma$ [273]. For the remaining terms once more (4.19) has been utilized.

In practice, the forms (4.302), (4.304) and (4.305) have not been applied very often, in spite of their simple form.

4.8 Meta-GGA (MGGA)

Given the limitations of standard GGA-type functionals discussed in Sect. 4.5.5, an extension of the GGA's rather restricted set of density variables (i.e. n and ∇n) by other semi-local functionals of the density offers itself quite naturally.[8] The most obvious variable for such an extension is the Laplacian of the density, $\nabla^2 n$, which is the key quantity in the second order gradient correction for E_{x}, Eq. (4.201), as well as in the corresponding expansion for E_{c}. The simplest extension of the GGA (4.281) for exchange is thus obtained by inclusion of $\nabla^2 n$ in an extended kernel $F_{\mathrm{x}}(\xi,\eta)$. Functionals of this type have been suggested quite early [197, 199]. However, the potential corresponding to expressions as (4.201), Eq. (4.208), involves the fourth gradient of the density, which diverges severely at the positions of nuclei and is very difficult to evaluate numerically. In fact, already the evaluation of $\nabla^2 n$ is

[8] For a completely different approach based on the properties of surfaces see [274].

numerically much more demanding than that of the first gradient. For this reason, GGAs depending directly on $\nabla^2 n$ have not been pursued intensively (for one such functional see [275]).

In addition to $\nabla^2 n$, the kinetic energy density[9]

$$t_s(\boldsymbol{r}) := \frac{\hbar^2}{2m}\sum_k \Theta_k \left|\nabla\phi_k(\boldsymbol{r})\right|^2 , \tag{4.306}$$

has been used for the representation of $E_{xc}[n]$ [276–280, 231, 232]. In fact, t_s can serve as a substitute for $\nabla^2 n$ [231]. In order to verify this statement, one has to consider the gradient expansion of t_s. The lowest order terms of the gradient expansion of the total kinetic energy T_s have already been given in Eqs. (4.117) and (4.177),

$$T_s^{GE}[n] = \frac{\hbar^2}{m}\int d^3r \left\{\frac{3(3\pi^2)^{2/3}}{10} n^{5/3} + \frac{1}{72}\frac{(\nabla n)^2}{n} + \ldots\right\}. \tag{4.307}$$

However, the total T_s defines the local $t_s(\boldsymbol{r})$ only up to terms which integrate to zero. One can easily demonstrate by use of Gauss' theorem that one consistent second order gradient term,

$$\frac{\hbar^2}{m}\int d^3r \nabla^2 n ,$$

vanishes for finite systems. As a result, the contribution of $\frac{\hbar^2}{m}\nabla^2 n$ to t_s is no longer visible in the total T_s.

A complete determination of the GE for $t_s(\boldsymbol{r})$ via the linear response approach of Sect. 4.4 is impossible. Rather one has to resort to a direct evaluation of t_s in real space, for which the commutator expansion of Kirzhnits [175] (or related techniques) are well suited. In this way one obtains as GE for the kinetic energy density (4.306) to second order [193–195]

$$t_s = \frac{\hbar^2}{m}\left\{\frac{3(3\pi^2)^{2/3}}{10} n^{5/3} + \frac{1}{72}\frac{(\nabla n)^2}{n} + \frac{1}{6}\nabla^2 n\right\}. \tag{4.308}$$

Combination of Eqs. (4.179), (4.182) and (4.308) then yields [231]

$$\eta = \frac{m}{\hbar^2}\frac{3t_s}{2(3\pi^2)^{2/3} n^{5/3}} - \frac{9}{20} - \frac{\xi}{12}. \tag{4.309}$$

One can therefore replace η by the right-hand side of this relation (with t_s evaluated by Eq. (4.306)), the result being correct to second order in the gradients. Abbreviating the right-hand side of (4.309) by $\overline{\eta}$, one can then consistently rewrite the fourth order GE for E_x, Eq. (4.202), as

---

[9] While any energy functional defines the corresponding energy density only up to partial integration, it has become standard to define the kinetic energy density by Eq. (4.306).

$$E_{\rm x}^{\rm GE}[n] = \int d^3r\, e_{\rm x}^{\rm HEG}(n) \left\{1+\frac{10}{81}\xi+\frac{146}{2025}\left[\overline{\eta}^2-\frac{5}{2}\overline{\eta}\xi+b\xi^2\right]+\ldots\right\}. \quad (4.310)$$

In the next step one constructs a GGA-type functional which reproduces the fourth order terms in (4.310). Such an extended GGA, termed *Meta-GGA* (MGGA), necessarily involves an extended kernel $F_{\rm x}(\xi,\overline{\eta})$,

$$E_{\rm x}^{\rm MGGA}[n] = \int d^3r\, e_{\rm x}^{\rm HEG}(n)\, F_{\rm x}(\xi,\overline{\eta}). \quad (4.311)$$

A simple kernel in the spirit of the PBE-GGA, which reproduces the GE (4.310) in the limit of slowly varying density, is [231]

$$F_{\rm x}^{\rm PKZB}(\xi,\overline{\eta}) = 1+\kappa-\frac{\kappa}{1+x/\kappa} \quad (4.312)$$

$$x = \frac{10}{81}\xi+\frac{146}{2025}\overline{\eta}^2-\frac{73}{405}\overline{\eta}\xi+\left[b+\frac{1}{\kappa}\left(\frac{10}{81}\right)^2\right]\xi^2. \quad (4.313)$$

The parameter κ drops out of the two leading gradient terms, so that the limit (4.310) is obtained for arbitrary values of κ. This parameter is then used to ensure that the MGGA (4.311) obeys the Lieb-Oxford bound (4.254) locally, which leads to a value of $\kappa = 0.804$. Lacking any rigorous information on the gradient coefficient b, Perdew et al. [231] fit this coefficient to the dissociation energies of a set of 20 molecules (which gives $b = 0.113$). As usual in the case of the exchange, the extension to spin-polarized systems is provided by Eq. (4.19).

Ideally, one would like to apply the same strategy for the extension of the correlation part of the GGA. However, no information on the fourth order gradient corrections for $E_{\rm c}[n_\uparrow,n_\downarrow]$ is available. In their MGGA Perdew et al. [231] therefore utilize the kinetic energy density only for the elimination of the inherent self-interaction in the GGA, while retaining the properties of the GGA for slowly varying densities. In order to avoid the unitarity problem of the conventional SIC-LDA approach (4.301), the MGGA starts from the SPP-SIC (4.302). This expression has the disadvantage that it does not preserve the limit of a uniform gas, i.e. the SPP-SIC is not restricted to the subtraction of the correlation energy of the individual particles. A quantity which allows to distinguish between a single particle and the HEG is the kinetic energy density. Splitting the total $t_{\rm s}$ into the two spin-channels,

$$t_{\rm s}(\boldsymbol{r}) = \sum_\sigma t_{{\rm s},\sigma}(\boldsymbol{r}), \quad (4.314)$$

the kinetic energy density of a single particle in the state ϕ_σ can be easily expressed in terms of its density [174], if one chooses ϕ_σ to be real,

$$t_{{\rm s},\sigma}(\boldsymbol{r}) \xrightarrow[n_\sigma=\phi_\sigma^2]{} t_{{\rm W},\sigma}(\boldsymbol{r}) = \frac{\hbar^2}{8m}\frac{[\boldsymbol{\nabla} n_\sigma(\boldsymbol{r})]^2}{n_\sigma(\boldsymbol{r})}. \quad (4.315)$$

While the so-called *von Weizsäcker functional* $t_{W,\sigma}$ coincides with the exact $t_{s,\sigma}$ for a single particle, it vanishes for uniform density. The ratio $t_{W,\sigma}/t_{s,\sigma}$ is therefore well suited to suppress contributions in the limit of slowly varying density (compare the discussion in Sect. 6.5.4). In fact, if $(t_{W,\sigma}/t_{s,\sigma})^2$ is used as switching factor, this factor vanishes like a fourth order gradient term for vanishing $\boldsymbol{\nabla} n_\sigma$, leaving the second order gradient correction unaffected. The final MGGA for correlation is thus given by

$$E_c^{\rm PKZB}[n_\uparrow,n_\downarrow] = \int d^3r \left\{ e_c^{\rm GGA}(n_\uparrow,n_\downarrow,\boldsymbol{\nabla} n_\uparrow,\boldsymbol{\nabla} n_\downarrow)\left[1+C\left(\frac{\sum_\sigma t_{W,\sigma}}{\sum_\sigma t_{s,\sigma}}\right)^2\right] \right.$$
$$\left. -(1+C)\sum_\sigma \left(\frac{t_{W,\sigma}}{t_{s,\sigma}}\right)^2 e_c^{\rm GGA}(n_\sigma,0,\boldsymbol{\nabla} n_\sigma,\mathbf{0})\right\}. \quad (4.316)$$

This expression vanishes for a single particle and reproduces the GGA without any SIC for slowly varying densities. For $e_c^{\rm GGA}$ the MGGA utilizes the combination of the LDA and the PBE gradient correction (4.280). The coefficient C has been chosen so that $E_c^{\rm PKZB}$ gives essentially the same correlation contribution to the surface energies of jellium (i.e. a semi-infinite electron gas experiencing the attractive potential of a positive background density which is constant in one half of space and vanishes in the other) as the PBE-GGA, which leads to $C = 0.53$.

The MGGA has later been revised [232], in response to the inaccurate equilibrium bond lengths which the PKZB form predicts [281]. In the case of the exchange the revision is primarily based on the non-uniqueness of the representation (4.312), (4.313). The construction of an optimized kernel F_x starts with an alternative representation of η. One first defines a suitable ratio of kinetic energies densities,

$$\alpha = \frac{t_s - t_W}{t_s^{\rm HEG}} = \frac{t_s}{t_s^{\rm HEG}} - \frac{5}{3}\xi, \quad (4.317)$$

where t_W denotes the von Weizsäcker functional (4.315) for a spin-saturated many-electron system,

$$t_W = \sum_\sigma t_{W,\sigma} = \frac{\hbar^2}{m}\frac{(\boldsymbol{\nabla} n)^2}{8n}. \quad (4.318)$$

If expanded to second order in the gradients, α reduces to

$$\alpha = 1 + \frac{20}{9}\eta - \frac{40}{27}\xi, \quad (4.319)$$

which is easily verified by insertion of Eq. (4.308) into (4.317). Using this limit, it is straightforward to show that the quantity

$$\tilde{\eta} = \frac{9}{20}\frac{\alpha - 1}{[1+\beta\alpha(\alpha-1)]^{1/2}} + \frac{2}{3}\xi \quad (4.320)$$

approaches η for slowly varying density, irrespective of the value of β: this component of the denominator does not enter the second order GE, it merely introduces some freedom to satisfy some constraint yet to be chosen (see below).

Now the form of $F_x(\xi,\tilde{\eta})$ needs to be specified. In order to remove some of the ambiguity in this expression one has to incorporate additional conditions on F_x. To this aim Tao et al. [232] consider a spin-saturated 2-electron system. For this system the exact t_s reduces to the von Weizsäcker expression (4.318), so that α vanishes. This implies a fixed relation between $\tilde{\eta}$ and ξ,

$$\tilde{\eta} = -\frac{9}{20}+\frac{2}{3}\xi\,.$$

If this limit is inserted into F_x, a purely ξ-dependent kernel is obtained, i.e. an expression like the kernel f_x of standard GGAs,

$$F_x\left(\xi,\tilde{\eta}=-\frac{9}{20}+\frac{2}{3}\xi\right) \xrightarrow[\text{2-electron}]{} f_x(\xi)\,. \tag{4.321}$$

In general, the corresponding potential (4.290) diverges at the positions of the nucleus, due to its η-dependence (in complete analogy to the GE2 potential, Eq. (4.206)). This divergence can be avoided by requiring

$$\left.\frac{df_x(\xi)}{d\xi}\right|_{\xi=\xi_0} = 0\,, \tag{4.322}$$

with ξ_0 denoting the value of ξ at the origin of the 2-electron atom. In addition, Tao et al. assumed the parameter b in (4.310) to be zero. A kernel which satisfies these two additional conditions is

$$F_x^{\mathrm{TPSS}}(\xi,\tilde{\eta}) = 1+\kappa-\frac{\kappa}{1+\tilde{x}/\kappa}\,, \tag{4.323}$$

with

$$\tilde{x} = \left\{\left[\frac{10}{81}+\gamma\frac{z^2}{(1+z^2)^2}\right]\xi+\frac{146}{2025}\tilde{\eta}^2-\frac{73}{405}\tilde{\eta}\left[\frac{1}{2}\left(\frac{3}{5}z\right)^2+\frac{1}{2}\xi^2\right]^{1/2} + \frac{1}{\kappa}\left(\frac{10}{81}\right)^2\xi^2+2\sqrt{\delta}\,\frac{10}{81}\left(\frac{3}{5}z\right)^2+\delta\mu\xi^3\right\}\frac{1}{(1+\sqrt{\delta}\,\xi)^2} \tag{4.324}$$

$$z = \frac{t_W}{t_s}\,. \tag{4.325}$$

The value of κ is the same as for the PKZB-MGGA. The parameters $\gamma = 1.59096$ and $\delta = 1.537$ are chosen to enforce the condition (4.322) and to yield the correct exchange energy of the hydrogen atom (for H one has $E_x = -E_H = -0.3125$ Hartree). Moreover, β is chosen to have the smallest value which ensures that F_x^{TPSS} is a monotonically increasing function of ξ for any fixed value of α (which gives

$\beta = 0.4$). Finally, $\mu = 0.21951$ restores the behavior of the PKZB-MGGA for large ξ.

The revision of the correlation part of the MGGA aims at an improved description of spin-polarization. The TPSS-MGGA for correlation is given by [232]

$$E_{\rm c}^{\rm TPSS}[n_\uparrow, n_\downarrow] = \int d^3r\, e_{\rm c}^{\rm revPKZB} \left[1 + \frac{D}{n}\left(\frac{t_{\rm W}}{t_{\rm s}}\right)^3 e_{\rm c}^{\rm revPKZB}\right] \tag{4.326}$$

$$\begin{aligned} e_{\rm c}^{\rm revPKZB} &= e_{\rm c}^{\rm PBE}(n_\uparrow, n_\downarrow, \nabla n_\uparrow, \nabla n_\downarrow)\left[1 + C(\zeta,\rho)\left(\frac{t_{\rm W}}{t_{\rm s}}\right)^2\right] \\ &\quad - [1 + C(\zeta,\rho)]\left(\frac{t_{\rm W}}{t_{\rm s}}\right)^2 \\ &\quad \times \sum_\sigma \max\left[e_{\rm c}^{\rm PBE}(n_\sigma, 0, \nabla n_\sigma, \mathbf{0}),\right. \\ &\qquad\qquad \left.\frac{n_\sigma}{n} e_{\rm c}^{\rm PBE}(n_\uparrow, n_\downarrow, \nabla n_\uparrow, \nabla n_\downarrow)\right] \end{aligned} \tag{4.327}$$

$$C(\zeta,0) = 0.53 + 0.87\zeta^2 + 0.50\zeta^4 + 2.26\zeta^6 \tag{4.328}$$

$$C(\zeta,\rho) = \frac{C(\zeta,0)}{\{1 + \rho^2[(1+\zeta)^{-4/3} + (1-\zeta)^{-4/3}]/2\}^4}\,. \tag{4.329}$$

Here ζ denotes the spin-polarization (4.119) and

$$\rho = \frac{|\nabla\zeta|}{2(3\pi^2 n)^{1/3}}\,. \tag{4.330}$$

One first notes that the core of the SIC term involves the maximum of the local values of $e_{\rm c}^{\rm PBE}(n_\sigma, 0, \nabla n_\sigma, \mathbf{0})$ and $e_{\rm c}^{\rm PBE}(n_\uparrow, n_\downarrow, \nabla n_\uparrow, \nabla n_\downarrow)$, unlike the PKZB-MGGA. This modification implements the rigorous constraint that the value of $E_{\rm c}[n_\uparrow, n_\downarrow]$ has to be negative for all possible spin-densities—in the PKZB-MGGA the positive SIC term can become slightly larger than the negative main term in regions in which only a single spin-orbital is non-vanishing, as in the asymptotic region of the Li atom.

It remains to discuss the treatment of spin in the SIC components. One first observes that the energy density (4.327) vanishes for a system containing only a single particle,

$$n_\uparrow = \phi^2\,, \quad n_\downarrow = 0 \qquad \Longrightarrow \qquad t_{\rm s} = t_{\rm W}\,, \quad \zeta = \pm 1\,,$$

irrespective of the value of $C(\zeta,\rho)$. On the other hand, for a spin-saturated many-particle system one has

$$e_c^{\mathrm{revPKZB}} = e_c^{\mathrm{PBE}}\left(\frac{n}{2},\frac{n}{2},\frac{\nabla n}{2},\frac{\nabla n}{2}\right)\left[1+C(0,0)\left(\frac{t_W}{t_s}\right)^2\right]$$
$$-2[1+C(0,0)]\left(\frac{t_W}{t_s}\right)^2 e_c^{\mathrm{PBE}}\left(\frac{n}{2},0,\frac{\nabla n}{2},\mathbf{0}\right).$$

If applied to the completely delocalized states of the jellium problem, the SIC should give no contribution to the surface correlation energy. This can be achieved for the standard range of valence electron densities ($2 \leq r_s \leq 6$—see Table 4.2) by setting $C(0,0) = 0.53$ (as in the PKZB-form) and by introducing the additional factor $1+(D/n)(t_W/t_s)^3 e_c^{\mathrm{revPKZB}}$ in the TPSS-MGGA (4.326) with $D = -2.8\hbar^2/(me^4)$.

The form of $C(\zeta,0)$ is motivated by the limit of very strong interaction. In this limit the Coulomb repulsion completely dominates over quantum effects, so that the correlation energy becomes independent of the spin-polarization of the system. This behavior is implemented via the form (4.328), which ensures that the TPSS correlation energy obtained from non-uniform spin-densities with uniform ζ,

$$\nabla\zeta = \mathbf{0} \qquad \Longrightarrow \qquad (1-\zeta)\nabla n_\uparrow = (1+\zeta)\nabla n_\downarrow ,$$

is independent of ζ in the range $0 \leq |\zeta| \leq 0.7$.

Finally, the SIC-term should not affect the transition region between the valence and the core regime of a monovalent atom like Li. While in the valence regime one has $\zeta = \pm 1$ and thus $\rho = 0$, ζ and ρ are close to zero in the core regime. In the region in which core and valence densities overlap ζ switches between these two values, so that ρ is large. The ρ-dependence of $C(\zeta,\rho)$ lets this quantity become small in the overlap zone, in this way suppressing the SIC ingredient of the first term on the right-hand side of Eq. (4.327) (compare [282], for an extensive discussion of the MGGA see [283]).

The functionals (4.311), (4.316) and (4.323), (4.326) avoid the singularities resulting from the potential (4.208). The price to be paid, however, is the dependence of the MGGA on the KS orbitals. As in the case of the SIC-LDA the corresponding potential has, as a matter of principle, to be evaluated via the optimized potential method discussed in Chap. 6. This procedure is computationally much more demanding than a standard GGA calculation. However, the additional gradient terms in the MGGA are expected to give only a limited correction to the GGA potential. In practice, the MGGA is therefore often applied *a posteriori*, using the solutions obtained by a self-consistent GGA calculation (Post-GGA approach). Alternatively, selfconsistent calculations with the MGGA are based on the HF-type equations which emerge from a minimization of the total MGGA energy with respect to the individual orbitals, rather than the density—this procedure is sometimes termed generalized KS (GKS) approach (see also Sect. 6.5.4).

Some reference data for atomic exchange energies obtained with the MGGA are provided in Table 4.13, atomic MGGA correlation energies are included in Table 4.5. Table 4.8 lists the mean absolute errors in the spectroscopic constants of the molecules in Table 4.7 resulting from the two MGGAs via the Post-GGA treatment. One finds no improvement over standard GGAs for bond lengths and only

Table 4.13 Exchange energies of spin-saturated, closed-subshell atoms: PKZB-MGGA, TPSS-MGGA, B3LYP-hybrid and PBE0-hybrid energies obtained by insertion of exact exchange-only densities into the functionals (4.311), (4.323), (6.127) and (6.134) in comparison with exact values (all energies are in Hartree). Also given is the corresponding percentage error Δ. The hybrid data are discussed in Sect. 6.5.4.

Atom	Exact	PKZB		TPSS		B3LYP		PBE0	
	$-E_x$	$-E_x$	Δ	$-E_x$	Δ	$-E_x$	Δ	$-E_x$	Δ
He	1.026	1.020	−0.54	1.030	0.44	1.014	−1.13	1.017	−0.89
Be	2.666	2.648	−0.66	2.677	0.43	2.632	−1.28	2.643	−0.84
Ne	12.105	11.951	−1.27	12.180	0.62	12.043	−0.51	12.076	−0.24
Mg	15.988	15.745	−1.52	16.050	0.38	15.887	−0.63	15.933	−0.35
Ar	30.175	29.644	−1.76	30.216	0.14	29.974	−0.66	30.041	−0.44
Ca	35.199	34.587	−1.74	35.249	0.14	34.986	−0.61	35.061	−0.39
Zn	69.619	68.651	−1.39	69.798	0.26	69.480	−0.20	69.553	−0.09
Kr	93.833	92.295	−1.64	93.719	−0.12	93.444	−0.41	93.527	−0.33
Sr	101.926	100.250	−1.64	101.761	−0.16	101.502	−0.42	101.591	−0.33
Cd	148.880	146.558	−1.56	148.493	−0.26	148.329	−0.37	148.415	−0.31
Xe	179.064	176.257	−1.57	178.449	−0.34	178.369	−0.39	178.450	−0.34

a very moderate advance for vibrational frequencies, but much better atomization energies. A more detailed inspection of the individual data shows that the TPSS functional also provides an improved description of hydrogen bonds [282]. In addition, the TPSS-MGGA often significantly reduces the error in the lattice constants of solids in comparison with the PBE-GGA [232].

4.9 LDA+U

It is a long-standing problem for effective single-particle methods to deal with the simultaneous presence of highly delocalized (itinerant) band states and of strongly localized, atomic-like states in many compounds containing rare-earth or late transition metal elements. While the d- and, in particular, the f-states essentially retain their atomic character in solids, the valence s- and p-states tend to form bands. The preceding discussion of the properties of the LDA indicates that the LDA has difficulties with the description of highly localized states, primarily due to the insufficient treatment of the SIC. As a result the LDA predicts itinerant d-states and a metallic ground state for many transition metal oxides, for which a sizable energy gap between occupied and unoccupied subbands is observed in experiment (an example is given in Fig. 6.3). Moreover, in spite of its overall improvement over the LDA, the GGA does not really provide a satisfactory SIC either (see also Fig. 6.3).

The fact that the band picture is not adequate for rare-earth and transition metal compounds suggests a decomposition of the complete Hilbert space into two subsystems, following the Anderson model [284]: (i) the localized d- or f-states for which a more explicit, orbital-dependent treatment of all Coulomb effects is required, and (ii) the s- and p-states which are well described by the LDA (or GGA). The technical implementation of this concept in DFT is the LDA+U method [285–288]. For its derivation the only assumption required is that the d- or f-states are localized within well-separated atomic spheres, so that the bulk states are well represented by a superposition of the corresponding atomic states only.

The starting point of the discussion is the spin-resolved form of the sum of the Hartree and the exact exchange energy, Eq. (4.16),

$$\begin{aligned} E_{\mathrm{H}}+E_{\mathrm{x}} &= \frac{1}{2}\sum_{\sigma,\sigma'}\sum_{\alpha,\alpha'}\Theta_{\alpha\sigma}\Theta_{\alpha'\sigma'}\int d^3r d^3r' \,|\phi_{\alpha\sigma}(\boldsymbol{r})|^2 w(\boldsymbol{r},\boldsymbol{r}')\,|\phi_{\alpha'\sigma}(\boldsymbol{r}')|^2 \\ &\quad -\frac{1}{2}\sum_{\sigma}\sum_{\alpha,\alpha'}\Theta_{\alpha\sigma}\Theta_{\alpha'\sigma}\int d^3r d^3r' \\ &\quad \times \phi^*_{\alpha\sigma}(\boldsymbol{r})\phi_{\alpha'\sigma}(\boldsymbol{r})w(\boldsymbol{r},\boldsymbol{r}')\phi^*_{\alpha'\sigma}(\boldsymbol{r}')\phi_{\alpha\sigma}(\boldsymbol{r}'). \end{aligned} \tag{4.331}$$

The states $\phi_{\alpha\sigma}$ can e.g. be Bloch states, for which α corresponds to the combination of the crystal momentum $\boldsymbol{k}$ and a band index (in this case one of the spatial integrations has to be restricted to one unit cell). The intra-shell contribution to $E_{\mathrm{H}}+E_{\mathrm{x}}$, i.e. the complete Hartree plus exchange energy associated with a single atomic shell nl of an atom at site a, is obtained by projection of the actual KS states $\phi_{\alpha\sigma}$ onto the standard set of atomic orbitals φ_{anlm} with principal quantum number n, angular momentum quantum number l and corresponding z-projection m at site a,

$$c_{\alpha\sigma;anlm} = \int d^3r\, \varphi^*_{anlm}(\boldsymbol{r})\phi_{\alpha\sigma}(\boldsymbol{r})\,. \tag{4.332}$$

For given nl and a this overlap vanishes for all $\phi_{\alpha\sigma}$, except the few localized states which emerge from the φ_{anlm} in the solid (or molecule). The bulk states formed by the φ_{anlm}, on the other hand, can be expanded in terms of the orthonormal set of atomic orbitals: within the atomic sphere in which $\phi_{\alpha\sigma}$ is localized at site a one simply has

$$\int d^3r\, \varphi^*_{anlm}(\boldsymbol{r})\varphi_{an'l'm'}(\boldsymbol{r}) = \delta_{nn'}\,\delta_{ll'}\,\delta_{mm'} \tag{4.333}$$

$$\Longrightarrow \qquad \phi_{\alpha\sigma}(\boldsymbol{r}) = \sum_m c_{\alpha\sigma;anlm}\,\varphi_{anlm}(\boldsymbol{r})\,. \tag{4.334}$$

The energy associated with the shell nl at site a is thus obtained by insertion of (4.334) into Eq. (4.331),

$$\begin{aligned} [E_{\mathrm{H}}+E_{\mathrm{x}}]_{anl} &= \frac{1}{2}\sum_{\sigma,\sigma'}\sum_{\boldsymbol{m}}\Lambda^{\sigma;anl}_{mm''}\Lambda^{\sigma';anl}_{m'm'''}(mm'|w|m''m''')_{anl} \\ &\quad -\frac{1}{2}\sum_{\sigma}\sum_{\boldsymbol{m}}\Lambda^{\sigma;anl}_{mm'''}\Lambda^{\sigma;anl}_{m'm''}(mm'|w|m''m''')_{anl}\,, \end{aligned} \tag{4.335}$$

where the matrix elements of the basis functions

$$(mm'|w|m''m''')_{anl} = \int d^3r\,d^3r'\,\varphi^*_{anlm}(\boldsymbol{r})\varphi_{anlm''}(\boldsymbol{r})\,w(\boldsymbol{r},\boldsymbol{r}')\,\varphi^*_{anlm'}(\boldsymbol{r}')\varphi_{anlm'''}(\boldsymbol{r}')\,, \tag{4.336}$$

the occupation matrix

$$\Lambda^{\sigma;anl}_{mm''} = \sum_{\alpha} \Theta_{\alpha\sigma}\, c^*_{\alpha\sigma;anlm}\, c_{\alpha\sigma;anlm''}\,, \tag{4.337}$$

and the abbreviation $\boldsymbol{m} \equiv m, m', m'', m'''$ have been introduced. At this point the reference to the shell nl as well as to the site a is dropped, direct and exchange matrix elements are combined in an alternative fashion, and the intra-shell direct plus exchange energy is given its standard name,

$$\begin{aligned} E^U[\underline{\underline{\Lambda}}^\sigma] &:= [E_{\rm H} + E_{\rm x}]_{anl} \\ &= \frac{1}{2}\sum_{\sigma}\sum_{\boldsymbol{m}} \Lambda^{\sigma}_{mm''}\Lambda^{-\sigma}_{m'm'''}(mm'|w|m''m''') \\ &\quad +\frac{1}{2}\sum_{\sigma}\sum_{\boldsymbol{m}} \Lambda^{\sigma}_{mm''}\Lambda^{\sigma}_{m'm'''}\left[(mm'|w|m''m''') - (mm'|w|m'''m'')\right]. \end{aligned} \tag{4.338}$$

The derivation shows that this expression can be directly extended to the case of more than one shell nl and/or more than one site a.

In the next step the Slater integrals (4.336) have to be evaluated. Insertion of the atomic orbitals

$$\varphi_{anlm}(\boldsymbol{r}) = \frac{P_{nl}(r)}{r}\, Y_{lm}(\Omega) \tag{4.339}$$

(the center of the coordinate system is chosen to be at the location of site a) into (4.336) and expansion of $w(\boldsymbol{r},\boldsymbol{r}')$ in spherical harmonics allows a straightforward calculation of these matrix elements (for all details of the expansion and the angular integrations involved see [289]),

$$(mm'|w|m''m''') = \sum_{L=0}^{2l} a_L(m,m',m'',m''')\,F_L \tag{4.340}$$

$$a_L(m,m',m'',m''') = (L0\,l0|l0)^2 \sum_{M=-L}^{L} (LM\,lm''|lm)\,(LM\,lm'|lm''') \tag{4.341}$$

$$F_L = \int_0^\infty dr \int_0^\infty dr'\, P_{nl}(r)^2 w_L(r,r') P_{nl}(r')^2\,, \tag{4.342}$$

where $(LM\,lm|l'm')$ denotes the Clebsch-Gordan coefficients (in the definition of Rose [289]). If w was the pure Coulomb interaction, one would have

$$w_L(r,r') = e^2\,\frac{r_<^L}{r_>^{L+1}} \qquad \text{with} \quad r_< = \min\{r,r'\},\;\; r_> = \max\{r,r'\}\,.$$

In reality, screening of the Coulomb interaction is important, so that usually an effective, screened interaction is used to determine the radial Slater integrals F_L.

Due to the exact cancellation of the self-interaction energy in the expression (4.338), E^U accounts for the intra-shell xc-energy much better than the LDA. In order to correct the LDA one would therefore like to combine E^U with the LDA functional. However, when adding E^U to $E^{\rm LDA}$, one has to make sure that double counting of the intra-shell Hartree and xc-energy is avoided. The form of this double counting correction could be determined by different strategies, depending on the implementation of the combined functional. As long as no shape approximation is used, the Hartree energy is handled exactly in $E^{\rm LDA}$, so that no modification of this component is necessary. On the other hand, often a spherical approximation (atomic sphere approximation—ASA) is used for the total KS potential experienced by all localized states. The average Hartree energy per particle of the atomic shell nl utilized in the ASA,

$$\frac{1}{(2l+1)^2}\sum_{m,m'}(mm'|w|mm') = F_0$$

(compare the discussion below) differs significantly from the actual Hartree energy resulting for the individual states φ_{anlm}. For instance, the self-interaction energy of a single d-state with $m=0$ is obtained as

$$(00|w|00) = F_0 + \frac{4}{49}(F_2+F_4)\ .$$

Thus, if the ASA is applied, not only the xc-energy, but also the Hartree term in $E^{\rm LDA}$ needs to be corrected for open-shell effects. In addition, one wants to include screening effects, as emphasized before (concerning screening, compare, however, the discussion of the tadpole contributions to the KS perturbation expansion in Sect. 6.4.1). In the LDA+U approach the double counting (dc) correction is therefore chosen so that E^U is cancelled by the dc correction, if the shell (of bands) emerging from the atomic shell nl is completely filled, since in the case of filled shells $E_{\rm H}$ is handled correctly by $E^{\rm LDA}$ even in the ASA.

In order to evaluate E^U for a completely filled shell one starts by considering the occupation matrix $\Lambda^\sigma_{mm''}$, Eq. (4.337),

$$\Lambda^\sigma_{mm''} = \sum_\alpha \Theta_{\alpha\sigma}\int d^3r\int d^3r'\,\varphi^*_{anlm''}(\boldsymbol{r})\phi_{\alpha\sigma}(\boldsymbol{r})\phi^*_{\alpha\sigma}(\boldsymbol{r}')\varphi_{anlm}(\boldsymbol{r}')\ . \tag{4.343}$$

Since the states $\phi_{\alpha\sigma}$ are eigenstates of the KS Hamiltonian, they form a complete orthonormal basis. The completeness relation in the Hilbert space spanned by the occupied states may therefore be written as (for each spin-channel separately)

$$\sum_\alpha \Theta_{\alpha\sigma}\,\phi_{\alpha\sigma}(\boldsymbol{r})\phi^*_{\alpha\sigma}(\boldsymbol{r}') = \delta^{(3)}(\boldsymbol{r}-\boldsymbol{r}') \quad \text{(in the space of occupied states)}\ . \tag{4.344}$$

If the shell is completely filled, this completeness relation can be used in all matrix elements of the atomic states φ_{anlm}, as for instance[10] for (4.343). This allows the evaluation of the occupation matrix by use of the orthonormality relation (4.333),

$$\Lambda^{\sigma}_{mm''} = \delta_{m,m''} . \tag{4.345}$$

Insertion of this result into E^U gives

$$E^U_{\text{closed}} = \sum_{\sigma} \sum_{m,m'} (mm'|w|mm') - \frac{1}{2} \sum_{\sigma} \sum_{m,m'} (mm'|w|m'm) . \tag{4.346}$$

Use of the matrix elements (4.340) as well as of the properties of the Clebsch-Gordan coefficients then leads to

$$E^U_{\text{closed}} = 2(2l+1)^2 F_0 - (2l+1) \sum_{L=0}^{2l} (L0\,l0|l0)^2 F_L . \tag{4.347}$$

One now identifies the self-interaction correction contribution to the exchange energy (the monopole term) to rewrite E^U_{closed} as

$$E^U_{\text{closed}} = \frac{1}{2} N^{anl} (N^{anl} - 1) U - \frac{1}{2} \sum_{\sigma} N^{anl}_{\sigma} (N^{anl}_{\sigma} - 1) J , \tag{4.348}$$

where $N^{anl}_{\sigma} = 2l+1$ is the number of spin-σ electrons in the shell nl, the quantity $N^{anl} = N^{anl}_{\uparrow} + N^{anl}_{\downarrow}$ is the total number of electrons in the shell, $U \equiv F_0$ is the screened Coulomb repulsion energy per electron and J denotes the Stoner exchange parameter,

$$J = \frac{1}{2l} \sum_{L=2}^{2l} (L0\,l0|l0)^2 F_L . \tag{4.349}$$

If the complete shell is filled, subtraction of the energy (4.348) from the general expression (4.338) gives zero. In order to correct the expression (4.338) for double counting in the case of a partially filled shell, one therefore subtracts the energy[11]

---

[10] Note that the completeness relation (4.344) could be used for all sites of a lattice which are equivalent to the site a. On the contrary, use of the completeness relation for the φ_{anlm} is restricted to the atomic sphere around site a.

[11] The expression (4.350) is also obtained by averaging over all possible ways to distribute the N^{anl}_{σ} electrons with spin σ among the $2l+1$ available m-substates of an open-subshell atom [290],

$$E^U_{\text{av}} = \sum_{\boldsymbol{q}} E^U(\boldsymbol{q}) \qquad \text{with} \qquad q^{\sigma}_m = 0,1 \quad ; \quad \sum_{m=-l}^{l} q^{\sigma}_m = N^{anl}_{\sigma}$$

$$E^U(\boldsymbol{q}) = \frac{1}{2} \sum_{\sigma,\sigma'} \sum_{m,m'} q^{\sigma}_m q^{\sigma'}_{m'} (mm'|w|mm') - \frac{1}{2} \sum_{\sigma} \sum_{m,m'} q^{\sigma}_m q^{\sigma}_{m'} (mm'|w|m'm) ,$$

which, in turn is equivalent to a spherical average.

$$E^{\mathrm{dc}}[\underline{\underline{\Lambda}}^{\sigma}] = \frac{U}{2}N^{anl}(N^{anl}-1) - \frac{J}{2}\sum_{\sigma}N_{\sigma}^{anl}(N_{\sigma}^{anl}-1) \qquad (4.350)$$

from E^U, where, however, N_{σ}^{anl} now denotes the *actual* number of spin-σ electrons in the shell (and $N^{anl} = N_{\uparrow}^{anl} + N_{\downarrow}^{anl} \neq 2(2l+1)$). This number is given by the occupation factor $\Theta_{\alpha\sigma}$ weighted with the overlap of $\phi_{\alpha\sigma}$ and all φ_{anlm},

$$N_{\sigma}^{anl} = \sum_{\alpha}\Theta_{\alpha\sigma}\sum_{m}|c_{\alpha\sigma;anlm}|^2 = \sum_{m}\Lambda_{mm}^{\sigma}\,. \qquad (4.351)$$

In practice, the quantities U and J in Eq. (4.350) are often treated as parameters, rather than evaluated from the integrals F_L. In addition, sometimes modified double counting corrections are utilized [291, 292].

The final LDA+U functional is then defined as [285]

$$E^{\mathrm{LDA}+U}[n_{\sigma},\underline{\underline{\Lambda}}^{\sigma}] := E^{\mathrm{LDA}}[n_{\sigma}] + E^{U}[\underline{\underline{\Lambda}}^{\sigma}] - E^{\mathrm{dc}}[\underline{\underline{\Lambda}}^{\sigma}]\,. \qquad (4.352)$$

This functional is orbital-dependent since the occupation matrix depends on the KS states. Unlike the case of the SIC-LDA, however, invariance under unitary transformations among the KS states emerging from the shell nl is ensured.

Ignoring the exchange contributions beyond the SIC as well as the difference between the intra-shell Coulomb integrals for the actual bulk states and the U of the atomic orbitals, one can qualitatively express $E^{\mathrm{LDA}+U}$ as

$$E^{\mathrm{LDA}+U} \sim E^{\mathrm{LDA}} + \frac{U}{2}\sum_{\alpha\neq\beta}\Theta_{\alpha}\Theta_{\beta} - \frac{U}{2}N(N-1)\,,$$

where $\Theta_{\alpha} = 0,1$ represents the occupation of the state α (the spin is now included in α for brevity) and the sum over α,β only extends over all substates in the shell nl. The orbital energies corresponding to this functional can then be obtained via Janak's theorem (see Sect. 3.4),

$$\varepsilon_{\alpha} = \frac{\partial E^{\mathrm{LDA}+U}}{\partial\Theta_{\alpha}} \sim \varepsilon_{\alpha}^{\mathrm{LDA}} + U\left(\frac{1}{2} - \Theta_{\alpha}\right)\,.$$

Compared to the LDA eigenvalue, one finds a lowering of the occupied states by $-U/2$ and a raise of the unoccupied states by $+U/2$. The complete d- or f-band thus splits up into occupied lower Hubbard bands and unoccupied upper Hubbard bands, with an energy separation determined by the intra-shell Coulomb integral U. The LDA+U method therefore recovers the physics of Mott-Hubbard insulators.

Due to its orbital-dependence $E^{\mathrm{LDA}+U}$ basically belongs to the class of functionals for which the OPM of Chap. 6 should be used to determine the corresponding potential. As in the case of the SIC-LDA, this is, however, not the standard procedure. Rather one resorts to a nonlocal Hamiltonian which is obtained by minimization of $E^{\mathrm{LDA}+U}$ with respect to $\phi_{\alpha\sigma}$ [287],

$$
\begin{aligned}
[\hat{H}_\sigma^{\text{LDA}+U}\phi_{\alpha\sigma}](\boldsymbol{r}) &= \frac{\delta E^{\text{LDA}+U}}{\delta\phi^*_{\alpha\sigma}(\boldsymbol{r})} \\
&= \hat{H}_\sigma^{\text{LDA}}\phi_{\alpha\sigma}(\boldsymbol{r}) + \sum_{m,m''}\int d^3r'\,\varphi_{anlm}(\boldsymbol{r})u^\sigma_{mm''}\varphi^*_{anlm''}(\boldsymbol{r}')\phi_{\alpha\sigma}(\boldsymbol{r}') \qquad (4.353)
\end{aligned}
$$

$$
\begin{aligned}
u^\sigma_{mm''} &= \sum_{m',m'''}\left[\sum_{\sigma'}\Lambda^{\sigma'}_{m'm'''}(mm'|w|m''m''') - \Lambda^\sigma_{m'm'''}(mm'|w|m'''m'')\right] \\
&\quad -\delta_{m,m''}\left[U\left(N^{anl}-\frac{1}{2}\right) - J\left(N^{anl}_\sigma - \frac{1}{2}\right)\right]. \qquad (4.354)
\end{aligned}
$$

The additional term compared to the LDA has the form of a projection operator on the atomic subspace nl. As this operator acts on all KS states, the quality of the LDA+U approach is determined by the degree to which the localized states $\phi_{\alpha\sigma}$, for which the explicit treatment of the intra-shell Coulomb interaction is necessary, retain the character of the atomic states φ_{anlm}. A clear separation of the complete Hilbert space into the subspace emerging from the φ_{anlm} and a remainder must be possible.

The LDA+U approach has been successfully applied to a wide variety of transition metal and rare earth compounds (see e.g. [288]). In particular, the LDA+U provides an accurate account of the Mott-Hubbard character of the $3d$ transition metal oxides, most notably MnO, FeO, CoO and NiO. This success seems particularly noteworthy as, in spite of their localized nature, the $3d$ states strongly hybridize with the oxygen $2p$ states in these compounds.

Chapter 5
Virial Theorems

5.1 Scaling Behavior of Energy Contributions

Virial theorems are derived by considering the behavior of all relevant quantities under a *global scaling transformation* of the position vectors $\boldsymbol{r}_i$ of the electrons,

$$\boldsymbol{r}_i \longrightarrow \lambda\, \boldsymbol{r}_i\,, \tag{5.1}$$

while keeping all other length scales of the system (positions of the nuclei, box sizes etc) fixed. This transformation is thus not equivalent to a simple rescaling of the unit of length: in the latter case all(!) quantities with the dimension of a length are scaled in a homogeneous fashion.

In accordance with the transformation (5.1) one defines the scaled ground-state wavefunction of the interacting system as

$$(\boldsymbol{r}_1,\sigma_1;\ldots\boldsymbol{r}_N,\sigma_N|\Psi_{0,\lambda}\rangle := \lambda^{3N/2}\,(\lambda\boldsymbol{r}_1,\sigma_1;\ldots\lambda\boldsymbol{r}_N,\sigma_N|\Psi_0\rangle\,. \tag{5.2}$$

The prefactor $\lambda^{3N/2}$ ensures proper normalization to 1,

$$\begin{aligned}\langle\Psi_{0,\lambda}|\Psi_{0,\lambda}\rangle &= \lambda^{3N}\sum_{\sigma_1\ldots\sigma_N}\int d^3r_1\ldots\int d^3r_N\,|(\lambda\boldsymbol{r}_1,\sigma_1;\ldots\lambda\boldsymbol{r}_N,\sigma_N|\Psi_0\rangle|^2\\ &= \langle\Psi_0|\Psi_0\rangle = 1\,.\end{aligned} \tag{5.3}$$

Similarly, one can introduce the scaled KS ground state,

$$(\boldsymbol{r}_1,\sigma_1;\ldots\boldsymbol{r}_N,\sigma_N|\Phi_{0,\lambda}\rangle := \lambda^{3N/2}\,(\lambda\boldsymbol{r}_1,\sigma_1;\ldots\lambda\boldsymbol{r}_N,\sigma_N|\Phi_0\rangle\,, \tag{5.4}$$

which is a Slater determinant of scaled KS orbitals,

$$\phi_{i,\lambda}(\boldsymbol{r},\sigma) := \lambda^{3/2}\,\phi_i(\lambda\boldsymbol{r},\sigma)\,. \tag{5.5}$$

By construction, both (5.2) and (5.4) lead to the same scaled ground state density,

The original version of this chapter was revised. An erratum can be found at https://doi.org/10.1007/978-3-642-14090-7_12

E. Engel, R.M. Dreizler, Virial Theorems. In: E. Engel, R.M. Dreizler, Density Functional Theory, Theoretical and Mathematical Physics, pp. 219–226 (2011)
DOI 10.1007/978-3-642-14090-7_5

$$n_\lambda(\boldsymbol{r}) := \lambda^3 n(\lambda \boldsymbol{r}). \tag{5.6}$$

For $\lambda \to 0$ the initial density $n(\boldsymbol{r})$ is therefore smeared out in space, so that, on the fixed external length scale(s), $n_\lambda(\boldsymbol{r})$ becomes uniform and probes the low-density limit. On the other hand, $n_\lambda(\boldsymbol{r})$ approaches a δ-like peak for $\lambda \to \infty$, if $n(\boldsymbol{r})$ characterizes a finite system.

On the basis of these definitions one can analyze the scaling behavior of various ground state expectation values, starting with the kinetic energy. Taking into account the scaling of the momentum operator $\boldsymbol{\nabla} = \partial/\partial \boldsymbol{r}$, one finds the same quadratic scaling with λ for both the exact and the KS kinetic energy,

$$\langle \Psi_{0,\lambda}|\hat{T}|\Psi_{0,\lambda}\rangle = \lambda^2 \langle \Psi_0|\hat{T}|\Psi_0\rangle \tag{5.7}$$

$$\langle \Phi_{0,\lambda}|\hat{T}|\Phi_{0,\lambda}\rangle = \lambda^2 \langle \Phi_0|\hat{T}|\Phi_0\rangle . \tag{5.8}$$

Equation (5.8) can be reformulated as a scaling relation for $T_s[n]$, Eq. (3.21),

$$T_s[n_\lambda] = \langle \Phi_{0,\lambda}|\hat{T}|\Phi_{0,\lambda}\rangle = \lambda^2 \langle \Phi_0|\hat{T}|\Phi_0\rangle = \lambda^2 T_s[n] , \tag{5.9}$$

where the unique correspondence between $|\Phi_{0,\lambda}\rangle$ and n_λ as well as the universality of $T_s[n]$ have been used on the left-hand side (the one-to-one correspondence results from the Hohenberg-Kohn theorem for noninteracting particles, as soon as v-representability of n_λ is assumed). The next energy component of interest is the external potential energy,

$$\begin{aligned}\langle \Psi_{0,\lambda}|\hat{V}_{\text{ext}}|\Psi_{0,\lambda}\rangle &= \langle \Phi_{0,\lambda}|\hat{V}_{\text{ext}}|\Phi_{0,\lambda}\rangle \\ &= \int d^3r\, n_\lambda(\boldsymbol{r})\, v_{\text{ext}}(\boldsymbol{r}) = \int d^3r\, n(\boldsymbol{r})\, v_{\text{ext}}(\boldsymbol{r}/\lambda) . \end{aligned} \tag{5.10}$$

Of course, no statement is possible without further specification of v_{ext}. The best-known case is that of a single atom with the nuclear Coulomb potential centered at the origin,

$$\int d^3r\, n_\lambda(\boldsymbol{r})\, v_{\text{ext}}^{\text{atom}}(\boldsymbol{r}) = \lambda \int d^3r\, n(\boldsymbol{r})\, v_{\text{ext}}^{\text{atom}}(\boldsymbol{r}) . \tag{5.11}$$

The scaling factor can be absorbed into the positions of the nuclei in the more interesting case of molecular or crystalline systems with the external potential (1.6),

$$\int d^3r\, n_\lambda(\boldsymbol{r})\, v_{\text{ext}}(\boldsymbol{r}) = -\lambda \int d^3r\, n(\boldsymbol{r}) \sum_\alpha \frac{Z_\alpha e^2}{|\boldsymbol{r} - \lambda \boldsymbol{R}_\alpha|} . \tag{5.12}$$

Finally, it remains to examine the interaction energy. For the Coulomb interaction one has

$$\langle \Psi_{0,\lambda}|\hat{W}|\Psi_{0,\lambda}\rangle = \lambda \langle \Psi_0|\hat{W}|\Psi_0\rangle \tag{5.13}$$

$$\langle \Phi_{0,\lambda}|\hat{W}|\Phi_{0,\lambda}\rangle = \lambda \langle \Phi_0|\hat{W}|\Phi_0\rangle , \tag{5.14}$$

with $\hat{W}$ given by Eq. (2.5) and $w(\boldsymbol{r},\boldsymbol{r}') = e^2/|\boldsymbol{r}-\boldsymbol{r}'|$. By definition the expectation value (5.14) consists of the Hartree and the exchange contribution,

$$E_{\rm H}[n_\lambda] + E_{\rm x}[n_\lambda] = \langle \Phi_{0,\lambda}|\hat{W}|\Phi_{0,\lambda}\rangle = \lambda\,\langle \Phi_0|\hat{W}|\Phi_0\rangle = \lambda\left\{E_{\rm H}[n] + E_{\rm x}[n]\right\}. \quad (5.15)$$

The scaling behavior of $E_{\rm H}[n]$ is easily obtained from (3.25),

$$E_{\rm H}[n_\lambda] = \lambda\, E_{\rm H}[n]\,, \quad (5.16)$$

so that one finds for the exchange functional [293],

$$E_{\rm x}[n_\lambda] = \lambda\, E_{\rm x}[n]\,. \quad (5.17)$$

The scaling properties of $E_{\rm c}[n]$ are more involved. It is clear that $E_{\rm c}[n]$ can not obey the simple power law observed for the exact interaction energy, Eq. (5.13), since $E_{\rm c}[n]$ contains the difference between the exact and the KS kinetic energy. In order to derive a relation for $E_{\rm c}[n]$ one uses the fact that $|\Psi_{0,\lambda}\rangle$ minimizes $\langle\Psi_\lambda|\hat{T}+\lambda\hat{W}|\Psi_\lambda\rangle$ [293]. On the other hand, the Levy-Lieb functional $F_{\rm LL}[n]$, Eq. (2.59), minimizes $\langle\Psi|\hat{T}+\hat{W}|\Psi\rangle$ for given constraint $|\Psi\rangle \to n$, so that one finds [293]

$$F_{\rm LL}[n_\lambda] \;\le\; \langle \Psi_{0,\lambda}|\hat{T}+\hat{W}|\Psi_{0,\lambda}\rangle\,.$$

Decomposing $F_{\rm LL}$ as usual, $F_{\rm LL} = T_{\rm s,LL} + E_{\rm H} + E_{\rm xc,LL}$, then allows one to establish an inequality for $E_{\rm xc,LL}[n]$ under scaling [293],

$$E_{\rm c,LL}[n_\lambda] \;\le\; \lambda^2\big(T - T_{\rm s,LL}[n]\big) + \lambda\big(W - E_{\rm H}[n] - E_{\rm x,LL}[n]\big)\,, \quad (5.18)$$

with the obvious definitions

$$T = \langle \Psi_0|\hat{T}|\Psi_0\rangle \quad (5.19)$$
$$W = \langle \Psi_0|\hat{W}|\Psi_0\rangle\,. \quad (5.20)$$

What is the use of these scaling relations?[1] The answer is given under the heading of *virial theorems*.

5.2 Conventional Virial Theorem

We first consider the virial theorem of standard quantum many-body theory. The crucial observation for its derivation is the fact that, for given, unscaled Hamiltonian, the total energy of the system has its minimum for $\lambda = 1$, as $|\Psi_{0,\lambda}\rangle$ approaches the

---

[1] A large variety of additional scaling relations for $E_{\rm xc}$, its components and the corresponding functional derivatives have been established in the literature [294–306] (for a compact compilation of the most important results see e.g. [307, 308]). In addition to the uniform coordinate scaling (5.6), non-uniform scaling, such as $n^x_\lambda(x,y,z) = \lambda n(\lambda x,y,z)$, and scaling of the interaction strength (either with or without simultaneous coordinate scaling) have been considered. These investigations have led to an overwhelming number of identities and inequalities for various parts and limits of $E_{\rm xc}$, which have been extensively used as constraints in the construction of xc-functionals [309, 206, 211, 310, 207, 231, 232].

true ground state $|\Psi_0\rangle$ in this limit. This corresponds to the relation

$$\frac{d}{d\lambda}\langle\Psi_{0,\lambda}|\hat{H}|\Psi_{0,\lambda}\rangle\bigg|_{\lambda=1} = 0\,. \tag{5.21}$$

Insertion of (5.7), (5.12) and (5.13) yields

$$\begin{aligned} 0 &= 2\langle\Psi_0|\hat{T}|\Psi_0\rangle + \langle\Psi_0|\hat{W}|\Psi_0\rangle - \frac{d}{d\lambda}\left[\lambda\int d^3r n(\boldsymbol{r})\sum_\alpha \frac{Z_\alpha e^2}{|\boldsymbol{r}-\lambda\boldsymbol{R}_\alpha|}\right]_{\lambda=1} \\ &= 2T + W - \int d^3r n(\boldsymbol{r})\sum_\alpha \frac{Z_\alpha e^2}{|\boldsymbol{r}-\boldsymbol{R}_\alpha|} \\ &\quad -\sum_\beta \boldsymbol{R}_\beta\cdot\frac{\partial}{\partial\boldsymbol{R}_\beta}\left[\int d^3r n(\boldsymbol{r})\sum_\alpha \frac{Z_\alpha e^2}{|\boldsymbol{r}-\boldsymbol{R}_\alpha|}\right]_{\text{expl.}}. \end{aligned}$$

Here the relation

$$\frac{\partial}{\partial\boldsymbol{R}_{\beta,j}}\boldsymbol{R}_{\alpha,i} = \delta_{\alpha\beta}\delta_{ij} \tag{5.22}$$

has been utilized and the index "expl" indicates that only the explicit $\boldsymbol{R}_\alpha$-dependence has to be taken into account in the evaluation of the partial derivative (but not the implicit dependence of $n(\boldsymbol{r})$ on $\boldsymbol{R}_\alpha$). Identification of $\langle\Psi_0|\hat{V}_{\text{ext}}|\Psi_0\rangle$ leads to the desired virial theorem

$$0 = 2T + W + \langle\Psi_0|\hat{V}_{\text{ext}}|\Psi_0\rangle + \sum_\alpha \boldsymbol{R}_\alpha\cdot\frac{\partial}{\partial\boldsymbol{R}_\alpha}\langle\Psi_0|\hat{V}_{\text{ext}}|\Psi_0\rangle\bigg|_{\text{expl.}}. \tag{5.23}$$

In the special case of atoms Eq. (5.23) simplifies to the famous relation

$$0 \;=\; 2T + W + \langle\Psi_0|\hat{V}_{\text{ext}}^{\text{atom}}|\Psi_0\rangle \;=\; T + E_{\text{tot}}^{\text{atom}}\,. \tag{5.24}$$

5.3 DFT Virial Theorem

In order to derive virial relations for density functionals, one first considers the generic situation

$$A[n_\lambda] = \lambda^k A[n]\,. \tag{5.25}$$

Using the fact that $A[n_\lambda]$ depends on λ only via the scaled density n_λ (all other length scales remain fixed), one obtains quite generally [311],

$$\frac{d}{d\lambda}A[n_\lambda]\bigg|_{\lambda=1} = kA[n] = \int d^3r\frac{dn_\lambda(\boldsymbol{r})}{d\lambda}\frac{\delta A[n_\lambda]}{\delta n_\lambda(\boldsymbol{r})}\bigg|_{\lambda=1}. \tag{5.26}$$

One can now insert (5.6),

$$kA[n] = \int d^3r \left[3\lambda^2 n(\lambda\boldsymbol{r}) + \lambda^3 \boldsymbol{r} \cdot \frac{\partial}{\partial(\lambda\boldsymbol{r})} n(\lambda\boldsymbol{r})\right]_{\lambda=1} \frac{\delta A[n]}{\delta n(\boldsymbol{r})}$$
$$= \int d^3r \left[3n(\boldsymbol{r}) + \boldsymbol{r} \cdot \boldsymbol{\nabla} n(\boldsymbol{r})\right] \frac{\delta A[n]}{\delta n(\boldsymbol{r})}. \tag{5.27}$$

If $n(\boldsymbol{r})$ decays sufficiently fast for large $|\boldsymbol{r}|$, i.e. for finite systems, one can integrate by parts,

$$kA[n] = -\int d^3r n(\boldsymbol{r}) \boldsymbol{r} \cdot \boldsymbol{\nabla} \frac{\delta A[n]}{\delta n(\boldsymbol{r})}. \tag{5.28}$$

Either the form (5.27) or (5.28) has to be used, depending on the type of system considered. For brevity, the subsequent discussion is restricted to the case of finite systems.

One can now apply (5.28) to the individual energy components. For the kinetic energy (5.9) the variational equation (3.54) (for the KS system) leads to

$$2T_s[n] = -\int d^3r n(\boldsymbol{r}) \boldsymbol{r} \cdot \boldsymbol{\nabla} \frac{\delta T_s[n]}{\delta n(\boldsymbol{r})} = \int d^3r n(\boldsymbol{r}) \boldsymbol{r} \cdot \boldsymbol{\nabla} v_s(\boldsymbol{r}). \tag{5.29}$$

Similarly, Eqs. (5.16) and (5.17) lead to virial relations for the Hartree and the exchange energy functional [293],

$$E_H[n] = -\int d^3r n(\boldsymbol{r}) \boldsymbol{r} \cdot \boldsymbol{\nabla} v_H(\boldsymbol{r}) \tag{5.30}$$

$$E_x[n] = -\int d^3r n(\boldsymbol{r}) \boldsymbol{r} \cdot \boldsymbol{\nabla} v_x(\boldsymbol{r}). \tag{5.31}$$

The most complicated component is the correlation energy, as it does not exhibit the straight power law behavior assumed for the derivation of (5.28). Nevertheless, the steps which lead from Eq. (5.26) to the right-hand side of Eq. (5.28) can also be applied to E_c [293],

$$\left.\frac{d}{d\lambda} E_c[n_\lambda]\right|_{\lambda=1} = -\int d^3r n(\boldsymbol{r}) \boldsymbol{r} \cdot \boldsymbol{\nabla} v_c(\boldsymbol{r}). \tag{5.32}$$

Note that the relations (5.29)–(5.32) are valid for all kinds of external potentials v_{ext}. Only the particle–particle interaction has been chosen to be Coulombic. These relations thus allow a check of numerical procedures for a very general class of problems.

On the basis of (5.9), (5.16), (5.17) and (5.32) one can derive the DFT analog of Eq. (5.23). Starting from the minimum property of the total energy functional,

$$\frac{d}{d\lambda}E[n_\lambda]\bigg|_{\lambda=1} = 0, \tag{5.33}$$

one obtains with the decomposition (3.24) (see [312]),

$$\begin{aligned} 0 = {} & 2T_s[n] + E_H[n] + E_x[n] - \int d^3r\, n(\boldsymbol{r})\, \boldsymbol{r}\cdot\nabla v_c(\boldsymbol{r}) \\ & + \langle\Psi_0|\hat{V}_{ext}|\Psi_0\rangle + \sum_\alpha \boldsymbol{R}_\alpha \cdot \frac{\partial}{\partial \boldsymbol{R}_\alpha}\langle\Psi_0|\hat{V}_{ext}|\Psi_0\rangle\bigg|_{expl.}, \end{aligned} \tag{5.34}$$

where (5.32) has been utilized for the evaluation of $dE_c[n_\lambda]/d\lambda$ and the treatment of $E_{ext}[n]$ follows that in Eq. (5.23).

Combining (5.34) with (5.23), one finally obtains for the specific external potential (1.6) [312]

$$E_c[n] = -\int d^3r\, n(\boldsymbol{r})\, \boldsymbol{r}\cdot\nabla v_c(\boldsymbol{r}) - T + T_s[n]. \tag{5.35}$$

5.4 Hellmann-Feynman Theorem

The interaction between the external sources, the nuclei, has been ignored in the discussion of the virial theorems (5.23) and (5.34). We now consider the complete Born-Oppenheimer (BO) energy surface, including the energy (1.4). In order to establish a virial relation, one first evaluates the gradient of the total BO energy with respect to the position of the nucleus α, $\boldsymbol{R}_\alpha$,

$$\frac{d}{d\boldsymbol{R}_\alpha}E_{BO} = \frac{d}{d\boldsymbol{R}_\alpha}\left[E[n] + \frac{1}{2}\sum_{\beta\neq\gamma}\frac{Z_\beta Z_\gamma e^2}{|\boldsymbol{R}_\beta - \boldsymbol{R}_\gamma|}\right]. \tag{5.36}$$

The electronic part of this gradient is given by

$$\frac{d}{d\boldsymbol{R}_\alpha}E[n] = \frac{\partial}{\partial\boldsymbol{R}_\alpha}E[n]\bigg|_{expl.} + \int d^3r\, \frac{\delta E[n]}{\delta n(\boldsymbol{r})}\frac{\partial n(\boldsymbol{r})}{\partial \boldsymbol{R}_\alpha}.$$

However, as $F[n]$ is universal, an explicit $\boldsymbol{R}_\alpha$-dependence of $E[n]$ is only found in $E_{ext}[n]$. In addition, the variational equation (2.38) is valid on the BO surface, i.e. as soon as the electronic ground state for given positions of the nuclei is reached,

$$\frac{d}{d\boldsymbol{R}_\alpha}E[n] = \frac{\partial}{\partial\boldsymbol{R}_\alpha}\langle\Psi_0|\hat{V}_{ext}|\Psi_0\rangle\bigg|_{expl.} - \mu\frac{\partial}{\partial\boldsymbol{R}_\alpha}\int d^3r\, n(\boldsymbol{r})$$

$$
\begin{aligned}
&= \frac{\partial}{\partial \boldsymbol{R}_\alpha} \langle \Psi_0 | \hat{V}_{\text{ext}} | \Psi_0 \rangle \Big|_{\text{expl.}} \\
&= \int d^3 r\, n(\boldsymbol{r}) \frac{Z_\alpha e^2 (\boldsymbol{R}_\alpha - \boldsymbol{r})}{|\boldsymbol{R}_\alpha - \boldsymbol{r}|^3}
\end{aligned}
\tag{5.37}
$$

$$
\Longrightarrow \quad \frac{d}{d\boldsymbol{R}_\alpha} E_{\text{BO}} = \int d^3 r\, n(\boldsymbol{r}) \frac{Z_\alpha e^2 (\boldsymbol{R}_\alpha - \boldsymbol{r})}{|\boldsymbol{R}_\alpha - \boldsymbol{r}|^3} - \sum_{\beta \neq \alpha} \frac{Z_\alpha Z_\beta e^2 (\boldsymbol{R}_\alpha - \boldsymbol{R}_\beta)}{|\boldsymbol{R}_\alpha - \boldsymbol{R}_\beta|^3} . \tag{5.38}
$$

Equation (5.38) is an important result in itself, as it shows that the gradients of the total BO energy, i.e. the forces on the nuclei (within a classical description), are completely determined by the explicit $\boldsymbol{R}_\alpha$-dependence of the electronic energy. This result simplifies the calculation of forces in actual applications considerably.

Equations (5.37) and (5.38) represent the original form of the *Hellmann-Feynman theorem* [313, 314], formulated in terms of the total energy functional. They are special cases of a rather general result which can be proven for parameter-dependent expectation values of the form

$$
\langle \Psi_{0,\lambda} | \hat{H}_\lambda | \Psi_{0,\lambda} \rangle \qquad \text{with} \qquad \hat{H}_\lambda | \Psi_{0,\lambda} \rangle = E_\lambda | \Psi_{0,\lambda} \rangle . \tag{5.39}
$$

The derivative of $\langle \Psi_{0,\lambda} | \hat{H}_\lambda | \Psi_{0,\lambda} \rangle$ with respect to the parameter λ reduces to the derivative of the operator $\hat{H}_\lambda$, as long as the eigenstates $|\Psi_{0,\lambda}\rangle$ of $\hat{H}_\lambda$ are normalized to one for all values of λ,

$$
\begin{aligned}
&\frac{d}{d\lambda} \langle \Psi_{0,\lambda} | \hat{H}_\lambda | \Psi_{0,\lambda} \rangle \\
&= \left\langle \frac{d\Psi_{0,\lambda}}{d\lambda} \middle| \hat{H}_\lambda \middle| \Psi_{0,\lambda} \right\rangle + \left\langle \Psi_{0,\lambda} \middle| \frac{d\hat{H}_\lambda}{d\lambda} \middle| \Psi_{0,\lambda} \right\rangle + \left\langle \Psi_{0,\lambda} \middle| \hat{H}_\lambda \middle| \frac{d\Psi_{0,\lambda}}{d\lambda} \right\rangle \\
&= E_\lambda \frac{d}{d\lambda} \langle \Psi_{0,\lambda} | \Psi_{0,\lambda} \rangle + \left\langle \Psi_{0,\lambda} \middle| \frac{d\hat{H}_\lambda}{d\lambda} \middle| \Psi_{0,\lambda} \right\rangle \\
&= \left\langle \Psi_{0,\lambda} \middle| \frac{d\hat{H}_\lambda}{d\lambda} \middle| \Psi_{0,\lambda} \right\rangle .
\end{aligned}
\tag{5.40}
$$

In (5.37) and (5.38) the positions of the nuclei play the role of the parameter λ.

Equation (5.38) leads directly to a virial relation, if one considers a local or global minimum of the BO energy. At a minimum the forces on the nuclei vanish,

$$
\begin{aligned}
\frac{\partial}{\partial \boldsymbol{R}_\alpha} \langle \Psi_0 | \hat{V}_{\text{ext}} | \Psi_0 \rangle \Big|_{\text{expl.}} &= \int d^3 r\, n(\boldsymbol{r}) \frac{Z_\alpha e^2 (\boldsymbol{R}_\alpha - \boldsymbol{r})}{|\boldsymbol{R}_\alpha - \boldsymbol{r}|^3} \\
&= \sum_{\beta \neq \alpha} \frac{Z_\alpha Z_\beta e^2 (\boldsymbol{R}_\alpha - \boldsymbol{R}_\beta)}{|\boldsymbol{R}_\alpha - \boldsymbol{R}_\beta|^3} .
\end{aligned}
\tag{5.41}
$$

This allows an evaluation of the most complicated contribution to the general virial theorems (5.23) and (5.34),

$$\sum_\alpha \boldsymbol{R}_\alpha \cdot \frac{\partial}{\partial \boldsymbol{R}_\alpha} \langle \Psi_0 | \hat{V}_{\text{ext}} | \Psi_0 \rangle \bigg|_{\text{expl.}} = \frac{1}{2} \sum_{\alpha \neq \beta} \frac{Z_\alpha Z_\beta e^2}{|\boldsymbol{R}_\alpha - \boldsymbol{R}_\beta|} . \tag{5.42}$$

Insertion of (5.42) into the conventional virial relation (5.23) then yields a relation between the total BO energy and the (full) kinetic energy of the electrons,

$$0 = 2T + W + \langle \Psi_0 | \hat{V}_{\text{ext}} | \Psi_0 \rangle + \frac{1}{2} \sum_{\alpha \neq \beta} \frac{Z_\alpha Z_\beta e^2}{|\boldsymbol{R}_\alpha - \boldsymbol{R}_\beta|} = T + E_{\text{BO}}$$
$$\Longrightarrow \qquad E_{\text{BO}} = -T . \tag{5.43}$$

Finally, combination of (5.42) with (5.34) yields the DFT variant of (5.43) (see [312]),

$$0 = 2T_s[n] + E_{\text{ext}}[n] + E_{\text{H}}[n] + E_{\text{x}}[n] - \int d^3r\, n(\boldsymbol{r})\, \boldsymbol{r} \cdot \nabla v_{\text{c}}(\boldsymbol{r}) + \frac{1}{2} \sum_{\alpha \neq \beta} \frac{Z_\alpha Z_\beta e^2}{|\boldsymbol{R}_\alpha - \boldsymbol{R}_\beta|} \tag{5.44}$$

$$\Longrightarrow \qquad E_{\text{BO}} = -T_s[n] + E_{\text{c}}[n] + \int d^3r\, n(\boldsymbol{r})\, \boldsymbol{r} \cdot \nabla v_{\text{c}}(\boldsymbol{r}) . \tag{5.45}$$

Equation (5.45) provides a consistency check for actual calculations.

Chapter 6
Orbital Functionals: Optimized Potential Method

6.1 Motivation

The illustrative results for (M)GGAs which have been presented in Chap. 4 demonstrate the success of the GGA concept. Studies of the applicability and the accuracy of different forms of the (M)GGA for a large variety of systems support this statement [5, 200, 211, 224, 238, 315–326]. A completely different concept [113, 114, 127] is introduced in this chapter. In this approach E_{xc} is expressed in terms of the KS orbitals and eigenvalues,[1] rather than the density itself, so that E_{xc} becomes an implicit functional of n. The prototype functional of this class is the exact exchange, Eq. (4.5).

Given the success of the GGA, one may ask why such orbital-dependent xc-functionals are of any interest? The answer to this question necessarily consists in a list of systems for which the GGA (and, of course, also the LDA) is not just inaccurate, but rather fails completely.

6.1.1 Atomic Negative Ions

A first fundamental failure of the GGA is observed for atomic negative ions [327, 328]. This failure originates from the semi-local density-dependence of the GGA exchange potential. The situation is more easily analyzed in the case of the LDA, to which we restrict the explicit discussion. The LDA for the exchange potential is given by Eq. (4.111). As the density of finite systems decays exponentially in the asymptotic region, one finds an exponential decay of $v_{\mathrm{x}}^{\mathrm{LDA}}$,

$$n(\boldsymbol{r}) \xrightarrow[|\boldsymbol{r}|\to\infty]{} e^{-\alpha r} \qquad \Longrightarrow \qquad v_{\mathrm{x}}^{\mathrm{LDA}}(\boldsymbol{r}) \xrightarrow[|\boldsymbol{r}|\to\infty]{} e^{-\alpha r/3},$$

[1] In the following the term *orbital-dependent* is always understood to include a possible eigenvalue-dependence.

E. Engel, R.M. Dreizler, Orbital Functionals: Optimized Potential Method. In: E. Engel, R.M. Dreizler, Density Functional Theory, Theoretical and Mathematical Physics, pp. 227–306 (2011)
DOI 10.1007/978-3-642-14090-7_6

in obvious contradiction to the exact result (4.20). An equivalent exponential decay is also obtained for the LDA correlation potential ($1/r_s \sim n^{1/3}$). Moreover, for neutral atoms the electrostatic potential of the nucleus cancels with the monopole term in v_H, Eq. (3.42). Consequently, the total v_s decays faster than $1/r$. This implies that a neutral atom does not exhibit a Rydberg series of excited states within the framework of the LDA and is therefore not able to bind an additional electron, i.e. to form a negative ion.[2]

The same deficiency is found for all GGAs of the form (4.281), for which the potential, Eq. (4.290), depends on the first two gradients of the density. While for some forms of the kernel f_x, Eq. (4.281), a power-law behavior of v_x^{GGA} for large r can emerge from (4.290), this standard form of the GGA potential necessarily decays faster than $-1/r$ [220] (see Sect. 6.3 for an explicit example of this behavior). As a consequence, GGAs do not predict the existence of atomic negative ions either. The same problem is observed for many negatively charged small molecules.

The $-1/r$-behavior of the exact v_x results from the self-interaction contribution to E_x, which is required to cancel the self-interaction included in E_H, as discussed in Sect. 4.1. This indicates that one needs a rather nonlocal exchange functional to reproduce the $-1/r$-behavior: the self-interaction component of v_x has to be as nonlocal as the Coulomb integral in v_H. In fact, an exact cancellation of the self-interaction in E_H is only possible if the same Coulomb matrix elements are present in E_x. This requires an orbital-dependent representation of E_x.

6.1.2 Dispersion Forces

The LDA and GGA also fail to reproduce the London dispersion force (one of the van der Waals forces) [329–331]. In this case the problem results from the short-ranged nature of the LDA/GGA correlation functional. In the LDA the correlation energy density is given by the energy density e_c^{HEG} of the homogeneous electron gas (HEG), evaluated with the local density, Eq. (4.109). As $e_c^{HEG}(n_0 \to 0) = 0$ (see Eq. (4.107)), one realizes that only regions in space with a non-vanishing density contribute to the correlation energy. Now consider two neutral, closed-subshell atoms, which are so far apart that their densities do not overlap. As indicated in Fig. 6.1 the density of this system corresponds to the sum of the two atomic densities. In this situation dispersion forces become important, as there is neither an electrostatic interaction between the two atoms nor can any bonding orbitals be formed. Only the attraction between virtual (dipole) excitations of the two atoms can lead to binding via the London dispersion force. In the LDA, however, any inter-atomic interaction provided by E_c requires an overlap of the atomic densities, as the binding energy E_b must result from the nonlinear density dependence of E_c^{LDA},

[2] Note, however, that (4.20) is only a necessary, but not a sufficient criterion for the stability of a negative ion. Ultimately, the stability depends on the relative size of the total energies of the N and the $N+1$ electron system.

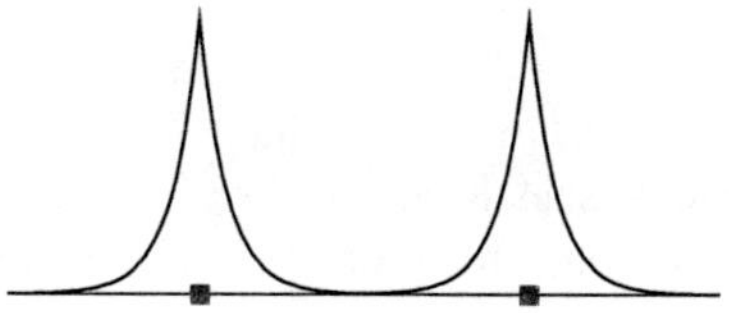

Fig. 6.1 Electronic density of two atoms at large separation.

$$E_{\mathrm{b}} = E_{\mathrm{c}}^{\mathrm{LDA}}[n_A + n_B] - E_{\mathrm{c}}^{\mathrm{LDA}}[n_A] - E_{\mathrm{c}}^{\mathrm{LDA}}[n_B] \, .$$

For this reason dispersion forces can not be reproduced by the LDA.

The behavior of the LDA in the case of dispersive bonds is illustrated in Fig. 6.2, in which the BO surface of the He dimer is shown. He_2 is a particularly sensitive

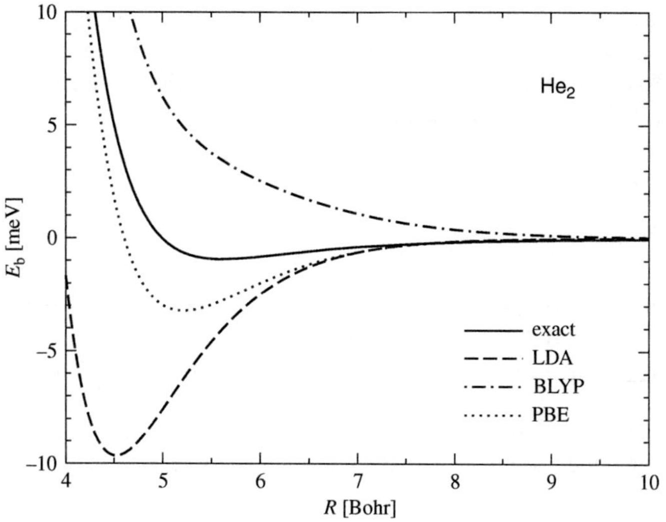

Fig. 6.2 Energy surface E_{b} of He_2: Selfconsistent LDA [134], BLYP [219, 221] and PBE [207] data versus exact result [332].

system, as manifest in the meV scale of Fig. 6.2: the dispersive bond is extremely weak, which leads to a highly delocalized ground state wavefunction [333]. He_2 thus provides an ideal testing ground for approximate correlation functionals.

As is obvious from Fig. 6.2, the LDA predicts the minimum of E_{b} to be at much smaller separation than the true E_{b} [332] and overestimates the corresponding well depth by an order of magnitude. For large internuclear separations the LDA surface vanishes exponentially (which is not apparent due to the scale of Fig. 6.2). The figure confirms the argument that the LDA requires the densities of the two atoms to overlap, in order to produce binding. As a consequence, the dimer is contracted far too much in the LDA. The LDA is not suitable to describe dispersive bonds, in spite of the fact that it generates an attractive E_b.

As in the case of negative ions, this problem is not resolved by use of the GGA, in which the energy density depends on the first two gradients of the density,

$$E_c^{GGA}[n] = \int d^3r\, e_c^{GGA}(n, (\nabla n)^2, \nabla^2 n) .$$

The ∇n-dependence of $e_c^{GGA}(\boldsymbol{r})$ only accounts for the density in the immediate vicinity of $\boldsymbol{r}$. $e_c^{GGA}(\boldsymbol{r})$ therefore vanishes wherever $n(\boldsymbol{r})$ vanishes. Figure 6.2 explicitly illustrates this statement for two standard versions of the GGA. While the PBE-GGA basically shows the same picture as the LDA, the BLYP-GGA remains repulsive for all internuclear separations. The discrepancy between the two GGA results already indicates the inadequacy of the GGA for this system. In spite of the quantitative improvement of the PBE-form over the LDA, the attractive potential generated by the PBE-GGA relies on the same mechanism as the potential produced by the LDA. Neither the LDA nor the GGA can mediate the long-range force resulting from virtual excitations.[3] Not only the exact exchange functional is highly nonlocal, but also the exact correlation functional.

6.1.3 Strongly Correlated Systems

The third class of systems for which both the LDA and the GGA meet with problems are strongly correlated systems [338, 339, 237]. The most prominent examples of this type of solids are the $3d$ transition metal monoxides MnO, FeO, CoO and NiO. These systems, which crystallize in the rock salt structure,[4] are insulating antiferromagnets of type II.[5] Both the LDA and the GGA predict, however, FeO and CoO to be metallic and underestimate the band gap in MnO and NiO dramatically [338, 339, 237]. This is illustrated in Fig. 6.3 in which the LDA and PBE-GGA band

[3] The dispersion force between two well-separated perturbations in a uniform electron gas is, however, included in the linear response contribution to the xc-functional, Eq. (4.159) [160, 161]: for large separation R of the two localized perturbations one obtains the required attractive $1/R^6$-interaction. This opens a route to account for the dispersion interaction by nonlocal, but still explicitly density-dependent functionals [162–170]. In an alternative approach the polarizability of the interacting fragments is modeled in terms of multipole moments constructed from the exchange hole on the two centers [334–337]. This model is then inserted into the closure approximated expression for the standard second order energy correction. The resulting functional depends, however, on the occupied KS orbitals.

[4] One finds a minor rhombohedral distortion of the unit cell, which, however, appears to be irrelevant for the insulating nature of these compounds.

[5] In the type II antiferromagnets the magnetic moments of all transition metal atoms in planes orthogonal to the body-diagonal are aligned, with the direction of the spins alternating from plane to plane.

structures for FeO are plotted—the band structures obtained with most other GGAs are rather similar [339, 237].

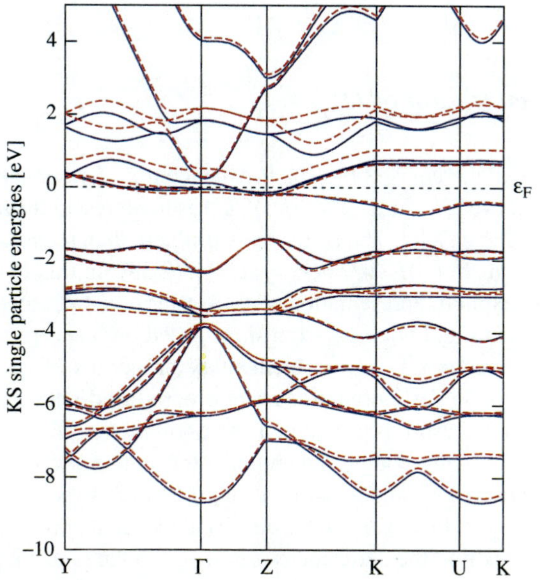

Fig. 6.3 Band structure of antiferromagnetic (type II) FeO obtained by plane-wave-pseudopotential calculation with the LDA (*solid lines*) and PBE-GGA (*dashed lines*). The valence space of Fe includes the $3s$, $3p$, $3d$ and $4s$ states, the cut-off energy of the plane-wave basis is $E_{\rm cut} = 300$ Ry, 20 special k-points have been used for the integration over the 1st Brillouin zone.

The origin of this problem is not *a priori* clear from the structure of the LDA and GGA. There are, however, definite indications of the reason for these difficulties:

- On the one hand, there exists one parameterization of the GGA which predicts FeO and CoO to be antiferromagnetic insulators [237]. This functional is the only GGA whose kernel was explicitly optimized to reproduce the exact atomic exchange potentials as accurately as possible [244]. Although the results obtained with this GGA are not really convincing (in particular, the corresponding gaps are still much too small), this suggests that the quality of the exchange potential plays an important role in the description of Mott insulators.
- The SIC-LDA, Eq. (4.301), also leads to insulating ground states [270, 271].
- Similarly, partial inclusion of exact exchange (i.e. for the $3d$ states at the iron sites) in the spirit of a hybrid functional (see Sect. 6.5.4) opens a gap [340].
- Finally, an LDA+U treatment solves the problem [285, 288].

All these results indicate that once again the inappropriate handling of the self-interaction (in this case of the $3d$ states) is a major reason for the failure of the LDA and GGA.

6.1.4 Third Generation of DFT

Let us finally look at the concept of orbital-dependent xc-functionals from a somewhat wider perspective. The initial idea of DFT was to represent the total energy as a functional of the density, $E[n]$. The many-body problem then reduces to the solution of the variational equation $\delta E[n]/\delta n(\boldsymbol{r}) = \mu$, Eq. (2.38). In this first generation of DFT one needs an explicit representation of T_s in terms of n, for which only Thomas-Fermi-type functionals and nonlocal extensions based on linear response are available to date (see Appendix D). Even elementary properties of quantum systems, most notably the atomic shell structure, are not reproduced with these functionals. This was the reason to resort to an implicit representation of T_s in terms of the KS orbitals, recasting the many-body problem in the form of the KS equations. Only an explicit representation of E_{xc} in terms of n is required in this variant of DFT (second generation). Unfortunately, the explicitly density-dependent approximations for E_{xc} currently available exhibit the deficiencies discussed in the preceding sections. The obvious next step for an improvement is a representation of E_{xc}, or at least of E_x, in terms of the KS orbitals. The formally correct way of implementing this third generation of DFT is the so-called *Optimized (Effective) Potential Method* (OPM or OEP), in which the many-body problem is approached by the simultaneous solution of the KS equations and an integral equation which determines v_{xc} (to be discussed below).

Unfortunately, the OPM turns out to be computationally very demanding even for rather simple orbital-dependent expressions. For this reason applications of orbital-dependent xc-functionals often rely on the so-called *generalized Kohn-Sham* (GKS) *approach*, in which the KS equations are replaced by the HF-type orbital-dependent equations which result from a minimization of the total energy functional with respect to the orbitals.[6] Alternatively, orbital-dependent expressions are simply applied *a posteriori* by insertion of the solutions of a self-consistent GGA (or even LDA) calculation (Post-GGA/LDA treatment), rather than implemented self-consistently. The GKS or Post-GGA schemes are used in particular for the model-based[7] functionals of Sect. 6.5. In the following, however, we will focus completely on the rigorous way of handling orbital-dependent functionals, i.e. the OPM. In fact,

---

[6] If applied to the exact exchange without any correlation, the GKS scheme coincides with the HF approach.

[7] Lacking any better characterization, we call all those functionals model-based which either can not be derived by a systematic sequence of approximations from the exact E_{xc}, or contain parameters which can be determined in more than one way. The class of model-based functionals includes both non-empirical functionals whose parameters are adjusted to theoretical conditions, as well as semi-empirical forms whose parameters are fitted to experimental observations.

this is the standard approach in the case of the exact exchange and the first-principles correlation functionals of Sect. 6.4.

In summary, there exist three basic variants of DFT, which differ both in quality and in computational efficiency. Depending on the physical phenomenon under investigation, one can choose the method which is most appropriate. However, it seems fair to say that, given the present computer power, the OPM implementation of orbital-dependent functionals primarily serves to provide benchmark results and proof-of-principle solutions.

6.2 Derivation of OPM Integral Equation

The most important question to be answered at this point is how to calculate the multiplicative potential (3.43), which corresponds to an xc-functional of the type (4.5). There are three different ways for the derivation of the basic equation, which determines this potential. As each of these derivations is instructive by itself, all three will be outlined below.[8] However, before starting the discussion of the OPM equations, we first introduce a notation which allows dealing with both spin-saturated (DFT) and spin-polarized (SDFT) systems in a more coherent fashion.

6.2.1 Compact Notation

In Chaps. 1–4 the spin degree of freedom has always been specified explicitly as an argument of the KS orbitals, $\phi_k(\boldsymbol{r}\sigma)$, and related quantities such as Green's functions. This explicit notation allows a more ready access to the relations used in DFT, be it at the price of somewhat lengthier equations. From this point on, we will use the standard notation, in which the orbitals are understood as bispinors,

$$\phi_k(\boldsymbol{r}) \equiv \begin{pmatrix} \phi_k(\boldsymbol{r}\uparrow) \\ \phi_k(\boldsymbol{r}\downarrow) \end{pmatrix} . \tag{6.1}$$

In the framework of both DFT and SDFT, to which we restrict the discussion in this chapter, the quantum number k consists of an orbital quantum number α and a spin quantum number s, according to either Eq. (3.9) for the case of spin-saturated systems or to Eq. (3.93) for spin-polarized systems. As the spinor character of ϕ_k is, in this case, determined by the Pauli spinor χ_s, either its lower or its upper component vanishes,

$$\phi_{k=\alpha\uparrow}(\boldsymbol{r}) = \begin{pmatrix} \phi_{\alpha\uparrow}(\boldsymbol{r}) \\ 0 \end{pmatrix} \qquad \phi_{k=\alpha\downarrow}(\boldsymbol{r}) = \begin{pmatrix} 0 \\ \phi_{\alpha\downarrow}(\boldsymbol{r}) \end{pmatrix} . \tag{6.2}$$

[8] For an extension of the OPM to excited states see [341], to time-dependent DFT see [342].

The scalar orbitals $\phi_{\alpha s}$ of the right-hand side are defined by either (3.9) or (3.93). In order to distinguish between the bispinor on the left-hand sides of Eq. (6.2) and the scalar orbitals on the right-hand sides, we will always label the bispinor with the compact index k (etc) in the following, with the understanding that $k \equiv \alpha s$.

This allows a unified representation of the KS equations, covering both DFT and SDFT. If the KS potential v_s is understood as the diagonal matrix of the spin-dependent potentials (3.99),

$$v_s(\boldsymbol{r}) = \begin{pmatrix} v_s^{\uparrow}(\boldsymbol{r}) & 0 \\ 0 & v_s^{\downarrow}(\boldsymbol{r}) \end{pmatrix}, \tag{6.3}$$

and ϕ_k is the bispinor (6.1), the KS equations of both DFT and SDFT can be written as

$$\left\{ -\frac{\hbar^2 \boldsymbol{\nabla}^2}{2m} + v_s(\boldsymbol{r}) \right\} \phi_k(\boldsymbol{r}) = \varepsilon_k \phi_k(\boldsymbol{r}) \,. \tag{6.4}$$

If the spin-component of k corresponds to spin-up, only the upper component of ϕ_k is nonzero, according to (6.2). Consequently, the matrix-spinor product $v_s\phi_k$ reduces to

$$v_s \, \phi_k = v_s^{\uparrow} \, \phi_{\alpha\uparrow}(\boldsymbol{r}) \,,$$

so that the KS equations for spin-up emerge (similarly for spin-down).

The nonlocal operator v_s may be decomposed in the form

$$v_s(\boldsymbol{r}) = v_s^{\uparrow}(\boldsymbol{r}) P_{\uparrow} + v_s^{\downarrow}(\boldsymbol{r}) P_{\downarrow} \,, \tag{6.5}$$

with the aid of the projection operators

$$P_{\uparrow} = \begin{pmatrix} 1\,0 \\ 0\,0 \end{pmatrix} = \chi_{\uparrow}\chi_{\uparrow}^{\dagger} \qquad P_{\downarrow} = \begin{pmatrix} 0\,0 \\ 0\,1 \end{pmatrix} = \chi_{\downarrow}\chi_{\downarrow}^{\dagger} \,. \tag{6.6}$$

Using P_σ, the spin-density can be expressed as

$$n_\sigma(\boldsymbol{r}) = \sum_k \Theta_k \phi_k^{\dagger}(\boldsymbol{r}) P_\sigma \phi_k(\boldsymbol{r}) \,. \tag{6.7}$$

6.2.2 Direct Functional Derivative

The simplest way to derive the basic equation of the OPM is the transformation of the functional derivative (3.43) into derivatives with respect to ϕ_k and ε_k, using the chain rule for functional differentiation [343, 344] (for a general discussion of the chain rule see Appendix A). In the following we formulate this approach directly for the more general situation of spin-density functional theory, assuming E_{xc} to be a functional of the KS orbitals ϕ_k and the corresponding eigenvalues ε_k. At the end of the discussion we will then reduce the spin-dependent equations to the limit of spin-saturated systems.

Relying on the one-to-one correspondence between the set of spin-densities $(n_\uparrow, n_\downarrow)$ and the corresponding set of KS potentials $(v_\mathrm{s}^\uparrow, v_\mathrm{s}^\downarrow)$, one first obtains

$$v_\mathrm{xc}^\sigma(\boldsymbol{r}) = \frac{\delta E_\mathrm{xc}}{\delta n_\sigma(\boldsymbol{r})} = \sum_{\sigma'} \int d^3r' \frac{\delta v_\mathrm{s}^{\sigma'}(\boldsymbol{r}')}{\delta n_\sigma(\boldsymbol{r})} \frac{\delta E_\mathrm{xc}}{\delta v_\mathrm{s}^{\sigma'}(\boldsymbol{r}')} . \tag{6.8}$$

However, for given number of spin-up/down electrons, $N_\sigma = \int d^3r n_\sigma(\boldsymbol{r})$, the spin-density n_σ is uniquely determined by v_s^σ only, and vice versa. On the one hand, the solutions $\phi_{\alpha\sigma}$ of the KS equation (3.98) are completely determined for given v_s^σ, irrespective of the form of the KS potential for the opposite spin. Therefore, as soon as N_σ is fixed, n_σ is fixed. On the other hand, for given n_σ the associated KS potential is fully determined (up to a constant) by virtue of the HK-theorem for noninteracting systems, $v_\mathrm{s}^\sigma = -\delta T_\mathrm{s,L}[n_\uparrow, n_\downarrow]/\delta n_\sigma + \mu_\sigma$, since $T_\mathrm{s,L}[n_\uparrow, n_\downarrow]$, Eq. (3.96), is additive with respect to spin. In other words: for given N_σ the two spin channels are completely decoupled in the noninteracting KS system.[9,10] As a consequence, one has

$$\frac{\delta v_\mathrm{s}^{\sigma'}(\boldsymbol{r}')}{\delta n_\sigma(\boldsymbol{r})} = \delta_{\sigma\sigma'} \frac{\delta v_\mathrm{s}^\sigma(\boldsymbol{r}')}{\delta n_\sigma(\boldsymbol{r})} , \tag{6.9}$$

which simplifies Eq. (6.8). In a second step one utilizes the one-to-one correspondence between v_s^σ and the associated single-particle states (compare the discussion following Eq. (3.21)),

$$v_\mathrm{xc}^\sigma(\boldsymbol{r}) = \int d^3r' \frac{\delta v_\mathrm{s}^\sigma(\boldsymbol{r}')}{\delta n_\sigma(\boldsymbol{r})} \sum_k \left\{ \int d^3r'' \left[\frac{\delta \phi_k^\dagger(\boldsymbol{r}'')}{\delta v_\mathrm{s}^\sigma(\boldsymbol{r}')} \frac{\delta E_\mathrm{xc}}{\delta \phi_k^\dagger(\boldsymbol{r}'')} + c.c. \right] \right.$$
$$\left. + \frac{\delta \varepsilon_k}{\delta v_\mathrm{s}^\sigma(\boldsymbol{r}')} \frac{\partial E_\mathrm{xc}}{\partial \varepsilon_k} \right\} . \tag{6.10}$$

The sum over k is not restricted to occupied states and includes both spin channels. However, the variational derivative $\delta\phi_k^\dagger/\delta v_\mathrm{s}^\sigma$ vanishes if the spin-quantum number in k is opposite to σ, as will be shown explicitly below. The functional derivative $\delta E_\mathrm{xc}/\delta\phi_k^\dagger$ is a bispinor. It has the simple form

---

[9] Note that this statement is no longer correct, if only the total number of electrons is specified, while N_σ is allowed to vary in accordance with $N = N_\uparrow + N_\downarrow$. This is typically the case for infinite systems for which the magnetic moment results from a minimization of the total energy with respect to $N_\uparrow, N_\downarrow$. In this situation n_σ depends on both $v_\mathrm{s}^\uparrow$ and $v_\mathrm{s}^\downarrow$. Nevertheless, v_s^σ is still uniquely determined by n_σ only (up to some constant, as usual).

[10] This statement should not be mixed up with the fact that the total KS potentials in the KS equations for the interacting system are constructed from both spin-channels: for the present task of replacing functional derivatives via the chain rule only the one-to-one correspondence between the spin-densities and potentials of the KS system is relevant.

$$\frac{\delta E_{\rm xc}}{\delta \phi^\dagger_{k=\alpha\uparrow}(\boldsymbol{r})} = \begin{pmatrix} \frac{\delta E_{\rm xc}}{\delta \phi^*_{\alpha\uparrow}(\boldsymbol{r})} \\ 0 \end{pmatrix} ; \qquad \frac{\delta E_{\rm xc}}{\delta \phi^\dagger_{k=\alpha\downarrow}(\boldsymbol{r})} = \begin{pmatrix} 0 \\ \frac{\delta E_{\rm xc}}{\delta \phi^*_{\alpha\downarrow}(\boldsymbol{r})} \end{pmatrix}$$

in the present situation.

The functional derivative $\delta E_{\rm xc}/\delta n_\sigma$ has now been expressed in terms of quantities which can be evaluated: the derivatives $\delta E_{\rm xc}/\delta \phi_k$ and $\partial E_{\rm xc}/\partial \varepsilon_k$ can be calculated for any explicit expression at hand. For instance, for the exact exchange (4.5) one finds

$$\frac{\delta E_{\rm x}}{\delta \phi_k^\dagger(\boldsymbol{r}')} = -e^2 \Theta_k \sum_l \Theta_l \phi_l(\boldsymbol{r}') \int d^3r \, \frac{\phi_l^\dagger(\boldsymbol{r})\, \phi_k(\boldsymbol{r})}{|\boldsymbol{r}-\boldsymbol{r}'|} \tag{6.11}$$

and $\partial E_{\rm x}/\partial \varepsilon_k = 0$. The additivity of $E_{\rm x}$ with respect to spin shows up in (6.11) as soon as Eq. (6.2) is taken into account.

The derivatives $\delta \phi_k^\dagger/\delta v_{\rm s}^\sigma$ and $\delta \varepsilon_k/\delta v_{\rm s}^\sigma$ can be evaluated by use of the KS equation (6.4). An infinitesimal variation $\delta v_{\rm s}$ of the total KS potential (which corresponds to a diagonal 2×2 matrix in the case of SDFT) induces infinitesimal changes $\delta \phi_k$ and $\delta \varepsilon_k$ of the KS states and eigenvalues, respectively,

$$\left\{ -\frac{\hbar^2 \boldsymbol{\nabla}^2}{2m} + v_{\rm s}(\boldsymbol{r}) + \delta v_{\rm s}(\boldsymbol{r}) \right\} [\phi_k(\boldsymbol{r}) + \delta\phi_k(\boldsymbol{r})] = [\varepsilon_k + \delta\varepsilon_k]\, [\phi_k(\boldsymbol{r}) + \delta\phi_k(\boldsymbol{r})] \; .$$

To first order one thus finds

$$\left\{ -\frac{\hbar^2 \boldsymbol{\nabla}^2}{2m} + v_{\rm s}(\boldsymbol{r}) - \varepsilon_k \right\} \delta\phi_k(\boldsymbol{r}) \;=\; [\delta\varepsilon_k - \delta v_{\rm s}(\boldsymbol{r})]\, \phi_k(\boldsymbol{r}) \; . \tag{6.12}$$

Multiplication by $\phi_k^\dagger(\boldsymbol{r})$, integration over $\boldsymbol{r}$ and use of the unperturbed KS equation then yield

$$\delta\varepsilon_k = \int d^3r\, \phi_k^\dagger(\boldsymbol{r})\, \delta v_{\rm s}(\boldsymbol{r})\, \phi_k(\boldsymbol{r}) \; . \tag{6.13}$$

Upon insertion into (6.12) one arrives at

$$\left\{ -\frac{\hbar^2 \boldsymbol{\nabla}^2}{2m} + v_{\rm s}(\boldsymbol{r}) - \varepsilon_k \right\} \delta\phi_k(\boldsymbol{r}) = \left[\int d^3r'\, \phi_k^\dagger(\boldsymbol{r}')\delta v_{\rm s}(\boldsymbol{r}')\phi_k(\boldsymbol{r}') - \delta v_{\rm s}(\boldsymbol{r}) \right] \phi_k(\boldsymbol{r}) \; .$$

This inhomogeneous differential equation can be solved with the aid of the associated Green's function G_k,

$$\left\{ -\frac{\hbar^2 \boldsymbol{\nabla}^2}{2m} + v_{\rm s}(\boldsymbol{r}) - \varepsilon_k \right\} G_k(\boldsymbol{r},\boldsymbol{r}') \;=\; \delta^{(3)}(\boldsymbol{r}-\boldsymbol{r}') - \phi_k(\boldsymbol{r})\, \phi_k^\dagger(\boldsymbol{r}') \, , \tag{6.14}$$

which can be identified as[11]

$$G_k(\boldsymbol{r},\boldsymbol{r}') = \sum_{l\neq k} \frac{\phi_l(\boldsymbol{r})\,\phi_l^\dagger(\boldsymbol{r}')}{\varepsilon_l - \varepsilon_k}\,. \tag{6.15}$$

G_k has the obvious symmetry

$$G_k^\dagger(\boldsymbol{r},\boldsymbol{r}') = G_k(\boldsymbol{r}',\boldsymbol{r})\,. \tag{6.16}$$

In terms of the Green's function, $\delta\phi_k$ is given by

$$\delta\phi_k(\boldsymbol{r}) = -\int d^3r'\, G_k(\boldsymbol{r},\boldsymbol{r}')\,\delta v_{\rm s}(\boldsymbol{r}')\,\phi_k(\boldsymbol{r}')\,. \tag{6.17}$$

It should be no surprise that Eqs. (6.13) and (6.17) are quite familiar from first order perturbation theory. By insertion of (6.5) into (6.13), (6.17) one obtains

$$\frac{\delta\phi_k^\dagger(\boldsymbol{r})}{\delta v_{\rm s}^\sigma(\boldsymbol{r}')} = -\,\phi_k^\dagger(\boldsymbol{r}')\,P_\sigma\,G_k(\boldsymbol{r}',\boldsymbol{r}) \tag{6.18}$$

$$\frac{\delta\varepsilon_k}{\delta v_{\rm s}^\sigma(\boldsymbol{r})} = \phi_k^\dagger(\boldsymbol{r})\,P_\sigma\,\phi_k(\boldsymbol{r})\,. \tag{6.19}$$

It remains to deal with the factor $\delta v_{\rm s}^\sigma/\delta n^\sigma$. The inverse of this quantity is the static response function of the KS system,

$$\frac{\delta n^\sigma(\boldsymbol{r})}{\delta v_{\rm s}^{\sigma'}(\boldsymbol{r}')} \equiv \chi_{\rm s}^{\sigma\sigma'}(\boldsymbol{r},\boldsymbol{r}')\,. \tag{6.20}$$

As indicated by its name and notation, $\chi_{\rm s}^{\sigma\sigma'}$ is related to the density–density response function (4.67). The discussion of this point is, however, relegated to Sect. 6.2.4. $\chi_{\rm s}^{\sigma\sigma'}$ is easily calculated by insertion of (6.18) and its hermitian con-

---

[11] A closer look at the expression (6.15) exhibits that G_k contains contributions from both spins. With the explicit form of the quantum number k, $k=\alpha s$, one finds in the case of SDFT,

$$\begin{aligned} G_{\alpha s}(\boldsymbol{r},\boldsymbol{r}') &= \sum_{s'} \sum_{\substack{\beta \\ \beta\neq\alpha \text{ if } s'=s}} \frac{\phi_{\beta s'}(\boldsymbol{r})\,\phi_{\beta s'}^\dagger(\boldsymbol{r}')}{\varepsilon_{\beta s'}-\varepsilon_{\alpha s}} \\ &= \begin{pmatrix} \displaystyle\sum_{\substack{\beta \\ \beta\neq\alpha \text{ if } s=\uparrow}} \frac{\phi_{\beta\uparrow}(\boldsymbol{r})\,\phi_{\beta\uparrow}^*(\boldsymbol{r}')}{\varepsilon_{\beta\uparrow}-\varepsilon_{\alpha s}} & 0 \\ 0 & \displaystyle\sum_{\substack{\beta \\ \beta\neq\alpha \text{ if } s=\downarrow}} \frac{\phi_{\beta\downarrow}(\boldsymbol{r})\,\phi_{\beta\downarrow}^*(\boldsymbol{r}')}{\varepsilon_{\beta\downarrow}-\varepsilon_{\alpha s}} \end{pmatrix} . \end{aligned}$$

The contribution to $G_{\alpha s}$ combining eigenvalues from both spin channels is a result of the particular notation chosen, which covers spin-saturated and spin-polarized systems at the same time. It drops out of all relevant formulae.

jugate into (6.7),

$$\chi_s^{\sigma\sigma'}(\boldsymbol{r},\boldsymbol{r}') = -\sum_k \Theta_k \phi_k^\dagger(\boldsymbol{r}) P_\sigma G_k(\boldsymbol{r},\boldsymbol{r}') P_{\sigma'} \phi_k(\boldsymbol{r}') + c.c.. \tag{6.21}$$

$\chi_s^{\sigma\sigma'}$ is diagonal in spin, as can be shown by use of Eqs. (6.2) and (6.6),

$$\chi_s^{\sigma\sigma'}(\boldsymbol{r},\boldsymbol{r}') = -\delta_{\sigma\sigma'} \sum_\alpha \Theta_{\alpha\sigma} \phi_{\alpha\sigma}^*(\boldsymbol{r}) G_{\alpha\sigma}(\boldsymbol{r},\boldsymbol{r}') \phi_{\alpha\sigma}(\boldsymbol{r}') + c.c.. \tag{6.22}$$

Insertion of G_k into (6.21) allows the derivation of an alternative form for $\chi_s^{\sigma\sigma'}$. Splitting the summation over all states $l \neq k$ in G_k into contributions from occupied and unoccupied states, one arrives at

$$\begin{aligned}
\chi_s^{\sigma\sigma'}(\boldsymbol{r},\boldsymbol{r}') = &- \sum_{kl;l\neq k} \Theta_k \Theta_l \, \phi_k^\dagger(\boldsymbol{r}) P_\sigma \frac{\phi_l(\boldsymbol{r})\,\phi_l^\dagger(\boldsymbol{r}')}{\varepsilon_l - \varepsilon_k} P_{\sigma'} \phi_k(\boldsymbol{r}') + c.c. \\
&- \sum_k \Theta_k \sum_l (1-\Theta_l)\, \phi_k^\dagger(\boldsymbol{r}) P_\sigma \frac{\phi_l(\boldsymbol{r})\,\phi_l^\dagger(\boldsymbol{r}')}{\varepsilon_l - \varepsilon_k} P_{\sigma'} \phi_k(\boldsymbol{r}') + c.c..
\end{aligned}$$

The first term on the right-hand side vanishes, as can be shown by interchanging the indices k and l in the $c.c.$ term,

$$c.c. = - \sum_{kl;l\neq k} \Theta_k \Theta_l \phi_l^\dagger(\boldsymbol{r}') P_{\sigma'} \frac{\phi_k(\boldsymbol{r}')\,\phi_k^\dagger(\boldsymbol{r})}{\varepsilon_k - \varepsilon_l} P_\sigma \phi_l(\boldsymbol{r}).$$

One thus finds

$$\chi_s^{\sigma\sigma'}(\boldsymbol{r},\boldsymbol{r}') = -\sum_{kl} \Theta_k (1-\Theta_l) \frac{\phi_k^\dagger(\boldsymbol{r}) P_\sigma \phi_l(\boldsymbol{r})\,\phi_l^\dagger(\boldsymbol{r}') P_{\sigma'} \phi_k(\boldsymbol{r}')}{\varepsilon_l - \varepsilon_k} + c.c., \tag{6.23}$$

which is often more useful than (6.21) in actual calculations.

If one multiplies Eq. (6.10) by $\chi_s^{\sigma\sigma}(\boldsymbol{r},\boldsymbol{r}''')$, integrates over $\boldsymbol{r}$, and uses

$$\int d^3r \, \frac{\delta n_\sigma(\boldsymbol{r})}{\delta v_s^\sigma(\boldsymbol{r}''')} \, \frac{\delta v_s^\sigma(\boldsymbol{r}')}{\delta n_\sigma(\boldsymbol{r})} = \delta^{(3)}(\boldsymbol{r}''' - \boldsymbol{r}')$$

as well as the symmetry of the response function (6.21),

$$\chi_s^{\sigma\sigma}(\boldsymbol{r}''',\boldsymbol{r}) = \chi_s^{\sigma\sigma}(\boldsymbol{r},\boldsymbol{r}'''),$$

one ends up with an integral equation for v_{xc}^σ,

$$\int d^3r' \, \chi_s^{\sigma\sigma}(\boldsymbol{r},\boldsymbol{r}') \, v_{xc}^\sigma(\boldsymbol{r}') = \Lambda_{xc}^\sigma(\boldsymbol{r}), \tag{6.24}$$

with the inhomogeneity given by

$$\Lambda_{\rm xc}^{\sigma}(\boldsymbol{r}) = -\sum_k \int d^3r' \left[\phi_k^\dagger(\boldsymbol{r}) P_\sigma G_k(\boldsymbol{r},\boldsymbol{r}') \frac{\delta E_{\rm xc}}{\delta \phi_k^\dagger(\boldsymbol{r}')} + c.c.\right]$$
$$+ \sum_k \phi_k^\dagger(\boldsymbol{r}) P_\sigma \phi_k(\boldsymbol{r}) \frac{\partial E_{\rm xc}}{\partial \varepsilon_k}. \tag{6.25}$$

The xc-potential is spin-independent in the case of spin-saturated systems. Summing (6.24) over both spins, one finds

$$\int d^3r' \, \chi_{\rm s}(\boldsymbol{r},\boldsymbol{r}') v_{\rm xc}(\boldsymbol{r}') = \Lambda_{\rm xc}(\boldsymbol{r}), \tag{6.26}$$

where $\chi_{\rm s}$ denotes the complete static KS response function,

$$\chi_{\rm s}(\boldsymbol{r},\boldsymbol{r}') = \sum_\sigma \chi_{\rm s}^{\sigma\sigma}(\boldsymbol{r},\boldsymbol{r}') = -\sum_k \Theta_k \phi_k^\dagger(\boldsymbol{r}) G_k(\boldsymbol{r},\boldsymbol{r}') \phi_k(\boldsymbol{r}') + c.c., \tag{6.27}$$

and the inhomogeneity is given by

$$\Lambda_{\rm xc}(\boldsymbol{r}) = \sum_k \left\{ -\int d^3r' \left[\phi_k^\dagger(\boldsymbol{r}) G_k(\boldsymbol{r},\boldsymbol{r}') \frac{\delta E_{\rm xc}}{\delta \phi_k^\dagger(\boldsymbol{r}')} + c.c.\right] + |\phi_k(\boldsymbol{r})|^2 \frac{\partial E_{\rm xc}}{\partial \varepsilon_k} \right\}. \tag{6.28}$$

From now on we will no longer distinguish between Eq. (6.26) and its spin-dependent counterpart (6.24)—for brevity, the subsequent discussion will be restricted to the spin-saturated formalism.

Equation (6.26), termed the *OPM integral equation*, is the central equation of the OPM. It allows the calculation of the multiplicative xc-potential for a given orbital- and eigenvalue-dependent functional $E_{\rm xc}$, as soon as all ϕ_k and ε_k are available. As Eq. (6.26) is linear in $E_{\rm xc}$, each of the components of $E_{\rm xc}$ can be treated separately.

Each cycle of a selfconsistent KS calculation consists of two steps, (i) the solution of the differential equations (3.44) for given $v_{\rm s}$, and (ii) the determination of this $v_{\rm s}$ for given KS orbitals. Step (ii) includes the evaluation of the xc-potential. In the case of an LDA or GGA potential one calculates the density and its derivatives and inserts these quantities into a given analytical formula. In the OPM the solution of Eq. (6.26) replaces this simple insertion of n into a functional.

6.2.3 Total Energy Minimization

The physics behind the OPM integral equation becomes more transparent in the second derivation of Eq. (6.26). This alternative approach, which constitutes the original derivation of the OPM [124, 125], relies on energy minimization. The starting point is a total energy functional expressed in terms of the KS orbitals and eigenvalues, $E[\phi_k, \varepsilon_k]$. As the Hohenberg-Kohn theorem for noninteracting particles guarantees that there is a unique relation between n and $v_{\rm s}$, the standard minimization of the total energy E with respect to n can be substituted by a minimization with respect to v_s,

$$0 = \frac{\delta E[\phi_k, \varepsilon_k]}{\delta v_s(\boldsymbol{r})} \tag{6.29}$$

(for fixed particle number). The derivative (6.29) can be handled in the same fashion as in (6.10),

$$\frac{\delta E[\phi_k, \varepsilon_k]}{\delta v_s(\boldsymbol{r})} = \sum_k \left\{ \int d^3r' \left[\frac{\delta \phi_k^\dagger(\boldsymbol{r}')}{\delta v_s(\boldsymbol{r})} \frac{\delta E}{\delta \phi_k^\dagger(\boldsymbol{r}')} + c.c. \right] + \frac{\delta \varepsilon_k}{\delta v_s(\boldsymbol{r})} \frac{\partial E}{\partial \varepsilon_k} \right\}. \tag{6.30}$$

In addition to the ingredients which are already known from the previous section, Eq. (6.30) contains the functional derivatives of E with respect to ϕ_k and ε_k, which can be evaluated from Eqs. (3.21), (3.24)–(3.26),

$$\frac{\delta E}{\delta \phi_k^\dagger(\boldsymbol{r})} = \Theta_k \left[-\frac{\hbar^2 \nabla^2}{2m} + v_{\rm ext}(\boldsymbol{r}) + v_{\rm H}(\boldsymbol{r}) \right] \phi_k(\boldsymbol{r}) + \frac{\delta E_{\rm xc}}{\delta \phi_k^\dagger(\boldsymbol{r})} \tag{6.31}$$

$$\frac{\partial E}{\partial \varepsilon_k} = \frac{\partial E_{\rm xc}}{\partial \varepsilon_k}. \tag{6.32}$$

One can now use the KS equations to rewrite $\delta E / \delta \phi_k^\dagger$,

$$\frac{\delta E}{\delta \phi_k^\dagger(\boldsymbol{r})} = \Theta_k \left[\varepsilon_k - v_{\rm xc}(\boldsymbol{r}) \right] \phi_k(\boldsymbol{r}) + \frac{\delta E_{\rm xc}}{\delta \phi_k^\dagger(\boldsymbol{r})}. \tag{6.33}$$

Insertion of the unpolarized forms of (6.18), (6.19) as well as of (6.32), (6.33) into (6.30) leads to

$$\begin{aligned} 0 = \sum_k \int d^3r' \left[\phi_k^\dagger(\boldsymbol{r}) G_k(\boldsymbol{r}, \boldsymbol{r}') \left(\Theta_k \phi_k(\boldsymbol{r}') \left(v_{\rm xc}(\boldsymbol{r}') - \varepsilon_k \right) + \frac{\delta E_{\rm xc}}{\delta \phi_k^\dagger(\boldsymbol{r}')} \right) + c.c. \right] \\ + \sum_k |\phi_k(\boldsymbol{r})|^2 \frac{\partial E_{\rm xc}}{\partial \varepsilon_k}. \end{aligned} \tag{6.34}$$

After identification of the various components of χ_s and $\Lambda_{\rm xc}$, which show up in Eq. (6.34), as well as use of the orthogonality relation

$$\int d^3r \, \phi_k^\dagger(\boldsymbol{r}) G_k(\boldsymbol{r}, \boldsymbol{r}') = \int d^3r' \, G_k(\boldsymbol{r}, \boldsymbol{r}') \, \phi_k(\boldsymbol{r}') = 0 \tag{6.35}$$

(which is easily proven by insertion of Eq. (6.15)) one again obtains the OPM integral equation (6.26).

This derivation seems to indicate that the x-only limit of the OPM is identical with the HF approach, as the total energy functional $E[\phi_k, \varepsilon_k]$ formally agrees with the HF energy functional in this limit. The HF approach corresponds, however, to a free minimization of the total energy functional with respect to ϕ_k and ε_k. Equation

(6.29), on the other hand, is not equivalent to a free minimization of E: rather the ϕ_k and ε_k have to satisfy the KS equations with their multiplicative total potential. This requirement represents a subsidiary condition for the minimization of E, which is actually implemented into the OPM equation via Eq. (6.33). This point will be addressed further from a quantitative point of view in Sect. 6.3.

6.2.4 Invariance of Density

The starting point of the third derivation of the OPM integral equation is the equality of the KS density n_s with the density n of the interacting system [102, 345], Eq. (3.23),

$$0 = n_s(\boldsymbol{r}) - n(\boldsymbol{r}) . \tag{6.36}$$

Note that the relation (6.36) relies on the complete framework of the Hohenberg-Kohn and KS formalism. In particular, it implies the application of the minimum principle for the total ground state energy. This statement provides the link between (6.36) and the arguments of Sects. 6.2.2 and 6.2.3.

The interacting density n can be expressed in terms of the 1-particle Green's function (3.118) of the interacting system. Similarly, n_s can be expressed in terms of the Green's function G_s of the noninteracting KS system, Eq. (4.71). G_s has the form (3.124), if ϕ_k and ε_k are understood as the KS states and eigenvalues. In both cases the densities are obtained by taking the appropriate limit $t' \to t$ (compare Eqs. (3.106), (3.122)),

$$0 = -i\,\mathrm{tr}\big\{G_s(\boldsymbol{r}t, \boldsymbol{r}t^+) - G(\boldsymbol{r}t, \boldsymbol{r}t^+)\big\} . \tag{6.37}$$

As usual, t^+ indicates an infinitesimal positive time-shift of t, i.e.

$$f(t^+) \equiv \lim_{\varepsilon \to 0} f(t + |\varepsilon|) .$$

The interacting Green's function obeys the Dyson equation (3.123),

$$G(1,2) = G_0(1,2) + \int d3d4\, G_0(1,3)\Sigma(3,4)G(4,2) , \tag{6.38}$$

where the (Harvard) notation

$$k \equiv (\boldsymbol{r}_k t_k) \tag{6.39}$$

$$\int d3 \equiv \int d^3r_3 \int dt_3 \tag{6.40}$$

$$\delta(3,4) \equiv \delta^{(3)}(\boldsymbol{r}_3 - \boldsymbol{r}_4)\,\delta(t_3 - t_4) \tag{6.41}$$

has been used. G_0 represents the Green's function of electrons which experience the external potential v_{ext}, but do not interact, Σ is the full self-energy of the interacting system, which may be split into an xc- and a Hartree contribution,

$$\Sigma(3,4) = \Sigma_{xc}(3,4) + \frac{1}{\hbar}\delta(3,4)v_H(\boldsymbol{r}_3). \tag{6.42}$$

On the other hand, the KS (single-particle) Green's function satisfies a Dyson equation in which the self-energy is simply given by $v_H + v_{xc}$,

$$G_s(1,2) = G_0(1,2) + \frac{1}{\hbar}\int d3d4\, G_0(1,3)\delta(3,4)[v_H(\boldsymbol{r}_3) + v_{xc}(\boldsymbol{r}_3)]G_s(4,2). \tag{6.43}$$

If one now subtracts Eqs. (6.38) and (6.43) from each other, one finds a relation between G and G_s,

$$G(1,2) = G_s(1,2) + \int d3d4\, G_s(1,3)\left\{\Sigma_{xc}(3,4) - \frac{1}{\hbar}\delta(3,4)v_{xc}(\boldsymbol{r}_3)\right\}G(4,2). \tag{6.44}$$

Equation (6.44) is a Dyson equation, whose irreducible kernel is given by the difference between the full self-energy and the KS self-energy. Upon insertion of (6.44) into (6.37) one obtains

$$0 = -i\,\mathrm{tr}\int d3d4\, G_s(1,3)\left[\Sigma_{xc}(3,4) - \frac{1}{\hbar}\delta(3,4)v_{xc}(\boldsymbol{r}_3)\right]G(4,1^+). \tag{6.45}$$

Equation (6.45), referred to as the *Sham–Schlüter equation*, is a complicated integral equation connecting the KS Green's function, the xc-component of the full self-energy, the xc-potential and the full Green's function. Does this relation have anything to do with the OPM equation (6.26)?

The first step towards an answer to this question is provided by repeated application of the Dyson equation (6.44). After insertion of (6.44) the leading term in (6.45) contains the product of $G_s(1,3)$ with $G_s(4,1)$. Partial evaluation of the 4-integration then yields,

$$\begin{aligned}
&\int d^3r_3 \int dt_3\, \chi_s(1,3)v_{xc}(\boldsymbol{r}_3) \\
&= -i\,\mathrm{tr}\int d3d4\, G_s(1,3)\Sigma_{xc}(3,4)G_s(4,1^+) \\
&\quad -i\,\mathrm{tr}\int d3d4\, G_s(1,3)\left[\Sigma_{xc}(3,4) - \frac{1}{\hbar}\delta(3,4)v_{xc}(\boldsymbol{r}_3)\right] \\
&\quad\quad \times \int d5d6\, G_s(4,5)\left[\Sigma_{xc}(5,6) - \frac{1}{\hbar}\delta(5,6)v_{xc}(\boldsymbol{r}_5)\right]G(6,1^+),
\end{aligned} \tag{6.46}$$

where the time-ordered KS response function (4.67),

$$\chi_s(1,3) = -\frac{i}{\hbar}\,\mathrm{tr}\big[G_s(1,3)G_s(3,1)\big] \tag{6.47}$$

has been introduced in order to make the similarity of (6.46) with (6.26) more apparent. In fact, the left-hand side of Eq. (6.46) is identical with the corresponding term of the OPM equation, if one performs the dt_3 integration and identifies the static response function (zero-frequency limit of the Fourier transform of $\chi_s(1,3)$),

$$\int dt_3\, \chi_s(1,3) = \int dt_3\, \chi_s(\boldsymbol{r}_1,\boldsymbol{r}_3,t_1-t_3) = \chi_s(\boldsymbol{r}_1,\boldsymbol{r}_3,\omega=0) \equiv \chi_s(\boldsymbol{r}_1,\boldsymbol{r}_3)\,. \quad (6.48)$$

The identity of (6.48) with (6.27) can be verified explicitly via the Lehmann representation (4.68). For $\omega = 0$ the pole shifts in (4.68) are no longer relevant, so that one ends up exactly with the form of χ_s given in Eqs. (6.27), (6.23).

On the other hand, the right-hand side of Eq. (6.46) is still different from the inhomogeneity (6.28). In fact, the right-hand side depends on v_{xc} itself, so that (6.46) represents a nonlinear integral equation for v_{xc}.

In order to explain the difference between Eqs. (6.46) and (6.26), one has to remember that an arbitrary, orbital-dependent E_{xc} has been assumed in the discussion of Sects. 6.2.2 and 6.2.3, i.e. the form of E_{xc} has not been specified beyond the fact that it depends on the ϕ_k and ε_k. The starting point (6.36) of the present approach, together with the application of the Dyson equation for both the KS and the interacting system, automatically implies the use of the exact E_{xc}. In order to make closer contact between the first two and the third derivation of the OPM equation, one has to insert the exact E_{xc}, studied in Sect. 4.2.1, into the OPM equation (6.26).

The comparison of Eq. (6.46) with Eq. (6.26) is rather straightforward in the x-only limit, which corresponds to a lowest order expansion of E_{xc} in powers of the coupling constant e^2. In this limit Eq. (6.46) reduces to

$$\int d^3r_3\, \chi_s(\boldsymbol{r}_1,\boldsymbol{r}_3)v_x(\boldsymbol{r}_3) = -i\,\mathrm{tr}\int d3d4\, G_s(1,3)\Sigma_x(3,4)G_s(4,1^+)\,,$$

since each factor of Σ_{xc} or v_{xc} introduces an additional factor of e^2 on the right-hand side of (6.46). Insertion of the exchange contribution Σ_x to the full self-energy, i.e. the standard 1-loop self-energy (3.125), leads to the exchange-only limit of Eq. (6.26), which is obtained by insertion of (6.11) into the inhomogeneity (6.28). This demonstrates explicitly that the Sham-Schlüter equation (6.46) agrees with the standard OPM equation in the x-only limit.

The situation is much more complicated for the correlation contribution. A detailed verification of the identity of (6.46) and (6.26) to all orders in e^2 is not available to date. Such a comparison is hampered by the fact that the exact E_{xc} in Eq. (4.65) not only depends on ϕ_k and ε_k via the Green's function (4.71), but also on its functional derivative v_{xc}. On the one hand, this is exactly what is required in view of the structure of the right-hand side of (6.46), which also depends on v_{xc}. On the other hand, the v_{xc}-dependence implies that the derivative $\delta E_{xc}/\delta\phi_k^\dagger$ on the right-hand side of (6.28) includes contributions from $\delta v_{xc}/\delta\phi_k^\dagger$, which can only be handled within an expansion in powers of e^2 (see Sect. 6.4.1). Nevertheless, there can be little doubt about the general agreement of (6.46) and (6.26).

6.2.5 Exact Relations Based on OPM Integral Equation

Before investigating the properties of orbital-dependent xc-functionals from a quantitative point of view, it seems worthwhile to list a few exact relations, which emerge from an analysis of the OPM integral equation. One first of all recognizes that the OPM equation determines v_{xc} only up to an additive constant. In fact, as norm-conservation, i.e. Eq. (6.35), requires

$$\int d^3r\,\chi_{\mathrm{s}}(\boldsymbol{r},\boldsymbol{r}') = \int d^3r'\,\chi_{\mathrm{s}}(\boldsymbol{r},\boldsymbol{r}') = 0\,, \tag{6.49}$$

one can add any constant to v_{xc} without altering the left-hand side of Eq. (6.26). The same is true for the individual spin channels in the case of SDFT, i.e. of Eq. (6.24). One therefore has to ensure the explicit normalization of v_{xc} in the process of solving the OPM equation.

For finite systems one usually requires v_{xc} to vanish asymptotically,

$$\lim_{|\boldsymbol{r}|\to\infty} v_{\mathrm{xc}}(\boldsymbol{r}) = 0\,. \tag{6.50}$$

Let us therefore examine the solution of (6.26) for finite systems, with the aim to extract the behavior of v_{xc} for large $|\boldsymbol{r}|$.

One starts with (6.34),

$$\begin{aligned} 0 = \sum_k \int d^3r' \left\{ \phi_k^\dagger(\boldsymbol{r}) G_k(\boldsymbol{r},\boldsymbol{r}') \left[\Theta_k v_{\mathrm{xc}}(\boldsymbol{r}')\phi_k(\boldsymbol{r}') - \frac{\delta E_{\mathrm{xc}}}{\delta \phi_k^\dagger(\boldsymbol{r}')} \right] + c.c. \right\} \\ + \sum_k |\phi_k(\boldsymbol{r})|^2 \frac{\partial E_{\mathrm{xc}}}{\partial \varepsilon_k}\,. \end{aligned} \tag{6.51}$$

Further analysis of (6.51) requires a specification of the eigenvalue-dependence of E_{xc} and thus of the form of the correlation contribution to E_{xc}. The discussion of (6.51) for large $|\boldsymbol{r}|$ will, therefore, be restricted to the x-only limit. In addition, we assume the energetically highest occupied KS state ϕ_h to be nondegenerate. This implies that all other occupied states decay exponentially faster than ϕ_h, while ϕ_h decays exponentially faster than all unoccupied states.[12] Insertion of G_k and (6.11) into (6.51) gives

$$0 = \sum_{kl;l\neq k} \Theta_k \int d^3r'\,\phi_k^\dagger(\boldsymbol{r}) \frac{\phi_l(\boldsymbol{r})\,\phi_l^\dagger(\boldsymbol{r}')}{\varepsilon_l - \varepsilon_k} F_k(\boldsymbol{r}') + c.c.\,, \tag{6.52}$$

with the abbreviation

---

[12] It is sometimes helpful to think of finite systems as being enclosed in a large box or spherical cavity of radius R, with hard-wall boundary conditions applied at the surface of the box or cavity [346], in order to simplify the discussion of positive energy states. In this situation all KS states vanish on the surface, so that at first glance the following arguments, based on exponential suppression, appear to be incorrect. However, the behavior of bound, i.e. negative energy, states is essentially unaffected by the boundary conditions at R in an extended region between the finite system and the surface, as long as R is sufficiently large. For systems enclosed in a box the arguments given in this section therefore apply to this intermediate range of large $\boldsymbol{r}$, which are not yet close to the surface.

$$F_k(\mathbf{r}') = v_{\mathrm{x}}(\mathbf{r}')\phi_k(\mathbf{r}') + e^2 \sum_j \Theta_j \phi_j(\mathbf{r}') \int d^3r'' \frac{\phi_j^\dagger(\mathbf{r}'')\,\phi_k(\mathbf{r}'')}{|\mathbf{r}'' - \mathbf{r}'|} \,. \tag{6.53}$$

In analogy to the derivation of Eq. (6.23), the sum over l can be split into two parts,

$$\begin{aligned} 0 = &\sum_{kl;l\neq k} \Theta_k \Theta_l \frac{\phi_k^\dagger(\mathbf{r})\phi_l(\mathbf{r})}{\varepsilon_l - \varepsilon_k} \int d^3r'\, \phi_l^\dagger(\mathbf{r}') F_k(\mathbf{r}') + c.c. \\ &+ \sum_k \Theta_k \sum_l (1-\Theta_l) \frac{\phi_k^\dagger(\mathbf{r})\phi_l(\mathbf{r})}{\varepsilon_l - \varepsilon_k} \int d^3r'\, \phi_l^\dagger(\mathbf{r}') F_k(\mathbf{r}') + c.c.\,. \end{aligned} \tag{6.54}$$

For very large $|\mathbf{r}|$ both expressions are dominated by the contributions of the highest occupied orbital. This statement is easily verified for the first line. In the first expression the sum over l is restricted to occupied states, so that the most weakly decaying pair $\phi_k^\dagger(\mathbf{r})\phi_l(\mathbf{r})$ is obtained if either $k = h$ and $l = h - 1$ or vice versa. The analysis is somewhat more involved for the second line. In this expression all $\phi_k^\dagger(\mathbf{r})$ are multiplied by an infinite series of unoccupied $\phi_l(\mathbf{r})$. In order to conclude that the term $k = h$ dominates the second line one thus has to make sure that the summation over all unoccupied $\phi_l(\mathbf{r})$ does not lead to an exponential suppression of the term with $k = h$ relative to energetically lower states k. To this aim, let us examine the coefficient associated with an individual pair $\phi_k^\dagger(\mathbf{r})\phi_l(\mathbf{r})$,

$$a_{kl} = \frac{1}{\varepsilon_l - \varepsilon_k} \int d^3r' \left[\phi_l^\dagger(\mathbf{r}') v_{\mathrm{x}}(\mathbf{r}')\phi_k(\mathbf{r}') + e^2 \sum_j \Theta_j \int d^3r'' \frac{\phi_l^\dagger(\mathbf{r}')\phi_j(\mathbf{r}')\,\phi_j^\dagger(\mathbf{r}'')\phi_k(\mathbf{r}'')}{|\mathbf{r}'' - \mathbf{r}'|} \right] .$$

For a given l, a_{kl} becomes the larger, the less localized $\phi_k(\mathbf{r})$ is. This is immediately obvious for the first contribution, as the degree of localization is reduced the higher the energy of a state is—unoccupied states are always less localized than occupied ones, so that the overlap of $\phi_l^\dagger(\mathbf{r}')$ and $\phi_k(\mathbf{r}')$ is maximized for the least localized ϕ_k, i.e. for $k = h$. In the second term the overlap between $\phi_l^\dagger(\mathbf{r}')$ and $\phi_j(\mathbf{r}')$ is maximal for $j = h$. Simultaneous maximization of the overlap of the second pair involved, $\phi_j^\dagger(\mathbf{r}'')$ and $\phi_k(\mathbf{r}'')$, then requires $k = h$. This overlap argument is particularly applicable to the completely delocalized high energy states. Consequently, in the second line of Eq. (6.54) the term with $k = h$ dominates asymptotically for each individual l, so that the summation over l cannot compensate differences in the asymptotic decay of the $\phi_k(\mathbf{r})$. The most weakly decaying contribution to (6.54) is obtained for $k = h$—all other terms are suppressed exponentially.

Moreover, the combination $\phi_h^\dagger(\mathbf{r})\phi_{h-1}(\mathbf{r})$, which dominates the first line of Eq. (6.54), always decays faster than $\phi_h^\dagger(\mathbf{r})\phi_l(\mathbf{r})$, if l is an unoccupied level. For large $|\mathbf{r}|$, Eq. (6.54) thus reduces to

$$0 = \phi_h^\dagger(\mathbf{r}) \int d^3r' \sum_l (1-\Theta_l) \frac{\phi_l(\mathbf{r})\,\phi_l^\dagger(\mathbf{r}')}{\varepsilon_l - \varepsilon_h} F_h(\mathbf{r}') + c.c.\,. \tag{6.55}$$

At this point one can use the fact that all discrete states can be chosen to be real. In addition, all the bispinors ϕ_k have only one non-vanishing component,

$$\phi_h^\dagger \phi_l = \phi_{\alpha s}^* \phi_{\beta s'} \delta_{ss'} \qquad \text{if } h = \alpha s,\ l = \beta s' .$$

One can thus divide Eq. (6.55) by the non-vanishing component of $\phi_h(\mathbf{r})$,

$$0 = \sum_l (1-\Theta_l) \frac{\phi_l(\mathbf{r})}{\varepsilon_l - \varepsilon_h} \int d^3r'\, \phi_l^\dagger(\mathbf{r}') F_h(\mathbf{r}') + c.c.\,,$$

with the understanding that this relation only holds for that bispinor component of ϕ_l which is projected out by ϕ_h. If one applies the operator

$$-\frac{\hbar^2\nabla^2}{2m}+v_s(\boldsymbol{r})-\varepsilon_h$$

to this equation, one arrives at

$$\begin{aligned}0 &= \sum_l (1-\Theta_l)\phi_l(\boldsymbol{r})\int d^3r'\,\phi_l^\dagger(\boldsymbol{r}')F_h(\boldsymbol{r}')+c.c.\\ &= F_h(\boldsymbol{r})-\sum_l \Theta_l\phi_l(\boldsymbol{r})\int d^3r'\,\phi_l^\dagger(\boldsymbol{r}')F_h(\boldsymbol{r}')+c.c.\,,\end{aligned}$$

where the completeness relation has been used to obtain the second line. Both contributions of the second line are real, so that the $c.c.$ term can be dropped. Asymptotically, the second expression is dominated by the state $l=h$, all other terms are suppressed exponentially,

$$0 = F_h(\boldsymbol{r})-\phi_h(\boldsymbol{r})\int d^3r'\,\phi_h^\dagger(\boldsymbol{r}')F_h(\boldsymbol{r}')\,. \tag{6.56}$$

Equation (6.56) allows to establish two important relations. As $F_h(\boldsymbol{r})$ decays faster than $\phi_h(\boldsymbol{r})$ (provided the normalization (6.50) is used), the asymptotically leading term in (6.56) yields an identity for the highest occupied KS state (the *Krieger-Li-Iafrate identity*) [347],

$$\int d^3r\,v_x(\boldsymbol{r})|\phi_h(\boldsymbol{r})|^2 = -e^2\sum_j \Theta_j\int d^3r\int d^3r'\,\frac{\phi_h^\dagger(\boldsymbol{r})\phi_j(\boldsymbol{r})\phi_j^\dagger(\boldsymbol{r}')\phi_h(\boldsymbol{r}')}{|\boldsymbol{r}-\boldsymbol{r}'|}\,. \tag{6.57}$$

The next to leading order term in (6.56) constrains the asymptotic behavior of v_x,

$$F_h(\boldsymbol{r})\xrightarrow[|\boldsymbol{r}|\to\infty]{}0 \quad\Longrightarrow\quad v_x(\boldsymbol{r})\xrightarrow[|\boldsymbol{r}|\to\infty]{}-e^2\int d^3r'\,\frac{|\phi_h(\boldsymbol{r}')|^2}{|\boldsymbol{r}-\boldsymbol{r}'|}\xrightarrow[|\boldsymbol{r}|\to\infty]{}-\frac{e^2}{|\boldsymbol{r}|}\,, \tag{6.58}$$

so that one ultimately ends up with Eq. (4.20).[13] Equation (6.58) explicitly confirms the earlier statements concerning the physical origin of the limit (4.20) (compare Sect. 4.1). The present discussion is easily extended to Eq. (6.24), so that Eq. (6.58) also applies to the individual spin channels.

Both Eqs. (6.57) and (6.58) allow an unambiguous normalization of v_x in the case of finite, non-degenerate systems. An analogous, though more complicated, statement is available for v_c [348]. It is clear from the physical background of (6.58) that (4.20) also applies to degenerate KS systems. For solids, on the other hand, it is more convenient to fix the average of v_{xc} in the unit cell. The precise value of this average is irrelevant for non-magnetic systems. In the case of spin-polarized solids, however, the normalization of v_{xc} is intrinsically related to the magnetic moment in the unit cell, i.e. to the relative occupation of spin-up and spin-down bands. In this case, the correct ground state is obtained by minimization of the total energy with respect to the pair of spin-up and spin-down averages of v_{xc}.

One further limit of (6.26) appears to be worth a comment. If there is only one occupied orbital ($k=h$) for each spin, the x-only limit of Eq. (6.26) reads

[13] Equation (6.58) has originally been proven for closed-subshell (spherical) atoms, using an alternative approach for the discussion of G_k [125].

$$\phi_h^\dagger(\boldsymbol{r}) \int d^3r'\, G_h(\boldsymbol{r},\boldsymbol{r}')\phi_h(\boldsymbol{r}') \left[v_{\rm x}(\boldsymbol{r}') + e^2 \int d^3r'' \frac{\phi_h^\dagger(\boldsymbol{r}'')\phi_h(\boldsymbol{r}'')}{|\boldsymbol{r}''-\boldsymbol{r}'|} \right] + c.c. = 0\,.$$

One easily identifies

$$v_{\rm x}(\boldsymbol{r}) = -e^2 \int d^3r' \frac{|\phi_h(\boldsymbol{r}')|^2}{|\boldsymbol{r}'-\boldsymbol{r}|} = -\frac{e^2}{2} \int d^3r' \frac{n(\boldsymbol{r}')}{|\boldsymbol{r}'-\boldsymbol{r}|}$$

as the solution of the OPM integral equation for spin-saturated two-electron systems, in perfect agreement with Eq. (4.22). For these systems the exchange potential just eliminates the self-interaction of the electrons, but does not include any Pauli repulsion among equal spins.

The next point to be discussed at this stage is an exact property of $E_{\rm c}$. In order to derive this relation, we assume the KS system to have a completely discrete spectrum, i.e. all eigenstates ϕ_k to be normalizable. If required, this may be ensured by enclosing the finite system in a very large box [346, 349]. Integration of (6.51) over $\boldsymbol{r}$ and use of (6.35) then yields [350]

$$\sum_k \frac{\delta E_{\rm xc}}{\delta \varepsilon_k} = 0. \tag{6.59}$$

The sum rule (6.59) is automatically satisfied as long as $E_{\rm xc}$ only depends on eigenvalue differences,

$$E_{\rm xc} = \sum_{ln} E_{ln}(\varepsilon_l - \varepsilon_n)$$

$$\Longrightarrow \quad \sum_k \frac{\partial E_{\rm xc}}{\partial \varepsilon_k} = \sum_k \sum_{ln} \frac{\partial E_{ln}(\varepsilon)}{\partial \varepsilon}\bigg|_{\varepsilon=\varepsilon_l-\varepsilon_n} (\delta_{kl} - \delta_{kn}) = 0\,. \tag{6.60}$$

All first-principles approximations for $E_{\rm c}$ discussed so far in the literature (see below), have the form (6.60).

Finally, it seems worthwhile to demonstrate that the OPM yields the conventional functional derivative $v_{\rm xc} = \delta E_{\rm xc}/\delta n$, if $E_{\rm xc}$ depends explicitly on the density. In this case Eq. (6.26) reduces to

$$\begin{aligned}
\int d^3r'\, \chi_{\rm s}(\boldsymbol{r},\boldsymbol{r}')\, v_{\rm xc}(\boldsymbol{r}') &= -\sum_k \int d^3r'\, \phi_k^\dagger(\boldsymbol{r}) G_k(\boldsymbol{r},\boldsymbol{r}') \frac{\delta E_{\rm xc}[n]}{\delta \phi_k^\dagger(\boldsymbol{r}')} + c.c. \\
&= -\sum_k \int d^3r'\, \phi_k^\dagger(\boldsymbol{r}) G_k(\boldsymbol{r},\boldsymbol{r}')\phi_k(\boldsymbol{r}') \frac{\delta E_{\rm xc}[n]}{\delta n(\boldsymbol{r}')} + c.c. \\
&= \int d^3r'\, \chi_{\rm s}(\boldsymbol{r},\boldsymbol{r}') \frac{\delta E_{\rm xc}[n]}{\delta n(\boldsymbol{r}')}\,.
\end{aligned}$$

If one now multiplies both sides by $\chi_{\rm s}^{-1}$, one recovers the original definition of $v_{\rm xc}$.

6.2.6 Krieger-Li-Iafrate Approximation (KLI)

One has now reached the point at which it is clear that, as a matter of principle, orbital-dependent functionals can be handled in a fashion consistent with the spirit of DFT. Moreover, the subsequent sections will show that the OPM integral equation can also be solved in practice. However, in view of the complicated structure of Eqs. (6.26), (6.28) and, in particular, of Eq. (6.45), the question concerning the efficiency of the OPM scheme arises quite naturally. The answer obviously depends on the system under consideration and on the numerical implementation of the OPM chosen. Nevertheless, as a rule of thumb, one can say that OPM calculations are essentially one or two orders of magnitude more involved than corresponding GGA calculations. Consequently, an approximate (semi-analytical) solution of the OPM integral equation is of definitive interest.

The main reason for the demanding nature of the OPM is the presence of the Green's function (6.15) both in the response function (6.27) and in the inhomogeneity (6.28). The Green's function depends on the complete KS spectrum and not just on the occupied states. A full solution of (6.26) therefore requires the evaluation and, perhaps, the storage of all occupied and unoccupied KS states.

One way to circumvent the explicit evaluation of unoccupied states is an approximate procedure that has been suggested by Krieger, Li and Iafrate (KLI) [351]. The idea is to use a *closure approximation* (also called *common energy denominator approximation*) for the Green's function, i.e. to approximate the eigenvalue difference in the denominator of (6.15) by some average $\Delta\overline{\varepsilon}$ [124, 351],

$$G_k(\boldsymbol{r},\boldsymbol{r}') \approx \sum_{l\neq k} \frac{\phi_l(\boldsymbol{r})\phi_l^\dagger(\boldsymbol{r}')}{\Delta\overline{\varepsilon}} = \frac{\delta^{(3)}(\boldsymbol{r}-\boldsymbol{r}') - \phi_k(\boldsymbol{r})\phi_k^\dagger(\boldsymbol{r}')}{\Delta\overline{\varepsilon}} . \tag{6.61}$$

Insertion into the OPM integral equation leads to

$$\begin{aligned} v_{\rm xc}^{\rm av}(\boldsymbol{r}) &= \frac{1}{2n(\boldsymbol{r})} \sum_k \left\{ \left[\phi_k^\dagger(\boldsymbol{r}) \frac{\delta E_{\rm xc}}{\delta \phi_k^\dagger(\boldsymbol{r})} + c.c. \right] + |\phi_k(\boldsymbol{r})|^2 \left[\Delta v_k - \Delta\overline{\varepsilon} \frac{\partial E_{\rm xc}}{\partial \varepsilon_k} \right] \right\} \\ \Delta v_k &= \int d^3r \left\{ \Theta_k |\phi_k(\boldsymbol{r})|^2 v_{\rm xc}^{\rm av}(\boldsymbol{r}) - \phi_k^\dagger(\boldsymbol{r}) \frac{\delta E_{\rm xc}}{\delta \phi_k^\dagger(\boldsymbol{r})} \right\} + c.c.. \end{aligned} \tag{6.62}$$

This approximation is completely unambiguous as soon as $E_{\rm xc}$ is independent of ε_k. On the other hand, the presence of $\partial E_{\rm xc}/\partial \varepsilon_k$ introduces a new energy scale in the case of eigenvalue-dependent $E_{\rm xc}$ (via $\Delta\overline{\varepsilon}$). Given the basic concept of the closure approximation, it is obvious that this term should be neglected. The consequences of this step have been investigated in detail for the relativistic exchange [350] (see Chap. 8). In this case neglect of the $\partial E_{\rm xc}/\partial \varepsilon_k$-contribution represents an excellent approximation. One should note, however, that the relativistic exchange depends only weakly on the eigenvalues ε_k, unlike orbital-dependent correlation functionals (compare Sect. 6.5). In any case, no systematic scheme for the specification of $\Delta\overline{\varepsilon}$ has been suggested so far. The KLI approximation is thus always understood to imply neglect of the $\partial E_{xc}/\partial \varepsilon_k$-term,

$$v_{xc}^{KLI}(\boldsymbol{r}) = \frac{1}{2n(\boldsymbol{r})}\sum_k \left\{ \left[\phi_k^\dagger(\boldsymbol{r}) \frac{\delta E_{xc}}{\delta \phi_k^\dagger(\boldsymbol{r})} + c.c. \right] + |\phi_k(\boldsymbol{r})|^2 \Delta v_k^{KLI} \right\} . \tag{6.63}$$

A careful look at Eqs. (6.62), (6.63) shows that one has not yet found a full resolution of the problem, as v_{xc}^{KLI} appears both on the left-hand and on the right-hand side of (6.63). Fortunately, one can recast Eqs. (6.62), (6.63) as a set of linear equations which allow the determination of Δv_k^{KLI} without prior knowledge of v_{xc}^{KLI} [351]. This provides an analytical solution of the integral equation (6.63). Alternatively, one can iterate Eqs. (6.62), (6.63) until self-consistency, starting with some approximation for Δv_k^{KLI}, e.g. obtained from the LDA.

When applied to the exact exchange, a calculation with the KLI method is as efficient as a Hartree-Fock calculation, and often only slightly less efficient than a GGA calculation. At this point one should nevertheless keep in mind that the KLI approximation only speeds up the calculation of G_k, but not that of the other ingredients of the OPM equation. The most time-consuming step in a KLI calculation is usually the evaluation of $\delta E_{xc}/\delta \phi_k^\dagger$: as soon as the exact exchange is used, the evaluation of Slater integrals is required, which is generally more time consuming than the calculation of density gradients.

The KLI approximation preserves both the identity (6.57) and the exact asymptotic behavior of v_x for finite systems, Eq. (6.58). It is exact for spin-saturated two-electron systems, i.e. it also satisfies Eq. (4.22).

A variant of the KLI approximation within the x-only limit is the so-called *localized HF approximation* (LHF) [352, 328]. Its derivation is based on the assumption that the HF and the x-only DFT ground state orbitals (resulting from the multiplicative LHF potential) are identical. Subtraction of the total HF and x-only DFT energies, using this assumption, leads to the following expression for the exchange potential,

$$\begin{aligned} v_x^{LHF}(\boldsymbol{r}) = \frac{1}{2n(\boldsymbol{r})} \Bigg\{ \sum_k \left[\int d^3r'\, \phi_k^\dagger(\boldsymbol{r}) v_x^{HF}(\boldsymbol{r},\boldsymbol{r}') \phi_k(\boldsymbol{r}') + c.c. \right] \\ + \sum_{kl} \Theta_k \Theta_l\, \phi_k^\dagger(\boldsymbol{r}) \phi_l(\boldsymbol{r}) \Delta v_{kl} \Bigg\} \end{aligned} \tag{6.64}$$

$$\Delta v_{kl} = \int d^3r \int d^3r'\, \phi_k^\dagger(\boldsymbol{r}) \left\{ v_x^{LHF}(\boldsymbol{r})\, \delta^{(3)}(\boldsymbol{r}-\boldsymbol{r}') - v_x^{HF}(\boldsymbol{r},\boldsymbol{r}') \right\} \phi_l(\boldsymbol{r}') + c.c.$$

$$v_x^{HF}(\boldsymbol{r},\boldsymbol{r}') = -e^2 \sum_m \Theta_m \frac{\phi_m(\boldsymbol{r})\, \phi_m^\dagger(\boldsymbol{r}')}{|\boldsymbol{r}-\boldsymbol{r}'|} . \tag{6.65}$$

The KLI approximation is obtained from this result by neglect of the off-diagonal terms,

$$\Delta v_{kl} \longrightarrow \Delta v_{kk}\, \delta_{kl} .$$

Quantitatively one finds only limited differences between KLI and LHF results [352].

6.2.7 OPM in Case of Degeneracy

In the derivation of the OPM integral equation we have so far assumed that the KS states are non-degenerate. We now consider the situation in which some symmetry of the system leads to degenerate KS states. These states will be denoted by $\phi_{k\nu}$ in the following, where k identifies all KS energy levels and ν distinguishes the KS states within the degenerate subspaces,

$$\left\{-\frac{\hbar^2\nabla^2}{2m}+v_s(\boldsymbol{r})\right\}\phi_{k\nu}(\boldsymbol{r})=\varepsilon_k\phi_{k\nu}(\boldsymbol{r}). \tag{6.66}$$

The functions $\phi_{k\nu}$ are assumed to be orthonormal,

$$\int d^3r\,\phi_{k\nu}^\dagger(\boldsymbol{r})\phi_{l\rho}(\boldsymbol{r})=\delta_{kl}\,\delta_{\nu\rho}\,. \tag{6.67}$$

Our derivation of the associated OPM equation follows the line of Sect. 6.2.2. The discussion is restricted to the spin-saturated situation.

In order to evaluate $\delta\phi_{k\nu}/\delta v_s$, one introduces a perturbation, in analogy to Eq. (6.12). In general, the perturbation δv_s will lift the degeneracy of the unperturbed states. In the present situation, however, one is only interested in symmetry-preserving perturbations, as the resulting potential is to be used in a self-consistent calculation. The perturbed states can therefore be characterized by the same quantum numbers as the unperturbed states. To first order in the symmetry-preserving δv_s one has

$$\left\{-\frac{\hbar^2\nabla^2}{2m}+v_s(\boldsymbol{r})-\varepsilon_k\right\}\delta\phi_{k\nu}(\boldsymbol{r})=\left[\delta\varepsilon_k-\delta v_s(\boldsymbol{r})\right]\phi_{k\nu}(\boldsymbol{r})\,. \tag{6.68}$$

Multiplication by $\phi_{k\rho}^\dagger(\boldsymbol{r})$, integration over $\boldsymbol{r}$ and use of (6.66) allow the determination of the eigenvalue shifts,

$$\delta\varepsilon_k\,\delta_{\nu\rho}=\int d^3r\,\phi_{k\rho}^\dagger(\boldsymbol{r})\,\delta v_s(\boldsymbol{r})\,\phi_{k\nu}(\boldsymbol{r})\,. \tag{6.69}$$

Upon insertion into (6.68) one arrives at

$$\begin{aligned}&\left\{-\frac{\hbar^2\nabla^2}{2m}+v_s(\boldsymbol{r})-\varepsilon_k\right\}\delta\phi_{k\nu}(\boldsymbol{r})\\&=\left[\int d^3r'\,\phi_{k\nu}^\dagger(\boldsymbol{r}')\delta v_s(\boldsymbol{r}')\phi_{k\nu}(\boldsymbol{r}')-\delta v_s(\boldsymbol{r})\right]\phi_{k\nu}(\boldsymbol{r})\,.\end{aligned}$$

This differential equation can be solved via a suitably redefined Green's function G_k,

$$\left\{-\frac{\hbar^2\nabla^2}{2m}+v_s(\boldsymbol{r})-\varepsilon_k\right\}G_k(\boldsymbol{r},\boldsymbol{r}')=\delta^{(3)}(\boldsymbol{r}-\boldsymbol{r}')-\sum_\nu \phi_{k\nu}(\boldsymbol{r})\,\phi_{k\nu}^\dagger(\boldsymbol{r}') \tag{6.70}$$

$$G_k(\boldsymbol{r},\boldsymbol{r}')=\sum_{l\neq k}\sum_\rho \frac{\phi_{l\rho}(\boldsymbol{r})\,\phi_{l\rho}^\dagger(\boldsymbol{r}')}{\varepsilon_l-\varepsilon_k}\,. \tag{6.71}$$

Utilizing (6.69) and (6.70), $\delta\phi_{k\nu}$ is given by

$$\delta\phi_{k\nu}(\boldsymbol{r})=-\int d^3r'\,G_k(\boldsymbol{r},\boldsymbol{r}')\,\delta v_s(\boldsymbol{r}')\,\phi_{k\nu}(\boldsymbol{r}')\,. \tag{6.72}$$

At first glance, Eqs. (6.69), (6.72) are very similar to their non-degenerate counterparts (6.13), (6.17). However, the point to be noted is that $\delta v_s(\boldsymbol{r})$ can no longer be varied arbitrarily: rather, the variation has to preserve the symmetry, which has to be taken into account when performing the functional derivative.

One has basically two options to implement this constraint. The first is the direct use of a form of $\delta v_s(\boldsymbol{r})$ relevant for the problem under investigation. For instance, in the case of spherically symmetric systems $\delta v_s(\boldsymbol{r})$ just depends on $|\boldsymbol{r}|$, so that the angular coordinates can be integrated out in both (6.69) and (6.72). In this case the functional derivative determining the OPM equation is a derivative with respect to the spherically symmetric potential $\delta v_s(|\boldsymbol{r}|)$, leading to an integral equation in the variable $|\boldsymbol{r}|$.

Alternatively, one may resort to an explicit symmetrization of an unconstrained variation,

$$\frac{\delta F[v_s]}{\delta v_s(\boldsymbol{r})}=\lim_{\varepsilon\to 0}\frac{1}{\varepsilon}\left\{F\left[v_s(\boldsymbol{r}')+\frac{\varepsilon}{h}\sum_R \hat{P}_R(\boldsymbol{r})\,\delta^{(3)}(\boldsymbol{r}-\boldsymbol{r}')\right]-F\left[v_s(\boldsymbol{r}')\right]\right\}, \tag{6.73}$$

where $\hat{P}_R$ denotes the symmetry operators which commute with the KS Hamiltonian,

$$\left[\hat{P}_R,\hat{H}_{\rm KS}\right]=0 \qquad \forall\,R\,, \tag{6.74}$$

and the sum over R runs over all h elements of the symmetry group of $\hat{H}_{\rm KS}$. Using (6.73) in Eqs. (6.69), (6.72), the functional derivative can be implemented as an unconstrained 3-dimensional variation of $\delta v_s(\boldsymbol{r})$, which leads to

$$\frac{\delta\phi_{k\nu}^\dagger(\boldsymbol{r}')}{\delta v_s(\boldsymbol{r})}=-\frac{1}{h}\sum_R \hat{P}_R(\boldsymbol{r})\phi_{k\nu}^\dagger(\boldsymbol{r})\,G_k(\boldsymbol{r},\boldsymbol{r}') \tag{6.75}$$

$$\frac{\delta\varepsilon_k}{\delta v_s(\boldsymbol{r})}=\frac{1}{h}\sum_R \hat{P}_R(\boldsymbol{r})\phi_{k\nu}^\dagger(\boldsymbol{r})\,\phi_{k\nu}(\boldsymbol{r})\,. \tag{6.76}$$

The ν-dependence of the right-hand side of Eq. (6.76) is eliminated automatically by the sum over R (see below). One can now proceed as in Sect. 6.2.2 and derive an

OPM equation of the form (6.26) with the ingredients[14]

$$\chi_s(\boldsymbol{r},\boldsymbol{r}') = -\frac{1}{h}\sum_R \hat{P}_R(\boldsymbol{r})\sum_{kv}\Theta_k\phi_{kv}^\dagger(\boldsymbol{r})G_k(\boldsymbol{r},\boldsymbol{r}')\phi_{kv}(\boldsymbol{r}') + c.c. \tag{6.77}$$

$$\Lambda_{xc}(\boldsymbol{r}) = \frac{1}{h}\sum_R \hat{P}_R(\boldsymbol{r})\sum_{kv}\Bigg\{ -\int d^3r'\left[\phi_{kv}^\dagger(\boldsymbol{r})G_k(\boldsymbol{r},\boldsymbol{r}')\frac{\delta E_{xc}}{\delta\phi_{kv}^\dagger(\boldsymbol{r}')} + c.c.\right] + |\phi_{kv}(\boldsymbol{r})|^2\frac{\partial E_{xc}}{\partial\varepsilon_k}\Bigg\}. \tag{6.78}$$

In both expressions the symmetrization is applied to a product of two KS orbitals, once Eq. (6.71) is utilized for G_k. The symmetrization can be carried through more explicitly, if the quantum numbers of the KS states are chosen according to the irreducible representations of the group of $\hat{H}_{KS}$,

$$\phi_{kv}(\boldsymbol{r}) \longrightarrow \phi_{a\alpha v}(\boldsymbol{r}),$$

where α denotes the irreducible representations and a represents the remaining quantum number(s). One can now utilize the fact that any symmetry operator $\hat{P}_R$ only couples states belonging to the same energy level and the same irreducible representation,

$$\hat{P}_R(\boldsymbol{r})\,\phi_{a\alpha v}(\boldsymbol{r}) = \sum_{i=1}^{l_\alpha} D_{iv}^{(\alpha)}(R)\,\phi_{a\alpha i}(\boldsymbol{r}). \tag{6.79}$$

The function $D_{iv}^{(\alpha)}(R)$ denotes the matrix representation of the group element R in the irreducible representation α and l_α is the dimension of this representation. With this choice the summation over all group elements present in Eqs. (6.77) and (6.78) can be performed via the orthogonality theorem for irreducible representations,

$$\begin{aligned}
\frac{1}{h}\sum_R \hat{P}_R(\boldsymbol{r})\phi_{a\alpha v}^\dagger(\boldsymbol{r})\phi_{b\beta\rho}(\boldsymbol{r}) &= \frac{1}{h}\sum_R\sum_{i=1}^{l_\alpha} D_{iv}^{(\alpha)}(R)^*\,\phi_{a\alpha i}^\dagger(\boldsymbol{r})\sum_{j=1}^{l_\beta} D_{j\rho}^{(\beta)}(R)\,\phi_{b\beta j}(\boldsymbol{r}) \\
&= \sum_{i=1}^{l_\alpha}\sum_{j=1}^{l_\beta}\left[\frac{1}{h}\sum_R D_{iv}^{(\alpha)}(R)^* D_{j\rho}^{(\beta)}(R)\right]\phi_{a\alpha i}^\dagger(\boldsymbol{r})\phi_{b\beta j}(\boldsymbol{r}) \\
&= \delta_{\alpha\beta}\,\delta_{\mu\rho}\,\frac{1}{l_\alpha}\sum_{i=1}^{l_\alpha}\phi_{a\alpha i}^\dagger(\boldsymbol{r})\phi_{b\alpha i}(\boldsymbol{r}).
\end{aligned} \tag{6.80}$$

The summations over states in Eqs. (6.77) and (6.78) then reduce to

[14] Note that the general form of $\delta E_{xc}/\delta\phi_{kv}^\dagger$ automatically accounts for the correct symmetry, if the appropriate form of the orbitals, i.e. ϕ_{kv}, is used for its evaluation.

$$\frac{1}{h}\sum_R \hat{P}_R(\boldsymbol{r}) \sum_{a\alpha\nu} \sum_{b\beta\rho;b\beta\neq a\alpha} \phi^\dagger_{a\alpha\nu}(\boldsymbol{r})\phi_{b\beta\rho}(\boldsymbol{r})A_{a\alpha\nu,b\beta\rho}$$
$$= \sum_{ab\alpha;b\neq a} \frac{1}{l_\alpha}\sum_{i=1}^{l_\alpha} \phi^\dagger_{a\alpha i}(\boldsymbol{r})\phi_{b\alpha i}(\boldsymbol{r})\sum_\nu A_{a\alpha\nu,b\alpha\nu} \tag{6.81}$$

(the summation over $b\beta\rho$ results from the sums over l and ρ in Eq. (6.71)).

As an explicit illustration, we consider a spin-saturated, spherically symmetric system, for which

$$\phi_{a\alpha\nu}(\boldsymbol{r}) \equiv \frac{P_{nl}(r)}{r} Y_{lm}(\Omega_r)\,\chi_s\,, \tag{6.82}$$

i.e. $a \equiv n$, $\alpha \equiv l$, $\nu \equiv m$ and $l_\alpha = 2l+1$. After insertion of (6.81), the symmetry-preserving response function (6.77) reduces to

$$\chi_{\mathrm{s}}(\boldsymbol{r},\boldsymbol{r}') = -\sum_{nn'l;n'\neq n} \frac{\Theta_{nl}}{2l+1} \sum_{m'=-l}^{l}\sum_{ss'} \frac{P_{nl}(r)}{r} Y^*_{lm'}(\Omega_r)\,\chi^\dagger_s \frac{P_{n'l}(r)}{r} Y_{lm'}(\Omega_r)\,\chi_{s'}$$
$$\times \sum_{m=-l}^{l} \frac{1}{\varepsilon_{n'l}-\varepsilon_{nl}} \frac{P_{n'l}(r')}{r'} Y^*_{lm}(\Omega_{r'})\,\chi^\dagger_{s'} \frac{P_{nl}(r')}{r'} Y_{lm}(\Omega_{r'})\,\chi_s + c.c.$$
$$= -\frac{4}{(4\pi r r')^2}\sum_{nl}\Theta_{nl}(2l+1)P_{nl}(r)\sum_{n';n'\neq n}\frac{P_{n'l}(r)P_{n'l}(r')}{\varepsilon_{n'l}-\varepsilon_{nl}}P_{nl}(r')\,. \tag{6.83}$$

Similarly, using (6.11), one finds for the inhomogeneity (6.78) in the x-only limit

$$\Lambda_{\mathrm{x}}(\boldsymbol{r}) = \frac{2}{4\pi r^2}\sum_{nn'l;n'\neq n}\Theta_{nl}\frac{P_{nl}(r)P_{n'l}(r)}{\varepsilon_{n'l}-\varepsilon_{nl}}\sum_{m=-l}^{l}\int\frac{d^3r'}{r'^2}P_{n'l}(r')Y^*_{lm}(\Omega_{r'})$$
$$\times \sum_{n''l''m''}\Theta_{n''l''}P_{n''l''}(r')Y_{l''m''}(\Omega_{r'})$$
$$\times \int\frac{d^3r''}{r''^2}\frac{P_{n''l''}(r'')Y^*_{l''m''}(\Omega_{r''})P_{nl}(r'')Y_{lm}(\Omega_{r''})}{|\boldsymbol{r}'-\boldsymbol{r}''|}$$
$$+c.c.\,. \tag{6.84}$$

Identifying

$$\frac{\delta E_{\mathrm{x}}}{\delta P_{nl}(r')} = -4\sum_m\sum_{n''l''m''}\Theta_{nl}\Theta_{n''l''}\int d\Omega_{r'}\,Y^*_{lm}(\Omega_{r'})\,P_{n''l''}(r')\,Y_{l''m''}(\Omega_{r'})$$
$$\times\int\frac{d^3r''}{r''^2}\frac{P_{n''l''}(r'')Y^*_{l''m''}(\Omega_{r''})P_{nl}(r'')Y_{lm}(\Omega_{r''})}{|\boldsymbol{r}'-\boldsymbol{r}''|}\,, \tag{6.85}$$

one can rewrite (6.84) as

$$\Lambda_{\mathrm{x}}(\boldsymbol{r}) = -\frac{1}{4\pi r^2}\sum_{nl}P_{nl}(r)\int_0^\infty dr'\sum_{n';n'\neq n}\frac{P_{n'l}(r)P_{n'l}(r')}{\varepsilon_{n'l}-\varepsilon_{nl}}\frac{\delta E_{\mathrm{x}}}{\delta P_{nl}(r')}\,. \tag{6.86}$$

Insertion of (6.84) and (6.86) into (6.26) leads to exactly the same OPM equation as obtained by variation of the exchange energy of a spherical system with respect to a manifestly spherical density [125, 223].

It remains to discuss the KLI approximation in the presence of degeneracy. Averaging the eigenvalue denominator of (6.71) as in Eq. (6.61) gives

$$G_{a\alpha}(\boldsymbol{r},\boldsymbol{r}') \approx \frac{\delta^{(3)}(\boldsymbol{r}-\boldsymbol{r}') - \sum_{\nu=1}^{l_\alpha} \phi_{a\alpha\nu}(\boldsymbol{r})\,\phi^\dagger_{a\alpha\nu}(\boldsymbol{r}')}{\Delta\bar{\varepsilon}}. \tag{6.87}$$

Insertion into the OPM integral equation with kernel (6.77) and inhomogeneity (6.78) then yields

$$\begin{aligned} v_{\rm xc}^{\rm KLI}(\boldsymbol{r}) &= \frac{1}{2n(\boldsymbol{r})}\left\{\frac{1}{h}\sum_R \hat{P}_R(\boldsymbol{r})\sum_{a\alpha\nu}\left[\phi^\dagger_{a\alpha\nu}(\boldsymbol{r})\frac{\delta E_{\rm xc}}{\delta\phi^\dagger_{a\alpha\nu}(\boldsymbol{r})} + c.c.\right]\right. \\ &\qquad\left. + \sum_{a\alpha\nu}\frac{1}{l_\alpha}\phi^\dagger_{a\alpha\nu}(\boldsymbol{r})\phi_{a\alpha\nu}(\boldsymbol{r})\sum_\rho \Delta v^{\rm KLI}_{a\alpha\rho}\right\} \qquad (6.88) \\ \Delta v^{\rm KLI}_{a\alpha\rho} &= \int d^3r\left\{\Theta_{a\alpha}\phi^\dagger_{a\alpha\rho}(\boldsymbol{r})v_{\rm xc}^{\rm KLI}(\boldsymbol{r})\phi_{a\alpha\rho}(\boldsymbol{r}) - \phi^\dagger_{a\alpha\rho}(\boldsymbol{r})\frac{\delta E_{\rm xc}}{\delta\phi^\dagger_{a\alpha\rho}(\boldsymbol{r})}\right\} + c.c.. \end{aligned}$$

In the derivation of Eq. (6.88) the fact that (6.88) generates a manifestly symmetric $v_{\rm xc}^{\rm KLI}$ has been used, so that an explicit symmetrization is not necessary. The result (6.88) is invariant under a unitary transformation of the KS states within the degenerate subspaces.

6.3 Exchange-Only Results

Before addressing the issue of correlation in more detail, it is instructive to study the x-only limit from a quantitative point of view. Quite a number of exact x-only results have already been used for the analysis of the LDA and GGA (in Sects. 4.3.5 and 4.5.5). It is clear from this discussion that use of the exact exchange (often termed *EXX approach*) is worth the increased computational effort in many situations. The present section supports this statement with additional results, focussing on two aspects: the first is an assessment of the quality of the KLI approximation. It has been pointed out in Sect. 6.2.6 that large-scale applications of orbital-dependent functionals would benefit substantially from the efficiency of the KLI approximation. Before using the KLI approach in such applications one has to make sure that it yields reasonable results at least for the simplest orbital-dependent functional, the exact $E_{\rm x}$. The second purpose of this section is to demonstrate the role of the exact exchange in the electronic structure of solids. For additional results the reader is referred to the extensive literature on practical realizations and applications of the exact exchange [103, 125, 223, 241, 243, 244, 344, 351, 353–382].

The x-only ground state energies of closed-subshell atoms resulting from the exact DFT exchange are compared with the corresponding HF values in Table 6.1 (all calculations were performed fully numerically with finite differences methods). The energies obtained by solution of the full OPM equation (6.26) for the exchange

Table 6.1 Exchange-only ground state energies of closed-subshell atoms: Selfconsistent OPM results [223] versus KLI, LDA, PW91-GGA [211] and HF [383] energies (all energies in mHartree).

Atom	$-E_{\mathrm{tot}}$	$E_{\mathrm{tot}} - E_{\mathrm{tot}}^{\mathrm{OPM}}$			
	OPM	KLI	LDA	GGA	HF
He	2861.7	0.0	138.0	6.5	0.0
Be	14572.4	0.1	349.1	18.2	−0.6
Ne	128545.4	0.6	1054.7	−23.5	−1.7
Mg	199611.6	0.9	1362.8	−0.5	−3.1
Ar	526812.2	1.7	2294.8	41.2	−5.3
Ca	676751.9	2.2	2591.8	25.7	−6.3
Zn	1777834.4	3.7	3924.5	−252.6	−13.8
Kr	2752042.9	3.2	5176.8	−18.4	−12.0
Sr	3131533.4	3.6	5535.4	−8.8	−12.2
Pd	4937906.0	4.5	6896.0	−65.2	−15.0
Cd	5465114.4	6.0	7292.6	−31.9	−18.7
Xe	7232121.1	6.1	8463.8	54.9	−17.3
Ba	7883526.6	6.5	8792.5	15.7	−17.3
Yb	13391416.3	10.0	10505.6	−852.4	−39.9
Hg	18408960.5	9.1	13040.4	−221.5	−31.0
Rn	21866745.7	8.5	14424.3	8.3	−26.5
Ra	23094277.9	8.7	14807.2	0.5	−25.8
No	32789472.7	12.9	17202.9	−373.1	−39.5

(4.5) are given in the first column. This rigorous handling of the exact DFT exchange functional provides the reference data for the x-only limit. The energies for all other methods are given relative to this reference standard. The KLI approximation for the exact exchange is of primary interest here—the LDA and GGA data are only included for completeness. One observes that the KLI energies are extremely close to the correct OPM energies. For helium the KLI approximation is, as explained in Sect. 6.2.6, exact. All other KLI energies are higher than their full OPM counterparts. This is consistent with the fact that the full OPM generates that potential which minimizes the energy expression at hand. The deviation of the KLI approximation systematically increases with atomic size, i.e. the number of shells present. Nevertheless, the deviation is still no larger than 15 mHartree even for very heavy atoms. An indication of the relevance of this error is obtained by a comparison with the corresponding error of LDA and GGA energies: even the GGA results differ much more from the exact data than the KLI numbers.

We next compare OPM and Hartree-Fock results. It has been emphasized in Sect. 6.2.3 that the x-only OPM represents a restricted HF energy minimization: the same energy expression is minimized, but under the subsidiary condition of generating a multiplicative exchange potential in the case of the OPM. The consequences of this subsidiary condition can be gleaned from Table 6.1. The differences are rather small. The OPM energy for He is identical with the HF value, as the HF equation for the occupied state can be trivially recast as a KS equation with the OPM exchange potential (4.22) in this case (this is only true for the occupied state, but not for the rest of the spectrum). Even for the heaviest elements the differences between OPM and HF energies are below 40 mHartree. The additional variational freedom of the HF approach appears to be of limited importance. This observation is made in many situations (see below). The x-only OPM is in many respects physically equivalent to the HF approximation.

This statement is corroborated by Table 6.2, in which the x-only ground state energies of a number of diatomic molecules are presented (evaluated at the experimental bond lengths). For a reason that will become clear in a moment, the KLI

Table 6.2 Exchange-only ground state energies of diatomic molecules: Selfconsistent OPM [372] results versus KLI [243], LDA and HF [384] energies at the experimental bond lengths (all energies in mHartree).

Molecule	State	R_e	$-E_{tot}$	$E_{tot} - E_{tot}^{KLI}$		
		[Bohr]	KLI	OPM	LDA	HF
H_2	$^1\Sigma$	1.400	1133.6	0.5	89.9	0.0
Li_2	$^1\Sigma$	5.046	14870.5		473.7	−1.1
Be_2	$^1\Sigma$	4.600	29127.4		666.2	−6.3
B_2	$^3\Sigma$	3.003	49085.2		823.6	
C_2	$^1\Sigma$	2.348	75394.0		956.3	−12.6
N_2	$^1\Sigma$	2.075	108985.1	5.6	1229.0	−8.0
O_2	$^3\Sigma$	2.281	149681.3	11.5	1447.0	
F_2	$^1\Sigma$	2.678	198760.2	16.3	1703.3	−12.2
LiH	$^1\Sigma$	3.014	7986.8		282.6	−0.5
BH	$^1\Sigma$	2.336	25129.0		499.1	−2.6
NH	$^3\Sigma$	2.047	54982.9	3.4	711.3	
FH	$^1\Sigma$	1.733	100067.5	11.0	916.3	−3.3
BF	$^1\Sigma$	2.386	124162.1		1312.1	−6.7
CO	$^1\Sigma$	2.132	112783.3	6.7	1252.5	−7.6
NO	$^2\Pi$	2.175	129295.5		1336.5	

energies [243] are used as a reference in Table 6.2. All other energies are given relative to the KLI values. If one compares the KLI and HF energies one finds, as expected, that the HF energies are always lower—with the exception of H_2. For this spin-saturated two-electron system both energies must be identical.

On the other hand, the full OPM results [372] are energetically higher than the KLI data, even though the OPM produces, by construction, the optimum exchange potential. The reason for this unexpected ordering can be found in the technical details of the calculations. The HF results are obtained fully numerically, using large real-space grids [384]. All DFT calculations rely on basis set expansions. Extremely large two-center basis sets have been used [243] in the case of the KLI (and LDA) calculations, so that the KLI numbers are essentially converged with respect to the basis set size. On the other hand, the OPM results are obtained with standard Gaussian basis sets of more modest size, so that the basis set limit is not yet reached. Clearly, converged OPM energies must lie between KLI and HF results. In other words: the ordering $E^{\rm HF} \leq E^{\rm OPM} \leq E^{\rm KLI}$ must only be preserved if the same computational basis is applied for all three methods. Table 6.3 demonstrates that this ordering is actually maintained, if the OPM, KLI and HF energies are obtained with the same, highly accurate basis set [371].[15] The results allow the conclusion that the

Table 6.3 Exchange-only ground state energies of small molecules: Selfconsistent OPM, KLI and HF energies obtained with the same, high-accuracy basis set [371] at the experimental bond geometry (all energies in mHartree).

Molecule	$-E_{\rm tot}$	$E_{\rm tot}-E_{\rm tot}^{\rm OPM}$	
	OPM	KLI	HF
CO	112784.9	2.0	−5.8
H_2O	76064.8	0.8	−2.5
C_2H_2	76850.9	1.1	−4.7

error of the KLI approximation is smaller than the variation resulting from use of different basis sets: an appropriate choice of the basis set is more important than the exact implementation of the OPM integral equation. Compared with the full OPM, the KLI approximation either allows a speed-up of molecular calculations (keeping the basis set fixed) or a gain in accuracy by enlarging the basis set.

It has been emphasized before that the physical and chemical properties usually depend on energy differences, rather than on total energies. In Table 6.4 the simplest energy difference, the (first) ionization potential (IP), is studied for atoms. Again the KLI results are extremely close to the OPM data, which in turn agree very well with the HF IPs. Moreover, the exact x-only eigenvalues of the highest occupied (molecular) orbitals (HOMO) are rather close to the corresponding IPs, consistent with the statement in Sect. 3.6.1. This is a direct consequence of the correct asymptotic $-1/r$-behavior of the exact x-only potential, which dominates the total KS potential in the case of neutral atoms and molecules. In view of the correct asymptotic behavior of the KLI potential, $v_{\rm x}^{\rm KLI}(r\longrightarrow\infty)\sim -1/r$, it should be no surprise that the HOMO eigenvalues obtained with the KLI approximation are generally close to their OPM counterparts. Similar results for molecules [371] are given in Table 6.5.

A more critical energy difference is the electron affinity (EA). The EAs of the prototype negative ions F^- and Na^- are listed in Table 6.6, together with the HOMO eigenvalues obtained with the full OPM and the KLI approximation. One should first note the mere existence of these systems within the OPM [354], in contrast to the situation encountered in the LDA and the GGA. This deficiency of conventional

[15] Note that the basis set used for the representation of the exchange potential has to be in balance with that used for the KS orbitals [385, 386], i.e. the variational freedom in the exchange potential must not be larger than that in the orbitals. Otherwise spurious results may be obtained [387].

Table 6.4 Exchange-only ionization potentials of atoms: Selfconsistent OPM results versus KLI, LDA, PW91-GGA and HF data. Also given is the eigenvalue $\varepsilon_{\mathrm{HOMO}}$ of the highest occupied (molecular) orbital obtained in the OPM (all energies in mHartree).

Atom	$-\varepsilon_{\mathrm{HOMO}}$	IP	$\mathrm{IP-IP^{OPM}}$			
	OPM	OPM	KLI	LDA	GGA	HF
He	918	862	0	−51	4	0
Be	309	295	0	−14	6	1
Mg	253	242	0	−4	12	1
Ca	196	188	0	1	12	0
Sr	179	171	0	3	13	0
Cu	240	231	−2	47	54	5
Ag	222	215	−1	36	41	3
Au	223	216	−2	38	42	2
Li	196	196	0	−11	4	0
Na	182	181	0	−2	10	1
K	148	147	0	2	10	0
Rb	138	137	0	4	12	1
Cs	124	123	0	5	11	0
Zn	293	276	0	34	44	5

Table 6.5 Exchange-only eigenvalues of highest occupied molecular orbital of small molecules: OPM, KLI and HF energies obtained with the same, high-accuracy basis set [371] at the experimental bond geometry (all energies in mHartree).

Molecule	OPM	KLI	HF
CO	553	550	555
H_2O	509	507	511
C_2H_2	411	410	411

Table 6.6 Exchange-only electron affinities of atoms: Selfconsistent KLI versus OPM results. Also given is the highest occupied eigenvalue $\varepsilon_{\mathrm{HOMO}}$ (all energies in mHartree).

Atom	Method	$-\varepsilon_{\mathrm{HOMO}}$	EA
F^-	OPM	181.0	48.5
	KLI	180.4	48.5
Na^-	OPM	13.3	58.4
	KLI	13.2	58.3

density functionals, which was an important motivation for studying implicit functionals (Sect. 6.1.1), is automatically resolved by use of the exact exchange. The existence of atomic negative ions is a direct consequence of the $-1/r$-behavior of the exact exchange potential in the large-r regime. Since the KLI approximation preserves this feature, the KLI-EAs are almost identical to their OPM counterparts.

One next observes the huge difference between the EA and the HOMO eigenvalue $\varepsilon_{\mathrm{HOMO}}$. This discrepancy is somewhat surprising, given the facts that (i) the

IPs of neutral atoms are in reasonable agreement with the corresponding $\varepsilon_{\rm HOMO}$ (see Table 6.4) and that (ii) the exact EA is identical with the exact $\varepsilon_{\rm HOMO}$ [92] (see Sect. 3.6.1). One has to keep in mind, however, that the data in Table 6.6 correspond to the x-only limit. The difference between the EA and $\varepsilon_{\rm HOMO}$ reflects the important role which correlation plays for negative ions. In the case of singly charged negative ions the otherwise asymptotically dominating $-1/r$-term in $v_{\rm ext}+v_{\rm H}+v_{\rm x}$ cancels out completely, so that the total $v_{\rm s}$ is particularly sensitive to its correlation component. For that reason the x-only EAs should not be expected to be close to the experimental EAs.

The most interesting quantities in quantum chemistry are the spectroscopic constants. A set of data for some diatomic molecules is given in Table 6.7. As full OPM

Table 6.7 Exchange-only spectroscopic constants of diatomic molecules: Selfconsistent KLI [243] versus HF [388, 384, 225] results.

Molecule	Method	R_e [Bohr]	D_e [eV]	ω [cm^{-1}]
H_2	KLI	1.386	3.638	4603
	HF	1.386	3.631	4583
Li_2	KLI	5.266	0.168	338
	HF	5.259	0.176	337
B_2	KLI	3.068	0.608	972
	HF	3.096	0.75	939
C_2	KLI	2.332	0.281	1933
	HF	2.341	0.38	1898
N_2	KLI	2.011	4.972	2736
	HF	2.013	4.952	2713
O_2	KLI	2.184	1.441	1981
	HF	2.21	1.455	2002
F_2	KLI	2.496	−1.607	1283
	HF	2.507	−1.627	1276
LiH	KLI	3.037	1.483	1427
	HF	3.035	1.487	1430
FH	KLI	1.694	4.203	4501
	HF	1.695	4.197	4472
CO	KLI	2.080	7.530	2444
	HF	2.082	7.534	2416
Cl_2	KLI	3.727		613
	HF	3.726		618

results for these quantities are not yet available, the KLI numbers can only be compared with HF data. However, the exact OPM energy must be somewhere between the KLI and the HF energy for each individual molecular geometry (for fixed basis set). As long as the KLI and HF energy surfaces are very close, one can be sure that KLI and OPM results are equally close. This is exactly what one finds: the KLI and HF spectroscopic constants (as a measure for the energy surface) show very good agreement—in particular, if one takes into account that not all HF results in Table 6.7 might be fully converged with respect to the size of the basis set. In conclusion, one can state that KLI results for finite systems are generally very close to the corresponding OPM values in the x-only limit.

Does this statement also apply to extended systems? The x-only lattice constants, cohesive energies and bulk moduli of aluminum (fcc lattice) and silicon (diamond structure) obtained with both the full OPM and the KLI approximation are given in Table 6.8. Again the KLI results are found to be very close to the OPM data.

Table 6.8 Lattice constants a, cohesive energies $E_{\rm coh}$ and bulk moduli B of prototype metals and semiconductors: Selfconsistent OPM versus KLI results [380]. All calculations use the plane-wave pseudopotential scheme [389, 103] in combination with self-consistent exact exchange pseudopotentials [380] (for technical details see Figs. 6.6 and 6.7). The same pseudopotential is employed for both the full OPM and the KLI calculation for a given system.

Solid	Method	a [Bohr]	$E_{\rm coh}$ [eV]	B [GPa]
Al	OPM	7.814	1.662	63.9
	KLI	7.816	1.658	63.2
Si	OPM	10.409	6.213	103.8
	KLI	10.406	6.120	104.2

Nevertheless, it is worthwhile to note that the cohesive energies differ by about 0.1 eV in the case of silicon. This is somewhat larger than the deviations obtained for aluminum and most diatomic molecules.

Similar agreement is observed for the corresponding single-particle energies of aluminum, as shown in Fig. 6.4a. However, the same is not true for the band structure of silicon, as illustrated in Fig. 6.4b. The KLI and OPM band energies of the valence and conduction bands differ by about 0.3 eV, with the KLI approximation producing less bound valence states and more strongly bound conduction band states. As a consequence the band gap is underestimated by the KLI scheme by almost a factor of 2, $E_{\rm g}^{\rm OPM} = 1.22\,{\rm eV}$ versus $E_{\rm g}^{\rm KLI} = 0.69\,{\rm eV}$. One concludes that the KLI spectrum of extended systems—in particular, of the unoccupied KS states—can be seriously in error.

In order to understand this result, one has to analyze local quantities as the xc-potential (compare the corresponding discussion for the LDA and GGA in Sect. 4.5.5). The selfconsistent $v_{\rm x}^{\rm KLI}$ is compared with the exact $v_{\rm x}$ for the case of the neon atom in Fig. 6.5.[16] The only difference between the full OPM result and the KLI potential is found in the transition region from the K- to the L-shell, where

[16] Note that the differences between a selfconsistent potential and the potential obtained by insertion of a given density (as e.g. the exact KS density) into a functional are very small in the case of atoms [390]. The same is true for the solution of the OPM equation, if different sets of orbitals are utilized: the origin of the orbitals (and thus of the density) is much less important for the structure of atomic $v_{\rm xc}$ than the functional form of $E_{\rm xc}$.

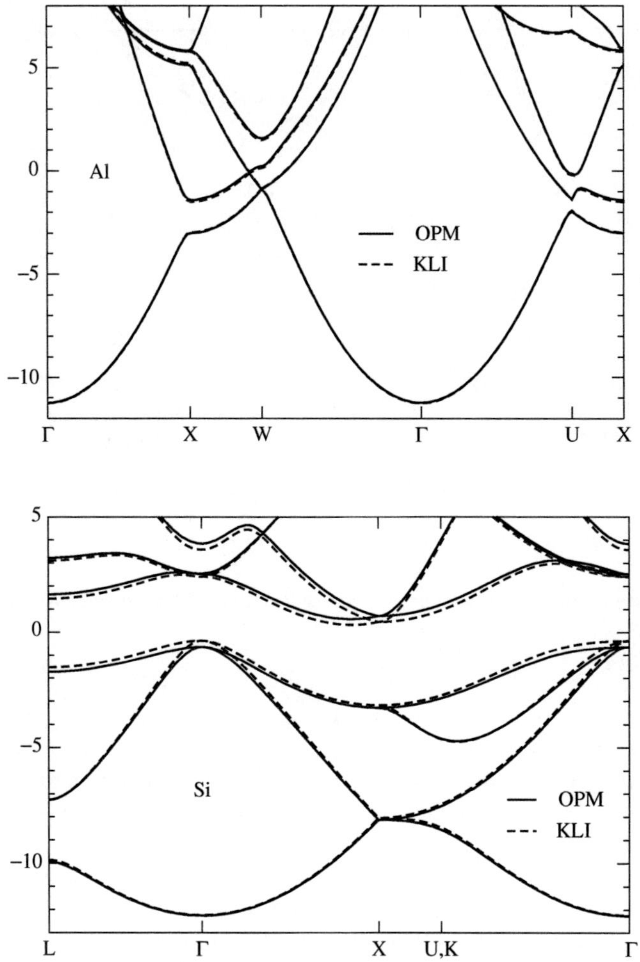

Fig. 6.4 Band structure of (**a**) aluminum (fcc structure) and (**b**) silicon (diamond structure): Self-consistent x-only KLI versus full OPM result (for technical details see Figs. 6.6 and 6.7).

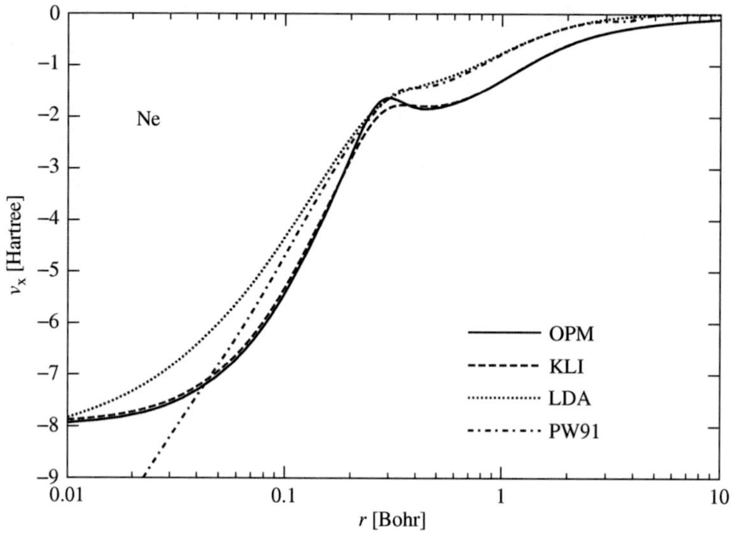

Fig. 6.5 Exchange potential of Ne atom: Selfconsistent OPM, KLI, LDA and PW91-GGA results.

the shell oscillation of v_x^{KLI} is not as pronounced as that of the exact potential. This smoothing of shell oscillations is a general feature of the KLI approximation. For large r both potentials explicitly show the $-1/r$-behavior discussed before, i.e. both potentials are self-interaction free. Figure 6.5 explains the findings in Tables 6.2–6.8 from a microscopic perspective.

Figure 6.6 provides a comparison similar to Fig. 6.5 for the case of a metal. The exchange potential of bulk aluminum is plotted along the [110] direction. As all potentials originate from pseudopotential calculations, the attractive part of v_x associated with the core electrons is missing in Fig. 6.6—the comparison focuses completely on the delocalized valence states of the metal (the corresponding valence densities resulting from selfconsistent calculations with the same functionals are shown in Fig. 6.6b). Again the KLI approximation is reasonably close to the OPM potential. The agreement is particularly convincing in comparison with the GGA result: the gradient corrections to the LDA even go into the wrong direction. In fact, the dependence of the GGA on the local density gradients introduces some completely artificial structures in the low density region.

The same overall picture can be observed in Fig. 6.7, which shows the exchange potential of bulk silicon along the [111] direction. The only difference worth noting in this case is an improvement over v_x^{LDA} by the GGA in high density regions, i.e. in the bonding regime between the nearest neighbor atoms. This result is corroborated by the corresponding self-consistent valence densities, plotted in Fig. 6.7b.

However, as the 2D representations in Figs. 6.8 and 6.9 show, the KLI approximation misses some directional information present in the full OPM potential. This

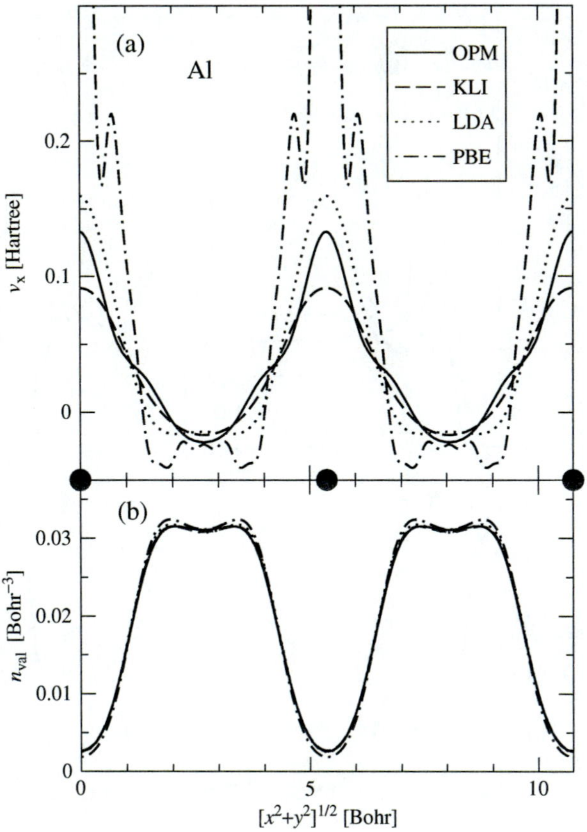

Fig. 6.6 (**a**) Exchange potential of fcc aluminum in [110] direction (surface diagonal of conventional cubic unit cell): Full OPM versus KLI approximation, LDA and PBE-GGA. All v_x have been evaluated from the KS states resulting from a self-consistent x-only OPM calculation within the plane-wave pseudopotential scheme (60 special k-points for integration over 1st Brillouin zone, $E_{cut} = 40$ Ry—this leads to about 460 states per k-point in G_k, $a = 7.6$ Bohr, • = positions of atoms). (**b**) Self-consistent valence densities obtained by full OPM, KLI, LDA and PBE-GGA calculations.

effect is particularly pronounced for silicon, as demonstrated in Fig. 6.9. Obviously, the closure approximation (6.61), on which the KLI scheme is based, leads to an averaging over directions, as all states in the response of the system are treated as being degenerate (similar to the fact that summation over the m-quantum number of atomic orbitals of type $P_{nl}(r)Y_{lm}(\theta,\varphi)$ is equivalent to spherical averaging). The KLI approximation thus has particular problems with the treatment of localized states which are energetically close, but not degenerate.

This can lead to rather fundamental failures, as demonstrated in Figs. 6.10 and 6.11. Figure 6.10 shows the band structure of FeO in the AF II phase which is

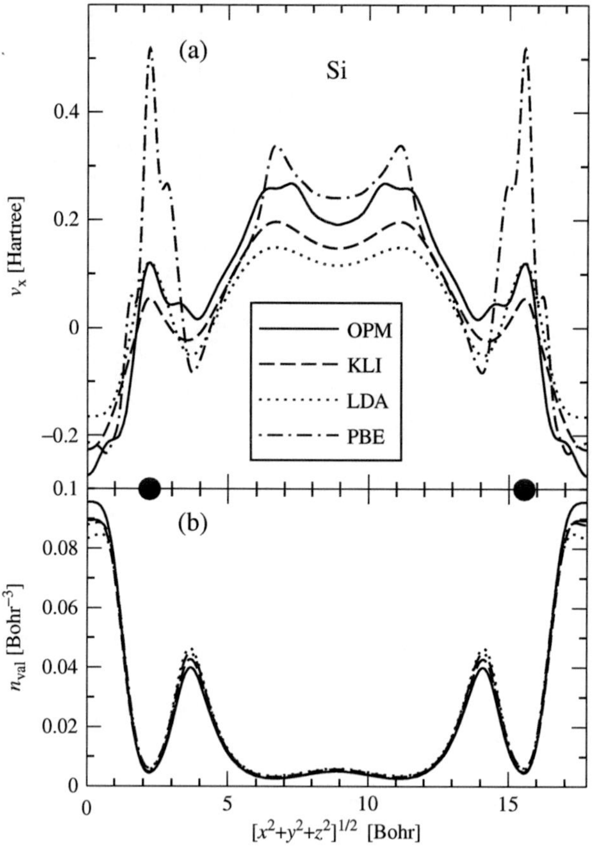

Fig. 6.7 (**a**) Exchange potential of Si (diamond structure) along [111] direction (body diagonal): Full OPM versus KLI approximation, LDA and PBE-GGA. All v_x have been evaluated with the KS states resulting from a self-consistent x-only OPM calculation within the plane-wave pseudopotential scheme (28 special k-points for integration over 1st Brillouin zone, $E_{cut} = 40$ Ry—this leads to about 1150 states per k-point in G_k, $a = 10.26$ Bohr, • = positions of atoms). (**b**) Self-consistent valence densities obtained by full OPM, KLI, LDA and PBE-GGA calculations.

obtained by application of the exact exchange within the KLI approximation (compare Sect. 6.1.3). Correlation is included via the LDA. The correlation contribution has, however, only a minor impact on the resulting bands. One notices a substantial rearrangement of essentially all bands in comparison with the LDA and GGA band structures in Fig. 6.3. However, the EXX/KLI calculation still predicts FeO to be a metal, in conflict with reality. On the other hand, a full OPM calculation with the exact exchange and LDA correlation [391] leads to the bands plotted in Fig. 6.11. One now observes a fundamental gap of 1.7 eV, in very rough agreement with the experimental gap of 2.4 eV [392, 393]. The key factor for this success is the

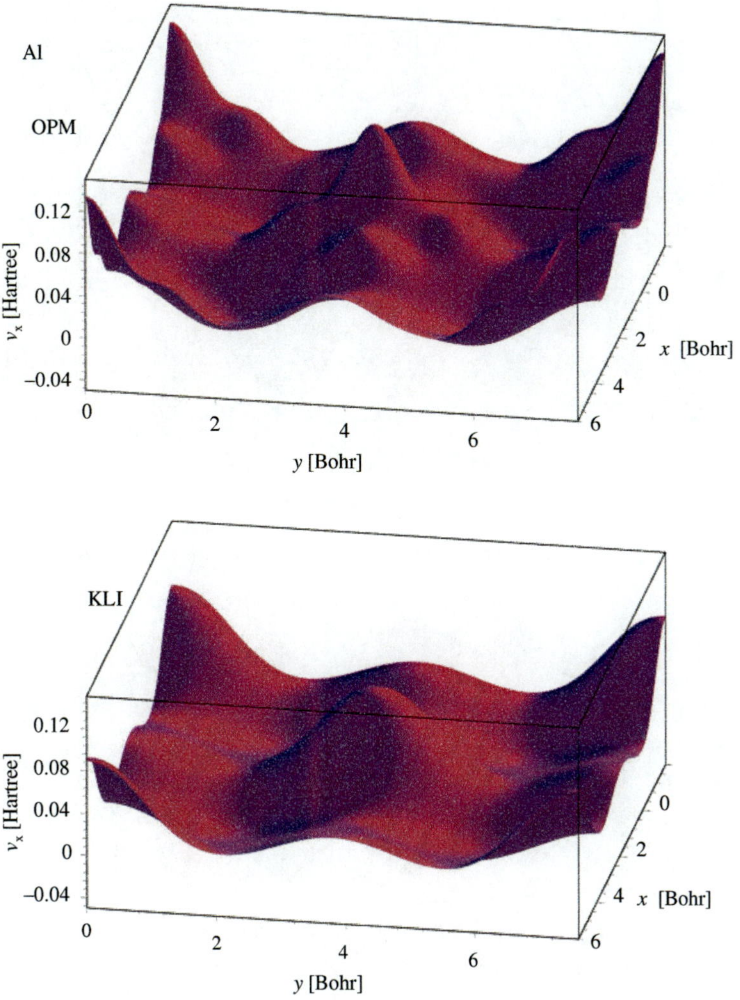

Fig. 6.8 Exchange potential of fcc aluminum in (100) plane (surface of conventional cubic unit cell): Self-consistent OPM and KLI potentials resulting from exact exchange (technical details as in Fig. 6.6).

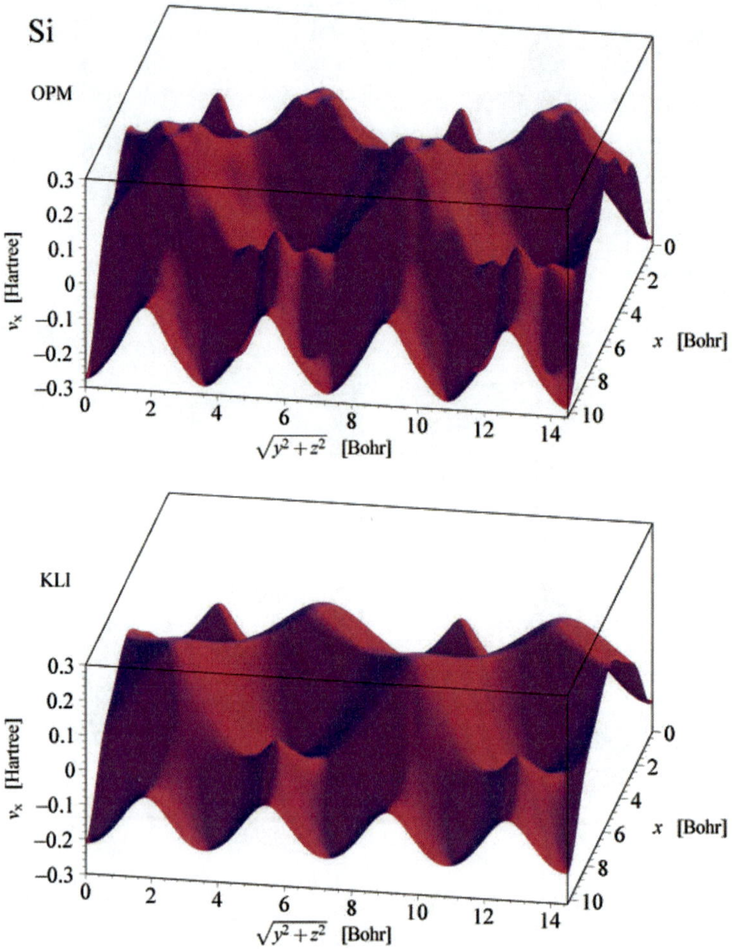

Fig. 6.9 Exchange potential of silicon (diamond structure) in (1,1,0) plane (diagonal plane of conventional cubic unit cell): Self-consistent OPM and KLI potentials resulting from exact exchange (technical details as in Fig. 6.7).

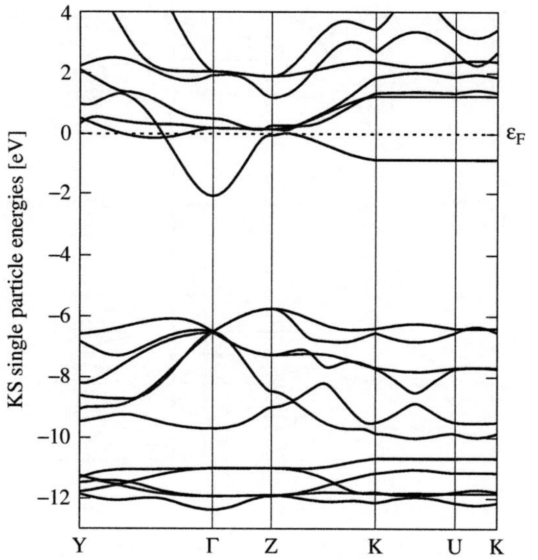

Fig. 6.10 As Fig. 6.3 for exact exchange within the KLI approximation in combination with LDA correlation.

complete cancellation of the self-interaction of the highly localized and oriented $3d$ states: proper elimination of the self-interaction stabilizes one of the minority spin t_{2g} states (the so-called a_{1g} state in the rhombohedral nomenclature—this state has $3(z')^2 - r^2$ character in the coordinate system with z'-axis perpendicular to the AF II planes) so much that a population imbalance of three t_{2g} states is energetically favorable and a gap emerges. The full OPM exchange allows the exchange potential to develop the very localized, attractive pockets, which are required to build up this population imbalance, while these pockets are smeared out in the KLI approximation. The same effect is found for CoO [391], for which the minority spin t_{2g} states are split into two occupied and one unoccupied band. Figure 6.11 should, however, not be interpreted as the final result, as replacement of the LDA correlation by some appropriate orbital-dependent correlation functional is expected to lead to modifications of the band structure. In addition, a serious comparison with experimental gaps requires the inclusion of the derivative discontinuity, Eq. (3.186), which is non-zero for orbital-dependent functionals as the exact exchange.

Some further band gaps obtained with the exact exchange are listed in Table 6.9, as a final illustration of the role of self-interaction in $v_{\rm x}$. All-electron OPM results based on (i) the Korringa-Kohn-Rostoker (KKR) method and the atomic sphere approximation (ASA) [360], (ii) the Linear-Muffin-Tin-Orbital method plus ASA [358–360], and (iii) the Full-Potential Linearized-Augmented-Plane-Wave (FP-LAPW) approach [394] are compared with full potential plane-wave pseudopotential (PW-PP) [103, 241] data for C, Si, Ge and GaAs. All calculations use the full

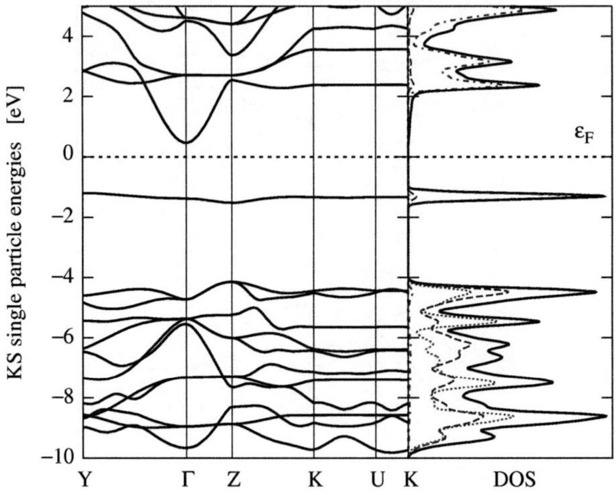

Fig. 6.11 As Fig. 6.3 for full OPM calculation with the exact exchange in combination with LDA correlation [391] ($E_{\rm cut}$ =160 Ry). Also shown are the total (*solid line*) and partial densities of states (DOS): O2*p*—*dashed line*; Fe3*d* at sites with majority spin ↑—*dotted line*; Fe3*d* at sites with majority spin ↓—*dash-dotted line.*

Table 6.9 Fundamental band gap $E_{\rm g}$ and band gap at Γ-point $E_{\rm g}(\Gamma)$ of semiconductors and insulators: FP-LAPW [394], KKR-ASA [360], LMTO-ASA [360] and plane-wave pseudopotential (PW-PP—in the spirit of [103, 377]) OPM results versus HF [395] and experimental data (all energies in eV—experimental values for $E_{\rm g}$ from [396–400], for $E_{\rm g}(\Gamma)$ from [401–404]; for $E_{\rm g}(\Gamma)$ of diamond one also finds 6.5±1.0 eV [405] and 7.3 eV [406]). The direct gap $\Delta_{\rm s}$ and the exact exchange contribution $\Delta_{\rm x}$ [103, 241] to the derivative discontinuity of $E_{\rm xc}$, Eq. (3.185), are given separately in the case of the EXX calculations.

	$E_{\rm x}$	$E_{\rm c}$	Method	C	Si	Ge	GaAs	Ne
$\Delta_{\rm s}(\Gamma)$	Exact	—	FP-LAPW	6.67	3.58	1.42	2.42	16.3
	Exact	—	PW-PP	6.24	3.17	1.46	2.06	14.72
	LDA	LDA	PW-PP	5.56	2.56	0.0	0.0	11.32
$E_{\rm g}(\Gamma)$			Expt. ($T>0$)	6.0±0.2	3.35	0.89	1.63	21.69
$\Delta_{\rm s}$	Exact	LDA	KKR-ASA	4.58	1.12	1.03		
	Exact	LDA	LMTO-ASA	4.65	1.25	1.12		
	Exact	LDA	PW-PP	4.81	1.35	1.22	2.11	14.76
	Exact	PBE	PW-PP	4.32	0.94	0.84	1.80	
	Exact	—	PW-PP	4.67	1.21	1.08	2.03	14.15
$\Delta_{\rm x}$	Exact	—	PW-PP	8.70	5.62	4.81	5.28	
$E_{\rm g}$	HF	—	LAPW/PW-PP	12.4	6.3	6.4	7.7	
			Expt. ($T=0$)	5.48	1.17	0.79	1.52	21.69

OPM, rather than the KLI approximation with its inherent limitations, illustrated in Fig. 6.4b. Values resulting from the combination of the exact E_{x} with either LDA or PBE-GGA correlation are listed in addition to the x-only data.

Results for both the fundamental band gap E_{g} (which is an indirect gap in the case of C and Si) and for the gap $E_{\mathrm{g}}(\Gamma)$ at the Γ-point, i.e. the origin of the first Brillouin zone ($\boldsymbol{k} = (0,0,0)$) are given.[17] The single-particle contribution Δ_{s}, i.e. the KS eigenvalue gap, is separated from the contribution Δ_{x} originating from the derivative discontinuity of the exact exchange, according to Eqs. (3.185), (3.186). The corresponding HF band gaps are listed for comparison.

Before comparing OPM results with experiment, one has to make sure that the OPM data are not affected by technical limitations. Several aspects of the computation can be critical:

- The atomic sphere approximation (ASA) has been applied in the case of the KKR and LMTO results. However, as shown in Fig. 6.9, the exchange potential in these crystals with a diamond structure is not really spherical inside the ASA sphere around the atomic sites (which implies that v_{x} is essentially constant between the sites).
- The PW-PP data, on the other hand, could suffer from an inadequate treatment of the core–valence interaction: While the exact exchange interaction among the valence states is included explicitly, the exchange component of the core–valence interaction is frozen to have the form of the core–valence interaction of the corresponding atom (via the pseudopotential—compare, however, [407, 408]).
- In the case of all methods a restricted KS spectrum resulting from a small basis set can lead to an inadequate representation of both the response function and the inhomogeneity of the OPM equation. This point is illustrated in Fig. 6.12 which shows the convergence of Δ_{s} with the resolution of the plane-wave basis used for v_{x} in case of the PW-PP scheme (for fixed cut-off energy for the KS states). As potentials resulting from Coulomb integrals are smoother than the underlying (orbital) densities, one would expect the representation of v_{x} to require fewer plane-waves than that of the KS states (whose spatial variation is determined by the dominating components of the total KS potential, the pseudopotential and v_{H}). On the other hand, the solution of the OPM integral equation on the reciprocal lattice is the most costly step in PW-PP calculations with the exact exchange, so that a truncation of the plane-wave expansion of v_{x} below the value used for the KS states suggests itself [241, 381, 407]. Figure 6.12 basically confirms this procedure. However, even for silicon an accurate evaluation of the gap requires a cut-off energy of more than 15 Ry.
- Finally, insufficient k-point sampling can spoil results. Use of a dense grid for the integration over the first Brillouin zone is particularly costly in EXX calculations, since the exact exchange scales quadratically with the number of grid points.

[17] Note that, as a matter of principle, $E_{\mathrm{g}}(\Gamma)$ is not a density functional, in contrast to the fundamental gap, Eq. (3.181). It is nevertheless usual to compare the KS eigenvalue gap at $\boldsymbol{k} = (0,0,0)$ with the experimental $E_{\mathrm{g}}(\Gamma)$.

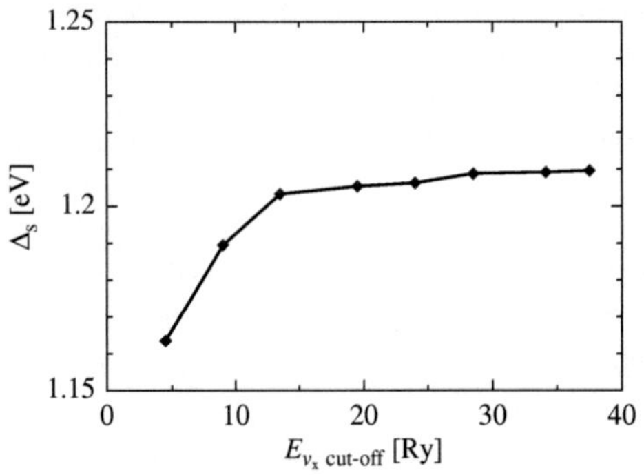

Fig. 6.12 Convergence of Δ_s with the energy cut-off used in the plane-wave representation of v_x within the PW-PP scheme for silicon. All results rely on a cut-off energy of 40 Ry for the KS states and 28 special k-points for the integration over the 1st Brillouin zone.

As a result of the various limitations, the agreement of the gaps obtained with different methods is not fully satisfactory. Nevertheless, it is obvious from Table 6.9 that Δ_s is consistently enlarged, when replacing the LDA exchange by the exact E_x (compare Table 4.11). In fact, the Δ_s obtained with the exact E_x are in much better agreement with the experimental data than the LDA gaps, irrespective of the correlation functional used. Surprisingly, one finds that the inclusion of correlation on the LDA level increases Δ_s compared with the x-only result, while use of the PBE-GGA for E_c leads to a reduced gap. As a consequence, inclusion of GGA correlation improves the agreement with experiment for Ge and GaAs, while it worsens the agreement for Si and the insulator C.

The picture becomes even less clear as soon as the derivative discontinuity of E_x is taken into account. The corresponding contribution Δ_x is much larger than Δ_s [103], so that the agreement with experiment is completely lost. In fact, the sum of Δ_s and Δ_x obtained in the x-only OPM calculation is almost equal to the very large band gap which one finds in the Hartree-Fock approximation [395]. In this sense the x-only OPM and the HF scheme are once more equivalent. Obviously, the correlation contribution to Δ_{xc} must cancel most of the large Δ_x.

Indeed, the good agreement of EXX KS gaps with experimental data led to the expectation that Δ_x and Δ_c cancel each other almost completely. Δ_x has therefore often been ignored in discussions of EXX results. Recently, however, a first application of the orbital-dependent RPA functional (see Sect. 6.4.2) demonstrated that Δ_x^{EXX} and Δ_c^{RPA} do not cancel completely [409] (see also [410]). Rather, their sum is of the same order of magnitude as the corresponding Δ_s, at least for the systems considered in [409] (Si, LiF and Ar).

In view of all these contradicting results it seems to be too early to draw definitive conclusions about the role of the exact exchange for band gaps. First some technical issues of the various implementations need to be clarified and an appropriate functional for correlation has to be fully established. The only reliable information which can be extracted from Table 6.9 is the fact that the transition from LDA to exact exchange leads to an increase of Δ_s (for fixed $E_c[n]$).

6.4 First-Principles Implicit Correlation Functionals

Some of the examples considered in the previous section already indicated that the exact exchange, while providing obvious progress compared with LDA and GGA exchange, has to be combined with an appropriate correlation functional, in order to be really useful in practice. This statement is corroborated by the accuracy of the spectroscopic constants resulting from the combination of the exact exchange (EXX) and PBE-GGA correlation. As demonstrated in Table 6.10 for the diatomic molecules of the G2 test set, the mean absolute errors obtained with the EXX-PBE functional are worse than the corresponding deviations of the LDA data (compare Table 4.8). In particular, the EXX-PBE approach leads to rather unsatisfactory results for bond lengths (see Table 6.16 for an explicit example). GGA and LDA correlation functionals are not suitable for use with the exact exchange, so that one is led to consider orbital-dependent correlation functionals. Given the first-principles nature of the exact E_x, a systematic derivation of a corresponding functional from the exact E_c suggests itself.

Suitable starting points for this endeavor have been established in Sect. 4.2: both KS-based many-body theory and the adiabatic connection can be employed [343, 361, 344, 362, 411–415]. It is beyond the scope of the present summary to review the various lines of attack that have been followed during the last years—the construction of first-principles orbital-dependent correlation functionals is still a matter of current-day research. The discussion is therefore restricted to two basic functionals of this type. The first of these functionals results from a perturbation expansion of the exact E_c [343], for which Eq. (4.65) provides the most transparent starting point (see Sect. 6.4.1). Due to its roots in perturbation theory, this functional is not applicable to systems without an energy gap between the highest occupied and lowest unoccupied KS state. The simplest functional available for gapless systems is obtained from the RPA, applied within the framework of KS-based many-body theory [361, 362, 411, 412]. The adiabatic connection formula (4.85) allows the most direct derivation in this case (see Sect. 6.4.2). Other variants of partial resummation of the KS perturbation expansion can be found in [416–421].

Both examples discussed below indicate that the structure of first-principles orbital-dependent correlation functionals is much more involved than that of conventional, density-based approximations. In particular, inclusion of higher order correlation complicates applications dramatically. It is therefore tempting to combine lowest order perturbation theory with a model-based representation of all higher order contributions. The first functional implementing this concept, the interaction strength interpolation (ISI) [422], is briefly discussed in Sect. 6.4.3.

6.4.1 Kohn-Sham Perturbation Theory

Equation (4.65) provides an exact representation of $E_{\rm xc}$ in terms of the KS orbitals and eigenvalues (which enter via the KS Green's function emerging from the application of e.g. Wick's theorem) as well as of its own functional derivative $v_{\rm xc}$. In fact, the power series in $\hat{H}_1$ on the right-hand side of Eq. (4.65) includes arbitrary powers of $v_{\rm xc}$, so that Eq. (4.65) and the resulting OPM equation are highly nonlinear.

How can one deal with this nonlinearity? Given the genesis of Eq. (4.65), an expansion of $E_{\rm xc}$ (and thus of $v_{\rm xc}$) in powers of e^2 is the natural first approach to the linearization of the OPM equation for the exact $E_{\rm xc}$ [343]. This approach is usually referred to as *Kohn-Sham* or *Görling-Levy perturbation theory*. The lowest order term in this expansion is the exchange energy. All higher orders correspond to $E_{\rm c}$,

$$E_{\rm xc} = \sum_{l=1}^{\infty} e^{2l}\, E_{\rm xc}^{(l)}[n] = E_{\rm x} + E_{\rm c}^{(2)} + \ldots \tag{6.89}$$

$$v_{\rm xc} = \sum_{l=1}^{\infty} e^{2l}\, v_{\rm xc}^{(l)}[n] = v_{\rm x} + v_{\rm c}^{(2)} + \ldots . \tag{6.90}$$

After insertion of (6.89), (6.90) into the OPM equation (6.26) both its right-hand and its left-hand side are given as power series with respect to e^2. Identity of both sides is then required order by order.

Unfortunately, the expression (4.65) does not lend itself to a direct expansion in powers of e^2: many-body perturbation theory for the energy (4.65) necessarily has to start with an expansion in powers of $\hat{H}_1$, Eq. (4.25). The basic elements of this expansion are the Green's function $G_{\rm s}$, the Coulomb interaction $\hat{W}$ and $v_{\rm xc}$. While $\hat{W}$ is linear in e^2, $v_{\rm xc}$ itself involves a complete power series in e^2. The lowest order term in this series is $v_{\rm x}$ (which is linear in e^2), so that each power of $v_{\rm xc}$ introduces at least one order of e^2. An expansion of (4.65) in powers of e^2 thus consists of two steps, first an expansion in powers of $\hat{W}$ and $v_{\rm xc}$, followed by an expansion of $v_{\rm xc}$ in powers of e^2.

In lowest order (e^2) the left-hand side of the OPM integral equation (6.26) just contains $v_{\rm x}$, while the inhomogeneity is determined by (6.11). This simply reflects the fact that $E_{\rm x}$ is a well-defined functional of only the ϕ_k. In lowest order one therefore ends up with the standard OPM equation for the exact exchange,

$$\int d^3r'\, \chi_{\rm s}(\boldsymbol{r},\boldsymbol{r}')v_{\rm x}(\boldsymbol{r}') = \Lambda_{\rm x}(\boldsymbol{r}) \tag{6.91}$$

$$\Lambda_{\rm x}(\boldsymbol{r}) = -\sum_k \int d^3r'\, \phi_k^\dagger(\boldsymbol{r})G_k(\boldsymbol{r},\boldsymbol{r}')\frac{\delta E_{\rm x}}{\delta\phi_k^\dagger(\boldsymbol{r}')} + c.c.. \tag{6.92}$$

The first time that the nonlinearity comes into play is in the order e^4. This lowest order correlation contribution (often referred to as *second order Görling-Levy functional* [343]) can be written as

$$E_c^{(2)} = E_c^{\mathrm{MP2}} + E_c^{\Delta\mathrm{HF}} . \tag{6.93}$$

The first of these terms is an expression which basically looks like the standard *second order Møller-Plesset* (MP2) *correction* to the HF energy,

$$E_c^{\mathrm{MP2}} = \frac{e^4}{2}\sum_{ijkl}\Theta_i\Theta_j(1-\Theta_k)(1-\Theta_l)\frac{(ij||kl)[(kl||ij)-(kl||ji)]}{\varepsilon_i+\varepsilon_j-\varepsilon_k-\varepsilon_l} . \tag{6.94}$$

However, the Slater integrals $(ij||kl)$ in (6.94),

$$(ij||kl) = \int d^3r_1 \int d^3r_2 \frac{\phi_i^\dagger(\boldsymbol{r}_1)\phi_k(\boldsymbol{r}_1)\phi_j^\dagger(\boldsymbol{r}_2)\phi_l(\boldsymbol{r}_2)}{|\boldsymbol{r}_1-\boldsymbol{r}_2|} , \tag{6.95}$$

are calculated with the KS orbitals ϕ_k, and the denominator of (6.94) is determined by the KS eigenvalues ε_k, so that E_c^{MP2} can give results which are quite different from standard MP2 data (see below). As usual, E_c^{MP2} can be split into a direct contribution (DIR) and an exchange term (second order exchange—SOX),

$$E_c^{\mathrm{MP2,DIR}} = \frac{e^4}{2}\sum_{ijkl}\Theta_i\Theta_j(1-\Theta_k)(1-\Theta_l)\frac{(ij||kl)(kl||ij)}{\varepsilon_i+\varepsilon_j-\varepsilon_k-\varepsilon_l} \tag{6.96}$$

$$E_c^{\mathrm{MP2,SOX}} = -\frac{e^4}{2}\sum_{ijkl}\Theta_i\Theta_j(1-\Theta_k)(1-\Theta_l)\frac{(ij||kl)(kl||ji)}{\varepsilon_i+\varepsilon_j-\varepsilon_k-\varepsilon_l} . \tag{6.97}$$

In the diagrammatic evaluation of (4.65) the direct term emerges from

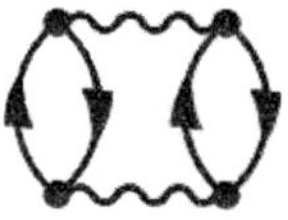

, (6.98)

with the solid line representing the KS Green's function (4.71) and the wavy line being the Coulomb interaction (for the Feynman rules required to translate diagrams into formulas and vice versa see Appendix L). The SOX term corresponds to

. (6.99)

The second contribution to (6.93) involves the difference between the orbital expectation values of the nonlocal HF-type exchange potential and of v_x,

$$E_c^{\Delta\mathrm{HF}} = \sum_{il}\frac{\Theta_i(1-\Theta_l)}{\varepsilon_i-\varepsilon_l}\left|\langle i|v_x|l\rangle + e^2\sum_j\Theta_j(ij||jl)\right|^2 \tag{6.100}$$

$$\langle i|v_{\rm x}|l\rangle = \int d^3r\,\phi_i^\dagger(\boldsymbol{r})v_{\rm x}(\boldsymbol{r})\phi_l(\boldsymbol{r}). \tag{6.101}$$

$E_{\rm c}^{\Delta\rm HF}$ accounts for the fact that the present perturbation expansion is not based on the HF Hamiltonian, but rather on the KS Hamiltonian. Diagrammatically, $E_{\rm c}^{\Delta\rm HF}$ results from the remaining second order terms in (4.65), including the interaction of the KS particles with the "external" potential $v_{\rm Hxc}$, Eq. (4.26):

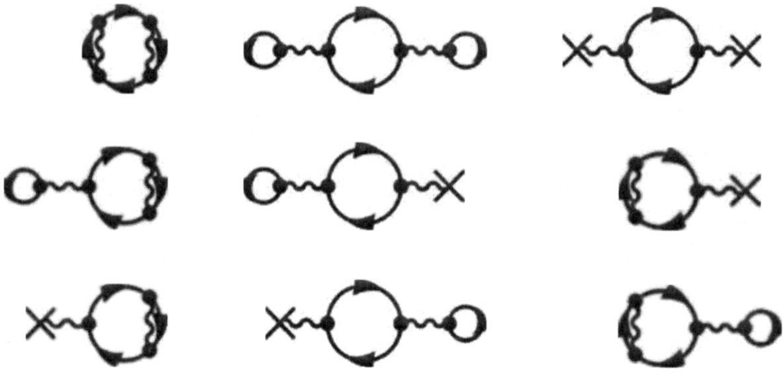

In these diagrams the wavy line with the cross denotes $v_{\rm Hxc}$. After decomposition of $v_{\rm Hxc}$ into $v_{\rm H}$ and $v_{\rm xc}$, the Hartree contributions cancel with the tadpole diagrams. The remaining terms can be added up to the expression (6.100) with $v_{\rm x}$ replaced by the full $v_{\rm xc}$. After this first step of the expansion the energy correction therefore features the full difference between the single-particle KS and HF Hamiltonians,

$$\langle\boldsymbol{r}|\hat{h}_{\rm s}|\boldsymbol{r}'\rangle = \delta^{(3)}(\boldsymbol{r}-\boldsymbol{r}')\left[\frac{(-i\hbar\boldsymbol{\nabla})^2}{2m}+v_{\rm ext}(\boldsymbol{r})+v_{\rm H}(\boldsymbol{r})+v_{\rm xc}(\boldsymbol{r})\right]$$

$$\langle\boldsymbol{r}|\hat{h}_{\rm HF}|\boldsymbol{r}'\rangle = \delta^{(3)}(\boldsymbol{r}-\boldsymbol{r}')\left[\frac{(-i\hbar\boldsymbol{\nabla})^2}{2m}+v_{\rm ext}(\boldsymbol{r})+v_{\rm H}(\boldsymbol{r})\right]+e^2\sum_j\Theta_j\frac{\phi_j(\boldsymbol{r})\phi_j^\dagger(\boldsymbol{r}')}{|\boldsymbol{r}-\boldsymbol{r}'|}.$$

However, due to the quadratic structure of (6.100) only the leading term $v_{\rm x}$ is relevant in the order e^4. As net result one finds a well-defined, linear expression for the correlation functional $E_{\rm c}^{(2)}$.

The contribution $E_{\rm c}^{\Delta\rm HF}$ once again illustrates the relation between the x-only OPM and the standard HF approach. In the x-only limit the OPM, i.e. the EXX approach, corresponds to a minimization of the HF energy expression under the subsidiary condition that the orbitals satisfy the KS equations (see Sect. 6.2.2). The EXX ground state energy is thus slightly higher than the HF value, which results from a free minimization of the same energy expression. The difference between the two energies can be evaluated order by order, using the difference between the HF and the EXX Hamiltonian as perturbation. In lowest order this procedure leads to the energy (6.100). This expression is always negative, consistent with the fact that the HF energy must be below the x-only OPM value. On the other hand, if one examines (6.100) quantitatively, one usually finds it to be rather small, which is immediately clear from Tables 6.1 and 6.2 (see also Sect. 6.6).

As already emphasized, $E_c^{(2)}$ is well-defined as soon as v_x is known. This implies, that the first step of a selfconsistent application of (6.93) is a solution of the x-only OPM equation (6.91) for given orbitals. Once v_x is available, it remains to evaluate

$$v_c^{(2)}(\boldsymbol{r}) = \frac{\delta E_c^{(2)}[\phi_k, \varepsilon_k, v_x]}{\delta n(\boldsymbol{r})}. \tag{6.102}$$

The calculation of the functional derivative of E_c^{MP2} with respect to n can be handled in analogy to Eq. (6.10), as this term does not depend on v_x. The same applies to the ϕ_k- and ε_k-dependence of $E_c^{\Delta\text{HF}}$. The subsequent discussion therefore focuses on the handling of the v_x-dependence of $E_c^{\Delta\text{HF}}$.

One starts by realizing that the explicit v_x-dependence of $E_c^{\Delta\text{HF}}$ is not fundamentally different from the ϕ_k- and ε_k-dependence. One thus has to include a functional derivative with respect to this additional variable, when eliminating the original derivative $\delta/\delta n$ via the chain rule (as in (6.10)). This leads to an additional contribution to the inhomogeneity (6.28),

$$\Delta\Lambda_c^{(2)}(\boldsymbol{r}) = \int d^3r' \, \frac{\delta v_x(\boldsymbol{r}')}{\delta v_s(\boldsymbol{r})} \frac{\delta E_c^{\Delta\text{HF}}}{\delta v_x(\boldsymbol{r}')}. \tag{6.103}$$

The first factor, the functional derivative of v_x with respect to v_s, is accessible via the x-only OPM equation. If one differentiates Eq. (6.91) with respect to v_s and isolates the desired derivative one finds

$$\frac{\delta v_x(\boldsymbol{r}_2)}{\delta v_s(\boldsymbol{r}_1)} = \int d^3r_3 \, \chi_s^{-1}(\boldsymbol{r}_2, \boldsymbol{r}_3) \left\{ \frac{\delta \Lambda_x(\boldsymbol{r}_3)}{\delta v_s(\boldsymbol{r}_1)} - \int d^3r_4 \, \frac{\delta \chi_s(\boldsymbol{r}_3, \boldsymbol{r}_4)}{\delta v_s(\boldsymbol{r}_1)} v_x(\boldsymbol{r}_4) \right\}. \tag{6.104}$$

The functional derivative of Λ_x with respect to the KS potential can again be obtained by use of the chain rule and the relations (6.18), (6.19),

$$\begin{aligned} \frac{\delta \Lambda_x(\boldsymbol{r}_3)}{\delta v_s(\boldsymbol{r}_1)} &= -\sum_k \int d^3r_4 \left\{ \phi_k^\dagger(\boldsymbol{r}_1) G_k(\boldsymbol{r}_1, \boldsymbol{r}_4) \frac{\delta \Lambda_x(\boldsymbol{r}_3)}{\delta \phi_k^\dagger(\boldsymbol{r}_4)} + c.c. \right\} \\ &\quad + \sum_k |\phi_k(\boldsymbol{r}_1)|^2 \frac{\partial \Lambda_x(\boldsymbol{r}_3)}{\partial \varepsilon_k}. \end{aligned} \tag{6.105}$$

The second new expression in (6.104) is the derivative of the linear response function with respect to the KS potential. Using the definition (6.27) of the linear KS response function, one can rewrite this quantity as the second functional derivative of n with respect to v_s, i.e. the quadratic response function of the KS system,

$$\frac{\delta \chi_s(\boldsymbol{r}_3, \boldsymbol{r}_4)}{\delta v_s(\boldsymbol{r}_1)} = \frac{\delta^2 n(\boldsymbol{r}_3)}{\delta v_s(\boldsymbol{r}_4) \delta v_s(\boldsymbol{r}_1)}. \tag{6.106}$$

This function can be evaluated in the same fashion as $\chi_{\rm s}$ (compare [350]). As the derivative $\delta E_{\rm c}^{\Delta \rm HF}/\delta v_{\rm x}$ can be taken directly, all ingredients of (6.103) are known, so that $v_{\rm c}^{(2)}$ can be calculated.

In principle, one can proceed in the same fashion to approach higher order contributions to $E_{\rm c}$. However, Eqs. (6.103)–(6.106) indicate that even a self-consistent application of $E_{\rm c}^{(2)}$ represents quite a demanding computational task. For this reason $E_{\rm c}^{(2)}$ is often applied in post-EXX fashion, i.e. evaluated *a posteriori* with the KS orbitals resulting from a self-consistent EXX calculation. Alternatively, the complete $E_{\rm c}^{\Delta \rm HF}$ or at least its contribution to $v_{\rm c}^{(2)}$ are neglected, as one expects this energy contribution to be small quite generally.

6.4.2 Kohn-Sham-Based Random Phase Approximation

The most straightforward way to go beyond $E_{\rm c}^{(2)}$ is the resummation of certain diagrammatic classes of the KS perturbation expansion. The simplest and, at the same time, most important of these classes are the so-called ring diagrams, which are known to be crucial for the description of metals (see e.g. [94]). As in the case of the homogeneous electron gas, the resulting correlation energy is termed random phase approximation (RPA) [361, 362, 411–414, 423–425]. Its derivation starts with the adiabatic connection formula (4.85). The heart of (4.85) is the time-ordered response function χ_λ of the interacting system with coupling strength λ and external potential u_λ, which has the same ground state density as the actual system of interest (obtained for $\lambda = 1$, $u_\lambda = v_{\rm ext}$). As in the case of any other interacting system, one can formulate a Dyson equation for χ_λ. Its derivation is discussed in detail in Sect. 7.4. This Dyson equation allows a systematic analysis of approximations for χ_λ. At this point, however, we will introduce the RPA for χ_λ in an *ad hoc* fashion, relying on the analogy with the RPA for the homogeneous electron gas, Eq. (4.100). Translated to the inhomogeneous situation, this equation reads[18]

$$\begin{aligned}\chi_\lambda^{\rm RPA}(\boldsymbol{r},\boldsymbol{r}',\omega) &= \chi_{\rm s}(\boldsymbol{r},\boldsymbol{r}',\omega) \\ &+ \int d^3r'' \int d^3r''' \, \chi_{\rm s}(\boldsymbol{r},\boldsymbol{r}'',\omega)\frac{\lambda e^2}{|\boldsymbol{r}''-\boldsymbol{r}'''|}\chi_\lambda^{\rm RPA}(\boldsymbol{r}''',\boldsymbol{r}',\omega) \,. \end{aligned} \quad (6.107)$$

Physically, this approximation corresponds to a screening of the Coulomb interaction by repeated, sequential excitation of a single virtual particle–hole pair, which, in the present situation, is a KS particle–hole pair. Insertion of the iterated equation (6.107) into (4.85) allows the evaluation of the coupling constant integral,

---

[18] Comparison with the Dyson equation (7.127) shows that (6.107) amounts to a complete neglect of the xc-kernel $f_{\rm xc}$.

$$\begin{aligned}
E_{\mathrm{c}}^{\mathrm{RPA}} &= -\frac{\hbar}{2}\int_0^\infty \frac{d\omega}{\pi}\sum_{n=2}^{\infty}\frac{1}{n}\int d^3r_1\int d^3r_1'\cdots\int d^3r_n\int d^3r_n' \\
&\quad\times\frac{e^2}{|\boldsymbol{r}_1-\boldsymbol{r}_1'|}\chi_{\mathrm{s}}(\boldsymbol{r}_1',\boldsymbol{r}_2,i\omega)\cdots\frac{e^2}{|\boldsymbol{r}_n-\boldsymbol{r}_n'|}\chi_{\mathrm{s}}(\boldsymbol{r}_n',\boldsymbol{r}_1,i\omega) \\
&= \frac{\hbar}{2}\int_0^\infty\frac{d\omega}{\pi}\,\mathrm{Tr}\left\{\ln\left[1-w\chi_{\mathrm{s}}(i\omega)\right]+w\chi_{\mathrm{s}}(i\omega)\right\},
\end{aligned}\tag{6.108}$$

where the trace in the last line indicates integration over all spatial coordinates. In order to arrive at a practically useful form of $E_{\mathrm{c}}^{\mathrm{RPA}}$ one has to specify χ_{s} in more detail. Starting from (4.68), one first obtains

$$\chi_{\mathrm{s}}(\boldsymbol{r},\boldsymbol{r}',i\omega) = \sum_{kl}\phi_k^\dagger(\boldsymbol{r})\phi_l(\boldsymbol{r})\phi_l^\dagger(\boldsymbol{r}')\phi_k(\boldsymbol{r}')\,C_{kl}(i\omega) \tag{6.109}$$

$$C_{kl}(i\omega) = (\Theta_k-\Theta_l)\frac{(\varepsilon_k-\varepsilon_l)-i\hbar\omega}{(\hbar\omega)^2+(\varepsilon_k-\varepsilon_l)^2}\,. \tag{6.110}$$

Insertion into (6.108) gives

$$\begin{aligned}
E_{\mathrm{c}}^{\mathrm{RPA}} &= -\frac{\hbar}{2}\int_0^\infty\frac{d\omega}{\pi}\sum_{n=2}^{\infty}\frac{1}{n}\sum_{k_1l_1}\cdots\sum_{k_nl_n} \\
&\quad\times(l_nk_1||k_nl_1)C_{k_1l_1}(i\omega)\cdots(l_{n-1}k_n||k_{n-1}l_n)C_{k_nl_n}(i\omega)\,,
\end{aligned}\tag{6.111}$$

where the Slater integrals (6.95) contain all spatial integrations. Comparison with (6.96) shows that the lowest order contribution to $E_{\mathrm{c}}^{\mathrm{RPA}}$ is exactly the second order direct term, $E_{\mathrm{c}}^{\mathrm{MP2,DIR}}$. On the other hand, the SOX contribution (6.97) is not included in the RPA.

Equation (6.111) can be brought into a more compact form by definition of the matrix

$$S_{ik,jl}(i\omega) := (kj||il)\,C_{jl}(i\omega) \tag{6.112}$$

$$\begin{aligned}
\Longrightarrow\quad E_{\mathrm{c}}^{\mathrm{RPA}} &= -\frac{\hbar}{2}\int_0^\infty\frac{d\omega}{\pi}\sum_{n=2}^{\infty}\frac{1}{n}\sum_{k_1l_1}\cdots\sum_{k_nl_n}S_{k_nl_n,k_1l_1}(i\omega)\cdots S_{k_{n-1}l_{n-1},k_nl_n}(i\omega) \\
&= \frac{\hbar}{2}\int_0^\infty\frac{d\omega}{\pi}\,\mathrm{Tr}\left\{\ln\left[\underline{\underline{1}}-\underline{\underline{S}}(i\omega)\right]+\underline{\underline{S}}(i\omega)\right\}.
\end{aligned}\tag{6.113}$$

where the trace runs over the multi-index ik of the matrix $\underline{\underline{S}}$. Finally, one can use

$$\mathrm{Tr}[\ln(\underline{\underline{A}})] = \ln|\det(\underline{\underline{A}})|$$

to rewrite (6.113) as

$$E_{\mathrm{c}}^{\mathrm{RPA}} = \frac{\hbar}{2}\int_0^\infty\frac{d\omega}{\pi}\left\{\ln\left|\det\left(\underline{\underline{1}}-\underline{\underline{S}}(i\omega)\right)\right|+\mathrm{Tr}\underline{\underline{S}}(i\omega)\right\}. \tag{6.114}$$

The corresponding potential can be calculated via the standard OPM integral equation (6.26).

As it stands, Eq. (6.114) is still completely general. A particularly useful form is obtained by choosing real KS orbitals (which is always possible for discrete states, as long as the spin-projection in z-direction is a good quantum number). In this case one of the contributions to (6.109),

$$-i\hbar\omega\sum_{kl}\phi_k^\dagger(\boldsymbol{r})\phi_l(\boldsymbol{r})\phi_l^\dagger(\boldsymbol{r}')\phi_k(\boldsymbol{r}')\frac{(\Theta_k-\Theta_l)}{(\hbar\omega)^2+(\varepsilon_k-\varepsilon_l)^2},$$

vanishes identically due to the symmetry of $\phi_k^\dagger(\boldsymbol{r})\phi_l(\boldsymbol{r})\phi_l^\dagger(\boldsymbol{r}')\phi_k(\boldsymbol{r}')$ under exchange of k and l. In addition, the remaining term can be reformulated as

$$\chi_s(\boldsymbol{r},\boldsymbol{r}',i\omega) = \sum_{kl}\phi_k^\dagger(\boldsymbol{r})\phi_l(\boldsymbol{r})\phi_l^\dagger(\boldsymbol{r}')\phi_k(\boldsymbol{r}')\overline{C}_{kl}(i\omega) \tag{6.115}$$

$$\overline{C}_{kl}(i\omega) = \Theta_k(1-\Theta_l)\frac{2(\varepsilon_k-\varepsilon_l)}{(\hbar\omega)^2+(\varepsilon_k-\varepsilon_l)^2}. \tag{6.116}$$

The form (6.116) of $\overline{C}_{kl}$ simplifies the evaluation of the most critical ingredient of E_c^{RPA}, the determinant det $\left(\underline{\underline{1}}-\underline{\underline{S}}(i\omega)\right)$, as the multi-index of $\underline{\underline{S}}$ now splits into two components—one index (k) is restricted to occupied states, while the second (l) involves only unoccupied states.

6.4.3 Interaction Strength Interpolation (ISI)

Given the extremely complicated form of the correlation contributions beyond $E_c^{(2)}$, one might ask whether it is possible to account for these higher order terms in a more efficient fashion? This is the aim of the interaction strength interpolation (ISI) [422], in which one attempts to express the higher order terms in the form of an explicit density functional. The starting point for the derivation of the ISI is the adiabatic connection (4.75). Insertion of χ_λ, Eq. (4.35), and use of the commutator (4.33) leads to a formula similar to (4.32),

$$E_{\mathrm{xc}} = \int_0^1 d\lambda\, W_\lambda[n], \tag{6.117}$$

with[19]

$$W_\lambda[n] = \langle\Psi_0(\lambda)[n]|\hat{W}|\Psi_0(\lambda)[n]\rangle - E_{\mathrm{H}}[n]. \tag{6.118}$$

The basic idea of the ISI is to obtain W_λ in the interesting regime $\lambda\approx 1$ from an interpolation between the weak ($\lambda\ll 1$) and the strong ($\lambda\gg 1$) interaction limit. The former limit is well-known by now: for weak interactions one can expand E_{xc} in powers of the coupling constant λ. This automatically yields a corresponding expansion for W_λ,

[19] $|\Psi_0(\lambda)\rangle$ denotes the ground state which results from (4.74)—see Sect. 4.2.2.

$$E_{xc} = \int_0^1 d\lambda \left\{ E_x + \sum_{l=2}^{\infty} l\lambda^{l-1} E_c^{(l)} \right\}. \tag{6.119}$$

The lowest two orders of W_λ are therefore determined by E_x and $E_c^{(2)}$.

The strong interaction limit requires a different approach. As the electrostatic forces completely dominate over kinetic effects for $\lambda \to \infty$, a simple model system (*point charge plus continuum model*) should be sufficient to extract W_λ [426]. In this way one finds the leading two orders of the expansion of W_λ in powers of $1/\sqrt{\lambda}$,

$$\lim_{\lambda \to \infty} W_\lambda[n] = W_\infty[n] + W'_\infty[n]\lambda^{-1/2} + \ldots, \tag{6.120}$$

with

$$W_\infty[n] = \int d^3r \left\{ An^{4/3} + B\frac{(\nabla n)^2}{n^{4/3}} \right\} \tag{6.121}$$

$$W'_\infty[n] = \int d^3r \left\{ Cn^{3/2} + D\frac{(\nabla n)^2}{n^{7/6}} \right\} \tag{6.122}$$

(for the values of the coefficients $A - D$ see [422]). Interpolation between E_x, $E_c^{(2)}$ on the one hand and W_∞, W'_∞ on the other leads to

$$E_{xc}^{ISI} = W_\infty + \frac{2X}{Y}\left[(1+Y)^{1/2} - 1 - Z\ln\left(\frac{(1+Y)^{1/2}+Z}{1+Z} \right) \right], \tag{6.123}$$

with the abbreviations

$$\begin{aligned} X &= \frac{xy^2}{z^2} & Y &= \frac{x^2y^2}{z^4} & Z &= \frac{xy^2}{z^3} - 1 \\ x &= -4E_c^{(2)} & y &= W'_\infty & z &= E_x - W_\infty . \end{aligned} \tag{6.124}$$

Equations (6.123), (6.124) should be understood as an effective resummation of the KS perturbation series. The correlation part of E_{xc}^{ISI} is obtained by subtraction of the exact E_x from (6.123).

6.5 Model-Based Orbital-Dependent Exchange-Correlation Functionals

In view of the complexity of the first-principles implicit correlation functionals of Sect. 6.4, one is automatically led to look for more accessible alternatives. Two functionals of this type have been combined with the OPM quite early, two additional options only very recently.

All these functionals have the advantage that they only depend on the occupied KS states, so that their application is much less time-consuming than that of the functionals of Sect. 6.4. On the other hand, the correlation components of these model-based forms are relatively local: their nonlocality is restricted to the first gradient of the KS orbitals, with obvious limitations. As a consequence, none of these functionals can deal with the London dispersion force. The argument given in Sect. 6.1.2 for the LDA and GGA applies equally well to these types of functionals.

6.5.1 Self-Interaction Corrected LDA

The first functional of this type is the self-interaction corrected LDA (SIC-LDA) [143, 262]. The SIC-LDA has already been introduced in Sect. 4.7 (to which we refer the reader for all details). Applications of this functional usually do not rely on the OPM, but rather on the use of orbital-dependent KS potentials. Few results combining the SIC-LDA with the OPM have been reported in the literature [262, 353].

6.5.2 Colle-Salvetti Functional

A second model-based orbital-dependent expression which was originally introduced in a different context is the Colle-Salvetti (CS) correlation functional [222]. It was initially developed for use within the Hartree-Fock scheme. The starting point for the derivation of the CS functional is an approximation for the correlated wavefunction $\Psi_0(\boldsymbol{r}_1\sigma_1,\ldots\boldsymbol{r}_N\sigma_N)$. The ansatz for Ψ_0 consists of a product of the HF Slater determinant and so-called *Jastrow factors*, which reflect the correlation between all pairs of particles,

$$\Psi_0(\boldsymbol{r}_1\sigma_1,\ldots\boldsymbol{r}_N\sigma_N) = \Phi_0^{\mathrm{HF}}(\boldsymbol{r}_1\sigma_1,\ldots\boldsymbol{r}_N\sigma_N)\prod_{i<j}\Big[1-\varphi(\boldsymbol{r}_i,\boldsymbol{r}_j)\Big]\,. \tag{6.125}$$

CS use a model for the correlation functions $\varphi(\boldsymbol{r}_i,\boldsymbol{r}_j)$ which satisfies the electron–electron cusp condition at $\boldsymbol{r}_i=\boldsymbol{r}_j$ [198]. The free parameter in this model is adjusted to the correlation energy of the He atom [222]. The final functional reads

$$\begin{aligned} E_{\mathrm{c}}^{\mathrm{CS}} = & -\frac{ab}{4}\int d^3r\,\beta\gamma\left[4\sum_{k\sigma}\Theta_{k\sigma}n_\sigma(\boldsymbol{\nabla}|\phi_{k\sigma}|)^2-|\boldsymbol{\nabla}n|^2-\sum_\sigma n_\sigma\boldsymbol{\nabla}^2n_\sigma+n\boldsymbol{\nabla}^2n\right] \\ & -a\int d^3r\,\gamma\frac{n}{\alpha}, \end{aligned} \tag{6.126}$$

with the abbreviations

$$\alpha(\boldsymbol{r}) = 1+dn(\boldsymbol{r})^{-1/3}$$

$$\beta(\boldsymbol{r}) = \frac{n(\boldsymbol{r})^{-5/3}\exp[-cn(\boldsymbol{r})^{-1/3}]}{\alpha(\boldsymbol{r})}$$
$$\gamma(\boldsymbol{r}) = 4\frac{n_\uparrow(\boldsymbol{r})n_\downarrow(\boldsymbol{r})}{n(\boldsymbol{r})^2}$$

(for the values of the coefficients *a–d* see after Eq. (4.289)). This functional depends on the spin-density n_σ and the orbital kinetic energy density, $|\boldsymbol{\nabla}\phi_{k\sigma}|^2$. In the context of DFT, this dependence makes E_c^{CS} an implicit functional for which the OPM has to be utilized. Only a few OPM-based applications of the CS functional are documented in the literature [427–429]. The CS functional is, however, the basis for the LYP-GGA, Eq. (4.289), which is one of the most often used correlation functionals.

6.5.3 Meta-GGA

A third model-based orbital-dependent xc-functional is the Meta-GGA [231, 232] already introduced in Sect. 4.8 (to which we refer the reader for all details). Similar to the CS-functional, the Meta-GGA is usually not applied within the OPM. Rather applications often utilize an *a posteriori* evaluation of the Meta-GGA energy by insertion of selfconsistent GGA densities/orbitals. Alternatively, selfconsistent calculations with the Meta-GGA rely on orbital-dependent potentials, i.e. on a minimization of the total energy in the spirit of the HF minimization (see also next section).

6.5.4 Global, Screened and Local Hybrid Functionals

It has been realized quite early [215, 430] that an admixture of the exact exchange to GGA-type functionals improves thermochemical results substantially, as compared to pure GGAs. The standard form of such a global *hybrid* functional is given by

$$E_x^{hyb} = a_0 E_x^{exact} + a_1 E_x^{GGA} + (1 - a_0 - a_1) E_x^{LDA} \tag{6.127}$$
$$E_c^{hyb} = b_1 E_c^{GGA} + (1 - b_1) E_c^{LDA}. \tag{6.128}$$

The form of the GGA and the mixing coefficients were originally chosen to optimally reproduce sets of thermochemical data. In this way one finds the B3LYP functional [216], which relies on the B88-GGA for exchange [219], the LYP-GGA for correlation [221] and $a_0 = 0.2$, $a_1 = 0.72$, $b_1 = 0.81$. For E_c^{LDA} the VWN parameterization is employed, with, however, (i) a set of parameters different from those provided in Sect. 4.3, and (ii) the spin-dependence of the exchange: in B3LYP E_c^{LDA} is replaced by (4.122), with $e_c^{HEG,RPA}(n,0)$ and $e_c^{HEG,RPA}(n,1)$ parameterized in the VWN form (4.108), using $x_0 = -0.409286$, $b = 13.0720$ and $c = 42.7198$

for the paramagnetic case and $x_0 = -0.743294$, $b = 20.1231$ and $c = 101.578$ for the ferromagnetic limit—this set of data corresponds to the RPA for the correlation energy of the HEG [134] (compare the discussion in [431]).

The mean average errors obtained with this hybrid for the spectroscopic constants of the molecules listed in Table 4.7 are given in Table 6.10 (atomic reference data obtained with the B3LYP functional are included in Tables 4.13 and 4.5). The improvement of these thermochemical data by the B3LYP functional is obvious. A good number of similar results can be found in the literature (see e.g. [430, 216, 435, 436]).

Table 6.10 Bond lengths R_e, dissociation energies D_e (including the zero-point energies [226]) and harmonic frequencies ω_e: Mean absolute deviation of global and local hybrid results from experimental data (taken from [227–229]—see also http://cccbdb.nist.gov/) for the set of diatomic molecules listed in Table 4.7. B3LYP: Eqs. (6.127), (6.128) [430, 216]; PBE0: Eqs. (6.134), (6.135) [217]); LH-BPW91: Eqs. (6.141), (6.146), (6.147) with B88-GGA exchange and PW91-GGA correlation [432]; LH-LDA: Eqs. (6.141), (6.148), (6.147) with the LDA for exchange and correlation [433]). The corresponding deviations of the combination of the exact exchange with the PBE-GGA for correlation (EXX+PBE) [434] are also given (see Sect. 6.4).

Method	R_e [Bohr]	D_e [eV]	ω_e [cm^{-1}]
B3LYP	0.018	0.148	33
PBE0	0.020	0.195	41
LH-BPW91	0.046	0.842	109
LH-LDA	0.026	0.215	35
EXX+PBE	0.088	1.187	213

The success of this construction immediately prompts the question concerning its justification. A simple argument can be given on the basis of the adiabatic connection (4.75) [215]. The crucial point is the decomposition of the total $E_{xc}[n]$ in terms of the coupling strength λ,

$$E_{xc}[n] = \int_0^1 d\lambda \, E_{xc,\lambda}[n] \,. \tag{6.129}$$

$E_{xc,\lambda}$ is known for small λ, for which a power series expansion in the coupling constant is legitimate,

$$E_{xc,\lambda}[n] = A^{(0)}[n] + \lambda A^{(1)}[n] + \lambda^2 A^{(2)}[n] + \lambda^3 A^{(3)}[n] + \ldots \,. \tag{6.130}$$

The lowest order coefficients of (6.130) are given by the series (6.89), $A^{(0)} = E_x$, $A^{(1)} = 2E_c^{(2)}$. The simplest approximation for (6.129) relies on the representation of the λ-dependence of $E_{xc,\lambda}$ by a straight line,

$$E_{xc,\lambda} = E_{xc,\lambda=0} + \lambda \left[E_{xc,\lambda=1} - E_{xc,\lambda=0}\right] \tag{6.131}$$

$$\Longrightarrow \quad E_{\mathrm{xc}}[n] = \frac{1}{2}\left[E_{\mathrm{x}} + E_{\mathrm{xc},\lambda=1}\right].$$

One can then argue that the LDA or GGA provide a reasonably accurate representation of $E_{\mathrm{xc},\lambda}$ at $\lambda = 1$ (as the xc-hole is deeper and thus more localized around its electron for full coupling strength [217]) to end up with [215]

$$E_{\mathrm{xc}}[n] = \frac{1}{2}\left[E_{\mathrm{x}} + E_{\mathrm{xc}}^{\mathrm{GGA}}\right]. \tag{6.132}$$

This qualitative argument indicates why functionals of the form (6.127) could be of interest. The quantitative performance of the functional (6.132) is less satisfying. The argument given can, however, be refined by a more detailed analysis of the shape of $E_{\mathrm{xc},\lambda}$ as a function of λ [217]. One starts with the ansatz

$$E_{\mathrm{xc},\lambda} = E_{\mathrm{xc},\lambda}^{\mathrm{GGA}} + \left[E_{\mathrm{x}} - E_{\mathrm{x}}^{\mathrm{GGA}}\right](1-\lambda)^{n-1}, \tag{6.133}$$

which (i) is exact for $\lambda = 0$ (as $E_{\mathrm{xc},\lambda=0}^{\mathrm{GGA}} = E_{\mathrm{x}}^{\mathrm{GGA}}$), and (ii) approaches $E_{\mathrm{xc},\lambda=1}^{\mathrm{GGA}}$ for $\lambda = 1$, similar to (6.132). The form (6.133) allows, however, any power-law shape for intermediate values of λ, not just a straight line. The appropriate power n depends on the system under consideration. For molecules one can refer to the success of many-body perturbation theory to fourth order in the coupling constant (the so-called *MP4 approximation*). If one assumes this observation to be also characteristic for the expansion of E_{xc}, it suggests that the optimum shape of (6.133) is obtained for $n = 4$.

If the GGA contribution to (6.133) is chosen to be of first-principles form, the HEG limit is also correctly reproduced. The resulting hybrid functional reads [217]

$$E_{\mathrm{x}}^{\mathrm{PBE0}} = \frac{1}{4}E_{\mathrm{x}}^{\mathrm{exact}} + \frac{3}{4}E_{\mathrm{x}}^{\mathrm{PBE}} \tag{6.134}$$

$$E_{\mathrm{c}}^{\mathrm{PBE0}} = E_{\mathrm{c}}^{\mathrm{PBE}}. \tag{6.135}$$

Some prototype results obtained with this so-called PBE0 functional are given in Table 6.11. The mean average errors found for the spectroscopic constants of the diatomic molecules in the G2 set are included in Table 6.10. A comparison with the PBE-GGA data in Tables 4.7 and 4.8 shows that the PBE0 hybrid improves the accuracy of the dissociation energy. Additional results may be found in Sect. 6.6 and [437, 435, 438, 439].

Applications of hybrid functionals usually rely on a HF-type implementation of their exact exchange component (often termed *generalized Kohn-Sham approach*—GKS), rather than on the solution of the OPM equation for the exact E_{x},

$$v_{\mathrm{x}}(\boldsymbol{r})\phi_k(\boldsymbol{r}) \longrightarrow [\hat{v}_{\mathrm{x}}\phi_k](\boldsymbol{r}) = \frac{\delta E_{\mathrm{x}}}{\delta \phi_k^{\dagger}(\boldsymbol{r})}.$$

This is uncritical for most thermochemical calculations in quantum chemistry, where the concept of hybrid functionals originated. However, as soon as one wants

Table 6.11 Bond lengths R_e, dissociation energies D_e (including the zero-point energies [226]) and harmonic frequencies ω_e of diatomic molecules: Selfconsistent PBE0 [217] and local hybrid (LH-LDA) results. The LH-LDA hybrid [433] combines Eqs. (6.141), (6.148), (6.147) with the LDA for exchange and correlation. It is evaluated with the orbitals resulting from a global hybrid in which 10% exact exchange is admixed to the LDA for exchange and correlation. For the atomic ground state non-spherical densities have been allowed.

Molecule	Method	R_e [Bohr]	D_e [eV]	ω_e [cm^{-1}]
H_2	PBE0	1.407	4.529	4406
	LH-LDA	1.403	4.972	4491
Li_2	PBE0	5.152	0.836	331
	LH-LDA	5.214	0.846	329
B_2	PBE0	3.047	2.848	1016
	LH-LDA	3.071	2.482	963
C_2	PBE0	2.351	5.219	1870
	LH-LDA	2.358	4.884	1846
N_2	PBE0	2.055	9.787	2464
	LH-LDA	2.061	9.445	2428
O_2	PBE0	2.251	5.406	1697
	LH-LDA	2.262	5.334	1673
F_2	PBE0	2.596	1.519	1067
	LH-LDA	2.615	1.633	1018
LiH	PBE0	3.014	2.296	1419
	LH-LDA	3.039	2.560	1402
FH	PBE0	1.734	5.939	4162
	LH-LDA	1.738	6.243	4156
CO	PBE0	2.120	11.090	2240
	LH-LDA	2.126	11.025	2227
NO	PBE0	2.150	6.603	2019
	LH-LDA	2.159	6.379	1982
OH	PBE0	1.833	4.414	3761
	LH-LDA	1.836	4.577	3758
NH	PBE0	1.960	3.697	3319
	LH-LDA	1.960	3.650	3318
CH	PBE0	2.124	3.501	2850
	LH-LDA	2.120	3.548	2878
CN	PBE0	2.186	7.754	2146
	LH-LDA	2.195	7.356	2144
LiF	PBE0	2.945	5.730	932
	LH-LDA	2.994	5.970	893
BeH	PBE0	2.544	2.422	2028
	LH-LDA	2.550	2.544	2043

to apply hybrid functionals to metallic systems, one has to face the logarithmic divergence of the derivative of the HF single particle (band) energies with respect to the crystal momentum at the Fermi surface (see Chap. 17 of [96]). This divergence results from the combination of (i) the long range of the unscreened Coulomb interaction, and (ii) the nonlocal structure of the HF exchange potential. While an OPM treatment of the exact exchange avoids the divergence, the GKS approach suffers from it, just as the pure HF scheme.

Quite generally, the infinite range of the Coulomb interaction complicates HF and also GKS calculations for metals substantially: while the range of the exchange interaction decays exponentially for systems with a gap [440] (with the decay constant being controlled by the size of the HOMO-LUMO or band gap), its decay is only algebraic for metallic systems [441] (at zero temperature). As a consequence, the real space methods typically applied in quantum chemistry become extremely demanding: for each center the exchange interaction with a large number of neighboring centers has to be taken into account.

In order to resolve this difficulty, the concept of *screened hybrid functionals* has been introduced [441]. The starting point for this class of functionals is a decomposition of the Coulomb interaction into a short-range and a long-range part. For technical reasons (i.e. the fact that integrals with Gaussian basis functions can be solved analytically) the decomposition is based on the error function,

$$\frac{1}{|\boldsymbol{r}|} = \underbrace{\frac{\mathrm{erfc}(\omega|\boldsymbol{r}|)}{|\boldsymbol{r}|}}_{\text{short-range}} + \underbrace{\frac{\mathrm{erf}(\omega|\boldsymbol{r}|)}{|\boldsymbol{r}|}}_{\text{long-range}} \tag{6.136}$$

$$\mathrm{erf}(x) = \frac{2}{\sqrt{\pi}}\int_0^x dt\, e^{-t^2} \qquad\qquad \mathrm{erfc}(x) = 1-\mathrm{erf}(x)$$

$$\Rightarrow \quad \mathrm{erf}(0)=0 \qquad \mathrm{erf}(\infty)=1\,,$$

where the empirical parameter ω remains to be determined. After insertion of (6.136) the exchange functional of the global PBE0 hybrid, Eq. (6.134), splits into four components,

$$E_{\mathrm{x}}^{\mathrm{PBE0}} = \frac{1}{4}\left[E_{\mathrm{x}}^{\mathrm{exact,sr}} + E_{\mathrm{x}}^{\mathrm{exact,lr}}\right] + \frac{3}{4}\left[E_{\mathrm{x}}^{\mathrm{PBE,sr}} + E_{\mathrm{x}}^{\mathrm{PBE,lr}}\right], \tag{6.137}$$

where $E_{\mathrm{x}}^{\mathrm{exact,sr}}$ denotes the Fock term with the interaction $\mathrm{erfc}(\omega|\boldsymbol{r}-\boldsymbol{r}'|)/|\boldsymbol{r}-\boldsymbol{r}'|$, and $E_{\mathrm{x}}^{\mathrm{exact,lr}}$ absorbs the remaining long-range part of (6.136). Similarly, $E_{\mathrm{x}}^{\mathrm{PBE,sr}}$ and $E_{\mathrm{x}}^{\mathrm{PBE,lr}}$ are the PBE-type GGAs for exchange which result from use of the short- and long-range Coulomb interactions inside the PBE exchange hole [441] (compare Sect. 4.5.2).

The critical term in (6.137), which complicates calculations for metallic systems, is $E_{\mathrm{x}}^{\mathrm{exact,lr}}$. However, for suitable values of ω this term is rather small. In addition, $E_{\mathrm{x}}^{\mathrm{exact,lr}}$ is often close to $E_{\mathrm{x}}^{\mathrm{PBE,lr}}$. This suggests the replacement of the PBE0 exchange by

$$E_{\rm x}^{\rm HSE} = \frac{1}{4}E_{\rm x}^{\rm exact,sr} + \frac{3}{4}E_{\rm x}^{\rm PBE,sr} + E_{\rm x}^{\rm PBE,lr} . \tag{6.138}$$

Obviously, $E_{\rm x}^{\rm HSE}$ approaches the PBE0 functional in the limit $\omega \to 0$. The correlation component of the PBE0 functional, on the other hand, remains unmodified in the screened hybrid,

$$E_{\rm c}^{\rm HSE} = E_{\rm c}^{\rm PBE} . \tag{6.139}$$

Extensive tests showed that results are not particularly sensitive to the value of ω, as long as this screening parameter is chosen sufficiently small. The value of ω=0.11 Bohr^{-1} turned out to be a good compromise between accuracy and computational efficiency [441–443]. With this choice the accuracy of the screened hybrid for standard sets of thermochemical data is comparable to that of its PBE0 parent functional. A detailed assessment of the screened hybrid for solids may be found in [444, 242, 443, 445, 446].

The major motivation for using the exact exchange functional is the complete cancellation of the self-interaction in $E_{\rm H}$. Unfortunately, this cancellation remains incomplete for hybrid functionals as (6.127) or (6.134). As a consequence, the corresponding $v_{\rm xc}$ does not satisfy (4.20), but rather decays like $-a/r$ (for finite systems). This local property of $v_{\rm xc}$, which is associated with the fact that asymptotically a single orbital dominates the density, raises the question whether one can improve hybrids by use of a local mixing procedure. Ideally, such a local mixing would allow the complete cancellation of the self-interaction in those regions in which the system is close to a single particle system. In order to distinguish between different regions of space a *local hybrid* [447] necessarily has to rely on the local admixture of the exact exchange energy density $e_{\rm x}^{\rm exact}(\boldsymbol{r})$ to some LDA or GGA energy density. Setting aside the issue of non-uniqueness of energy densities, the former quantity is most naturally defined as

$$e_{\rm x}^{\rm exact}(\boldsymbol{r}) = -\frac{e^2}{2}\sum_{kl}\Theta_k\Theta_l \int d^3r' \, \frac{\phi_k^\dagger(\boldsymbol{r})\phi_l(\boldsymbol{r})\phi_l^\dagger(\boldsymbol{r}')\phi_k(\boldsymbol{r}')}{|\boldsymbol{r}-\boldsymbol{r}'|} \tag{6.140}$$

(see also the discussion in [448]). The general form of a local hybrid [447] is then given by

$$E_{\rm x}^{\rm LH} = \int d^3r \left\{ f(\boldsymbol{r})e_{\rm x}^{\rm exact}(\boldsymbol{r}) + [1-f(\boldsymbol{r})]e_{\rm x}^{\rm LDA/GGA}(\boldsymbol{r}) \right\} . \tag{6.141}$$

Obviously, the crucial ingredient of (6.141) is the local mixing factor $f(\boldsymbol{r})$, for which some model is required.

Given the basic purpose of the local hybrid, the mixing factor has to distinguish between the one-electron regime, in which the exact exchange energy density is required for a complete cancellation of the self-interaction, and the electron gas regime, in which GGA-type functionals become exact. These two limits are also reflected in the kinetic energy density. Ignoring once more the non-uniqueness of any energy density, one usually defines the exact KS kinetic energy density as

$$t_s(\boldsymbol{r}) = \frac{\hbar^2}{2m} \sum_k \Theta_k \left[\boldsymbol{\nabla}\phi_k(\boldsymbol{r})\right]^2, \tag{6.142}$$

where all orbitals have been chosen to be real. In the case of a single particle t_s can be written in the form of the von Weizsäcker energy density [174],

$$t_s(\boldsymbol{r}) \xrightarrow[n=\phi_1^2]{} t_W(\boldsymbol{r}) = \frac{\hbar^2}{m} \frac{\left[\boldsymbol{\nabla} n(\boldsymbol{r})\right]^2}{8n(\boldsymbol{r})}. \tag{6.143}$$

On the other hand, t_W vanishes in the limit of a homogeneous electron gas. Both t_s and t_W are positive definite by construction and satisfy the inequality

$$t_s(\boldsymbol{r}) \geq t_W(\boldsymbol{r}). \tag{6.144}$$

Equation (6.144) is easily proven by considering

$$\begin{aligned}
0 &\leq \sum_{kl} \Theta_k \Theta_l \left[\phi_k \boldsymbol{\nabla}\phi_l - \phi_l \boldsymbol{\nabla}\phi_k\right]^2 \\
&= 2\left[\sum_l \Theta_l \phi_l^2\right]\left[\sum_k \Theta_k (\boldsymbol{\nabla}\phi_k)^2\right] - \frac{1}{2}\sum_{kl}\Theta_k\Theta_l \left(\boldsymbol{\nabla}\phi_k^2\right)\left(\boldsymbol{\nabla}\phi_l^2\right) \\
&= \frac{4m}{\hbar^2} n(\boldsymbol{r}) \left[t_s(\boldsymbol{r}) - t_W(\boldsymbol{r})\right].
\end{aligned}$$

One therefore finds

$$0 \leq \frac{t_W(\boldsymbol{r})}{t_s(\boldsymbol{r})} \leq 1, \tag{6.145}$$

with the left-hand bound being approached in regions in which the KS states are close to plane waves, the right-hand bound in regions in which only a single orbital is non-vanishing. As a consequence, the mixing factor

$$f(\boldsymbol{r}) = \frac{t_W(\boldsymbol{r})}{t_s(\boldsymbol{r})} \tag{6.146}$$

offers itself. Clearly, this choice is in no way unique, many alternatives being conceivable (see [432, 449, 433]). The functional (6.141) is easily extended to spin-polarized systems, using (4.19) and the corresponding statement for the kinetic energy.

It remains to define the correlation component of the local hybrid. As the cancellation of self-interaction is much more important for exchange than for correlation, the correlation component of the local hybrid is chosen to be a conventional (semi)local density functional,

$$E_c^{\mathrm{LH}} = E_c^{\mathrm{LDA/GGA}}. \tag{6.147}$$

Both the LDA and various GGAs have been utilized in connection with local hybrids. Applications initially relied on an evaluation of E_{xc}^{LH} with LDA or GGA orbitals. Recently, also selfconsistent calculations, using either the OPM [432] or the GKS scheme [450], have been reported. The differences between selfconsistent and post-LDA/GGA results are, however, generally rather small, so that one may use the more efficient post-LDA/GGA treatment.

It turns out that the results obtained with E_{xc}^{LH} with the mixing factor (6.146) are not fully convincing, irrespective of the approximation chosen for its semi-local ingredient and of its treatment via the OPM, the GKS or the post-LDA/GGA procedure [447, 432, 449, 433, 450]. On the one hand, the local hybrid clearly constitutes an improvement over the global hybrid for systems in which the cancellation of the self-interaction plays a predominant role, as symmetric radical cations (as e.g. H_2^+, He_2^+) or reaction barriers [447]. On the other hand, the accuracy of E_{xc}^{LH} for standard thermochemical data is clearly lower than that of global hybrids [447, 432]. An illustration of this statement is provided in Table 6.10. This table includes results from a local hybrid, which combines Eq. (6.141) with the mixing factor (6.146) and the B88-GGA for exchange as well as Eq. (6.147) with the PW91-GGA for correlation.

This poor performance has prompted attempts to optimize $f(\boldsymbol{r})$ [432, 449, 433]. It was found that the choice

$$f(\boldsymbol{r}) = 0.48\,\frac{t_W(\boldsymbol{r})}{t_s(\boldsymbol{r})} \tag{6.148}$$

gives much improved thermochemical results, even if the exact exchange is combined with the LDA [449, 433, 450]. Some prototype results obtained from a local hybrid using the mixing factor (6.148) and the LDA [134] for the density-dependent component of E_{xc}^{LH} are included in Tables 6.11 and 6.10. One finds almost the same average absolute error as with the best global hybrids.

6.6 Analysis of Orbital-Dependent Correlation Functionals

6.6.1 Dispersion Force

In view of the motivation for implicit correlation functionals given in Sect. 6.1.2, first the question of dispersion forces has to be addressed. As none of the model-based functionals of Sect. 6.5 can rigorously deal with these long-range forces, the present discussion focuses on the second order correlation functional (6.93) as the simplest first-principles functional. Some numerical results obtained with global hybrid functionals are provided at the end of this section.

Consider two neutral atoms A and B, separated by a sufficiently large distance R, so that no molecular orbitals are formed. For this system the overlap between the atomic orbitals $\phi_{k,A}$ centered on atom A and $\phi_{l,B}$ centered on atom B vanishes exponentially with increasing R. Even if A and B represent the same type of atom, one can therefore choose the KS states of the two-center system to be atomic orbitals, rather than molecular orbitals (which are trivial linear combinations of the atomic orbitals). The sums over all KS states in (6.94) and (6.100) then effectively split up into sums over the atomic states (compare e.g. [96]),

$$\sum_i \longrightarrow \sum_{i_A} + \sum_{i_B} .$$

This implies that only those Slater integrals in $E_c^{(2)}$ contribute which do not involve orbital pairs of type $\phi_{i_A}^\dagger(\boldsymbol{r})\phi_{j_B}(\boldsymbol{r})$, i.e. which do not link orbitals from different atomic centers at the same point $\boldsymbol{r}$. Consequently, for large R neither $E_c^{\Delta\mathrm{HF}}$, Eq. (6.100), nor the SOX component of E_c^{MP2}, Eq. (6.97), contributes to the interaction between the two atoms, as for these terms all sums over states only couple states belonging to the same atom,

$$\begin{aligned} E_c^{\mathrm{MP2,SOX}} &= -\frac{e^4}{2} \sum_{\alpha=A,B} \sum_{i_\alpha j_\alpha k_\alpha l_\alpha} \Theta_{i_\alpha}\Theta_{j_\alpha}(1-\Theta_{k_\alpha})(1-\Theta_{l_\alpha}) \\ &\qquad \times \frac{(i_\alpha j_\alpha||k_\alpha l_\alpha)(k_\alpha l_\alpha||j_\alpha i_\alpha)}{\varepsilon_{i_\alpha}+\varepsilon_{j_\alpha}-\varepsilon_{k_\alpha}-\varepsilon_{l_\alpha}} \\ E_c^{\Delta\mathrm{HF}} &= \sum_{\alpha=A,B} \sum_{i_\alpha l_\alpha} \frac{\Theta_{i_\alpha}(1-\Theta_{l_\alpha})}{\varepsilon_{i_\alpha}-\varepsilon_{l_\alpha}} \left| \langle i_\alpha | v_{\mathrm{x}} | l_\alpha \rangle + e^2 \sum_{j_\alpha} \Theta_{j_\alpha} (i_\alpha j_\alpha || j_\alpha l_\alpha) \right|^2 . \end{aligned}$$

Only the direct matrix elements in $E_c^{\mathrm{MP2,DIR}}$, which corresponds to the ring diagram (6.98), couple both centers for large R. In the present situation the complete ring diagram decomposes into four terms, as the electron-hole pair in any of the rings can either belong to atom A or atom B. The combinations AA and BB,

$$\begin{aligned} E_{\mathrm{c,AA+BB}}^{\mathrm{MP2,DIR}} &= \frac{e^4}{2} \sum_{\alpha=A,B} \sum_{i_\alpha j_\alpha k_\alpha l_\alpha} \Theta_{i_\alpha}\Theta_{j_\alpha}(1-\Theta_{k_\alpha})(1-\Theta_{l_\alpha}) \\ &\qquad \times \frac{(i_\alpha j_\alpha||k_\alpha l_\alpha)(k_\alpha l_\alpha||i_\alpha j_\alpha)}{\varepsilon_{i_\alpha}+\varepsilon_{j_\alpha}-\varepsilon_{k_\alpha}-\varepsilon_{l_\alpha}} , \end{aligned}$$

contribute to the atomic (on-site) correlation energies of A and B, just as $E_c^{\Delta\mathrm{HF}}$ and $E_c^{\mathrm{MP2,SOX}}$. Only the combinations AB and BA, which represent the interaction of virtual particle–hole excitations on two different centers, lead to molecular binding.

No other component of the total energy functional contributes to the interaction between the two atoms, provided that A and B are closed-subshell atoms (so that no static multipole moments are present). The interaction energy between A and B therefore reduces to

$$\begin{aligned} E_{\mathrm{int}} &\equiv E_{\mathrm{c,AB+BA}}^{\mathrm{MP2,DIR}} \\ &= e^4 \sum_{i_A k_A} \Theta_{i_A}(1-\Theta_{k_A}) \sum_{j_B l_B} \Theta_{j_B}(1-\Theta_{l_B}) \frac{(i_A j_B||k_A l_B)(k_A l_B||i_A j_B)}{\varepsilon_{i_A}+\varepsilon_{j_B}-\varepsilon_{k_A}-\varepsilon_{l_B}} . \end{aligned} \tag{6.149}$$

If one expands (6.149) in powers of $1/R$ and reintroduces the frequency integration inherent in the ring diagram, one ends up with an expression [350] which is much more familiar,

$$E_{\mathrm{int}} = -\frac{C_6}{R^6}\,; \qquad C_6 = -3e^4 \int_0^\infty \frac{du}{\pi}\, \alpha_{\mathrm{s},A}(iu)\alpha_{\mathrm{s},B}(iu)\,. \tag{6.150}$$

Here $\alpha_{\mathrm{s},A/B}(iu)$ denotes the atomic KS polarizability (for the case of closed subshells), evaluated at imaginary frequency,

$$\alpha_{\mathrm{s}}(\omega) \;=\; -\int d^3r_1 \int d^3r_2\; z_1\, z_2\, \chi_{\mathrm{s,R}}(\boldsymbol{r}_1,\boldsymbol{r}_2,\omega)\,. \tag{6.151}$$

The basic ingredient of $\alpha(\omega)$ is the frequency-dependent, retarded KS response function,

$$\chi_{\mathrm{s,R}}(\boldsymbol{r}_1,\boldsymbol{r}_2,\omega) \;=\; \sum_{k,l}[\Theta_k - \Theta_l]\frac{\phi_k^\dagger(\boldsymbol{r}_1)\phi_l(\boldsymbol{r}_1)\phi_l^\dagger(\boldsymbol{r}_2)\phi_k(\boldsymbol{r}_2)}{\hbar\omega - \varepsilon_l + \varepsilon_k + i\eta}\,. \tag{6.152}$$

Equation (6.150) has the standard form of a dispersion force. Obviously, $E_{\mathrm{c}}^{(2)}$ is able to reproduce the correct long-range behavior proportional to $1/R^6$. The same is true for $E_{\mathrm{c}}^{\mathrm{RPA}}$ as the diagram (6.98) is included in the RPA (see Eq. (6.108)).

However, the exact result for the coefficient C_6 involves the full atomic polarizabilities, while the present DFT-variant of C_6 is determined by the KS polarizabilities (as a consequence of second order perturbation theory). So, the next question is how well do the KS coefficients reproduce the exact C_6? Calculations [451] show that Eq. (6.150) yields reasonably accurate values for light atoms: as indicated by Table 6.12, the lowest order KS results overestimate the full C_6 by 10–20%. On the other hand, for heavier atoms higher order correlation becomes important, so that the C_6 obtained from Eq. (6.150) differ substantially from empirical results.

Table 6.12 Van der Waals coefficient C_6: Eq. (6.150) versus accurate empirical results.

Atoms	$E_{\mathrm{c}}^{(2)}$ [451]	Empirical [452, 453]
He–He	1.66	1.46
He–Ne	3.49	3.03
Ne–Ne	7.45	6.38
Xe–Xe	730.7	285.9
H–He	3.02	2.82 ± 0.02
H–Na	81.14	71.8 ± 0.3

At this point it is has been verified that the functional (6.93) reproduces the long-range behavior of the dispersion force. It is not yet clear, however, how it performs in the intermediate (bonding) regime. In order to predict the equilibrium geometry of a van der Waals bonded molecule, it is not sufficient to obtain the correct asymptotic

$1/R^6$-attraction. Rather the complete energy surface must be accurate. This point is illustrated in Fig. 6.13, in which the energy surface $E_b(R)$ of the He dimer is shown [454]. In Fig. 6.13 four different DFT results are compared with HF data [455]

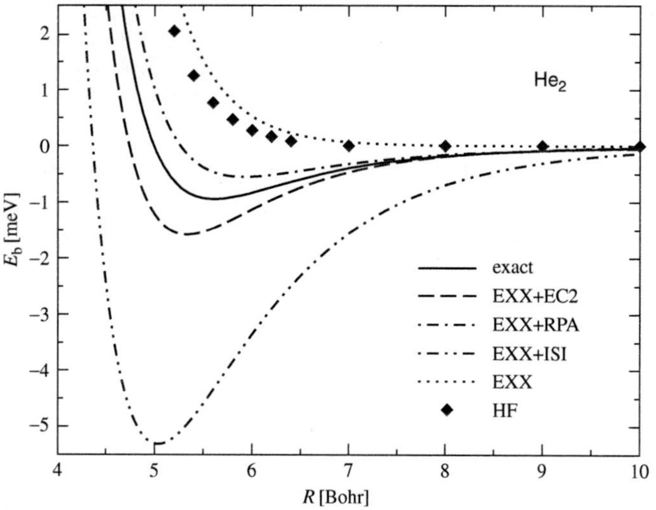

Fig. 6.13 Energy surface E_b of He_2: Exact E_x in combination with $E_c^{(2)}$ (EXX+EC2) [454], E_c^{RPA} (EXX+RPA) and E_c^{ISI} (EXX+ISI) versus EXX, HF [455] and exact data [332].

and the exact E_b, obtained by variational calculations with correlated wavefunctions [332] (all E_b are strictly nonrelativistic). The x-only results demonstrate explicitly that dispersive bonding is a pure correlation effect—both the EXX calculation and its HF counterpart predict a repulsive energy surface (as in many other situations the EXX data are rather close to the HF numbers—see Sect. 6.3).

Three "correlated" DFT results are also shown in Fig. 6.13. The first one is based on the combination of the exact E_x with $E_c^{(2)}$ (EXX+EC2—after a selfconsistent EXX calculation with the KLI approximation, $E_c^{(2)}$ is evaluated from the EXX orbitals and added to the EXX ground state energy). The agreement of the resulting $E_b^{EXX+EC2}(R)$ with the exact surface is far from perfect, but $E_b^{EXX+EC2}(R)$ is at least qualitatively correct. One can conclude that $E_c^{(2)}$ does not only give the desired asymptotic $1/R^6$ behavior, but also provides a realistic description of $E_b(R)$ for non-asymptotic R.

As pointed out before, the RPA contains the asymptotically leading term of $E_c^{(2)}$. It is therefore no surprise that the energy surfaces of the two functionals agree for large R. However, the RPA also leads to quite realistic results in the vicinity of the minimum of $E_b(R)$ [456] (see Fig. 6.13). The same is not true for the ISI energy surface, which is also plotted in Fig. 6.13. The He_2 molecule is an extremely sensitive system, so that a model-based treatment of higher order correlations is particularly difficult.

Since $E_c^{(2)}$ originates from second order perturbation theory, one might be tempted to consider its success for He_2 a trivial result: the arguments given in the beginning of this section apply to any second order perturbative energy functional, so that the $1/R^6$-behavior is common to all of them. However, the correct long-range behavior does not imply that the full function $E_b(R)$ is accurate. This is demonstrated explicitly in Fig. 6.14, in which four second order results for $E_b(R)$ are compared. In addition to the EXX+EC2-result, already shown in Fig. 6.13, the surface obtained

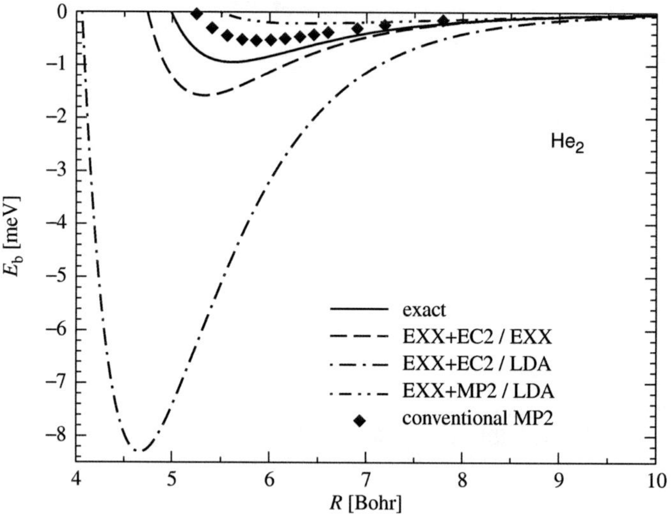

Fig. 6.14 Energy surface $E_b(R)$ of He_2 obtained from different perturbation expansions to second order: $E_c^{(2)}$ in combination with exact E_x evaluated with either self-consistent EXX (EXX+EC2/EXX) or LDA (EXX+EC2/LDA) orbitals versus conventional MP2 [457] and exact [332] result. Also shown is the result for LDA orbitals, if $E_c^{\Delta HF}$, Eq. (6.101), is left out (EXX+MP2/LDA).

in the conventional MP2 approach (second order perturbation theory on the basis of HF) is given. The EXX+EC2-result overestimates the exact well depth to roughly the same extent as the MP2 surface underestimates it.

In the third second order approach the EXX+EC2 functional is evaluated with self-consistent LDA orbitals, i.e. the difference between E_{xc}^{LDA} and $E_x + E_c^{(2)}$ is added perturbatively to the LDA surface. This is exactly the energy functional obtained from the perturbation expansion of Sect. 4.2.1, if the self-consistent LDA Hamiltonian is used as reference Hamiltonian. In this case the perturbing Hamiltonian is given by

$$\hat{H}_1 = \hat{W} + \int d^3r \hat{n}(\boldsymbol{r}) \left[v_{ext}(\boldsymbol{r}) - v_s^{LDA}(\boldsymbol{r}) \right] ,$$

where v_s^{LDA} denotes the self-consistent total LDA potential. One can then follow the derivation of Sect. 4.2.1 with this $\hat{H}_1$ until one arrives at Eq. (4.64) with $E_s = T_s^{LDA} + \int d^3r\, n^{LDA} v_s^{LDA}$. Insertion of the present $\hat{H}_1$ and expansion to second order finally leads to the ground state energy

$$T_s^{\text{LDA}} + \int d^3r\, v_{\text{ext}} n^{\text{LDA}} + E_{\text{H}}[n^{\text{LDA}}] + E_{\text{x}}[\phi_i^{\text{LDA}}] + E_c^{(2)}[\phi_i^{\text{LDA}}, \varepsilon_i^{\text{LDA}}, v_{\text{xc}}^{\text{LDA}}] \,,$$

where $v_{\text{xc}}^{\text{LDA}}$ enters into $E_c^{(2)}$ via $E_c^{\Delta\text{HF}}$. By construction this functional yields the $1/R^6$-asymptotics. However, the C_6-coefficient obtained from (6.150) with LDA orbitals [451] is much larger than the C_6 resulting from EXX orbitals (which is already too large). This is reflected by the large-R behavior of the corresponding energy surface (see Fig. 6.14). Even more important is the complete failure of the LDA-based second order expansion in the vicinity of the minimum of E_b. It either yields by far too much bonding, if $E_c^{(2)}[\phi_i^{\text{LDA}}, \varepsilon_i^{\text{LDA}}, v_{\text{xc}}^{\text{LDA}}]$ is included completely, or by far too little, if the component $E_c^{\Delta\text{HF}}$, Eq. (6.101) is left out. This demonstrates that application of second order perturbation theory does not automatically guarantee a realistic energy surface for dispersive bonds. A suitable noninteracting reference Hamiltonian, which provides the starting point for the expansion, is required. The EXX Hamiltonian offers itself as a particularly consistent choice (compare also [458, 459]).

For completeness, the energy surfaces obtained with two standard hybrid functionals, B3LYP [430, 216] and PBE0 [217], are shown in Fig. 6.15. While B3LYP

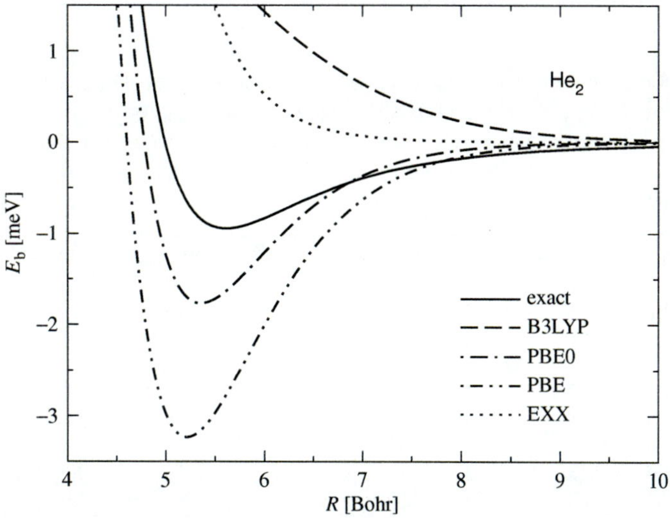

Fig. 6.15 Energy surface E_b of He_2: Self-consistent B3LYP and PBE0 results versus EXX, PBE data as well as exact surface [332].

generates a strictly repulsive energy surface, PBE0 leads to an attractive potential well whose depth and minimum position are reasonably close to the exact values [437]. This success results from a suitable superposition of the much too attractive PBE surface and the repulsive EXX surface (both are included in Fig. 6.15). As a consequence of the subtle balance between the GGA and EXX components, the energy surfaces obtained from hybrid functionals are quite sensitive to the precise mixing ratio between exact and GGA exchange. This fact emphasizes the importance of the ratio of 1/4 derived from the adiabatic connection. Asymptotically, the

PBE0 surface falls off much faster than $1/R^6$, consistent with the short-range character of its components.

6.6.2 Correlation Energy

After the demonstration that $E_c^{(2)}$ and related first-principles functionals are able to deal with dispersion forces, the next step is a quantitative study of more conventional properties. Correlation energies of closed-subshell atoms, obtained in various approximations, are listed in Table 6.13 and compared with exact quantum chemical correlation energies [115] (which have been extracted by combining variational results for two- and three-electron systems with experimental data for the ionization energies of the remaining electrons). The difference (4.11) between the exact DFT correlation energy and the quantum chemical correlation energy (4.9) is rather small, so that E_c^{QC} can serve as reference standard for the present comparison [116] (compare the discussion of Table 4.1 in Sect. 4.1). $E_c^{(2)}$ clearly overestimates the

Table 6.13 Correlation energies ($-E_c$) of closed-subshell atoms: LDA [134], PW91-GGA [211], CS [222], $E_c^{(2)}$ [343], E_c^{RPA} [414, 421], E_c^{RPA+} [460, 461, 421] and ISI [422] results (all DFT energies have been obtained by insertion of EXX densities) in comparison with conventional MP2 [145, 144] and exact E_c^{QC} [115] energies (in mHartree). The contribution (6.100) to $E_c^{(2)}$ is also listed separately.

Atom	MP2	$-E_c^{QC}$	ISI	$-E_c^{(2)}$	$-E_c^{\Delta HF}$	$-E_c^{RPA}$	$-E_c^{RPA+}$	LDA	GGA	CS
He	37	42	39	48	0.0	83	47	113	46	42
Be	76	94	101	124	0.6	179	108	225	94	93
Ne	388	391	411	478	1.7	597	400	746	382	375
Mg	428	438	462	522	3.2	687	453	892	450	451
Ar	709	722	777	866	5.4	1101	742	1431	771	743
Ca	798		890	996	6.4			1581	847	824
Zn	1678		1789	2016	14.9			2668	1526	1426
Cd	2618		2847	3104	19.5			4571	2739	2412
Xe	3088		3214	3487	17.7			5199	3149	2732

exact atomic correlation energies, consistent with the result for the energy surface of He_2. The accuracy of $E_c^{(2)}$ is lower than that of the conventional MP2 scheme, in particular for heavier atoms. The deviations are also much larger than those observed for standard GGAs. On the other hand, the ISI extension of $E_c^{(2)}$ eliminates most of the error produced by $E_c^{(2)}$.

The component $E_c^{\Delta HF}$ of $E_c^{(2)}$ is listed separately in Table 6.13. It vanishes for two-electron systems and is more than 2 orders of magnitude smaller than E_c^{MP2} for all other atoms. This suggests that $E_c^{\Delta HF}$ can be neglected in most situations, which definitively simplifies the application of $E_c^{(2)}$.

A comparison of $E_c^{(2)}$ with E_c^{RPA}, Eq. (6.114), illustrates the importance of the SOX contribution (6.97) for finite systems [460, 461, 411, 421]. The RPA is not particularly useful for atoms and molecules without inclusion of some SOX correction. However, as is clear from the discussion of dispersion forces, the SOX term is of short-range nature. For this reason one would expect the LDA to give a reasonably accurate account of the SOX contribution. The SOX energy, as well as other higher order corrections, may thus be included in the form [460, 461]

$$E_c^{RPA+} = E_c^{RPA} + E_c^{LDA} - E_c^{LDA-RPA}, \tag{6.153}$$

where $E_c^{LDA-RPA}$ denotes the LDA for E_c^{RPA}. The RPA+ approach has been shown to give rather accurate correlation energies for atoms [421]. Some examples are listed in Table 6.13.

For completeness, Table 6.13 also includes the energies resulting from the CS-functional (6.126). This functional leads to rather accurate values for light atoms, but underestimates E_c for the heavier elements.

Comparable data for the He isoelectronic series are given in Table 6.14 [344]. These numbers demonstrate that $E_c^{(2)}$ obeys the correct scaling law with respect

Table 6.14 Correlation energies ($-E_c$) of the He isoelectronic series: LDA [134], PW91-GGA [211], CS [222], $E_c^{(2)}$ [343], E_c^{RPA} [421] and ISI [422] results (all energies obtained by insertion of EXX densities) in comparison with conventional MP2 [462] and exact [81] energies (in mHartree).

Ion	LDA	GGA	CS	$-E_c^{RPA}$	$-E_c^{(2)}$	ISI	Exact	MP2
He	112.8	45.9	41.6	83.1	48.21	39.4	42.04	37.1
Ne^{8+}	203.0	61.7	40.6	90.8	46.81	45.0	45.69	44.4
Ca^{18+}	243.3	67.7	35.9	91.4	46.69	45.8	46.18	45.4
Zn^{28+}	267.2	71.3	33.2	91.5	46.67		46.34	45.7
Zr^{38+}	284.4	74.0	31.4	91.5	46.66	46.3	46.42	45.9
Sn^{48+}	297.7	76.0	30.0		46.65		46.47	46.0
Nd^{58+}	308.7	77.8	29.0		46.64	46.3	46.51	
Yb^{68+}	318.0	79.3	28.2		46.63		46.53	
Hg^{78+}	326.1	80.6	27.6		46.62	46.4	46.55	
Th^{88+}	333.2	81.7	27.0		46.62		46.56	
Fm^{98+}	339.6	82.8	26.0		46.62	46.4	46.57	

to the nuclear charge Z: $E_c^{(2)}$ becomes exact in the limit of large Z, in which the correlation energy of two-electron ions approaches a constant. The GGA energies, on the other hand, exhibit a systematic increase with increasing Z. The opposite behavior is found for the CS functional, with an error of about 50% for heavy ions. The ISI functional, whose dominant component is $E_c^{(2)}$, also approaches a constant value for large Z.

A very sensitive test for correlation functionals is provided by atomic EAs. Table 6.15 lists the EAs for H^- obtained with various functionals. The exact exchange

Table 6.15 Correlation energy and electron affinity of H^-: Results obtained by combination of the exact exchange with different correlation functionals (LDA [134], PW91-GGA [211], CS [222], EC2 ($E_c^{(2)}$) [343], RPA (E_c^{RPA}) [361] and ISI [422]) in comparison with conventional MP2 [463] and exact [464] energies (in mHartree). All correlation functionals have been evaluated with self-consistent EXX orbitals.

Method	$-E_c$	EA
EXX	—	−12.1
EXX+LDA	75.7	62.6
EXX+GGA	35.5	22.8
EXX+CS	31.2	18.9
EXX+EC2	54.6	42.6
EXX+RPA	74.2	42.4
EXX+ISI	34.3	22.2
Exact	39.8	27.8
MP2	27.3	15.2

is used in all cases, only the correlation functional varies. The EXX calculation predicts H^- to be unbound, which emphasizes once more the importance of correlation.[20] Inclusion of LDA correlation produces an EA which is more than a factor of 2 too large, since the EA of H^- directly reflects the accuracy of the correlation energy. $E_c^{(2)}$ also clearly overestimates the correlation energy of H^-, so that the same is true for the EA. $E_c^{(2)}$ and the conventional MP2 energy bracket the exact EA, in the same fashion as they bracket the correlation energies of neutral atoms and positive ions. The inclusion of higher order terms in $E_c^{(2)}$ via the ISI definitively improves the agreement with the exact EA, although one notices a tendency to overcorrect the error of $E_c^{(2)}$.

The next topic of this quantitative investigation of orbital-dependent correlation functionals are covalently bound molecules. Since the results for N_2 are quite characteristic for many diatomic molecules, the spectroscopic constants of N_2 obtained with a variety of approximations are listed in Table 6.16. One first of all observes that the EXX approach yields a reasonably accurate bond length, but substantially underestimates the experimental dissociation energy. In comparison to the EXX results, the combination of exact exchange with LDA or GGA correlation improves D_e, even though the remaining error is quite large. At the same time R_e is reduced and ω_e is increased, which worsens the agreement with experiment. Similar statements apply if the CS functional is added to the exact exchange.

Turning to the first-principles orbital-dependent correlation functionals, one finds that $E_c^{(2)}$ predicts N_2 to be unbound. In order to understand this result one has to go back to Eq. (6.94) and examine the structure of this expression. If the separation of the two nitrogen atoms in N_2 is increased, the highest occupied and the lowest unoccupied KS levels in the molecule approach each other further and further. So,

[20] Note that the EXX method yields a stable solution for the H^- ion. However, the EXX ground energy of the ion is higher than that of the neutral hydrogen atom.

Table 6.16 Spectroscopic constants of N_2: Results obtained by combination of the exact exchange (EXX) with different correlation functionals (LDA [134], PW91-GGA [211], CS [222], EC2 ($E_c^{(2)}$) [343] ISI [422] and RPA [361, 411]) in comparison with HF, MP2, quadratic configuration interaction with single and double excitations (QCISD) [225], SIC-LDA [143], conventional LDA and PW91-GGA, as well as experimental [227] results (all EXX calculations rely on the KLI approximation; EC2, ISI and RPA correlation have been evaluated with EXX orbitals; EXX+CS-data from [428]).

Method	R_e [Bohr]	D_e [eV]	ω_e [cm^{-1}]
Expt.	2.074	9.905	2359
HF	2.037	4.952	2738
MP2	2.135	9.333	2180
QCISD	2.105	8.488	2400
EXX	2.011	4.973	2736
EXX+LDA	1.994	7.381	2780
EXX+GGA	1.997	7.574	2804
EXX+CS	1.998	7.818	
EXX+EC2		unbound	
EXX+ISI	2.235	12.225	1401
EXX+RPA	2.077	9.603	2354
LDA	2.068	11.601	2393
PBE	2.082	10.583	2344
BLYP	2.082	10.440	2327
B3LYP	2.059	9.910	2416
PBE0	2.055	9.787	2464
SIC-LDA(x+c)	1.876	−49.490	3245

with increasing R the energy gap in the denominator of (6.94) shrinks more and more. As this divergence is not compensated by vanishing Slater integrals in the numerator, E_c^{MP2} becomes larger and larger when the atoms are taken apart. This effect sets in already in the vicinity of the equilibrium separation in the case of N_2. As a consequence, one does not even find a local minimum in the energy surface.

This problem is intrinsically related to (i) the perturbative nature of E_c^{MP2}, and (ii) the existence of a Rydberg series in the EXX spectrum. The same effect shows up in other instances: one finds, for example, that the correlation energy of the beryllium atom is particularly overestimated by $E_c^{(2)}$, which is due to the presence of the low-lying unoccupied $2p$ states (compare Table 6.13). The presence of the Rydberg series is useful for the calculation of many atomic properties, most notably for the description of negative ions or of excited states. On the other hand, it is not very helpful if the treatment of correlation is based on perturbation theory. This statement is corroborated by the fact that the conventional MP2 calculation gives quite reasonable results for N_2, as the underlying HF Hamiltonian does not yield a Rydberg series.

On the other hand, the Rydberg states do no longer pose a problem, as soon as a suitable partial resummation of the KS perturbation series is applied. This is demonstrated in Table 6.13 for the case of the RPA, which yields quite accurate results for N_2 (compare, however, [411]).

This statement raises the question whether the model-based resummation of the perturbation series via the ISI can resolve this fundamental problem of a straight perturbation expansion as well? The ISI functional indeed leads to a bound N_2. However, it does not perform particularly well quantitatively.

For completeness, Table 6.16 also contains the spectroscopic constants obtained with the SIC-LDA for both exchange and correlation. In this calculation the standard molecular orbitals have been used for the evaluation of the xc-energy and potential, without applying any additional localization scheme. The importance of the unitarity problem discussed in Sect. 4.7 becomes obvious. In particular, the dissociation energy is completely misrepresented. This problem can be traced to the SIC energies of the molecular core states [267]: the Coulomb contribution to (4.301), resulting from the two-center molecular core states ($1\sigma_g$, $1\sigma_u$), differs substantially from the one obtained with the one-center atomic core states ($1s$), as

$$\phi_{1\sigma_{g/u}} \approx \frac{1}{\sqrt{2}}\left[\phi_{A,1s} \pm \phi_{B,1s}\right] .$$

Realistic SIC-LDA results for molecules or solids can only be obtained on the basis of some localization prescription for quasi-degenerate states [267]. Such a scheme essentially consists of using localized linear combinations of $\phi_{1\sigma_{g/u}}$ for the evaluation of the SIC functional and has to be applied to all core and semi-core states.

6.6.3 Correlation Potential

In the applications discussed in Sects. 6.6.1 and 6.6.2 all orbital-dependent functionals have been evaluated with orbitals obtained by selfconsistent calculations with the exact E_x, i.e. in post-EXX form. However, ultimately one would like to apply any approximation for E_c in a selfconsistent fashion, so that the effective potential includes correlation effects. The correlation potentials obtained from the LDA and GGA have, as illustrated in Sect. 4.5.5, little in common with the exact atomic correlation potential [83]. This raises the question whether orbital-dependent functionals provide any improvement, in particular, in view of the first-principles nature of functionals like $E_c^{(2)}$.

Unfortunately, in the case of functionals which depend on unoccupied KS states the solution of the OPM integral equation (6.26) is not as straightforward as for the exact exchange. The underlying problem has already surfaced in Sect. 6.2.5, where the derivation of Eq. (6.59) required a discrete, normalizable KS spectrum. Quite generally, the OPM integral equation does not have a solution, which satisfies the boundary condition (6.50), as soon as positive energy continuum KS states are present [465, 466, 346]. This problem can be resolved by working with a completely discrete KS spectrum [467, 346], which is most elegantly introduced via a suitable basis set [417, 468]. Alternatively, one can embed finite systems in a large box or spherical cavity, requiring hard-wall boundary conditions at the surface of the box [349]. If the surface of the box is sufficiently far from the system of interest, the boundary conditions do not affect results in any significant way. Most potentials shown below have been generated with this approach. Note that in the case of periodic systems the standard Born-von Karman boundary conditions automatically ensure a discrete positive energy spectrum.

The question whether $E_c^{(2)}$ provides an improvement over LDA and GGA is addressed in Figs. 6.16–6.20, in which the correlation potentials of He, Be, Ne, Li and N are shown. Exact reference potentials are available for He, Be and Ne. These (numerically) exact v_c have been obtained by inversion of the KS equations

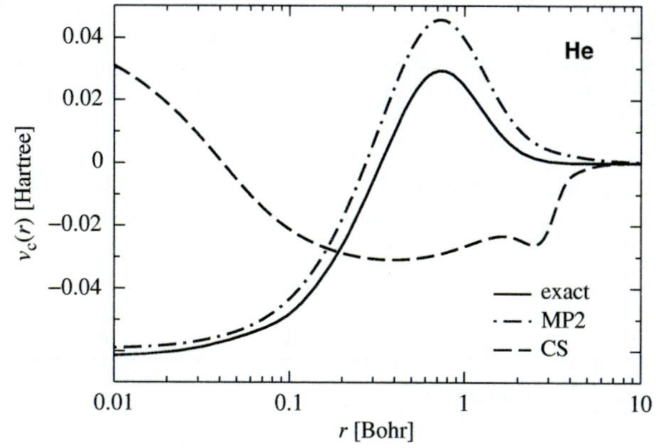

Fig. 6.16 Correlation potential of helium: $v_c^{MP2} = \delta E_c^{MP2}/\delta n$, Eq. (6.94), and $v_c^{CS} = \delta E_c^{CS}/\delta n$, Eq. (6.126), versus exact potential [83]. All approximations have been evaluated with the exact KS density. Note that $E_c^{\Delta HF}$ vanishes for spin-saturated 2-electron systems, so that $v_c^{MP2} = v_c^{(2)}$.

[83] (see also [117–122]): In this approach first a Monte Carlo calculation is performed in order to generate a highly accurate representation of the atomic ground state density (on a numerical grid). This exact density is then used as input for the actual inversion of the KS equations, which leads to the exact total KS potential and the exact KS orbitals. Finally, subtraction of the exact v_x, which is obtained by solution of the x-only OPM integral equation resulting from the exact KS orbitals, and the exact v_H yields the exact v_c. For consistency, all approximate correlation potentials have also been evaluated with the exact ground state densities, which corresponds to the use of the exact KS orbitals in case of $v_c^{(2)}$. Only the MP2 component (6.94) has been included in the evaluation of $v_c^{(2)}$, as E_c^{MP2} is much larger than $E_c^{\Delta HF}$ for the atoms discussed (compare Table 6.13) ($E_c^{\Delta HF}$ even vanishes for He).

Several features are obvious from Figs. 6.16–6.18:

- $v_c^{(2)}$ reproduces the shell-structure of the exact v_c, unlike all available explicitly density-dependent approximations (compare Fig. 4.19). However, the amplitudes of the individual structures are considerably overestimated.

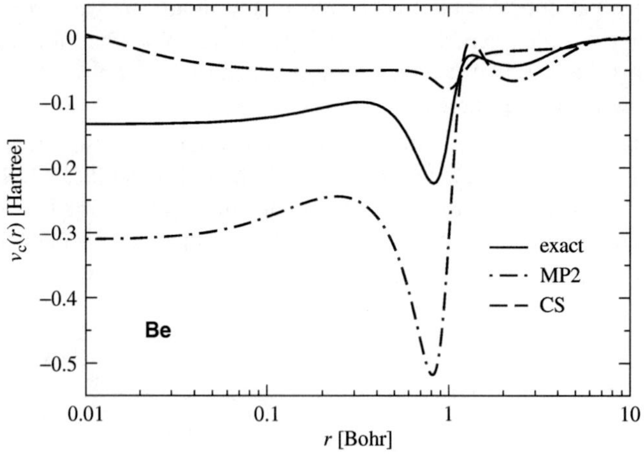

Fig. 6.17 Correlation potential of beryllium: $v_c^{MP2} = \delta E_c^{MP2}/\delta n$, Eq. (6.94), and $v_c^{CS} = \delta E_c^{CS}/\delta n$, Eq. (6.126), versus exact potential [83]. All approximations have been evaluated with the exact KS density.

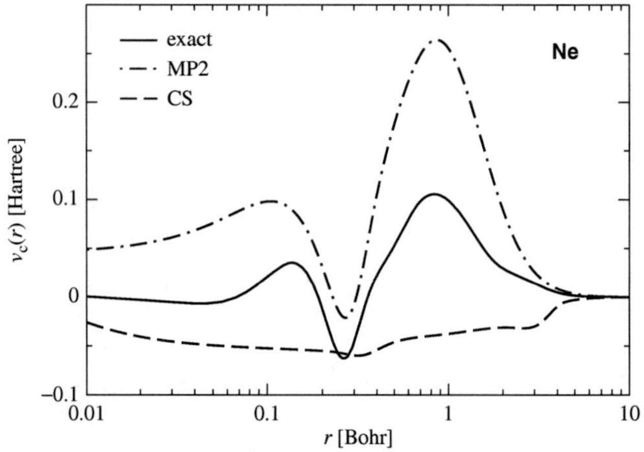

Fig. 6.18 Correlation potential of neon: $v_c^{MP2} = \delta E_c^{MP2}/\delta n$, Eq. (6.94), and $v_c^{CS} = \delta E_c^{CS}/\delta n$, Eq. (6.126), versus exact potential [83]. All approximations have been evaluated with the exact KS density.

- $v_c^{(2)}$ is asymptotically attractive on the left-hand side of the periodic table (Be) and repulsive on the right-hand side (Ne), as required by the error in EXX ionization potentials [469].
- $v_c^{(2)}$ reproduces the $-\alpha/(2r^4)$-behavior of the exact v_c in the asymptotic regime [92, 467]. However, the atomic polarizability α is substantially overestimated by $v_c^{(2)}$ (as is already clear from the energy surface of the He dimer).

$v_c^{(2)}$ can equally well be evaluated for spin-polarized systems. Examples for open-shell atoms are given in Figs. 6.19 and 6.20. As no Monte Carlo-based reference

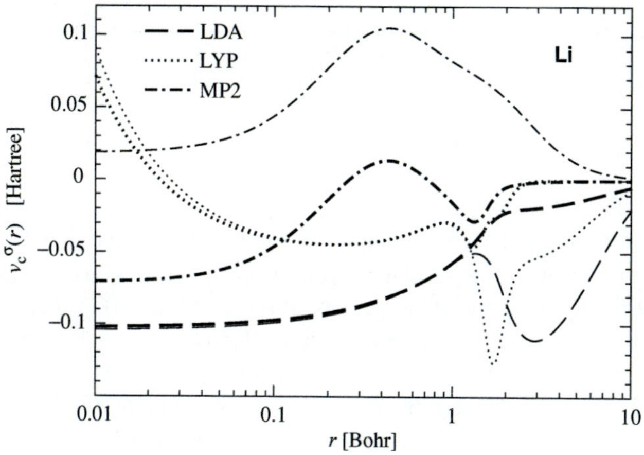

Fig. 6.19 Correlation potential of lithium: $v_c^{MP2} = \delta E_c^{MP2}/\delta n$, Eq. (6.94) versus LDA and LYP-GGA potential. *Thick lines* correspond to the majority spin channel, *thin lines* provide the minority spin potential. All approximations have been evaluated with the exact x-only spin-densities.

results are available for these spin-polarized atoms, the potentials in Figs. 6.19 and 6.20 have been evaluated with KS orbitals obtained from self-consistent calculations with the exact exchange. For comparison the correlation potentials obtained with the LDA and the LYP-GGA are also shown.

Besides the structures already observed for closed-subshell atoms, two new features show up:

- In the core region, in which the densities of the two spin-channels are essentially identical, the MP2 correlation potentials of the two channels differ by an almost constant shift, while the corresponding LDA and GGA potentials are very close (compare Fig. 4.16).
- In the case of Li the MP2 correlation potential of the majority-spin channel, $v_c^{\uparrow}$, is attractive, while the minority-spin potential, $v_c^{\downarrow}$, is repulsive. This reflects

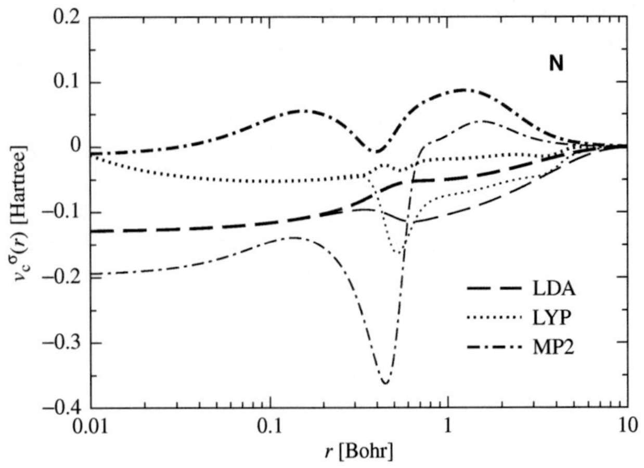

Fig. 6.20 As Fig. 6.19 for nitrogen.

the fact that the majority-spin channel is dominated by the $2s$ orbital, so that the resulting $v_c^\uparrow$ has some similarity with the correlation potential found for Be. The minority-spin channel consists only of the $1s$ orbital. This channel therefore resembles the closed-shell He configuration. On the other hand, the majority-spin potential is repulsive in the case of N, while the minority-spin potential remains close to zero in the valence regime and becomes attractive in the core region. While the majority-spin channel with its completely filled $2p$ orbital has the same basic structure as the correlation potential of Ne, the minority-spin channel is qualitatively close to the Be potential.

Both features are also observed for the correlation potentials obtained by inversion of accurate CI spin-densities [470].

So far, all orbital-dependent potentials have been evaluated for a given set of KS orbitals. Ultimately, however, fully self-consistent calculations are desired. Unfortunately, self-consistent application of E_c^{MP2} is hampered by its perturbative origin: E_c^{MP2} is found to be variationally instable for systems with a small HOMO-LUMO gap [349, 468], as, for instance, the beryllium atom. This behavior is easily understood if one recalls the fact that the OPM scheme is equivalent to a total energy minimization. As E_c^{MP2} is strictly negative, any increase of $|E_c^{\mathrm{MP2}}|$, which does not seriously affect other energy components, leads to a lower total energy. If the HOMO-LUMO gap is small, v_c^{MP2} can reduce this gap substantially without modifying the KS orbitals too much (which would change the remaining energy components) and thereby increase the absolute value of E_c^{MP2}. Once the gap shrinks more and more, E_c^{MP2} starts to dominate the total energy and v_c^{MP2} controls the total KS potential, which further eases closing of the gap. This deficiency of E_c^{MP2} reflects its origin in KS-based perturbation theory, which relies on a clear separation of occupied and unoccupied states.

Even for systems for which the HOMO-LUMO gap is too large to be closed by v_c^{MP2} one observes the same basic effect: the difference between the self-consistent v_c^{MP2} and the v_c^{MP2} obtained from the exact KS states is larger than the corresponding difference for any other known functional. This is illustrated in Fig. 6.21 for neon. The shell oscillations of the self-consistent v_c^{MP2} are clearly larger than those

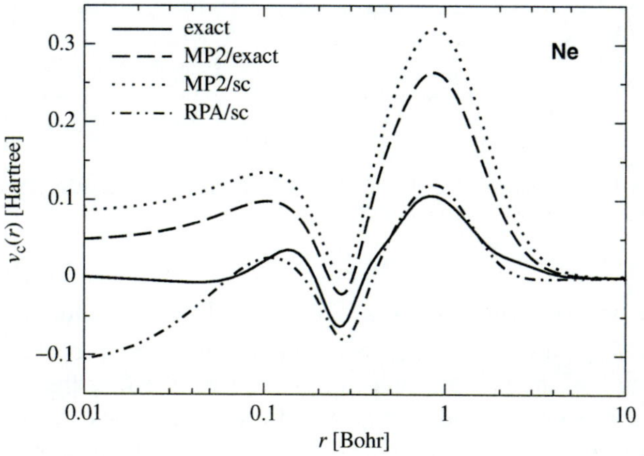

Fig. 6.21 Correlation potential of neon: v_c^{MP2} obtained from the exact KS density versus self-consistent v_c^{MP2} [349], self-consistent v_c^{RPA} [415], and exact result [83].

obtained by use of the exact KS orbitals in v_c^{MP2}. In this way additional correlation energy is gained, which lowers the total energy.

This observation points again at the importance of higher order correlation contributions. However, a *brute-force* inclusion of all third and fourth order terms of KS perturbation theory (in analogy to the MP4 level of quantum chemistry), is neither desirable (due to the resulting computational cost) nor can it be expected to resolve any problem originating from the perturbative nature of $E_c^{(2)}$. One is therefore led to consider some partial resummation of the KS perturbation series, such as the RPA.

It turns out that the RPA functional (6.114) is variationally stable for beryllium [415]. The correlation potentials resulting from self-consistent calculations with the combination of the exact E_x and E_c^{RPA} for neon and beryllium [415] are included in Figs. 6.21 and 6.22. In addition to the variational stability, one observes a clear quantitative improvement over v_c^{MP2}. Similar improvements have been obtained with other resummation schemes [418–420, 471].

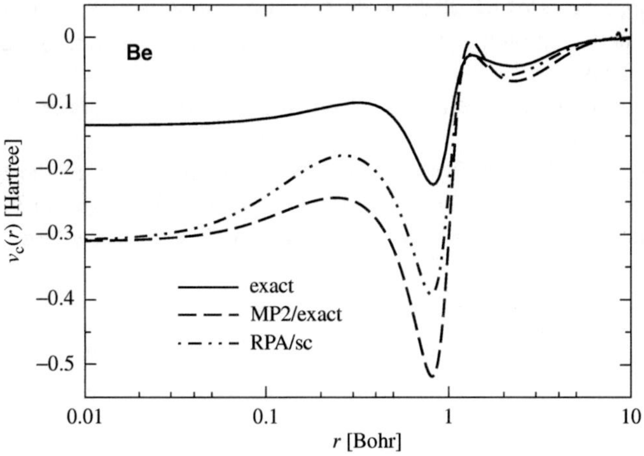

Fig. 6.22 Correlation potential of beryllium: v_c^{MP2} [420] obtained from the exact KS density versus self-consistent v_c^{RPA} [415], and exact result [83].

In summary: The results of Sects. 6.6.1–6.6.3 demonstrate the first-principles background of $E_c^{(2)}$ and related functionals. In fact, these results may be considered as a verification of the first-principles nature of DFT itself, as the formalism behind these expressions provides a truly systematic route to the exact xc-functional. On the other hand, the orbital-dependent treatment of correlation represents a much more serious challenge than that of exchange. The systematic derivation of such functionals via many-body theory leads to rather complicated expressions, in particular, if a resummation of the KS perturbation series is considered. Their rigorous implementation within the OPM not only requires the evaluation of Coulomb matrix elements of the complete set of KS states, but, in principle, also knowledge of higher order response functions. In practical calculations these first-principles functionals necessarily turn out to be rather difficult to handle, even if they are only applied non-self-consistently on the basis of some self-consistent EXX or GGA calculation.

6.7 Orbital-Dependent Representation of 2-Particle Density

Many-body techniques based on the KS Hamiltonian can not only be utilized for the derivation of first-principles xc-functionals, but also for the construction of density functionals for other ground state expectation values. This is of particular interest for quantities, which explicitly depend on the correlated motion of the electrons. A prototype of these quantities is the 2-particle density $n^{(2)}(\boldsymbol{r}_1,\boldsymbol{r}_2)$, Eq. (3.45). Using the perturbative approach of Sect. 6.4.1 to first order in e^2, one finds

$$\begin{aligned}
&n^{(2),(1)}(\boldsymbol{r}_1,\boldsymbol{r}_2)\\
&= -\frac{1}{2}\sum_i \Theta_i \sum_k (1-\Theta_k)\frac{\sum_j \Theta_j (kj||ji) + \langle k|v_{\rm x}|i\rangle}{\varepsilon_i - \varepsilon_k}\\
&\quad\times\Big\{ \phi_i^\dagger(\boldsymbol{r}_1)\phi_k(\boldsymbol{r}_1)n(\boldsymbol{r}_2) + n(\boldsymbol{r}_1)\phi_i^\dagger(\boldsymbol{r}_2)\phi_k(\boldsymbol{r}_2)\\
&\qquad -\sum_l \Theta_l\Big[\phi_l^\dagger(\boldsymbol{r}_1)\phi_k(\boldsymbol{r}_1)\phi_i^\dagger(\boldsymbol{r}_2)\phi_l(\boldsymbol{r}_2) + \phi_i^\dagger(\boldsymbol{r}_1)\phi_l(\boldsymbol{r}_1)\phi_l^\dagger(\boldsymbol{r}_2)\phi_k(\boldsymbol{r}_2)\Big]\Big\}\\
&\quad+\frac{1}{2}\sum_{i,j}\Theta_i\Theta_j\sum_{k,l}(1-\Theta_k)(1-\Theta_l)\frac{(ij||kl)-(ij||lk)}{\varepsilon_i+\varepsilon_j-\varepsilon_k-\varepsilon_l}\\
&\qquad\qquad \times\,\phi_i^\dagger(\boldsymbol{r}_1)\phi_k(\boldsymbol{r}_1)\phi_j^\dagger(\boldsymbol{r}_2)\phi_l(\boldsymbol{r}_2) + c.c. \quad .
\end{aligned} \tag{6.154}$$

This expression represents an implicit density functional for $n^{(2)}(\boldsymbol{r}_1,\boldsymbol{r}_2)$ in the same sense as (4.5) is an implicit density functional for the exchange energy.

The question of the quality of the description of the Coulomb correlation by $n^{(2),(1)}$ is addressed in Fig. 6.23, which shows the 2-particle density of the He ground state. In case of the He atom one has only three relevant coordinates, which charac-

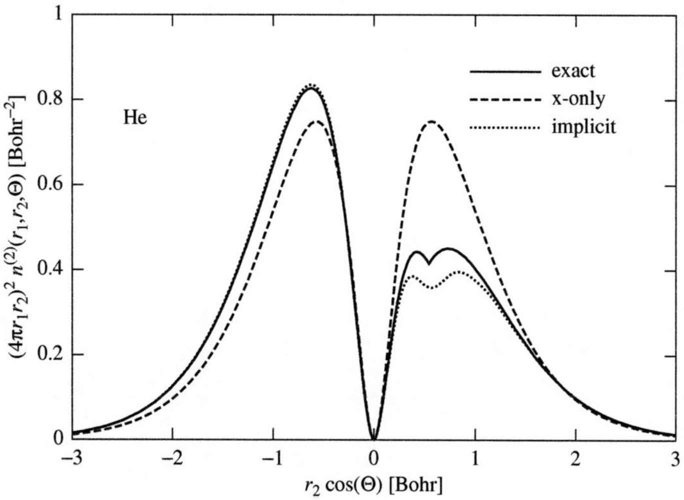

Fig. 6.23 2-particle-density $n^{(2)}(r_1,r_2,\Theta)$ of helium as a function of r_2 for r_1=0.543 Bohr and $\Theta=\pi,0$ (the *left half* corresponds to $\Theta=\pi$, the *right half* to $\Theta=0$). First order perturbative implicit functional versus x-only and exact result [80].

terize the positions of the two electrons relative to the nucleus in the plane spanned by the three particles (compare the discussion of Figs. 3.1 and 3.2 in Sect. 3.1). If the nucleus defines, as usual, the origin of the coordinate system, suitable coordinates are the radial distances of the two electrons from the origin, r_1, r_2, and the angle Θ between the position vectors of the electrons, $\boldsymbol{r}_1 \cdot \boldsymbol{r}_2 = r_1 r_2 \cos\Theta$. In Fig. 6.23 r_1 has been chosen as 0.543 Bohr, which is the radius at which radial density $r_1^2 n(r_1)$ has its maximum. Together with the nucleus at the origin the position of the first electron defines a straight line along which the second electron is moved, i.e. Fig. 6.23 shows the r_2-dependence for $\Theta = 0$ and $\Theta = \pi$. For r_2 =+0.543 Bohr and $\Theta = 0$ the two electrons are on top of each other.

The solid line represents the exact result, obtained from an accurate variational wavefunction [80], the dashed line the pure KS 2-particle density (compare Figs. 3.1 and 3.2 and the associated discussion). As this comparison shows, the implicit functional (6.154) incorporates most of the Coulomb correlation in $n^{(2)}$. It slightly overestimates the suppression of the on-top position, but, overall, is quite realistic (the somewhat less pronounced cusp is a basis set, i.e. numerical, effect). This result demonstrates explicitly that the concept of implicit functionals can not only be used for a representation of xc-energies, but also for the description of local correlation, which could be of interest in the context of scattering and multiple excitation processes.

Chapter 7
Time-Dependent Density Functional Theory

Time-dependent density functional theory (TDDFT) has developed rapidly since its beginnings in 1984 [472, 473]. TDDFT opens a direct access to excited states, but also to truly time-dependent processes. As a topical example of such processes, one may mention transport phenomena, like conduction through single molecules or atomic wires [474–476] (i.e. conduction in the quantum regime). A review of the large variety of applications of TDDFT is, however, not the aim of the present chapter. For this we refer to a recent set of lecture notes [477] (see also [478]). The present discussion is restricted to the basic elements of TDDFT, that is (i) the theorem of Runge and Gross, (ii) the time-dependent Kohn-Sham equations, (iii) the adiabatic local density approximation (ALDA) (sometimes called time-dependent LDA) plus direct extensions, (iv) linear response within TDDFT, and (v) a TDDFT approach to excitation energies.

7.1 Runge-Gross Theorem

The analog of the HK-theorem for time-dependent (td) systems is the Runge-Gross (RG) theorem [472]. It establishes a map between the td many-particle state and the corresponding td density. Since the reasoning behind the RG-theorem is quite different from the argumentation leading to the HK-theorem, it is worthwhile to discuss the proof of the RG-theorem in some detail.

The simplest Hamiltonian, which can be used to characterize a td system of interacting spin-$1/2$ particles, is given by

$$\hat{H}(t) = \hat{T} + \hat{V}_{\text{ext}}(t) + \hat{W}\,, \tag{7.1}$$

with an interaction between particles and external sources which couples the particle density to a time-dependent potential $v_{\text{ext}}(\boldsymbol{r}t)$,

E. Engel, R.M. Dreizler, Time-Dependent Density Functional Theory. In: E. Engel, R.M. Dreizler, Density Functional Theory, Theoretical and Mathematical Physics, pp. 307–349 (2011)
DOI 10.1007/978-3-642-14090-7_7

$$\hat{V}_{\text{ext}}(t) = \sum_{i=1}^{N} v_{\text{ext}}(\boldsymbol{r}_i t) = \int d^3r\, v_{\text{ext}}(\boldsymbol{r}t)\,\hat{n}(\boldsymbol{r}). \tag{7.2}$$

As in the stationary case, the particle–particle interaction $\hat{W}$ is assumed to be symmetric and independent of spin (and, of course, independent of time). For all systems of interest here $\hat{W}$ is the Coulomb interaction.

The many-body state $|\Psi(t)\rangle$ corresponding to the Hamiltonian (7.1) satisfies the td Schrödinger equation,

$$i\hbar\partial_t|\Psi(t)\rangle = \hat{H}(t)|\Psi(t)\rangle\,. \tag{7.3}$$

In addition to the equation of motion, the complete characterization of a td problem requires the specification of some initial value. In the present case the Schrödinger equation (7.3) has to be solved for a given initial state $|\Psi_0\rangle$ at time t_0,

$$|\Psi(t_0)\rangle = |\Psi_0\rangle\,. \tag{7.4}$$

The external potential (7.2) can describe many interesting td phenomena, as for instance the adiabatic motion of nuclei which governs (i) the rotational and vibrational excitations in molecules and solids, and (ii) low-energy ion-atom collisions. It also represents a good approximation for many other situations, as the interaction of atoms with applied laser fields. The explicit proof of the existence theorem of TDDFT will therefore be restricted to the Hamiltonian (7.1).

It is nevertheless advantageous to start the discussion with a detour, in order to prepare the actual proof. Let us for a moment consider a more general Hamiltonian, including the coupling to a td magnetic field,

$$\hat{H} = \sum_{\sigma}\int d^3r\,\hat{\psi}^\dagger(\boldsymbol{r}\sigma)\left\{\frac{1}{2m}\left[-i\hbar\boldsymbol{\nabla}+\frac{e}{c}\boldsymbol{A}_{\text{ext}}(\boldsymbol{r}t)\right]^2 + v_{\text{ext}}(\boldsymbol{r}t)\right\}\hat{\psi}(\boldsymbol{r}\sigma)+\hat{W}\,, \tag{7.5}$$

where the standard relations between the external electro-magnetic fields and potentials,

$$\boldsymbol{E}_{\text{ext}}(\boldsymbol{r}t) = \frac{1}{e}\boldsymbol{\nabla}v_{\text{ext}}(\boldsymbol{r}t) - \frac{1}{c}\frac{\partial\boldsymbol{A}_{\text{ext}}(\boldsymbol{r}t)}{\partial t} \tag{7.6}$$

$$\boldsymbol{B}_{\text{ext}}(\boldsymbol{r}t) = \boldsymbol{\nabla}\times\boldsymbol{A}_{\text{ext}}(\boldsymbol{r}t)\,, \tag{7.7}$$

are supposed to hold. A point to be noted here is the fact that the vector potential $\boldsymbol{A}_{\text{ext}}$ can be non-zero even for a vanishing magnetic field $\boldsymbol{B}_{\text{ext}}$. One and the same electric field $\boldsymbol{E}_{\text{ext}}$ (and thus the same physics) can either be represented in terms of v_{ext} (with $\boldsymbol{A}_{\text{ext}}$ set to zero) or in terms of $\boldsymbol{A}'_{\text{ext}}$ (with $v'_{\text{ext}}=0$), if $\boldsymbol{A}'_{\text{ext}}$ is chosen as

$$\boldsymbol{A}'_{\text{ext}}(\boldsymbol{r}t) = -\frac{c}{e}\int^t dt'\,\boldsymbol{\nabla}v_{\text{ext}}(\boldsymbol{r}t')\,. \tag{7.8}$$

Equation (7.8) is a special case of a td gauge transformation, for which the most general form is given by

$$v'_{\text{ext}}(\boldsymbol{r}t) = v_{\text{ext}}(\boldsymbol{r}t) - \frac{e}{c}\frac{\partial\Lambda(\boldsymbol{r}t)}{\partial t} \tag{7.9}$$

$$\boldsymbol{A}'_{\text{ext}}(\boldsymbol{r}t) = \boldsymbol{A}_{\text{ext}}(\boldsymbol{r}t) - \boldsymbol{\nabla}\Lambda(\boldsymbol{r}t)\,. \tag{7.10}$$

While for the stationary Hamiltonian (2.143) the electric (v_{ext}) and magnetic ($\boldsymbol{A}_{\text{ext}}$) potentials are not coupled by the gauge transformation (2.145)–(2.147), the situation is no longer as simple for td Hamiltonians.

It is therefore necessary to study the relation between the td potentials $v_{\text{ext}}(\boldsymbol{r}t)$, $\boldsymbol{A}_{\text{ext}}(\boldsymbol{r}t)$ and the corresponding td state $|\Psi(t)\rangle$ in more detail, before discussing the RG-theorem for the system (7.1). In view of the gauge freedom the obvious question is: when does one consider two potentials to be physically non-equivalent? In order to answer this question, one first notes that the gauge-transformed state

$$(\boldsymbol{r}_1\sigma_1\ldots\boldsymbol{r}_N\sigma_N|\Psi'(t)\rangle = \exp\left[\frac{ie}{\hbar c}\sum_{k=1}^{N}\Lambda(\boldsymbol{r}_k t)\right](\boldsymbol{r}_1\sigma_1\ldots\boldsymbol{r}_N\sigma_N|\Psi(t)\rangle \tag{7.11}$$

is a solution of the td Schrödinger equation (7.3) with the potentials $v'_{\text{ext}}(\boldsymbol{r}t), \boldsymbol{A}'_{\text{ext}}(\boldsymbol{r}t)$, provided that $|\Psi(t)\rangle$ is the solution of (7.3) for the Hamiltonian (7.5) with the unprimed potentials. One can, without loss of generality, choose $\Lambda(\boldsymbol{r}t_0) = 0$, so that the initial states coincide, $|\Psi'(t_0)\rangle = |\Psi(t_0)\rangle = |\Psi_0\rangle$. It is obvious that sets of potentials $v_{\text{ext}}, \boldsymbol{A}_{\text{ext}}$ and $v'_{\text{ext}}, \boldsymbol{A}'_{\text{ext}}$ which are related by Eqs. (7.9), (7.10) are physically equivalent. In turn, the uniqueness of the solution of the td Schrödinger equation leads to the statement: two states which differ by more than the phase transformation (7.11) can never emerge from two potentials which are related by Eqs. (7.9), (7.10).

The crucial question, which has to be answered, is: can $|\Psi(t)\rangle$ and $|\Psi'(t)\rangle$ be related by a phase transformation of the form (7.11), if the potentials differ by more than the gauge transformation (7.9), (7.10)? Obviously, an affirmative answer would exclude any density functional approach from the very outset, as the densities corresponding to $|\Psi'(t)\rangle$ and $|\Psi(t)\rangle$ are identical in this case. Fortunately, the question can be answered rigorously with "no". For the case $\boldsymbol{A}_{\text{ext}} = \boldsymbol{0}$, i.e. for the Hamiltonian (7.1), the answer is provided by the original RG-theorem [472]. This limit not only excludes the presence of magnetic fields, but also implies a particular choice of gauge. For the general Hamiltonian (7.5) a corresponding proof has been given by Ghosh and Dhara [479] (see also [480]). The scheme of this proof follows closely the line of RG. Our explicit discussion will therefore be restricted to the RG theorem.

Before proceeding to the RG theorem for the Hamiltonian (7.1), it is instructive to consider the reduced class of phase transformations allowed by the choice $\boldsymbol{A}_{\text{ext}} = \boldsymbol{0}$, i.e. transformations by a *global* td phase $\alpha(t)$,

$$|\Psi'(t)\rangle = e^{-iN\alpha(t)/\hbar}|\Psi(t)\rangle\,, \tag{7.12}$$

in more detail. The additional phase can, as in the general situation discussed above, be absorbed into a gauge-transformed potential v'_{ext},

$$v'_{\text{ext}}(\boldsymbol{r}t) = v_{\text{ext}}(\boldsymbol{r}t) + \dot{\alpha}(t) \tag{7.13}$$

($\dot{\alpha} = \partial_t \alpha$). On the other hand, as soon as the potentials v_{ext} and v'_{ext} differ by more than the trivial gauge transformation (7.13),

$$v'_{\text{ext}}(\boldsymbol{r}t) \neq v_{\text{ext}}(\boldsymbol{r}t) + c(t) \qquad \text{for } t_0 \leq t < t_1 , \tag{7.14}$$

the solutions $|\Psi(t)\rangle$ and $|\Psi'(t)\rangle$ of (7.3) differ by more than a td global phase,

$$|\Psi'(t)\rangle \neq e^{-iN\alpha(t)/\hbar}|\Psi(t)\rangle \qquad \text{for } t_0 \leq t < t_1 , \tag{7.15}$$

even if the same initial condition is chosen for both states.

The proof of this statement is based on *reductio ad absurdum*. Assume that, in spite of the condition (7.14), Eq. (7.12) is valid. Now consider the corresponding Schrödinger equations,

$$\begin{aligned} i\hbar\partial_t|\Psi(t)\rangle &= \hat{H}(t)|\Psi(t)\rangle \qquad &(7.16)\\ i\hbar\partial_t|\Psi'(t)\rangle &= \hat{H}'(t)|\Psi'(t)\rangle \\ \Longrightarrow \quad e^{-iN\alpha(t)/\hbar}[N\dot{\alpha}(t) + i\hbar\partial_t]|\Psi(t)\rangle &= e^{-iN\alpha(t)/\hbar}\hat{H}'(t)|\Psi(t)\rangle . \qquad &(7.17) \end{aligned}$$

Upon subtraction of (7.17) from (7.16) one obtains

$$0 = \left[\hat{V}_{\text{ext}}(t) - \hat{V}'_{\text{ext}}(t) + \dot{\alpha}(t)\hat{N}\right]|\Psi(t)\rangle , \tag{7.18}$$

($\hat{N}|\Psi(t)\rangle = N|\Psi(t)\rangle$) or, in first quantized form,

$$0 = \sum_{i=1}^{N}\Big(v_{\text{ext}}(\boldsymbol{r}_i t) - v'_{\text{ext}}(\boldsymbol{r}_i t) + \dot{\alpha}(t)\Big)\Psi(\boldsymbol{r}_1\sigma_1 \ldots \boldsymbol{r}_N\sigma_N, t) . \tag{7.19}$$

As in the stationary situation, one can now keep $N-1$ of the $\boldsymbol{r}_i$ fixed and consider (7.19) as a function of time and the N-th position. As the wavefunction in (7.19) equals zero only for selected values of its arguments, the relation can only be satisfied, if

$$0 = v_{\text{ext}}(\boldsymbol{r}t) - v'_{\text{ext}}(\boldsymbol{r}t) + \dot{\alpha}(t) \qquad \text{for } t_0 \leq t < t_1 ,$$

in obvious contradiction to the condition (7.14). One can therefore state: as soon as $v'_{\text{ext}}(\boldsymbol{r}t)$ differs from $v_{\text{ext}}(\boldsymbol{r}t)$ by more than a td function $c(t)$ the corresponding many-particle states differ by more than a global td phase. Of course, all these statements have to be understood as statements for functions of time over a finite period $t_0 < t < t_1$, not just for a single instant of time.

Now consider the set of all Hamiltonians of the form (7.1), (7.2) for which

(a) the Taylor expansion of the potential about the initial time t_0 exists,

$$v_{\text{ext}}(\boldsymbol{r}t) = \sum_{k=0}^{\infty} \frac{1}{k!} \partial_t^k v_{\text{ext}}(\boldsymbol{r}t)\Big|_{t=t_0} (t-t_0)^k , \tag{7.20}$$

(b) only one representative v_{ext} is included for each class of physically equivalent potentials according to Eq. (7.14).

The last condition is an obvious extension of the situation for stationary systems for which external potentials are only considered to be different if they differ by more than a global constant. The mathematical role of condition (7.20) will become clear later on. This requirement excludes the possibility that the td potential is switched on at t_0 in a step-like fashion. As realistic fields necessarily build up continuously, this is not a fundamental restriction, but merely a requirement for the time scale on which a process is considered.

The statement of the RG-theorem then is [472]:

For a given initial state $|\Psi(t_0)\rangle$ there exists a one-to-one correspondence between the set of all potentials, which satisfy (7.14) and (7.20),

$$\mathscr{V} = \Big\{ v_{\text{ext}}(\boldsymbol{r}t) \,\Big|\, v_{\text{ext}}(\boldsymbol{r}t) = \sum_{k=0}^{\infty} \frac{1}{k!} \partial_t^k v_{\text{ext}}(\boldsymbol{r}t)\Big|_{t=t_0} (t-t_0)^k ; \\ v'_{\text{ext}}(\boldsymbol{r}t) \neq v_{\text{ext}}(\boldsymbol{r}t) + c(t) \Big\}, \tag{7.21}$$

and the set of associated td densities,

$$\mathscr{N} = \Big\{ n(\boldsymbol{r}t) \,\Big|\, n(\boldsymbol{r}t) = \langle\Psi(t)|\hat{n}(\boldsymbol{r})|\Psi(t)\rangle ;\ i\hbar\partial_t|\Psi(t)\rangle = \hat{H}(t)|\Psi(t)\rangle ; \\ \hat{H} = \hat{T} + \int d^3r\, v_{\text{ext}}(\boldsymbol{r}t)\,\hat{n}(\boldsymbol{r}) + \hat{W} ;\ v_{\text{ext}}(\boldsymbol{r}t) \in \mathscr{V} \Big\}, \tag{7.22}$$

within the finite time interval $t_0 < t < t_1$.

Note that it is not sufficient to specify the initial density

$$n_0(\boldsymbol{r}) = n(\boldsymbol{r}t_0) = \langle\Psi(t_0)|\hat{n}(\boldsymbol{r})|\Psi(t_0)\rangle \tag{7.23}$$

instead of $|\Psi(t_0)\rangle$. This signifies that the RG-theorem does not, in general, define a pure density functional method. There is, however, an important class of situations in which the initial state $|\Psi(t_0)\rangle$ is uniquely determined by n_0, i.e. all those time-dependent systems which are in their ground state until $t = t_0$. In this case the stationary HK-theorem applies at t_0, so that $|\Psi(t_0)\rangle$ is a functional of the ground state density $n(\boldsymbol{r}t_0)$.

The proof of the RG-theorem proceeds in two steps. In the first step a corresponding statement for the current density is verified. This statement is extended to the particle density in the second step.

Step 1(a): Consider the (paramagnetic) current density

$$\boldsymbol{j}_{\text{p}}(\boldsymbol{r}t) = \langle\Psi(t)|\hat{\boldsymbol{j}}_{\text{p}}(\boldsymbol{r})|\Psi(t)\rangle \tag{7.24}$$

$$\hat{\boldsymbol{j}}_{\mathrm{p}}(\boldsymbol{r}) = \frac{-i\hbar}{2m}\sum_{i=1}^{N}\left[\boldsymbol{\nabla}_i\delta^{(3)}(\boldsymbol{r}-\boldsymbol{r}_i)+\delta^{(3)}(\boldsymbol{r}-\boldsymbol{r}_i)\boldsymbol{\nabla}_i\right]$$
$$= \frac{-i\hbar}{2m}\sum_{\sigma}\left[\hat{\psi}^{\dagger}(\boldsymbol{r}\sigma)\left(\boldsymbol{\nabla}\hat{\psi}(\boldsymbol{r}\sigma)\right)-\left(\boldsymbol{\nabla}\hat{\psi}^{\dagger}(\boldsymbol{r}\sigma)\right)\hat{\psi}(\boldsymbol{r}\sigma)\right]. \qquad (7.25)$$

The first time-derivative of $\boldsymbol{j}_{\mathrm{p}}(\boldsymbol{r}t)$ can be analyzed via Heisenberg's equation of motion,

$$\frac{d}{dt}\langle\Psi(t)|\hat{O}(t)|\Psi(t)\rangle = \left\langle\Psi(t)\left|\frac{d\hat{O}(t)}{dt}-\frac{i}{\hbar}[\hat{O}(t),\hat{H}(t)]\right|\Psi(t)\right\rangle, \qquad (7.26)$$

which is valid for the expectation value of any operator $\hat{O}(t)$. In Eq. (7.26) the term with $d\hat{O}(t)/dt$ accounts for the time-dependence of the operator itself (in the Schrödinger representation), while the time-derivative of the state $|\Psi(t)\rangle$ has been reformulated using the Schrödinger equation (7.1). Application of (7.26) to the current operator (7.25), which does not depend explicitly on time, leads to

$$\partial_t\boldsymbol{j}_{\mathrm{p}}(\boldsymbol{r}t) = -\frac{i}{\hbar}\langle\Psi(t)|\left[\hat{\boldsymbol{j}}_{\mathrm{p}}(\boldsymbol{r}),\hat{H}(t)\right]|\Psi(t)\rangle. \qquad (7.27)$$

With the aid of (7.27) one can compare the time-derivatives of the current densities resulting from two different potentials $v'_{\mathrm{ext}}(\boldsymbol{r}t)$ and $v_{\mathrm{ext}}(\boldsymbol{r}t)$ of the set $\mathcal{V}$ at the initial time t_0. Utilizing the basic commutation relation

$$[\hat{\boldsymbol{j}}_{\mathrm{p}}(\boldsymbol{r}),\hat{n}(\boldsymbol{r}')] = \frac{-i\hbar}{2m}\left\{\left[\boldsymbol{\nabla}\delta^{(3)}(\boldsymbol{r}-\boldsymbol{r}')\right]\sum_{\sigma}\left[\hat{\psi}^{\dagger}(\boldsymbol{r}\sigma)\hat{\psi}(\boldsymbol{r}'\sigma)+\hat{\psi}^{\dagger}(\boldsymbol{r}'\sigma)\hat{\psi}(\boldsymbol{r}\sigma)\right]\right.$$
$$\left.-\delta^{(3)}(\boldsymbol{r}-\boldsymbol{r}')\boldsymbol{\nabla}\hat{n}(\boldsymbol{r})\right\}, \qquad (7.28)$$

one obtains for the Hamiltonian (7.1)

$$\partial_t\left[\boldsymbol{j}_{\mathrm{p}}(\boldsymbol{r}t)-\boldsymbol{j}'_{\mathrm{p}}(\boldsymbol{r}t)\right]\Big|_{t=t_0} = -\frac{1}{m}n(\boldsymbol{r}t_0)\boldsymbol{\nabla}\left[v_{\mathrm{ext}}(\boldsymbol{r}t_0)-v'_{\mathrm{ext}}(\boldsymbol{r}t_0)\right]. \qquad (7.29)$$

Consequently, as long as the two potentials differ by more than a global constant at t_0, $v_{\mathrm{ext}}(\boldsymbol{r}t_0)\neq v'_{\mathrm{ext}}(\boldsymbol{r}t_0)+c_0$, the currents $\boldsymbol{j}_{\mathrm{p}}(\boldsymbol{r}t)$ and $\boldsymbol{j}'_{\mathrm{p}}(\boldsymbol{r}t)$ differ at a time t which is infinitesimally larger than t_0.

However, the two potentials will often be identical at $t=t_0$, most notably, if their time-dependent components are switched on smoothly for $t\geq t_0$. One therefore has to deal with the case that $v_{\mathrm{ext}}(\boldsymbol{r}t_0)=v'_{\mathrm{ext}}(\boldsymbol{r}t_0)+c_0$, which requires an examination of the second time-derivative of the current. It turns out that $\boldsymbol{j}_{\mathrm{p}}$ and $\boldsymbol{j}'_{\mathrm{p}}$ become different immediately after t_0 if $\partial_t[v_{\mathrm{ext}}(\boldsymbol{r}t)-v'_{\mathrm{ext}}(\boldsymbol{r}t)]_{t=t_0}$ is not identical with some global constant. As this need not always be the case, one is led to consider the next time-derivative of the current.

Step 1(b): It is thus necessary to discuss the general situation, in which a finite number, say $k-1$, of time-derivatives of $v'_{\mathrm{ext}}(\boldsymbol{r}t)$ and $v_{\mathrm{ext}}(\boldsymbol{r}t)$ agree at $t=t_0$ (up to global constants),

$$\partial_t^l v'_{\text{ext}}(\boldsymbol{r}t)\Big|_{t=t_0} = \partial_t^l v_{\text{ext}}(\boldsymbol{r}t)\Big|_{t=t_0} + c_l \qquad l = 0,\dots k-1 \tag{7.30}$$

(which includes the special value $k = 1$). In this case one has to consider the $(k+1)$-th derivative of the current. Quite generally, one finds by recursive use of (7.26),

$$\begin{aligned}
&\partial_t^{k+1} \boldsymbol{j}_{\text{p}}(\boldsymbol{r}t) \\
&= \partial_t^{k-1} \Big\langle \Psi(t) \Big| \Big(-\frac{i}{\hbar}\Big) \partial_t \big[\hat{\boldsymbol{j}}_{\text{p}}(\boldsymbol{r}), \hat{H}(t)\big] + \Big(-\frac{i}{\hbar}\Big)^2 \Big[\big[\hat{\boldsymbol{j}}_{\text{p}}(\boldsymbol{r}), \hat{H}(t)\big], \hat{H}(t)\Big] \Big| \Psi(t) \Big\rangle \\
&= \partial_t^{k-2} \Big\langle \Psi(t) \Big| \; \Big(-\frac{i}{\hbar}\Big) \partial_t^2 \big[\hat{\boldsymbol{j}}_{\text{p}}(\boldsymbol{r}), \hat{H}(t)\big] \\
&\qquad + \Big(-\frac{i}{\hbar}\Big)^2 \partial_t \Big[\big[\hat{\boldsymbol{j}}_{\text{p}}(\boldsymbol{r}), \hat{H}(t)\big], \hat{H}(t)\Big] \\
&\qquad + \Big(-\frac{i}{\hbar}\Big)^2 \Big[\partial_t \big[\hat{\boldsymbol{j}}_{\text{p}}(\boldsymbol{r}), \hat{H}(t)\big], \hat{H}(t)\Big] \\
&\qquad + \Big(-\frac{i}{\hbar}\Big)^3 \Big[\Big[\big[\hat{\boldsymbol{j}}_{\text{p}}(\boldsymbol{r}), \hat{H}(t)\big], \hat{H}(t)\Big], \hat{H}(t)\Big] \Big| \Psi(t) \Big\rangle .
\end{aligned}$$

From the lowest two orders indicated, one can already glean how order by order more complicated combinations of time-derivatives and multiple commutators emerge. In fact, all possible sequences of differentiation and commutation appear. Ordered with respect to the number of derivatives to be taken, one finally obtains

$$\begin{aligned}
&\partial_t^{k+1} \boldsymbol{j}_{\text{p}}(\boldsymbol{r}t) \\
&= \Big\langle \Psi(t) \Big| \; \Big(-\frac{i}{\hbar}\Big) \partial_t^k \big[\hat{\boldsymbol{j}}_{\text{p}}(\boldsymbol{r}), \hat{H}(t)\big] \\
&\qquad + \Big(-\frac{i}{\hbar}\Big)^2 \sum_{j=0}^{k-1} \partial_t^{k-1-j} \Big[\partial_t^j \big[\hat{\boldsymbol{j}}_{\text{p}}(\boldsymbol{r}), \hat{H}(t)\big], \hat{H}(t)\Big] \\
&\qquad + \cdots \\
&\qquad + \Big(-\frac{i}{\hbar}\Big)^{k+1} \underbrace{\Big[\dots\Big[\big[\hat{\boldsymbol{j}}_{\text{p}}(\boldsymbol{r}), \hat{H}(t)\big], \hat{H}(t)\Big], \dots \hat{H}(t)\Big]}_{(k+1)-times} \Big| \Psi(t) \Big\rangle ,
\end{aligned} \tag{7.31}$$

where the last line denotes a $(k+1)$-fold commutator. Once all time-derivatives are applied to the Hamiltonians in the individual expressions, all terms in (7.31) consist of multiple commutators of $\hat{\boldsymbol{j}}_{\text{p}}$ with the Hamiltonian and its derivatives. The general form of these expressions is given by

$$D_n^{\boldsymbol{\alpha}}(\hat{H},t) := \underbrace{\Big[\dots\Big[\big[\hat{\boldsymbol{j}}_{\text{p}}(\boldsymbol{r}), \partial_t^{\alpha_1}\hat{H}(t)\big], \partial_t^{\alpha_2}\hat{H}(t)\Big], \dots \partial_t^{\alpha_n}\hat{H}(t)\Big]}_{n-times} , \tag{7.32}$$

with the total number of derivatives and commutators being fixed by the order of the derivative on the left-hand side of (7.31),

$$\sum_{i=1}^{n} \alpha_i + n = k+1\,. \tag{7.33}$$

An analogous relation holds for $\boldsymbol{j}'_{\mathrm{p}}$ in terms of $\hat{H}'$. For the present proof one has to evaluate the difference between $\partial_t^{k+1}\boldsymbol{j}_{\mathrm{p}}$ and $\partial_t^{k+1}\boldsymbol{j}'_{\mathrm{p}}$ at $t=t_0$. However, at $t=t_0$ the first $k-1$ time-derivatives of $\hat{H}'$ can, in view of assumption (7.30), be expressed in terms of the corresponding derivatives of $\hat{H}$,

$$\partial_t^l \hat{H}'(t)\Big|_{t=t_0} = \partial_t^l [\hat{H}(t) + c(t)\hat{N}]\Big|_{t=t_0} \qquad \text{for } l=0,\ldots k-1\,, \tag{7.34}$$

with

$$c(t) := \sum_{l=0}^{k-1} \frac{c_l}{l!}(t-t_0)^l \tag{7.35}$$

and $\hat{N}$ denoting the particle number operator. This allows the replacement of $\hat{H}'$ in $D_n^{\boldsymbol{\alpha}}(\hat{H}',t_0)$ by $\hat{H}+c\hat{N}$ for all $n>1$, as none of the $\alpha_i<k$ can be larger than $k-1$ in this case,

$$\begin{aligned} &D_{n>1}^{\boldsymbol{\alpha}}(\hat{H}',t_0) \\ &= \underbrace{\Big[\cdots\Big[\Big[\hat{\boldsymbol{j}}_{\mathrm{p}}, \partial_t^{\alpha_1}(\hat{H}+c\hat{N})\Big], \partial_t^{\alpha_2}(\hat{H}+c\hat{N})\Big], \ldots \partial_t^{\alpha_n}(\hat{H}+c\hat{N})\Big]}_{n-times}\Big|_{t=t_0}\,. \end{aligned} \tag{7.36}$$

Now one can use the fact that both $\hat{\boldsymbol{j}}_{\mathrm{p}}$ and $\hat{H}$ commute with $\hat{N}$,

$$\big[\hat{\boldsymbol{j}}_{\mathrm{p}}(\boldsymbol{r}),\hat{N}\big] = 0 \qquad ; \qquad \big[\hat{H}(t),\hat{N}\big] = 0\,. \tag{7.37}$$

As a consequence of (7.37) the contributions of all the terms $\partial_t^{\alpha_i}c(t)\hat{N}$ in (7.36) to the multiple commutator vanish, irrespective of the order of the time-derivatives involved,

$$\Big[\partial_t^{\alpha}\hat{H}(t), \partial_t^{\beta}c(t)\hat{N}\Big] = \Big(\partial_t^{\beta}c(t)\Big)\,\partial_t^{\alpha}\,\big[\hat{H}(t),\hat{N}\big] = 0\,.$$

One therefore finds

$$D_n^{\boldsymbol{\alpha}}(\hat{H}',t_0) = D_n^{\boldsymbol{\alpha}}(\hat{H},t_0) \qquad \text{for } n>1\,. \tag{7.38}$$

As a result only the first term on the right-hand side of (7.31) contributes to the difference between $\partial_t^{k+1}\boldsymbol{j}_{\mathrm{p}}$ and $\partial_t^{k+1}\boldsymbol{j}'_{\mathrm{p}}$ at $t=t_0$,

$$\partial_t^{k+1}\left[\boldsymbol{j}_\mathrm{p}(\boldsymbol{r}t)-\boldsymbol{j}'_\mathrm{p}(\boldsymbol{r}t)\right]_{t=t_0} = -\frac{i}{\hbar}\left\langle\Psi(t)\middle|\partial_t^k\left[\hat{\boldsymbol{j}}_\mathrm{p}(\boldsymbol{r}),\hat{H}(t)-\hat{H}'(t)\right]\middle|\Psi(t)\right\rangle\Big|_{t=t_0}. \tag{7.39}$$

The evaluation of this expression again relies on Eq. (7.28),

$$\partial_t^{k+1}\left[\boldsymbol{j}_\mathrm{p}(\boldsymbol{r}t)-\boldsymbol{j}'_\mathrm{p}(\boldsymbol{r}t)\right]_{t=t_0} = -\frac{1}{m}n(\boldsymbol{r}t_0)\,\partial_t^k\boldsymbol{\nabla}\left[v_\mathrm{ext}(\boldsymbol{r}t)-v'_\mathrm{ext}(\boldsymbol{r}t)\right]_{t=t_0}. \tag{7.40}$$

Equation (7.40) shows why the existence of the Taylor expansion (7.20) had to be assumed.

In order to complete Step 1 of the proof of the RG-theorem, one argues as follows. By assumption the potentials $v_\mathrm{ext}(\boldsymbol{r}t)$ and $v'_\mathrm{ext}(\boldsymbol{r}t)$ differ by an $\boldsymbol{r}$-dependent function for all $t>t_0$. This implies that at least one of the time-derivatives of the two potentials at t_0 differs by more than a global constant,

$$\boldsymbol{\nabla}\partial_t^k\left[v_\mathrm{ext}(\boldsymbol{r}t)-v'_\mathrm{ext}(\boldsymbol{r}t)\right]\Big|_{t=t_0}\neq 0 \qquad \text{for at least one } k. \tag{7.41}$$

As a consequence of (7.40) and (7.41) the current densities differ immediately after t_0,

$$\boldsymbol{j}_\mathrm{p}(\boldsymbol{r},t_0+\varepsilon)\neq\boldsymbol{j}'_\mathrm{p}(\boldsymbol{r},t_0+\varepsilon). \tag{7.42}$$

This is the final result of Step 1 of the proof and at the same time an important result in itself.

Step 2: The continuity equation,

$$\partial_t n(\boldsymbol{r}t)+\boldsymbol{\nabla}\cdot\boldsymbol{j}_\mathrm{p}(\boldsymbol{r}t)=0, \tag{7.43}$$

allows the extension of the statement (7.42) to the td densities n and n', which correspond to $\boldsymbol{j}_\mathrm{p}$ and $\boldsymbol{j}'_\mathrm{p}$, respectively. Let k again denote the lowest order time-derivative for which the two potentials under consideration differ by more than a constant, so that (7.30) and (7.41) apply. One can then rewrite the time-derivative of the density as

$$\begin{aligned}\partial_t^{k+2}\Big[n(\boldsymbol{r}t)-n'(\boldsymbol{r}t)\Big]_{t=t_0} &= -\boldsymbol{\nabla}\cdot\partial_t^{k+1}\Big[\boldsymbol{j}_\mathrm{p}(\boldsymbol{r}t)-\boldsymbol{j}'_\mathrm{p}(\boldsymbol{r}t)\Big]_{t=t_0}\\ &= \frac{1}{m}\boldsymbol{\nabla}\cdot\Big[n(\boldsymbol{r}t_0)\,\boldsymbol{\nabla}\Delta_k(\boldsymbol{r})\Big],\end{aligned} \tag{7.44}$$

with the abbreviation

$$\Delta_k(\boldsymbol{r}) := \partial_t^k\Big[v_\mathrm{ext}(\boldsymbol{r}t)-v'_\mathrm{ext}(\boldsymbol{r}t)\Big]_{t=t_0}. \tag{7.45}$$

In order to prove that $n(\boldsymbol{r}t)$ and $n'(\boldsymbol{r}t)$ differ for $t>t_0$ one has to show that

$$\boldsymbol{\nabla}\cdot\left[n(\boldsymbol{r}t_0)\,\boldsymbol{\nabla}\Delta_k(\boldsymbol{r})\right]$$

does not vanish for at least one k and $\boldsymbol{r}$. The proof proceeds by *reductio ad absurdum*. Assume that for all k

$$\boldsymbol{\nabla}\cdot\Big[n(\boldsymbol{r}t_0)\,\boldsymbol{\nabla}\Delta_k(\boldsymbol{r})\Big]=0 \qquad \forall\,\boldsymbol{r} \tag{7.46}$$

and consider the integral

$$\int d^3r\,n(\boldsymbol{r}t_0)\,\Big[\boldsymbol{\nabla}\Delta_k(\boldsymbol{r})\Big]^2 = -\int d^3r\,\Delta_k(\boldsymbol{r})\boldsymbol{\nabla}\Big[n(\boldsymbol{r}t_0)\,\boldsymbol{\nabla}\Delta_k(\boldsymbol{r})\Big] + \oint d\boldsymbol{S}\cdot\Big[n(\boldsymbol{r}t_0)\,\Delta_k(\boldsymbol{r})\boldsymbol{\nabla}\Delta_k(\boldsymbol{r})\Big]. \tag{7.47}$$

The first contribution on the right-hand side of (7.47) vanishes by assumption. The second term, on the other hand, is a surface integral over a surface at infinity. For all systems for which $n(\boldsymbol{r}t_0)\Delta_k(\boldsymbol{r})\boldsymbol{\nabla}\Delta_k(\boldsymbol{r})$ vanishes for $|\boldsymbol{r}|\to\infty$ one has

$$\oint d\boldsymbol{S}\cdot\Big[n(\boldsymbol{r}t_0)\,\Delta_k(\boldsymbol{r})\boldsymbol{\nabla}\Delta_k(\boldsymbol{r})\Big]=0\,.$$

This relation is valid in particular for all *finite* systems, as long as the td potential does not increase faster than $n(\boldsymbol{r}t_0)$ falls off. On the other hand, in the case of extended systems the condition $n(\boldsymbol{r}t_0)\Delta_k(\boldsymbol{r})\boldsymbol{\nabla}\Delta_k(\boldsymbol{r}) \xrightarrow{|\boldsymbol{r}|\to\infty} 0$ requires Δ_k (or $\boldsymbol{\nabla}\Delta_k$) to decay for $|\boldsymbol{r}|\to\infty$ (compare [481, 482]). In both cases the assumption (7.46) leads to

$$\int d^3r\,n(\boldsymbol{r}t_0)\,\Big[\boldsymbol{\nabla}\Delta_k(\boldsymbol{r})\Big]^2=0\,.$$

In view of (7.41), which implies that

$$\Big[\boldsymbol{\nabla}\Delta_k(\boldsymbol{r})\Big]^2>0\,,$$

for some $\boldsymbol{r}$, one thus ends up with a contradiction, provided, one excludes the case that $n(\boldsymbol{r}t_0)$ and $\boldsymbol{\nabla}\Delta_k(\boldsymbol{r})$ are nonzero in different regions of space.[1] As a consequence one has

$$n(\boldsymbol{r},t_0+\varepsilon)\neq n'(\boldsymbol{r},t_0+\varepsilon)\,. \tag{7.48}$$

Of course, if the densities n and n' differ directly after t_0, they differ as functions of time in the time interval $t_0<t<t_1$. Note, however, that this does not automatically exclude the possibility that n and n' coincide for some intermediate instant of time.

This concludes the proof of the RG-theorem. As a side-remark one may mention that n and n' already differ to first order in $v_{\rm ext}$, i.e. on the linear response level, as the right-hand side of Eq. (7.44) is of first order in $v_{\rm ext}$.

---

[1] This is a trivial situation as it means that the potentials differ in a region of space which is not accessible to the particles, so that they can not experience any difference.

The combination of the RG-theorem with the arguments given at the beginning of this section establishes a unique relation between $n(\boldsymbol{r}t)$ and the td state $|\Psi(t)\rangle$. Let us collect the various pieces of information. Provided that

(i) $v'_{\rm ext}(\boldsymbol{r}t) \neq v_{\rm ext}(\boldsymbol{r}t) + c(t)$,
(ii) both $v'_{\rm ext}(\boldsymbol{r}t)$ and $v_{\rm ext}(\boldsymbol{r}t)$ can be expanded according to Eq. (7.20),
(iii) the gauge $\boldsymbol{A}'_{\rm ext} = \boldsymbol{A}_{\rm ext} = \boldsymbol{0}$ is chosen,

both the current densities $\boldsymbol{j}'_{\rm p}(\boldsymbol{r}t)$ and $\boldsymbol{j}_{\rm p}(\boldsymbol{r}t)$ and the charge densities $n'(\boldsymbol{r}t)$ and $n(\boldsymbol{r}t)$ resulting from $v'_{\rm ext}(\boldsymbol{r}t)$ and $v_{\rm ext}(\boldsymbol{r}t)$, respectively, differ, even if the time-evolution of the primed and the unprimed system is started with the same initial state. However, $n'(\boldsymbol{r}t) \neq n(\boldsymbol{r}t)$ requires that the corresponding states $|\Psi'(t)\rangle$ and $|\Psi(t)\rangle$ differ by more than the phase transformation (7.11): as long as $\boldsymbol{A}'_{\rm ext} = \boldsymbol{A}_{\rm ext} = \boldsymbol{0}$, potentials $v'_{\rm ext}$ and $v_{\rm ext}$ which differ by more than a global td function lead to states which differ by more than a local(!) td phase. In other words: the relation (7.11) between two td states can only result from the gauge transformation (7.9), (7.10) of the potentials $v_{\rm ext}(\boldsymbol{r}t)$ and $\boldsymbol{A}_{\rm ext}(\boldsymbol{r}t)$, at least as long as $\boldsymbol{B}_{\rm ext} = \boldsymbol{0}$.

As $n'(\boldsymbol{r}t) \neq n(\boldsymbol{r}t)$ automatically implies that $v'_{\rm ext}(\boldsymbol{r}t) \neq v_{\rm ext}(\boldsymbol{r}t) + c(t)$, one can conclude: for a given initial state $|\Psi(t_0)\rangle = |\Psi_0\rangle$, the td density $n(\boldsymbol{r}t)$ determines $|\Psi(t)\rangle$ uniquely up to a global td phase. In mathematical terms this means that the td state is a unique functional $|\Psi[n,\Psi_0](t)\rangle$ of $n(\boldsymbol{r}t)$, up to some phase,

$$|\Psi(t)\rangle = e^{-iN\alpha(t)/\hbar}\,|\Psi[n,\Psi_0](t)\rangle\,. \tag{7.49}$$

There is a unique $|\Psi[n,\Psi_0](t)\rangle$ for each set of physically equivalent potentials. The phase $\alpha(t)$, on the other hand, cannot be determined at all by knowledge of $n(\boldsymbol{r}t)$ and $|\Psi_0\rangle$. Rather, the choice of the gauge for which $\alpha(t) \equiv 0$ is a matter of convenience: it simply depends on the representative $v_{\rm ext}$ of each class of equivalent potentials which is used to define $|\Psi[n,\Psi_0](t)\rangle$.

Equation (7.49) allows the formulation of arbitrary expectation values as (unique) density functionals. An important expectation value is the quantum mechanical action,

$$A(t_1,t_0) := \int_{t_0}^{t_1} dt\,\langle\Psi(t)|i\hbar\partial_t - \hat{H}(t)|\Psi(t)\rangle\,. \tag{7.50}$$

Variation of A with respect to the state $\langle\Psi(t)|$ under the constraint of norm conservation[2] [483] leads to the Schrödinger operator (7.3),

$$\frac{\delta A}{\delta\langle\Psi(t)|} = \left[i\hbar\partial_t - \hat{H}(t)\right]|\Psi(t)\rangle\,, \tag{7.51}$$

i.e. A is stationary for the correct state $|\Psi(t)\rangle$ which develops from a given initial state $|\Psi(t_0)\rangle$ (Frenkel's variational principle). One can therefore solve the

[2] Note that the variations $|\delta\Psi(t)\rangle$ used in Eq. (7.51) need not necessarily satisfy any boundary conditions. The only assumption required is that with $|\delta\Psi(t)\rangle$ also $|i\delta\Psi(t)\rangle$ is an allowed variation [483].

Schrödinger equation (7.3) by searching for the state $|\Psi(t)\rangle$ which makes A stationary.

The existence of $|\Psi[n,\Psi_0](t)\rangle$ makes it possible to define an action density functional,[3]

$$A[n] := \int_{t_0}^{t_1} dt\,\langle\Psi[n,\Psi_0](t)|i\hbar\partial_t - \hat{H}(t)|\Psi[n,\Psi_0](t)\rangle + N\left[\alpha(t_1)-\alpha(t_0)\right], \quad (7.52)$$

which, at first glance, seems to suffer from the non-uniqueness of gauge. The phase contribution can, however, be easily absorbed into the external potential,

$$A[n] = \int_{t_0}^{t_1} dt\,\langle\Psi[n,\Psi_0](t)|i\hbar\partial_t - \hat{H}'(t)|\Psi[n,\Psi_0](t)\rangle, \quad (7.53)$$

where $\hat{H}'$ is obtained from $\hat{H}$ with the gauge transformation (7.13). $\hat{H}'$ and $\hat{H}$ are physically equivalent, so that the stationarity of (7.53) is equivalent to that of (7.52). In other words: once the gauge is fixed for all possible elements of $\mathscr{V}$ and both $|\Psi[n,\Psi_0](t)\rangle$ and the set of associated Hamiltonians is defined accordingly, the gauge arbitrariness is eliminated. This gauge fixing could, for instance, be implemented by keeping some spatial average of $v_{\mathrm{ext}}(\boldsymbol{r}t)$ constant throughout time. From now on the gauge is assumed to be chosen so that $\alpha(t)\equiv 0$ for all t.

The functional (7.53) becomes stationary for the correct td density $n(\boldsymbol{r}t)$, due to the one-to-one correspondence between $n(\boldsymbol{r}t)$ and the class of equivalent $|\Psi(t)\rangle$ (the global phase is irrelevant at this point). This suggests that the variational equation[4] [472]

$$\frac{\delta A[n]}{\delta n(\boldsymbol{r}t)} = 0 \quad (7.54)$$

is equivalent to the td Schrödinger equation, in analogy to the equivalence of the variational equation (2.38) with the stationary Schrödinger equation. However, the td situation is complicated by the fact that the action (7.50) *per se* not only incorporates the time-evolution in positive direction of time, but also backward propagation: variation of the action (7.50) with respect to the state $|\Psi(t)\rangle$ leads to the time-reversed Schrödinger equation, as time-reversal simply amounts to complex conjugation for the Hamiltonian (7.1). Therefore A also becomes stationary for the state $|\Psi(t)\rangle$ which evolves backward in time from a given final state $|\Psi(t_1)\rangle$. As a result, stationarity of (7.53) with respect to the density cannot *a priori* distinguish between propagation in positive and in negative direction of time[5] [484].

---

[3] As the functional $|\Psi[n,\Psi_0](t)\rangle$ depends on the initial state $|\Psi_0\rangle$, the same applies to all density functionals constructed from $|\Psi[n,\Psi_0](t)\rangle$—for brevity, we do not indicate this dependence explicitly, $A[n,\Psi_0]\equiv A[n]$.

[4] In Eq. (7.54) the variation with respect to $n(\boldsymbol{r}t)$ is performed for fixed $v_{\mathrm{ext}}(\boldsymbol{r}t)$.

[5] By contrast, the equations resulting from variation of the ground state energy expectation value of stationary systems with respect to $\langle\Psi|$ and $|\Psi\rangle$ coincide, as one is dealing with a boundary value problem, rather than the initial value problem (7.3).

As it stands, Eq. (7.54) therefore leads to problems with causality [485, 486]. This problem is most easily identified by an analysis of the second functional derivative of A with respect to the external potential. For this analysis the total external potential is split into a stationary part v_0 and a td component v_1, with a corresponding decomposition of the total Hamiltonian,

$$v_{ext}(\boldsymbol{r}t) = v_0(\boldsymbol{r}) + v_1(\boldsymbol{r}t) \qquad \Longleftrightarrow \qquad \hat{H}(t) = \hat{H}_0 + \hat{H}_1(t)\,. \tag{7.55}$$

In addition, v_1 is assumed to have all the properties required for the existence of the functional Taylor expansion of the expectation value (7.50) about $v_1 = 0$,

$$\begin{aligned} A[v_1] &= A[v_1=0] + \int_{t_0}^{t_1} dt \int d^3r \, \frac{\delta A}{\delta v_1(\boldsymbol{r}t)}\bigg|_{v_1=0} v_1(\boldsymbol{r}t) \\ &+ \frac{1}{2}\int_{t_0}^{t_1} dt \int d^3r \int_{t_0}^{t_1} dt' \int d^3r' \, \frac{\delta^2 A}{\delta v_1(\boldsymbol{r}t)\delta v_1(\boldsymbol{r}'t')}\bigg|_{v_1=0} v_1(\boldsymbol{r}t) v_1(\boldsymbol{r}'t') \\ &+ \ldots\,. \end{aligned} \tag{7.56}$$

For the true wavefunction corresponding to $\hat{H}(t)$, i.e. the wavefunction for which A is stationary, A vanishes, so that the same holds for each individual term in the expansion (7.56). In particular, one has

$$\frac{\delta^2 A}{\delta v_1(\boldsymbol{r}t)\delta v_1(\boldsymbol{r}'t')}\bigg|_{v_1=0} = 0\,. \tag{7.57}$$

It is instructive to verify Eq. (7.57) by an explicit evaluation. Functional differentiation of (7.50) with respect to v_1 at two times $t' \neq t''$ with $t_0 < t', t'' < t_1$ yields

$$\begin{aligned} &\frac{\delta^2 A}{\delta v_1(\boldsymbol{r}''t'')\delta v_1(\boldsymbol{r}'t')} \\ &= \int_{t_0}^{t_1} dt \left[\langle \frac{\delta^2\Psi(t)}{\delta v_1(\boldsymbol{r}''t'')\delta v_1(\boldsymbol{r}'t')} | i\hbar\partial_t - \hat{H}(t)|\Psi(t)\rangle + \langle\Psi(t)|i\hbar\partial_t - \hat{H}(t)| \frac{\delta^2\Psi(t)}{\delta v_1(\boldsymbol{r}''t'')\delta v_1(\boldsymbol{r}'t')}\rangle \right] \\ &+ \int_{t_0}^{t_1} dt \left[\langle \frac{\delta\Psi(t)}{\delta v_1(\boldsymbol{r}''t'')} | i\hbar\partial_t - \hat{H}(t)| \frac{\delta\Psi(t)}{\delta v_1(\boldsymbol{r}'t')}\rangle + \langle \frac{\delta\Psi(t)}{\delta v_1(\boldsymbol{r}'t')} | i\hbar\partial_t - \hat{H}(t)| \frac{\delta\Psi(t)}{\delta v_1(\boldsymbol{r}''t'')}\rangle \right] \\ &- \int_{t_0}^{t_1} dt \left[\langle \frac{\delta\Psi(t)}{\delta v_1(\boldsymbol{r}''t'')} | \frac{\delta\hat{H}_1(t)}{\delta v_1(\boldsymbol{r}'t')} |\Psi(t)\rangle + \langle \frac{\delta\Psi(t)}{\delta v_1(\boldsymbol{r}'t')} | \frac{\delta\hat{H}_1(t)}{\delta v_1(\boldsymbol{r}''t'')} |\Psi(t)\rangle \right] \\ &- \int_{t_0}^{t_1} dt \left[\langle\Psi(t)| \frac{\delta\hat{H}_1(t)}{\delta v_1(\boldsymbol{r}''t'')} | \frac{\delta\Psi(t)}{\delta v_1(\boldsymbol{r}'t')}\rangle + \langle\Psi(t)| \frac{\delta\hat{H}_1(t)}{\delta v_1(\boldsymbol{r}'t')} | \frac{\delta\Psi(t)}{\delta v_1(\boldsymbol{r}''t'')}\rangle \right]. \end{aligned} \tag{7.58}$$

Once the differentiation is executed, one can use the stationarity of A, so that the first term of the right-hand side vanishes. After partial integration the stationarity of A can also be employed for the second term of the right-hand side of (7.58),

$$\begin{aligned} &\int_{t_0}^{t_1} dt \, \langle\Psi(t)|i\hbar\partial_t - \hat{H}(t)| \frac{\delta^2\Psi(t)}{\delta v_1(\boldsymbol{r}''t'')\delta v_1(\boldsymbol{r}'t')}\rangle \\ &= i\hbar \, \langle\Psi(t)| \frac{\delta^2\Psi(t)}{\delta v_1(\boldsymbol{r}''t'')\delta v_1(\boldsymbol{r}'t')}\rangle\bigg|_{t_0}^{t_1}. \end{aligned} \tag{7.59}$$

An explicit representation of $|\Psi(t)\rangle$ in terms of v_1 is required for the evaluation of the remaining terms. However, $|\Psi(t)\rangle$ can be obtained from the initial state $|\Psi_0\rangle$ by application of the time-evolution operator in the Schrödinger representation,

$$|\Psi(t)\rangle = \hat{U}_S(t,t_0)|\Psi_0\rangle \tag{7.60}$$

$$\begin{aligned}\hat{U}_S(t,t_0) &= \sum_{n=0}^{\infty}\frac{(-i)^n}{\hbar^n n!}\int_{t_0}^{t}dt_1\cdots\int_{t_0}^{t}dt_n\,T\left[\hat{H}(t_1)\cdots\hat{H}(t_n)\right]\\ &= T\exp\left[-\frac{i}{\hbar}\int_{t_0}^{t}dt'\,\hat{H}(t')\right].\end{aligned} \tag{7.61}$$

By construction the operator $\hat{U}_S$ ensures proper normalization, i.e. $\langle\Psi(t)|\Psi(t)\rangle = 1$ if $\langle\Psi_0|\Psi_0\rangle = 1$,

$$\frac{\delta\langle\Psi(t)|\Psi(t)\rangle}{\delta v_1(\boldsymbol{r}'t')} = 0, \tag{7.62}$$

as is required by the form (7.50) of the action. Using Eqs. (7.60), (7.61), the required derivatives can be expressed as

$$\left|\frac{\delta\Psi(t)}{\delta v_1(\boldsymbol{r}'t')}\right\rangle = -\frac{i}{\hbar}\Theta(t-t')\hat{U}_S(t,t')\hat{n}(\boldsymbol{r}')\hat{U}_S(t',t_0)|\Psi_0\rangle \tag{7.63}$$

$$\begin{aligned}\left|\frac{\delta^2\Psi(t)}{\delta v_1(\boldsymbol{r}''t'')\delta v_1(\boldsymbol{r}'t')}\right\rangle &= -\frac{1}{\hbar^2}\Theta(t-t')\Theta(t'-t'')\\ &\quad\times\hat{U}_S(t,t')\hat{n}(\boldsymbol{r}')\hat{U}_S(t',t'')\hat{n}(\boldsymbol{r}'')\hat{U}_S(t'',t_0)|\Psi_0\rangle\\ &\quad-\frac{1}{\hbar^2}\Theta(t-t'')\Theta(t''-t')\\ &\quad\times\hat{U}_S(t,t'')\hat{n}(\boldsymbol{r}'')\hat{U}_S(t'',t')\hat{n}(\boldsymbol{r}')\hat{U}_S(t',t_0)|\Psi_0\rangle\end{aligned} \tag{7.64}$$

$$\frac{\delta\hat{H}_1(t)}{\delta v_1(\boldsymbol{r}'t')} = \delta(t-t')\hat{n}(\boldsymbol{r}'). \tag{7.65}$$

In the limit $v_1 \to 0$ Eqs. (7.63), (7.64) reduce to

$$\left|\frac{\delta\Psi(t)}{\delta v_1(\boldsymbol{r}'t')}\right\rangle = -\frac{i}{\hbar}\Theta(t-t')e^{-i\hat{H}_0(t-t_0)/\hbar}\hat{n}_0(\boldsymbol{r}'t')|\Psi_0\rangle \tag{7.66}$$

$$\begin{aligned}\left|\frac{\delta^2\Psi(t)}{\delta v_1(\boldsymbol{r}''t'')\delta v_1(\boldsymbol{r}'t')}\right\rangle &= -\frac{1}{\hbar^2}\Theta(t-t')\Theta(t'-t'')e^{-i\hat{H}_0(t-t_0)/\hbar}\hat{n}_0(\boldsymbol{r}'t')\hat{n}_0(\boldsymbol{r}''t'')|\Psi_0\rangle\\ &\quad-\frac{1}{\hbar^2}\Theta(t-t'')\Theta(t''-t')e^{-i\hat{H}_0(t-t_0)/\hbar}\hat{n}_0(\boldsymbol{r}''t'')\hat{n}_0(\boldsymbol{r}'t')|\Psi_0\rangle,\end{aligned} \tag{7.67}$$

where $\hat{n}_0$ denotes the density operator in the Heisenberg representation with respect to $\hat{H}_0$,

$$\hat{n}_0(\boldsymbol{r}t) = e^{i\hat{H}_0(t-t_0)/\hbar}\,\hat{n}(\boldsymbol{r})\,e^{-i\hat{H}_0(t-t_0)/\hbar}. \tag{7.68}$$

With the aid of Eqs. (7.65)–(7.67) one can evaluate (7.58) in the limit $v_1 \to 0$. Collecting all terms, one finds

$$\begin{aligned}&\left.\frac{\delta^2 A}{\delta v_1(\boldsymbol{r}''t'')\delta v_1(\boldsymbol{r}'t')}\right|_{v_1=0}\\ &= -\frac{i}{\hbar}\Theta(t'-t'')\langle\Psi_0|\hat{n}_0(\boldsymbol{r}'t')\hat{n}_0(\boldsymbol{r}''t'')|\Psi_0\rangle - \frac{i}{\hbar}\Theta(t''-t')\langle\Psi_0|\hat{n}_0(\boldsymbol{r}''t'')\hat{n}_0(\boldsymbol{r}'t')|\Psi_0\rangle\\ &\quad+\frac{1}{\hbar^2}\int_{t_0}^{t_1}dt\,\Theta(t-t'')\langle\Psi_0|\hat{n}_0(\boldsymbol{r}''t'')e^{i\hat{H}_0(t-t_0)/\hbar}\left[i\hbar\partial_t - \hat{H}_0\right]\Theta(t-t')e^{-i\hat{H}_0(t-t_0)/\hbar}\hat{n}_0(\boldsymbol{r}'t')|\Psi_0\rangle\end{aligned}$$

$$
\begin{aligned}
&+\frac{1}{\hbar^2}\int_{t_0}^{t_1} dt\,\Theta(t-t')\langle\Psi_0|\hat{n}_0(\boldsymbol{r}'t')e^{i\hat{H}_0(t-t_0)/\hbar}\left[i\hbar\partial_t-\hat{H}_0\right]\Theta(t-t'')e^{-i\hat{H}_0(t-t_0)/\hbar}\hat{n}_0(\boldsymbol{r}''t'')|\Psi_0\rangle \\
&-\frac{i}{\hbar}\Theta(t'-t'')\langle\Psi_0|\hat{n}_0(\boldsymbol{r}''t'')\hat{n}_0(\boldsymbol{r}'t')|\Psi_0\rangle-\frac{i}{\hbar}\Theta(t''-t')\langle\Psi_0|\hat{n}_0(\boldsymbol{r}'t')\hat{n}_0(\boldsymbol{r}''t'')|\Psi_0\rangle \\
&+\frac{i}{\hbar}\Theta(t''-t')\langle\Psi_0|\hat{n}_0(\boldsymbol{r}''t'')\hat{n}_0(\boldsymbol{r}'t')|\Psi_0\rangle+\frac{i}{\hbar}\Theta(t'-t'')\langle\Psi_0|\hat{n}_0(\boldsymbol{r}'t')\hat{n}_0(\boldsymbol{r}''t'')|\Psi_0\rangle \\
&=0,
\end{aligned}
\tag{7.69}
$$

consistent with Eq. (7.57).

The relevance of this (obvious) result becomes more clear if A is split into two contributions according to Eq. (7.55),

$$
A = A_0-\int_{t_0}^{t_1} dt\,\langle\Psi(t)|\hat{H}_1(t)|\Psi(t)\rangle\,. \tag{7.70}
$$

The functional derivative of the second term on the right-hand side can be directly extracted from the preceding discussion. Combination with Eq. (7.69) gives

$$
\begin{aligned}
\left.\frac{\delta^2 A_0}{\delta v_1(\boldsymbol{r}''t'')\delta v_1(\boldsymbol{r}'t')}\right|_{v_1=0} &= -\frac{i}{\hbar}\Theta(t'-t'')\langle\Psi_0|\left[\hat{n}_0(\boldsymbol{r}'t'),\hat{n}_0(\boldsymbol{r}''t'')\right]|\Psi_0\rangle \\
&\quad-\frac{i}{\hbar}\Theta(t''-t')\langle\Psi_0|\left[\hat{n}_0(\boldsymbol{r}''t''),\hat{n}_0(\boldsymbol{r}'t')\right]|\Psi_0\rangle\,.
\end{aligned}
\tag{7.71}
$$

The expression on the right-hand side of Eq. (7.71) is easily identified as the retarded response function of the $\hat{H}_0$-system, Eq. (4.39),

$$
\left.\frac{\delta^2 A_0}{\delta v_1(\boldsymbol{r}''t'')\delta v_1(\boldsymbol{r}'t')}\right|_{v_1=0} = \chi_{\mathrm{R}}(\boldsymbol{r}'t',\boldsymbol{r}''t'')+\chi_{\mathrm{R}}(\boldsymbol{r}''t'',\boldsymbol{r}'t'), \tag{7.72}
$$

provided that $|\Psi_0\rangle$ is assumed to be the ground state corresponding to $\hat{H}_0$ (compare Sect. 4.2.1). The unsymmetric form of χ_{R} ensures that the response of the system always appears after the perturbation which causes the response (for an explicit derivation see e.g. [94], compare also Sect. 7.4). In the case of backward evolution the advanced response function plays the same role,

$$
\chi_{\mathrm{A}}(\boldsymbol{r}'t',\boldsymbol{r}''t'') = \chi_{\mathrm{R}}(\boldsymbol{r}''t'',\boldsymbol{r}'t')\,, \tag{7.73}
$$

so that (7.72) can also be expressed as

$$
\left.\frac{\delta^2 A_0}{\delta v_1(\boldsymbol{r}''t'')\delta v_1(\boldsymbol{r}'t')}\right|_{v_1=0} = \chi_{\mathrm{R}}(\boldsymbol{r}'t',\boldsymbol{r}''t'')+\chi_{\mathrm{A}}(\boldsymbol{r}'t',\boldsymbol{r}''t'')\,. \tag{7.74}
$$

Yet another form of (7.72) is obtained by use of the time-ordered response (4.35)

$$
\left.\frac{\delta^2 A_0}{\delta v_1(\boldsymbol{r}''t'')\delta v_1(\boldsymbol{r}'t')}\right|_{v_1=0} = \chi(\boldsymbol{r}'t',\boldsymbol{r}''t'')+\chi^*(\boldsymbol{r}'t',\boldsymbol{r}''t'')\,. \tag{7.75}
$$

Equations (7.72), (7.74), (7.75) are explicit manifestations of the fact that each individual component of A is symmetric with respect to the direction of time—both retarded and advanced contributions are included in the functional (7.53).

The same statements hold for the action density functional $A[n]$ [484]. In order to illustrate the relevance of these statements for TDDFT one has to consider the functional derivative of $A[n]$ with respect to the density. For this purpose it is convenient to express A_0 as a density functional

$$\begin{aligned} A_0[n] &:= A[n] + \int_{t_0}^{t_1} dt \int d^3r\, n(\boldsymbol{r}t) v_1(\boldsymbol{r}t) \\ &= \int_{t_0}^{t_1} dt\, \langle \Psi[n,\Psi_0](t) | i\hbar\partial_t - \hat{H}_0 | \Psi[n,\Psi_0](t) \rangle \,. \end{aligned} \tag{7.76}$$

Compared with $A[n]$, $A_0[n]$ has the advantage that it does not depend explicitly on v_1. The stationarity condition (7.54) then translates to

$$\frac{\delta A_0[n]}{\delta n(\boldsymbol{r}t)} = v_1(\boldsymbol{r}t)\,, \tag{7.77}$$

where v_0 and v_1 have to be kept fixed in the variational process on the left-hand side.

If one considers once more the second functional derivative of A with respect to the td potential and uses the fact that $A_0[n]$ does not *explicitly* depend on v_1, one obtains

$$\begin{aligned} &\frac{\delta^2 A}{\delta v_1(\boldsymbol{r}t)\delta v_1(\boldsymbol{r}'t')} \\ &= \int_{t_0}^{t_1} dt'' \int d^3r'' \int_{t_0}^{t_1} dt''' \int d^3r''' \frac{\delta n(\boldsymbol{r}''t'')}{\delta v_1(\boldsymbol{r}t)} \frac{\delta n(\boldsymbol{r}'''t''')}{\delta v_1(\boldsymbol{r}'t')} \frac{\delta^2 A_0[n]}{\delta n(\boldsymbol{r}''t'')\delta n(\boldsymbol{r}'''t''')} \\ &+ \int_{t_0}^{t_1} dt'' \int d^3r'' \left(\frac{\delta A_0[n]}{\delta n(\boldsymbol{r}''t'')} - v_1(\boldsymbol{r}''t'') \right) \frac{\delta^2 n(\boldsymbol{r}''t'')}{\delta v_1(\boldsymbol{r}t)\delta v_1(\boldsymbol{r}'t')} \\ &- \frac{\delta n(\boldsymbol{r}t)}{\delta v_1(\boldsymbol{r}'t')} - \frac{\delta n(\boldsymbol{r}'t')}{\delta v_1(\boldsymbol{r}t)}\,. \end{aligned} \tag{7.78}$$

This relation is now evaluated for the density which makes $A[n]$ stationary. One can then use Eq. (7.77) for the second term on the right-hand side. On the other hand, the left-hand side of Eq. (7.78) vanishes in the limit $v_1 \to 0$, Eq. (7.69). In this limit one thus obtains

$$\begin{aligned} &\int_{t_0}^{t_1} dt'' \int d^3r'' \int_{t_0}^{t_1} dt''' \int d^3r''' \\ &\times \left[\frac{\delta n(\boldsymbol{r}''t'')}{\delta v_1(\boldsymbol{r}t)} \frac{\delta n(\boldsymbol{r}'''t''')}{\delta v_1(\boldsymbol{r}'t')} \frac{\delta^2 A_0[n]}{\delta n(\boldsymbol{r}''t'')\delta n(\boldsymbol{r}'''t''')} \right]_{v_1=0} \\ &= \left.\frac{\delta n(\boldsymbol{r}t)}{\delta v_1(\boldsymbol{r}'t')}\right|_{v_1=0} + \left.\frac{\delta n(\boldsymbol{r}'t')}{\delta v_1(\boldsymbol{r}t)}\right|_{v_1=0}\,. \end{aligned} \tag{7.79}$$

The quantities on the right-hand side are identified as the response function of the $\hat{H}_0$-system. In the case of forward evolution this is the retarded response

$$\chi_{\rm R}(\boldsymbol{r}t,\boldsymbol{r}'t') = \left.\frac{\delta n(\boldsymbol{r}t)}{\delta v_1(\boldsymbol{r}'t')}\right|_{v_1=0} . \tag{7.80}$$

However, the right-hand side of Eq. (7.79) features a combination of two response functions, which is symmetric in t and t'. In fact, the left-hand side of Eq. (7.79) coincides with

$$\left.\frac{\delta^2 A_0}{\delta v_1(\boldsymbol{r}t)\delta v_1(\boldsymbol{r}'t')}\right|_{v_1=0} ,$$

so that one can use Eq. (7.72) to rewrite (7.79) as

$$\begin{aligned}&\int_{t_0}^{t_1} dt'' \int d^3r'' \int_{t_0}^{t_1} dt''' \int d^3r''' \\ &\times \left[\frac{\delta n(\boldsymbol{r}''t'')}{\delta v_1(\boldsymbol{r}t)} \frac{\delta^2 A_0[n]}{\delta n(\boldsymbol{r}''t'')\delta n(\boldsymbol{r}'''t''')} \frac{\delta n(\boldsymbol{r}'''t''')}{\delta v_1(\boldsymbol{r}'t')}\right]_{v_1=0} \\ &= \chi_{\rm R}(\boldsymbol{r}t,\boldsymbol{r}'t') + \chi_{\rm R}(\boldsymbol{r}'t',\boldsymbol{r}t) .\end{aligned} \tag{7.81}$$

With the inverse of the retarded response function [487],

$$\chi_{\rm R}^{-1}(\boldsymbol{r}t,\boldsymbol{r}'t') = \left.\frac{\delta v_1(\boldsymbol{r}t)}{\delta n(\boldsymbol{r}'t')}\right|_{v_1=0} \tag{7.82}$$

$$\int dt' \int d^3r'\, \chi_{\rm R}^{-1}(\boldsymbol{r}t,\boldsymbol{r}'t')\chi_{\rm R}(\boldsymbol{r}'t',\boldsymbol{r}''t'') = \delta(t-t'')\,\delta^{(3)}(\boldsymbol{r}-\boldsymbol{r}'') , \tag{7.83}$$

Eq. (7.81) may finally be resolved as

$$\left.\frac{\delta^2 A_0[n]}{\delta n(\boldsymbol{r}t)\delta n(\boldsymbol{r}'t')}\right|_{v_1=0} = \chi_{\rm R}^{-1}(\boldsymbol{r}t,\boldsymbol{r}'t') + \chi_{\rm R}^{-1}(\boldsymbol{r}'t',\boldsymbol{r}t) . \tag{7.84}$$

This result can equivalently be expressed in terms of the inverse of the advanced response function (7.73),

$$\left.\frac{\delta^2 A_0[n]}{\delta n(\boldsymbol{r}t)\delta n(\boldsymbol{r}'t')}\right|_{v_1=0} = \chi_{\rm R}^{-1}(\boldsymbol{r}t,\boldsymbol{r}'t') + \chi_{\rm A}^{-1}(\boldsymbol{r}t,\boldsymbol{r}'t') . \tag{7.85}$$

Irrespective of the response function chosen, the right-hand sides of (7.84), (7.85) are symmetric in t and t'. This symmetry reflects the fact that both retarded and advanced contributions are included in the functional (7.76). The result (7.84) is the TDDFT counterpart of Eq. (7.72). Both directions of time are covered by the *action density functional*.

As a consequence, the stationarity condition (7.54) is not fully equivalent to the td Schrödinger equation, whose form implies a specification of the direction of time. In other words: Eq. (7.54) is a necessary condition for finding the true td density, but not a sufficient condition. It must be supplemented by an explicit choice of the direction of time. When one uses Eq. (7.54) in the context of causal propagation, the appropriate retarded component of $A_0[n]$ has to be projected out [484].

The implications of the causality problem are still a matter of debate. For details the reader is referred to the original literature [484, 488, 489, 487, 490–495]. One option for a mathematically rigorous implementation of forward evolution is offered by the Keldysh formalism [488, 489]. In this approach the action functional is redefined in terms of the Keldysh pseudo-time, which then allows the derivation of a causal equation of motion. An alternative solution to the causality problem relies on the inclusion of the additional boundary term resulting from the fact that the variation of the wavefunction is not restricted at the end point t_1 [495]. This inclusion restores causality, but leads to expressions which depend on the Berry curvature. A detailed discussion of these two approaches is, however, beyond the scope of this text (for an introduction to the Keldysh formalism and its application to TDDFT see [477]). We will from now on apply the stationarity condition (7.54) with the understanding that only its causal component is to be used.

Some further remarks on the functional $A[n]$ are appropriate at this point:

- By construction, the domain of $A[n]$ is restricted to those densities which are obtained from the solution of the Schrödinger equation for some potential $v_{\mathrm{ext}}(\boldsymbol{r}t)$, i.e. to interacting v-representable densities. On the other hand, in order to guarantee that the functional derivative (7.54) exists, $A[n]$ needs to be defined on a sufficiently dense set of densities $n(\boldsymbol{r}t)$. Ideally, the domain of $A[n]$ would be the set of all normalizable $n(\boldsymbol{r}t)$. However, not even the stationary Hohenberg-Kohn functional $E[n]$ has such a large domain. One has to define $A[n]$ in a more general way, following the Levy-Lieb concept (see Sect. 2.3), in order to rigorously ensure functional differentiability (for further details see [496, 497]).
- The functional $A[n]$ is not only nonlocal in space, but also in time. All times between t_0 and t contribute via (7.54) to the time-evolved density at time t. $A[n]$ therefore involves "memory effects", that is the properties of the system at time t depend on the way the system has developed in the time interval from t_0 to time t.
- The functional $|\Psi[n,\Psi_0](t)\rangle$ not only allows the definition of $A[n]$, but also of other density functionals. A quantity of obvious interest is the cross section, which provides the essential information on excitation and collision processes. The probability for the system to end up in a particular eigenstate of the asymptotic Hamiltonian for $t \to \infty$ is given by the overlap of $|\Psi[n,\Psi_0](t \to \infty)\rangle$ with this eigenstate. Unfortunately, the density alone is not very well suited to extract information on these overlap matrix elements, as $n(\boldsymbol{r}t)$ does not convey any phase information. Only observables which involve some spatial separation can be obtained directly from the density. In this way one can e.g. differentiate electron transfer, excitation and ionization in the case of collision processes. As a way out of this restriction, the time-dependent Kohn-Sham state, discussed in Sect. 7.2, offers itself.
- The functional

$$B[n] := A[n] + \int_{t_0}^{t_1} dt \int d^3r\, n(\boldsymbol{r}t) v_{\text{ext}}(\boldsymbol{r}t)$$
$$= \int_{t_0}^{t_1} dt\, \langle \Psi[n,\Psi_0](t) | i\hbar\partial_t - \hat{T} - \hat{W} | \Psi[n,\Psi_0](t) \rangle \qquad (7.86)$$

is universal in the sense that it does not depend on v_{ext}—according to the RG-theorem the td state is completely determined by the td density. $B[n]$ coincides with $A_0[n]$, Eq. (7.76), if one chooses[6] $v_1 = v_{\text{ext}}$, $v_0 = 0$. All relations derived for $A_0[n]$ therefore also apply to $B[n]$, as soon as $\hat{H}_0$ is reduced to $\hat{T} + \hat{W}$. In particular, $B[n]$ requires projection on causal evolution in the same way as $A_0[n]$.

- The generalization of the RG-theorem to the Hamiltonian (7.5) [479, 480] requires the explicit use of the current density $\boldsymbol{j}_{\text{p}}$, Eq. (7.25), as basic DFT variable, as is to be expected from stationary CDFT. One can show that the td charge and current densities determine the td external potentials v_{ext} and $\boldsymbol{A}_{\text{ext}}$ up to the gauge transformation (7.9), (7.10). As a consequence, the td state is determined by n and $\boldsymbol{j}_{\text{p}}$ up to a phase transformation of type (7.11). This then allows the definition of a corresponding action density functional for the Hamiltonian (7.5) [479].
- The combination of Eq. (7.27) with the continuity equation (7.43) yields a direct hydrodynamic equation for $n(\boldsymbol{r}t)$,

$$\partial_t^2 n(\boldsymbol{r}t) = \frac{i}{\hbar} \boldsymbol{\nabla} \cdot \langle \Psi[n,\Psi_0](t) | \left[\hat{\boldsymbol{j}}_{\text{p}}(\boldsymbol{r}), \hat{H}(t) \right] | \Psi[n,\Psi_0](t) \rangle , \qquad (7.87)$$

which is equivalent to the variational equation (7.54).

7.2 Time-Dependent Kohn-Sham Equations

The stationarity of the action functional, Eq. (7.54), could be explicitly exploited as soon as a suitable approximation for $A[n]$ was available. In practice, however, the derivation of such an approximation is even more difficult than that of total energy functionals for stationary systems. For this reason the standard way to set up and apply a workable version of TDDFT is the use of time-dependent Kohn-Sham (TDKS) equations. This approach automatically implements a choice for the direction of time. The derivation of the TDKS equations closely follows the path taken in the stationary limit (see Chap. 3).

The RG-theorem not only applies to interacting many-particle systems, but also to noninteracting systems with td Hamiltonians of the type

$$\hat{H}_{\text{s}}(t) = \hat{T} + \hat{V}_{\text{s}}(t) \,; \qquad \hat{V}_{\text{s}}(t) = \int d^3r\, \hat{n}(\boldsymbol{r}) v_{\text{s}}(\boldsymbol{r}t) \,. \qquad (7.88)$$

[6] None of the results between Eqs. (7.55) and (7.85) depends on the presence of v_0.

The nature or presence of the interaction does not play any role in the proof given in the preceding section. For a given initial state $|\Phi(t_0)\rangle$ the density of a noninteracting system thus determines the external potential $v_s(\boldsymbol{r}t)$ of the system and the corresponding state $|\Phi(t)\rangle$,

$$i\hbar\partial_t|\Phi(t)\rangle = \hat{H}_s(t)|\Phi(t)\rangle\,, \tag{7.89}$$

uniquely for all $t \geq t_0$.

The many-body state $|\Phi(t)\rangle$ can be specified explicitly. If the initial state $|\Phi(t_0)\rangle$ is a Slater determinant constructed from N single-particle orbitals $\phi_i(\boldsymbol{r}\sigma t_0)$, $|\Phi(t)\rangle$ has a determinantal form for all t,

$$\begin{aligned}(\boldsymbol{r}_1\sigma_1,\ldots\boldsymbol{r}_N\sigma_N|\Phi(t)\rangle &\equiv \Phi(\boldsymbol{r}_1\sigma_1,\ldots\boldsymbol{r}_N\sigma_N,t)\\ &= \frac{1}{\sqrt{N!}}\det\begin{pmatrix}\phi_1(\boldsymbol{r}_1\sigma_1 t) & \cdots & \phi_N(\boldsymbol{r}_1\sigma_1 t)\\ \vdots & & \vdots\\ \phi_1(\boldsymbol{r}_N\sigma_N t) & \cdots & \phi_N(\boldsymbol{r}_N\sigma_N t)\end{pmatrix},\end{aligned} \tag{7.90}$$

with the orthonormal single-particle orbitals being solutions of

$$i\hbar\partial_t\phi_i(\boldsymbol{r}\sigma t) = \left\{-\frac{\hbar^2\boldsymbol{\nabla}^2}{2m} + v_s(\boldsymbol{r}t)\right\}\phi_i(\boldsymbol{r}\sigma t)\,. \tag{7.91}$$

This statement can be easily verified. As a consequence of Eq. (7.91) the state (7.90) manifestly satisfies the td Schrödinger equation for the Hamiltonian (7.88). As soon as (7.90) satisfies the given initial condition, the uniqueness of the solution of the noninteracting Schrödinger equation (7.89) guarantees that (7.90) is the correct td state for all t. The density of the noninteracting system is therefore given by

$$n(\boldsymbol{r}t) = \sum_\sigma\sum_{i=1}^{N}|\phi_i(\boldsymbol{r}\sigma t)|^2 \tag{7.92}$$

for all t (remember that the quantum number i includes the spin).

The RG-theorem ensures the existence of a DFT description of this noninteracting system. For instance, the corresponding action functional is

$$A_s[n] = B_s[n] - \int_{t_0}^{t_1} dt \int d^3r\, n(\boldsymbol{r}t)\, v_s(\boldsymbol{r}t) \tag{7.93}$$

$$B_s[n] = \int_{t_0}^{t_1} dt\,\langle\Phi[n](t)|i\hbar\partial_t - \hat{T}|\Phi[n](t)\rangle\,, \tag{7.94}$$

where $|\Phi[n](t)\rangle$ denotes the density functional obtained for the noninteracting state (7.90) (the dependence of $|\Phi[n](t)\rangle$ on the initial state $|\Phi(t_0)\rangle$ has been suppressed for brevity). The stationarity condition

$$\frac{\delta A_s[n]}{\delta n(\boldsymbol{r}t)} = \frac{\delta B_s[n]}{\delta n(\boldsymbol{r}t)} - v_s(\boldsymbol{r}t) = 0 \tag{7.95}$$

is a necessary condition for the solution of Eq. (7.89). As in the interacting situation, however, application of (7.95) requires proper projection on forward evolution.

Once the density functional description of noninteracting, time-dependent systems is established, one can return to the interacting system (7.1) with initial state (7.4). Let us assume that one can find a noninteracting system with potential $v_s(\boldsymbol{r}t)$ and determinantal state (7.90) for which the density (7.92) is identical with the density of the interacting system for all $t \geq t_0$. The existence of such an auxiliary noninteracting system is not guaranteed. Without a suitable redefinition of the action functional noninteracting v-representability is an assumption.

One may, however, convince oneself that at least the minimal requirements for this assumption are fulfilled. Obviously, noninteracting v-representability first of all requires the densities of the two systems to be identical at $t = t_0$. This initial density is fixed by the initial condition (7.4) via (7.23). An initial Slater determinant for the auxiliary noninteracting system which reproduces the given density can be explicitly constructed by the scheme discussed in Sect. 2.3. Of course, the state $|\Phi(t_0)\rangle$ differs from the initial state of the interacting system. In addition, the initial Slater determinant and hence the complete time-evolution of the noninteracting state are not even determined uniquely. This poses no problem, however, as all equivalent determinants (i) yield the same density and (ii) consist of single-particle orbitals which satisfy the same differential equation (7.91) (the potential $v_s(\boldsymbol{r}t)$ is uniquely determined by the interacting $n(\boldsymbol{r}t)$, provided it exists). Consequently, the equivalent determinants can only differ by unitary transformations.

The next step is a suitable decomposition of the interacting action functional, using (7.93), (7.94),

$$\begin{aligned} A[n] = B_s[n] &- \int_{t_0}^{t_1} dt \int d^3r\, n(\boldsymbol{r}t)\, v_{\mathrm{ext}}(\boldsymbol{r}t) \\ &- \frac{1}{2}\int_{t_0}^{t_1} dt \int d^3r \int d^3r'\, n(\boldsymbol{r}t)\, w(\boldsymbol{r},\boldsymbol{r}')\, n(\boldsymbol{r}'t) - A_{\mathrm{xc}}[n], \end{aligned} \tag{7.96}$$

which defines the xc-action functional,[7]

$$A_{\mathrm{xc}}[n] = B_s[n] - B[n] - \frac{1}{2}\int_{t_0}^{t_1} dt \int d^3r \int d^3r'\, n(\boldsymbol{r}t)\, w(\boldsymbol{r},\boldsymbol{r}')\, n(\boldsymbol{r}'t)\,. \tag{7.97}$$

Of course, $A_{\mathrm{xc}}[n]$ is subject to the same causality problem as $B[n]$ and its noninteracting limit $B_s[n]$. When using the exact $A_{\mathrm{xc}}[n]$ or approximations, causality has to be implemented explicitly.

In order to find the explicit form of $v_s(\boldsymbol{r}t)$, one relies, in analogy to the stationary situation, on the stationarity principle. The stationarity condition (7.54) implies

$$A[n+\delta n] = A[n] + \mathcal{O}(\delta n^2)\,, \tag{7.98}$$

[7] Note the sign convention for $A_{\mathrm{xc}}[n]$, which differs from that of the total $A[n]$.

provided n is the density which makes $A[n]$ stationary, i.e. the functional Taylor expansion of $A[n]$ about the stationary point does not contain a first order contribution. If both n and $n+\delta n$ are noninteracting v-representable, the functional form (7.92) applies to both n and $n+\delta n$,

$$n(\boldsymbol{r}t)+\delta n(\boldsymbol{r}t)=\sum_{\sigma}\sum_{i=1}^{N}\Big\{|\phi_i(\boldsymbol{r}\sigma t)|^2+\phi_i^\dagger(\boldsymbol{r}\sigma t)\delta\phi_i(\boldsymbol{r}\sigma t)+\delta\phi_i^\dagger(\boldsymbol{r}\sigma t)\phi_i(\boldsymbol{r}\sigma t)\Big\}, \tag{7.99}$$

with the initial conditions

$$\delta n(\boldsymbol{r}t_0)=0\,; \qquad \delta\phi_i(\boldsymbol{r}\sigma t_0)=0\,. \tag{7.100}$$

Without (7.100) one would compare completely different systems. In the same fashion as for (3.36) one can then derive

$$B_\mathrm{s}[n+\delta n]-B_\mathrm{s}[n]=\int_{t_0}^{t_1}dt\int d^3r\,\delta n(\boldsymbol{r}t)\,v_\mathrm{s}(\boldsymbol{r}t)\,. \tag{7.101}$$

Insertion of (7.101) and the functional Taylor expansion of the remaining components of $A[n]$ into (7.98) leads to

$$0=\int_{t_0}^{t_1}dt\int d^3r\,\delta n(\boldsymbol{r}t)\Big\{v_\mathrm{s}(\boldsymbol{r}t)-v_\mathrm{ext}(\boldsymbol{r}t)-v_\mathrm{H}(\boldsymbol{r}t)-v_\mathrm{xc}(\boldsymbol{r}t)\Big\}, \tag{7.102}$$

with the definitions

$$v_\mathrm{H}[n](\boldsymbol{r}t)=\int d^3r'\,w(\boldsymbol{r},\boldsymbol{r}')\,n(\boldsymbol{r}'t) \tag{7.103}$$

$$v_\mathrm{xc}[n](\boldsymbol{r}t)=\frac{\delta A_\mathrm{xc}[n]}{\delta n(\boldsymbol{r}t)}\,. \tag{7.104}$$

Equation (7.102) must be valid for arbitrary variation $\delta n(\boldsymbol{r}t)$, so that the desired td KS potential has the form

$$v_\mathrm{s}(\boldsymbol{r}t)=v_\mathrm{ext}(\boldsymbol{r}t)+v_\mathrm{H}(\boldsymbol{r}t)+v_\mathrm{xc}(\boldsymbol{r}t)\,. \tag{7.105}$$

v_s depends on the density via (7.103) and (7.104). The td KS equations [498, 499] read

$$i\hbar\partial_t\phi_i(\boldsymbol{r}\sigma t)=\Big\{-\frac{\hbar^2\boldsymbol{\nabla}^2}{2m}+v_\mathrm{ext}(\boldsymbol{r}t)+v_\mathrm{H}(\boldsymbol{r}t)+v_\mathrm{xc}(\boldsymbol{r}t)\Big\}\phi_i(\boldsymbol{r}\sigma t)\,. \tag{7.106}$$

A number of remarks are worthwhile at this point:

- Equation (7.106) is an initial value problem in which the time-evolution of the orbitals is controlled by the td density. The time development of the orbitals is coupled back to the orbitals themselves, so that indirectly the complete history of the system enters the time-derivative of the orbitals at time t. In addition, the xc-

action functional is nonlocal in time. The same is then true for the xc-potential, so that also direct memory effects are contained in (7.106).
A second, different kind of memory effect is introduced into the xc-action functional by the initial state dependence of $|\Psi[n,\Psi_0](t)\rangle$.

- The KS equations have to be solved for given initial orbitals $\phi_i(\boldsymbol{r}\sigma t_0)$ which yield the initial density

$$n(\boldsymbol{r}t_0) = \langle\Phi(t_0)|\hat{n}(\boldsymbol{r})|\Phi(t_0)\rangle = \langle\Psi(t_0)|\hat{n}(\boldsymbol{r})|\Psi(t_0)\rangle\,. \tag{7.107}$$

  As already emphasized there exists more than one set of $\phi_i(\boldsymbol{r}\sigma t_0)$ which satisfies this condition. On the other hand, there is no criterion which of these sets one should choose. Fortunately, all sets yield the same td density, so that all physical quantities which do not explicitly depend on $|\Phi(t_0)\rangle$ are invariant against the choice of the initial orbitals. The initial Slater determinant is unique if the system is in its ground state for $t \le t_0$. This removes any ambiguity in the situation which is encountered most frequently.
- Use of the td KS equations amounts to a causal implementation of $\delta B_{\rm s}[n]/\delta n$, i.e. of the leading component of the total action functional.
- The crucial quantity of TDDFT is the xc-action functional, which absorbs all complicated many-body effects. Unfortunately, $A_{\rm xc}[n]$ is plagued by the same problems as $A[n]$: both retarded and advanced contributions are included. When implementing approximations for the xc-action and/or the xc-potential, causality has to be incorporated by hand.
- An extension of the td KS equations to time-dependent current density functional theory, i.e. to the Hamiltonian (7.5), has also been formulated [479]. In this scheme the td xc-potential depends on both n and the current density $\boldsymbol{j}_{\rm p}$, Eq. (7.24).

7.3 Exchange-Correlation Action: Adiabatic Local Density Approximation and Beyond

The simplest concept for setting up approximations for $A_{\rm xc}[n]$ is the use of xc-functionals of stationary DFT with a replacement of the stationary density by the actual td density [500, 501, 499, 502]. It can directly be illustrated for the LDA,

$$A_{\rm xc}^{\rm ALDA}[n] = \int_{t_0}^{t_1} dt \int d^3r\, e_{\rm xc}^{\rm HEG}(n(\boldsymbol{r}t))\,, \tag{7.108}$$

where $e_{\rm xc}^{\rm HEG}(n_0)$ is the xc-energy density of the (stationary) homogeneous electron gas at gas density n_0. The functional (7.108) is usually called adiabatic LDA (ALDA), as the use of a stationary functional with the td density is particularly appropriate for adiabatic processes, in which the system has sufficient time to adjust to the td external potential. The ALDA is sometimes also referred to as time-dependent LDA.

The corresponding ALDA potential $v_{xc}^{ALDA}(\boldsymbol{r}t)$ is completely determined by the density at time t,

$$v_{xc}^{ALDA}(\boldsymbol{r}t) = \frac{\delta A_{xc}^{ALDA}[n]}{\delta n(\boldsymbol{r}t)} = \left.\frac{de_{xc}^{HEG}(n)}{dn}\right|_{n=n(\boldsymbol{r}t)} = v_{xc}^{LDA}(n(\boldsymbol{r}t)). \tag{7.109}$$

The ALDA is local both in space and in time. It therefore shares the limitations resulting from locality in space with the stationary LDA. As an example one may mention the failure of the ALDA for long molecular chains [503]. In these systems the axial polarizabilities and hyper-polarizabilities are by far overestimated by the ALDA, due to the insensitivity of the ALDA potential in the center of the chain to a rearrangement of the densities at the end points. The ALDA contains no memory of the time-evolution of the system beyond the information that is contained in $n(\boldsymbol{r}t)$ at time t. v_{xc}^{ALDA} automatically satisfies the causality requirement, if $n(\boldsymbol{r}t)$ results from causal evolution.

In the same way one can base approximations on stationary GGA or spin-dependent LDA functionals. The exact exchange, depending on the td KS orbitals $\phi_i(\boldsymbol{r}\sigma t)$, can be used in a td situation as well. The calculation of $\delta A_{xc}[n]/\delta n$ has to be performed by a time-dependent extension of the optimized potential method [342, 488] in this case. However, the resulting xc-potentials $v_{xc}(\boldsymbol{r}t)$ all depend on either the density or the td KS orbitals at time t only.

The derivation of xc-action functionals with explicit memory, on the other hand, turns out to be rather difficult. A simple functional with memory was developed quite early within the framework of td linear response [473] (for a derivation see Sect. 7.4). It was found, however, that this functional violates a fundamental constraint on the nonlocality of $A_{xc}[n]$ with respect to time [504]. Consider a finite number of interacting electrons confined by a parabolic quantum well (rather than by the Coulomb potential of the nuclei). The ground state density of this system is denoted by $n_0(\boldsymbol{r})$. If now a uniform td electric field is switched on, the electrons perform a collective motion, in which the shape of the td density remains identical with that of $n_0(\boldsymbol{r})$, $n(\boldsymbol{r}t) = n_0(\boldsymbol{r} - \boldsymbol{R}(t))$. The electron density oscillates as a rigid object (this result is often referred to as *harmonic potential theorem* [504]—see also [505–507]). One can show that the exact td xc-potential is compatible with the harmonic potential theorem. Quite generally, the exact td xc-potential satisfies the relation [508]

$$v_{xc}[n'](\boldsymbol{r},t) = v_{xc}[n](\boldsymbol{r} - \boldsymbol{R}(t),t), \tag{7.110}$$

if the density n' is obtained from n by a rigid boost $\boldsymbol{R}(t)$,

$$n'(\boldsymbol{r},t) = n(\boldsymbol{r} - \boldsymbol{R}(t),t), \tag{7.111}$$

and the system was in its ground state at $t = 0$, with $\boldsymbol{R}(0) = d\boldsymbol{R}/dt(0) = \boldsymbol{0}$.

One can verify directly that the ALDA complies with Eq. (7.110). On the other hand, compatibility with (7.110) is impossible for any functional which is nonlocal in time, but local in space [509]. The combination of memory with Eq. (7.110) necessarily requires the xc-functional to be nonlocal in space, which quite naturally leads to a current dependence. A number of approximations for such a fully nonlocal xc-functional have been suggested—for further details see [509–514].

7.4 Time-Dependent Linear Response

Let us now consider the important case that a weak time-dependent perturbation is switched on at $t = t_0$, while the system is in its ground state for $t \leq t_0$. This is, for example, the situation encountered in photo-absorption processes. In the context of TDDFT this special case has the advantage that all functionals are pure density functionals: they do no longer depend on the initial state, as this state is uniquely determined by the initial density.

The external potential is then given by

$$v_{\text{ext}}(\boldsymbol{r}t) = \begin{cases} v_0(\boldsymbol{r}) & \text{for } t \leq t_0 \\ v_0(\boldsymbol{r}) + \delta v(\boldsymbol{r}t) & \text{for } t > t_0 \end{cases} . \tag{7.112}$$

The initial state at t_0 is the ground state $|\Psi_0\rangle$ with ground state density n_0,

$$n(\boldsymbol{r}t) = \begin{cases} n_0(\boldsymbol{r}) = \langle\Psi_0|\hat{n}(\boldsymbol{r})|\Psi_0\rangle & \text{for } t \leq t_0 \\ n_0(\boldsymbol{r}) + \delta n(\boldsymbol{r}t) & \text{for } t > t_0 \end{cases} . \tag{7.113}$$

n_0 is obtained from the stationary KS orbitals corresponding to v_0,

$$\left\{ -\frac{\hbar^2\boldsymbol{\nabla}^2}{2m} + v_0(\boldsymbol{r}) + v_{\text{H}}[n_0](\boldsymbol{r}) + v_{\text{xc}}[n_0](\boldsymbol{r}) \right\} \phi_i(\boldsymbol{r}\sigma) = \varepsilon_i \phi_i(\boldsymbol{r}\sigma), \tag{7.114}$$

in the standard form,

$$n_0(\boldsymbol{r}) = \sum_i \Theta_i \sum_\sigma |\phi_i(\boldsymbol{r}\sigma)|^2 . \tag{7.115}$$

For the examination of the linear (first order) response of the system to the perturbation δv one applies the linear response formalism to the induced density shift δn,

$$\delta n(\boldsymbol{r}t) = \delta n^{(1)}(\boldsymbol{r}t) + \mathscr{O}(\delta v^2) \tag{7.116}$$

$$\delta n^{(1)}(\boldsymbol{r}t) = \int d^3r' dt' \, \chi_{\text{R}}(\boldsymbol{r}, \boldsymbol{r}', t - t') \, \delta v(\boldsymbol{r}'t') \tag{7.117}$$

(for a derivation of Eq. (7.117) see e.g. Chap. 13 of [94]). Equation (7.117) represents the time-dependent extension of the density response of stationary systems,

Eq. (4.135). The quantity χ_R is the retarded response function of the unperturbed interacting system, Eq. (4.39),

$$\chi_R(\boldsymbol{r},\boldsymbol{r}',t-t') = -\frac{i}{\hbar}\Theta(t-t')\,\langle\Psi_0|[\hat{n}(\boldsymbol{r}t),\hat{n}(\boldsymbol{r}'t')]|\Psi_0\rangle\,. \tag{7.118}$$

This definition is equivalent to

$$\chi_R(\boldsymbol{r},\boldsymbol{r}',t-t') = \frac{\delta n(\boldsymbol{r}t)}{\delta v_{ext}(\boldsymbol{r}'t')}\,, \tag{7.119}$$

which, in turn, is equivalent to Eq. (7.117). The equivalence of (7.118) and (7.119) has already been apparent from Eqs. (7.79) and (7.72).

Alternatively, $\delta n^{(1)}$ can be evaluated within TDDFT. As the exact $n(\boldsymbol{r}t)$ is reproduced by the KS system, the same is true for its component $\delta n^{(1)}(\boldsymbol{r}t)$ to first order in δv. However, within the KS system $\delta n^{(1)}(\boldsymbol{r}t)$ is obtained as linear response to the total KS potential,

$$\delta n^{(1)}(\boldsymbol{r}t) = \int d^3r'dt'\,\chi_{R,s}(\boldsymbol{r},\boldsymbol{r}',t-t')\,\delta v_s(\boldsymbol{r}'t')\,, \tag{7.120}$$

where $\chi_{R,s}$ is the retarded KS response function,

$$\chi_{R,s}(\boldsymbol{r},\boldsymbol{r}',t-t') := -\frac{i}{\hbar}\Theta(t-t')\,\langle\Phi_0|[\hat{n}(\boldsymbol{r}t),\hat{n}(\boldsymbol{r}'t')]|\Phi_0\rangle\,, \tag{7.121}$$

and δv_s is the change of the total KS potential to first order in δv. The precise form of δv_s is obtained from Eqs. (7.103)–(7.105), using a functional Taylor expansion about the unperturbed density n_0,

$$\begin{aligned}
v_s(\boldsymbol{r}t) &= v_0(\boldsymbol{r}t) + \delta v(\boldsymbol{r}t) + \int d^3r'\,w(\boldsymbol{r},\boldsymbol{r}')\,\left[n_0(\boldsymbol{r}') + \delta n(\boldsymbol{r}'t)\right] \\
&\quad + \left.\frac{\delta A_{xc}[n]}{\delta n(\boldsymbol{r}t)}\right|_{n=n_0(\boldsymbol{r})+\delta n(\boldsymbol{r}t)} \\
&= v_0(\boldsymbol{r}t) + \delta v(\boldsymbol{r}t) + v_H[n_0](\boldsymbol{r}) + \int d^3r'\,w(\boldsymbol{r},\boldsymbol{r}')\,\delta n^{(1)}(\boldsymbol{r}'t) + v_{xc}[n_0](\boldsymbol{r}) \\
&\quad + \int d^3r' \int dt'\,\left.\frac{\delta^2 A_{xc}[n]}{\delta n(\boldsymbol{r}t)\delta n(\boldsymbol{r}'t')}\right|_{n=n_0(\boldsymbol{r})} \delta n^{(1)}(\boldsymbol{r}'t') + \mathcal{O}(\delta v^2)\,.
\end{aligned} \tag{7.122}$$

The kernel of the last term is usually abbreviated as

$$f_{xc}[n_0](\boldsymbol{r},\boldsymbol{r}',t-t') := \left.\frac{\delta^2 A_{xc}[n]}{\delta n(\boldsymbol{r}t)\delta n(\boldsymbol{r}'t')}\right|_{n=n_0(\boldsymbol{r})}\,. \tag{7.123}$$

Again projection on the causal component of $A_{xc}[n]$ is implied. With the help of (7.117) one identifies

$$\begin{aligned}\delta v_s(\boldsymbol{r}t) &= \delta v(\boldsymbol{r}t) + \int d^3r'\, w(\boldsymbol{r},\boldsymbol{r}')\,\delta n^{(1)}(\boldsymbol{r}'t) \\ &\quad + \int d^3r'dt'\, f_{\mathrm{xc}}[n_0](\boldsymbol{r},\boldsymbol{r}',t-t')\,\delta n^{(1)}(\boldsymbol{r}'t') \\ &= \int d^3r'dt' \Big\{ \delta^{(3)}(\boldsymbol{r}-\boldsymbol{r}')\delta(t-t') + \int d^3r''\, w(\boldsymbol{r},\boldsymbol{r}'')\,\chi_{\mathrm{R}}(\boldsymbol{r}'',\boldsymbol{r}',t-t') \\ &\qquad + \int d^3r''dt''\, f_{\mathrm{xc}}[n_0](\boldsymbol{r},\boldsymbol{r}'',t-t'')\,\chi_{\mathrm{R}}(\boldsymbol{r}'',\boldsymbol{r}',t''-t') \Big\} \\ &\qquad \times \delta v(\boldsymbol{r}'t') . \end{aligned} \tag{7.124}$$

If one equates (7.117) and (7.120), one obtains a Dyson-type equation, relating the full response with the KS response function,

$$\begin{aligned}\chi_{\mathrm{R}}(\boldsymbol{r},\boldsymbol{r}',t-t') &= \chi_{\mathrm{R,s}}(\boldsymbol{r},\boldsymbol{r}',t-t') + \int d^3r''dt'' \int d^3r'''dt'''\, \chi_{\mathrm{R,s}}(\boldsymbol{r},\boldsymbol{r}'',t-t'') \\ &\quad \times \Big\{ \delta(t''-t''')w(\boldsymbol{r}'',\boldsymbol{r}''') + f_{\mathrm{xc}}[n_0](\boldsymbol{r}'',\boldsymbol{r}''',t''-t''') \Big\} \\ &\quad \times \chi_{\mathrm{R}}(\boldsymbol{r}''',\boldsymbol{r}',t'''-t') . \end{aligned} \tag{7.125}$$

This relation allows the evaluation of an approximation for the full response function from the much simpler KS response function for any given approximation for f_{xc}. Even if the ALDA is employed for f_{xc},

$$\begin{aligned} f_{\mathrm{xc}}^{\mathrm{ALDA}}[n_0](\boldsymbol{r},\boldsymbol{r}',t-t') &= \left.\frac{\delta v_{\mathrm{xc}}^{\mathrm{ALDA}}[n](\boldsymbol{r}t)}{\delta n(\boldsymbol{r}'t')}\right|_{n=n_0(\boldsymbol{r})} \\ &= \left.\frac{d^2 e_{\mathrm{xc}}^{\mathrm{HEG}}(n)}{dn^2}\right|_{n=n_0(\boldsymbol{r})} \delta^{(3)}(\boldsymbol{r}-\boldsymbol{r}')\delta(t-t') , \end{aligned} \tag{7.126}$$

the resulting χ_{R} goes beyond the random phase approximation, which corresponds to $f_{\mathrm{xc}} = 0$. However, within the ALDA the xc-interaction, represented by f_{xc}, is instantaneous.

For the derivation of more refined approximations for f_{xc} it is advantageous to formulate Eq. (7.125) in frequency space,

$$\begin{aligned}\chi_{\mathrm{R}}(\boldsymbol{r},\boldsymbol{r}',\omega) &= \chi_{\mathrm{R,s}}(\boldsymbol{r},\boldsymbol{r}',\omega) + \int d^3r'' \int d^3r'''\, \chi_{\mathrm{R,s}}(\boldsymbol{r},\boldsymbol{r}'',\omega) \\ &\quad \times \Big\{ w(\boldsymbol{r}'',\boldsymbol{r}''') + f_{\mathrm{xc}}[n_0](\boldsymbol{r}'',\boldsymbol{r}''',\omega) \Big\} \chi_{\mathrm{R}}(\boldsymbol{r}''',\boldsymbol{r}',\omega) . \end{aligned} \tag{7.127}$$

The Fourier transforms follow the convention (4.41). In frequency space the instantaneous character of the ALDA for f_{xc} leads to a function which does not depend on the frequency

$$f_{\mathrm{xc}}^{\mathrm{ALDA}}[n_0](\boldsymbol{r},\boldsymbol{r}',\omega) = \left.\frac{d^2 e_{\mathrm{xc}}^{\mathrm{HEG}}(n)}{dn^2}\right|_{n=n_0(\boldsymbol{r})} \delta^{(3)}(\boldsymbol{r}-\boldsymbol{r}') . \tag{7.128}$$

Equation (7.127) reminds one of the Dyson equation for the time-ordered response (4.35), which relates the full response function with its irreducible (proper) kernel Π [94, 95],

$$\chi(\boldsymbol{r},\boldsymbol{r}',\omega) = \Pi(\boldsymbol{r},\boldsymbol{r}',\omega) + \int d^3r'' \int d^3r''' \,\Pi(\boldsymbol{r},\boldsymbol{r}'',\omega)w(\boldsymbol{r}'',\boldsymbol{r}''')\chi(\boldsymbol{r}''',\boldsymbol{r}',\omega) \tag{7.129}$$

(compare the discussion of Eq. (4.92)). The relation between (7.127) and (7.129) can be exploited for the construction of a simple frequency-dependent extension of the ALDA for f_{xc}. To this aim one restricts both equations to the homogeneous situation, i.e. to the homogeneous electron gas, for which a Fourier representation in momentum space leads to algebraic equations,

$$\begin{aligned}
\chi_{R}(\boldsymbol{q},\omega) &= \chi_{R,s}(\boldsymbol{q},\omega) + \chi_{R,s}(\boldsymbol{q},\omega)\left[w(\boldsymbol{q}) + f_{xc}[n_0](\boldsymbol{q},\omega)\right]\chi_{R}(\boldsymbol{q},\omega) \\
\chi(\boldsymbol{q},\omega) &= \Pi(\boldsymbol{q},\omega) + \Pi(\boldsymbol{q},\omega)w(\boldsymbol{q})\chi(\boldsymbol{q},\omega) \\
&= \Pi^{(0)}(\boldsymbol{q},\omega) + \Pi^{(0)}(\boldsymbol{q},\omega)\left[w(\boldsymbol{q}) + \frac{1}{\Pi^{(0)}(\boldsymbol{q},\omega)} - \frac{1}{\Pi(\boldsymbol{q},\omega)}\right]\chi(\boldsymbol{q},\omega) .
\end{aligned}$$

The connection between (7.127) and (7.129) is fully established, if one uses the relation between the retarded and the time-ordered response function, which can be derived on the basis of the Lehmann representation, Eq. (4.45),

$$f_{xc}[n_0](\boldsymbol{q},\omega) = \begin{cases} \dfrac{1}{\Pi^{(0)}(\boldsymbol{q},\omega)} - \dfrac{1}{\Pi(\boldsymbol{q},\omega)} & \text{for } \omega \geq 0 \\[2ex] \dfrac{1}{\Pi^{(0)}(\boldsymbol{q},\omega)^*} - \dfrac{1}{\Pi(\boldsymbol{q},\omega)^*} & \text{for } \omega \leq 0 \end{cases} \tag{7.130}$$

($\Pi^{(0)} = \chi^{(0)} = \chi_s$ for the HEG). Any approximation for Π defines, via (7.130), an approximation for f_{xc}. Note that causality has now been implemented in f_{xc} by treating f_{xc} as an integral part of a causal equation.

On the basis of the compressibility sum rule (4.154) the ALDA can be characterized in an alternative fashion. To this aim one extracts the xc-contribution from (4.154),

$$\frac{1}{\Pi(\boldsymbol{q},\omega)} - \frac{1}{\Pi^{(0)}(\boldsymbol{q},\omega)} = -\left.\frac{d^2 e_{xc}^{HEG}(n)}{dn^2}\right|_{n=n_0} , \tag{7.131}$$

and verifies that the ALDA, Eq. (7.128), corresponds to the approximation

$$f_{xc}^{ALDA}[n_0](\boldsymbol{q},\omega) = \frac{1}{\Pi^{(0)}(\boldsymbol{q}=\boldsymbol{0},\omega=0)} - \frac{1}{\Pi(\boldsymbol{q}=\boldsymbol{0},\omega=0)} . \tag{7.132}$$

The simplest extension of the ALDA is a smooth interpolation between the static ($\omega = 0$) and the high-frequency limit of Π. This is the concept behind the Gross-Kohn (GK) approximation [473]

$$f_{\rm xc}^{\rm GK}[n_0](\boldsymbol{q},\omega) = \frac{1}{\Pi^{(0)}(\boldsymbol{q}=\boldsymbol{0},\omega)} - \frac{1}{\Pi(\boldsymbol{q}=\boldsymbol{0},\omega)} \tag{7.133}$$

$$f_{\rm xc}^{\rm GK}[n_0](\boldsymbol{r},\boldsymbol{r}',\omega) = \left[\frac{1}{\Pi^{(0)}(\boldsymbol{q}=\boldsymbol{0},\omega)} - \frac{1}{\Pi(\boldsymbol{q}=\boldsymbol{0},\omega)}\right]_{n=n_0(\boldsymbol{r})} \delta^{(3)}(\boldsymbol{r}-\boldsymbol{r}').$$

In real space the frequency-dependence of $f_{\rm xc}^{\rm GK}$ turns into nonlocality in time. The GK approximation thus is the simplest xc-functional with memory. It turns out, however, that for intermediate and particularly for high densities the frequency-dependence of $f_{\rm xc}^{\rm GK}$ is rather weak. This explains the somewhat surprising success of the ALDA in the description of photo-absorption processes for atoms. On the other hand, it also implies that the memory effects in the GK functional are only weak. In addition, as indicated earlier, the GK approximation does not satisfy the exact relation (7.110).

7.5 Spin-Polarized Time-Dependent Density Functional Theory

In general, spin-polarized systems are characterized by an external potential and an external magnetic field which are both simultaneously spin- and time-dependent. Restricting the discussion to the case of a magnetic field with only a z-component, $B_{\rm ext}(\boldsymbol{r}t)$, one can express the external fields in terms of spin-up and -down potentials,

$$v_{\rm ext}^{\sigma}(\boldsymbol{r}t) = v_{\rm ext}(\boldsymbol{r}t) - {\rm sign}(\sigma)\mu_B B_{\rm ext}(\boldsymbol{r}t). \tag{7.134}$$

In this limit the spin-densities $n_\sigma(\boldsymbol{r}t)$ are the fundamental quantities of TDDFT, so that all components of the action become functionals of $n_\sigma(\boldsymbol{r}t)$ [515]. No new concepts are involved in the derivation of the formalism, so that its details can be suppressed. Only a brief list of the most important results is given. We start with the corresponding spin-dependent TDKS equations, [515],

$$i\hbar\partial_t\phi_i(\boldsymbol{r}\sigma t) = \left\{-\frac{\hbar^2\boldsymbol{\nabla}^2}{2m} + v_{\rm ext}^{\sigma}(\boldsymbol{r}t) + v_{\rm H}(\boldsymbol{r}t) + v_{\rm xc}^{\sigma}(\boldsymbol{r}t)\right\}\phi_i(\boldsymbol{r}\sigma t) \tag{7.135}$$

$$v_{\rm xc}^{\sigma}(\boldsymbol{r}t) = \frac{\delta A_{\rm xc}[n_\uparrow,n_\downarrow]}{\delta n_\sigma(\boldsymbol{r}t)} \tag{7.136}$$

$$n_\sigma(\boldsymbol{r}t) = \sum_{i=1}^{N} |\phi_i(\boldsymbol{r}\sigma t)|^2. \tag{7.137}$$

The spin-dependent ALDA is based on the corresponding stationary LSDA functional,

$$A_{\rm xc}^{\rm ALDA}[n_\uparrow,n_\downarrow] = \int_{t_0}^{t_1} dt \int d^3r\, e_{\rm xc}^{\rm HEG}(n_\uparrow(\boldsymbol{r}t),n_\downarrow(\boldsymbol{r}t)) \tag{7.138}$$

$$\frac{\delta A_{\text{xc}}^{\text{ALDA}}[n_\uparrow,n_\downarrow]}{\delta n_\sigma(\boldsymbol{r}t)} = \left.\frac{de_{\text{xc}}^{\text{HEG}}(n_\uparrow,n_\downarrow)}{dn_\sigma}\right|_{n_\sigma=n_\sigma(\boldsymbol{r}t)} . \tag{7.139}$$

The linear response approach [516] assumes a potential of the form

$$v_{\text{ext}}^{\sigma}(\boldsymbol{r}t) = \begin{cases} v_0^{\sigma}(\boldsymbol{r}) & \text{for } t \le t_0 \\ v_0^{\sigma}(\boldsymbol{r}) + \delta v^{\sigma}(\boldsymbol{r}t) & \text{for } t > t_0 \end{cases} , \tag{7.140}$$

with associated densities

$$n_\sigma(\boldsymbol{r}t) = \begin{cases} n_{\sigma,0}(\boldsymbol{r}) = \langle\Psi_0|\hat{n}_\sigma(\boldsymbol{r})|\Psi_0\rangle & \text{for } t \le t_0 \\ n_{\sigma,0}(\boldsymbol{r}) + \delta n_\sigma(\boldsymbol{r}t) & \text{for } t > t_0 \end{cases} . \tag{7.141}$$

The resulting spin-dependent Dyson equation [516] then reads

$$\begin{aligned} \chi_{\text{R}}(\boldsymbol{r}\sigma,\boldsymbol{r}'\sigma',\omega) = \chi_{\text{R,s}}(\boldsymbol{r}\sigma,\boldsymbol{r}'\sigma',\omega) + \sum_{\sigma''\sigma'''} \int d^3r''\, d^3r'''\, \chi_{\text{R,s}}(\boldsymbol{r}\sigma,\boldsymbol{r}''\sigma'',\omega) \\ \times \left\{ w(\boldsymbol{r}'',\boldsymbol{r}''') + f_{\text{xc}}(\boldsymbol{r}''\sigma'',\boldsymbol{r}'''\sigma''',\omega) \right\} \\ \times \chi_{\text{R}}(\boldsymbol{r}'''\sigma''',\boldsymbol{r}'\sigma',\omega) , \end{aligned} \tag{7.142}$$

with the retarded spin–spin response functions

$$\chi_{\text{R}}(\boldsymbol{r}\sigma,\boldsymbol{r}'\sigma',t-t') := -\frac{i}{\hbar}\Theta(t-t')\,\langle\Psi_0|\left[\hat{n}_\sigma(\boldsymbol{r}t),\hat{n}_{\sigma'}(\boldsymbol{r}'t')\right]|\Psi_0\rangle \tag{7.143}$$

$$\chi_{\text{R,s}}(\boldsymbol{r}\sigma,\boldsymbol{r}'\sigma',t-t') := -\frac{i}{\hbar}\Theta(t-t')\,\langle\Phi_0|\left[\hat{n}_\sigma(\boldsymbol{r}t),\hat{n}_{\sigma'}(\boldsymbol{r}'t')\right]|\Phi_0\rangle \tag{7.144}$$

and the spin-dependent xc-kernel

$$f_{\text{xc}}(\boldsymbol{r}\sigma,\boldsymbol{r}'\sigma',t-t') := \left.\frac{\delta^2 A_{\text{xc}}[n_\uparrow,n_\downarrow]}{\delta n_\sigma(\boldsymbol{r}t)\,\delta n_{\sigma'}(\boldsymbol{r}'t')}\right|_{n_\sigma=n_{\sigma,0}(\boldsymbol{r})} . \tag{7.145}$$

7.6 Excited States: Part II

Why should one be interested in the solution of the Dyson equations (7.127) or (7.142) for some approximate f_{xc}? The answer is provided by a look at the Lehmann representation of the full response function. Straightforward extension of the result (4.44) for the complete response function to the spin-resolved response gives[8]

[8] For the basic properties of the retarded response function the reader is referred to the discussion in Sect. 4.2.1.

$$\chi_{\mathrm{R}}(\boldsymbol{r}\sigma,\boldsymbol{r}'\sigma',\omega) = \sum_{n\neq 0} \frac{\langle\Psi_0|\hat{n}_\sigma(\boldsymbol{r})|\Psi_n\rangle\langle\Psi_n|\hat{n}_{\sigma'}(\boldsymbol{r}')|\Psi_0\rangle}{\hbar\omega-\Omega_n+i\eta} - \sum_{n\neq 0} \frac{\langle\Psi_0|\hat{n}_{\sigma'}(\boldsymbol{r}')|\Psi_n\rangle\langle\Psi_n|\hat{n}_\sigma(\boldsymbol{r})|\Psi_0\rangle}{\hbar\omega+\Omega_n+i\eta}, \tag{7.146}$$

where Ω_n is the excitation energy of the N-particle state $|\Psi_n\rangle$, Eq. (4.42),

$$\Omega_n := E_n - E_0 \tag{7.147}$$

(the ground state is assumed to be non-degenerate, as usual). All excitation energies of the stationary interacting system appear as poles of χ_{R} in the complex frequency-plane. Consequently, one way to calculate the excitation energies is to identify the poles of χ_{R}. For this reason the solution of the Dyson equation (7.142) is not only of interest for the description of time-dependent phenomena, but also as a vehicle for the discussion of excited states of stationary systems.

The obvious question that arises in this context concerns the relation between the exact excitation energies Ω_n and the KS excitation energies

$$\omega_{\beta\alpha\sigma} := \varepsilon_{\beta\sigma} - \varepsilon_{\alpha\sigma}\,, \tag{7.148}$$

which show up in the Lehmann representation of the spin-resolved KS response function (7.144),

$$\chi_{\mathrm{R,s}}(\boldsymbol{r}\sigma,\boldsymbol{r}'\sigma',\omega) = \delta_{\sigma,\sigma'}\sum_{\alpha\beta}(\Theta_{\alpha\sigma}-\Theta_{\beta\sigma})\frac{\phi^*_{\alpha\sigma}(\boldsymbol{r})\phi_{\beta\sigma}(\boldsymbol{r})\phi^*_{\beta\sigma}(\boldsymbol{r}')\phi_{\alpha\sigma}(\boldsymbol{r}')}{\hbar\omega-\omega_{\beta\alpha\sigma}+i\eta} \tag{7.149}$$

(the detailed form of the spin-dependent KS orbitals, Eq. (3.93), has been used to arrive at this result). Note that only eigenvalue differences between excited single-particle states (not occupied in the KS ground state) and the holes left in the ground state determinant are relevant in (7.149), as

$$\Theta_{\alpha\sigma} - \Theta_{\beta\sigma} = \Theta_{\alpha\sigma}(1-\Theta_{\beta\sigma}) - \Theta_{\beta\sigma}(1-\Theta_{\alpha\sigma})\,.$$

If one compares (7.149) with (7.146) two points come to mind: the first concerns the nature of the excitations. Only one particular class of excitations manifests itself as poles in the noninteracting response, i.e. the single-particle excitations. In the noninteracting case double excitations require more than one interaction of the particles with the perturbing potential and therefore do not show up in the linear response function. One has to go to nonlinear (quadratic) response to generate such excitations in a noninteracting system (for an attempt to utilize the linear KS response for the description of double excitations see [517–519]). The second point is: how can the complete spectrum of excitations present in (7.146) be generated by the right-hand side of the Dyson equation (7.142), if the first term in this expression has poles at quite different values of ω? And how can the poles at ω_{kl} be suppressed to reproduce the fact that the full function χ_{R} is finite at these values of ω?

In order to answer these questions [520, 515, 516] one considers the Dyson equation (7.142) for frequencies in the vicinity of one of the true excitation energies, i.e. in the limit $\hbar\omega \to \Omega_n$. More precisely, one performs a Laurent expansion of both sides of this equation about $\hbar\omega = \Omega_n$ and then compares the residues of the pole obtained on both sides. As long as there is no KS excitation energy $\omega_{\beta\alpha\sigma}$ which coincides with Ω_n, the KS response function is finite at $\hbar\omega = \Omega_n$, so that only the second term on the right-hand side of Eq. (7.142) contributes to the residue. Since all $\Omega_n > 0$, the second sum of the Lehmann representation of the full response function, Eq. (7.146), is also finite at this pole. Thus, if the excited state $|\Psi_n\rangle$ is nondegenerate, only a single term of the first sum over all states in (7.146) determines the residue on both sides. After elimination of the common factor of $\langle\Psi_n|\hat{n}_{\sigma'}(\boldsymbol{r}')|\Psi_0\rangle$ from the residue one obtains

$$\begin{aligned}\langle\Psi_0|\hat{n}_\sigma(\boldsymbol{r})|\Psi_n\rangle &= \sum_{\sigma'''}\int d^3r''d^3r''' \sum_{\alpha\beta}(\Theta_{\alpha\sigma}-\Theta_{\beta\sigma})\frac{\rho_{\alpha\beta\sigma}(\boldsymbol{r})\rho_{\beta\alpha\sigma}(\boldsymbol{r}'')}{\Omega_n-\omega_{\beta\alpha\sigma}+i\eta} \\ &\quad\times\Big[w(\boldsymbol{r}'',\boldsymbol{r}''')+f_{\rm xc}(\boldsymbol{r}''\sigma,\boldsymbol{r}'''\sigma''',\Omega_n/\hbar)\Big] \\ &\quad\times\langle\Psi_0|\hat{n}_{\sigma'''}(\boldsymbol{r}''')|\Psi_n\rangle\,, \end{aligned} \tag{7.150}$$

where the abbreviation

$$\rho_{\alpha\beta\sigma}(\boldsymbol{r}) := \phi^*_{\alpha\sigma}(\boldsymbol{r})\phi_{\beta\sigma}(\boldsymbol{r}) \qquad \rho^*_{\alpha\beta\sigma}(\boldsymbol{r}) = \rho_{\beta\alpha\sigma}(\boldsymbol{r}) \tag{7.151}$$

has been introduced for the Lehmann amplitudes of the KS response function. The pole shift η can now be dropped, as the validity of Eq. (7.150) rests on the assumption that $\Omega_n \neq \omega_{\beta\alpha\sigma}$ for arbitrary occupied α and unoccupied β. Next one defines the amplitudes

$$\begin{aligned}\xi_{n\alpha\beta\sigma} &:= \frac{1}{\Omega_n-\omega_{\beta\alpha\sigma}}\sum_{\sigma'}\int d^3r\int d^3r'\,\rho_{\beta\alpha\sigma}(\boldsymbol{r}) \\ &\quad\times\Big[w(\boldsymbol{r},\boldsymbol{r}')+f_{\rm xc}(\boldsymbol{r}\sigma,\boldsymbol{r}'\sigma',\Omega_n/\hbar)\Big]\langle\Psi_0|\hat{n}_{\sigma'}(\boldsymbol{r}')|\Psi_n\rangle\,,\end{aligned} \tag{7.152}$$

to rewrite Eq. (7.150) as

$$\langle\Psi_0|\hat{n}_\sigma(\boldsymbol{r})|\Psi_n\rangle = \sum_{\alpha\beta}(\Theta_{\alpha\sigma}-\Theta_{\beta\sigma})\,\rho_{\alpha\beta\sigma}(\boldsymbol{r})\,\xi_{n\alpha\beta\sigma}\,. \tag{7.153}$$

Insertion of (7.153) into (7.152) then leads to [516]

$$\big(\Omega_n-\omega_{\beta\alpha\sigma}\big)\,\xi_{n\alpha\beta\sigma} = \sum_{\mu\nu\sigma'} K_{\alpha\beta\sigma,\mu\nu\sigma'}(\Omega_n/\hbar)(\Theta_{\mu\sigma'}-\Theta_{\nu\sigma'})\,\xi_{n\mu\nu\sigma'}\,, \tag{7.154}$$

where the matrix element

$$K_{\alpha\beta\sigma,\mu\nu\sigma'}(\omega) := \int d^3r\int d^3r'\,\rho_{\beta\alpha\sigma}(\boldsymbol{r})\Big[w(\boldsymbol{r},\boldsymbol{r}')+f_{\rm xc}(\boldsymbol{r}\sigma,\boldsymbol{r}'\sigma',\omega)\Big]\rho_{\mu\nu\sigma'}(\boldsymbol{r}') \tag{7.155}$$

has been introduced. At this point it is important to realize that only the $\xi_{n\alpha\beta\sigma}$ for which either α denotes an occupied and β an unoccupied KS state or vice versa contribute to the actual Lehmann amplitudes (7.153). All other amplitudes, though properly defined via Eq. (7.152), are irrelevant for the excitation problem. Accordingly, the single-particle indices in the set of linear equations (7.154) only include all particle–hole pairs $\alpha\beta$ and $\mu\nu$, but no particle–particle or hole–hole pairs.

Equation (7.153) can be loosely interpreted as a "basis set expansion" of the Lehmann amplitudes in terms of the KS particle–hole amplitudes $\rho_{\beta\alpha\sigma}(\boldsymbol{r})$. One should note, however, the dimensionalities involved. If one uses an N-particle Hilbert space spanned by the determinants generated by M linearly independent single-particle basis functions, one finds $\binom{M}{N}$ many-body states $|\Psi_n\rangle$ and thus $\binom{M}{N}$ excitation energies Ω_n (ignoring any symmetries). On the other hand, the number of KS particle–hole amplitudes is only $N(M-N)$. Consequently, Eq. (7.153) implies that only $N(M-N)$ Lehmann amplitudes vary independently (ignoring complex conjugation). In other words: although the interacting response function contains contributions from all $\binom{M}{N}$ states of the Hilbert space, effectively only $N(M-N)$ independent excitations are described.

So far, the excited state $|\Psi_n\rangle$ has been assumed to be non-degenerate. However, an equivalent result can also be derived if there are several states degenerate with $|\Psi_n\rangle$. In this situation the Lehmann amplitudes of all degenerate states contribute to the residue, so that a superposition of the corresponding factors $\langle\Psi_0|\hat{n}_\sigma(\boldsymbol{r})|\Psi_n\rangle\langle\Psi_n|\hat{n}_{\sigma'}(\boldsymbol{r}')|\Psi_0\rangle$ is obtained both on the left- and on the right-hand side of Eq. (7.150). Nevertheless, if the coefficients $\xi_{n\alpha\beta\sigma}$ are redefined accordingly,

$$\begin{aligned}\xi_{n\alpha\beta\sigma}(\boldsymbol{r}''\sigma'') := \frac{1}{\Omega_n - \omega_{\beta\alpha\sigma}} \sum_{\sigma'} \int d^3r \int d^3r' \rho_{\beta\alpha\sigma}(\boldsymbol{r}) \\ \times \Big[w(\boldsymbol{r},\boldsymbol{r}') + f_{\rm xc}(\boldsymbol{r}\sigma,\boldsymbol{r}'\sigma',\Omega_n/\hbar)\Big] \\ \times \sum_{deg.n} \langle\Psi_0|\hat{n}_{\sigma'}(\boldsymbol{r}')|\Psi_n\rangle\langle\Psi_n|\hat{n}_{\sigma''}(\boldsymbol{r}'')|\Psi_0\rangle\,, \end{aligned}\qquad (7.156)$$

Eq. (7.154) emerges again.

Equation (7.154) can be recast in a slightly different form, in order to show that one is dealing with a pseudo-eigenvalue problem [520, 515, 516],

$$\begin{aligned}\sum_{\mu\nu\sigma'} \big[\omega_{\beta\alpha\sigma}\,\delta_{\alpha\mu}\,\delta_{\beta\nu}\,\delta_{\sigma\sigma'} + K_{\alpha\beta\sigma,\mu\nu\sigma'}(\Omega_n/\hbar)(\Theta_{\mu\sigma'}-\Theta_{\nu\sigma'})\big]\,\xi_{n\mu\nu\sigma'} \\ = \Omega_n\,\xi_{n\alpha\beta\sigma}\,. \end{aligned}\qquad (7.157)$$

The true excitation energies are therefore given by the zeros of the characteristic polynomial,

$$\det\big[(\omega_{\beta\alpha\sigma}-\Omega_n)\,\delta_{\alpha\mu}\,\delta_{\beta\nu}\,\delta_{\sigma\sigma'} + K_{\alpha\beta\sigma,\mu\nu\sigma'}(\Omega_n/\hbar)(\Theta_{\mu\sigma'}-\Theta_{\nu\sigma'})\big] = 0\,. \qquad (7.158)$$

As in Eq. (7.154), the matrix indices are understood to run over all particle–hole pairs $\alpha\beta$ and $\mu\nu$. Unlike ordinary eigenvalue problems, however, Eq. (7.157) has more solutions for Ω_n than linearly independent "eigenvectors" $(\xi_{n\alpha\beta\sigma})$, as long as the matrix $(K_{\alpha\beta\sigma,\mu\nu\sigma'})$ depends on Ω_n.

An alternative, more symmetric pseudo-eigenvalue equation for the true excitation energies can be derived, if the KS orbitals are chosen to be real. One starts with the transformation $\alpha \leftrightarrow \beta$ in half of the right-hand side of Eq. (7.150) and uses the symmetry $\rho_{\alpha\beta\sigma}(\boldsymbol{r}) = \rho_{\beta\alpha\sigma}(\boldsymbol{r})$ as well as $\omega_{\alpha\beta\sigma} = -\omega_{\beta\alpha\sigma}$ to express Eq. (7.150) as

$$\begin{aligned}\langle\Psi_0|\hat{n}_\sigma(\boldsymbol{r})|\Psi_n\rangle = \sum_{\sigma'''}\int d^3r''d^3r''' \sum_{\alpha\beta}(\Theta_{\alpha\sigma}-\Theta_{\beta\sigma})\,\omega_{\beta\alpha\sigma}\frac{\rho_{\alpha\beta\sigma}(\boldsymbol{r})\rho_{\beta\alpha\sigma}(\boldsymbol{r}'')}{\Omega_n^2-\omega_{\beta\alpha\sigma}^2}\\ \times\Big[w(\boldsymbol{r}'',\boldsymbol{r}''')+f_{\mathrm{xc}}(\boldsymbol{r}''\sigma,\boldsymbol{r}'''\sigma''',\Omega_n/\hbar)\Big]\\ \times\langle\Psi_0|\hat{n}_{\sigma'''}(\boldsymbol{r}''')|\Psi_n\rangle\,.\end{aligned} \tag{7.159}$$

In this case the amplitudes are defined as

$$\begin{aligned}\overline{\xi}_{n\alpha\beta\sigma} :=& \frac{\big[(\Theta_{\alpha\sigma}-\Theta_{\beta\sigma})\omega_{\beta\alpha\sigma}\big]^{1/2}}{\Omega_n^2-\omega_{\beta\alpha\sigma}^2}\sum_{\sigma'}\int d^3r\int d^3r'\,\rho_{\beta\alpha\sigma}(\boldsymbol{r})\\ &\times\Big[w(\boldsymbol{r},\boldsymbol{r}')+f_{\mathrm{xc}}(\boldsymbol{r}\sigma,\boldsymbol{r}'\sigma',\Omega_n/\hbar)\Big]\langle\Psi_0|\hat{n}_{\sigma'}(\boldsymbol{r}')|\Psi_n\rangle \qquad (7.160)\\ =& \frac{\big[(\Theta_{\alpha\sigma}-\Theta_{\beta\sigma})\omega_{\beta\alpha\sigma}\big]^{1/2}}{\Omega_n+\omega_{\beta\alpha\sigma}}\xi_{n\alpha\beta\sigma}\,.\end{aligned}$$

Insertion into Eq. (7.159) yields

$$\langle\Psi_0|\hat{n}_\sigma(\boldsymbol{r})|\Psi_n\rangle = \sum_{\alpha\beta}\big[(\Theta_{\alpha\sigma}-\Theta_{\beta\sigma})\omega_{\beta\alpha\sigma}\big]^{1/2}\rho_{\alpha\beta\sigma}(\boldsymbol{r})\,\overline{\xi}_{n\alpha\beta\sigma}\,. \tag{7.161}$$

The result (7.161) can now be used to eliminate the Lehmann amplitude on the right-hand side of the definition (7.160). With the kernel

$$\begin{aligned}S_{\alpha\beta\sigma,\mu\nu\sigma'}(\omega) := \omega_{\beta\alpha\sigma}^2\,\delta_{\alpha\mu}\,\delta_{\beta\nu}\,\delta_{\sigma\sigma'}+\big[(\Theta_{\alpha\sigma}-\Theta_{\beta\sigma})\omega_{\beta\alpha\sigma}\big]^{1/2}\\ \times K_{\alpha\beta\sigma,\mu\nu\sigma'}(\omega)\big[(\Theta_{\mu\sigma'}-\Theta_{\nu\sigma'})\omega_{\nu\mu\sigma'}\big]^{1/2},\end{aligned} \tag{7.162}$$

Eq. (7.160) then becomes Casida's equation [521, 522],

$$\Omega_n^2\,\overline{\xi}_{n\alpha\beta\sigma} = \sum_{\mu\nu\sigma'} S_{\alpha\beta\sigma,\mu\nu\sigma'}(\Omega_n/\hbar)\,\overline{\xi}_{n\mu\nu\sigma'}\,. \tag{7.163}$$

Once again, the matrix indices are understood to run over all particle–hole pairs $\alpha\beta$ and $\mu\nu$. Compared with the initial pseudo-eigenvalue problem (7.157), Eq. (7.163) has the advantage that the coefficient matrix $(S_{\alpha\beta\sigma,\mu\nu\sigma'})$ is hermitian.[9] As in the

[9] Without any assumption on the causal structure of f_{xc} one easily verifies that

$$f_{\mathrm{xc}}(\boldsymbol{r}\sigma,\boldsymbol{r}'\sigma',\omega) = f_{\mathrm{xc}}^*(\boldsymbol{r}\sigma,\boldsymbol{r}'\sigma',-\omega)\,.$$

case of Eq. (7.157) there must exist more solutions Ω_n than there are linearly independent "eigenvectors" $(\xi_{n\alpha\beta\sigma})$. However, if $f_{\rm xc}$ is approximated by a frequency-independent expression (as in the ALDA), Eq. (7.163) becomes a regular hermitian eigenvalue problem, so that the number of eigenvalues can not be larger than the number of eigenvectors. In this approximation only a subset of the true excitation energies is accessible.

The solution of the pseudo-eigenvalue equations (7.157) or (7.163) not only determines the true excitation energies of the interacting system, it also provides information about the Lehmann amplitudes via Eqs. (7.153) or (7.161). One of the central quantities of interest is the dynamic polarizability, which determines the dipole moment $\boldsymbol{\mu}(t)$ induced by a homogeneous electric field[10] $\boldsymbol{E}(t) = \boldsymbol{E}_0 e^{-i\omega_0 t}$,

$$\mu_k(t) = -e\int d^3r\, r_k\, \delta n(\boldsymbol{r}t) \tag{7.164}$$

$$= e^2 \sum_{l=1}^{3} \alpha_{kl}(\omega_0) E_{0,l} e^{-i\omega_0 t}\,. \tag{7.165}$$

The relation between α_{kl} and the density–density response function is easily established after identification of the perturbing potential which corresponds to the homogeneous electric field,

$$\begin{aligned} \delta v(\boldsymbol{r}t) &= e\boldsymbol{r}\cdot\boldsymbol{E}_0\, e^{-i\omega_0 t} \\ \Longrightarrow\quad \delta v(\boldsymbol{r}\omega) &= e\boldsymbol{r}\cdot\boldsymbol{E}_0\, 2\pi\,\delta(\omega-\omega_0)\,, \end{aligned}$$

and use of the linear response relation (7.117). Similar to the KS polarizability (6.151), the exact dynamic polarizability is given in terms of the full retarded response function as

$$\alpha_{kl}(\omega) = -\sum_{\sigma}\int d^3r \sum_{\sigma'}\int d^3r'\, r_k r'_l\, \chi_{\rm R}(\boldsymbol{r}\sigma,\boldsymbol{r}'\sigma',\omega)\,. \tag{7.166}$$

Use of the Lehmann representation (7.146) and of the relation (7.161) leads to

---

If, in addition, the symmetry of the functional derivative (7.123) is used (i.e. if the projection on the causal contribution to (7.123) is not implemented), one has

$$f_{\rm xc}(\boldsymbol{r}\sigma,\boldsymbol{r}'\sigma',t-t') = f_{\rm xc}(\boldsymbol{r}'\sigma',\boldsymbol{r}\sigma,t'-t)$$

and thus

$$f_{\rm xc}(\boldsymbol{r}\sigma,\boldsymbol{r}'\sigma',\omega) = f_{\rm xc}(\boldsymbol{r}'\sigma',\boldsymbol{r}\sigma,-\omega)\,.$$

[10] Equations (7.164), (7.165) define the dipole moment and thus the polarizability in terms of a complex electric field. The transition to a real field $\boldsymbol{E}_0\cos(\omega_0 t)$ by superposition of a second field with frequency $-\omega_0$ is, however, straightforward via Eq. (4.46). Note also, that we have extracted a factor of e^2 from the polarizability tensor α_{kl} compared to the standard definition of the polarizability in classical electrodynamics (see e.g. [36]).

$$\alpha_{kl}(\omega) = -\sum_{\alpha\beta\sigma}\sum_{\mu\nu\sigma'} R_{k\alpha\beta\sigma} \sum_{n\neq 0}\left[\frac{\overline{\xi}_{n\alpha\beta\sigma}\overline{\xi}^{*}_{n\mu\nu\sigma'}}{\hbar\omega-\Omega_n+i\eta} - \frac{\overline{\xi}^{*}_{n\alpha\beta\sigma}\overline{\xi}_{n\mu\nu\sigma'}}{\hbar\omega+\Omega_n+i\eta}\right] R_{l\mu\nu\sigma'}\,, \quad (7.167)$$

with

$$R_{k\alpha\beta\sigma} := \left[(\Theta_{\alpha\sigma}-\Theta_{\beta\sigma})\omega_{\beta\alpha\sigma}\right]^{1/2}\int d^3r\, r_k\, \rho_{\alpha\beta\sigma}(\boldsymbol{r})\,. \quad (7.168)$$

A particularly simple result is obtained for the diagonal elements of α_{kl} and thus for the mean polarizability [521]. Denoting the vectors $R_{k\alpha\beta\sigma}$ and $\overline{\xi}_{n\alpha\beta\sigma}$ in the space of KS particle–hole excitations $\alpha\beta$ with spin σ by $\underline{R}_k$ and $\underline{\overline{\xi}}_n$, respectively, and their contraction by $\sum_{\alpha\beta\sigma}$ in form of the matrix notation $\underline{R}_k^\dagger\underline{\overline{\xi}}_n$, one has

$$\alpha_{kk}(\omega) = \underline{R}_k^\dagger\left[\sum_{n\neq 0}\frac{2\Omega_n}{\Omega_n^2-(\hbar\omega+i\eta)^2}\,\underline{\overline{\xi}}_n\underline{\overline{\xi}}_n^\dagger\right]\underline{R}_k\,. \quad (7.169)$$

Unfortunately, however, the pseudo-eigenvalue problems (7.157) and (7.163) only determine the direction of the eigenvectors in the space of KS single-particle excitations, but not the length of the eigenvectors. It therefore remains to discuss the question of normalization. As a matter of principle, the normalization of the eigenvectors is controlled by the normalization of the true eigenstates, $\langle\Psi_n|\Psi_n\rangle = 1$. This condition, however, is not easily implemented in the present pseudo-eigenvalue scheme.

The normalization is also restricted by various sum rules which the Lehmann amplitudes satisfy. The most important of these exact relations is the Thomas-Reiche-Kuhn (TRK) sum rule (f-sum rule). Starting from the Eq. (7.166), one arrives at

$$\begin{aligned}
&-i\int_{-\infty}^{\infty}\frac{d\omega}{2\pi}\,\omega\,\alpha_{kk}(\omega)\\
&= -i\sum_{n\neq 0}\left|\langle\Psi_0|\int d^3r\, r_k\,\hat{n}(\boldsymbol{r})|\Psi_n\rangle\right|^2\int_{-\infty}^{\infty}\frac{d\omega}{\pi}\,\frac{\omega\Omega_n}{(\hbar\omega+i\eta)^2-\Omega_n^2}\\
&= \frac{1}{\hbar^2}\int d^3r\int d^3r'\, r_k r'_k\sum_n \Omega_n\,\langle\Psi_0|\hat{n}(\boldsymbol{r})|\Psi_n\rangle\langle\Psi_n|\hat{n}(\boldsymbol{r}')|\Psi_0\rangle
\end{aligned} \quad (7.170)$$

by contour integration (note that $\Omega_0 = 0$). The excitation energies Ω_n can be represented in terms of the full Hamiltonian by use of the Schrödinger equation, which then allows to exploit the completeness of the $|\Psi_n\rangle$ and the fact that the right-hand side of Eq. (7.170) is real,

$$-i\int_{-\infty}^{\infty}\frac{d\omega}{2\pi}\,\omega\,\alpha_{kk}(\omega) = -\frac{1}{2\hbar^2}\int d^3r\int d^3r'\, r_k r'_k\,\langle\Psi_0|\left[\left[\hat{H},\hat{n}(\boldsymbol{r})\right],\hat{n}(\boldsymbol{r}')\right]|\Psi_0\rangle\,.$$

Use of the commutators (L.17) and (L.18) plus subsequent partial integration then leads to the desired sum rule,

$$-i\int_{-\infty}^{\infty}\frac{d\omega}{2\pi}\,\omega\,\alpha_{kk}(\omega) = -\frac{i}{2\hbar}\int d^3r\int d^3r'\,r_k r'_k\,\nabla\cdot\langle\Psi_0|\left[\hat{\boldsymbol{j}}_{\rm p}(\boldsymbol{r}),\hat{n}(\boldsymbol{r}')\right]|\Psi_0\rangle$$
$$= \frac{N}{2m}\,. \tag{7.171}$$

Application of the TRK sum rule to (7.169) restricts the overall normalization of all eigenvectors,

$$\underline{R}_k^\dagger\left[\sum_{n\neq 0}\Omega_n\,\overline{\underline{\xi}}_n\,\overline{\underline{\xi}}_n^\dagger\right]\underline{R}_k = \frac{N\hbar^2}{2m}\,. \tag{7.172}$$

While further conditions of this type can be derived, none of them allows to pin down the normalization of individual eigenvectors. Sum rules can only be used as a check on the basis sets used in applications.

One thus has to resort to a more indirect argument for the normalization of the $\overline{\underline{\xi}}_n$ [521]. To this aim one solves the Dyson equation (7.142) by iteration, so that on the right-hand side an infinite sum of the type

$$\chi_{\rm R,s}\sum_{n=0}^{\infty}\left[(w+f_{\rm xc})\chi_{\rm R,s}\right]^n$$

emerges. In order to express this sum in compact form, consider the basic element

$$B = \sum_{\sigma''}\int d^3r''\sum_{\sigma'''}\int d^3r'''\,\chi_{\rm R,s}(\boldsymbol{r}\sigma,\boldsymbol{r}''\sigma'',\omega)$$
$$\times\Big\{w(\boldsymbol{r}'',\boldsymbol{r}''')+f_{\rm xc}(\boldsymbol{r}''\sigma'',\boldsymbol{r}'''\sigma''',\omega)\Big\}\chi_{\rm R,s}(\boldsymbol{r}'''\sigma''',\boldsymbol{r}'\sigma',\omega)$$

in more detail. Insertion of the KS response function (7.149) for real orbitals leads to

$$B = \int d^3r''\int d^3r'''\sum_{\alpha\beta\sigma''}\delta_{\sigma\sigma''}\frac{(\Theta_{\alpha\sigma''}-\Theta_{\beta\sigma''})\omega_{\beta\alpha\sigma''}}{(\hbar\omega+i\eta)^2-\omega_{\beta\alpha\sigma}^2}\rho_{\alpha\beta\sigma}(\boldsymbol{r})\rho_{\beta\alpha\sigma''}(\boldsymbol{r}'')$$
$$\times\Big\{w(\boldsymbol{r}'',\boldsymbol{r}''')+f_{\rm xc}(\boldsymbol{r}''\sigma'',\boldsymbol{r}'''\sigma''',\omega)\Big\}$$
$$\times\sum_{\mu\nu\sigma'''}\delta_{\sigma'''\sigma'}\frac{(\Theta_{\mu\sigma'''}-\Theta_{\nu\sigma'''})\omega_{\nu\mu\sigma'''}}{(\hbar\omega+i\eta)^2-\omega_{\nu\mu\sigma'}^2}\rho_{\mu\nu\sigma'''}(\boldsymbol{r}''')\rho_{\nu\mu\sigma'}(\boldsymbol{r}')\,.$$

With the matrices

$$\Lambda_{\alpha\beta\sigma,\mu\nu\sigma'}(\omega) := \frac{\delta_{\alpha\mu}\,\delta_{\beta\nu}\,\delta_{\sigma\sigma'}}{(\hbar\omega+i\eta)^2-\omega_{\beta\alpha\sigma}^2} \tag{7.173}$$

$$M_{\alpha\beta\sigma,\mu\nu\sigma'}(\omega) := \left[(\Theta_{\alpha\sigma}-\Theta_{\beta\sigma})\omega_{\beta\alpha\sigma}\right]^{1/2}K_{\alpha\beta\sigma,\mu\nu\sigma'}(\omega)$$
$$\times\left[(\Theta_{\mu\sigma'}-\Theta_{\nu\sigma'})\omega_{\nu\mu\sigma'}\right]^{1/2} \tag{7.174}$$

B can be rewritten as

$$\begin{aligned}B = &\sum_{\alpha\beta}\sum_{\mu\nu}\left[(\Theta_{\alpha\sigma}-\Theta_{\beta\sigma})\omega_{\beta\alpha\sigma}\right]^{1/2}\rho_{\alpha\beta\sigma}(\boldsymbol{r})\\ &\times\sum_{\alpha_1\beta_1\sigma_1}\sum_{\alpha_2\beta_2\sigma_2}\Lambda_{\alpha\beta\sigma,\alpha_1\beta_1\sigma_1}(\omega)M_{\alpha_1\beta_1\sigma_1,\alpha_2\beta_2\sigma_2}(\omega)\Lambda_{\alpha_2\beta_2\sigma_2,\mu\nu\sigma'}(\omega)\\ &\times\left[(\Theta_{\mu\sigma'}-\Theta_{\nu\sigma'})\omega_{\nu\mu\sigma'}\right]^{1/2}\rho_{\nu\mu\sigma'}(\boldsymbol{r}')\,. \end{aligned}\tag{7.175}$$

In the Dyson equation this structure repeats itself to all orders. In addition, the lowest order contribution, $\chi_{\mathrm{R,s}}$, is easily expressed in terms of $\Lambda_{\alpha\beta\sigma,\mu\nu\sigma'}(\omega)$. Insertion of the resulting Dyson series into Eq. (7.166) and use of the definition (7.168) allows to express the polarizability in matrix notation as

$$\alpha_{kk}(\omega) = -\underline{R}_k^{\dagger}\left[\underline{\underline{\Lambda}}\sum_{n=0}^{\infty}\left[\underline{\underline{M}}\,\underline{\underline{\Lambda}}\right]^n\right]\underline{R}_k \tag{7.176}$$

($[\underline{\underline{A}}]^0$ is the unit matrix). The geometric series of matrices can be resummed,

$$\alpha_{kk}(\omega) = \underline{R}_k^{\dagger}\left[\left(\underline{\underline{M}}-\underline{\underline{\Lambda}}^{-1}\right)^{-1}\right]\underline{R}_k\,, \tag{7.177}$$

where $\underline{\underline{\Lambda}}^{-1}$ and $\left(\underline{\underline{\Lambda}}^{-1}-\underline{\underline{M}}\right)^{-1}$ denote the corresponding inverse matrices. Due to the diagonal structure of $\underline{\underline{\Lambda}}$ the inverse is trivially given by

$$\Lambda^{-1}_{\alpha\beta\sigma,\mu\nu\sigma'}(\omega) = \left[(\hbar\omega+i\eta)^2-\omega^2_{\beta\alpha\sigma}\right]\delta_{\alpha\mu}\,\delta_{\beta\nu}\,\delta_{\sigma\sigma'}\,, \tag{7.178}$$

so that

$$M_{\alpha\beta\sigma,\mu\nu\sigma'}(\omega)-\Lambda^{-1}_{\alpha\beta\sigma,\mu\nu\sigma'}(\omega) = S_{\alpha\beta\sigma,\mu\nu\sigma'}(\omega)-(\hbar\omega+i\eta)^2\,\delta_{\alpha\mu}\,\delta_{\beta\nu}\,\delta_{\sigma\sigma'}\,, \tag{7.179}$$

with $\underline{\underline{S}}$ defined by Eq. (7.162). One thus finally arrives at

$$\alpha_{kk}(\omega) = \underline{R}_k^{\dagger}\left[\left(\underline{\underline{S}}(\omega)-(\hbar\omega+i\eta)^2\underline{\underline{1}}\right)^{-1}\right]\underline{R}_k\,. \tag{7.180}$$

A comparison of this result with Eq. (7.169) reminds one of the decomposition of the operator $\left(\underline{\underline{S}}(\omega)-(\hbar\omega+i\eta)^2\underline{\underline{1}}\right)^{-1}$ via the spectral theorem. In fact, if the $\overline{\underline{\xi}}_n$ were regular, orthogonal eigenvectors, i.e. if f_{xc} was frequency-independent, a normalization of

$$\overline{\underline{\xi}}_n^{\dagger}\,\overline{\underline{\xi}}_n = \frac{1}{2\Omega_n} \qquad (n\neq 0) \tag{7.181}$$

would lead to the completeness relation

$$\sum_{n\neq 0} 2\Omega_n \overline{\underline{\xi}}_n \overline{\underline{\xi}}_n^\dagger = \underline{\underline{1}} \,. \tag{7.182}$$

Any operator in the space of single-particle excitations could then be represented in terms of the eigenvectors $\overline{\underline{\xi}}_n$ and eigenvalues Ω_n [521], in particular

$$\left(\underline{\underline{S}} - (\hbar\omega + i\eta)^2 \underline{\underline{1}}\right)^{-1} = \sum_{n\neq 0} \frac{2\Omega_n}{\Omega_n^2 - (\hbar\omega + i\eta)^2} \overline{\underline{\xi}}_n \overline{\underline{\xi}}_n^\dagger \,. \tag{7.183}$$

In other words: if a frequency-independent approximation for $f_{\rm xc}$ is used, the identity of (7.169) and (7.180) together with the orthogonality and completeness of the eigenvectors requires the normalization (7.181). For the true $f_{\rm xc}$, the normalization (7.181) is an approximation. Lowest order corrections for the frequency-dependence can be obtained by a Taylor expansion of (7.183) about the eigenvalues Ω_n [521]. A variety of results for molecular excitation energies and dynamic polarizabilities obtained with the adiabatic approximation may be found in [523, 524].

The pseudo-eigenvalue equations (7.157) and (7.163) can be applied in a straightforward fashion, as soon as a basis set expansion of the KS orbitals is utilized. However, they also offer themselves for a perturbative evaluation. In fact, if one assumes that there is only one single KS excitation energy $\omega_{\beta\alpha\sigma}$ which is close to the true excitation energy Ω_n, the sum over all KS particle–hole pairs on the right-hand side of Eq. (7.150) is completely dominated by a single term. If the sum is approximated by this dominant contribution, i.e. if only the leading term of a Taylor expansion of the right-hand side of (7.150) in powers of $\Omega_n - \omega_{\beta\alpha\sigma}$ is kept, the pseudo-eigenvalue problem (7.157) reduces to a single algebraic equation, which determines Ω_n,

$$\Omega_n = \omega_{\beta\alpha\sigma} + K_{\alpha\beta\sigma,\alpha\beta\sigma}(\omega_{\beta\alpha\sigma}/\hbar) \,. \tag{7.184}$$

Moreover, if there are several (q) degenerate KS excitations which are close to Ω_n,

$$\omega_{\beta_1\alpha_1\sigma} = \ldots \omega_{\beta_q\alpha_q\sigma} \equiv \overline{\omega} \approx \Omega_n > 0 \,, \tag{7.185}$$

only the sum over these q particle–hole pairs shows up in Eq. (7.157). In addition, the non-linearity of (7.157) has disappeared, as $K_{\alpha\beta\sigma,\mu\nu\sigma'}(\omega)$ has to be evaluated at $\omega = \overline{\omega}/\hbar$. The corresponding true excitation energies are determined by the finite-order characteristic polynomial (7.158), with $K_{\alpha\beta\sigma,\mu\nu\sigma'}(\Omega_n/\hbar)$ replaced by $K_{\alpha\beta\sigma,\mu\nu\sigma'}(\overline{\omega}/\hbar)$.

As a result of this first order perturbative approach one can deal with nearly degenerate states $|\Psi_n\rangle$ more easily, as the corresponding excitation energies are obtained simultaneously by solution of a single eigenvalue problem. Nearly degenerate excited states are often present in photo-absorption processes. As an example, consider the excitation of an electron from a closed-shell ground state with total spin 0. Even if only a single spatial configuration is available to the excited electron, the spin of the electron can be either parallel or antiparallel to the spin of the hole left in the ground state. The excited state can thus either be a spin singlet or a spin triplet state. The characteristic polynomial then has the form

$$\det \begin{pmatrix} \overline{\omega} - \Omega + K_{\uparrow,\uparrow} & K_{\uparrow,\downarrow} \\ K_{\downarrow,\uparrow} & \overline{\omega} - \Omega + K_{\downarrow,\downarrow} \end{pmatrix} \quad \text{with} \quad K_{\sigma,\sigma'} \equiv K_{\alpha\beta\sigma,\alpha\beta\sigma'}(\overline{\omega}/\hbar), \quad (7.186)$$

since the KS excitation energies do not depend on the spin,

$$\omega_{\alpha\beta\uparrow} = \omega_{\alpha\beta\downarrow} = \overline{\omega}$$

(due to the spin-saturated ground state). The corresponding singlet and triplet excitation energies are obtained as

$$\Omega = \overline{\omega} + \frac{K_{\uparrow,\uparrow} + K_{\downarrow,\downarrow}}{2} \pm \left[\left(\frac{K_{\uparrow,\uparrow} + K_{\downarrow,\downarrow}}{2} \right)^2 - K_{\uparrow,\uparrow} K_{\downarrow,\downarrow} + K_{\uparrow,\downarrow} K_{\downarrow,\uparrow} \right]^{1/2} . \quad (7.187)$$

This result can be simplified by use of the symmetries of $K_{\sigma,\sigma'}$. As the initial state is spin-saturated, $n_{\uparrow,0}(\boldsymbol{r}) = n_{\downarrow,0}(\boldsymbol{r})$, the KS orbitals do not depend on spin,

$$\phi_{\alpha\uparrow}(\boldsymbol{r}) = \phi_{\alpha\downarrow}(\boldsymbol{r}) \equiv \phi_{\alpha}(\boldsymbol{r}) ,$$

and the kernel $f_{xc}(\boldsymbol{r}\sigma, \boldsymbol{r}'\sigma', \omega)$, Eq. (7.145), is symmetric with respect to spin,

$$f_{xc}(\boldsymbol{r}\sigma, \boldsymbol{r}'\sigma', \omega) = f_{xc}(\boldsymbol{r}\sigma', \boldsymbol{r}'\sigma, \omega) . \quad (7.188)$$

These symmetries are transferred to $K_{\sigma,\sigma'}$,

$$K_{\sigma,\sigma'} = K_{\sigma',\sigma} ; \qquad K_{\sigma,\sigma} = K_{-\sigma,-\sigma} ; \qquad K^*_{\sigma,\sigma'} = K_{\sigma,\sigma'} , \quad (7.189)$$

so that the excitations energies can be expressed as

$$\Omega = \overline{\omega} + K_{\uparrow,\uparrow} \pm K_{\uparrow,\downarrow} . \quad (7.190)$$

It remains to identify the true states $|\Psi_n\rangle$ to which the solutions of Eq. (7.158) correspond. For this identification necessarily some information on the $|\Psi_n\rangle$ is required. Fortunately, it is often sufficient to know the noninteracting states which are approached by $|\Psi_0\rangle$ and $|\Psi_n\rangle$ in the limit of vanishing interaction. If the eigenvectors are available, Eq. (7.153) then allows to establish a connection between the solution Ω_n and $|\Psi_n\rangle$.

Concerning the identification of the $|\Psi_n\rangle$, it is nevertheless instructive to reconsider the original residue equation (7.150). If there are q degenerate KS excitation energies, as specified in Eq. (7.185), and if one aims at a first order perturbative evaluation of the excitation energies, Eq. (7.150) reduces to

$$\begin{aligned} &(\Omega_n - \overline{\omega}) \langle \Psi_0 | \hat{n}_\sigma(\boldsymbol{r}) | \Psi_n \rangle \\ &= \sum_{a=1}^{q} \rho_{\alpha_a\beta_a\sigma}(\boldsymbol{r}) \sum_{\sigma'} \int d^3r'' d^3r''' \rho_{\beta_a\alpha_a\sigma}(\boldsymbol{r}'') \\ &\quad \times \left[w(\boldsymbol{r}'', \boldsymbol{r}''') + f_{xc}(\boldsymbol{r}''\sigma, \boldsymbol{r}'''\sigma', \overline{\omega}/\hbar) \right] \langle \Psi_0 | \hat{n}_{\sigma'}(\boldsymbol{r}''') | \Psi_n \rangle . \end{aligned} \quad (7.191)$$

In order to proceed one has to specify suitable approximations for $|\Psi_0\rangle$ and $|\Psi_n\rangle$, in accordance with the forms of $|\Psi_0\rangle$ and $|\Psi_n\rangle$ obtained in the noninteracting limit. Let us again consider the excitation of a single electron from a spin-saturated, closed-shell ground state to a spin singlet and a spin triplet excited state—degeneracies resulting from spatial symmetries are suppressed for brevity. One could, for example, think of the $1s^2 \to 1s2p$ excitation of the helium atom. A lowest order approximation for the true ground state of the helium atom is provided by the corresponding KS ground state

$$(\boldsymbol{r}_1\sigma_1\boldsymbol{r}_2\sigma_2|\Psi_0\rangle \approx \phi_\alpha(\boldsymbol{r}_1)\phi_\alpha(\boldsymbol{r}_2)\chi_{00}(\sigma_1,\sigma_2)\,, \tag{7.192}$$

with χ_{00} given by (3.49). An approximation in terms of the KS orbitals is also possible for the excited states. In the case of the spin singlet one has

$$(\boldsymbol{r}_1\sigma_1\boldsymbol{r}_2\sigma_2|\Psi^{\mathrm{S}}\rangle \approx \frac{1}{\sqrt{2}}\Big[\phi_\alpha(\boldsymbol{r}_1)\phi_\beta(\boldsymbol{r}_2)+\phi_\alpha(\boldsymbol{r}_2)\phi_\beta(\boldsymbol{r}_1)\Big]\chi_{00}(\sigma_1,\sigma_2)\,, \tag{7.193}$$

while the spin triplet is given by

$$(\boldsymbol{r}_1\sigma_1\boldsymbol{r}_2\sigma_2|\Psi^{\mathrm{T},M}\rangle \approx \frac{1}{\sqrt{2}}\Big[\phi_\alpha(\boldsymbol{r}_1)\phi_\beta(\boldsymbol{r}_2)-\phi_\alpha(\boldsymbol{r}_2)\phi_\beta(\boldsymbol{r}_1)\Big]\chi_{1M}(\sigma_1,\sigma_2) \tag{7.194}$$

$$\chi_{10}(\sigma_1,\sigma_2) = \frac{1}{\sqrt{2}}\Big\{\chi_+(\sigma_1)\chi_-(\sigma_2)+\chi_-(\sigma_1)\chi_+(\sigma_2)\Big\} \tag{7.195}$$

$$\chi_{1,\pm1}(\sigma_1,\sigma_2) = \chi_\pm(\sigma_1)\chi_\pm(\sigma_2)\,. \tag{7.196}$$

The Lehmann amplitudes of the true states can therefore to lowest order be expressed as

$$\langle\Psi_0|\hat{n}_\sigma(\boldsymbol{r})|\Psi^{\mathrm{S}}\rangle \approx \frac{1}{\sqrt{2}}\phi_\alpha^*(\boldsymbol{r})\phi_\beta(\boldsymbol{r}) \tag{7.197}$$

$$\langle\Psi_0|\hat{n}_\sigma(\boldsymbol{r})|\Psi^{\mathrm{T},M}\rangle \approx \frac{1}{\sqrt{2}}\phi_\alpha^*(\boldsymbol{r})\phi_\beta(\boldsymbol{r})\Big[\delta_{\sigma\downarrow}-\delta_{\sigma\uparrow}\Big]\delta_{M,0}\,. \tag{7.198}$$

These results are also correct for comparable systems with more than 2 electrons. Insertion of Eqs. (7.197) and (7.198) into (7.191) leads to

$$\Omega^{\mathrm{S}}-\overline{\omega} = \sum_{\sigma'} K_{\sigma,\sigma'} \tag{7.199}$$

$$\Omega^{\mathrm{T},0}-\overline{\omega} = \sum_{\sigma'} K_{\sigma,\sigma'}\left(\delta_{\sigma',\sigma}-\delta_{\sigma',\ \sigma}\right), \tag{7.200}$$

with $K_{\sigma,\sigma'}$ defined as in (7.186). The final solutions for the excitation energies are therefore

$$\Omega^{\mathrm{S}} = \overline{\omega}+K_{\uparrow,\uparrow}+K_{\uparrow,\downarrow} \tag{7.201}$$

$$\Omega^{\mathrm{T},0} = \overline{\omega}+K_{\uparrow,\uparrow}-K_{\uparrow,\downarrow}\,, \tag{7.202}$$

in agreement with the earlier result (7.190). The lowest order approximations (7.192)–(7.194) allow an unambiguous assignment of the solutions of the eigenvalue problem to the actual states.

Some illustrative results obtained with this approach are listed in Table 7.1 [486] (many further results may be found in [516, 525]). In this table the excitation of the

Table 7.1 Energetically lowest $s \to p$ excitation energies of closed-subshell atoms: comparison of spin-singlet and spin-triplet excitation energies obtained from first order perturbative approach, Eqs. (7.201), (7.202), with the eigenvalue difference $\overline{\omega}$ and experimental results. Also given is the difference between the total energies obtained by self-consistent calculations for the excited and the ground state (Δ_{SCF}). The LDA has been used both in the KS calculation for the ground state as well as for the xc-kernel in (7.188). All data are taken from [486].

Atom	State	$\overline{\omega}$	Ω_n [Ry]		
		[Ry]	1st order	Expt.	Δ_{SCF}
Be	$^1P_1$	0.257	0.399	0.388	0.331
	$^3P_0$			0.200	
	$^3P_1$	0.257	0.192	0.200	0.181
	$^3P_2$			0.200	
Mg	$^1P_1$	0.249	0.351	0.319	0.299
	$^3P_0$			0.199	
	$^3P_1$	0.249	0.209	0.199	0.206
	$^3P_2$			0.200	
Ca	$^1P_1$	0.176	0.263	0.216	0.211
	$^3P_0$			0.138	
	$^3P_1$	0.176	0.145	0.139	0.144
	$^3P_2$			0.140	
Zn	$^1P_1$	0.352	0.477	0.426	0.403
	$^3P_0$			0.294	
	$^3P_1$	0.352	0.314	0.296	0.316
	$^3P_2$			0.300	
Sr	$^1P_1$	0.163	0.241	0.198	0.193
	$^3P_0$			0.130	
	$^3P_1$	0.163	0.136	0.132	0.135
	$^3P_2$			0.136	
Cd	$^1P_1$	0.303	0.427	0.398	0.346
	$^3P_0$			0.274	
	$^3P_1$	0.303	0.269	0.279	0.272
	$^3P_2$			0.290	

most weakly bound s-electron of the alkaline earth elements and the zinc series is considered, so that the noninteracting limit of the ground state is given by a single Slater determinant as in the example of helium. The s-electron is excited into the lowest available p-level. Strictly speaking, one has to take into account an orbital degeneracy in addition to the spin degeneracy leading to Eqs. (7.201), (7.202). For simplicity, however, the threefold degeneracy associated with the z-projection of the angular momentum has been ignored in the evaluation of the first order excitations energies. The average value of the excitation energy calculated in this way has to be compared with the average of the experimental energies for the three levels $^3P_{0,1,2}$. The calculations are based on the LDA, both for the solution of the stationary KS equations as well as for the evaluation of the matrix element (7.188). It is obvious that the first order results are much closer to reality than the KS eigenvalue differences.

For comparison the total energy differences between the KS ground and excited state are also given (following the approach discussed in Sect. 2.7). Again the LDA has been utilized in the calculations. Over the small set of atoms considered here this so-called Δ_{SCF}-method gives excitation energies of similar quality as the first order energies from TDDFT.

Chapter 8
Relativistic Density Functional Theory

Given the success of nonrelativistic DFT, the question concerning a suitable relativistic extension arises quite naturally.[1] Unfortunately, relativistic DFT (RDFT) is not as easily established as its nonrelativistic counterpart. The difficulties originate from the underlying many-body theory, that is quantum electrodynamics (QED). A direct evaluation of expectation values, such as ground state energies or Green's functions is not possible because of divergences. These divergences result from the presence of anti-particle (negative energy) states, which in particular complicates the handling of electron-hole pairs. It needs an elaborate formalism, the so-called renormalization program of QED, in order to deal with these divergences properly. Without the renormalization procedure it is neither possible to prove the relativistic equivalent of the Hohenberg-Kohn theorem nor can one derive simple density functionals as e.g. the relativistic LDA. Once the existence theorem has been put onto a proper foundation, the assembly of the relativistic Kohn-Sham equations (selfconsistent Dirac equations with multiplicative potentials incorporating radiative corrections) is reasonably straightforward, even though the handling of magnetic interactions turns out to be rather involved.

Clearly, an exhaustive discussion of QED and the associated renormalization scheme is beyond the scope of this book. On the other hand, it is not possible to give convincing arguments in favor of an existence theorem of RDFT without addressing this issue. We therefore briefly review the relevant aspects of QED of inhomogeneous systems (bound state QED) in Sect. 8.2. Some background information on the renormalization program of QED is provided in the Appendices F–I. This material may serve as a reminder and reference for notation, but is not intended to be a substitute for a text on QED. We nevertheless hope that with this information the reader is able to follow the steps taken in Sect. 8.2 and is thus well-prepared for the discussion of the existence theorem of RDFT in Sect. 8.3.[2]

---

[1] In this text we will not address the issue of time-dependent RDFT. A corresponding existence theorem has been put forward in [526] (see also [527, 528]).

[2] The present text basically follows the line of [529]. For an alternative path through the QED background see [28]).

The original version of this chapter was revised. An erratum can be found at
https://doi.org/10.1007/978-3-642-14090-7_12

E. Engel, R.M. Dreizler, Relativistic Density Functional Theory. In: E. Engel, R.M. Dreizler, Density Functional Theory, Theoretical and Mathematical Physics, pp. 351–399 (2011)
DOI 10.1007/978-3-642-14090-7_8

On the other hand, a reader not interested in the details of the existence theorem and the various field theoretical facets of RDFT may skip Sects. 8.2–8.5 and proceed directly in Sect. 8.6, in which RDFT within the *no-pair* approximation is introduced. In this limit, which is the form of RDFT used in practice, all contributions originating from the creation of particle-antiparticle pairs are neglected.

8.1 Notation

Let us introduce our notation, before starting with the actual discussion of QED. Space-time points are denoted by four vectors

$$x \equiv (x^\mu) = (ct, \boldsymbol{r}) = (ct, r^1, r^2, r^3) = (ct, r^i) . \tag{8.1}$$

Greek (Minkowski) indices always extend from 0 to 3, Roman indices always denote spatial components, $i, j, \ldots = 1,2,3$. The associated metric tensor reads

$$g_{\mu\nu} \equiv \begin{pmatrix} 1 & 0 & 0 & 0 \\ 0 & -1 & 0 & 0 \\ 0 & 0 & -1 & 0 \\ 0 & 0 & 0 & -1 \end{pmatrix} = \begin{pmatrix} 1 & 0 \\ 0 & -\delta_{ij} \end{pmatrix} . \tag{8.2}$$

Contraction of any four vector with the metric tensor transforms covariant into contravariant components and vice versa,

$$a_\mu = g_{\mu\nu} a^\nu ; \qquad a^\mu = g^{\mu\nu} a_\nu . \tag{8.3}$$

The spatial vector $\boldsymbol{a}$, which characterizes the actual physical three vector, consists of the contravariant components $\boldsymbol{a} = (a^1, a^2, a^3)$, from which the covariant components a_i differ by a minus sign, $a_i = -a^i$. Note that the distinction between contravariant and covariant components is not used in all previous chapters of this book. In the nonrelativistic situation only contravariant vectors show up and are often labelled as a_i. The four gradient is abbreviated by

$$\partial_\mu \equiv \frac{\partial}{\partial x^\mu} = \left(\frac{1}{c}\frac{\partial}{\partial t}, \boldsymbol{\nabla} \right) . \tag{8.4}$$

The summation convention is used throughout,

$$a^\mu b_\mu \equiv \sum_{\mu=0}^{3} a^\mu b_\mu ; \qquad a^i b_i \equiv \sum_{i=1}^{3} a^i b_i = -\sum_{i=1}^{3} a^i b^i = -\boldsymbol{a}\cdot\boldsymbol{b} . \tag{8.5}$$

γ^μ, $\boldsymbol{\alpha}$ and β are the standard Dirac matrices,

$$\alpha^\mu = \gamma^0 \gamma^\mu = (1, \boldsymbol{\alpha}) \qquad \beta = \gamma^0 , \tag{8.6}$$

which satisfy the commutation relations

$$\{\gamma^\mu,\gamma^\nu\} = \gamma^\mu\gamma^\nu + \gamma^\nu\gamma^\mu \ = \ 2g^{\mu\nu} \tag{8.7}$$

$$\{\alpha^i,\alpha^j\} = -2g^{ij} \ = \ 2\delta_{ij} \tag{8.8}$$

$$\{\alpha^i,\beta\} = 0\,. \tag{8.9}$$

Finally, it is often convenient to work with the forms

$$\hat{\overline{\psi}} \equiv \hat{\psi}^\dagger\gamma^0 \qquad \text{and} \qquad \overline{\phi}_k \equiv \phi_k^\dagger\gamma^0 \tag{8.10}$$

of the hermitian conjugate field operator $\hat{\psi}^\dagger$ and wavefunction $\phi_k^\dagger$, respectively. As in all other chapters of this text we use $e = |e|$.

8.2 Field Theoretical Background

As already indicated, the appropriate starting point for a relativistic description of the electronic structure of atoms, molecules and solids is QED. In a truly covariant QED-approach to these systems both the electrons and the nuclei would have to be treated as dynamical degrees of freedom (possibly on a classical level in the case of the nuclei). However, in view of the large difference between the electron mass and the nuclear mass (in particular for heavy nuclei) the Born-Oppenheimer approximation is usually applied, at least for the discussion of ground state properties. The nuclei are treated as fixed external sources, assumed to be at rest in a common Lorentz frame (which introduces a preferred reference frame). As external sources, the nuclei may either be represented by classical charge distributions which interact with the quantized photon field or in the form of a classical potential which couples to the electron field. Both viewpoints are completely equivalent. In this contribution the second approach is chosen: the electrons interact with each other by the exchange of photons and experience a static external potential. In order to keep the discussion as general as possible, the external potential is represented by a complete four vector, $V^\mu(\boldsymbol{x})$. In addition to the Coulomb field of the nuclei, V^μ may also include external electromagnetic fields or nuclear magnetic moments.

The system is therefore characterized by the Lagrangian [530]

$$\mathscr{L} = \mathscr{L}_\mathrm{e} + \mathscr{L}_\gamma + \mathscr{L}_\mathrm{int} + \mathscr{L}_\mathrm{ext}\,. \tag{8.11}$$

Here $\mathscr{L}_\mathrm{e}$ represents the Lagrangian of noninteracting fermions (i.e. of noninteracting electrons plus their antiparticles, the positrons),

$$\mathscr{L}_\mathrm{e}(x) = \frac{1}{4}\Big\{ \Big[\hat{\overline{\psi}}(x), \Big(i\hbar c\gamma^\mu \overrightarrow{\partial}_\mu - mc^2\Big)\hat{\psi}(x)\Big] + \Big[\hat{\overline{\psi}}(x)\Big(-i\hbar c\gamma^\mu \overleftarrow{\partial}_\mu - mc^2\Big), \hat{\psi}(x)\Big]\Big\}, \tag{8.12}$$

where $\hat{\psi}(x)$ denotes the fermion field operator,

$$\left\{\hat{\psi}_a(t,\boldsymbol{x}),\hat{\psi}_b(t,\boldsymbol{y})\right\} = \left\{\hat{\psi}_a^\dagger(t,\boldsymbol{x}),\hat{\psi}_b^\dagger(t,\boldsymbol{y})\right\} = 0 \tag{8.13}$$

$$\left\{\hat{\psi}_a(t,\boldsymbol{x}),\hat{\psi}_b^\dagger(t,\boldsymbol{y})\right\} = \delta_{ab}\,\delta^{(3)}(\boldsymbol{x}-\boldsymbol{y}) \tag{8.14}$$

(a,b are the spinor indices), and the vector bars on top of the partial derivatives indicate the direction in which the derivative has to be taken—in the second term of $\mathscr{L}_e$ the partial derivatives act on $\hat{\overline{\psi}}$. $\mathscr{L}_\gamma$ is the Lagrangian of noninteracting photons,

$$\mathscr{L}_\gamma(x) = -\frac{1}{16\pi}\hat{F}_{\mu\nu}(x)\hat{F}^{\mu\nu}(x) - \frac{\lambda}{8\pi}(\partial_\mu\hat{A}^\mu(x))^2\,. \tag{8.15}$$

The operator $\hat{A}_\mu(x)$ denotes the photon field operator,

$$\left[\hat{A}^\mu(t,\boldsymbol{x}),\hat{A}^\nu(t,\boldsymbol{y})\right] = \left[\hat{\pi}^\mu(t,\boldsymbol{x}),\hat{\pi}^\nu(t,\boldsymbol{y})\right] = 0 \tag{8.16}$$

$$\left[\hat{A}^\mu(t,\boldsymbol{x}),\hat{\pi}^\nu(t,\boldsymbol{y})\right] = i g^{\mu\nu}\,\delta^{(3)}(\boldsymbol{x}-\boldsymbol{y}) \tag{8.17}$$

$$\hat{\pi}^\mu(x) = \frac{\partial\mathscr{L}_\gamma}{\partial\left(\partial_0\hat{A}_\mu(x)\right)} = \frac{1}{4\pi}\left[\hat{F}^{\mu 0}(x) - \lambda g^{\mu 0}\partial_\nu\hat{A}^\nu(x)\right]\,, \tag{8.18}$$

and $\hat{F}_{\mu\nu}(x)$ is the associated field tensor,

$$\hat{F}_{\mu\nu}(x) = \partial_\mu\hat{A}_\nu(x) - \partial_\nu\hat{A}_\mu(x)\,. \tag{8.19}$$

Finally, $\mathscr{L}_{\rm int}$ and $\mathscr{L}_{\rm ext}$ represent the interaction between fermions and photons and between fermions and the static external potential V_μ,[3] respectively,

$$\mathscr{L}_{\rm int}(x) = e\,\hat{j}^\mu(x)\,\hat{A}_\mu(x) \tag{8.20}$$

$$\mathscr{L}_{\rm ext}(x) = e\,\hat{j}^\mu(x)\,V_\mu(\boldsymbol{x}) \tag{8.21}$$

($e = |e|$). Here $\hat{j}^\mu$ is the fermion four current density,

$$\hat{j}^\mu(x) = \frac{1}{2}\left[\hat{\overline{\psi}}(x),\gamma^\mu\hat{\psi}(x)\right]. \tag{8.22}$$

Both the fermion Lagrangian and the four current have been written in a form which ensures their correct behavior under charge conjugation [530] (for details see Appendix F). The four current (as a charge current rather than a probability current) must change its sign,

$$\hat{\mathscr{C}}\,\hat{j}^\mu(x)\,\hat{\mathscr{C}}^\dagger = -\hat{j}^\mu(x)\,, \tag{8.23}$$

[3] In Sects. 8.2–8.4 and the Appendices F–J the external four potential is denoted by V_μ in order to allow an easy distinction from the photon field operator $\hat{A}_\mu$. The transition to the standard notation used in nonrelativistic (C)DFT will be made in Sect. 8.6.

under the charge conjugation $\hat{\mathscr{C}}$, which transforms electrons into positrons and vice versa. As a potential which attracts electrons repels positrons, the photon field behaves as $\hat{\mathscr{C}}\hat{A}^\mu\hat{\mathscr{C}}^\dagger = -\hat{A}^\mu$. Moreover, the Lagrangian $\mathscr{L}_e$ is invariant under transformation with $\hat{\mathscr{C}}$ (note the minus sign in front of $ic\gamma^\mu \overleftarrow{\partial}_\mu$ in the second term). As a result, the total Lagrangian transforms as

$$\hat{\mathscr{C}}\ \mathscr{L}[V^\mu]\ \hat{\mathscr{C}}^\dagger = \mathscr{L}[-V^\mu]\,. \tag{8.24}$$

As expected, the fermion charge only manifests itself in the coupling to the external potential V^μ, which is not affected by $\hat{\mathscr{C}}$.

For the photon field we work in the covariant gauge [531], which relies on the Gupta-Bleuler indefinite metric quantization. Compared to classical electrodynamics, this approach requires the additional term

$$-\frac{\lambda}{8\pi}(\partial_\mu\hat{A}^\mu(x))^2$$

in the Lagrangian. In the indefinite metric quantization the classical gauge condition $\partial_\mu A^\mu = 0$ is not applied directly to the quantized photon field. Rather $\partial_\mu\hat{A}^\mu \neq 0$ is required in order to set up a covariant quantization scheme, which necessarily involves four independent fields. In addition to the two real transverse modes one introduces two artificial fields in the first step. The gauge condition is then restored and the two artificial degrees of freedom are eliminated by constraining the allowed set of physical states, so that $\langle\Psi|\partial_\mu\hat{A}^\mu|\Psi\rangle = 0$. Explicit formulae will subsequently be given for some particular choice of λ, which is, by misuse of language, also called a particular gauge. The choice of this gauge primarily determines the form of one of the building blocks of the theory characterized by (8.11), namely the propagator of noninteracting photons. In Sects. 8.2–8.4 Feynman gauge ($\lambda = 1$) will be used. On the other hand, the renormalization procedure of QED presented in the appendices is more easily discussed in Landau gauge ($\lambda \to \infty$).

The freedom to choose any of the covariant gauges for the photon field results from the gauge invariance of the Lagrangian (8.11): a gauge transformation of the photon field,

$$\hat{A}_\mu(x) \longrightarrow \hat{A}'_\mu(x) = \hat{A}_\mu(x) + \partial_\mu\Lambda(x)\,; \qquad \partial_\mu\partial^\mu\Lambda(x) = 0\,, \tag{8.25}$$

can be absorbed by an appropriate phase transformation of the fermion field

$$\hat{\psi}(x) \longrightarrow \hat{\psi}'(x) = \exp[ie\Lambda(x)/(\hbar c)]\ \hat{\psi}(x)\,, \tag{8.26}$$

leaving the Lagrangian (8.11) invariant,

$$\mathscr{L}[\hat{\psi}',\hat{A}'] = \mathscr{L}[\hat{\psi},\hat{A}]\,. \tag{8.27}$$

On the other hand, due to the choice of a particular Lorentz frame, only static gauge transformations are admitted for the external potential in order to remain within the common rest frame of the nuclei,

$$V'_\mu(\boldsymbol{x}) = V_\mu(\boldsymbol{x}) + \partial_\mu \Lambda(x) \tag{8.28}$$
$$\hat{\psi}'(x) = \exp[ie\Lambda(x)/(\hbar c)]\ \hat{\psi}(x) \tag{8.29}$$
$$\Lambda(x) = Ct + \lambda(\boldsymbol{x})\,; \qquad \Delta\lambda(\boldsymbol{x}) = 0 \tag{8.30}$$
$$\Longrightarrow \quad \mathscr{L}[\hat{\psi}', V'] = \mathscr{L}[\hat{\psi}, V]\,. \tag{8.31}$$

Not only the Lagrangian, but also the four current $\hat{j}^\mu(x)$, Eq. (8.22), is invariant under the transformations (8.25), (8.26) and (8.28)–(8.30). The invariances of $\mathscr{L}$ and $\hat{j}^\mu$ under the combination of a gauge transformation of the internal or external photon fields and a corresponding phase transformation of the fermion field operator are manifestations of the gauge invariance of the expectation values $\langle\Psi|\mathscr{L}[\hat{\psi},\hat{A}]|\Psi\rangle$ and $\langle\Psi|\hat{j}^\mu|\Psi\rangle$. In the following, Coulomb gauge,

$$\partial_i V^i(\boldsymbol{x}) = \nabla \cdot \boldsymbol{V}(\boldsymbol{x}) = 0\,,$$

will be applied consistently for the external potential, as already indicated by the restricted set of gauge transformations in (8.30).

The field equations for the basic operators of QED follow from the Lagrangian (8.11) via the principle of stationary action [531],

$$\left[i\hbar c\gamma^\mu \partial_\mu - mc^2 + e\gamma^\mu \hat{A}_\mu(x) + e\gamma^\mu V_\mu(\boldsymbol{x})\right] \hat{\psi}(x) = 0 \tag{8.32}$$
$$\partial_\mu \partial^\mu \hat{A}^\nu(x) - (1-\lambda)\partial^\nu \partial_\mu \hat{A}^\mu(x) + 4\pi e\, \hat{j}^\nu(x) = 0\,. \tag{8.33}$$

Their invariance under the transformations (8.25), (8.26) and (8.28)–(8.30) is directly obvious.

An immediate consequence of the local gauge invariance of the Lagrangian is current conservation,[4]

$$\partial_\mu \hat{j}^\mu(x) = 0\,, \tag{8.34}$$

which implies the conservation of total charge,

$$\hat{Q} = \int d^3x\, \hat{j}^0(x) = \frac{1}{2}\int d^3x \left[\hat{\psi}^\dagger(x), \hat{\psi}(x)\right]. \tag{8.35}$$

Any eigenstate of the system characterized by (8.11) can therefore be classified with respect to its charge (but not particle number).

Energy conservation can be directly deduced from the "continuity" equation for the energy momentum tensor $\hat{T}^{\mu\nu}$, utilizing Noether's theorem [530]. For the Lagrangian (8.11) $\hat{T}^{\mu\nu}$ reads[5]

---

[4] Equation (8.34) may be derived directly by combining the field equation (8.32) with its hermitian conjugate.

[5] We base our considerations on the "symmetric" energy momentum tensor $\hat{T}^{\mu\nu}$, rather than the canonical tensor $\hat{\Theta}^{\mu\nu}$. Both versions of the energy momentum tensor satisfy identical "continuity" equations, i.e. all physical results are independent of this choice. $\hat{T}^{\mu\nu}$ represents a covariant combination of the energy, momentum and stress densities of the system.

$$
\begin{aligned}
\hat{T}^{\mu\nu}(x) = \frac{i\hbar c}{8}\Big[\hat{\bar{\psi}}(x), \Big(\gamma^{\mu}\overset{\rightarrow\nu}{\partial} + \gamma^{\nu}\overset{\rightarrow\mu}{\partial} - \gamma^{\mu}\overset{\leftarrow\nu}{\partial} - \gamma^{\nu}\overset{\leftarrow\mu}{\partial}\Big)\hat{\psi}(x)\Big] \\
+\frac{1}{4\pi}\Big\{\hat{F}^{\mu\rho}(x)\hat{F}_{\rho}{}^{\nu}(x) + \frac{1}{4}g^{\mu\nu}\hat{F}(x)^2 - \frac{\lambda}{2}g^{\mu\nu}\Big(\partial_{\rho}\hat{A}^{\rho}(x)\Big)^2 \\
-\lambda\Big(\partial_{\rho}\partial_{\tau}\hat{A}^{\tau}(x)\Big)\Big(g^{\mu\nu}\hat{A}^{\rho}(x) - g^{\mu\rho}\hat{A}^{\nu}(x) - g^{\nu\rho}\hat{A}^{\mu}(x)\Big)\Big\} \\
+\frac{e}{2}\Big(\hat{j}^{\mu}(x)\hat{A}^{\nu}(x) + \hat{j}^{\nu}(x)\hat{A}^{\mu}(x)\Big) \\
-\frac{e}{2}\Big(\hat{j}^{\mu}(x)V^{\nu}(\boldsymbol{x}) - \hat{j}^{\nu}(x)V^{\mu}(\boldsymbol{x})\Big). \qquad (8.36)
\end{aligned}
$$

The last line of Eq. (8.36) indicates that one is dealing with an open system: the source field breaks the symmetry of $\hat{T}^{\mu\nu}$. For this reason the “continuity” equation for $\hat{T}^{\mu\nu}$ contains a source term,

$$\partial_{\mu}\hat{T}^{\mu\nu}(x) = -e\,\hat{j}_{\mu}(x)\,\partial^{\nu}V^{\mu}(\boldsymbol{x}). \qquad (8.37)$$

However, as $V^{\mu}(\boldsymbol{x})$ does not depend on time, one finds

$$\partial_{\mu}\hat{T}^{\mu 0}(x) = 0 \qquad \Longrightarrow \qquad \partial_0\int d^3x\,\hat{T}^{00}(x) = 0. \qquad (8.38)$$

This implies conservation of energy in the rest frame of the sources and allows the identification of the Hamiltonian,

$$
\begin{aligned}
\hat{H} &= \int d^3x\,\hat{T}^{00} = \hat{H}_{\mathrm{e}} + \hat{H}_{\gamma} + \hat{H}_{\mathrm{int}} + \hat{H}_{\mathrm{ext}} && (8.39)\\
\hat{H}_{\mathrm{e}}(x^0) &= \frac{1}{2}\int d^3x\,\big[\hat{\psi}^{\dagger}(x), \big(-i\hbar c\boldsymbol{\alpha}\cdot\boldsymbol{\nabla} + \beta mc^2\big)\,\hat{\psi}(x)\big] && (8.40)\\
\hat{H}_{\gamma}(x^0) &= -\frac{1}{8\pi}\int d^3x\Big\{[\partial^0\hat{A}_{\mu}(x)][\partial^0\hat{A}^{\mu}(x)] + \boldsymbol{\nabla}\hat{A}_{\mu}(x)\cdot\boldsymbol{\nabla}\hat{A}^{\mu}(x)\Big\} && (8.41)\\
\hat{H}_{\mathrm{int}}(x^0) &= -e\int d^3x\,\hat{j}^{\mu}(x)\,\hat{A}_{\mu}(x) && (8.42)\\
\hat{H}_{\mathrm{ext}}(x^0) &= -e\int d^3x\,\hat{j}^{\mu}(x)\,V_{\mu}(\boldsymbol{x}) && (8.43)
\end{aligned}
$$

(Feynman gauge $\lambda = 1$ is used here).

The ground state $|\Psi_0\rangle$ of the Hamiltonian $\hat{H}$ is in general nondegenerate, i.e. as long as $V_{\mu}(\boldsymbol{x})$ does not exhibit some spatial symmetries. Not only continuous symmetries, but also the discrete symmetries usually considered in QED [531] depend on the special form of V_{μ} (compare [532]). For parity to be a good quantum number reflection symmetry of the potential is required. Under charge conjugation the Hamiltonian shows the same transformation behavior as the Lagrangian, $\hat{\mathscr{C}}\hat{H}[V^{\mu}]\hat{\mathscr{C}}^{\dagger} = \hat{H}[-V^{\mu}]$. Finally, time reversal symmetry requires purely electrostatic potentials of the form $V^{\mu} = (V^0, \mathbf{0})$. While the twofold degeneracy, which results for such potentials, leads to an additional conserved quantum number, this does not cause any problems as the Fock space can be decomposed accordingly. In the subsequent discussion it will always be assumed that $|\Psi_0\rangle$ is nondegenerate.

Unfortunately, a straightforward calculation of the ground state expectation value $\langle\Psi_0|\hat{H}|\Psi_0\rangle$ of the Hamiltonian (8.39) is not possible: without further prescriptions the theory based on the Lagrangian (8.11) is not well-defined, but suffers from three types of divergences.[6] These divergences show up both in the Green's functions of the theory as well as in the expectation values characterizing physical observables. This is true in particular for the ground state energy and the ground state four current density, which are the basic ingredients of the RDFT formalism. A physically consistent removal of these divergences requires a renormalization of the fundamental parameters of the theory (the electron charge e and mass m as well as the normalization of the field operators). The renormalization procedure, which is usually formulated within the framework of perturbation theory, is described in detail in the Appendices F–I, which address in particular the case of inhomogeneous systems. Here only a brief summary is given.

The first type of divergence results directly from the presence of negative energy states, whose contribution leads to divergent vacuum expectation values (e.g. for the energy and the charge). This difficulty already exists for noninteracting fermions and can be resolved by explicit subtraction of vacuum expectation values (or, alternatively, by normal-ordering). For instance, if one considers noninteracting electrons subject to some external potential, i.e. a Hamiltonian of the form

$$\hat{H}_{\rm s} = \hat{H}_{\rm e} + \hat{H}_{\rm ext}\,, \tag{8.44}$$

the renormalized Hamiltonian can be given as

$$\hat{H}'_{\rm s,R} = \hat{H}_{\rm s} - \langle 0_{\rm s}|\hat{H}_{\rm s}|0_{\rm s}\rangle\,, \tag{8.45}$$

where $|0_{\rm s}\rangle$ represents the vacuum in the presence of the external potential. The Hamiltonian (8.45) leads to a finite ground state energy by construction (for details see Appendix F). However, the prescription (8.45) introduces different reference energies for different external potentials, i.e. it ignores the energy difference between different vacua (the so-called Casimir energy [533]). This poses no problem as long as only processes are considered in which V^μ does not change (as e.g. the ionization of an electron). In the context of RDFT, however, a universal energy standard is required, as one wants to compare the energy values associated with different external potentials. A suitable definition of the renormalized Hamiltonian is in this case

$$\hat{H}_{\rm s,R} = \hat{H}_{\rm s} - \langle 0_0|\hat{H}_{\rm e}|0_0\rangle\,, \tag{8.46}$$

where $|0_0\rangle$ is the homogeneous vacuum corresponding to the noninteracting Hamiltonian $\hat{H}_{\rm e}$ without any external potential. This universal choice of the energy zero reintroduces, however, a divergence into the expectation values of $\hat{H}_{\rm s,R}$: while all expectation values $\langle\hat{H}_{\rm s,R}\rangle$ remain finite if V^μ vanishes, one encounters a divergence proportional to V^μ for non-vanishing potentials (see Appendix F). This singularity

6 The discussion of infrared divergences is not necessary for the present purpose and will be omitted. Infrared divergences can be suppressed by enclosing the system in a large (cubic) box and requiring periodic boundary conditions on the surface of the cube (compare [28]).

has exactly the same form as the second type of divergences present in standard QED without external potentials, the high-energy or ultraviolet (UV) divergences.

The removal of UV divergences constitutes the core of the renormalization program of QED [531]. In this procedure the divergences are first suppressed at all intermediate steps of calculations by introducing a suitable regularization of the divergent integrals and then absorbed into a redefinition of the fundamental parameters in the Lagrangian. This redefinition leads to additional contributions to the Green's functions, the so-called counterterms, when the Green's functions are expressed in terms of the actual physical parameters. The precise form of the counterterms is controlled by a few unique, elementary requirements on the vacuum Green's functions (called normalization conditions: for instance, one requires the Lehmann representation of the electron single-particle Green's function to have a simple pole at $E^2 = \boldsymbol{p}^2c^2 + m^2c^4$ with m being the true mass of the electron—see Appendix G). The complete program is usually set up for the vacuum state, proceeding order by order in the electron–electron coupling constant α (i.e. in a perturbative manner). As a result all vacuum Green's functions of QED without external potential are finite.

Since this procedure associates well-defined counterterms with each individual element (subdiagram) of the perturbation expansion, it also leads to finite Green's functions in the case of the relativistic homogeneous electron gas (RHEG), i.e. the system characterized by the Hamiltonian

$$\hat{H}^{\text{hom}} = \hat{H}_{\text{e}} + \hat{H}_{\gamma} + \hat{H}_{\text{int}} , \tag{8.47}$$

and a non-vanishing density $n_0 = \langle \hat{j}^0 \rangle$. The individual terms in the perturbation expansions of the RHEG Green's functions can be split into vacuum and electron gas contributions. The renormalization of the former follows the pattern applied in vacuum QED, the latter are finite without renormalization (Appendix H).

As final step of the UV renormalization program for inhomogeneous systems characterized by the full Lagrangian (8.11) it remains to deal with the external potential, not included in (8.47). The renormalization of Green's functions and observables corresponding to (8.11) is straightforward, if a perturbation expansion with respect to V^μ is used. For instance, one obtains for the ground state four current density (see Appendix I),

$$\begin{aligned} j_\mu(\boldsymbol{r}) &= \langle \Psi_0^{\text{RHEG}} | \hat{j}_\mu(x) | \Psi_0^{\text{RHEG}} \rangle + \Delta j_\mu(\boldsymbol{r}) \\ &\quad + \sum_{n=1}^{\infty} \frac{(-e)^n}{n!} \int d^3r_1 \ldots \int d^3r_n \, \chi^{(n+1)}_{\text{c},\mu\mu_1\ldots\mu_n}(\boldsymbol{r}, \boldsymbol{r}_1, \ldots \boldsymbol{r}_n) \\ &\qquad \times V^{\mu_1}(\boldsymbol{r}_1) \ldots V^{\mu_n}(\boldsymbol{r}_n) \ , \end{aligned} \tag{8.48}$$

and an analogous perturbation expansion for the total counterterm Δj_μ,

$$\Delta j_\mu(\boldsymbol{r}) = \Delta j_\mu^{\text{hom}} + \Delta j_\mu^{\text{inhom}}(\boldsymbol{r}) \tag{8.49}$$

$$\begin{aligned} \Delta j_\mu^{\text{inhom}}(\boldsymbol{r}) &= \sum_{n=1}^{\infty} \frac{(-e)^n}{n!} \int d^3r_1 \ldots \int d^3r_n \, \Delta\chi^{(n+1)}_{\mu\mu_1\ldots\mu_n}(\boldsymbol{r}, \boldsymbol{r}_1, \ldots \boldsymbol{r}_n) \\ &\qquad \times V^{\mu_1}(\boldsymbol{r}_1) \ldots V^{\mu_n}(\boldsymbol{r}_n) \ . \end{aligned} \tag{8.50}$$

Here $\Delta j_\mu^{\text{hom}}$ represents the counterterms which keep the current expectation value of the interacting RHEG finite,

$$\langle\Psi_0^{\text{RHEG}}|\hat{j}^\mu(x)|\Psi_0^{\text{RHEG}}\rangle + \Delta j_\mu^{\text{hom}} = g^{\mu 0} n_0 \,, \tag{8.51}$$

and $\Delta\chi^{(n)}_{\mu_1\ldots\mu_n}$ denotes the counterterms required for the renormalization of the connected, n-th order response function $\chi^{(n)}_{\text{c},\mu_1\ldots\mu_n}$ of the RHEG (where all individual terms are understood to be regularized—for the precise definition of $\chi^{(n)}_{\text{c},\mu_1\ldots\mu_n}$ see Appendix H; note that Eq. (8.48) may also be used to define the $\chi^{(n)}_{\text{c},\mu_1\ldots\mu_n}$). Both $\Delta j_\mu^{\text{hom}}$ and all $\Delta\chi^{(n)}_{\mu_1\ldots\mu_n}$ are uniquely determined by the density n_0. Similarly, the ground state energy is given by

$$\begin{aligned} E = {} & \langle\Psi_0^{\text{RHEG}}|\hat{H}^{\text{hom}}|\Psi_0^{\text{RHEG}}\rangle - \langle 0|\hat{H}^{\text{hom}}|0\rangle + \Delta E \\ & - e\int d^3r\, \langle\Psi_0^{\text{RHEG}}|\hat{j}^\mu(x)|\Psi_0^{\text{RHEG}}\rangle V_\mu(\boldsymbol{r}) \\ & + \sum_{n=2}^{\infty} \frac{(-e)^n}{n!} \int d^3r_1 \ldots \int d^3r_n\, \chi^{(n)}_{\text{c},\mu_1\ldots\mu_n}(\boldsymbol{r}_1,\ldots\boldsymbol{r}_n) \\ & \qquad \times V^{\mu_1}(\boldsymbol{r}_1)\ldots V^{\mu_n}(\boldsymbol{r}_n)\,, \end{aligned} \tag{8.52}$$

where $|0\rangle$ denotes the vacuum of interacting QED without external potential. Again the counterterm ΔE can only be specified explicitly by an expansion,

$$\Delta E = \Delta E^{\text{hom}} + \Delta E^{\text{inhom}} \tag{8.53}$$

$$\begin{aligned} \Delta E^{\text{inhom}} = {} & -e\int d^3r\, \Delta j_\mu^{\text{hom}} V^\mu(\boldsymbol{r}) \\ & + \sum_{n=2}^{\infty} \frac{(-e)^n}{n!} \int d^3r_1 \ldots \int d^3r_n\, \Delta\chi^{(n)}_{\mu_1\ldots\mu_n}(\boldsymbol{r}_1,\ldots\boldsymbol{r}_n) \\ & \times V^{\mu_1}(\boldsymbol{r}_1)\ldots V^{\mu_n}(\boldsymbol{r}_n)\,. \end{aligned} \tag{8.54}$$

The quantity ΔE^{hom}, which is independent of V^μ, provides the counterterms for the ground state energy of the interacting RHEG,

$$\langle\Psi_0^{\text{RHEG}}|\hat{H}^{\text{hom}}|\Psi_0^{\text{RHEG}}\rangle - \langle 0|\hat{H}^{\text{hom}}|0\rangle + \Delta E^{\text{hom}} = E^{\text{RHEG}}\,. \tag{8.55}$$

With the aid of the perturbation expansion the renormalization of the inhomogeneous system is therefore reduced to the renormalization of the Green's functions of the RHEG. All counterterms can be explicitly specified order by order in the fine structure constant and the external potential.

Expressing the perturbation expansions of the unrenormalized energy and current in terms of the corresponding original expectation values, the renormalized ground state energy E and ground state four current density $j^\mu(x)$ can finally be written as

$$E = \langle\Psi_0|\hat{H}|\Psi_0\rangle - \langle 0|\hat{H}^{\text{hom}}|0\rangle + \Delta E \tag{8.56}$$

$$j^{\mu}(\boldsymbol{r}) = \langle\Psi_0|\hat{j}^{\mu}(x)|\Psi_0\rangle + \Delta j^{\mu}(\boldsymbol{r}). \tag{8.57}$$

As the renormalization procedure by construction preserves current conservation, Eq. (8.34), the current density (8.57) satisfies the relations

$$\int d^3r\, j^0(\boldsymbol{r}) = N \tag{8.58}$$

$$\boldsymbol{\nabla}\cdot\boldsymbol{j}(\boldsymbol{r}) = 0\,, \tag{8.59}$$

with N being the number of (excess) electrons. Equations (8.48)–(8.59) form a suitable starting point for the discussion of the relativistic extension of the Hohenberg-Kohn theorem, as one is now dealing only with finite quantities.

8.3 Existence Theorem

The extension of the HK-theorem to relativistic systems was first formulated by Rajagopal and Callaway [91] (see also [534–536]), utilizing a QED-based Hamiltonian and four current. These authors applied the standard *reductio ad absurdum* of HK to show that the ground state energy is a unique functional of the ground state four current. As questions related to zero-point energies, radiative corrections and, in particular, UV-divergences were not addressed, the argumentation was open to criticism. The HK argument relies on the comparison of energy values, so that a rigorous proof has to be based on the renormalized ground state energies and four currents (8.56) and (8.57), respectively. In other words: one has to make sure that the structure of the counterterms ΔE and Δj^{μ} is compatible with the logic of the HK argument. The compatibility can be demonstrated explicitly within a perturbative framework [72, 529], using the structure of the counterterms (8.49) and (8.53).

The argument of HK proceeds in two steps. In the first step the relation between external potentials and ground states is considered. Let us assume that two different potentials V_{μ} and V'_{μ} yield the same (nondegenerate) ground state $|\Psi_0\rangle$. The aim of the argument is to find a statement which contradicts this assumption. One first writes

$$\hat{H}|\Psi_0\rangle = E|\Psi_0\rangle \tag{8.60}$$

$$\hat{H}'|\Psi_0\rangle = E'|\Psi_0\rangle\,, \tag{8.61}$$

where $\hat{H}$ denotes the Hamiltonian (8.39) in the Schrödinger picture and $\hat{H}'$ is the corresponding Hamiltonian with the potential V'_{μ}. The issue of renormalization need not yet be addressed at this point as no expectation values are taken. For the present purpose a suitable regularization is sufficient. Upon subtraction of both eigenvalue equations,

$$-e\int d^3x\,\hat{j}^{\nu}(\boldsymbol{x})\left[V_{\nu}(\boldsymbol{x}) - V'_{\nu}(\boldsymbol{x})\right]|\Psi_0\rangle = \left[E - E'\right]|\Psi_0\rangle\,, \tag{8.62}$$

one is left with the question, whether the state on the left-hand side of Eq. (8.62) can be collinear with that on the right-hand side. If not, the desired contradiction would have been found. While the operator $\int d^3x\, \hat{j}^\nu [V_\nu - V'_\nu]$ in general does not commute with the Hamiltonian $\hat{H}$, this does not exclude the existence of a single common eigenstate $|\Psi_0\rangle$. An obvious contradiction only arises for multiplicative potentials of the form $V^\mu = (V^0, \mathbf{0})$. For purely electrostatic potentials one can therefore conclude that two potentials V_0 and V'_0 lead to different ground states, as long as V_0 and V'_0 differ by more than an additive constant ($V'_0 \neq V_0 + const$—as the total charge operator commutes with the Hamiltonian). One is thus led to the question whether the inclusion of a magnetic field can compensate the difference which results from two different electrostatic components V_0 and V'_0, or whether two different magnetic fields can yield the same ground state.

No rigorous answer to this question can be given to date. As discussed in Sect. 2.5, one can demonstrate explicitly that in nonrelativistic spin-density functional theory different magnetic fields can lead to the same ground state [37, 38]. However, the example given for this non-uniqueness relies crucially on the particular structure and the particular coupling of the magnetic field: the field simply adds spin-dependent constants to the spin-up and spin-down potentials of collinear SDFT. This counterexample can not be extended directly to the relativistic situation with its $j_\mu V^\mu$-coupling, so that a unique relation between V^μ and the corresponding ground state may or may not exist.

Fortunately, such a unique map between the space of four potentials and that of the corresponding ground states is not required for the existence of a ground state density functional. It is sufficient that the renormalized ground state four current j^μ, Eq. (8.57), determines the ground state $|\Psi_0\rangle$ uniquely (the second step of the HK argument). In order to prove this statement, let us compare two weakly inhomogeneous systems obtained by perturbing an electron gas with density n_0 by two external potentials V_μ and V'_μ which differ by more than a gauge transformation (the fact that the two systems must have the same average density just reflects the requirement that their charge has to be identical). The resulting ground states, assumed to be nondegenerate, are denoted by $|\Psi_0\rangle$ and $|\Psi'_0\rangle$, the ground state four current densities by j^μ and $j^{\mu\prime}$. Since V_μ and V'_μ differ by more than a gauge transformation, the same statement applies to the $|\Psi_0\rangle$ and $|\Psi'_0\rangle$. According to Eq. (8.56), the corresponding renormalized ground state energies are given by

$$E = \langle\Psi_0\Big|\hat{H}^{\rm hom} - e\int d^3x\, \hat{j}^\mu V_\mu\Big|\Psi_0\rangle - \langle 0|\hat{H}^{\rm hom}|0\rangle + \Delta E[V] \tag{8.63}$$

$$E' = \langle\Psi'_0\Big|\hat{H}^{\rm hom} - e\int d^3x\, \hat{j}^\mu V'_\mu\Big|\Psi'_0\rangle - \langle 0|\hat{H}^{\rm hom}|0\rangle + \Delta E[V'] . \tag{8.64}$$

Note that the counterterms do not depend on the state under consideration: within the perturbative framework ΔE is completely determined by the external potential and the average density of the underlying homogeneous electron gas (see Eqs. (8.48)–(8.55)).

As soon as $|\Psi'_0\rangle$ differs from $|\Psi_0\rangle$, the state $|\Psi'_0\rangle$ has some overlap with at least one of the excited states of the Hamiltonian with potential V_μ. In order to proceed with the usual HK-proof, one thus has to clarify how the expression $\langle\Psi_0'|\hat{H}^{\rm hom} - e\int d^3x\,\hat{j}^\mu V_\mu|\Psi_0'\rangle$, i.e. the energy corresponding to the state $|\Psi_0'\rangle$ in the unprimed system, is to be renormalized. To this aim one uses the fact that the counterterms only depend on the density n_0 and the associated potential, but not on the state itself. Ultimately, the form of the counterterms is determined by the normalization conditions for the Green's functions of vacuum QED and there is only one unique way to include these normalization conditions in perturbation expansions with respect to the external potential, i.e. in Eqs. (8.48)–(8.55): as long as the representation of the counterterms relies on the ground state response functions of the RHEG, their precise form is defined by that potential for which $|\Psi_0'\rangle$ is the ground state, i.e. V_μ'. One therefore has to rewrite the excited state energy $\langle\Psi_0'|\hat{H}^{\rm hom} - e\int d^3x\,\hat{j}^\mu V_\mu|\Psi_0'\rangle$ in such a way, that the ground state energy E', Eq. (8.64), can be extracted,

$$\begin{aligned}\langle\Psi_0'|\hat{H}^{\rm hom} - e\int d^3x\,\hat{j}^\mu V_\mu|\Psi_0'\rangle &= \langle\Psi_0'|\hat{H}^{\rm hom} - e\int d^3x\,\hat{j}^\mu V_\mu'|\Psi_0'\rangle \\ &\quad -e\int d^3x\,\langle\Psi_0'|\hat{j}^\mu|\Psi_0'\rangle(V_\mu - V_\mu')\,.\end{aligned}$$

Now the counterterms for the first term on the right-hand side are uniquely determined by (8.52), the counterterms for the second term by the unique renormalization of $\langle\Psi_0'|\hat{j}^\mu|\Psi_0'\rangle$ as in (8.48). The renormalized energy $E_{\rm es}$ associated with $|\Psi_0'\rangle$ in the unprimed system is given thus by

$$\begin{aligned}E_{\rm es} = \ &\langle\Psi_0'\Big|\hat{H}^{\rm hom} - e\int d^3x\,\hat{j}^\mu V_\mu'\Big|\Psi_0'\rangle - \langle 0|\hat{H}^{\rm hom}|0\rangle + \Delta E[V'] \\ &-e\int d^3x\Big[\langle\Psi_0'|\hat{j}^\mu|\Psi_0'\rangle + \Delta j^\mu[V']\Big](V_\mu - V_\mu')\,.\end{aligned} \tag{8.65}$$

This expression is easily rewritten as

$$E_{\rm es} = E' - e\int d^3x\,j^{\mu\prime}(V_\mu - V_\mu')\,, \tag{8.66}$$

where $j^{\mu\prime}$ now denotes the renormalized current. At this point, all quantities involved are finite.

As $|\Psi_0\rangle$ is assumed to be nondegenerate, the energy (8.66) associated with $|\Psi_0'\rangle$ in the unprimed system must be higher than the ground state energy,

$$E < E' - e\int d^3x\,j^{\mu\prime}(V_\mu - V_\mu')\,. \tag{8.67}$$

One can now interchange all primed and unprimed quantities in this argument to arrive at

$$E' < E - e\int d^3x\,j^{\mu}(V_\mu' - V_\mu)\,. \tag{8.68}$$

Upon addition of (8.67) and (8.68),

$$E + E' < E + E' - e\int d^3x (j^\mu - j^{\mu\prime})(V_\mu - V'_\mu), \tag{8.69}$$

one finally realizes that a contradiction arises for $j^{\mu\prime} = j^\mu$. We have shown that for states $|\Psi'_0\rangle$ and $|\Psi_0\rangle$ which differ by more than a gauge transformation one also has $j^{\mu\prime} \neq j^\mu$, so that the ground state of any such system is uniquely determined by the ground state four current.

On the other hand, if the two potentials only differ by a (static) gauge transformation,

$$V'_\mu(\boldsymbol{x}) = V_\mu(\boldsymbol{x}) + \partial_\mu \lambda(\boldsymbol{x}) , \qquad \Delta\lambda(\boldsymbol{x}) = 0,$$

the four currents obtained from (8.48) are identical due to the transversality of the response functions, Eq. (H.14). The same is then true for the counterterms (8.49) and (8.53), so that the inequalities (8.67)–(8.69) become equalities,

$$\int d^3x\, j^\mu (V_\mu - V'_\mu) = -\int d^3x\, \boldsymbol{j}\cdot\boldsymbol{\nabla}\lambda = \int d^3x\, \lambda \boldsymbol{\nabla}\cdot\boldsymbol{j} = 0$$

(due to Eq. (8.59)). In other words: there is no Ritz principle for states which differ by no more than a gauge transformation.[7]

Consequently, there exists a one-to-one correspondence between the class of all ground states which just differ by gauge transformations and the associated ground state four current. In mathematical terms one can state that $|\Psi_0\rangle$ is a unique functional of j^μ once the gauge has been fixed universally,

$$\Big\{ |\Psi_0\rangle \Big|\, |\Psi_0\rangle \text{ from } V_\mu + \partial_\mu \Lambda \Big\} \Longleftrightarrow j^\mu(\boldsymbol{x}) \quad \Longrightarrow \quad |\Psi_0\rangle = |\Psi_0[j]\rangle . \tag{8.70}$$

This situation is illustrated in Fig. 8.1. Note that for finite systems a unique gauge for the spatial components $\boldsymbol{V}$ can be obtained by combining Coulomb gauge, $\boldsymbol{\nabla}\cdot\boldsymbol{V} = 0$, with the requirement $\boldsymbol{V}(\boldsymbol{r}) \xrightarrow{|\boldsymbol{r}|\to\infty} 0$ (Helmholtz's theorem).

The proof given relies on a perturbation expansion with respect to both the electron–electron interaction and V^μ. The necessity for these expansions originates from the recursive nature of the renormalization scheme which proceeds order by order in the fine-structure constant and from the fact that the treatment of inhomogeneous systems has to be associated with the renormalization procedure for the homogeneous QED vacuum. Only in this framework is it possible to extract the structure of the required counterterms, which is the first important building block

[7] In view of this result the reader might wonder how the inequality (2.168) of CSDFT can emerge in the nonrelativistic limit. The point to be noted here is the fact that the Pauli Hamiltonian is not a consistent expansion of the QED Hamiltonian in the order of $1/c$: while all terms of the order $1/c$ are consistently included, there are additional terms of the order $1/c^2$ which are left out of the Pauli Hamiltonian. It is exactly the order $1/c^2$ treated inconsistently which ensures the validity of the Ritz inequality for states which only differ by a gauge transformation.

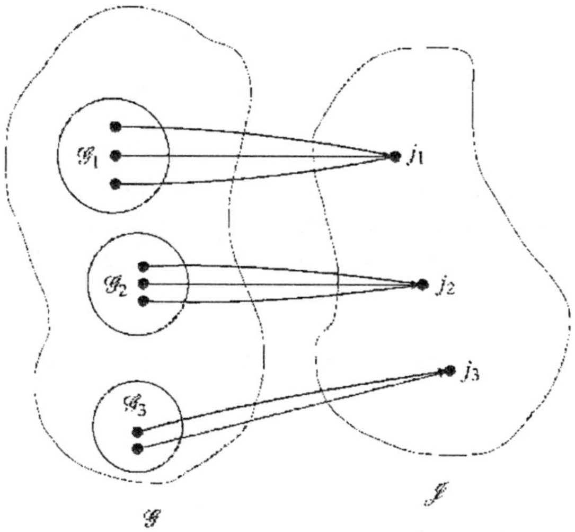

Fig. 8.1 Correspondence between subsets $\mathcal{G}_i$ of ground states which differ by no more than a gauge transformation and associated ground state four currents j_i.

for establishing the inequality (8.69). The discussion is formally valid to all orders, independent of the fact that the QED perturbation series is an asymptotic expansion. In addition, the case of finite systems is covered, at least in principle, by the limit $n_0 \longrightarrow 0$. This limit is particularly transparent for noninteracting systems as the corresponding counterterms (I.5) and (I.11) are independent of n_0 and are therefore directly applicable to arbitrary inhomogeneous systems, as e.g. the KS system.

The second important ingredient of the inequality (8.69) is the existence of a minimum principle for the ground state energy (8.56). While the Ritz variational principle is well established in the nonrelativistic context, no rigorous proof of a minimum principle for the renormalized ground state energies of bound state QED seems to be available. On the other hand, it is exactly the requirement that the energy spectrum must have a lower bound which is the main motivation for the first step of the renormalization program, the elimination of the divergent zero-point energy. This is not only true for standard QED without external potential, but also for noninteracting electrons subject to some V^{μ} (see Appendix F). Turning to interacting systems, the mere assumption of a nondegenerate ground state $|\Psi_0\rangle$ implies that all other states lead to energies higher than that of $|\Psi_0\rangle$. The perturbative approach used here preserves this ordering and it does so in a unique fashion: it provides a unique answer for the renormalization of the energy expectation value of any given state, be it the ground state or any other (as discussed above). In fact, if the renormalized energies would not reflect the minimum principle for the energy which is observed in nature, this would question the renormalization program, rather than the minimum principle.

Fixing the gauge once and for all, the relation (8.70) allows an interpretation of all ground state observables as unique functionals of j^{μ}. The most important functional of this type is the ground state energy itself,

$$E[j] = \langle \Psi_0[j] | \hat{H} | \Psi_0[j] \rangle - \langle 0 | \hat{H}^{\mathrm{hom}} | 0 \rangle + \Delta E \,. \tag{8.71}$$

In fact, in the case of the ground state energy functional the gauge problem is particularly irrelevant: similar to the situation for degenerate ground states discussed in Sect. 2.2, the energy is uniquely determined by j, since the same ground state energy is obtained for all states which only differ by gauge transformations.

The energy functional (8.71) not only contains all relativistic kinetic effects for both electrons and photons, but also all radiative effects. Utilizing once again the energy minimum principle, one may then formulate the basic variational principle of RDFT as

$$\frac{\delta}{\delta j^{\nu}(\boldsymbol{r})} \left\{ E[j] - \mu \int d^3x \, j^0(\boldsymbol{x}) \right\} = 0 \,. \tag{8.72}$$

The subsidiary condition ensures charge conservation, Eq. (8.58), and all quantities involved are supposed to be fully renormalized. Solution of (8.72) with the exact functional $E[j]$ yields the exact ground state four current j^{μ} and, upon insertion of j^{μ} into $E[j]$, the exact ground state energy.

As it stands the functional (8.71) is well defined for all j^{μ} which result as ground state four currents from some external potential V^{μ}. Strictly speaking, this does not yet guarantee the existence of the functional derivative $\delta E[j]/\delta j^{\mu}$ on the set of ground state four currents, which is a prerequisite for the applicability of the variational equation (8.72) (compare the discussion in Sect. 2.3). The question of the existence of the functional $E[j]$ for a sufficiently dense set of j^{μ}, i.e. the interacting v-representability of $E[j]$, has been investigated by Eschrig and collaborators [61, 28]. Interacting v-representability can be ensured on a formal level by a redefinition of $E[j]$ via the functional Legendre transform technique, which generalizes the Lieb functional (2.106) to the relativistic domain and, at the same time, to arbitrary non-integer particle number (see also [537]). For details the reader is referred to [28], which exposes the Legendre transform approach in detail for the nonrelativistic situation.

In view of the difficulties associated with the renormalization procedure one may ask whether it is possible to base RDFT on an approximate relativistic many-body approach, as e.g. the Dirac-Coulomb (DC) Hamiltonian,

$$\hat{H}^{\mathrm{DC}} = \hat{H}_{\mathrm{e}} + \hat{H}_{\mathrm{ext}} + \hat{H}_{\mathrm{e-e}} \tag{8.73}$$

$$\hat{H}_{\mathrm{e-e}} = \frac{e^2}{2} \int d^3r \int d^3r' \, \frac{\hat{\psi}^{\dagger}(\boldsymbol{r}) \hat{\psi}^{\dagger}(\boldsymbol{r}') \hat{\psi}(\boldsymbol{r}') \hat{\psi}(\boldsymbol{r})}{|\boldsymbol{r} - \boldsymbol{r}'|} \,, \tag{8.74}$$

or its Dirac-Coulomb-Breit (DCB) extension, rather than on QED. In this case the so-called *no-pair approximation* plays the role of the renormalization scheme. In this formalism a projection operator $\hat{\Lambda}_+$ is used to project out all negative energy states, in order to avoid that electrons access these states,

$$\hat{H}^{\mathrm{DC}}_{\mathrm{np}} = \hat{\Lambda}_+ \hat{H}^{\mathrm{DC}} \hat{\Lambda}_+ \,, \qquad \hat{j}^{\mu}_{\mathrm{np}} = \hat{\Lambda}_+ \hat{j}^{\mu} \hat{\Lambda}_+ \,. \tag{8.75}$$

However, $\hat{\Lambda}_+$ can be specified unambiguously only within some well-defined single-particle scheme (as the HF or KS approach). Even in this case $\hat{\Lambda}_+$ depends on the actual single-particle potential and in consequence on the external potential, $\hat{\Lambda}_+[V^\mu]$. For this reason one finds that $\hat{H}^{\rm DC}_{\rm np}$ is a nonlinear functional of V^μ, which does not allow the application of the usual *reductio ad absurdum* strategy of the HK-proof. In addition, the *no-pair* approximation introduces a gauge dependence of the ground state energy [350], so that an unambiguous comparison of two ground state energies is only possible if one neglects the Breit interaction and restricts oneself to an external potential of the form $V^\mu = (V^0, \mathbf{0})$. This demonstrates that the existence theorem of RDFT has to be based on a field theoretical formalism. The *no-pair* approximation, which is used in all applications, is much more easily introduced at a later stage, i.e. in the context of the single-particle equations of RDFT.

8.4 Relativistic Kohn-Sham Equations

The starting point for the derivation of the relativistic KS (RKS) scheme is the assumption that there exists a relativistic system of noninteracting particles with exactly the same ground state four current $j^\mu(\boldsymbol{r})$ as the interacting system that one is actually interested in. The question whether such a noninteracting system always exists, i.e. the question of noninteracting v-representability, has not been examined in the relativistic case. One would, however, expect similar statements as in the nonrelativistic situation (compare Sect. 3.2). Since the auxiliary system is noninteracting, its Hamiltonian is of the type (8.44), for which the renormalization procedure is discussed in detail in the Appendices F and I. The ground state corresponding to (8.44), i.e. to the Hamiltonian (F.23), is given by (F.13) and its vacuum four current has the form (F.26). Both the energy and the four current have to be renormalized. Subtraction of the appropriate zero-point energy is required according to Eq. (F.29), as well as the elimination of the lowest order UV-divergences, as indicated in Eqs. (I.5) and (I.11). The ground state four current j^μ of the auxiliary system and hence, by assumption, of the interacting system can be written as

$$j^\mu(\boldsymbol{r}) = \sum_k \Theta_k \phi_k^\dagger(\boldsymbol{r})\alpha^\mu\phi_k(\boldsymbol{r}) + j_{\rm V}^\mu(\boldsymbol{r}) \tag{8.76}$$

$$j_{\rm V}^\mu(\boldsymbol{r}) = \frac{1}{2}\Bigg\{\sum_{\varepsilon_k \le -mc^2} \phi_k^\dagger(\boldsymbol{r})\alpha^\mu\phi_k(\boldsymbol{r}) - \sum_{-mc^2<\varepsilon_k} \phi_k^\dagger(\boldsymbol{r})\alpha^\mu\phi_k(\boldsymbol{r})\Bigg\} + \Delta j^{\mu,(0)}(\boldsymbol{r}) \tag{8.77}$$

$$\Theta_k = \begin{cases} 0 \text{ for } \varepsilon_k \le -mc^2 \\ 1 \text{ for } -mc^2 < \varepsilon_k \le \varepsilon_{\rm F} \\ 0 \text{ for } \varepsilon_{\rm F} < \varepsilon_k \end{cases} . \tag{8.78}$$

The quantities ϕ_k denote the single-particle spinors of the auxiliary system and $\varepsilon_{\rm F}$ represents the Fermi level which separates occupied from unoccupied auxiliary states. In (8.76) the total current has been decomposed into a vacuum contribution $j_{\rm V}^\mu$ and the contribution of the actual electronic states. $\Delta j^{\mu,(0)}(\boldsymbol{r})$ is the lowest order counterterm which keeps the vacuum current $j_{\rm v}^{\mu}$ UV-finite. Using dimensional regularization, $\Delta j^{\mu,(0)}(\boldsymbol{r})$ is given by Eq. (I.5), with the total RKS-potential—specified below—replacing V^{μ} on the right-hand side.

In the next step one decomposes the ground state energy functional (8.71) in the form

$$E = T_{\rm s} + E_{\rm ext} + E_{\rm H} + E_{\rm xc}\,. \tag{8.79}$$

The counterterms for E given in Eq. (8.56) are understood to be included in the individual energy components. These are defined as follows: the noninteracting kinetic energy functional $T_{\rm s}$, i.e. the kinetic energy of the auxiliary system, is given by

$$T_{\rm s} = \sum_k \Theta_k \int d^3r\, \phi_k^{\dagger}(\boldsymbol{r})\big[-i\hbar c\boldsymbol{\alpha}\cdot\nabla + \beta mc^2\big]\phi_k(\boldsymbol{r}) + T_{\rm s,v} \tag{8.80}$$

$$\begin{aligned} T_{\rm s,v} = \frac{1}{2}\int d^3r \Big\{ &\sum_{\varepsilon_k \le -mc^2} \phi_k^{\dagger}(\boldsymbol{r})\big[-i\hbar c\boldsymbol{\alpha}\cdot\nabla + \beta mc^2\big]\phi_k(\boldsymbol{r}) \\ &- \sum_{-mc^2<\varepsilon_k} \phi_k^{\dagger}(\boldsymbol{r})\big[-i\hbar c\boldsymbol{\alpha}\cdot\nabla + \beta mc^2\big]\phi_k(\boldsymbol{r})\Big\} \\ &-\langle 0_0|\hat{H}_{\rm e}|0_0\rangle + \Delta T_{\rm s}^{\rm inhom}\,, \end{aligned} \tag{8.81}$$

with counterterms as specified in Eqs. (F.29) and (I.14). The interaction of the electrons with the external sources contributes the energy

$$E_{\rm ext}[j] = -e\int d^3r\, j_{\mu}(\boldsymbol{r})\, V^{\mu}(\boldsymbol{r})\,. \tag{8.82}$$

The counterterm (I.13) has already been absorbed into the renormalized current j_{μ}. The (direct) Hartree energy $E_{\rm H}$ is defined in a "covariant" fashion with the free photon propagator $D^0_{\mu\nu}$,

$$D^{0,\mu\nu}(x-y) = -i\frac{e^2}{\hbar c}\langle 0_0|T\hat{A}_0^{\mu}(x)\hat{A}_0^{\nu}(y)|0_0\rangle\,, \tag{8.83}$$

where $\hat{A}_0^{\mu}$ denotes the noninteracting photon field operator, and $|0_0\rangle$ is the corresponding noninteracting vacuum state (compare Appendix G). The prefactors have been chosen so that $D^{0,\mu\nu}(x-y)$ approaches the Coulomb interaction in the limit $c\to\infty$,

$$\lim_{c\to\infty} D^{0,\mu\nu}(x-y) = \delta(x^0-y^0)\frac{e^2}{|\boldsymbol{x}-\boldsymbol{y}|}g^{\mu\nu}$$

(in Feynman gauge). With this choice the Hartree energy is defined as

$$E_{\rm H}[j] = \frac{1}{2}\int d^3x \int d^4y\, j^{\mu}(\boldsymbol{x})\, D^0_{\mu\nu}(x-y)\, j^{\nu}(\boldsymbol{y})\,. \tag{8.84}$$

It includes all direct matrix elements of the electron–electron interaction (via exchange of photons) of the order e^2. $E_{\rm H}$ does not depend on the gauge chosen for

$D^0_{\mu\nu}$, as the propagator is contracted with a conserved current. Moreover, the integration over the time-like component y^0 can be performed for the full $D^0_{\mu\nu}$ with the result

$$E_{\rm H}[j] = \frac{e^2}{2}\int d^3x \int d^3y \frac{j_\mu(\boldsymbol{x})\, j^\mu(\boldsymbol{y})}{|\boldsymbol{x}-\boldsymbol{y}|}\,. \tag{8.85}$$

Finally, the xc-energy functional contains all remaining contributions to $E[j]$, i.e. (8.79) defines $E_{\rm xc}$. As the HK argument is also valid for noninteracting systems for which $E = T_{\rm s} + E_{\rm ext}$ one concludes that $T_{\rm s}$ is a density functional, $T_{\rm s}[j]$, in spite of its explicit orbital-dependent form. From this information and the fact that $E_{\rm ext}$ and $E_{\rm H}$ are obvious density functionals follows the statement that $E_{\rm xc}$ is also a unique functional of j^μ, $E_{\rm xc}[j]$.

As the ground state of the auxiliary system and therefore its ingredients ϕ_k are uniquely determined by the four current of the interacting system, the basic variational principle may be exploited by minimizing E with respect to the ϕ_k, rather than j^μ (a more careful argument can be given along the lines of Sect. 3.1, see also [537]). This minimization leads to the relativistic KS equations,

$$\left\{-i\hbar c\boldsymbol{\alpha}\cdot\boldsymbol{\nabla} + \beta mc^2 + \alpha_\mu v_{\rm s}^\mu(\boldsymbol{r})\right\}\phi_k(\boldsymbol{r}) = \varepsilon_k\phi_k(\boldsymbol{r})\,. \tag{8.86}$$

The multiplicative KS four potential $v_{\rm s}^\mu$ consists of the sum of the external potential V^μ, the Hartree potential $v_{\rm H}^\mu$ and the *xc*-potential $v_{\rm xc}^\mu$,[8]

$$v_{\rm s}^\mu(\boldsymbol{r}) = -eV^\mu(\boldsymbol{r}) + v_{\rm H}^\mu(\boldsymbol{r}) + v_{\rm xc}^\mu(\boldsymbol{r}) \tag{8.87}$$

$$v_{\rm H}^\mu(\boldsymbol{r}) = e^2\int d^3r' \frac{j^\mu(\boldsymbol{r}')}{|\boldsymbol{r}-\boldsymbol{r}'|} \tag{8.88}$$

$$v_{\rm xc}^\mu(\boldsymbol{r}) = \frac{\delta E_{\rm xc}[j]}{\delta j_\mu(\boldsymbol{r})}\,. \tag{8.89}$$

Equations (8.76), (8.86)–(8.89) have to be solved selfconsistently, of course.

As in nonrelativistic DFT, the xc-functional is the key ingredient of the KS equations. Exact representations for the relativistic $E_{\rm xc}$ can be derived along the lines of Sects. 4.2.1 and 4.2.2 [534, 350]. For instance, a relativistic extension of Eq. (4.65) is obtained by application of the coupling constant integration scheme to the difference between the total Hamiltonian (8.39) and the relativistic KS Hamiltonian. The result is [350]

---

[8] Note that the spatial components of $v_{\rm xc}^\mu$ have been defined in a way which ensures consistency with the Minkowski notation, but differs from the convention in nonrelativistic CDFT by a minus sign, as the contravariant component j^k corresponds to the physical three vector,

$$v_{\rm xc}^k(\boldsymbol{r}) = \frac{\delta E_{\rm xc}[j]}{\delta j_k(\boldsymbol{r})} = -\frac{\delta E_{\rm xc}[j]}{\delta j^k(\boldsymbol{r})}\,.$$

The transition to the standard CDFT notation will be made in the context of the *no-pair* approximation in Sect. 8.6.

$$E_{\mathrm{xc}} = \frac{1}{2}\lim_{\varepsilon\to 0}\int d^4x\,\delta(x^0)\int d^4y\,e^{-\varepsilon|y^0|}D^0_{\mu\nu}(x-y)$$
$$\times\left\{\langle\Phi_0|T\hat{j}^\mu_0(x)\hat{j}^\nu_0(y)|\Phi_0\rangle - j^\mu(\boldsymbol{x})j^\nu(\boldsymbol{y})\right\}$$
$$+\lim_{\varepsilon\to 0}\sum_{n=1}^{\infty}\frac{(-i)^n}{(n+1)!}\int_{-\infty}^{\infty}dt_1\cdots\int_{-\infty}^{\infty}dt_n$$
$$\times\langle\Phi_0|T\hat{W}(0)\hat{W}(t_1)\cdots\hat{W}(t_n)|\Phi_0\rangle_l\,, \tag{8.90}$$

where $\hat{W}$ is defined as

$$\hat{W}(x^0) = \frac{1}{2}\int d^3x\int d^4y\,e^{-\varepsilon(|x^0|+|y^0|)}\,\hat{j}^\mu_0(x)D^0_{\mu\nu}(x-y)\hat{j}^\nu_0(y)$$
$$-\int d^3x\,e^{-\varepsilon|x^0|}\left\{v_{\mathrm{H},\mu}(\boldsymbol{x})+v_{\mathrm{xc},\mu}(\boldsymbol{x})\right\}\hat{j}^\mu_0(x)\,, \tag{8.91}$$

$\hat{j}^\mu_0(x)$ denotes the current operator of the KS system (in the Heisenberg representation), $|\Phi_0\rangle$ is the KS ground state and the subscript l indicates that only contributions linked to $\hat{W}(0)$ are to be included in the evaluation of (8.90) via Wick's theorem. The counterterms required to keep E_{xc} UV-finite are not noted explicitly. A relativistic version of the adiabatic connection is also available [538].

Just as the kinetic energy (8.80), the xc-energy contains vacuum contributions. The modification of E_{xc} which results from these corrections is most readily illustrated for the exchange component. In direct extension of the nonrelativistic limit, the exact exchange E_{x} is defined as the contribution to E_{xc}, which is linear in e^2 in a perturbation expansion of E_{xc} with respect to the KS Hamiltonian [350], i.e. by the first term on the right-hand side of Eq. (8.90). Including the associated counterterms, E_{x} can be expressed as

$$E_{\mathrm{x}} = \frac{1}{2}\int d^3x\int d^4y\,D^0_{\mu\nu}(x-y)\,tr\left[G^{\mathrm{s}}(x,y)\,\gamma^\nu\,G^{\mathrm{s}}(y,x)\,\gamma^\mu\right]$$
$$-\frac{1}{2}\int d^3x\int d^4y\,D^0_{\mu\nu}(x-y)\,tr\left[G^0_{\mathrm{V}}(x,y)\,\gamma^\nu\,G^0_{\mathrm{V}}(y,x)\,\gamma^\mu\right]$$
$$+\Delta E^{\mathrm{hom}}_{\mathrm{x}}+\Delta E^{\mathrm{inhom}}_{\mathrm{x}}\,. \tag{8.92}$$

Here $G^{\mathrm{s}}(x,y)$ is the propagator of the KS system,

$$iG^{\mathrm{s}}(x,y) = \langle\Phi_0|T\hat{\psi}_0(x)\hat{\overline{\psi}}_0(y)|\Phi_0\rangle \tag{8.93}$$
$$= \Theta(x^0-y^0)\sum_{\varepsilon_{\mathrm{F}}<\varepsilon_k}\phi_k(\boldsymbol{x})\overline{\phi}_k(\boldsymbol{y})e^{-i\varepsilon_k(x^0-y^0)/(\hbar c)}$$
$$-\Theta(y^0-x^0)\sum_{\varepsilon_k\leq\varepsilon_{\mathrm{F}}}\phi_k(\boldsymbol{x})\overline{\phi}_k(\boldsymbol{y})e^{-i\varepsilon_k(x^0-y^0)/(\hbar c)}\,, \tag{8.94}$$

and $G^0_{\mathrm{V}}(x,y)$ denotes the noninteracting vacuum propagator (F.33). The second term in (8.92) represents the exchange contribution to $\langle 0|\,\hat{H}_{\mathrm{e}}+\hat{H}_\gamma+\hat{H}_{\mathrm{int}}\,|0\rangle$. $\Delta E^{\mathrm{hom}}_{\mathrm{x}}$ is the counterterm of the exchange energy of the RHEG, as specified in the second line of Eq. (H.50). $\Delta E^{\mathrm{inhom}}_{\mathrm{x}}$ is the exchange component of $\Delta E^{\mathrm{inhom}}$, determined by the first order vertex correction counterterm. In the case of the RHEG the expression (8.92) reduces to (H.50). On the other hand, the energy (8.92) constitutes the exchange contribution to the Casimir energy, if $G^{s}(x,y)$ is replaced by its vacuum limit (Eq. (8.94) with $\varepsilon_F = -mc^2$).

8.5 Towards a Workable RDFT Scheme: No-pair Approximation

The previous sections indicate the procedure for mapping the field theoretical many-body problem (8.11) to the auxiliary KS problem. In practice, however, Eqs. (8.76), (8.86)–(8.92) represent a computational problem of enormous complexity. The evaluation of essentially all quantities requires summation over all negative and positive energy solutions and appropriate renormalization in each step of the iterative KS procedure.[9] As the solution of this selfconsistency problem is extremely involved even in the case of atoms, one is bound to resort to approximations.

If one aims at electronic structure calculations for molecules or solids, neglect of all effects related to the existence of antiparticle (negative energy) states is legitimate, as these vacuum corrections only affect the innermost core states significantly. The technical implementation of this neglect is immediately clear for all single-particle quantities of RDFT such as the four current j^μ, Eq. (8.76), and the kinetic energy of non-interacting particles T_s, Eq. (8.80),

$$j^\mu_{\rm v}(\boldsymbol{r}) = 0\,; \qquad T_{\rm s,v} = 0\,. \tag{8.95}$$

In this way, all contributions to j^μ and T_s resulting from the creation of virtual particle-antiparticle pairs defined by the relativistic KS system, i.e. via the eigenstates of the relativistic KS equations (8.86), are suppressed.

In the case of many-body terms the treatment of vacuum corrections can only be specified explicitly in a perturbative framework. The noninteracting reference Hamiltonian then defines the type of virtual particle-antiparticle pairs which one deals with. In the present situation, consistency with Eq. (8.95) requires the suppression of KS particle-antiparticle pairs, i.e. the use of a perturbation expansion in terms of the KS Hamiltonian. All vacuum corrections in the xc-energy E_{xc} are consistently eliminated, if the contributions of the negative energy solutions to all intermediate sums over states in the expansion (8.90) are omitted. This so-called *no-pair* approximation can be implemented by neglect of the negative energy states in the KS propagator G^s, which is used for the perturbative evaluation of (8.90),

$$\begin{aligned} iG^{\rm s}_{\rm np}(x,y) = \; & \Theta(x^0-y^0) \sum_{\varepsilon_F<\varepsilon_k} \phi_k(\boldsymbol{x})\overline{\phi}_k(\boldsymbol{y}) e^{-i\varepsilon_k(x^0-y^0)/(\hbar c)} \\ & -\Theta(y^0-x^0) \sum_{-mc^2<\varepsilon_k\le\varepsilon_F} \phi_k(\boldsymbol{x})\overline{\phi}_k(\boldsymbol{y}) e^{-i\varepsilon_k(x^0-y^0)/(\hbar c)}\,. \end{aligned} \tag{8.96}$$

[9] It seems worthwhile to emphasize the fact that the KS vacuum $|0_s\rangle$ depends on v^μ_s and changes during the iterative solution of (8.86).

The *no-pair* approximation has to go hand in hand with an appropriate reordering of the field operators in the interaction (8.91), so that creation operators are all left of the annihilation operators. The correct form of $\hat{W}$ is obtained by normal-ordering the product of current operators in the first line of (8.91),

$$\hat{j}_0^{\mu}(x)\hat{j}_0^{\nu}(y) \longrightarrow \hat{\psi}_0^{\dagger}(y)\left[\hat{\psi}_0^{\dagger}(x)\alpha^{\mu}\hat{\psi}_0(x)\right]\alpha^{\nu}\hat{\psi}_0(y)\,. \tag{8.97}$$

This reordering has to be incorporated in the Feynman rules used for the evaluation of (8.90). Whenever the end points of a KS Green's function are linked by a single photon propagator $D^0_{\mu\nu}$, only the second term in Eq. (8.96) has to be included, as the first term results from the ordering $\hat{\psi}_0(x)\hat{\psi}_0^{\dagger}(y)$. This restriction can be easily implemented in frequency space, using the Lehmann representation of (8.96),

$$G^{\mathrm{s}}_{\mathrm{np}}(x,y) = \int \frac{d\omega}{2\pi}\, e^{-i\omega(x^0-y^0)/(\hbar c)}\, G^{\mathrm{s}}_{\mathrm{np}}(\boldsymbol{x},\boldsymbol{y},\omega) \tag{8.98}$$

$$G^{\mathrm{s}}_{\mathrm{np}}(\boldsymbol{x},\boldsymbol{y},\omega) = \sum_{\varepsilon_{\mathrm{F}}<\varepsilon_k} \frac{\phi_k(\boldsymbol{x})\overline{\phi}_k(\boldsymbol{y})}{\omega-\varepsilon_k+i\eta} + \sum_{-mc^2<\varepsilon_k\leq\varepsilon_{\mathrm{F}}} \frac{\phi_k(\boldsymbol{x})\overline{\phi}_k(\boldsymbol{y})}{\omega-\varepsilon_k-i\eta}\,. \tag{8.99}$$

If $\boldsymbol{x}$ and $\boldsymbol{y}$ are linked by a single photon propagator $D^0_{\mu\nu}(x,y)$, one simply associates a factor of $e^{i\omega\varepsilon}$ (with an infinitesimal $\varepsilon>0$) to $G^{\mathrm{s}}_{\mathrm{np}}(\boldsymbol{x},\boldsymbol{y},\omega)$, so that the contour of the subsequent ω-integration has to be closed in the upper half of the complex ω-plane. For instance, in the case of the exchange energy the *no-pair* approximation plus reordering leads to

$$\begin{aligned} E^{\mathrm{np}}_{\mathrm{x}} &= \frac{1}{2}\int d^3x \int d^4y\, D^0_{\mu\nu}(x-y) \int \frac{d\omega}{2\pi}\int \frac{d\omega'}{2\pi}\, e^{-i(\omega-\omega')(x^0-y^0)/(\hbar c)} \\ &\quad\times e^{i(\omega+\omega')\varepsilon}\, tr\left[G^{\mathrm{s}}_{\mathrm{np}}(\boldsymbol{x},\boldsymbol{y},\omega)\,\gamma^{\nu}\, G^{\mathrm{s}}_{\mathrm{np}}(\boldsymbol{y},\boldsymbol{x},\omega')\,\gamma^{\mu}\right]. \end{aligned} \tag{8.100}$$

Quite generally, the *no-pair* approximation has to be incorporated in E_{xc} during the derivation of specific approximations.

Even within the *no-pair* approximation the physical background of the RDFT formalism is not yet identical with that of the many-body methods based on the *no-pair* DC- or DCB-Hamiltonian. In *no-pair* RDFT still the complete electron–electron interaction mediated by the photon propagator is taken into account both in the Hartree energy (8.85) and in the construction of approximations for E_{xc} (see Appendix H). On the other hand, RDFT can be readily restricted to the Coulomb or Coulomb-Breit level. Choosing Feynman gauge, the full $D^0_{\mu\nu}$, Eq. (8.83),

$$D^0_{\mu\nu}(x-y) = \int \frac{d^4q}{(2\pi)^4}\, e^{-iq\cdot(x-y)}\, D^0_{\mu\nu}(q) \tag{8.101}$$

$$D^0_{\mu\nu}(q) = \frac{-4\pi e^2}{q^2+i\eta}\, g_{\mu\nu}\,, \tag{8.102}$$

reduces to

$$D^{0,\mathrm{CB}}_{\mu\nu}(q) = \frac{4\pi e^2}{\boldsymbol{q}^2}\begin{pmatrix} 1+\frac{(q_0)^2}{\boldsymbol{q}^2} & 0 \\ 0 & g_{ij} \end{pmatrix} \tag{8.103}$$

in the Coulomb-Breit limit and to

$$D^{0,\mathrm{C}}_{\mu\nu}(q) = \frac{4\pi e^2}{\boldsymbol{q}^2} g_{\mu 0} g_{\nu 0} \tag{8.104}$$

on the Coulomb level. As soon as E_{H} and the construction of E_{xc} are based on one of these forms the RDFT analog of the DC- or DCB-approach is obtained. It must be emphasized, however, that the full propagator (8.102) and its Coulomb-Breit limit (8.103) are equivalent to their Coulomb gauge counterparts (usually applied in quantum chemistry),

$$D^{0}_{\mu\nu}(q) = \frac{4\pi e^2}{\boldsymbol{q}^2} \begin{pmatrix} 1 & 0 \\ 0 & \frac{-\boldsymbol{q}^2}{q^2+i\eta}\left(g_{ij} + \frac{q_i q_j}{\boldsymbol{q}^2}\right) \end{pmatrix} \tag{8.105}$$

$$D^{0,\mathrm{CB}}_{\mu\nu}(q) = \frac{4\pi e^2}{\boldsymbol{q}^2} \begin{pmatrix} 1 & 0 \\ 0 & g_{ij} + \frac{q_i q_j}{\boldsymbol{q}^2} \end{pmatrix}, \tag{8.106}$$

only in gauge invariant expressions.

In general, gauge invariance can only be ensured by the inclusion of the negative energy continuum states in all intermediate sums over states. As soon as the *no-pair* approximation is applied a gauge dependence is introduced (with the exception of the *no-pair* exchange of RDFT, Eq. (8.100)—see [350]).

8.6 No-pair RDFT

Applications of the relativistic KS equations always rely on the *no-pair* approximation, in which all effects due the creation of particle-antiparticle pairs are neglected. Let us therefore summarize the basic relations of RDFT in this approximation. This also allows us to switch to the standard notation of nonrelativistic (C)DFT.

We start by recapitulating the existence theorem of RDFT:

There exists a one-to-one correspondence between the class of all ground states which just differ by gauge transformations and the associated ground state four current

$$j^{\mu}(\boldsymbol{r}) = \left(n(\boldsymbol{r}), \boldsymbol{j}(\boldsymbol{r})/c\right). \tag{8.107}$$

This correspondence is visualized in Fig. 8.1. The ground state $|\Psi_0\rangle$ is thus a unique functional of $(n, \boldsymbol{j})$, as soon as the gauge is fixed universally,

$$|\Psi_0\rangle = |\Psi_0[j]\rangle \equiv |\Psi_0[n, \boldsymbol{j}]\rangle .$$

As a result, all gauge invariant observables are unique functionals of $(n, \boldsymbol{j})$, in particular the total ground state energy,

$$E[j] \equiv E[n, \boldsymbol{j}] = \langle \Psi_0[n, \boldsymbol{j}] | \hat{H} | \Psi_0[n, \boldsymbol{j}] \rangle ,$$

where $\hat{H}$ is the QED Hamiltonian (8.39). The uniqueness of $E[n, \boldsymbol{j}]$ results from the fact that all $|\Psi_0\rangle$ which only differ by gauge transformations yield the same ground state energy (similar to the situation encountered for degenerate ground states—compare Sect. 2.2). The functional $E[n, \boldsymbol{j}]$ has a minimum for the ground state four current $(n, \boldsymbol{j})$ which corresponds to the actual external four potential V^{μ} at hand,

$$E[n,\boldsymbol{j}] < E[n',\boldsymbol{j}'] \qquad \forall\ (n',\boldsymbol{j}') \neq (n,\boldsymbol{j})\,. \tag{8.108}$$

The minimum principle (8.108) allows one to recast the relativistic many-body problem as a minimization procedure for $E[n, \boldsymbol{j}]$. Assuming that there exists a non-interacting system whose *no-pair* ground state four current is identical with the ground state four current of the interacting system, one can set up a corresponding RKS scheme, which performs the minimization in practice. In this scheme the components of four current are given by

$$n(\boldsymbol{r}) = \sum_k \Theta_k \phi_k^{\dagger}(\boldsymbol{r})\phi_k(\boldsymbol{r}) \tag{8.109}$$

$$\boldsymbol{j}(\boldsymbol{r}) = c\sum_k \Theta_k \phi_k^{\dagger}(\boldsymbol{r})\boldsymbol{\alpha}\phi_k(\boldsymbol{r})\,, \tag{8.110}$$

with the *no-pair* approximation implemented via the occupation factor Θ_k,

$$\Theta_k = \begin{cases} 0 \text{ for } \varepsilon_k \leq -2mc^2 \\ 1 \text{ for } -2mc^2 < \varepsilon_k \leq \varepsilon_{\mathrm{F}} \\ 0 \text{ for } \varepsilon_{\mathrm{F}} < \varepsilon_k \end{cases}. \tag{8.111}$$

Current conservation implies that the *no-pair* density integrates up to the number of electrons N,

$$\int d^3r\, n(\boldsymbol{r}) = N\,, \tag{8.112}$$

while the *no-pair* current satisfies the relation

$$\boldsymbol{\nabla}\cdot\boldsymbol{j}(\boldsymbol{r}) = 0\,, \tag{8.113}$$

just as the complete current, Eq. (8.59). The total *no-pair* energy and its components are obtained as

$$E = T_{\mathrm{s}} + E_{\mathrm{ext}} + E_{\mathrm{H}}^{\mathrm{C}} + E_{\mathrm{H}}^{\mathrm{T}} + E_{\mathrm{xc}} \tag{8.114}$$

$$T_{\mathrm{s}} = \sum_k \Theta_k \int d^3r\, \phi_k^{\dagger}(\boldsymbol{r})\left[-i\hbar c\boldsymbol{\alpha}\cdot\boldsymbol{\nabla} + (\beta-1)mc^2\right]\phi_k(\boldsymbol{r}) \tag{8.115}$$

$$E_{\mathrm{ext}} = \int d^3r \left\{ n(\boldsymbol{r})\, v_{\mathrm{ext}}(\boldsymbol{r}) + \frac{e}{c}\boldsymbol{j}(\boldsymbol{r})\cdot\boldsymbol{A}_{\mathrm{ext}}(\boldsymbol{r}) \right\} \tag{8.116}$$

$$E_{\mathrm{H}}^{\mathrm{C}} = \frac{e^2}{2}\int d^3r \int d^3r'\, \frac{n(\boldsymbol{r})\, n(\boldsymbol{r}')}{|\boldsymbol{r}-\boldsymbol{r}'|} \tag{8.117}$$

$$E_{\mathrm{H}}^{\mathrm{T}} = -\frac{e^2}{2c^2}\int d^3r \int d^3r' \frac{\boldsymbol{j}(\boldsymbol{r})\cdot\boldsymbol{j}(\boldsymbol{r}')}{|\boldsymbol{r}-\boldsymbol{r}'|}, \tag{8.118}$$

where the rest mass has now been subtracted from the kinetic energy. In addition, the Hartree energy E_{H} has been split into a Coulomb contribution $E_{\mathrm{H}}^{\mathrm{C}}$ and a transverse component $E_{\mathrm{H}}^{\mathrm{T}}$ which reflects the presence of the magnetic (Breit) interaction. The RKS equations have the same form as in the general situation. Splitting the total KS potential v_{s}^{μ} into its time-like and three vector components and switching to the nonrelativistic notation,

$$v_{\mathrm{s}}^{\mu} = \left(v_{\mathrm{s}}, -e\boldsymbol{A}_{\mathrm{s}}\right), \tag{8.119}$$

the RKS equations have the form

$$\left\{-i\hbar c\boldsymbol{\alpha}\cdot\boldsymbol{\nabla} + (\beta-1)mc^2 + v_{\mathrm{s}}(\boldsymbol{r}) + e\boldsymbol{\alpha}\cdot\boldsymbol{A}_{\mathrm{s}}(\boldsymbol{r})\right\}\phi_k(\boldsymbol{r}) = \varepsilon_k\phi_k(\boldsymbol{r}). \tag{8.120}$$

If the standard notation is also used for the components of v_{s}^{μ}, the external potential V^{μ}, the Hartree potential v_{H}^{μ} and the *xc*-potential v_{xc}^{μ},

$$V^{\mu} = \left(-v_{\mathrm{ext}}/e, \boldsymbol{A}_{\mathrm{ext}}\right), \quad v_{\mathrm{H}}^{\mu} = \left(v_{\mathrm{H}}, -e\boldsymbol{A}_{\mathrm{H}}\right), \quad v_{\mathrm{xc}}^{\mu} = \left(v_{\mathrm{xc}}, -e\boldsymbol{A}_{\mathrm{xc}}\right), \tag{8.121}$$

the relations between the potentials are given by

$$v_{\mathrm{s}}(\boldsymbol{r}) = v_{\mathrm{ext}}(\boldsymbol{r}) + v_{\mathrm{H}}(\boldsymbol{r}) + v_{\mathrm{xc}}(\boldsymbol{r}) \tag{8.122}$$

$$v_{\mathrm{H}}(\boldsymbol{r}) = e^2\int d^3r' \frac{n(\boldsymbol{r}')}{|\boldsymbol{r}-\boldsymbol{r}'|} \tag{8.123}$$

$$v_{\mathrm{xc}}(\boldsymbol{r}) = \frac{\delta E_{\mathrm{xc}}[n,\boldsymbol{j}]}{\delta n(\boldsymbol{r})} \tag{8.124}$$

$$\boldsymbol{A}_{\mathrm{s}}(\boldsymbol{r}) = \boldsymbol{A}_{\mathrm{ext}}(\boldsymbol{r}) + \boldsymbol{A}_{\mathrm{H}}(\boldsymbol{r}) + \boldsymbol{A}_{\mathrm{xc}}(\boldsymbol{r}) \tag{8.125}$$

$$\boldsymbol{A}_{\mathrm{H}}(\boldsymbol{r}) = -\frac{e}{c}\int d^3r' \frac{\boldsymbol{j}(\boldsymbol{r}')}{|\boldsymbol{r}-\boldsymbol{r}'|} \tag{8.126}$$

$$\boldsymbol{A}_{\mathrm{xc}}(\boldsymbol{r}) = \frac{c}{e}\frac{\delta E_{\mathrm{xc}}[n,\boldsymbol{j}]}{\delta \boldsymbol{j}(\boldsymbol{r})}. \tag{8.127}$$

The actual solution of Eq. (8.120) can now be restricted to the positive energy spectrum—usually one just needs the N energetically lowest positive energy states. The restriction to positive energy states has be ensured technically by imposing appropriate boundary conditions or using suitable basis sets. If desired, an *a posteriori* perturbative evaluation of vacuum corrections is possible on the basis of the *no-pair* KS ground state $|\Phi_0\rangle$ [539]. While this state differs from the true ground state, the inner core electrons for which radiative corrections become sizable are well described within the effective single-particle picture.

8.7 Variants of RDFT

Even within the *no-pair* approximation the RKS equations (8.120) are more involved than the nonrelativistic KS equations, primarily due to the fact that the RKS potential is a four vector. This raises the question whether one can find simplified forms in which the RKS potential reduces to one or two components. Fortunately, in most applications there is no external magnetic field

$$\boldsymbol{B}_{\text{ext}}(\boldsymbol{r}) = \boldsymbol{\nabla} \times \boldsymbol{A}_{\text{ext}}(\boldsymbol{r}) . \tag{8.128}$$

The external Hamiltonian reduces to

$$\hat{H}'_{\text{ext}} = \int d^3x \hat{n}(x) v_{\text{ext}}(\boldsymbol{x}) ; \qquad \hat{n}(x) = \hat{j}^0(x) \tag{8.129}$$

in this case. One can, as discussed in Sect. 8.3, then prove that there exists a one-to-one mapping between v_{ext}, the ground state and the ground state density [535],

$$\left\{ v_{\text{ext}} \middle| v_{\text{ext}} + const \right\} \Longleftrightarrow \left\{ |\Psi_0\rangle \middle| \, |\Psi_0\rangle \text{ from } v_{\text{ext}} + const \right\} \Longleftrightarrow n(\boldsymbol{r}) . \tag{8.130}$$

The ground state can be interpreted as a unique functional of the density only, $|\Psi_0[n]\rangle$. The same is true for ground state observables as the total energy, $E[n]$, and its various components. As a consequence, one obtains only one single variational equation, identical with the time-like component of (8.72), and the RKS equations (8.109)–(8.127) reduce to

$$\left\{ -i\hbar c \boldsymbol{\alpha} \cdot \boldsymbol{\nabla} + (\beta - 1)mc^2 + v_{\text{s}}(\boldsymbol{r}) \right\} \phi_k(\boldsymbol{r}) = \varepsilon_k \phi_k(\boldsymbol{r}) \tag{8.131}$$

$$n(\boldsymbol{r}) = \sum_k \Theta_k \phi_k^\dagger(\boldsymbol{r}) \phi_k(\boldsymbol{r}) \tag{8.132}$$

$$v_{\text{s}}(\boldsymbol{r}) = v_{\text{ext}}(\boldsymbol{r}) + v_{\text{H}}(\boldsymbol{r}) + \frac{\delta E_{\text{xc}}[n]}{\delta n(\boldsymbol{r})} + \frac{\delta E_{\text{H}}^{\text{T}}[\boldsymbol{j}[n]]}{\delta n(\boldsymbol{r})} . \tag{8.133}$$

It must be emphasized that a system with an external potential of the type $V^\mu = (V^0, \boldsymbol{0})$ can have a magnetic moment. In the present situation this implies that the spatial components of the four current must be viewed as functionals of the density, $\boldsymbol{j}[n] = \langle \Psi_0[n] | \hat{\boldsymbol{j}} | \Psi_0[n] \rangle \neq 0$. If this functional were known, one could include E_{H}^{T} exactly and reformulate the j^μ-dependent E_{xc} of Eq. (8.114) as a purely n-dependent functional, $E_{\text{xc}}[n] \equiv E_{\text{xc}}[n, \boldsymbol{j}[n]]$. This dependence is, however, not known, so that one usually simply neglects E_{H}^{T} and the $\boldsymbol{j}$-dependence of $E_{\text{xc}}[n, \boldsymbol{j}[n]]$ at this point. Moreover, for the large class of time-reversal invariant systems (closed subshells) $\boldsymbol{j}$ simply vanishes, so that $\boldsymbol{j}$-dependent terms do not contribute anyway.

As a matter of principle, the density is sufficient for an exact RDFT treatment of magnetic systems as long as $\boldsymbol{B}_{\text{ext}} = \boldsymbol{0}$, similar to the situation in nonrelativistic DFT. In practice, spin-density functional theory turned out to be necessary for the treatment of spin-polarized ground states in the nonrelativistic case. In the relativistic case full inclusion of magnetic effects is possible via the four current version of

RDFT. However, the standard energy functionals of (R)DFT are based on the electron gas (see Chap. 4 and Sect. 8.8), for which $\boldsymbol{j}$ vanishes. Consequently, the derivation of explicitly $\boldsymbol{j}$-dependent approximations for the energy has to be based on the response of the electron gas to some perturbing field, for which, however, only little information is available. Thus a relativistic extension of spin-density functional theory is desirable, whose basic variables are suitably generalized spin-densities.

The starting point for this generalization is the Gordon decomposition, in which the total current is split into the paramagnetic (orbital) component $\boldsymbol{j}_{\mathrm{p}}$, a gauge term proportional to the scalar density ρ_{s} and the curl of the magnetization density $\boldsymbol{m}$ (a derivation of the Gordon decomposition is given in Appendix K),

$$\boldsymbol{j}(\boldsymbol{r}) = \boldsymbol{j}_{\mathrm{p}}(\boldsymbol{r}) + \frac{e}{mc}\boldsymbol{A}_{\mathrm{ext}}(\boldsymbol{r})\rho_{\mathrm{s}}(\boldsymbol{r}) + \frac{c}{e}\boldsymbol{\nabla}\times\boldsymbol{m}(\boldsymbol{r}) \tag{8.134}$$

$$\boldsymbol{j}_{\mathrm{p}}(\boldsymbol{r}) = -\frac{i\hbar}{2m}\langle\Psi_0|\hat{\psi}^\dagger(x)\beta\left[\boldsymbol{\nabla}\hat{\psi}(x)\right] - \left[\boldsymbol{\nabla}\hat{\psi}^\dagger(x)\right]\beta\hat{\psi}(x)|\Psi_0\rangle \tag{8.135}$$

$$\rho_{\mathrm{s}}(\boldsymbol{r}) = \langle\Psi_0|\hat{\psi}^\dagger(x)\beta\hat{\psi}(x)|\Psi_0\rangle \tag{8.136}$$

$$\boldsymbol{m}(\boldsymbol{r}) = \mu_B\langle\Psi_0|\hat{\psi}^\dagger(x)\beta\boldsymbol{\Sigma}\hat{\psi}(x)|\Psi_0\rangle\,, \tag{8.137}$$

with

$$\boldsymbol{\Sigma} = \begin{pmatrix}\boldsymbol{\sigma} & \boldsymbol{0}\\ \boldsymbol{0} & \boldsymbol{\sigma}\end{pmatrix} \tag{8.138}$$

(the commutator form for operators is not used here, as the *no-pair* approximation is implied; $\mu_B = e\hbar/(2mc)$). With the aid of this decomposition one can introduce a relativistic "spin-density" functional formalism in the following fashion: on the basis of the Hamiltonian

$$\hat{H}'' = \hat{H}_{\mathrm{e}} + \hat{H}_{\gamma} + \hat{H}_{\mathrm{int}} + \int d^3r\left\{\hat{n}(\boldsymbol{r})v_{\mathrm{ext}}(\boldsymbol{r}) + \hat{\boldsymbol{m}}(\boldsymbol{r})\cdot\boldsymbol{B}_{\mathrm{ext}}(\boldsymbol{r})\right\} \tag{8.139}$$

one can establish an existence theorem, which states that the ground state $|\Psi_0\rangle$ is uniquely determined by the ground state charge and magnetization densities $n,\boldsymbol{m}$ [535, 540],

$$|\Psi_0\rangle \Longleftrightarrow (n,\boldsymbol{m}) \qquad \Longrightarrow \qquad |\Psi_0\rangle = |\Psi_0[n,\boldsymbol{m}]\rangle\,. \tag{8.140}$$

The expectation value $\langle\Psi_0[n,\boldsymbol{m}]|\hat{H}''|\Psi_0[n,\boldsymbol{m}]\rangle$ then defines the ground state energy functional $E[n,\boldsymbol{m}]$. As $E[n,\boldsymbol{m}]$ depends on $\boldsymbol{B}_{\mathrm{ext}}$ only via

$$\langle\Psi_0[n,\boldsymbol{m}]|\int\hat{\boldsymbol{m}}\cdot\boldsymbol{B}_{\mathrm{ext}}|\Psi_0[n,\boldsymbol{m}]\rangle$$

the limit $\boldsymbol{B}_{\mathrm{ext}}\to\boldsymbol{0}$ can be taken readily. If $\boldsymbol{B}_{\mathrm{ext}}\neq\boldsymbol{0}$, the Hamiltonian (8.139) can be viewed as an approximation, neglecting the coupling between $\boldsymbol{B}_{\mathrm{ext}}$ and the orbital current. On the other hand, for $\boldsymbol{B}_{\mathrm{ext}}=\boldsymbol{0}$ the Hamiltonian (8.139) provides an exact representation of the system. As a consequence also the relativistic "spin" density functional formalism (often abbreviated as RSDFT) based on (8.139), i.e. the map indicated in Eq. (8.140), becomes exact for $\boldsymbol{B}_{\text{ext}} = \boldsymbol{0}$—in this limit RSDFT simply covers a more general, though artificial, class of systems than required.

The RKS equations corresponding to (8.139) rely on a single-particle representation of the charge and the magnetization density [535, 541, 536, 542],

$$\Big\{ -i\hbar c\boldsymbol{\alpha}\cdot\boldsymbol{\nabla} + (\beta-1)mc^2 + v_{\text{s}} + \mu_B\beta\boldsymbol{\Sigma}\cdot\boldsymbol{B}_{\text{s}} \Big\}\phi_k = \varepsilon_k\phi_k \tag{8.141}$$

$$n(\boldsymbol{r}) = \sum_k \Theta_k \phi_k^\dagger(\boldsymbol{r})\phi_k(\boldsymbol{r}) \tag{8.142}$$

$$\boldsymbol{m}(\boldsymbol{r}) = \mu_B \sum_k \Theta_k \phi_k^\dagger(\boldsymbol{r})\beta\boldsymbol{\Sigma}\phi_k(\boldsymbol{r}) \tag{8.143}$$

$$v_{\text{s}}(\boldsymbol{r}) = v_{\text{ext}}(\boldsymbol{r}) + v_{\text{H}}(\boldsymbol{r}) + \frac{\delta E_{\text{xc}}[n,\boldsymbol{m}]}{\delta n(\boldsymbol{r})} + \frac{\delta E_{\text{H}}^{\text{T}}[\boldsymbol{j}[n,\boldsymbol{m}]]}{\delta n(\boldsymbol{r})} \tag{8.144}$$

$$\boldsymbol{B}_{\text{s}}(\boldsymbol{r}) = \boldsymbol{B}_{\text{ext}}(\boldsymbol{r}) + \frac{\delta E_{\text{xc}}[n,\boldsymbol{m}]}{\delta \boldsymbol{m}(\boldsymbol{r})} + \frac{\delta E_{\text{H}}^{\text{T}}[\boldsymbol{j}[n,\boldsymbol{m}]]}{\delta \boldsymbol{m}(\boldsymbol{r})} . \tag{8.145}$$

In Eqs. (8.141)–(8.145) E_{H}^{T} has been included consistently, with the understanding that $\boldsymbol{j}$ is a unique functional of n and $\boldsymbol{m}$, due to (8.140). However, as in Eqs. (8.131)–(8.133) E_{H}^{T} is usually neglected at this point. Equations (8.141)–(8.145) provide an appropriate starting point for density functional studies of magnetic systems with vanishing or small orbital currents. They can, in particular, be used to discuss the magnetic anisotropy of solids [543]. They have nevertheless not yet found widespread use.[10] The reasons are their rather intricate structure and the lack of practical approximations for $E_{\text{xc}}[n,\boldsymbol{m}]$. Only the relativistic LDA (RLDA) for $E_{\text{x}}[n,\boldsymbol{m}]$ is available [544–546, 541]. Some prototype results obtained with Eqs. (8.141)–(8.145) can be found in [547–549, 61].

An additional simplification offers itself, if the variation of the orientation of $\boldsymbol{m}$ with $\boldsymbol{r}$ (i.e. the *non-collinearity* of $\boldsymbol{m}$) is not relevant for the system. One can then restrict the (artificial) coupling between the electrons and the magnetic field to its z-component,

$$\hat{H}''' = \hat{H}_{\text{e}} + \hat{H}_\gamma + \hat{H}_{\text{int}} + \int d^3r\, \{\hat{n}(\boldsymbol{r})v_{\text{ext}}(\boldsymbol{r}) + \hat{m}_z(\boldsymbol{r})B_{\text{ext},z}(\boldsymbol{r})\} , \tag{8.146}$$

so that the ground state is uniquely determined by n and m_z,

$$|\Psi_0\rangle \Longleftrightarrow (n, m_z) \qquad \Longrightarrow \qquad |\Psi_0\rangle = |\Psi_0[n,m_z]\rangle , \tag{8.147}$$

or, alternatively, by the generalized spin-densities $n_\pm$,

$$n_\pm(\boldsymbol{r}) = \frac{1}{2}\left[n(\boldsymbol{r}) \pm \frac{1}{\mu_B} m_z(\boldsymbol{r}) \right] . \tag{8.148}$$

[10] Compare, however, the discussion of the weakly relativistic limit of the $\boldsymbol{m}$-dependent RDFT formalism in Sect. 2.5.

In terms of these quantities the corresponding RKS equations read

$$\left\{-i\hbar c\boldsymbol{\alpha}\cdot\boldsymbol{\nabla}+(\beta-1)mc^2+\sum_{\sigma=\pm}P_\sigma v_{\mathrm{s},\sigma}\right\}\phi_k=\varepsilon_k\phi_k \tag{8.149}$$

$$n_\pm(\boldsymbol{r})=\sum_k\Theta_k\phi_k^\dagger(\boldsymbol{r})P_\pm\phi_k(\boldsymbol{r}) \tag{8.150}$$

$$v_{\mathrm{s},\sigma}(\boldsymbol{r})=v_{\mathrm{ext}}(\boldsymbol{r})+v_{\mathrm{H}}(\boldsymbol{r})+v_{\mathrm{xc},\sigma}(\boldsymbol{r}) \tag{8.151}$$

$$v_{\mathrm{xc},\sigma}(\boldsymbol{r})=\frac{\delta E_{\mathrm{xc}}[n_+,n_-]}{\delta n_\sigma(\boldsymbol{r})} \tag{8.152}$$

$$P_\pm=\frac{1\pm\beta\Sigma_z}{2}\,. \tag{8.153}$$

Here $B_{\mathrm{ext},z}$ has already been set to zero and $E_{\mathrm{H}}^{\mathrm{T}}$ has been neglected. The reason for using the densities $n_\pm$, rather than n and m_z, becomes obvious as soon as one realizes the explicit form of the projection matrices (8.153),

$$P_+=\begin{pmatrix}1&0&0&0\\0&0&0&0\\0&0&0&0\\0&0&0&1\end{pmatrix}\quad,\qquad P_-=\begin{pmatrix}0&0&0&0\\0&1&0&0\\0&0&1&0\\0&0&0&0\end{pmatrix}. \tag{8.154}$$

After insertion of (8.154), Eqs. (8.149)–(8.152) are immediately identified as the direct relativistic extension of the standard two-component form of nonrelativistic spin-density functional theory. This suggests the use of nonrelativistic spin-density functionals $E_{\mathrm{xc}}[n_\uparrow,n_\downarrow]$ in Eq. (8.152), with $n_\uparrow,n_\downarrow$ replaced by n_+,n_-. With this approximation, which neglects relativistic contributions to $E_{\mathrm{xc}}[n_+,n_-]$, Eqs. (8.149)–(8.152) represent the standard RDFT approach to magnetic systems. Some prototype results for atoms obtained with Eqs. (8.149)–(8.152) can be found in [550–556]. A comparison of collinear with non-collinear results for open-subshell atoms and molecules is presented in [547, 548, 61].

It remains to discuss an alternative to the full RKS equations (8.120)–(8.127) for systems with a non-vanishing orbital current. Such an approach necessarily has to incorporate all components of the Gordon decomposition (8.134). The question then is: can one recast the orbital current

$$\boldsymbol{j}_{\mathrm{orb}}(\boldsymbol{r})=\boldsymbol{j}_{\mathrm{p}}(\boldsymbol{r})+\frac{e}{mc}\boldsymbol{A}_{\mathrm{ext}}(\boldsymbol{r})\rho_{\mathrm{s}}(\boldsymbol{r}) \tag{8.155}$$

in a way which allows a simpler combination with the magnetization density $\boldsymbol{m}$ than Eq. (8.134)? The answer is *yes* [536, 537]. The starting point of this reformulation is current conservation, Eq. (8.113). Applied to the Gordon decomposition of $\boldsymbol{j}$, current conservation leads to

$$\boldsymbol{\nabla}\cdot\boldsymbol{j}_{\mathrm{orb}}(\boldsymbol{r})=0\,. \tag{8.156}$$

However, according to Helmholtz's theorem [557] any vector field with vanishing divergence can be expressed as [536]

$$\boldsymbol{j}_{\rm orb}(\boldsymbol{r}) = \frac{c}{e}\nabla \times \boldsymbol{m}_l(\boldsymbol{r}), \tag{8.157}$$

with

$$\boldsymbol{m}_l(\boldsymbol{r}) = \frac{e}{c}\int d^3r' \frac{\nabla' \times \boldsymbol{j}_{\rm orb}(\boldsymbol{r}')}{4\pi|\boldsymbol{r}-\boldsymbol{r}'|} = \frac{e}{c}\int d^3r' \frac{\boldsymbol{j}_{\rm orb}(\boldsymbol{r}') \times (\boldsymbol{r}-\boldsymbol{r}')}{4\pi|\boldsymbol{r}-\boldsymbol{r}'|^3} \tag{8.158}$$

(provided this integral exists, which implies that $\boldsymbol{j}_{\rm orb}$ vanishes sufficiently fast in the limit $|\boldsymbol{r}| \to \infty$). Obviously, $\boldsymbol{m}_l$ determines $\boldsymbol{j}_{\rm orb}$ uniquely via Eq. (8.157). On the other hand, one can add an arbitrary gradient field to $\boldsymbol{m}_l$ without that $\boldsymbol{j}_{\rm orb}$ changes. However, as soon as one requires $\nabla \cdot \boldsymbol{m}_l = 0$ as well as $\boldsymbol{m}_l(\boldsymbol{r}) \xrightarrow{|\boldsymbol{r}|\to\infty} 0$ (which is automatically legitimate for finite systems and may be implemented for extended systems via a suitable limiting procedure), the representation (8.157), (8.158) becomes unique. This then establishes a one-to-one correspondence between $\boldsymbol{j}$ and the sum of $\boldsymbol{m}$ and $\boldsymbol{m}_l$ [536],

$$\boldsymbol{j}(\boldsymbol{r}) = \frac{c}{e}\nabla \times \boldsymbol{M}(\boldsymbol{r}) \tag{8.159}$$

$$\boldsymbol{M}(\boldsymbol{r}) = \boldsymbol{m}_l(\boldsymbol{r}) + \boldsymbol{m}(\boldsymbol{r}). \tag{8.160}$$

The fact that $\boldsymbol{M}$ uniquely determines $\boldsymbol{j}$ can be exploited in two different ways.[11] In the first approach one uses $\boldsymbol{M}$ for an indirect evaluation of the spatial components $\boldsymbol{A}_{\rm xc}$ of the original xc-potential (8.127). One starts with the observation that, in view of Eq. (8.159), the complete xc-functional $E_{\rm xc}[n,\boldsymbol{j}]$ can be directly expressed in terms of $\boldsymbol{M}$,

$$E_{\rm xc}[n,\boldsymbol{j}] \equiv \bar{E}_{\rm xc}[n,\boldsymbol{M}]. \tag{8.161}$$

Now consider the relation between $\boldsymbol{A}_{\rm xc}$ and the "magnetic field" resulting from $\bar{E}_{\rm xc}[n,\boldsymbol{M}]$,

$$\boldsymbol{B}_{\rm xc}(\boldsymbol{r}) := \frac{\delta \bar{E}_{\rm xc}[n,\boldsymbol{M}]}{\delta \boldsymbol{M}(\boldsymbol{r})}. \tag{8.162}$$

Use of the unique correspondence between $\boldsymbol{M}$ and $\boldsymbol{j}$ as well as of the chain rule allows to rewrite $\boldsymbol{B}_{\rm xc}$ as (note the conventions (8.107) and (8.127))

---

[11] In addition, one can, of course, rewrite the magnetic contribution to the external energy (8.116) as

$$\frac{e}{c}\int d^3r \boldsymbol{A}_{\rm ext}(\boldsymbol{r}) \cdot \boldsymbol{j}(\boldsymbol{r}) = \int d^3r \boldsymbol{B}_{\rm ext}(\boldsymbol{r}) \cdot \boldsymbol{M}(\boldsymbol{r}),$$

in order to avoid the problems resulting from the lacking periodicity of $\boldsymbol{A}_{\rm ext}$ in the case of solids.

$$
\begin{aligned}
B^k_{\mathrm{xc}}(\boldsymbol{r}) &= \sum_l \int d^3r' \frac{\delta E_{\mathrm{xc}}[n,\boldsymbol{j}]}{\delta j^l(\boldsymbol{r}')} \frac{\delta j^l(\boldsymbol{r}')}{\delta M^k(\boldsymbol{r})} \\
&= \sum_{lmn} \varepsilon_{lmn} \int d^3r' A^l_{\mathrm{xc}}(\boldsymbol{r}') \frac{\delta}{\delta M^k(\boldsymbol{r})} \frac{\partial}{\partial r'^m} M^n(\boldsymbol{r}') \\
\Longrightarrow \quad \boldsymbol{B}_{\mathrm{xc}}(\boldsymbol{r}) &= \boldsymbol{\nabla} \times \boldsymbol{A}_{\mathrm{xc}}(\boldsymbol{r}) \,.
\end{aligned} \tag{8.163}
$$

As a direct consequence one finds

$$
\boldsymbol{\nabla} \cdot \boldsymbol{B}_{\mathrm{xc}}(\boldsymbol{r}) = 0 \,. \tag{8.164}
$$

Given this relation, however, one can apply Helmholtz's theorem to $\boldsymbol{B}_{\mathrm{xc}}$ to obtain the relation

$$
\boldsymbol{B}_{\mathrm{xc}}(\boldsymbol{r}) \;=\; \boldsymbol{\nabla} \times \int d^3r' \frac{\boldsymbol{\nabla}' \times \boldsymbol{B}_{\mathrm{xc}}(\boldsymbol{r}')}{4\pi|\boldsymbol{r}-\boldsymbol{r}'|} \;=\; \boldsymbol{\nabla} \times \int d^3r' \frac{\boldsymbol{B}_{\mathrm{xc}}(\boldsymbol{r}') \times (\boldsymbol{r}-\boldsymbol{r}')}{4\pi|\boldsymbol{r}-\boldsymbol{r}'|^3} \,.
$$

In view of the identity (8.163) one thus arrives at

$$
\boldsymbol{A}_{\mathrm{xc}}(\boldsymbol{r}) = \int d^3r' \frac{\boldsymbol{B}_{\mathrm{xc}}(\boldsymbol{r}') \times (\boldsymbol{r}-\boldsymbol{r}')}{4\pi|\boldsymbol{r}-\boldsymbol{r}'|^3} \,, \tag{8.165}
$$

provided one ensures the uniqueness of $\boldsymbol{A}_{\mathrm{xc}}$ by choosing Coulomb gauge and the boundary condition $\boldsymbol{A}_{\mathrm{xc}}(\boldsymbol{r}) \xrightarrow{|\boldsymbol{r}|\to\infty} 0$.

Equation (8.165) can replace the original determination of $\boldsymbol{A}_{\mathrm{xc}}$ by direct functional differentiation of $E_{\mathrm{xc}}[n,\boldsymbol{j}]$. In this alternative evaluation of $\boldsymbol{A}_{\mathrm{xc}}$ [536] one first calculates the complete KS current, utilizing the Gordon decomposition,

$$
\boldsymbol{j}(\boldsymbol{r}) = \boldsymbol{j}_{\mathrm{orb}}(\boldsymbol{r}) + \frac{c}{e} \boldsymbol{\nabla} \times \boldsymbol{m}(\boldsymbol{r}) \,, \tag{8.166}
$$

with

$$
\begin{aligned}
\boldsymbol{j}_{\mathrm{orb}}(\boldsymbol{r}) &= -\frac{i\hbar}{2m} \sum_k \Theta_k \left[\phi_k^\dagger(\boldsymbol{r}) [\boldsymbol{\nabla} \phi_k(\boldsymbol{r})] - [\boldsymbol{\nabla} \phi_k^\dagger(\boldsymbol{r})] \phi_k(\boldsymbol{r}) \right] \\
&\quad + \frac{e}{mc} \boldsymbol{A}_{\mathrm{s}}(\boldsymbol{r}) \sum_k \Theta_k \phi_k^\dagger(\boldsymbol{r}) \beta \phi_k(\boldsymbol{r})
\end{aligned} \tag{8.167}
$$

and the KS magnetization density given as in (8.143). Only the total $\boldsymbol{j}$ is exactly reproduced by the full KS equations, so that the quantity (8.167) need not be identical with the exact orbital current (8.155) (for brevity we will nevertheless use the same notation for the KS currents as for their "exact" counterparts). However, no error is introduced, as long as both contributions to (8.166) are handled consistently. Once $\boldsymbol{j}_{\mathrm{orb}}$ and $\boldsymbol{m}$ are available, one can evaluate $\boldsymbol{M}$ via Eqs. (8.160) and (8.158). Insertion into Eq. (8.162) then yields $\boldsymbol{B}_{\mathrm{xc}}$. Finally, $\boldsymbol{A}_{\mathrm{xc}}$ is obtained from Eq. (8.165).

Why could this rather involved determination of $\boldsymbol{A}_{\mathrm{xc}}$ be of any interest? The answer is based on the fact that the functional dependence of $\bar{E}_{\mathrm{xc}}[n,\boldsymbol{M}]$ on $\boldsymbol{M}$ is the same as that of the purely spin-dependent functional $E_{xc}[n,\boldsymbol{m}]$ (defined by the mapping (8.140) on $\boldsymbol{m}$), as the exact xc-functional only depends on the sum of $\boldsymbol{m}$ and $\boldsymbol{m}_l$ and $E_{xc}[n,\boldsymbol{m}]$ is exact for all systems with $\boldsymbol{m}_l \equiv \boldsymbol{0}$. In actual applications of the above approach one could thus e.g. use the nonrelativistic LDA for $\bar{E}_{xc}[n,\boldsymbol{M}]$.

In the second way to exploit the unique relation between $\boldsymbol{j}$ and $\boldsymbol{M}$ one gives up the original form of the KS potential and rewrites the magnetic contribution to the xc-potential in (8.120),

$$\Theta_k e\boldsymbol{\alpha}\cdot\boldsymbol{A}_{xc}(\boldsymbol{r})\phi_k(\boldsymbol{r}) = \int d^3r' \frac{\delta E_{xc}[n,\boldsymbol{j}]}{\delta \boldsymbol{j}(\boldsymbol{r}')}\cdot\frac{\delta \boldsymbol{j}(\boldsymbol{r}')}{\delta\phi_k^\dagger(\boldsymbol{r})}$$
$$= \int d^3r' \frac{\delta \bar{E}_{xc}[n,\boldsymbol{M}]}{\delta \boldsymbol{M}(\boldsymbol{r}')}\cdot\frac{\delta \boldsymbol{M}(\boldsymbol{r}')}{\delta\phi_k^\dagger(\boldsymbol{r})}\,. \tag{8.168}$$

Using the decomposition (8.159), (8.160) for the KS current (8.166), one arrives at

$$\Theta_k e\boldsymbol{\alpha}\cdot\boldsymbol{A}_{xc}(\boldsymbol{r})\phi_k(\boldsymbol{r}) = \int d^3r'\,\boldsymbol{B}_{xc}(\boldsymbol{r}')\cdot\frac{\delta \boldsymbol{m}_l(\boldsymbol{r}')}{\delta\phi_k^\dagger(\boldsymbol{r})} + \mu_B\Theta_k\boldsymbol{B}_{xc}(\boldsymbol{r})\cdot\beta\boldsymbol{\Sigma}\phi_k(\boldsymbol{r})\,. \tag{8.169}$$

The second term on the right-hand side is immediately identified as the spin contribution to the magnetic xc-potential, already included in the KS equation (8.141). The first term represents the associated orbital contribution. For any given functional $\bar{E}_{xc}[n,\boldsymbol{M}]$ Eq. (8.169) allows an inclusion of the complete orbital contribution to the xc-potential in the KS equations,[12]

$$\left\{-i\hbar c\boldsymbol{\alpha}\cdot\boldsymbol{\nabla} + (\beta-1)mc^2 + v_s + \mu_B\beta\boldsymbol{\Sigma}\cdot\boldsymbol{B}_{xc} + \hat{P}_{xc}^{orb}\right\}\phi_k = \varepsilon_k\phi_k\,, \tag{8.170}$$

with

$$\Theta_k\hat{P}_{xc}^{orb}\phi_k(\boldsymbol{r}) = \int d^3r'\,\boldsymbol{B}_{xc}(\boldsymbol{r}')\frac{\delta \boldsymbol{m}_l(\boldsymbol{r}')}{\delta\phi_k^\dagger(\boldsymbol{r})} \tag{8.171}$$
$$= \frac{e}{c}\int d^3r'\,\boldsymbol{B}_{xc}(\boldsymbol{r}')\cdot\int d^3r''\left[\frac{\delta \boldsymbol{j}_{orb}(\boldsymbol{r}'')}{\delta\phi_k^\dagger(\boldsymbol{r})}\times\frac{(\boldsymbol{r}'-\boldsymbol{r}'')}{4\pi|\boldsymbol{r}'-\boldsymbol{r}''|^3}\right] \tag{8.172}$$

---

[12] The Breit contributions have been neglected in (8.170) for brevity. However, the transverse Hartree potential (8.126) can also be rewritten following Eq. (8.168). Using Eq. (8.166), one obtains [537]

$$E_H^T = -\frac{1}{2}\int d^3r\int d^3r'\,\frac{[\boldsymbol{\nabla}\times\boldsymbol{M}(\boldsymbol{r})]\cdot[\boldsymbol{\nabla}'\times\boldsymbol{M}(\boldsymbol{r}')]}{|\boldsymbol{r}-\boldsymbol{r}'|}\,.$$

The resulting magnetic field is then given by

$$\boldsymbol{B}_H(\boldsymbol{r}) = \frac{\delta\bar{E}_H[\boldsymbol{M}]}{\delta\boldsymbol{M}(\boldsymbol{r})} = \frac{e}{c}\int d^3r'\,\frac{\boldsymbol{j}(\boldsymbol{r}')\times(\boldsymbol{r}-\boldsymbol{r}')}{|\boldsymbol{r}-\boldsymbol{r}'|^3}\,.$$

$\boldsymbol{B}_H$ adds to $\boldsymbol{B}_{xc}$ in the KS equation (8.170).

The orbital polarization term $\hat{P}_{\mathrm{xc}}^{\mathrm{orb}}\phi_k$ can be evaluated further, if one uses the KS representation of $\boldsymbol{j}_{\mathrm{orb}}$, Eq. (8.167),

$$\begin{aligned}
&\hat{P}_{\mathrm{xc}}^{\mathrm{orb}}\phi_k(\boldsymbol{r}) \\
&= \frac{e}{mc}\int d^3r'\,\boldsymbol{B}_{\mathrm{xc}}(\boldsymbol{r}')\cdot\left[\frac{(\boldsymbol{r}-\boldsymbol{r}')}{4\pi|\boldsymbol{r}-\boldsymbol{r}'|^3}\times\left[-i\hbar\boldsymbol{\nabla}+\frac{e}{c}\boldsymbol{A}_{\mathrm{s}}(\boldsymbol{r})\right]\beta\phi_k(\boldsymbol{r})\right] &(8.173)\\
&= \frac{e}{mc}\int d^3r'\left[\frac{\boldsymbol{B}_{\mathrm{xc}}(\boldsymbol{r}')\times(\boldsymbol{r}-\boldsymbol{r}')}{4\pi|\boldsymbol{r}-\boldsymbol{r}'|^3}\right]\cdot\left[-i\hbar\boldsymbol{\nabla}+\frac{e}{c}\boldsymbol{A}_{\mathrm{s}}(\boldsymbol{r})\right]\beta\phi_k(\boldsymbol{r})\,. &(8.174)
\end{aligned}$$

Use of (8.165) finally yields

$$\hat{P}_{\mathrm{xc}}^{\mathrm{orb}}\phi_k(\boldsymbol{r}) = \frac{e}{mc}\boldsymbol{A}_{\mathrm{xc}}(\boldsymbol{r})\cdot\left[-i\hbar\boldsymbol{\nabla}+\frac{e}{c}\boldsymbol{A}_{\mathrm{s}}(\boldsymbol{r})\right]\beta\phi_k(\boldsymbol{r})\,. \tag{8.175}$$

As they stand, Eqs. (8.171)–(8.175) are not yet very helpful for applications, as knowledge of $\boldsymbol{A}_{\mathrm{xc}}$ or related quantities is required (as in the original KS equations). However, these equations offer themselves for an approximate treatment. One can, for instance, establish a more direct relation between $\boldsymbol{m}_l$ and $\boldsymbol{j}_{\mathrm{orb}}$ than provided by Eq. (8.158) in the case of specific situations (symmetries) [558] and then proceed from Eq. (8.171). The same is possible if one uses some ansatz for $\boldsymbol{m}_l$ [537, 28]. In order to utilize the forms (8.172)–(8.175) one can insert a complete set of localized orbitals,

$$\sum_l \psi_l(\boldsymbol{r})\psi_l^\dagger(\boldsymbol{r}'') = \delta^{(3)}(\boldsymbol{r}-\boldsymbol{r}'')\,, \tag{8.176}$$

(neglecting the negative energy states in the *no-pair* approximation) to obtain

$$\hat{P}_{\mathrm{xc}}^{\mathrm{orb}}\phi_k(\boldsymbol{r}) = \beta\sum_l C_{lk}\,\psi_l(\boldsymbol{r}) \tag{8.177}$$

with C_{lk} e.g. given by

$$\begin{aligned}
C_{lk} &= \frac{e}{mc}\int d^3r'\,\boldsymbol{B}_{\mathrm{xc}}(\boldsymbol{r}') \\
&\quad\cdot\int d^3r''\,\psi_l^\dagger(\boldsymbol{r}'')\left[\frac{(\boldsymbol{r}''-\boldsymbol{r}')}{4\pi|\boldsymbol{r}''-\boldsymbol{r}'|^3}\times\left[-i\hbar\boldsymbol{\nabla}''+\frac{e}{c}\boldsymbol{A}_{\mathrm{s}}(\boldsymbol{r}'')\right]\phi_k(\boldsymbol{r}'')\right]\,. \qquad (8.178)
\end{aligned}$$

The coefficients C_{lk} may then be evaluated approximately, neglecting for instance all relativistic corrections. Once the term $\boldsymbol{A}_{\mathrm{s}}/c$ is ignored, the remaining operator has the structure of a nonlocal angular momentum, which suggests the approximation[13]

$$C_{lk} \approx \frac{e}{mc}\sum_n\int d^3r'\,\psi_l^\dagger(\boldsymbol{r}')\boldsymbol{B}_{\mathrm{xc}}(\boldsymbol{r}')\cdot[\boldsymbol{r}'\times(-i\hbar\boldsymbol{\nabla}')\psi_n(\boldsymbol{r}')]\int d^3r\,\psi_n^\dagger(\boldsymbol{r})\phi_k(\boldsymbol{r})\,. \tag{8.179}$$

[13] One can show rigorously that the right-hand side of Eq. (8.179) is one contribution to the original nonlocal angular momentum integral,

If the localized states ψ_n are chosen as eigenfunctions of the angular momentum $\psi_n \to \psi_{nlm_ls}$ and $\boldsymbol{B}_{\mathrm{xc}}$ is restricted to the collinear form $(0,0,B_{\mathrm{xc}})$, the expression (8.179) finally simplifies to

$$C_{n'l'm'_ls',k} \approx 2\mu_B \sum_{nlm_ls} \langle \psi_{nlm_ls} | \phi_k \rangle \, m_l \int d^3r \, B_{\mathrm{xc}}(\boldsymbol{r}) \psi^{\dagger}_{n'l'm'_ls'}(\boldsymbol{r}) \psi_{nlm_ls}(\boldsymbol{r}) \,, \tag{8.180}$$

which can serve as motivation for empirical orbital polarization terms as the Brooks correction [559] (see also [560] for a treatment of orbital polarization within non-relativistic CDFT).

8.8 Relativistic Exchange-Correlation Functional: Concepts and Illustrative Results

There is a large variety of applications of R(S)DFT in the literature (see [561–568] for reviews, more specific results may e.g. be found in [569–587, 322, 62] for solids and in [588–605] for molecules). As in the nonrelativistic context no attempt is made here to review this vast body of literature. Rather the presentation focusses on a brief introduction of the few inherently relativistic approximations for E_{xc} and a few prototype results. The latter primarily serve to illustrate the limited importance of the relativistic contributions to E_{xc}, in particular of the transverse corrections, for standard electronic structure properties.

8.8.1 Relativistic Implicit Functionals: Optimized Potential Method

Given the importance of a proper cancellation of self-interaction in nonrelativistic DFT, the question concerning an exact treatment of exchange in RDFT arises automatically (for a relativistic extension of the SIC-LDA see [606]). The exact exchange of RDFT is given by Eq. (8.100) (within the *no-pair* approximation). Explicit evaluation of (8.100) using Feynman gauge yields[14] [350]

$$\begin{aligned} &\int d^3r' \, \boldsymbol{B}_{\mathrm{xc}}(\boldsymbol{r}') \cdot \left[\int d^3r'' \, \psi_l^{\dagger}(\boldsymbol{r}'') \frac{(\boldsymbol{r}''-\boldsymbol{r}') \times \nabla'' \phi_k(\boldsymbol{r}'')}{4\pi|\boldsymbol{r}''-\boldsymbol{r}'|^3} \right] \\ &= \int d^3r' \, \psi_l^{\dagger}(\boldsymbol{r}') \boldsymbol{B}_{\mathrm{xc}}(\boldsymbol{r}') \cdot [\boldsymbol{r}' \times \nabla' \phi_k(\boldsymbol{r}')] \\ &\quad - \int d^3r' \int d^3r'' \, \psi_l^{\dagger}(\boldsymbol{r}'') \frac{[\boldsymbol{r}' \times \nabla'' \phi_k(\boldsymbol{r}'')] \cdot [(\boldsymbol{r}''-\boldsymbol{r}') \cdot \nabla'] \boldsymbol{B}_{\mathrm{xc}}(\boldsymbol{r}')}{4\pi|\boldsymbol{r}''-\boldsymbol{r}'|^3} \,. \end{aligned}$$

[14] The exchange energy defined in this fashion should not be confused with the relativistic HF (RHF) exchange energy (at least from a rigorous point of view), as the RKS-orbitals satisfy the RKS equations (8.120) with their multiplicative potential, rather than the nonlocal RHF-equations. In fact, compared with the RHF-approach the exact exchange of RDFT has the advantage of be-

$$E_{\rm x} = -\frac{e^2}{2}\sum_{k,l}\Theta_k\Theta_l \int d^3r \int d^3r' \frac{\cos(\omega_{kl}|\boldsymbol{r}-\boldsymbol{r}'|)}{|\boldsymbol{r}-\boldsymbol{r}'|} \times \phi_k^\dagger(\boldsymbol{r})\alpha_\mu\phi_l(\boldsymbol{r})\ \phi_l^\dagger(\boldsymbol{r}')\alpha^\mu\phi_k(\boldsymbol{r}') \tag{8.181}$$

($\omega_{kl} = |\varepsilon_k - \varepsilon_l|/(\hbar c)$), which can be decomposed into the standard Coulomb component,

$$E_{\rm x}^{\rm C} = -\frac{e^2}{2}\sum_{k,l}\Theta_k\Theta_l \int d^3r \int d^3r' \frac{\phi_k^\dagger(\boldsymbol{r})\phi_l(\boldsymbol{r})\ \phi_l^\dagger(\boldsymbol{r}')\phi_k(\boldsymbol{r}')}{|\boldsymbol{r}-\boldsymbol{r}'|}, \tag{8.182}$$

and a transverse remainder,

$$E_{\rm x}^{\rm T} = E_{\rm x} - E_{\rm x}^{\rm C}. \tag{8.183}$$

The expressions (8.181)–(8.183) represent implicit functionals of the four current in the same sense as $T_{\rm s}$ is a functional of j^μ: the RKS-spinors are unique functionals of j^μ, as the ground state Slater determinant of the KS system is a unique functional of j^μ by virtue of the relativistic existence theorem for noninteracting particles. Orbital-dependent functionals can also be derived for the relativistic correlation functional $E_{\rm c} = E_{\rm xc} - E_{\rm x}$, following the lines of Sect. 6.4 (see [350, 556] for approximations of this type).

The evaluation of the xc-potential $v_{\rm xc}^\mu = \delta E_{\rm xc}/\delta j_\mu$,[15] for functionals of the type (8.181)–(8.183), requires the extension of the OPM introduced in Chap. 6. For this relativistic generalization (ROPM) one can follow any of the methods of Sect. 6.2. Here we use the chain rule for functional differentiation in order to replace the derivative with respect to j^μ by derivatives with respect to ϕ_k and ε_k [607, 350],

$$\frac{\delta E_{\rm xc}[\phi_k^{(\dagger)},\varepsilon_k]}{\delta j^\mu(\boldsymbol{r})} = \int d^3r' \frac{\delta v_{\rm s}^\rho(\boldsymbol{r}')}{\delta j^\mu(\boldsymbol{r})}\sum_k\left\{\int d^3r'' \left[\frac{\delta\phi_k^\dagger(\boldsymbol{r}'')}{\delta v_{\rm s}^\rho(\boldsymbol{r}')}\frac{\delta E_{\rm xc}}{\delta\phi_k^\dagger(\boldsymbol{r}'')} + c.c.\right] + \frac{\delta\varepsilon_k}{\delta v_{\rm s}^\rho(\boldsymbol{r}')}\frac{\partial E_{\rm xc}}{\partial\varepsilon_k}\right\}. \tag{8.184}$$

The k-summation on the right-hand side of (8.184) runs over all KS-levels, including the negative continuum states. As soon as the *no-pair* approximation is used for $E_{\rm xc}$, it reduces, however, to states with $\varepsilon_k > -mc^2$. The derivatives of $\phi_k^{(\dagger)}$ and ε_k with respect to $v_{\rm s}^\rho$ on the right-hand side of (8.184) can be evaluated as in Sect. 6.2.2, using the RKS equation (8.86). Multiplication of Eq. (8.184) by the static KS current–current response function,

---

ing gauge invariant due to the multiplicative nature of the RKS exchange potential [350], which justifies the use of the more simple Feynman gauge in (8.181).

[15] For the discussion of the fully relativistic form of the ROPM it is advantageous to use the four vector notation (8.89), rather than the nonrelativistic convention (8.124), (8.127). However, the transition from the Minkowski form to the latter is easily possible via Eqs. (8.107) and (8.121).

$$\chi_s^{\mu\nu}(\boldsymbol{r},\boldsymbol{r}') = \frac{\delta j^\mu(\boldsymbol{r})}{\delta v_{s,\nu}(\boldsymbol{r}')} = -\sum_k \Theta_k \phi_k^\dagger(\boldsymbol{r})\,\alpha^\mu G_k(\boldsymbol{r},\boldsymbol{r}')\,\alpha^\nu \phi_k(\boldsymbol{r}') + c.c. \tag{8.185}$$

$$G_k(\boldsymbol{r},\boldsymbol{r}') = \sum_{l\neq k} \frac{\phi_l(\boldsymbol{r})\,\phi_l^\dagger(\boldsymbol{r}')}{\varepsilon_l - \varepsilon_k}, \tag{8.186}$$

and subsequent integration leads to the ROPM integral equation,

$$\int d^3r'\,\chi_s^{\mu\nu}(\boldsymbol{r},\boldsymbol{r}')\,v_{xc,\nu}(\boldsymbol{r}') = \Lambda_{xc}^\mu(\boldsymbol{r}), \tag{8.187}$$

with the inhomogeneity

$$\Lambda_{xc}^\mu(\boldsymbol{r}) = -\sum_k \int d^3r' \left[\phi_k^\dagger(\boldsymbol{r})\alpha^\mu G_k(\boldsymbol{r},\boldsymbol{r}')\frac{\delta E_{xc}}{\delta\phi_k^\dagger(\boldsymbol{r}')} + c.c.\right] + \sum_k \phi_k^\dagger(\boldsymbol{r})\alpha^\mu \phi_k(\boldsymbol{r})\frac{\partial E_{xc}}{\partial \varepsilon_k}. \tag{8.188}$$

The integral equation (8.187) has to be solved selfconsistently together with the RKS equations (8.86). Of course, the spatial components v_{xc}^i of the solution of Eq. (8.187) do not vanish in general, so that the use of orbital-dependent functionals automatically accounts for the current dependence of E_{xc}.

In this procedure one also has to fix the gauge of v_{xc}^μ. First of all, similar to the situation without magnetic fields, v_{xc}^0 is only defined up to a global constant, as

$$\int d^3r'\,\chi_s^{\mu 0}(\boldsymbol{r},\boldsymbol{r}') = \int d^3r\,\chi_s^{0\mu}(\boldsymbol{r},\boldsymbol{r}') = 0 \tag{8.189}$$

(due to charge conservation). This is, however, not a real problem. For finite systems the constant is usually defined by the requirement $v_{xc}^0(\boldsymbol{r}) \xrightarrow{|\boldsymbol{r}|\to\infty} 0$. For infinite systems the average of v_{xc}^0 in the unit cell may be set to zero. The handling of the spatial components of v_{xc}^μ is somewhat more complicated. In the *no-pair* approximation the l-summation in G_k, Eq. (8.186), is restricted to $\varepsilon_k > -mc^2$. If one uses this approximation for G_k in (8.185), the transversality of $\chi_s^{\mu\nu}$ is lost (for a detailed discussion of this issue see [350]),

$$\partial_i \chi_s^{i\nu}(\boldsymbol{r},\boldsymbol{r}') \neq 0. \tag{8.190}$$

Therefore three-vector potentials $\boldsymbol{v}_{xc} = -e\boldsymbol{A}_{xc}$ which differ by static gauge transformations lead to different results for the left-hand side of (8.187). As a result, the gauge of $\boldsymbol{v}_{xc}$ is indirectly fixed. On the other hand, without the *no-pair* approximation the gauge of $\boldsymbol{v}_{xc}$ must be chosen explicitly.

A closer analysis of the integral equation (8.187) reveals important properties of the xc-potential [350]. One finds that both the KLI identity for the highest occupied orbital, Eq. (6.57), and the asymptotic behavior of v_x^0, Eq. (6.58),

$$v_{\rm x}^{0}(\boldsymbol{r}) \xrightarrow[|\boldsymbol{r}|\to\infty]{} -\frac{e^2}{r}, \tag{8.191}$$

remain unchanged in the relativistic situation (for finite systems). Moreover, the condition (6.59) also applies to the relativistic $E_{\rm xc}$. Equation (6.59) is, for instance, manifestly satisfied by the exact exchange (8.181).

ROPM equations are also available for the variants of RDFT outlined in Sect. 8.7. The purely n-dependent ROPM formalism [608, 609] is given by the time-like components of Eqs. (8.185)–(8.188). On the other hand, for the $\boldsymbol{m}$-dependent formalism the set of coupled OPM equations [610, 556],

$$\begin{aligned}
&\int d^3r' \left\{\chi_{nn}(\boldsymbol{r},\boldsymbol{r}')v_{\rm xc}(\boldsymbol{r}') + \boldsymbol{\chi}_{nm}(\boldsymbol{r},\boldsymbol{r}')\cdot\boldsymbol{B}_{\rm xc}(\boldsymbol{r}')\right\} \\
&= -\sum_k \int d^3r'\, \phi_k^\dagger(\boldsymbol{r}) G_k(\boldsymbol{r},\boldsymbol{r}') \frac{\delta E_{\rm xc}}{\delta \phi_k^\dagger(\boldsymbol{r}')} + c.c.
\end{aligned} \tag{8.192}$$

$$\begin{aligned}
&\int d^3r' \left\{\boldsymbol{\chi}_{mn}(\boldsymbol{r},\boldsymbol{r}')v_{\rm xc}(\boldsymbol{r}') + \boldsymbol{\chi}_{mm}(\boldsymbol{r},\boldsymbol{r}')\cdot\boldsymbol{B}_{\rm xc}(\boldsymbol{r}')\right\} \\
&= -\mu_B \sum_k \int d^3r'\, \phi_k^\dagger(\boldsymbol{r}) \beta\boldsymbol{\Sigma} G_k(\boldsymbol{r},\boldsymbol{r}') \frac{\delta E_{\rm xc}}{\delta \phi_k^\dagger(\boldsymbol{r}')} + c.c. \quad ,
\end{aligned} \tag{8.193}$$

involves the response functions

$$\chi_{nn}(\boldsymbol{r},\boldsymbol{r}') = \frac{\delta n(\boldsymbol{r})}{\delta v_{\rm s}(\boldsymbol{r}')} = -\sum_k \Theta_k \phi_k^\dagger(\boldsymbol{r}) G_k(\boldsymbol{r},\boldsymbol{r}')\phi_k(\boldsymbol{r}') + c.c. \tag{8.194}$$

$$\boldsymbol{\chi}_{mn}(\boldsymbol{r},\boldsymbol{r}') = \frac{\delta \boldsymbol{m}(\boldsymbol{r})}{\delta v_{\rm s}(\boldsymbol{r}')} = -\mu_B \sum_k \Theta_k \phi_k^\dagger(\boldsymbol{r}) \beta\boldsymbol{\Sigma} G_k(\boldsymbol{r},\boldsymbol{r}')\phi_k(\boldsymbol{r}') + c.c. \tag{8.195}$$

$$\boldsymbol{\chi}_{nm}(\boldsymbol{r},\boldsymbol{r}') = \frac{\delta n(\boldsymbol{r})}{\delta \boldsymbol{B}_{\rm s}(\boldsymbol{r}')} = \boldsymbol{\chi}_{mn}^\dagger(\boldsymbol{r}',\boldsymbol{r}) \tag{8.196}$$

$$\boldsymbol{\chi}_{mm}(\boldsymbol{r},\boldsymbol{r}') = \frac{\delta \boldsymbol{m}(\boldsymbol{r})}{\delta \boldsymbol{B}_{\rm s}(\boldsymbol{r}')} = -\mu_B^2 \sum_k \Theta_k \phi_k^\dagger(\boldsymbol{r}) \beta\boldsymbol{\Sigma} G_k(\boldsymbol{r},\boldsymbol{r}')\beta\boldsymbol{\Sigma}\phi_k(\boldsymbol{r}') + c.c.\,, \tag{8.197}$$

with G_k given by (8.186). Contributions to (8.192), (8.193) arising from $\partial E_{\rm xc}/\partial \varepsilon_k$ have been suppressed for brevity. The derivation of Eqs. (8.192)–(8.197) follows Sect. 6.2.2.

Finally, in the limit of collinear magnetization density, $\boldsymbol{m} = (0,0,m_z)$, one can express the OPM equations in terms of the generalized spin-densities (8.148),

$$\begin{aligned}
&\sum_{\sigma'} \int d^3r'\, \chi_{\sigma\sigma'}(\boldsymbol{r},\boldsymbol{r}') v_{{\rm xc},\sigma'}(\boldsymbol{r}') \\
&= -\sum_k \int d^3r'\, \phi_k^\dagger(\boldsymbol{r}) P_\sigma G_k(\boldsymbol{r},\boldsymbol{r}') \frac{\delta E_{\rm xc}}{\delta \phi_k^\dagger(\boldsymbol{r}')} + c.c.
\end{aligned} \tag{8.198}$$

$$\chi_{\sigma\sigma'}(\boldsymbol{r},\boldsymbol{r}') = \frac{\delta n_\sigma(\boldsymbol{r})}{\delta v_{{\rm s},\sigma'}(\boldsymbol{r}')} = -\sum_k \Theta_k \phi_k^\dagger(\boldsymbol{r}) P_\sigma G_k(\boldsymbol{r},\boldsymbol{r}') P_{\sigma'} \phi_k(\boldsymbol{r}') + c.c.. \tag{8.199}$$

G_k again has the form (8.186), P_σ is given by (8.153). Each of the solutions $v_{x,\sigma}$ of (8.198) satisfies the relation (8.191) in the case of finite systems [556].

The numerical solution of the integral equations (8.185)–(8.188), (8.192)–(8.197) or (8.198), (8.199) is complicated by the fact that the evaluation of G_k requires a summation over the complete positive energy spectrum. In the nonrelativistic situation the semi-analytical KLI approximation [351] has been introduced to address this problem (see Sect. 6.2.6). The KLI approximation can be readily extended to all variants of the relativistic OPM by use of a closure approximation for G_k [350, 610], in analogy to Eq. (6.61) (for an alternative derivation see [607]). The relativistic KLI scheme shares all the properties of its nonrelativistic counterpart discussed in Sect. 6.2.6 (for further information see e.g. [529]).

8.8.2 Role of Relativistic Corrections in E_{xc}: I. Prototype Results for Atoms

The exact E_x allows an unambiguous investigation of the importance of relativistic corrections in the xc-functional and provides reference data for an analysis of approximate functionals. Some illustrative results obtained by solution of the ROPM equation (8.187) within the x-only limit (REXX) are given in Tables 8.1–8.5 and Figs. 8.2–8.4. Table 8.1 lists the x-only ground state energies of closed-subshell atoms obtained with three different forms of E_x: the Coulomb limit (8.182), the fully relativistic exchange (8.181) and its Coulomb-Breit approximation. In all three cases

Table 8.1 Exchange-only ground state energies from REXX and RHF calculations for closed subshell atoms: Coulomb(C)- and Coulomb-Breit(C+B)-limit in comparison with complete transverse exchange (C+T) [350]. For the RHF approximation the energy difference with respect to the REXX approach is given, $\Delta E = E(\text{RHF}) - E(\text{REXX})$. These results are obtained by (a) finite difference calculations [611] and (b) basis set expansion [462] (all energies in mHartree—nuclear model and c as in [462]).

Atom	$-E^{C}$ REXX	ΔE^{C} RHF[a]	ΔE^{C} RHF[b]	$-E^{C+B}$ REXX	ΔE^{C+B} RHF[b]	$-E^{C+T}$ REXX
He	2862	0	0	2862	0	2862
Ne	128690	−2	−2	128674	−2	128674
Ar	528678	−5	−5	528546	−5	528546
Zn	1794598	−14		1793838	−12	1793841
Kr	2788849	−13	−12	2787423	−12	2787431
Cd	5593299	−20		5589466	−15	5589500
Xe	7446882	−19	−6	7441115	−3	7441179
Yb	14067621	−48		14053517		14053764
Hg	19648826	−39		19626225	9	19626715
Rn	23601947	−35	−19	23572625	11	23573332
No	36740625	−57		36685157		36686790

standard finite difference methods have been applied in the REXX calculations, so that there are no basis set limitations. In the Coulomb limit an unambiguous comparison with relativistic Hartree-Fock (RHF) results is possible, as accurate finite difference RHF data are available in this limit [611]. The differences between the Coulomb-REXX and Coulomb-RHF energies are very similar to those observed in the nonrelativistic limit (compare Table 6.1): due to the multiplicative nature of the OPM exchange potential, the REXX energies are slightly higher than the RHF data, although the same energy expression is minimized. However, the actual differences are extremely small. In fact, they are as small as those resulting from usual basis set limitations: the differences between the fully numerical RHF energies and those obtained with a Gaussian basis set are of the same order of magnitude as the differences between REXX and RHF data. Table 8.1 also shows the fully relativistic x-only ground state energies obtained by selfconsistent treatment of the complete transverse exchange. It is obvious that retardation corrections to the Breit interaction are only relevant for truly heavy atoms.

A microscopic analysis of the transverse exchange is provided in Fig. 8.2. Three results for the relativistic v_x of Hg (Coulomb, Coulomb-Breit and complete interaction) are shown together with the nonrelativistic v_x. Taking into account that the

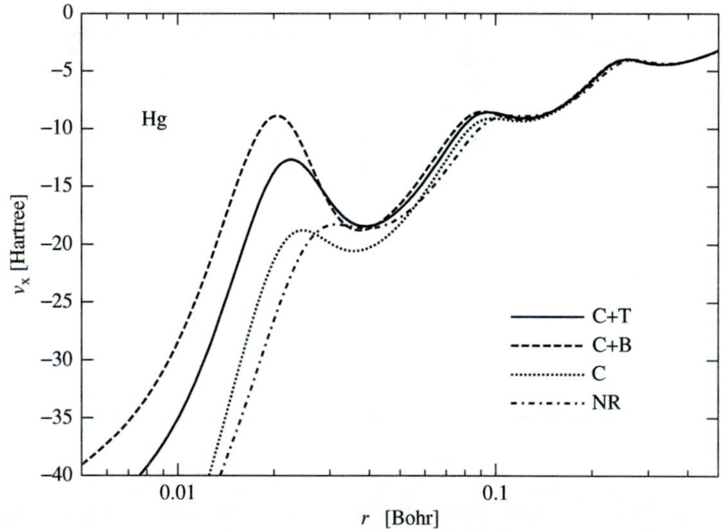

Fig. 8.2 REXX potentials for neutral Hg: Selfconsistent Coulomb (C), Coulomb-Breit (C+B) and fully transverse (C+T) results in comparison with nonrelativistic limit (NR).

r-expectation values of the $1s_{1/2}$ and $2s_{1/2}$ orbitals are 0.017 Bohr and 0.069 Bohr, respectively, one realizes that the transverse corrections in v_x are not only significant for the K-shell, but also for the L-shell. In fact, the difference between the complete relativistic v_x and its Coulomb limit is as large as the difference between the latter and the nonrelativistic v_x. The transverse exchange potential is repulsive, as is to be expected from the fact that the retarded exchange of photons reduces the interaction strength of the instantaneous Coulomb interaction.

The corrections shown in Fig. 8.2 are also visible in the single-particle spectrum of Hg listed in Table 8.2. The table demonstrates that not only the K- and L-shell

Table 8.2 Exchange-only single particle energies ($-\varepsilon_{nlj}$) for neutral Hg from selfconsistent REXX and RHF calculations, using (i) the complete relativistic exchange-only potential (C+T), (ii) its Coulomb-Breit approximation (C+B), and (iii) its Coulomb (C) limit. Also given are RGGA results which have been obtained with the relativistic extension [612] of the Becke functional [219], either for exchange-only (RB88) or including the Lee-Yang-Parr correlation GGA (RBLYP) [221] (all energies in Hartree [613]).

Level	REXX			RHF	RB88	RBLYP
	C+T	C+B	C	C	C+T	C+T
1*s*1/2	3036.871	3032.278	3047.431	3074.229	3036.453	3036.485
2*s*1/2	538.444	537.853	540.057	550.251	538.051	538.085
2*p*1/2	516.198	515.546	518.062	526.855	516.097	516.132
2*p*3/2	445.422	445.013	446.683	455.157	445.276	445.311
3*s*1/2	127.956	127.858	128.273	133.113	127.703	127.738
3*p*1/2	117.994	117.885	118.351	122.639	117.857	117.893
3*p*3/2	102.302	102.236	102.537	106.545	102.152	102.187
3*d*3/2	86.069	86.036	86.202	89.437	85.959	85.994
3*d*5/2	82.692	82.665	82.808	86.020	82.582	82.617
4*s*1/2	28.361	28.351	28.428	30.648	28.037	28.072
4*p*1/2	24.090	24.075	24.162	26.124	23.819	23.854
4*p*3/2	20.321	20.315	20.364	22.189	20.024	20.059
4*d*3/2	13.397	13.397	13.412	14.797	13.151	13.186
4*d*5/2	12.689	12.690	12.701	14.053	12.441	12.476
4*f*5/2	3.766	3.770	3.757	4.473	3.571	3.607
4*f*7/2	3.613	3.616	3.603	4.312	3.417	3.453
5*s*1/2	4.394	4.394	4.404	5.103	4.278	4.313
5*p*1/2	3.004	3.002	3.013	3.538	2.886	2.920
5*p*3/2	2.360	2.360	2.364	2.842	2.219	2.253
5*d*3/2	0.507	0.507	0.506	0.650	0.367	0.399
5*d*5/2	0.440	0.441	0.440	0.575	0.300	0.332
6*s*1/2	0.330	0.330	0.330	0.328	0.222	0.249

energies experience the transverse correction. Even in the case of the $4f$-levels the energy shift induced by v_x^T is as large as 25% of the shift resulting from inclusion of correlation. On the other hand, the valence levels remain unaffected by v_x^T.

The last statement is corroborated by the atomic ionization potentials (IPs) given in Table 8.3. Table 8.3 focusses on the ionization of s-electrons, which experience relativistic effects most strongly.[16] The table confirms that the transverse interaction

[16] The data in this table have been generated with the purely n-dependent version of RDFT, in which all KS particles experience only the total v_s, but no magnetic field. This is legitimate as long as only methods are compared and all methods are applied in the same fashion. In the case of the open s-shells studied in Table 8.3 the differences between the IPs of this approach and those

Table 8.3 Exchange-only ionization potentials of neutral atoms calculated from total energy differences between neutral and singly ionized states: for the REXX approach the selfconsistent inclusion of the transverse exchange (C+T) is compared with complete neglect of E_x^{T} (C). The (R)GGA data have been obtained with the B88-functional [219] and its relativistic extension [612] (all energies in mHartree [613]).

Atom	REXX		GGA	RGGA
	C+T	C	C	C+T
Li	196	196	179	179
Na	181	181	173	173
K	147	147	143	143
Rb	139	139	137	137
Cs	127	127	127	127
Fr	132	132	132	132
Cu	234	234	279	279
Ag	229	229	263	262
Au	279	280	321	319
Zn	284	284	339	339
Cd	269	269	316	315
Hg	312	313	365	363

is irrelevant for the outermost electrons and therefore for most quantities of interest in quantum chemistry and condensed matter physics. If desired at all, transverse corrections can be added perturbatively (for a detailed comparison of the selfconsistent with the perturbative treatment of $E_{\mathrm{x}}^{\mathrm{T}}$ see [350, 529]).

The $\boldsymbol{m}$-dependent REXX formalism allows the investigation of open-subshell atoms. The obvious question in this context concerns the balance between spin–orbit and exchange effects. The energetic ordering of the KS-levels and therefore of the ground state predicted by an RKS calculation depends on the relative size of the exchange interaction, favoring alignment of spins (more precisely, of magnetic moments), and of the spin–orbit splitting. This balance has been examined within the collinear REXX approach (8.198), (8.199), neglecting transverse exchange [556]. Some prototype results for the $6p$-elements are shown in Fig. 8.3. It is obvious that, even for these quite heavy elements, the strong spin alignment, that results from the exact exchange, breaks up the ordering of levels according to good total angular momentum j induced by spin–orbit coupling. This is particularly noticeable in Pb, for which the closing of a relativistic subshell is possible.

As a further illustration of the role of spin-alignment we consider the lowest-lying states of the chromium atom. The experimental excitation energies from the $3d^5(6S)4s\ {}^7S_3$ ground state to the first excited state, $3d^5(6S)4s\ {}^5S_2$, and to the multiplet $3d^44s^2\ {}^5D$ are shown in Fig. 8.4. Due to their different symmetries all three levels are basically accessible to DFT. The $3d^5(6S)4s\ {}^5S$ level involves the inversion of the $4s$ spin and the $3d^44s^2\ {}^5D$ multiplet the transfer of a d-electron to the $4s$-state. The KS representations of the ground state and the first excited state are

resulting from the magnetization-dependent version of RDFT are very small. For Fr, for instance, a total energy difference of 0.16 mHartree is found in the Coulomb-only limit.

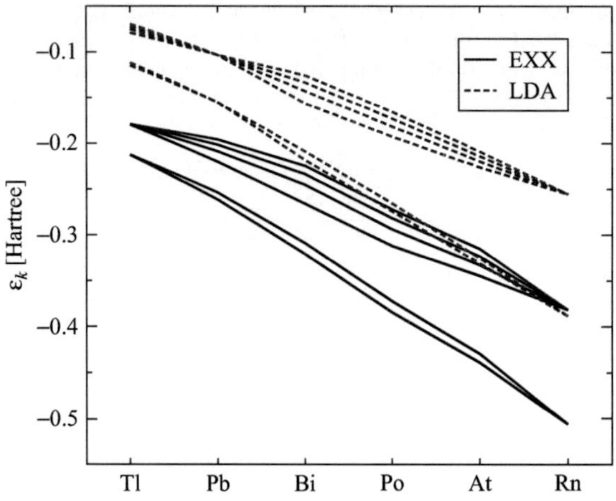

Fig. 8.3 KS eigenvalues of the $6p$-subshell for $6p$-elements with occupation $6p^n$: relativistic EXX versus LDA [134].

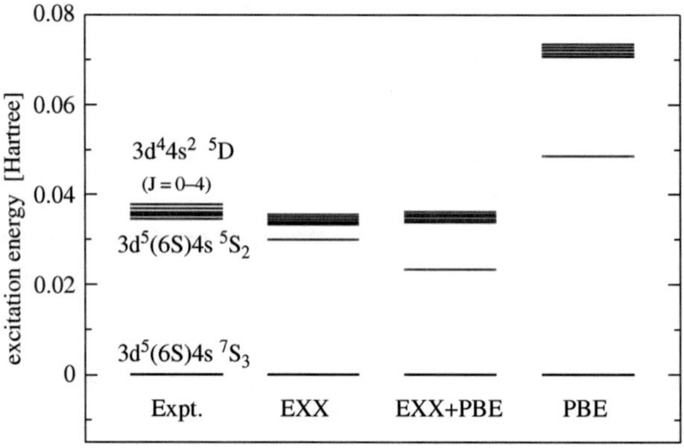

Fig. 8.4 Low-lying levels of the chromium atom: relativistic EXX results versus complete PBE-GGA [207], the combination of exact exchange with PBE correlation (EXX+PBE) and experimental [614] data. The experimental $3d^5(6S)4s\ ^5S_2$ state is only 0.7 mHartree lower than the lowest state ($J = 0$) of the $3d^4 4s^2\ ^5D$ multiplet and can therefore not be resolved on the scale used.

unambiguous, due to the filled $3d$-subshell. On the other hand, the $3d^4 4s^2$ configuration can be realized in different ways, corresponding to different occupations of the $3d$-substates. The energetically lowest of the corresponding KS states can be considered as a representation of the lowest state of the $3d^4 4s^2\ ^5D$ multiplet. The excitation energies resulting from the various occupations are indicated in Fig. 8.4, together with the excitation energy of the $3d^5(6S)4s\ ^5S$ state. An accurate reproduction of these excitation energies has been a long-standing challenge for (R)DFT [615–617], as is clear from the PBE-GGA results included in Fig. 8.4. Obviously, the right balance between spin-alignment and kinetic effects is only obtained if the exact exchange is used.

In fact, the REXX excitation energies are so close to their experimental counterparts that one might wonder whether the agreement still persists after inclusion of correlation. Figure 8.4 demonstrates that this is not the case if GGA correlation is added to the exact exchange. This result agrees with the observation that the combination of the exact exchange with LDA or GGA correlation does not give convincing results for many electronic structure properties (compare Tables 6.10 and 4.8). The cancellation of errors between exchange and correlation is no longer possible, if the exact exchange is utilized. A suitable orbital-dependent correlation functional has to be used in order to retain the quality of the pure REXX data [556].

8.8.3 Relativistic Local Density Approximation

In complete analogy to (4.109) the relativistic LDA (RLDA) for $E_{\mathrm{xc}}[j]$ is constructed from the xc-energy density $e_{\mathrm{xc}}^{\mathrm{RHEG}}$ of the RHEG. As $\boldsymbol{j}$ vanishes in the RHEG, the full j^μ-dependence of the exact xc-energy functional reduces to a pure density-dependence in the RLDA,

$$E_{\mathrm{xc}}^{\mathrm{RLDA}}[n] = \int d^3r\, e_{\mathrm{xc}}^{\mathrm{RHEG}}(n(\boldsymbol{r}))\,. \tag{8.200}$$

The derivation of its lowest order contribution, i.e. the exchange energy, is discussed in detail in Appendix H. It illustrates in particular the UV-renormalization required. The final result [618–621, 534, 535] can be expressed as the nonrelativistic exchange energy density $e_{\mathrm{x}}^{\mathrm{HEG}}$, Eq. (4.99), multiplied by a relativistic correction factor Φ_{x}. Separation into its Coulomb and transverse part according to (8.182), (8.183) yields

$$e_{\mathrm{x}}^{\mathrm{RHEG,C/T}}(n) = e_{\mathrm{x}}^{\mathrm{HEG}}(n)\; \Phi_{\mathrm{x}}^{\mathrm{C/T}}(\beta)\,, \tag{8.201}$$

where β represents the dimensionless relativistic density variable,

$$\beta = \frac{\hbar(3\pi^2 n)^{1/3}}{mc}\,. \tag{8.202}$$

The factors $\Phi_{\mathrm{x}}^{\mathrm{C/T}}$ are explicitly given in (H.56), (H.57). The variation of $\Phi_{\mathrm{x}}^{\mathrm{C/T}}$ as

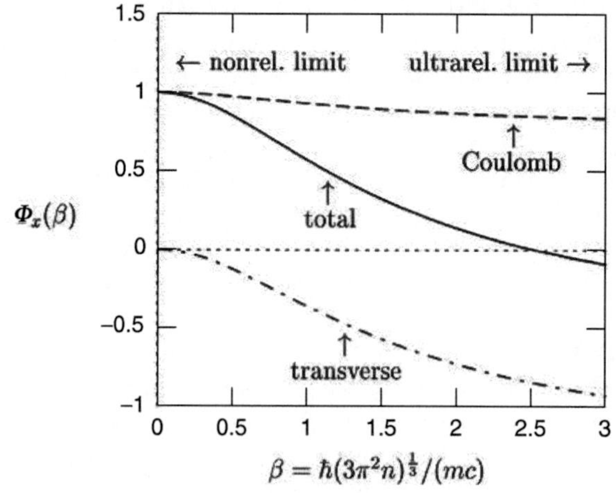

Fig. 8.5 Relativistic correction factor for the LDA exchange energy density: Coulomb contribution (H.56), transverse contribution (H.57) and total correction $\Phi_\mathrm{x}^\mathrm{C} + \Phi_\mathrm{x}^\mathrm{T}$.

well as of $\Phi_\mathrm{x}^\mathrm{C} + \Phi_\mathrm{x}^\mathrm{T}$ with β is shown in Fig. 8.5. One notices that the Coulomb contribution dominates in the low density limit and that it depends only weakly on β. $\Phi_\mathrm{x}^\mathrm{T}$ shows a stronger dependence on β and dominates in the high density regime, in which $e_\mathrm{x}^\mathrm{RHEG}$ even changes its sign.

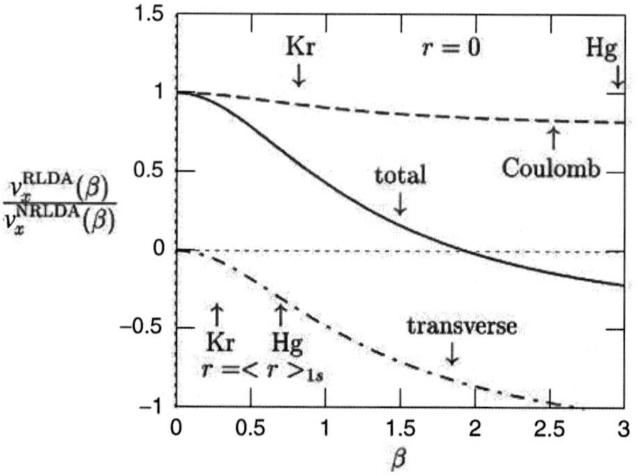

Fig. 8.6 Relativistic correction factor for the LDA exchange potential. The values of the densities of Kr and Hg at the origin ($r = 0$) and the r-expectation values of the $1s$-orbitals ($r = < r >_{1s}$) from RLDA-calculations are also indicated.

In Fig. 8.6 we plot the density dependence of the resulting exchange potentials. The range of density values relevant for electronic structure calculations is indicated by the β-values at the origin and at the r-expectation values of the $1s_{1/2}$-orbitals of the Kr and Hg atoms. As is obvious from Fig. 8.2 the exact v_x is negative in the vicinity of the origin even after inclusion of the transverse interaction. The fact that the transverse term leads to a sign change in e_x^{RLDA} and v_x^{RLDA} for high densities thus questions the usefulness of the RLDA from the very outset. The transverse interaction is much more important for an infinite system as the RHEG, in which the finite speed of light plays a role for electrons at large separation, than for atomic systems with their highly localized inner core electrons.

Some results obtained with the x-only RLDA are included in Tables 8.4 and 8.5. Table 8.4 shows that the RLDA produces ground state energies which are far from

Table 8.4 Selfconsistent exchange-only ground state energies of closed subshell atoms: RLDA, RGGA and GGA results versus REXX reference data (all energies in mHartree; the model for the form of the nuclei used here [613] differs from the one underlying Table 8.1). E_x^{T} has been treated selfconsistently in the RLDA, RGGA and REXX calculations. The PW91 form [206] has been applied for the GGA.

Atom	$-E$	$E - E^{\text{REXX}}$		
	REXX	RLDA	RGGA	GGA
He	2862	138	6	6
Ne	128674	1080	−24	−43
Ar	528546	2458	41	−111
Zn	1793840	4702	−263	−1146
Kr	2787429	6543	−22	−1683
Cd	5589496	10556	−35	−4537
Xe	7441173	13161	83	−6705
Yb	14053750	20888	−894	−17660
Hg	19626705	29161	−257	−27253
Rn	23573354	35207	−9	−35145
No	36687173	56937	−1344	−68097

the exact REXX values for heavy atoms. The origin of this failure can be traced, as to be expected from the preceding discussion, to the misrepresentation of E_x^{T} by the RLDA (see Table 8.5). Taking into account relativistic corrections to E_{xc} via the RLDA does not improve results compared to their complete neglect, i.e. the use of the nonrelativistic LDA in the RKS equations.

A magnetization dependent form of the RLDA for E_x has been derived from a polarized RHEG [544–546, 541]. Unfortunately, this functional, which has to be used together with the KS equations (8.141) or (8.149), is not directly obtained as a functional of n and $\boldsymbol{m}$, but rather in terms of spin-up and -down Fermi momenta, $k_{\text{F},\sigma}$. Its application requires an inversion of the function $\boldsymbol{m}(k_{\text{F},\uparrow}, k_{\text{F},\downarrow})$. Due to this complication and the numerically intricate form of the KS equations (8.141), (8.149), this functional has not found widespread use (results for atoms may be found in [550, 551, 555]). In addition, one can show that for atoms this functional yields

Table 8.5 Transverse exchange energies (E_x^T) for closed subshell atoms: selfconsistent REXX, RLDA and B88-RGGA results in comparison with perturbative RHF values (Coulomb gauge for E_x^T in the case of RHF—all energies in Hartree [613]).

Atom	REXX	RHF(p)	RLDA	RGGA
He	0.000064	0.000064	0.000147	0.000059
Ne	0.0167	0.0166	0.0350	0.0167
Ar	0.132	0.132	0.249	0.132
Zn	0.758	0.759	1.318	0.757
Kr	1.417	1.419	2.391	1.415
Cd	3.797	3.808	6.131	3.796
Xe	5.693	5.711	9.039	5.691
Yb	13.842	13.898	21.418	13.837
Hg	22.071	22.168	33.957	22.054
Rn	28.547	28.680	43.979	28.519
No	53.313	53.591	84.222	53.101

essentially the same results as the combination of the unpolarized RLDA (8.200) with the spin-dependence of the nonrelativistic E_x, Eq. (4.19), applied to $n_\pm$,

$$E_x^{\mathrm{RLSDA}}[n_+,n_-] = \frac{1}{2}\left\{E_x^{\mathrm{RLDA}}[2n_+] + E_x^{\mathrm{RLDA}}[2n_-]\right\}. \tag{8.203}$$

This statement is verified in Table 8.6, which lists some prototype ground state energies obtained with the two variants of treating the spin-dependence.[17] Of course,

Table 8.6 Dependence of atomic x-only ground state energies ($-E$) on the treatment of spin: correct magnetization-dependent form of the weakly relativistic LDA (XRR [546]) versus combination of the unpolarized, weakly relativistic LDA functional with the spin-dependence of the nonrelativistic E_x, Eq. (8.203) (all energies in mHartree).

Atom	XRR	Nonrel. spin
Cr	1045942.2	1045942.0
Fe	1267116.4	1267116.2
Eu	10814457.4	10814456.1
W	16101781.8	16101781.8
Au	18966363.2	18966363.2
U	27925396.0	27925395.9
Am	30335911.4	30335911.0

both versions of spin-dependent functionals lead to more realistic results for open-subshell atoms than the use of the unpolarized RLDA (8.200) (see e.g. [555]).

Correlation contributions to the RLDA have also been considered [538, 622]. Similar to the case of e_x^{RHEG}, one finds that the relativistic correction in e_c^{RHEG} is

[17] The two variants of treating the spin are consistently restricted to first order in $1/c^2$, as the magnetization-dependent functional of [546] is known analytically only in this weakly relativistic limit.

substantial. However, it turns out that the RLDA completely misrepresents atomic correlation energies [622]: in addition to the drastic overestimation of atomic E_c on the nonrelativistic level (see Table 4.5) the RLDA yields relativistic corrections which deviate by roughly 50% from the correct numbers (for a more extended discussion the reader is referred to [529]).

In summary, one can state that the RLDA does not give a realistic account of relativistic xc-effects, neither in the case of the exchange nor for correlation.

8.8.4 Relativistic Generalized Gradient Approximation

The direct use of nonrelativistic GGAs in RDFT calculations leads to substantial errors in total energies (see Table 8.4). Consequently, a relativistic extension of the GGA (RGGA) is desirable. The most important ingredient of any GGA is the lowest order gradient contribution proportional to $(\nabla n)^2$. The structure of this term is determined by the long-wavelength expansion of the linear response function of the electron gas. As shown in Appendix J, this approach automatically leads to a current-dependent functional in the relativistic situation. However, no result for the response function of the RHEG beyond the noninteracting limit can be found in the literature. Moreover, in addition to the term proportional to $(\nabla n)^2$, GGAs necessarily contain higher order gradient terms, whose form has to be fixed by additional (model-based or semi-empirical) requirements. As a result, only a semi-empirical form of the RGGA is available so far.

Neglecting any $\boldsymbol{j}$-dependence, an appropriate form of the RGGA for E_x is

$$E_x^{\mathrm{RGGA}} = \int d^3r\, e_x^{\mathrm{HEG}}(n) \left[\Phi_x(\beta) + g(\xi)\Phi_x^{\mathrm{GGA}}(\beta)\right], \qquad (8.204)$$

with $\xi = [\nabla n/(2(3\pi^2 n)^{1/3} n)]^2$ and $g(\xi)$ denoting the gradient part of a nonrelativistic GGA [612]. Following the strategy behind the B88-GGA [219], the relativistic correction factor Φ_x^{GGA} for the gradient term may be obtained by making a sufficiently flexible ansatz and fitting its coefficients to the exact relativistic E_x of a number of closed-subshell atoms, while keeping $g(\xi)$ fixed[18] (for details see [612, 623]). As demonstrated in Tables 8.4 and 8.5, the RGGA leads to much more accurate atomic E_x and ground state energies than both the RLDA and the GGA. The RGGA reaches the same level of accuracy as the nonrelativistic GGA does for light atoms. An improvement is also observed for v_x and the resulting single particle-energies (see Table 8.2). In the case of magnetic systems the RGGA (8.204) can be combined with the nonrelativistic spin-dependence (4.19), in analogy to Eq. (8.203).

Following the same strategy, a RGGA for correlation can be set up [623]. However, the absolute size of the contributions resulting from the relativistic corrections to the functional dependence of E_c on n is rather small. The accuracy of the RGGA

[18] Fits for different GGAs lead to very similar functions Φ_x^{GGA}, confirming the consistency of this approach.

for E_c is therefore completely dominated by the quality of its nonrelativistic basis. In other words: on the present level of sophistication the inclusion of relativistic corrections in GGAs for correlation is not necessary.

8.8.5 Role of Relativistic Corrections in E_{xc}: II. Prototype Results for Molecules and Solids

It remains to discuss the importance of relativistic corrections to the density-dependence of E_{xc} for standard electronic structure properties of molecules and solids. Given the results for atomic IPs in Sect. 8.8.1, one may suspect that these corrections can be safely ignored. This expectation has been confirmed by calculations both for noble metal compounds [624–626] and for bulk gold and platinum [627]. Prototype results for Cu_2 and Au_2 are given in Table 8.7. Even for Au, which

Table 8.7 Spectroscopic constants of noble metal dimers: LDA and BP86-GGA versus BP86-RGGA.

Method	Cu_2			Au_2		
	R_e [Bohr]	D_e [eV]	ω_e [cm^{-1}]	R_e [Bohr]	D_e [eV]	ω_e [cm^{-1}]
LDA[624, 626]	4.05	2.86	307	4.64	3.00	196
GGA[624]	4.16	2.28	287	4.75	2.30	179
RGGA[624]	4.17	2.27	285	4.76	2.27	177
Expt.	4.20	2.05	265	4.67	2.30	191

usually exhibits the effects of relativity most clearly [628], the impact of the correction factors $\Phi_{x/c}$ on the molecular binding properties is marginal. It should be noted, however, that in the case of the dissociation energy the similarity of GGA and RGGA results originates from the cancellation of large relativistic corrections to the individual ground state energies of the molecule and its constituents. The same observation is made for solids, as can be gleaned from Table 8.8. In practice, one can therefore resort to nonrelativistic xc-functionals, as long as an accurate description of the inner shells is irrelevant.

Table 8.8 Lattice constant a_0 and cohesive energy $E_{\rm coh}$ of Au and Pt obtained from LAPW calculations with relativistic and nonrelativistic LDA and PW91-GGA [627] in comparison to experiment [629, 238].

Method	Au				Pt	
	a_0 [Bohr]	$-E_{\rm coh}$ [eV]	$-E$(solid) [Ry]	$-E$(atom) [Ry]	a_0 [Bohr]	$-E_{\rm coh}$ [eV]
LDA	7.68	4.12	38075.445	38075.132	7.36	6.76
RLDA	7.68	4.09	37997.970	37997.669	7.37	6.73
GGA	7.87	2.91	38100.029	38099.815	7.51	5.34
RGGA	7.88	2.89	38049.253	38049.040	7.52	5.30
Expt.	7.67	3.78			7.40	5.85

Chapter 9
Further Reading

In the following we present a listing of topics which are not covered in the main text. For a detailed account of these areas of DFT the reader is referred to the original literature. The cited references can, however, only be indicative of the field, but in no way exhaustive.

- As already mentioned, a density functional description is also possible for systems at non-zero temperature [8]. In this case it is the grand potential and the intrinsic Helmholtz free energy which play the roles of the total energy functional and $F[n]$, respectively. Applications e.g. deal with plasmas [630–633] and the process of freezing [634, 255, 256].
- The foundations of DFT for bosons are addressed in [635–637], for mixtures of fermions and bosons in [638], TDDFT for bosons in [639]. It turns out that in the presence of a macroscopic Bose condensate, i.e. for Bose-condensed liquids like ^{4}He, two density variables are required for a characterization of the ground state: the standard single-particle density together with the condensate density.[1] The dependence of the energy functional on the condensate density was not always taken into account in early applications to ^{4}He (see e.g. [640–642]). For further applications see e.g. [643, 644].
- In the case of infinite, insulating solids subject to an electric field, i.e. dielectrics, the basis for the formulation of the HK-theorem breaks down, as, strictly speaking, there is no ground state[2] [646]: in addition to the density the macroscopic polarization is required for a complete description of insulators in the thermodynamic limit [646–650] (see also [651–654] for a discussion of subtleties involved). As a result, the total xc-potential contains an xc-electric field which, for a symmetric crystal and small applied field, is linear in the macroscopic electric field. A similar linear field dependence was observed for finite molecular chains [655, 503].

[1] Both densities are identical for noninteracting bosons at zero temperature.

[2] In general, an infinite system in the presence of an electric field is in a metastable state, which causes substantial difficulties in defining the polarization (see e.g. [645]).

E. Engel, R.M. Dreizler, Further Reading. In: E. Engel, R.M. Dreizler, Density Functional Theory, Theoretical and Mathematical Physics, pp. 401–402 (2011)
DOI 10.1007/978-3-642-14090-7_9

- It is impossible to handle very large systems, with say hundreds or thousands of atoms, with conventional implementations of the KS equations, even on the basis of the LDA. This class of problems can be approached in terms of order-N methods, which involve segmentation of the system, as first utilized in the divide-and-conquer technique [656–659]. The segmentation relies crucially on the *near-sightedness* (W. Kohn) of quantum systems, which manifests itself most clearly in the 1-particle density matrix $\gamma(\boldsymbol{r},\boldsymbol{r}')$ [660–663]. Even for metals an exponential decay of $\gamma(\boldsymbol{r},\boldsymbol{r}')$ with $|\boldsymbol{r}-\boldsymbol{r}'|$ is found, as soon as non-zero temperature is considered [662, 663] (at zero temperature γ decays as $\cos(k_F|\boldsymbol{r}-\boldsymbol{r}'|)/|\boldsymbol{r}-\boldsymbol{r}'|^2$). Further development of order-N methods was rapid and spawned a substantial number of techniques, which can be sampled from [664, 665] and some recent publications [666, 667] (for an order-N approach to exact exchange see [668]).
- A closely related aspect which is also not addressed in detail in this text is the technical implementation of the various DFT methods, and in particular their embedding in or combination with molecular dynamics or Green's function methods. Information on this aspect can be found in [669, 665, 107].
- DFT for multi-component systems (primarily aiming at electrons plus nuclei) has been formulated in [670, 671, 630, 672–675]. As can be expected, the densities of all components feature in the energy functional. An extension to time-dependent system can be found in [676, 677].
- A DFT approach to superconductivity based on the anomalous (or off-diagonal) density (the order parameter of BCS superconductors) as an additional density variable has been introduced in [678, 679], a time-dependent version in [680]. A local density approximation for the electronic interactions is also available [681]. The coupling of the electrons to the nuclei in the sense of a multicomponent DFT has been included in [682]. For applications see e.g. [683–694].
- The self-consistent mean field models, which have a longstanding tradition in nuclear physics, can also be interpreted as density functional approaches. In these models the interaction between the nucleons is basically described by phenomenological, density-dependent (Skyrme or Gogny) forces (for a recent review see [695]). An alternative approach to the discussion of nuclear properties is provided by meson exchange models. An existence theorem, the corresponding KS equations and an LDA for the xc-energy functional have been formulated for one of these models (quantum hadrodynamics) [696–698]. For more recent work on the foundations of DFT for nuclei see [699–701]. Further progress in the application of DFT in nuclear physics can be gleaned from [702] and [703].
- The existence theorem of (reduced) density matrix functional theory [11, 12] (see also [704, 22, 705, 706]) has already been mentioned in the context of nonlocal potentials. Density matrix functionals also promise a more accurate representation of the exchange-correlation energy [707] and have therefore recently attracted renewed interest [708–718] (for use of the density matrix in TDDFT see e.g. [719, 720]).
- Finally, it seems worthwhile to mention that there also exist density functional approaches based on the 2-particle density (3.45) (also called pair density) [721–725] and on the reduced 2-particle density matrix [726, 727].

Erratum to: Introduction

Erratum to:
Chapter 1 in: E. Engel, R.M. Dreizler, *Density of Functional Theory*, Theoretical and Mathematical Physics, DOI 10.1007/978-3-642-14090-7_1

1. Equation (1.24):

$$\sum_{l=1}^{M}\sum_{\sigma'}\left[\left\langle \eta_k \left| -\frac{\hbar^2\nabla^2}{2m}\delta_{\sigma\sigma'} + \hat{v}_{\text{eff},\sigma\sigma'} \right| \eta_l \right\rangle - \varepsilon_i \langle \eta_k | \eta_l \rangle \delta_{\sigma\sigma'}\right] b_{i,l\sigma'} = 0$$

E. Engel, R.M. Dreizler, Introduction. In: E. Engel, R.M. Dreizler, Density Functional Theory, Theoretical and Mathematical Physics, p. 8 (2011)
DOI 10.1007/978-3-642-14090-7_10

Erratum to: Density Functional Theory

Erratum to:
E. Engel, R.M. Dreizler, *Density of Functional Theory*, Theoretical and Mathematical Physics, DOI 10.1007/978-3-642-14090-7

Equation (2.91) on page 31 should read

$$\begin{aligned}
&\|fg\|_1 \leq \|f\|_p \, \|g\|_q \quad \text{with} \quad \frac{1}{p}+\frac{1}{q}=1 \\
\Rightarrow\ &\|nv_{\text{ext}}\|_1 \leq \|n\|_3 \|v_{\text{ext}}\|_{3/2} < \infty .
\end{aligned}$$

Equation (D.9) on page 438 should read

$$\begin{aligned}
n_0 &= \sum_{i=1}^{\infty} \Theta(\varepsilon_{\mathrm{F}} - \varepsilon_i) \sum_{\sigma=\uparrow,\downarrow} \phi_i^*(\boldsymbol{r}\sigma)\phi_i(\boldsymbol{r}\sigma) \\
&= \sum_{\alpha_1,\alpha_2,\alpha_3=-\infty}^{\infty} \Theta(\varepsilon_{\mathrm{F}} - \varepsilon_{\boldsymbol{k}}) \sum_{\sigma=\uparrow,\downarrow} \phi_{\boldsymbol{k}s}^*(\boldsymbol{r}\sigma)\phi_{\boldsymbol{k}s}(\boldsymbol{r}\sigma) \\
&= \sum_{\alpha_1,\alpha_2,\alpha_3=-\infty}^{\infty} \Theta\left(\varepsilon_{\mathrm{F}} - \frac{\hbar^2\boldsymbol{k}^2}{2m}\right)\frac{2}{L^3} .
\end{aligned}$$

Equation (D.10) on pages 438, 439 should read

$$\begin{aligned}
\frac{T_{\mathrm{S}}(V)}{V} &= \frac{1}{V}\sum_{i=1}^{\infty} \Theta(\varepsilon_{\mathrm{F}} - \varepsilon_i) \sum_{\sigma=\uparrow,\downarrow} \int_V d^3r \phi_i^*(\boldsymbol{r}\sigma)\frac{-\hbar^2\boldsymbol{\nabla}^2}{2m}\phi_i(\boldsymbol{r}\sigma) \\
&= \frac{1}{L^3}\sum_{\alpha_1,\alpha_2,\alpha_3=-\infty}^{\infty} \Theta(\varepsilon_{\mathrm{F}} - \varepsilon_{\boldsymbol{k}}) \sum_{\sigma=\uparrow,\downarrow} \int_0^L dx \int_0^L dy \int_0^L dz \phi_{\boldsymbol{k}s}^*(\boldsymbol{r}\sigma)\frac{\hbar^2\boldsymbol{k}^2}{2m}\phi_{\boldsymbol{k}s}(\boldsymbol{r}\sigma) \\
&= \sum_{\alpha_1,\alpha_2,\alpha_3=-\infty}^{\infty} \Theta\left(\varepsilon_{\mathrm{F}} - \frac{\hbar^2\boldsymbol{k}^2}{2m}\right)\frac{2}{L^3}\,\frac{\hbar^2\boldsymbol{k}^2}{2m} .
\end{aligned}$$

E. Engel, R.M. Dreizler, Density Functional Theory. In: E. Engel, R.M. Dreizler, Density Functional Theory, Theoretical and Mathematical Physics (2011)
DOI 10.1007/978-3-642-14090-7_11

Equation (D.11) on page 439 should read

$$\Delta k_i = \frac{2\pi}{L}\Delta\alpha_i \Rightarrow \Delta\alpha_1\Delta\alpha_2\Delta\alpha_3 = \left(\frac{L}{2\pi}\right)^3 \Delta^3 k$$

$$\sum_{\alpha_1,\alpha_2,\alpha_3=-\infty}^{\infty} \xrightarrow{L\to\infty} \left(\frac{L}{2\pi}\right)^3 \int d^3k.$$

Erratum to: Density Functional Theory

Eberhard Engel • Reiner M. Dreizler

E. Engel, R.M. Dreizler, *Density Functional Theory*, Theoretical and Mathematical Physics, https://doi.org/10.1007/978-3-642-14090-7

The original version of this book was inadvertently published with errors. Those errors have been corrected as follows:

Chapter 5: Virial Theorems

Page 221

Equation (5.18) was revised. The quantities involved satisfy an inequality, rather than an equality.
The derivation of Eq. (5.18) was erroneously based on the implicit assumption that the total energy obtained from the scaled full wavefunction $|\Psi_{0,\lambda}\rangle$, Eq. (5.2), is identical with

$$E[n_\lambda] \;=\; T_{\rm s}[n_\lambda]+E_{\rm ext}[n_\lambda]+E_{\rm H}[n_\lambda]+E_{\rm xc}[n_\lambda]$$

not only for $\lambda = 1$, but also for $\lambda \neq 1$. The correct relation between these two energies [293] follows from the fact that the unscaled ground state $|\Psi_0\rangle = |\Psi_{0,\lambda=1}\rangle$ minimizes the expectation value $\langle\Psi|\hat{T}+\hat{W}|\Psi\rangle$, if $|\Psi\rangle$ is restricted to that part of the Hilbert space which gives the correct ground state density,

$$E_0 \;=\; \min_n \left\{ \min_{|\Psi\rangle\to n} \langle\Psi|\hat{T}+\hat{W}|\Psi\rangle + \int d^3r\, n(\boldsymbol{r})\, v_{\rm ext}(\boldsymbol{r}) \right\} .$$

As a consequence, the scaled state $|\Psi_{0,\lambda}\rangle$ minimizes $\langle\Psi_\lambda|\hat{T}+\lambda\hat{W}|\Psi_\lambda\rangle$. In fact, the scaling behavior of the expection values involved, Eqs. (5.7) and (5.13), directly leads to the relation

$$\langle\Psi_\lambda|\hat{T}+\lambda\hat{W}|\Psi_\lambda\rangle \;=\; \lambda^2\langle\Psi|\hat{T}+\hat{W}|\Psi\rangle .$$

The updated online versions of these chapters can be found at
https://doi.org/10.1007/978-3-642-14090-7_5
https://doi.org/10.1007/978-3-642-14090-7_8

E. Engel, R.M. Dreizler, Density Functional Theory. In: E. Engel, R.M. Dreizler, Density Functional Theory, Theoretical and Mathematical Physics (2011)
DOI 10.1007/978-3-642-14090-7_12

However, restricting $|\Psi\rangle$ on the right-hand side of this equation to give the correct ground state density automatically implies that the density is restricted to n_λ on the left-hand side (by construction of $|\Psi_\lambda\rangle$). Therefore $|\Psi_{0,\lambda}\rangle$ minimizes the expectation value on left-hand side [293], instead of minimizing $\langle\Psi|\hat{T}+\hat{W}|\Psi\rangle$. This result can then be combined with the Levy-Lieb constrained search definition of the density functional $F_{\rm LL}[n]$, Eq. (2.59). Since $F_{\rm LL}[n]$ minimizes the expectation value $\langle\Psi|\hat{T}+\hat{W}|\Psi\rangle$ for given constraint $|\Psi\rangle \to n$, one finds [293]

$$F_{\rm LL}[n_\lambda] \;\leq\; \langle\Psi_{0,\lambda}|\hat{T}+\hat{W}|\Psi_{0,\lambda}\rangle\,.$$

Decomposing $F_{\rm LL}$ as

$$F_{\rm LL}[n_\lambda] \;=\; T_{\rm s,LL}[n_\lambda]+E_{\rm H}[n_\lambda]+E_{\rm xc,LL}[n_\lambda]\,,$$

finally allows to establish an inequality for the correlation functional $E_{\rm xc,LL}[n]$ under scaling [293],

$$E_{\rm c,LL}[n_\lambda] \;\leq\; \lambda^2\left[\langle\Psi_0|\hat{T}|\Psi_0\rangle - T_{\rm s,LL}[n]\right] + \lambda\left[\langle\Psi_0|\hat{W}|\Psi_0\rangle - E_{\rm H}[n] - E_{\rm x,LL}[n]\right]\,, \tag{5.18}$$

which has to replace the old Eq. (5.18). Note that this inequality is strict for $\lambda \neq 1$, while an identity is obtained for $\lambda = 1$. We thank Mel Levy for making us aware of this mistake.

Equation (5.18) was used to derive relation (5.35). However, Eq. (5.35) can also be established by combination of Eqs. (5.23) and (5.34), which are both independent of (5.18). Therefore Eq. (5.35) remains valid.

Page 224

The order of the equations (5.33), (5.34) and (5.35) was changed and all three equations were renumbered. The updated order is shown below.

Equation (5.33) (former 5.34):

$$\frac{d}{d\lambda}E[n_\lambda]\bigg|_{\lambda=1} = 0,$$

Equation (5.34) (former 5.35):

$$0 = 2T_{\rm s}[n] + E_{\rm H}[n] + E_{\rm x}[n] - \int d^3r\, n(\boldsymbol{r})\,\boldsymbol{r}\cdot\boldsymbol{\nabla} v_{\rm c}(\boldsymbol{r})$$
$$+\langle\Psi_0|\hat{V}_{\rm ext}|\Psi_0\rangle + \sum_\alpha \boldsymbol{R}_\alpha\cdot\frac{\partial}{\partial\boldsymbol{R}_\alpha}\langle\Psi_0|\hat{V}_{\rm ext}|\Psi_0\rangle\bigg|_{\rm expl.}\,,$$

Equation (5.35) (former 5.33):

$$E_{\rm c}[n] = -\int d^3r\, n(\boldsymbol{r})\,\boldsymbol{r}\cdot\boldsymbol{\nabla} v_{\rm c}(\boldsymbol{r}) - T + T_{\rm s}[n]\,.$$

Chapter 8: Relativistic Density Functional Theory

Page 374

Equation (8.111) was revised: Since in the *no-pair* RDFT formalism we have chosen to subtract the rest mass from the energy, the occupation function has to be adjusted accordingly.

$$\Theta_k = \begin{cases} 0 \text{ for } \varepsilon_k \leq -2mc^2 \\ 1 \text{ for } -2mc^2 < \varepsilon_k \leq \varepsilon_F \\ 0 \text{ for } \varepsilon_F < \varepsilon_k \end{cases} .$$

Page 396

Table 8.6 was revised. The data for Au have been corrected and the updated version is shown below.

Atom	XRR	Nonrel. spin
Cr	1045942.2	1045942.0
Fe	1267116.4	1267116.2
Eu	10814457.4	10814456.1
W	16101781.8	16101781.8
Au	18966363.2	18966363.2
U	27925395.7	27925395.6
Am	30335910.7	30335910.3

Appendix A
Functionals and the Functional Derivative

In this Appendix we provide a minimal introduction to the concept of functionals and the functional derivative. No attempt is made to maintain mathematical rigor. A more extended and mathematically more precise discussion of the material summarized here can be found in the books of Courant and Hilbert [728] and of Atkinson and Han [29] (for the special context of DFT see also [28]).

A.1 Definition of the Functional

A functional is defined by a rule, which associates a number (real or complex) with a function of one or several variables,

$$f(x) \text{ or } f(\boldsymbol{r}_1, \ldots) \quad \stackrel{\text{rule}}{\longrightarrow} \quad F[f], \tag{A.1}$$

or, more generally, which associates a number with a set of functions,

$$f_1, f_2, \ldots \quad \stackrel{\text{rule}}{\longrightarrow} \quad F[f_1, f_2, \ldots]. \tag{A.2}$$

This definition is quite well described by the designation as a function of a function. Some examples are:

- A definite integral over a continuous function $f(x)$

$$F[f] = \int_{x_1}^{x_2} f(x)dx \tag{A.3}$$

  (similarly one can have integrals with functions of several variables).
- A slightly more general form is

$$F_w[f] = \int_{x_1}^{x_2} w(x)f(x)dx, \tag{A.4}$$

E. Engel, R.M. Dreizler, Appendices. In: E. Engel, R.M. Dreizler, Density Functional Theory, Theoretical and Mathematical Physics, pp. 403–531 (2011)
DOI 10.1007/978-3-642-14090-7

that is an integral over the function f with a fixed weight function $w(x)$.

- A prescription which associates a function with the value of this function at a particular point in the interior of a given interval $[x_1, x_2]$

$$F[f] = f(x_0) \qquad x_0 \in (x_1, x_2)\,. \tag{A.5}$$

This functional can be represented in integral form with the aid of the δ-function,

$$F_\delta[f] = \int_{x_1}^{x_2} \delta(x - x_0) f(x) dx\,, \tag{A.6}$$

that is with a weight function in the form of a generalized function (a distribution).

The examples (A.3) and (A.5) directly show that a functional can itself be a function of a variable, i.e. of one of the parameters in its definition, as the boundaries in the integral (A.3) or the point x_0 in the functional (A.5). The dependence on such a parameter y is denoted as $F[f](y)$.

So far, all examples are characterized by the fact that they depend linearly on the function $f(x)$, so that they satisfy the relation

$$F[c_1 f_1 + c_2 f_2] = c_1 F[f_1] + c_2 F[f_2]\,, \tag{A.7}$$

with c_1, c_2 being complex numbers. Examples of nonlinear functionals are:

- The energy functional of the simplest DFT, the Thomas-Fermi kinetic energy,

$$F_{\rm TF}[n] \equiv T_{\rm s}^{\rm TF}[n] = C_{\rm TF} \int d^3r\, n^{5/3}(\boldsymbol{r})\,. \tag{A.8}$$

- A nonlocal functional of two functions,

$$F_w[f_1, f_2] = \int f_1(x_1) w(x_1, x_2) f_2(x_2) dx_1 dx_2\,. \tag{A.9}$$

- The action integral of classical mechanics,

$$F[\boldsymbol{q}] \equiv A[\boldsymbol{q}] = \int_{t_1}^{t_2} dt\, L(\boldsymbol{q}(t), \dot{\boldsymbol{q}}(t), t)\,. \tag{A.10}$$

The abbreviation $\boldsymbol{q}(t)$ stands for a set of generalized coordinates, which depend on time.

- Any matrix element of quantum mechanics, e.g. the ground state energy and the S-matrix element of potential scattering theory,

$$F[\Psi_0, \Psi_0^*] \equiv E[\Psi_0, \Psi_0^*] = \int d^3r\, \Psi_0^*(\boldsymbol{r}) \hat{H} \Psi_0(\boldsymbol{r})$$

$$F[\Psi_{\boldsymbol{k}}, \Psi_{\boldsymbol{q}}^*] \equiv S[\Psi_{\boldsymbol{k}}, \Psi_{\boldsymbol{q}}^*] = \int d^3r\, \Psi_{\boldsymbol{q}}^*(\boldsymbol{r}) \hat{S} \Psi_{\boldsymbol{k}}(\boldsymbol{r})\,.$$

It seems worthwhile to emphasize that the two functions Ψ_0 and Ψ_0^* have to be considered as being independent, so that one is dealing with a functional of two functions. Alternatively, a dependence on the real and the imaginary part of the wavefunctions can be used to characterize the functional.

A.2 Functional Derivative

Usually knowledge of the complete functional $F[f]$, as for example the classical action $A[\boldsymbol{q}]$ for all possible trajectories in phase space or the value of the integral (A.3) for all continuous functions, is not required. Rather it is the behavior of the functional in the vicinity of the function f_0, which makes $F[f]$ extremal or stationary, which is of interest.[1] The implementation of the search for f_0 involves the exploration of the space of functions in the vicinity of f_0 in a suitable fashion.

A variation of any function f by an infinitesimal but arbitrary amount can be represented in the form

$$\begin{aligned} \delta f(x) &= \varepsilon\,\eta(x) && \text{for one variable} \\ \delta f(\boldsymbol{r}_1,\boldsymbol{r}_2,\ldots) &= \varepsilon\,\eta(\boldsymbol{r}_1,\boldsymbol{r}_2,\ldots) && \text{for several variables}\,. \end{aligned} \tag{A.11}$$

The quantity ε is an infinitesimal number, η is an arbitrary function. In order to explore the properties of the functionals a generalization of the (ordinary or partial) derivative (of first and higher order)—the functional derivative—is required. It can be defined via the variation δF of the functional $F[f]$ which results from variation of f by δf,

$$\delta F := F[f+\delta f] - F[f]\,. \tag{A.12}$$

The technique used to evaluate δF is a Taylor expansion of the functional $F[f+\delta f] = F[f+\varepsilon\eta]$ in powers of δf, respectively of ε. The functional $F[f+\varepsilon\eta]$ is an ordinary function of ε. This implies that the expansion in terms of powers of ε is a standard Taylor expansion,

$$F[f+\varepsilon\,\eta] = F[f] + \left.\frac{dF[f+\varepsilon\eta]}{d\varepsilon}\right|_{\varepsilon=0}\varepsilon + \frac{1}{2}\left.\frac{d^2F[f+\varepsilon\eta]}{d\varepsilon^2}\right|_{\varepsilon=0}\varepsilon^2 + \ldots \tag{A.13}$$

$$= \sum_{n=0}^{N}\frac{1}{n!}\left.\frac{d^nF[f+\varepsilon\eta]}{d\varepsilon^n}\right|_{\varepsilon=0}\varepsilon^n + \mathcal{O}\left(\varepsilon^{N+1}\right)\,. \tag{A.14}$$

As indicated, the sum in (A.14) can be finite or infinite. In the latter case, it has to be assumed that the function $F(\varepsilon)$ can be differentiated with respect to ε any number of times.

[1] Often functionals are introduced to recast some equation(s) in the form of an extremum or stationarity principle.

The derivatives with respect to ε now have to be related to the functional derivatives. This is achieved by a suitable definition. The definition of the functional derivative (also called variational derivative) is

$$\left.\frac{dF[f+\varepsilon\eta]}{d\varepsilon}\right|_{\varepsilon=0} =: \int dx_1\, \frac{\delta F[f]}{\delta f(x_1)}\, \eta(x_1)\,. \tag{A.15}$$

This definition implies that the left-hand side can be brought into the form on the right-hand side, i.e. the form of a linear functional with kernel $\delta F[f]/\delta f$ acting on the test function η. This is by no means guaranteed for arbitrary functionals and arbitrary f. It is exactly this point where rigorous mathematics sets in. A functional for which (A.15) is valid is called *differentiable*.[2] We will, however, not go into any details concerning the existence of the functional derivative, nor will we make any attempt to characterize the space of (test) functions which are allowed in (A.15) (as usual, the existence of all integrals involved is assumed, of course).

The definition (A.15) can be thought of as an extension of the first total differential of a function of several variables,

$$f(x_1,x_2,\ldots) \quad\longrightarrow\quad df = \sum_{n=1}^{N} \frac{\partial f}{\partial x_n}\, dx_n\,,$$

to the case of an infinite set of variables $f(x_1)$. The definition of the second order functional derivative corresponds to the second order total differential,

---

[2] More precisely, a functional $F[f]$ which maps an open subset of some Banach space $\mathscr{X}$ (i.e. some complete normed vector space) of functions f onto another Banach space $\mathscr{Y}$ (which could be the set of real or complex numbers) is called *Fréchet differentiable*, if there exists a linear continuous operator $\delta F_f^{\mathrm{F}} : \mathscr{X} \to \mathscr{Y}$ with the property

$$\lim_{\|\eta\|\to 0} \frac{\|F[f+\eta] - F[f] - \delta F_f^{\mathrm{F}}[\eta]\|_{\mathscr{Y}}}{\|\eta\|_{\mathscr{X}}} = 0\,.$$

Here $\|F\|_{\mathscr{Y}}$ and $\|\eta\|_{\mathscr{X}}$ denote the norms in the two Banach spaces. The Fréchet derivative has to be distinguished from the Gâteaux derivative, which exists if there is a linear continuous operator $\delta F_f^{\mathrm{G}} : \mathscr{X} \to \mathscr{Y}$ such that

$$\delta F_f^{\mathrm{G}}[\eta] = \lim_{\lambda\to 0} \frac{\|F[f+\lambda\eta] - F[f]\|_{\mathscr{Y}}}{\lambda}\,.$$

If the right-hand side of this relation exists, but does not yield a linear continuous operator, it is called the Gâteaux differential,

$$F'[f,\eta] = \lim_{\lambda\to 0} \frac{\|F[f+\lambda\eta] - F[f]\|_{\mathscr{Y}}}{\lambda}\,.$$

Thus any Fréchet differentiable functional is also Gâteaux differentiable, but the converse is not true. The existence of the Fréchet derivative is only ensured, if the Gâteaux derivative is continuous or if the Gâteaux differential is uniform with respect to η with $\|\eta\| = 1$.

$$\left.\frac{d^2F[f+\varepsilon\eta]}{d\varepsilon^2}\right|_{\varepsilon=0} =: \int dx_1dx_2\,\frac{\delta^2F[f]}{\delta f(x_1)\delta f(x_2)}\,\eta(x_1)\eta(x_2)\,. \tag{A.16}$$

The definition of the general derivative can be guessed at this stage. The functional derivative of n-th order is given by

$$\left.\frac{d^nF[f+\varepsilon\eta]}{d\varepsilon^n}\right|_{\varepsilon=0} =: \int dx_1\ldots dx_n\,\frac{\delta^nF[f]}{\delta f(x_1)\ldots\delta f(x_n)}\,\eta(x_1)\ldots\eta(x_n)\,. \tag{A.17}$$

This derivative constitutes the kernel of the Taylor expansion of a functional F in terms of the variation $\delta f(x) = \varepsilon\eta(x)$,

$$\begin{aligned} F[f+\varepsilon\eta] = \sum_{n=0}^{N}\frac{1}{n!}\int dx_1\ldots dx_n\,&\frac{\delta^nF[f]}{\delta f(x_1)\ldots\delta f(x_n)}\,\delta f(x_1)\ldots\delta f(x_n)\\ &+\mathscr{O}\left(\varepsilon^{N+1}\right)\,, \end{aligned} \tag{A.18}$$

again with N being either finite or infinite.

The actual calculation of the functional derivative relies on the evaluation of the difference (A.12). This will be illustrated with the aid of a few examples.

- According to Eq. (A.12), the variation of the functional (A.6) is

$$\delta F_\delta = \int_{x_1}^{x_2}\delta(x-x_0)\varepsilon\eta(x)dx\,.$$

  Comparison with the definition (A.15) shows that

$$\frac{\delta F_\delta}{\delta f(x)} = \delta(x-x_0)\,, \tag{A.19}$$

  as $\eta(x)$ can vary freely. A very useful formula is obtained if the definition

$$F_\delta[f] = f(x_0)$$

  is used explicitly,

$$\frac{\delta F_\delta}{\delta f(x)} = \frac{\delta f(x_0)}{\delta f(x)} = \delta(x-x_0)\,. \tag{A.20}$$

  All higher order functional derivatives of F_δ vanish.
- This example is readily extended to the functional

$$f(x_0)^\alpha = \int dx\,\delta(x-x_0)\,f(x)^\alpha\,.$$

  Its variation can be evaluated by straightforward Taylor expansion,

$$\delta f(x_0)^\alpha = \int dx\,\delta(x-x_0)\,\left[(f(x)+\varepsilon\eta(x))^\alpha - f(x)^\alpha\right]$$

$$= \int dx\, \delta(x - x_0) \Big[\alpha f(x)^{\alpha-1} \varepsilon\eta(x) + \frac{\alpha(\alpha-1)}{2} f(x)^{\alpha-2} (\varepsilon\eta(x))^2 + \ldots \Big] .$$

The functional derivative is again identified by comparison with the definition (A.15),

$$\frac{\delta f(x_0)^\alpha}{\delta f(x)} = \delta(x - x_0)\, \alpha f(x)^{\alpha-1} . \tag{A.21}$$

In order to calculate the second functional derivative one can simply reuse Eq. (A.21),

$$\frac{\delta^2 f(x_0)^\alpha}{\delta f(x_1) \delta f(x_2)} = \delta(x_1 - x_0)\, \delta(x_2 - x_0)\, \alpha(\alpha - 1) f(x)^{\alpha-2} . \tag{A.22}$$

- The variation of the Thomas-Fermi functional (A.8) is obtained from

$$\delta F_{\rm TF} = C_{\rm TF} \int d^3r \left[(n(\boldsymbol{r}) + \varepsilon\eta(\boldsymbol{r}))^{5/3} - n(\boldsymbol{r})^{5/3} \right]$$

in the form of a binomial expansion

$$\delta F_{\rm TF} = C_{\rm TF} \int d^3r\, n(\boldsymbol{r})^{5/3} \sum_{k=1}^{\infty} \binom{5/3}{k} \left(\frac{\varepsilon\eta(\boldsymbol{r})}{n(\boldsymbol{r})} \right)^k .$$

The functional derivatives, which can be extracted from this expression, are

$$\frac{\delta F_{\rm TF}}{\delta n(\boldsymbol{r})} = \frac{5}{3} C_{\rm TF}\, n(\boldsymbol{r})^{2/3} \tag{A.23}$$

for the first derivative and, applying (A.21),

$$\frac{\delta^2 F_{\rm TF}}{\delta n(\boldsymbol{r}) \delta n(\boldsymbol{r}')} = \frac{10}{9} C_{\rm TF}\, n(\boldsymbol{r})^{-1/3}\, \delta^{(3)}(\boldsymbol{r} - \boldsymbol{r}')$$

for the second derivative.

- The variation of the nonlocal functional

$$F_w[f] = \int_{y_1}^{y_2} dx_1 \int_{y_1}^{y_2} dx_2\, f(x_1) w(x_1, x_2) f(x_2) \tag{A.24}$$

is

$$\delta F_w = \int_{y_1}^{y_2} dx_1 \int_{y_1}^{y_2} dx_2\, w(x_1, x_2) [f(x_1) \varepsilon\eta(x_2) + f(x_2) \varepsilon\eta(x_1) + \varepsilon\eta(x_1) \varepsilon\eta(x_2)] . \tag{A.25}$$

The variational derivatives are

$$\frac{\delta F_w}{\delta f(x)} = \int_{y_1}^{y_2} dx_2 \, [w(x,x_2) + w(x_2,x)] f(x_2) \tag{A.26}$$

and

$$\frac{\delta^2 F_w}{\delta f(x_1) f(x_2)} = w(x_1,x_2) + w(x_2,x_1) \,. \tag{A.27}$$

All derivatives with $n > 2$ vanish for this example.

A.3 Calculational Rules

The calculation of the functional derivative can be abbreviated using a variation in terms of the δ-function: for the functionals relevant in physics all local, δ-type variations of $f(x)$ are equivalent to probing the functional with arbitrary general variations $\eta(x)$. The functional derivative can therefore be recast in the form of the (almost familiar) limiting value

$$\frac{\delta F}{\delta f(x_1)} = \lim_{\varepsilon \to 0} \frac{F[f(x) + \varepsilon \delta(x - x_1)] - F[f(x)]}{\varepsilon} \,. \tag{A.28}$$

The reader may check that this form follows from the definition (A.15) with the replacement $\eta(x) \longrightarrow \delta(x - x_1)$ and that it reproduces the results of the examples. When using the form (A.28), one has to remember that the variation $\delta f = \varepsilon \delta(x - x_1)$ should always be understood in the sense of a representation of the δ-function via some sequence of regular functions, so that powers of the δ-function are uncritical.

As the functional derivatives constitute an extension of the concept of the ordinary derivative, most of the rules for ordinary derivatives can be taken over. For example, the product rule of functional differentiation can be obtained directly with the argument

$$\left[\frac{d(F_1[f+\varepsilon\eta]F_2[f+\varepsilon\eta])}{d\varepsilon}\right]_{\varepsilon=0} = \left[\frac{dF_1[f+\varepsilon\eta]}{d\varepsilon}F_2[f+\varepsilon\eta]\right]_{\varepsilon=0} + \left[F_1[f+\varepsilon\eta]\frac{dF_2[f+\varepsilon\eta]}{d\varepsilon}\right]_{\varepsilon=0} ,$$

which is valid as F_1 and F_2 are *functions* of ε. In the actual limit $\varepsilon \to 0$ there follows with (A.15)

$$\frac{\delta(F_1 F_2)}{\delta f(x)} = \frac{\delta F_1}{\delta f(x)} F_2 + F_1 \frac{\delta F_2}{\delta f(x)} \,. \tag{A.29}$$

Let us next extend the chain rule for functions to functionals. Consider a functional F which depends on some function $G(y)$, which itself is a functional of $f(x)$, $G[f](y)$. The functional F therefore is also a functional of $f(x)$. Its variation with f is then given by

$$\begin{aligned}\delta F_f &= F[G[f(x)+\varepsilon\eta(x)](y)] - F[G[f(x)](y)] \\ &= \left.\frac{dF[G[f(x)+\varepsilon\eta(x)](y)]}{d\varepsilon}\right|_{\varepsilon=0} \varepsilon + \mathcal{O}(\varepsilon^2) \qquad \text{(A.30)} \\ &= \int \frac{\delta F[f]}{\delta f(x)} \varepsilon\eta(x)\, dx + \mathcal{O}(\varepsilon^2)\,, \qquad \text{(A.31)}\end{aligned}$$

where the last line simply represents the definition of the functional derivative of F with respect to f, according to Eq. (A.15). Similarly, the variation of G with f is obtained as

$$\begin{aligned}\delta G(y) &= G[f(x)+\varepsilon\eta(x)](y) - G[f(x)](y) \\ &= \left.\frac{dG[f(x)+\varepsilon\eta(x)](y)}{d\varepsilon}\right|_{\varepsilon=0} \varepsilon + \mathcal{O}(\varepsilon^2) \qquad \text{(A.32)} \\ &= \int \frac{\delta G[f](y)}{\delta f(x)} \varepsilon\eta(x)\, dx + \mathcal{O}(\varepsilon^2)\,. \qquad \text{(A.33)}\end{aligned}$$

Now, to first order in ε one can express $G[f(x)+\varepsilon\eta(x)](y)$ via Eq. (A.33),

$$G[f(x)+\varepsilon\eta(x)](y) = G[f(x)](y) + \int \frac{\delta G[f](y)}{\delta f(x)} \varepsilon\eta(x)\, dx + \mathcal{O}(\varepsilon^2)\,,$$

to obtain

$$\int \frac{\delta F[f]}{\delta f(x)} \eta(x)\, dx = \left.\frac{dF[G[f(x)](y) + \int \frac{\delta G[f](y)}{\delta f(x)} \varepsilon\eta(x) dx + \mathcal{O}(\varepsilon^2)]}{d\varepsilon}\right|_{\varepsilon=0} + \mathcal{O}(\varepsilon)\,. \qquad \text{(A.34)}$$

However, the derivative on the right-hand side has exactly the form of the variation of F with G,

$$\begin{aligned}\delta F_G &= F[G(y)+\varepsilon\bar{\eta}(y)] - F[G(y)] \\ &= \left.\frac{dF[G(y)+\varepsilon\bar{\eta}(y)]}{d\varepsilon}\right|_{\varepsilon=0} \varepsilon + \mathcal{O}(\varepsilon^2)\,, \qquad \text{(A.35)}\end{aligned}$$

with $\bar{\eta}$ given by

$$\bar{\eta}(y) = \int \frac{\delta G[f](y)}{\delta f(x)} \eta(x) dx\,. \qquad \text{(A.36)}$$

Provided that $\bar{\eta}(y)$ probes the complete space around $G(y)$, in which $F[G]$ is defined, when $\eta(x)$ goes through all legitimate variations of $f(x)$, the expression (A.34) coincides with the corresponding functional derivative of F with respect to $G(y)$,

$$\left.\frac{dF[G(y)+\varepsilon\bar{\eta}(y)]}{d\varepsilon}\right|_{\varepsilon=0} = \int \frac{\delta F[G]}{\delta G(y)}\bar{\eta}(y)\,dy + \mathscr{O}(\varepsilon)\,. \tag{A.37}$$

Combination of Eqs. (A.34), (A.36) and (A.37) finally yields

$$\int \frac{\delta F[f]}{\delta f(x)}\eta(x)\,dx = \int \frac{\delta F[G]}{\delta G(y)}\frac{\delta G[f](y)}{\delta f(x)}\eta(x)\,dxdy\,,$$

and thus, due to the arbitrary form of $\eta(x)$,

$$\frac{\delta F[f]}{\delta f(x)} = \int \frac{\delta F[G]}{\delta G(y)}\frac{\delta G[f](y)}{\delta f(x)}\,dy\,. \tag{A.38}$$

Equation (A.38) represents the chain rule of functional differentiation. It is valid, if the variation $\eta(x)$ generates all possible variations $\bar{\eta}(y)$ in the neighborhood of $G[f](y)$. This is guaranteed if there is a one-to-one correspondence between the admissible functions $f(x)$ and the corresponding functions $G(y)$ (at least locally) and both spaces of functions are sufficiently dense to define a functional derivative. The condition of a unique correspondence is satisfied in particular, if the kernel $\frac{\delta G[f](y)}{\delta f(x)}$ is invertible.

It is worthwhile to note a special case of the rule (A.38). If there is a unique relation between $f(x)$ and $G(y)$, i.e. if the form of the complete function $G(y)$ is uniquely determined by $f(x)$ and vice versa, one can consider the functional $F[G[f(x)](y)] \equiv f(x_0)$. Application of the chain rule (A.38) then leads to

$$\begin{aligned}\delta(x-x_0) = \frac{\delta f(x_0)}{\delta f(x)} &= \frac{\delta F[f]}{\delta f(x)} = \int \frac{\delta F[G]}{\delta G(y)}\frac{\delta G[f](y)}{\delta f(x)}\,dy \\ &= \int \frac{\delta f(x_0)}{\delta G(y)}\frac{\delta G(y)}{\delta f(x)}\,dy\,.\end{aligned} \tag{A.39}$$

This relation shows that one can always insert a complete set of variations in a variational derivative (here $\delta f(x_0)/\delta f(x)$), as long as there exists a one-to-one correspondence between the functions involved.

A.4 Variational Principle

An apt example for the discussion of variational principles on the basis of functional calculus is the derivation of the Euler-Lagrange equations for the action functional (A.10). For the case of one degree of freedom,

$$A[q] = \int_{t_1}^{t_2} dt\, L(q,\dot{q},t)\,, \tag{A.40}$$

which suffices to point out the main features, extrema are characterized by setting the first variation equal to zero. This implies

$$\delta A = \int_{t_1}^{t_2} dt\, [L(q+\delta q,\dot{q}+\delta\dot{q},t) - L(q,\dot{q},t)] = 0 \tag{A.41}$$

to first order in the variation of the variable and its derivative. Taylor expansion of the first term to first order gives

$$\delta A = \int_{t_1}^{t_2} dt \left[\frac{\partial L}{\partial q}\delta q + \frac{\partial L}{\partial \dot{q}}\delta\dot{q}\right] = 0\,. \tag{A.42}$$

This is followed by partial integration of the second term with the result

$$\delta A = \int_{t_1}^{t_2} dt \left[\frac{\partial L}{\partial q} - \frac{d}{dt}\frac{\partial L}{\partial \dot{q}}\right]\delta q + \left[\frac{\partial L}{\partial \dot{q}}\delta q\right]_{t_1}^{t_2} = 0\,. \tag{A.43}$$

For arbitrary variations δq the Euler-Lagrange equations have to be satisfied,

$$\frac{\partial L}{\partial q} - \frac{d}{dt}\frac{\partial L}{\partial \dot{q}} = 0\,. \tag{A.44}$$

No further conditions apply, if the variation at the end points is restricted by boundary conditions,

$$\delta q(t_1) = \delta q(t_2) = 0\,. \tag{A.45}$$

This restriction does not apply to the case of a free boundary, for which arbitrary variations at the points t_1 and t_2 are permitted. Therefore it is necessary to demand in addition the "natural boundary conditions" (see [728])

$$\left[\frac{\partial L}{\partial \dot{q}}\right]_{t_1} = \left[\frac{\partial L}{\partial \dot{q}}\right]_{t_2} = 0 \tag{A.46}$$

in this case.

Appendix B
Second Quantization in Many-Body Theory

The language of second quantization allows a compact formulation of quantum many-particle problems. The name "second quantization" arose actually in the context of quantum field theory, where the need to accommodate particle creation or annihilation processes demanded the replacement of wavefunctions by operators. This Appendix provides an introduction to this tool tailored to the requirements of many-body theory.

B.1 *N*-Particle Hilbert Space

B.1.1 Realization in First Quantized Form

The basic elements of the discussion are square-integrable single-particle wavefunctions in configuration or configuration-spin space. These functions span the 1-particle Hilbert space $\mathscr{H}_1$. They will be denoted by

$$\phi_\alpha(x) \; . \tag{B.1}$$

The index α represents a set of quantum numbers, which characterize the state of the particle completely, as e.g. the quantum numbers of the nonrelativistic hydrogen problem,

$$\alpha \longrightarrow n, l, m, m_s \; . \tag{B.2}$$

For the present purpose it is most convenient to characterize the states by a single discrete label which orders all states in a well-defined sequence.

The variable x in Eq. (B.1) stands for the spatial coordinates $\boldsymbol{r}$ and, if applicable, additional internal degrees of freedom. A relevant example are the components of the bispinor wavefunction of a spin-1/2 fermion,

$$\phi_\alpha(x) = \phi_\alpha(\boldsymbol{r}\sigma) = \begin{cases} \phi_\alpha(\boldsymbol{r}, +\frac{1}{2}) & \text{if } \sigma = +\frac{1}{2} \equiv\uparrow \\ \phi_\alpha(\boldsymbol{r}, -\frac{1}{2}) & \text{if } \sigma = -\frac{1}{2} \equiv\downarrow \end{cases} . \tag{B.3}$$

For brevity the complex functions (B.1) are supposed to be orthonormal

$$\int dx\, \phi_\alpha^*(x)\phi_\beta(x) = \delta_{\alpha\beta} . \tag{B.4}$$

The integral $\int dx$ abbreviates integration over space and summation over all internal degrees of freedom, as e.g. in the case of spin-1/2 fermions

$$\int dx \equiv \sum_{\sigma=\uparrow,\downarrow} \int d^3r .$$

In addition, the functions $\phi_\alpha(x)$ are assumed to form a complete set,

$$\sum_\alpha \phi_\alpha(x)\phi_\alpha^*(x') = \delta^{(3)}(\boldsymbol{r}-\boldsymbol{r}')\,\delta_{\sigma,\sigma'} \equiv \delta(x,x') . \tag{B.5}$$

The Hilbert space $\mathscr{H}_N$ of N identical particles is the *tensor product* of N single-particle Hilbert spaces,

$$\mathscr{H}_N = \mathscr{H}_1 \otimes \mathscr{H}_1 \otimes \cdots \otimes \mathscr{H}_1 . \tag{B.6}$$

This space is spanned by the product wavefunctions

$$\Phi^{\mathrm{c}}_{\alpha_1\cdots\alpha_N}(x_1x_2\cdots x_N) = \phi_{\alpha_1}(x_1)\phi_{\alpha_2}(x_2)\cdots\phi_{\alpha_N}(x_N) . \tag{B.7}$$

In these *canonical* N-particle states particle number 1 with coordinates $\boldsymbol{r}_1$ and spin-projection σ_1 is in the single-particle state α_1, particle number 2 at x_2 in state α_2, and so on. Orthonormality and completeness of this basis of $\mathscr{H}_N$ follow from the corresponding properties of the 1-particle functions,

$$\begin{aligned} &\int dx_1\cdots dx_N\, \Phi^{\mathrm{c}\,*}_{\alpha_1\cdots\alpha_N}(x_1\cdots x_N)\Phi^{\mathrm{c}}_{\beta_1\cdots\beta_N}(x_1\cdots x_N) \\ &= \int dx_1\, \phi_{\alpha_1}^*(x_1)\phi_{\beta_1}(x_1)\cdots\int dx_N\, \phi_{\alpha_N}^*(x_N)\phi_{\beta_N}(x_N) \\ &= \delta_{\alpha_1\beta_1}\cdots\delta_{\alpha_N\beta_N} \end{aligned} \tag{B.8}$$

$$\begin{aligned} &\sum_{\alpha_1\cdots\alpha_N} \Phi^{\mathrm{c}}_{\alpha_1\cdots\alpha_N}(x_1\cdots x_N)\Phi^{\mathrm{c}\,*}_{\alpha_1\cdots\alpha_N}(y_1\cdots y_N) \\ &= \sum_{\alpha_1}\phi_{\alpha_1}(x_1)\phi_{\alpha_1}^*(y_1)\cdots\sum_{\alpha_N}\phi_{\alpha_N}(x_N)\phi_{\alpha_N}^*(y_N) \\ &= \delta(x_1,y_1)\cdots\delta(x_N,y_N) . \end{aligned} \tag{B.9}$$

The states $\Phi^{\mathrm{c}}_{\alpha_1\cdots\alpha_N}$ are ordered with respect to the particles and their labels, which is only possible if the individual particles can be distinguished. However, in the case of N identical quantum particles, their fermionic or bosonic nature has to be taken into account, so that only the corresponding subspaces $\mathscr{F}_N$ and $\mathscr{B}_N$ of $\mathscr{H}_N$ are of interest. The wavefunctions of the fermion sector, to which we restrict the discussion, are antisymmetric: *any* wavefunction Ψ describing N identical fermions satisfies the relation

$$\Psi(x_{p_1} \cdots x_{p_N}) = (-1)^P \Psi(x_1 \cdots x_N)\,, \tag{B.10}$$

where $p_1, \ldots p_N$ denotes an arbitrary permutation of the numbers $1, \ldots N$. The sign of the permutations $(-1)^P$ corresponds to the property even (+) or odd (−), according to the number P of pairwise transpositions necessary to restore the natural order.[1]

A basis in $\mathscr{F}_N$ can be constructed from the canonical basis by explicit antisymmetrization,

$$\begin{aligned}
\Phi_{\alpha_1 \cdots \alpha_N}(x_1 x_2 \cdots x_N) &= \frac{1}{\sqrt{N!}} \sum_{P \in S_N} (-1)^P \Phi^{\mathrm{c}}_{\alpha_{p_1} \cdots \alpha_{p_N}}(x_1 x_2 \cdots x_N) \\
&= \frac{1}{\sqrt{N!}} \sum_{P \in S_N} (-1)^P \phi_{\alpha_{p_1}}(x_1) \phi_{\alpha_{p_2}}(x_2) \cdots \phi_{\alpha_{p_N}}(x_N) \\
&= \frac{1}{\sqrt{N!}} \sum_{P \in S_N} (-1)^P \phi_{\alpha_1}(x_{p_1}) \phi_{\alpha_2}(x_{p_2}) \cdots \phi_{\alpha_N}(x_{p_N})\,.
\end{aligned} \tag{B.11}$$

The sum runs over all permutations of N ordered objects. This set of permutations constitutes the symmetric group S_N. The last lines indicate that the basis functions of $\mathscr{F}_N$ take the form of a determinant—a *Slater determinant*.

The antisymmetrized states $\Phi_{\alpha_1 \cdots \alpha_N}$ do no longer associate a particular single-particle quantum number with a given particle. Their determinantal structure is a direct manifestation of the Pauli principle which is expressed by Eq. (B.10). The function $\Phi_{\alpha_1 \cdots \alpha_N}$ vanishes if two of the labels $\alpha_1 \cdots \alpha_N$ are identical. This allows a definite (although arbitrary) ordering of the quantum numbers in $\Phi_{\alpha_1 \cdots \alpha_N}$ in the form $\alpha_1 < \alpha_2 \cdots < \alpha_N$.

The basis functions (B.11) are orthonormal and complete in $\mathscr{F}_N$, provided the 1-particle basis is orthonormal and complete in $\mathscr{H}_1$. For the illustration of orthonormality one simply uses the definition (B.11),

$$\begin{aligned}
&\int dx_1 \cdots dx_N\, \Phi^*_{\alpha_1 \cdots \alpha_N}(x_1 \cdots x_N) \Phi_{\beta_1 \cdots \beta_N}(x_1 \cdots x_N) \\
&= \frac{1}{N!} \sum_{P,P' \in S_N} (-1)^{P+P'} \int dx_1\, \phi^*_{\alpha_{p_1}}(x_1) \phi_{\beta_{p'_1}}(x_1) \cdots \int dx_N\, \phi^*_{\alpha_{p_N}}(x_N) \phi_{\beta_{p'_N}}(x_N) \\
&= \frac{1}{N!} \sum_{P,P' \in S_N} (-1)^{P+P'} \delta_{\alpha_{p_1} \beta_{p'_1}} \cdots \delta_{\alpha_{p_N} \beta_{p'_N}}\,.
\end{aligned}$$

[1] A given permutation can be generated by different sequences of pairwise transpositions. However, the number of pairwise transpositions required is uniquely either even or odd.

The sum over the permutations P' in the last line yields, for a given permutation P, exactly the same terms that are obtained with the regular ordering $1,2,\ldots N$. The sign that is required to bring a given permutation $p'_1,\ldots p'_N$ into the natural order is just $(-1)^{P'}$. One can therefore eliminate the multiplicity by arranging the first indices of the Kronecker symbols in regular order and sum only over all the permutations of the second,

$$\begin{aligned}
&\int dx_1 \cdots dx_N\, \Phi^*_{\alpha_1\cdots\alpha_N}(x_1\cdots x_N)\,\Phi_{\beta_1\cdots\beta_N}(x_1\cdots x_N) \\
&= \sum_{P\in S_N} (-1)^P \delta_{\alpha_1\beta_{p_1}} \cdots \delta_{\alpha_N\beta_{p_N}} \\
&= \det\begin{pmatrix} \langle\alpha_1|\beta_1\rangle & \cdots & \langle\alpha_1|\beta_N\rangle \\ \vdots & \vdots & \vdots \\ \langle\alpha_N|\beta_1\rangle & \cdots & \langle\alpha_N|\beta_N\rangle \end{pmatrix} . \qquad \text{(B.12)}
\end{aligned}$$

The final result takes the form of a determinant. It becomes simpler, if the state labels are arranged in a given order. For

$$\alpha_1 < \alpha_2 < \cdots < \alpha_N \qquad \text{and} \qquad \beta_1 < \beta_2 < \cdots < \beta_N \qquad \text{(B.13)}$$

one obtains

$$\int dx_1 \cdots dx_N\, \Phi^*_{\alpha_1\cdots\alpha_N}(x_1\cdots x_N)\,\Phi_{\beta_1\cdots\beta_N}(x_1\cdots x_N) = \delta_{\alpha_1\beta_1}\cdots\delta_{\alpha_N\beta_N} . \qquad \text{(B.14)}$$

The completeness relation follows in a similar fashion

$$\begin{aligned}
&\sum_{\alpha_1\cdots\alpha_N} \Phi_{\alpha_1\cdots\alpha_N}(x_1\cdots x_N)\Phi^*_{\alpha_1\cdots\alpha_N}(y_1\cdots y_N) \\
&= \frac{1}{N!}\sum_{P,P'\in S_N} (-1)^{P+P'} \left\{\sum_{\alpha_1} \phi_{\alpha_1}(x_{p_1})\phi^*_{\alpha_1}(y_{p'_1})\right\}\cdots\left\{\sum_{\alpha_N} \phi_{\alpha_N}(x_N)\phi^*_{\alpha_N}(y_{p'_N})\right\} \\
&= \frac{1}{N!}\sum_{P,P'\in S_N} (-1)^P(-1)^{P'}\delta(x_{p_1},y_{p'_1})\cdots\delta(x_{p_N},y_{p'_N}) \\
&= \sum_{P\in S_N} (-1)^P \delta(x_1,y_{p_1})\cdots\delta(x_N,y_{p_N}) . \qquad \text{(B.15)}
\end{aligned}$$

The transition from the second to last to the last line involves the same argument concerning the multiplicity of terms as in the case of the orthogonality relation.

The derivation of the completeness relation has to be augmented by one additional point: one has to take into account the fact that the operator

$$\sum_{\alpha_1\cdots\alpha_N} \Phi_{\alpha_1\cdots\alpha_N}(x_1\cdots)\Phi^*_{\alpha_1\cdots\alpha_N}(y_1\cdots)$$

acts only on the antisymmetric states in $\mathscr{F}_N$. If a product of Kronecker symbols, as in (B.15), is contracted with an arbitrary antisymmetric wavefunction Ψ, the permutation of the coordinates $y_1, \ldots y_N$ leads to

$$\int dy_1 \cdots dy_N \, \delta(x_1, y_{p_1}) \cdots \delta(x_N, y_{p_N}) \Psi(y_1 \cdots y_N) = (-1)^P \Psi(x_1 \cdots x_N) \, . \quad \text{(B.16)}$$

Within the space $\mathscr{F}_N$ one thus obtains

$$\sum_{\alpha_1 \cdots \alpha_N} \Phi_{\alpha_1 \cdots \alpha_N}(x_1 \cdots x_N) \Phi^*_{\alpha_1 \cdots \alpha_N}(y_1 \cdots y_N) = N! \, \delta(x_1, y_1) \cdots \delta(x_N, y_N) \, . \quad \text{(B.17)}$$

The factor $N!$ results from the overcompleteness of the basis set formed by the $\Phi_{\alpha_1 \cdots \alpha_N}$ in the space $\mathscr{F}_N$: since all states $\Phi_{\alpha_1 \cdots \alpha_N}$ which differ only by a permutation of the set $\alpha_1, \ldots \alpha_N$ coincide (up to an irrelevant sign), any basis state shows up $N!$ times in the sum on the left-hand side of Eq. (B.17). The factor is easily eliminated by use of an ordered sum,

$$\sum_{\alpha_1 < \alpha_2 \cdots < \alpha_N} \Phi_{\alpha_1 \cdots \alpha_N}(x_1 \cdots x_N) \Phi^*_{\alpha_1 \cdots \alpha_N}(y_1 \cdots y_N) = \delta(x_1, y_1) \cdots \delta(x_N, y_N) \, . \quad \text{(B.18)}$$

B.1.2 Formal Representation

The same statements can be made on a more formal level, if one adopts the *Dirac notation.*

The discussion of the formal representation also begins with a look at the Hilbert space of one particle, $\mathscr{H}_1$. The 1-particle wavefunctions are interpreted as a scalar product of two state vectors $|x\rangle$ and $|\alpha\rangle \equiv |\phi_\alpha\rangle$ which is written as

$$\phi_\alpha(x) \equiv \langle x|\alpha\rangle \, . \quad \text{(B.19)}$$

A state vector of the form $\langle x|$ is called a *bra-vector*, of the form $|\alpha\rangle$ a *ket-vector.* The scalar product itself is therefore often referred to as a *bra-ket*. Since Eq. (B.19) relates the wavefunction to a scalar product, one finds for the complex conjugate wavefunction,

$$\phi^*_\alpha(x) = \langle \alpha|x\rangle \, . \quad \text{(B.20)}$$

The state $\langle\alpha|$ is the adjoint of the state $|\alpha\rangle$.

The set of vectors $|\alpha\rangle$ and the set $|x\rangle$ are elements of different vector spaces. The notation implies that "factors" with the label α carry the information concerning the quantum labels, "factors" with x define the representation space of the particle as position and spin space. The separation of the wavefunction in terms of two abstract ingredients allows, for example, an easy transition to alternative representation spaces as the momentum-spin space.

The states $|\alpha\rangle$ span the familiar Hilbert space $\mathscr{H}_1$. They form a complete and orthonormal set. This is expressed by the relations[2]

$$\sum_{\alpha}|\alpha\rangle\langle\alpha| = \hat{1}_{\mathscr{H}_1}\,; \qquad \langle\alpha|\beta\rangle = \delta_{\alpha\beta}\,. \tag{B.21}$$

The consistency of these relations can be checked by considering

$$\langle x|\alpha\rangle = \sum_{\beta}\langle x|\beta\rangle\langle\beta|\alpha\rangle = \sum_{\beta}\langle x|\beta\rangle\delta_{\alpha\beta} = \langle x|\alpha\rangle\,.$$

The vector $|x\rangle \equiv |\boldsymbol{r}\sigma\rangle$ is an eigenstate of the position operator $\hat{\boldsymbol{r}}$ and the spin-projection operator $\hat{s}_z$. It is characterized by the corresponding eigenvalues $\boldsymbol{r}$ and $\sigma = \pm 1$,

$$\hat{\boldsymbol{r}}|x\rangle = \boldsymbol{r}\,|x\rangle \tag{B.22}$$

$$\hat{s}_z|x\rangle = \sigma\frac{\hbar}{2}\,|x\rangle\,. \tag{B.23}$$

The notation indicates that operators $\hat{o}$ are, as the state vectors, abstracted elements. The states $|x\rangle$ satisfy the improper orthogonality relation

$$\langle x|x'\rangle = \delta_{\sigma\sigma'}\,\delta(\boldsymbol{r}-\boldsymbol{r}') \equiv \delta(x,x')\,. \tag{B.24}$$

Alternatively one may interpret $\langle x|x'\rangle$ as a wavefunction, the representation of an eigenstate of the position/spin operator in position/spin space. Equation (B.24) then states that the probability to find a particle at any other point in space than its eigenvalue vanishes for eigenstates of $\hat{\boldsymbol{r}}$. Two different wavefunctions $\langle x|x'\rangle$ and $\langle x|x''\rangle$ are orthogonal, as required by their definition as eigenstates of $\hat{\boldsymbol{r}}$ and $\hat{s}_z$,

$$\int dx\,\langle x'|x\rangle\langle x|x''\rangle = \delta(x',x'') = \langle x'|x''\rangle\,. \tag{B.25}$$

Equation (B.25) also demonstrates that the state vectors $|x\rangle$ are not properly normalizable, so that they are not elements of the Hilbert space $\mathscr{H}_1$. They can nevertheless be used to represent the elements of $\mathscr{H}_1$ in the sense of a basis set expansion, as they form a complete basis in a vector space which contains $\mathscr{H}_1$. An example for such a representation is the Fourier representation of normalizable functions in terms of

[2] These relations are replaced by

$$\langle\alpha|\beta\rangle = S_{\alpha\beta} \qquad \sum_{\alpha\beta}|\alpha\rangle S^{-1}_{\alpha\beta}\langle\beta| = \hat{1}_{\mathscr{H}_1}\,,$$

in the case of a non-orthogonal basis. The matrix elements $S^{-1}_{\alpha\beta}$ are elements of the inverse overlap matrix, which is defined by

$$\mathsf{S}\mathsf{S}^{-1} = 1 \quad \rightarrow \quad \sum_{\beta} S_{\alpha\beta}S^{-1}_{\beta\gamma} = \delta_{\alpha\gamma}\,.$$

non-normalizable plane waves. The completeness relation for the states $|x\rangle$ can be extracted from Eq. (B.25), which is valid for arbitrary $|x'\rangle$, $|x''\rangle$,

$$\int dx|x\rangle\langle x| \equiv \sum_{\sigma}\int d^3r\,|\boldsymbol{r}\sigma\rangle\langle\boldsymbol{r}\sigma| = \hat{1}\,. \tag{B.26}$$

The quantity $\hat{1}$ stands for the unit operator in the space which contains $\mathscr{H}_1$.

With these basic elements of the Dirac notation the orthogonality and completeness relations of the one particle wavefunctions, (B.4) and (B.5) respectively, can be reproduced in a consistent fashion. The notation also opens access to all formal aspects of quantum mechanics.

In the next step the Dirac notation can be extended to deal with N-particle systems. The N-particle Hilbert space $\mathscr{H}_N$ is spanned by the product states

$$|\alpha_1 \cdots \alpha_N) = |\alpha_1\rangle \otimes \cdots \otimes |\alpha_N\rangle\,. \tag{B.27}$$

In these N-particle states the particle k is in the quantum state α_k, i.e. the position of a single-particle state in the tensor product on the right-hand side characterizes a particular particle of the system. It is usual to omit the product sign $\otimes$, when working with the states $|\alpha_1 \cdots \alpha_N)$. Nevertheless, the convention associating particle k with position k still applies.

The bra-ket combination of (B.27) with

$$|x_1 \cdots x_N) = |x_1\rangle \otimes \cdots \otimes |x_N\rangle \tag{B.28}$$

yields the product wavefunction (B.7),

$$\Phi^{\rm c}_{\alpha_1\cdots\alpha_N}(x_1\cdots x_N) = (x_1 \cdots x_N|\alpha_1 \cdots \alpha_N) = \langle x_1|\alpha_1\rangle \cdots \langle x_N|\alpha_N\rangle\,. \tag{B.29}$$

The N-particle states (B.27) constitute a basis of $\mathscr{H}_N$. They form a complete set provided the 1-particle basis is complete,

$$\sum_{\alpha_1\cdots\alpha_N} |\alpha_1 \cdots \alpha_N)(\alpha_1 \cdots \alpha_N| = \sum_{\alpha_1}|\alpha_1\rangle\langle\alpha_1|\cdots\sum_{\alpha_N}|\alpha_N\rangle\langle\alpha_N| = \hat{1}_{\mathscr{H}_N}\,, \tag{B.30}$$

where $\hat{1}_{\mathscr{H}_N}$ represents the unit operator in $\mathscr{H}_N$. Similarly, one has in the x-representation,

$$\int dx_1\cdots dx_N\,|x_1 \cdots x_N)(x_1 \cdots x_N| = \hat{1}_N\,, \tag{B.31}$$

where $\hat{1}_N$ is the unit operator of the (N-particle) space which contains $\mathscr{H}_N$.

The fermion and boson sectors of $\mathscr{H}_N$ are defined in the same fashion as before. The fermion sector is spanned by the antisymmetrized states[3]

---

[3] Many-body states in the form of products will be denoted by $|\cdots)$, antisymmetrized states by $|\cdots\rangle$.

$$\begin{aligned} |\alpha_1 \cdots \alpha_N\rangle &= \frac{1}{\sqrt{N!}} \sum_{P \in S_N} (-1)^P |\alpha_{p_1} \cdots \alpha_{p_N}) \\ &= \frac{1}{\sqrt{N!}} \sum_{P \in S_N} (-1)^P |\alpha_{p_1}) \cdots |\alpha_{p_N}) \,. \end{aligned} \tag{B.32}$$

In the states $|\alpha_1 \cdots \alpha_N\rangle$ the position of the quantum number is no longer related to a particular particle. A given particle is not in a particular single-particle state. The individual terms on the right-hand side of (B.32) are, however, product states of the form (B.27), so that the position k in the product characterizes a particular particle.

The N-fermion wavefunction (B.11) is given by the bra-ket combination

$$\begin{aligned} (x_1 \cdots x_N|\alpha_1 \cdots \alpha_N\rangle &= \frac{1}{\sqrt{N!}} \sum_{P \in S_N} (-1)^P (x_1 \cdots x_N|\alpha_{p_1} \cdots \alpha_{p_N}) \\ &= \frac{1}{\sqrt{N!}} \sum_{P \in S_N} (-1)^P (x_1|\alpha_{p_1}) \cdots (x_N|\alpha_{p_N}) \\ &= \frac{1}{\sqrt{N!}} \sum_{P \in S_N} (-1)^P (x_{p_1}|\alpha_1) \cdots (x_{p_N}|\alpha_N) \,. \end{aligned} \tag{B.33}$$

Note, that only one of the state vectors in the bra-ket scalar product is an antisymmetrized state, either the bra or the ket vector,

$$\Phi_{\alpha_1 \cdots \alpha_N}(x_1 \cdots x_N) = \langle x_1 \cdots x_N|\alpha_1 \cdots \alpha_N) = (x_1 \cdots x_N|\alpha_1 \cdots \alpha_N\rangle \,. \tag{B.34}$$

The other is a simple product state.

The antisymmetric N-fermion state vectors satisfy the orthonormality relation (B.12),

$$\langle \alpha_1 \cdots \alpha_N|\beta_1 \cdots \beta_N\rangle = \sum_{P \in S_N} (-1)^P \delta_{\alpha_1 \beta_{p_1}} \cdots \delta_{\alpha_N \beta_{p_N}} \,, \tag{B.35}$$

which may be verified by insertion of (B.32) and subsequent use of (B.21) for the individual particles. As a single-particle state can at most be occupied by one fermion, at most one of the possible permutations of the single-particle overlap matrices can be non-zero. If the state labels are arranged in a strict order, $\alpha_1 < \cdots < \alpha_N$, the result can be written as

$$\langle \alpha_1 \cdots \alpha_N|\beta_1 \cdots \beta_N\rangle = \delta_{\alpha_1 \beta_1} \cdots \delta_{\alpha_N \beta_N} \,. \tag{B.36}$$

Similarly, the completeness relations (B.17) and (B.18) have the form

$$\frac{1}{N!} \sum_{\alpha_1 \cdots \alpha_N} |\alpha_1 \cdots \alpha_N\rangle\langle \alpha_1 \cdots \alpha_N| = \hat{1}_{\mathscr{F}_N} \tag{B.37}$$

$$\sum_{\alpha_1 < \alpha_2 \cdots < \alpha_N} |\alpha_1 \cdots \alpha_N\rangle\langle \alpha_1 \cdots \alpha_N| = \hat{1}_{\mathscr{F}_N} \,. \tag{B.38}$$

B.2 Fock Space

There are several reasons to combine the Hilbert spaces for all possible particle numbers into a more general space, the *Fock space*. Particle numbers of a particular species are not necessarily conserved in quantum processes or there might be the need to describe a thermodynamical equilibrium without a fixed number of particles. The Fock space of fermions $\mathscr{F}$ is defined as the direct sum of the N-fermions spaces $\mathscr{F}_N$ for all particle numbers,

$$\mathscr{F} = \mathscr{F}_0 \oplus \mathscr{F}_1 \oplus \cdots \oplus \mathscr{F}_N \oplus \cdots . \tag{B.39}$$

In addition to the well-defined spaces $\mathscr{F}_N$, it includes a sector $\mathscr{F}_0$ containing no particle at all. The only state in $\mathscr{F}_0$ is the so-called *vacuum state*

$$|0\rangle \qquad \text{with} \qquad \langle 0|0\rangle = 1 . \tag{B.40}$$

The actual specification of this state requires the application of the creation and annihilation operators, which will be detailed in the next section. A complete and orthonormal basis of $\mathscr{F}$ is obtained by combining all N-fermion basis sets with $|0\rangle\langle 0|$, so that the completeness relation in $\mathscr{F}$ reads

$$|0\rangle\langle 0| + \sum_{N=1}^{\infty} \frac{1}{N!} \sum_{\alpha_1 \cdots \alpha_N} |\alpha_1 \cdots \alpha_N\rangle\langle \alpha_1 \cdots \alpha_N| = \hat{1}_{\mathscr{F}} . \tag{B.41}$$

B.2.1 Creation and Annihilation Operators

The action of a fermion creation operator $\hat{a}^{\dagger}_{\alpha}$ on a N-fermion basis state generates an $(N+1)$-fermion basis state

$$\hat{a}^{\dagger}_{\alpha}|\alpha_1 \cdots \alpha_N\rangle := |\alpha\, \alpha_1 \cdots \alpha_N\rangle . \tag{B.42}$$

The $(N+1)$-fermion state is properly normalized and antisymmetrized. The definition (B.42) of the operators $\hat{a}^{\dagger}_{\alpha}$ is unambiguous, as all states involved are well-defined. Extension of Eq. (B.42) to $N = 0$ defines the vacuum state as the state from which $\hat{a}^{\dagger}_{\alpha}$ generates the single-particle state $|\alpha\rangle$,

$$\hat{a}^{\dagger}_{\alpha}|0\rangle = |\alpha\rangle . \tag{B.43}$$

Combination of the definitions (B.42) and (B.43) allows a representation of any N-fermion basis state in terms of creation operators and the vacuum,

$$|\alpha_1 \cdots \alpha_N\rangle = \hat{a}^{\dagger}_{\alpha_1} \cdots \hat{a}^{\dagger}_{\alpha_N}|0\rangle . \tag{B.44}$$

The entire basis of Fock space can be generated by the repeated action of creation operators on the vacuum state.

The associated annihilation operator $\hat{a}_\alpha$ is defined by hermitian conjugation of $\hat{a}^\dagger_\alpha$

$$\hat{a}_\alpha := \left(\hat{a}^\dagger_\alpha\right)^\dagger . \tag{B.45}$$

Consequently one has

$$\begin{aligned} \langle\alpha_1| &= \langle 0|\,\hat{a}_{\alpha_1} \\ \langle\alpha_1\,\alpha_2\cdots\alpha_N| &= \langle\alpha_2\cdots\alpha_N|\,\hat{a}_{\alpha_1} = \langle 0|\,\hat{a}_{\alpha_N}\cdots\hat{a}_{\alpha_1} . \end{aligned}$$

The antisymmetry of fermion states of the form (B.44) is incorporated by demanding specific commutation relations for the creation operators. With the interchange of two quantum numbers in (B.44) one arrives at

$$\begin{aligned} |\alpha_1\,\alpha_2\,\alpha_3\cdots\alpha_N\rangle &= \hat{a}^\dagger_{\alpha_1}\hat{a}^\dagger_{\alpha_2}\hat{a}^\dagger_{\alpha_3}\cdots\hat{a}^\dagger_{\alpha_N}|0\rangle \\ &= -|\alpha_2\,\alpha_1\,\alpha_3\cdots\alpha_N\rangle \\ &= -\hat{a}^\dagger_{\alpha_2}\hat{a}^\dagger_{\alpha_1}\hat{a}^\dagger_{\alpha_3}\cdots\hat{a}^\dagger_{\alpha_N}|0\rangle . \end{aligned} \tag{B.46}$$

This relation requires that the creation operators (and hence the annihilation operators) satisfy *anticommutation relations*,

$$\left\{\hat{a}^\dagger_\alpha,\hat{a}^\dagger_\beta\right\} = \left\{\hat{a}_\alpha,\hat{a}_\beta\right\} = 0 \qquad \text{with} \quad \{\hat{A},\hat{B}\} = \hat{A}\hat{B}+\hat{B}\hat{A} , \tag{B.47}$$

as (B.46) must hold for arbitrary states $|\alpha_3,\ldots\alpha_N\rangle$.

The commutation relation between creation and annihilation operators can be derived in the following fashion: as first step consider the expectation value of an annihilation operator for arbitrary basis set states,

$$\langle\alpha_1\cdots\alpha_M|\hat{a}_\mu|\beta_1\cdots\beta_N\rangle = \langle\mu\,\alpha_1\cdots\alpha_M|\beta_1\cdots\beta_N\rangle . \tag{B.48}$$

The right-hand side of (B.48) necessarily vanishes if $M+1\neq N$, irrespective of the values of the quantum numbers involved. This shows that the state $\hat{a}_\mu|\beta_1\cdots\beta_N\rangle$ is a $(N-1)$-particle state—the operator $\hat{a}_\mu$ annihilates one particle. In particular, the expression $\langle 0|\hat{a}_\mu|0\rangle = 0$ requires

$$\hat{a}_\mu|0\rangle = 0 \qquad \left(\text{similarly, } \langle 0|\hat{a}^\dagger_\mu = 0\right) . \tag{B.49}$$

Particles can not be destroyed, if there are no particles.

The next step is an investigation of the action of the annihilation operator on an arbitrary basis state. With the completeness relation (B.41) one obtains

$$\hat{a}_\mu|\beta_1\cdots\beta_N\rangle = \sum_{M=1}^{\infty}\frac{1}{M!}\sum_{\alpha_1\cdots\alpha_M}|\alpha_1\cdots\alpha_M\rangle\langle\alpha_1\cdots\alpha_M|\hat{a}_\mu|\beta_1\cdots\beta_N\rangle$$
$$= \sum_{M=1}^{\infty}\frac{1}{M!}\sum_{\alpha_1\cdots\alpha_M}\langle\mu\,\alpha_1\cdots\alpha_M|\beta_1\cdots\beta_N\rangle\,|\alpha_1\cdots\alpha_M\rangle$$
$$= \frac{1}{(N-1)!}\sum_{\alpha_1\cdots\alpha_{N-1}}\langle\mu\,\alpha_1\cdots\alpha_{N-1}|\beta_1\cdots\beta_N\rangle\,|\alpha_1\cdots\alpha_{N-1}\rangle\,.$$

The last line can be processed further with the orthonormality relation (B.35),

$$\hat{a}_\mu|\beta_1\cdots\beta_N\rangle$$
$$= \frac{1}{(N-1)!}\sum_{\alpha_1\cdots\alpha_{N-1}}\sum_{P\in S_N}(-1)^P\delta_{\mu\beta_{p_1}}\,\delta_{\alpha_1\beta_{p_2}}\cdots\delta_{\alpha_{N-1}\beta_{p_N}}|\alpha_1\cdots\alpha_{N-1}\rangle$$
$$= \frac{1}{(N-1)!}\sum_{P\in S_N}(-1)^P\delta_{\mu\beta_{p_1}}|\beta_{p_2}\cdots\beta_{p_N}\rangle\,.$$

The sum over the $N!$ permutations P can be written more explicitly in terms of an expansion with respect to the entry with the index i as

$$\hat{a}_\mu|\beta_1\cdots\beta_N\rangle = \frac{1}{(N-1)!}\sum_{i=1}^{N}(-1)^{i-1}\delta_{\mu\beta_i}\sum_{P'\in S_{N-1}}(-1)^{P'}|\beta_{p'_1}\cdots\not{\beta_i}\cdots\beta_{p'_N}\rangle\,.$$

The sum over the permutations P' of the numbers $1,\ldots,i-1,i+1,\ldots N$ (the omission of i is indicated by $\not{\beta_i}$) represents $(N-1)!$ times the same $(N-1)$-particle state

$$|\beta_1\cdots\beta_{i-1}\,\beta_{i+1}\cdots\beta_N\rangle = \frac{1}{(N-1)!}\sum_{P'\in S_{N-1}}(-1)^{P'}|\beta_{p'_1}\cdots\not{\beta_i}\cdots\beta_{p'_N}\rangle\,.$$

The final result

$$\hat{a}_\mu|\beta_1\cdots\beta_N\rangle = \sum_{i=1}^{N}(-1)^{i-1}\delta_{\mu\beta_i}|\beta_1\cdots\beta_{i-1}\,\beta_{i+1}\cdots\beta_N\rangle \qquad \text{(B.50)}$$

shows: the right-hand is only non-zero, if the quantum number μ is identical with one of the β_i,

$$\hat{a}_\mu|\beta_1\cdots\beta_N\rangle = \begin{cases}(-1)^{i-1}|\beta_1\cdots\beta_{i-1}\,\beta_{i+1}\cdots\beta_N\rangle & \text{if } \mu=\beta_i\\ 0 & \text{otherwise}\end{cases}\,. \qquad \text{(B.51)}$$

Combination of (B.50) with (B.42) then yields

$$\hat{a}_\mu\hat{a}_\nu^\dagger|\alpha_1\cdots\alpha_N\rangle = \delta_{\mu\nu}|\alpha_1\cdots\alpha_N\rangle$$
$$+\sum_{i=1}^{N}(-1)^i\delta_{\mu\alpha_i}|\nu\alpha_1\cdots\alpha_{i-1}\,\alpha_{i+1}\cdots\alpha_N\rangle\,, \qquad \text{(B.52)}$$

as well as

$$\hat{a}_\nu^\dagger \hat{a}_\mu |\alpha_1 \cdots \alpha_N\rangle = \sum_{i=1}^{N} (-1)^{i-1} \delta_{\mu\alpha_i} |\nu\alpha_1 \cdots \alpha_{i-1}\, \alpha_{i+1} \cdots \alpha_N\rangle \,. \tag{B.53}$$

Both relations are valid for arbitrary $|\alpha_1 \cdots \alpha_N\rangle$, so that one can extract the anti-commutation relation

$$\{\hat{a}_\mu, \hat{a}_\nu^\dagger\} = \delta_{\mu\nu} \,. \tag{B.54}$$

With Eqs. (B.42)–(B.50) and (B.54) the set of basic relations for creation and annihilation operators is complete. *All* operations and manipulations in Fock space can be handled with these tools.

The creation or destruction of a particle has so far been associated with a basis labelled by an index α. A transition to an alternative basis can be achieved with the aid of completeness relations. For example, the relations (B.21) and (B.26) can be used to write down the identities (valid for any kind of particle)

$$|\alpha\rangle = \int dx |x\rangle\langle x|\alpha\rangle = \int dx\, \phi_\alpha(x)|x\rangle \tag{B.55}$$

$$|x\rangle = \sum_\alpha |\alpha\rangle\langle\alpha|x\rangle = \sum_\alpha \phi_\alpha^*(x)\,|\alpha\rangle \,, \tag{B.56}$$

which can be interpreted as a unitary basis transformation in $\mathscr{H}_1$. The second of these relations suggests the introduction of the operators $\hat{\psi}(x)$ and $\hat{\psi}^\dagger(x)$ with

$$|x\rangle = \hat{\psi}^\dagger(x)|0\rangle \qquad \text{and} \qquad \langle x| = \langle 0|\,\hat{\psi}(x) \,. \tag{B.57}$$

These operators describe the creation and the destruction of a particle at the "position x". For this reason they are usually referred to as *field operators*. In other words: the basis transformations (B.55) and (B.56) induce a corresponding transformation between the associated creation and annihilation operators,

$$\hat{\psi}^\dagger(x) = \sum_\alpha \phi_\alpha^*(x)\,\hat{a}_\alpha^\dagger = \sum_\alpha \langle\alpha|x\rangle\,\hat{a}_\alpha^\dagger \tag{B.58}$$

$$\hat{\psi}(x) = \sum_\alpha \phi_\alpha(x)\hat{a}_\alpha = \sum_\alpha \langle x|\alpha\rangle\hat{a}_\alpha \,, \tag{B.59}$$

with the inverse transformation

$$\hat{a}_\alpha = \int dx\, \phi_\alpha^*(x)\,\hat{\psi}(x) \tag{B.60}$$

$$\hat{a}_\alpha^\dagger = \int dx\, \phi_\alpha(x)\,\hat{\psi}^\dagger(x) \,. \tag{B.61}$$

The relations (B.58), (B.59) indicate directly that the field operators are objects with two components in the case of spin 1/2 fermions,

$$\hat{\psi}(x) = \hat{\psi}(\boldsymbol{r}\sigma) = \begin{cases} \hat{\psi}(\boldsymbol{r}, +\frac{1}{2}) \text{ if } \sigma = +\frac{1}{2} \\ \hat{\psi}(\boldsymbol{r}, -\frac{1}{2}) \text{ if } \sigma = -\frac{1}{2} \end{cases} . \tag{B.62}$$

The anticommutation relations (B.54) and (B.47) and the transformations (B.58)–(B.61) can only be consistent, if the field operators satisfy

$$\{\hat{\psi}(x), \hat{\psi}^{\dagger}(x')\} = \delta(x, x') \tag{B.63}$$

$$\{\hat{\psi}(x), \hat{\psi}(x')\} = \{\hat{\psi}^{\dagger}(x), \hat{\psi}^{\dagger}(x')\} = 0 . \tag{B.64}$$

The structure of the commutation relations is conserved under basis transformations. A transformation between the basis $|\alpha\rangle$ and any other (single-particle) basis proceeds in the same fashion.

B.2.2 1-Particle Operators

The Pauli principle requires that quantum particles are indistinguishable. Observables of many particle systems can, as a consequence, only be represented by operators which are symmetric under exchange of particles.

An important class of operators in N-particle space are those constructed by summation over terms acting on a single particle,

$$\hat{F} = \sum_{i=1}^{N} \hat{f}_i . \tag{B.65}$$

They are referred to as 1-particle (or single-particle) operators. More correctly they might be called 1-particle operators in an N-particle system. A second important type of operators is constructed by summation of terms linking two particles,

$$\hat{W} = \sum_{i,j=1;\, i<j}^{N} \hat{w}_{ij} . \tag{B.66}$$

These operators are therefore called 2-particle operators.

A 1-particle operator $\hat{f}$ can be specified in the x-, the α- or any other representation. In the Dirac notation one obtains for instance for the operator of the kinetic energy of a single particle in the x-representation

$$\langle x|\hat{t}|x'\rangle = \delta(x, x') \frac{(-i\hbar\boldsymbol{\nabla}')^2}{2m} . \tag{B.67}$$

The α- and the x-representation of an operator $\hat{f}$ can be related with the aid of the completeness relation

$$\langle\alpha|\hat{f}|\beta\rangle = \int dx dx' \, \langle\alpha|x'\rangle\langle x'|\hat{f}|x\rangle\langle x|\beta\rangle . \tag{B.68}$$

The action of $\hat{f}$ on a 1-particle state $|\gamma\rangle$ can also be rewritten with the completeness relation as

$$\hat{f}|\gamma\rangle = \sum_{\alpha} |\alpha\rangle\langle\alpha|\hat{f}|\gamma\rangle \,. \tag{B.69}$$

This implies that the representation of a 1-particle operator in terms of creation and annihilation operators must have the form

$$\hat{f} = \sum_{\alpha\beta} \langle\alpha|\hat{f}|\beta\rangle\, \hat{a}^{\dagger}_{\alpha}\hat{a}_{\beta} \,. \tag{B.70}$$

The operators $\hat{a}^{\dagger}_{\alpha}$ and $\hat{a}_{\beta}$ are specified in terms of the single-particle basis to which $|\gamma\rangle$ belongs. Equation (B.70) can be verified by insertion,

$$\hat{f}|\gamma\rangle = \sum_{\alpha\beta} \langle\alpha|\hat{f}|\beta\rangle\, \hat{a}^{\dagger}_{\alpha}\hat{a}_{\beta}\hat{a}^{\dagger}_{\gamma}|0\rangle = \sum_{\alpha} \langle\alpha|\hat{f}|\gamma\rangle\, \hat{a}^{\dagger}_{\alpha}|0\rangle \,, \tag{B.71}$$

and comparison with Eq. (B.69).

The 1-particle operator $\hat{F} = \sum_i \hat{f}_i$ in Fock space is completely characterized by the action of $\hat{f}$ within the 1-particle segment of this space. It follows that the operator (B.70) can also serve as a representation of the operator $\hat{F}$,

$$\hat{F} = \sum_{\alpha\beta} \langle\alpha|\hat{f}|\beta\rangle\, \hat{a}^{\dagger}_{\alpha}\hat{a}_{\beta} \,. \tag{B.72}$$

Due to the combination $\hat{a}^{\dagger}_{\alpha}\hat{a}_{\beta}$ the operator only connects states of the same segment of Fock space. In order to evaluate the action of this operator on a N-particle state one may use the commutation relation

$$\left[\hat{F}, \hat{a}^{\dagger}_{\alpha}\right] = \sum_{\beta\gamma} \langle\beta|\hat{f}|\gamma\rangle \left[\hat{a}^{\dagger}_{\beta}\hat{a}_{\gamma}, \hat{a}^{\dagger}_{\alpha}\right] = \sum_{\beta} \langle\beta|\hat{f}|\alpha\rangle \hat{a}^{\dagger}_{\beta} \,. \tag{B.73}$$

Use of (B.73) allows a direct evaluation of $\hat{F}|\alpha_1 \cdots \alpha_N\rangle$ as soon as $[\hat{F}, \hat{a}^{\dagger}_{\alpha}]$ is introduced by suitable addition and subtraction of terms,

$$\begin{aligned} \hat{F}\, \hat{a}^{\dagger}_{\alpha_1} \cdots \hat{a}^{\dagger}_{\alpha_N}|0\rangle &= [\hat{F}, \hat{a}^{\dagger}_{\alpha_1}]\, \hat{a}^{\dagger}_{\alpha_2} \cdots \hat{a}^{\dagger}_{\alpha_N}|0\rangle + \hat{a}^{\dagger}_{\alpha_1} [\hat{F}, \hat{a}^{\dagger}_{\alpha_2}]\, \hat{a}^{\dagger}_{\alpha_3} \cdots \hat{a}^{\dagger}_{\alpha_N}|0\rangle \\ &\quad + \cdots + \hat{a}^{\dagger}_{\alpha_1} \cdots \hat{a}^{\dagger}_{\alpha(N-1)} [\hat{F}, \hat{a}^{\dagger}_{\alpha_N}]\, |0\rangle \,. \end{aligned} \tag{B.74}$$

After replacement of the commutator one obtains

$$\begin{aligned} &= \sum_{\beta_1} \langle\beta_1|\hat{f}|\alpha_1\rangle\, \hat{a}^{\dagger}_{\beta_1}\hat{a}^{\dagger}_{\alpha_2} \cdots \hat{a}^{\dagger}_{\alpha_N}|0\rangle + \sum_{\beta_2} \langle\beta_2|\hat{f}|\alpha_2\rangle\, \hat{a}^{\dagger}_{\alpha_1}\hat{a}^{\dagger}_{\beta_2}\hat{a}^{\dagger}_{\alpha_3} \cdots \hat{a}^{\dagger}_{\alpha_N}|0\rangle \\ &\quad + \cdots + \sum_{\beta_N} \langle\beta_N|\hat{f}|\alpha_N\rangle\, \hat{a}^{\dagger}_{\alpha_1} \cdots \hat{a}^{\dagger}_{\alpha_{N-1}}\hat{a}^{\dagger}_{\beta_N}|0\rangle \,. \end{aligned} \tag{B.75}$$

This explicit result can be written in the compact form

$$\hat{F}|\alpha_1 \cdots \alpha_N\rangle = \sum_{i=1}^{N} \sum_{\beta_i} \langle \beta_i | \hat{f} | \alpha_i \rangle \, \hat{a}^\dagger_{\alpha_1} \cdots \hat{a}^\dagger_{\beta_i} \cdots \hat{a}^\dagger_{\alpha_N} |0\rangle \, . \tag{B.76}$$

The notation indicates that $\hat{a}^\dagger_{\beta_i}$ stands at position i in the sequence of creation operators. Each of the particles is transferred with a certain probability, determined by the matrix element $\langle \beta_i | \hat{f} | \alpha_i \rangle$, into a single-particle state which is not already present in $|\alpha_1 \cdots \alpha_N\rangle$. The result (B.75) can also be used to evaluate the only non-vanishing matrix elements of $\hat{F}$,

$$\langle \alpha_1 \cdots \alpha_N | \hat{F} | \alpha_1 \cdots \alpha_N \rangle = \sum_{i=1} \langle \alpha_i | \hat{f} | \alpha_i \rangle \tag{B.77}$$

$$\langle \alpha_1 \cdots \beta_k \cdots \alpha_N | \hat{F} | \alpha_1 \cdots \alpha_N \rangle = \langle \beta_k | \hat{f} | \alpha_k \rangle \, . \tag{B.78}$$

The label $\beta_k \neq \alpha_i, \; i = 1, \ldots N$ replaces α_k in the bra-state of Eq. (B.78). 1-particle operators can only connect states of Fock space with the same number of particles, which differ at most in one occupation.

The operator (B.72) in the second quantized representation can alternatively be written in terms of the field operators

$$\begin{aligned} \hat{F} = \sum_{\alpha\beta} \langle \alpha | \hat{f} | \beta \rangle \hat{a}^\dagger_\alpha \hat{a}_\beta &= \int dx dx' \sum_{\alpha\beta} \langle \alpha | x' \rangle \langle x' | \hat{f} | x \rangle \langle x | \beta \rangle \hat{a}^\dagger_\alpha \hat{a}_\beta \\ &= \int dx dx' \, \hat{\psi}^\dagger(x') \langle x' | \hat{f} | x \rangle \hat{\psi}(x) \, , \end{aligned} \tag{B.79}$$

or, for that matter, in terms of any other basis, which is related by a unitary transformation, as e.g.

$$\hat{b}^\dagger_k = \sum_\alpha C_{k,\alpha} \hat{a}^\dagger_\alpha \qquad \hat{b}_k = \sum_\alpha C^*_{k,\alpha} \hat{a}_\alpha \tag{B.80}$$

with the inverse

$$\hat{a}^\dagger_\alpha = \sum_k C^*_{k,\alpha} b^\dagger_k \qquad \hat{a}_\alpha = \sum_k C_{k,\alpha} \hat{b}_k \, . \tag{B.81}$$

In the basis with the creation and annihilation operators $\hat{b}_k, \hat{b}^\dagger_k$ one obtains

$$\hat{F} = \sum_{k_1, k_2} \langle k_1 | \hat{f} | k_2 \rangle \, \hat{b}^\dagger_{k_1} \hat{b}_{k_2} \, . \tag{B.82}$$

The form of the representation is independent of the basis chosen.

B.2.3 2-Particle Operators

Similar statements can be made for 2-particle operators, though detailed calculations and proofs are more involved. These operators are characterized by matrix elements in the 2-particle sector of the Fock space, e.g. in the x-representation by

$$(x_1'x_2'|\hat{w}|x_1x_2)\,.$$

The notation indicates that the matrix element under consideration is the canonical matrix element, obtained with the product states of the 2-particle Hilbert space. One example is the interaction between two particles, which is usually local with respect to the coordinates of the two particles involved,

$$(x_1'x_2'|\hat{w}|x_1x_2) = \delta(x_1,x_1')\,\delta(x_2,x_2')\,w(x_1,x_2)\,. \tag{B.83}$$

The function $w(x_1,x_2)$ has to be symmetric and real, as

- the corresponding force has to satisfy Newton's third axiom, and
- the operator $\hat{w}$ has to be hermitian.

It may either be spin-dependent, or not,

$$w(x_1,x_2) = w(\boldsymbol{r}_1,\boldsymbol{r}_2)\,. \tag{B.84}$$

Equation (B.84) applies in particular to the Coulomb interaction, which is of primary interest in the present context. In fact, the Coulomb force is a good example for an interaction which is, in addition, Galilei invariant. The function $w(\boldsymbol{r}_1,\boldsymbol{r}_2)$ depends only on the difference of the position vectors in this case

$$w(\boldsymbol{r}_1,\boldsymbol{r}_2) = w(\boldsymbol{r}_1-\boldsymbol{r}_2)\,. \tag{B.85}$$

The α-representation of the two-body interaction is again obtained with the aid of the completeness relation,

$$(\beta_1\beta_2|\hat{w}|\alpha_1\alpha_2) = \int dx_1'dx_2'dx_1dx_2\;(\beta_1\beta_2|x_1'x_2')(x_1'x_2'|\hat{w}|x_1x_2)(x_1x_2|\alpha_1\alpha_2)\,, \tag{B.86}$$

in detail for the case (B.83),

$$(\beta_1\beta_2|\hat{w}|\alpha_1\alpha_2) = \int dx_1dx_2\;\phi_{\beta_1}^*(x_1)\phi_{\beta_2}^*(x_2)w(x_1,x_2)\phi_{\alpha_1}(x_1)\phi_{\alpha_2}(x_2)\,. \tag{B.87}$$

The order of the quantum labels in the 2-particle bra- and ket-states is, as indicated explicitly in (B.87), of relevance. The first label in the bra- and in the ket-state is associated with the coordinate x_1, the second with x_2. The matrix element satisfies the symmetry relations

$$(\beta_1\beta_2|\hat{w}|\alpha_1\alpha_2) = (\beta_2\beta_1|\hat{w}|\alpha_2\alpha_1) = (\alpha_1\alpha_2|\hat{w}|\beta_1\beta_2)^*\,. \tag{B.88}$$

The action of a 2-particle operator on antisymmetric 2-particle states can be reformulated with the aid of the completeness relation (B.37),

$$\hat{w}|\alpha_1\alpha_2\rangle = \frac{1}{2!}\sum_{\beta_1\beta_2}\langle\beta_1\beta_2|\hat{w}|\alpha_1\alpha_2\rangle|\beta_1\beta_2\rangle\,. \tag{B.89}$$

The matrix element in (B.89) is the *antisymmetric* matrix element, which can be expressed in terms of canonical matrix elements by use of Eq. (B.32),

$$\begin{aligned}\langle\beta_1\beta_2|\hat{w}|\alpha_1\alpha_2\rangle &= \frac{1}{2}\{(\beta_1\beta_2| - (\beta_2\beta_1|\}\hat{w}\{|\alpha_1\alpha_2) - |\alpha_2\alpha_1)\}\\ &= (\beta_1\beta_2|\hat{w}|\alpha_1\alpha_2) - (\beta_1\beta_2|\hat{w}|\alpha_2\alpha_1)\,.\end{aligned} \tag{B.90}$$

The second line in Eq. (B.90) follows from the symmetry of the interaction against the interchange of the two particles. The properties of this matrix element,

$$\begin{aligned}\langle\beta_1\beta_2|\hat{w}|\alpha_1\alpha_2\rangle &= -\langle\beta_2\beta_1|\hat{w}|\alpha_1\alpha_2\rangle \;=\; -\langle\beta_1\beta_2|\hat{w}|\alpha_2\alpha_1\rangle\\ &= \langle\beta_2\beta_1|\hat{w}|\alpha_2\alpha_1\rangle \;=\; \langle\alpha_1\alpha_2|\hat{w}|\beta_1\beta_2\rangle^*\,,\end{aligned} \tag{B.91}$$

follow directly from the definition and the properties (B.88) of the direct matrix elements involved. Combination of Eqs. (B.88)–(B.91) yields the alternative form

$$\hat{w}|\alpha_1\alpha_2\rangle = \sum_{\beta_1\beta_2}(\beta_1\beta_2|\hat{w}|\alpha_1\alpha_2)\,|\beta_1\beta_2\rangle\,. \tag{B.92}$$

The second quantized form of a 2-particle operator in Fock space that reproduces (B.92) is

$$\hat{W} = \frac{1}{2}\sum_{\alpha\beta\gamma\delta}(\alpha\beta|\hat{w}|\gamma\delta)\,\hat{a}^\dagger_\alpha a^\dagger_\beta\hat{a}_\delta\hat{a}_\gamma\,. \tag{B.93}$$

One should take note of the sequence of the labels of the operators with respect to the sequence of the labels of the states. The calculation of the action of this operator on a 2-particle state of fermion Fock space involves the evaluation of

$$\hat{W}|\alpha_1\alpha_2\rangle = \frac{1}{2}\sum_{\beta_1\beta_2\gamma_1\gamma_2}(\beta_1\beta_2|\hat{w}|\gamma_1\gamma_2)\,\hat{a}^\dagger_{\beta_1}\hat{a}^\dagger_{\beta_2}\hat{a}_{\gamma_2}\hat{a}_{\gamma_1}\hat{a}^\dagger_{\alpha_1}\hat{a}^\dagger_{\alpha_2}|0\rangle\,. \tag{B.94}$$

Rearrangement of the creation and annihilation operators,

$$\hat{a}^\dagger_{\beta_1}\hat{a}^\dagger_{\beta_2}\hat{a}_{\gamma_2}\hat{a}_{\gamma_1}\hat{a}^\dagger_{\alpha_1}\hat{a}^\dagger_{\alpha_2}|0\rangle = (\delta_{\alpha_1\gamma_1}\delta_{\alpha_2\gamma_2} - \delta_{\alpha_1\gamma_2}\delta_{\alpha_2\gamma_1})\,\hat{a}^\dagger_{\beta_1}\hat{a}^\dagger_{\beta_2}|0\rangle \tag{B.95}$$

leads, with (B.91), to the same result as (B.92), namely

$$\hat{W}|\alpha_1\alpha_2\rangle = \sum_{\beta_1\beta_2}(\beta_1\beta_2|\hat{w}|\alpha_1\alpha_2)\,|\beta_1\beta_2\rangle\,. \tag{B.96}$$

The evaluation of the action of the operator $\hat{W}$ on a N-particle state also relies on the use of a suitable commutator,

$$\left[\hat{W},\hat{a}^\dagger_\alpha\right] = \sum_{\beta_1\beta_2\alpha_2} (\beta_1\beta_2|\hat{w}|\alpha\alpha_2)\,\hat{a}^\dagger_{\beta_1}\hat{a}^\dagger_{\beta_2}\hat{a}_{\alpha_2}\,. \tag{B.97}$$

Consequent permutation of the commutator through the string of creation operators of the N-particle state yields

$$\begin{aligned}
&\hat{W}|\alpha_1\cdots\alpha_N\rangle \\
&= \sum_{i=1}^{N} \hat{a}^\dagger_{\alpha_1}\cdots\left[\hat{W},\hat{a}^\dagger_{\alpha_i}\right]\cdots\hat{a}^\dagger_{\alpha_N}|0\rangle \\
&= \sum_{i=1}^{N-1}\sum_{\beta_1\beta_2\gamma_2} (\beta_1\beta_2|\hat{w}|\alpha_i\gamma_2)\,\hat{a}^\dagger_{\alpha_1}\cdots\hat{a}^\dagger_{\alpha_{i-1}}\hat{a}^\dagger_{\beta_1}\hat{a}^\dagger_{\beta_2}\hat{a}_{\gamma_2}\hat{a}^\dagger_{\alpha_{i+1}}\cdots\hat{a}^\dagger_{\alpha_N}|0\rangle\,.
\end{aligned} \tag{B.98}$$

The expression $\hat{a}^\dagger_{\beta_1}\hat{a}^\dagger_{\beta_2}\hat{a}_{\gamma_2}$ stands exactly at the position of $\hat{a}^\dagger_{\alpha_i}$ (the contribution with $i = N$ has been omitted, as it vanishes). The pair $\hat{a}^\dagger_{\beta_2}\hat{a}_{\gamma_2}$ now has to be commuted through the chain of operators to its right,

$$\begin{aligned}
\hat{W}|\alpha_1\cdots\alpha_N\rangle = &\sum_{i=1}^{N-1}\sum_{j=i+1}^{N}\sum_{\beta_1\beta_2} (\beta_1\beta_2|\hat{w}|\alpha_i\alpha_j) \\
&\times\hat{a}^\dagger_{\alpha_1}\cdots\hat{a}^\dagger_{\alpha_{i-1}}\hat{a}^\dagger_{\beta_1}\hat{a}^\dagger_{\alpha_{i+1}}\cdots\hat{a}^\dagger_{\alpha_{j-1}}\hat{a}^\dagger_{\beta_2}\hat{a}^\dagger_{\alpha_{j+1}}\cdots\hat{a}^\dagger_{\alpha_N}|0\rangle\,.
\end{aligned} \tag{B.99}$$

This expression can be symmetrized with respect to i and j by use of

$$\sum_{i=1}^{N-1}\sum_{j=i+1}^{N} A_{ij} = \sum_{j=2}^{N}\sum_{i=1}^{j-1} A_{ij}$$

and subsequent simultaneous relabelling $i \leftrightarrow j$, $\beta_1 \leftrightarrow \beta_2$ in half of the right-hand side,

$$\begin{aligned}
\hat{W}|\alpha_1\cdots\alpha_N\rangle = &\frac{1}{2}\sum_{i,j=1;i\neq j}^{N}\sum_{\beta_1\beta_2} (\beta_1\beta_2|\hat{w}|\alpha_i\alpha_j) \\
&\times\hat{a}^\dagger_{\alpha_1}\cdots\hat{a}^\dagger_{\alpha_{i-1}}\hat{a}^\dagger_{\beta_1}\hat{a}^\dagger_{\alpha_{i+1}}\cdots\hat{a}^\dagger_{\alpha_{j-1}}\hat{a}^\dagger_{\beta_2}\hat{a}^\dagger_{\alpha_{j+1}}\cdots\hat{a}^\dagger_{\alpha_N}|0\rangle\,.
\end{aligned}$$

The canonical matrix element can be replaced by its antisymmetric counterpart by using the commutation relations to interchange the positions of $\hat{a}^\dagger_{\beta_1}$ and $\hat{a}^\dagger_{\beta_2}$ in the sequence of creation operators,

$$\begin{aligned}
\hat{W}|\alpha_1\cdots\alpha_N\rangle = &\frac{1}{4}\sum_{i,j=1;i\neq j}^{N}\sum_{\beta_1\beta_2} \langle\beta_1\beta_2|\hat{w}|\alpha_i\alpha_j\rangle \\
&\times\hat{a}^\dagger_{\alpha_1}\cdots\hat{a}^\dagger_{\alpha_{i-1}}\hat{a}^\dagger_{\beta_1}\hat{a}^\dagger_{\alpha_{i+1}}\cdots\hat{a}^\dagger_{\alpha_{j-1}}\hat{a}^\dagger_{\beta_2}\hat{a}^\dagger_{\alpha_{j+1}}\cdots\hat{a}^\dagger_{\alpha_N}|0\rangle\,.
\end{aligned} \tag{B.100}$$

This result shows that at most two of the particles in $|\alpha_1 \cdots \alpha_N\rangle$ are promoted to different single-particle states by application of $\hat{W}$. The matrix elements, which are non-zero, are

- the expectation value,

$$\langle \alpha_1 \cdots \alpha_N | \hat{W} | \alpha_1 \cdots \alpha_N \rangle = \frac{1}{2} \sum_{i,j=1}^{N} \langle \alpha_i \alpha_j | \hat{w} | \alpha_i \alpha_j \rangle , \tag{B.101}$$

- matrix elements with N-particle states, which differ in one quantum number ($\beta_k \neq \alpha_i$, $i = 1 \dots N$),

$$\langle \alpha_1 \cdots \alpha_{k-1} \beta_k \alpha_{k+1} \cdots \alpha_N | \hat{W} | \alpha_1 \cdots \alpha_k \cdots \alpha_N \rangle = \sum_{i=1}^{N} \langle \beta_k \alpha_i | \hat{w} | \alpha_k \alpha_i \rangle , \tag{B.102}$$

- matrix elements with N-particle states, which differ in two quantum numbers ($\beta_k, \beta_l \neq \alpha_i$, $i = 1 \dots N$),

$$\begin{aligned} &\langle \alpha_1 \cdots \alpha_{k-1} \beta_k \alpha_{k+1} \cdots \alpha_{l-1} \beta_l \alpha_{l+1} \cdots \alpha_N | \hat{W} | \alpha_1 \cdots \alpha_k \cdots \alpha_l \cdots \alpha_N \rangle \\ &= \langle \beta_k \beta_l | \hat{w} | \alpha_k \alpha_l \rangle \end{aligned} \tag{B.103}$$

(with the understanding that β_k stands on position k etc.).

The second quantized form of a 2-particle operator in the x-representation can be obtained from the α-representation with the aid of completeness relation (B.31). Insertion of (B.31) into (B.93) and subsequent use of (B.29) leads to

$$\begin{aligned} \hat{W} &= \frac{1}{2} \sum_{\beta_1 \beta_2 \alpha_1 \alpha_2} \int dx_1 dx_2 \, (\beta_1 \beta_2 | x_1 x_2) w(x_1, x_2) (x_1 x_2 | \alpha_1 \alpha_2) \, \hat{a}^\dagger_{\beta_1} \hat{a}^\dagger_{\beta_2} \hat{a}_{\alpha_2} \hat{a}_{\alpha_1} \\ &= \frac{1}{2} \int dx_1 dx_2 \sum_{\beta_1 \beta_2 \alpha_1 \alpha_2} \phi^*_{\beta_1}(x_1) \phi^*_{\beta_2}(x_2) w(x_1, x_2) \phi_{\alpha_1}(x_1) \phi_{\alpha_2}(x_2) \\ &\qquad \times \hat{a}^\dagger_{\beta_1} \hat{a}^\dagger_{\beta_2} \hat{a}_{\alpha_2} \hat{a}_{\alpha_1} \, . \end{aligned}$$

With Eqs. (B.58), (B.59) one finally obtains

$$\hat{W} = \frac{1}{2} \int dx_1 dx_2 \, \hat{\psi}^\dagger(x_1) \hat{\psi}^\dagger(x_2) w(x_1, x_2) \hat{\psi}(x_2) \hat{\psi}(x_1) \, . \tag{B.104}$$

Once again, specific attention should be given to the order of the arguments of the annihilation operators.

Appendix C
Scaling Behavior of Many-Body Methods

In order to provide some background for the discussion of the scaling behavior of many-body methods with the basis set size M indicated in the Introduction, we explicitly consider the most relevant expressions which one has to deal with in this Appendix. The analysis is still quite simple for the class of matrix elements, which have to be evaluated in any of the *ab-initio* methods, i.e. the matrix elements of a single-particle operator. Let us thus first consider a multiplicative potential v as the prototype of such an operator.

In an algebraic eigenvalue problem of the type (1.24) usually two steps are involved. In order to determine the eigenvectors $b_{i,l\sigma}$ the Hamilton matrix has to be evaluated first. Once the $b_{i,l\sigma}$ are known, other quantities, like the energy of the system can be calculated in a second step.[1] In the case of a multiplicative potential v M^2 integrals

$$\langle \eta_k | \hat{v} | \eta_l \rangle = \int d^3r \, \eta_k^*(\boldsymbol{r}) \, v(\boldsymbol{r}) \, \eta_l(\boldsymbol{r}) \qquad k,l = 1,\ldots M\,, \tag{C.1}$$

have to be evaluated in the first step.[2] Three aspects are relevant in this context:

- If v is a given potential the M^2 integrals have to be evaluated only once. However, the single-particle potential is often determined during the calculation, rather than given *a priori*. This is the case, in particular, for the HF scheme, which represents the starting point for many of the more advanced approaches. In a selfconsistent scheme the evaluation of the matrix elements $\langle \eta_k | \hat{v} | \eta_l \rangle$ has to be repeated a number of times. This repetition introduces an additional factor into the total computational cost, which, however, is independent of M and will be ignored in the following.
- On the other hand, the construction of v itself usually depends on M. The associated scaling cannot be determined without specification of a particular method and will therefore be examined later.

[1] In practice, these two steps often go hand in hand, of course.

[2] For Hermitian operators the actual number is $M(M+1)/2$, which for large M corresponds to $\mathcal{O}(M^2)$.

- It remains to address the cost of handling the spatial integral in (C.1). If the integral is known analytically, one can simply store the M^2 coefficients $\langle\eta_k|\hat{v}|\eta_l\rangle$ for repeated use. However, this is rarely the case, so that a numerical evaluation of (C.1) is usually unavoidable.[3] The summation over a spatial grid introduces an additional scaling factor of M, as the number of grid points required to represent M linearly independent basis functions is proportional to M. In the following the number of grid points will therefore simply be identified with M. As a result M^3 operations are needed in order to set up the table of all $\langle\eta_k|\hat{v}|\eta_l\rangle$, if numerical integration is used in Eq. (C.1).

Once the matrix elements$\langle\eta_k|\hat{v}|\eta_l\rangle$ are available and the eigenvalue problem (1.24) is solved, the evaluation of the associated energy,

$$\sum_{i=1}^{N}\langle\phi_i|\hat{v}|\phi_i\rangle = \sum_{i=1}^{N}\sum_{k,l}^{M}\sum_{\sigma} b^*_{i,k\sigma} b_{i,l\sigma}\langle\eta_k|\hat{v}|\eta_l\rangle\,, \tag{C.2}$$

involves a summation over N terms for each of the M^2 matrix elements kl (the multiplicities associated with spin are irrelevant at this point). The scaling of N and M is, however, intrinsically related, i.e. M increases linearly with N. For the present discussion N can therefore simply be replaced by M, so that one ends up with a total scaling of M^3.

It is instructive to compare this procedure with an alternative possibility for the calculation of (C.2). The first step of this second path is the evaluation of all orbitals (1.23), for which a summation over M terms is required for all $N = M$ orbitals on all M grid points. Once all $\phi_i(\boldsymbol{r})$ are stored, it takes M^2 operations to calculate the density

$$n(\boldsymbol{r}) = \sum_{\sigma}\sum_{i=1}^{N}|\phi_i(\boldsymbol{r}\sigma)|^2\,. \tag{C.3}$$

The energy (C.2) can finally be evaluated by numerical integration over $n(\boldsymbol{r})v(\boldsymbol{r})$, which is linear in M. Again one ends up with an M^3 scaling. In the alternative approach the storage of the $M\times M$ array $\phi_i(\boldsymbol{r})$ replaces the storage of the $M\times M$ array $\langle\eta_k|\hat{v}|\eta_l\rangle$ necessary in the first approach, so that no additional memory is needed.

The second approach can easily be extended to nonlocal single-particle potentials. In this case one would pre-evaluate the 1-particle density matrix

$$\gamma(\boldsymbol{r}\sigma,\boldsymbol{r}'\sigma') = \sum_{i=1}^{N}\phi_i(\boldsymbol{r}\sigma)\,\phi_i^*(\boldsymbol{r}'\sigma')\,, \tag{C.4}$$

rather than the density. This step scales as M^3. The integration

[3] The numerical integration can be avoided if v is expanded in terms of a separate basis set which allows an analytical treatment of $\langle\eta_k|\hat{v}|\eta_l\rangle$. This point will, however, not be expanded here.

$$\sum_{\sigma,\sigma'}\int d^3r \int d^3r'\ \gamma(\boldsymbol{r}\sigma,\boldsymbol{r}'\sigma')v(\boldsymbol{r}\sigma,\boldsymbol{r}'\sigma')$$

then requires M^2 operations.

The same arguments can be applied to the kinetic energy. This is immediately clear if the gradients of η_k can be evaluated analytically. However, even if the partial derivatives of η_k (or ϕ_i) have to be calculated numerically, the total scaling is not affected, as differentiation is linear in M.

In summary: the numerical calculation of the Hamilton matrix scales like M^3 with the basis set size, as long as the Hamiltonian consists only of single-particle operators, whose evaluation does not introduce an additional M-dependence. The same scaling behavior is found for the actual diagonalization of the Hamilton matrix by standard techniques.[4] In practice, however, the diagonalization is less time consuming than the evaluation of the matrix elements.

The situation becomes more complicated as soon as the Coulomb interaction, a 2-particle operator, is taken into account, i.e. as soon as the determination of $\hat{v}_{\mathrm{eff},\sigma\sigma'}$ is addressed. Let us explicitly consider the HF approximation in which only very specific Coulomb matrix elements are required. In order to extract the scaling behavior it is sufficient to analyze the exchange contribution

$$E_{\mathrm{x}} = -\frac{e^2}{2}\sum_{i,j=1}^{N}\sum_{\sigma,\sigma'}\int d^3r \int d^3r'\ \frac{\phi_i^*(\boldsymbol{r}\sigma)\phi_j^*(\boldsymbol{r}'\sigma')\phi_j(\boldsymbol{r}\sigma)\phi_i(\boldsymbol{r}'\sigma')}{|\boldsymbol{r}-\boldsymbol{r}'|}\ , \qquad \text{(C.5)}$$

which is the most demanding term in the HF approach. One possible method for the evaluation of (C.5) consists of the following sequence of operations:

1. evaluate and store $\phi_i(\boldsymbol{r}\sigma)$ (scales as M^3)
2. evaluate and store $\gamma(\boldsymbol{r}\sigma,\boldsymbol{r}'\sigma')$ (scales as M^3)
3. evaluate $E_{\mathrm{x}} = -\frac{e^2}{2}\sum_{\sigma,\sigma'}\int d^3r \int d^3r'\ \frac{|\gamma(\boldsymbol{r}\sigma,\boldsymbol{r}'\sigma')|^2}{|\boldsymbol{r}-\boldsymbol{r}'|}$ (scales as M^2)

A net scaling of M^3 is found. The same is true for the exchange contribution to the effective single-particle Hamiltonian

$$\sum_{j=1}^{N}(\eta_k\phi_j|\frac{1}{|\boldsymbol{r}-\boldsymbol{r}'|}|\phi_j\eta_n) = \sum_{j=1}^{N}\int d^3r \int d^3r'\ \frac{\eta_k^*(\boldsymbol{r})\phi_j^*(\boldsymbol{r}'\sigma')\phi_j(\boldsymbol{r}\sigma)\eta_n(\boldsymbol{r}')}{|\boldsymbol{r}-\boldsymbol{r}'|} \qquad \text{(C.6)}$$

$$= \int d^3r\eta_k^*(\boldsymbol{r})\left[\int d^3r'\ \frac{\gamma(\boldsymbol{r}\sigma,\boldsymbol{r}'\sigma')}{|\boldsymbol{r}-\boldsymbol{r}'|}\eta_n(\boldsymbol{r}')\right]\ .$$

The calculation of the quantity in brackets scales as M^3. Once it is stored for all $\boldsymbol{r}$ and n, one can perform the $\boldsymbol{r}$-integration for all k,l, which again scales as M^3.

However, the numerical evaluation of matrix elements of the Coulomb interaction is complicated by the singularity at $\boldsymbol{r}=\boldsymbol{r}'$ and by the long range of the inter-

[4] Here we ignore advanced techniques as iterative diagonalization [729], the Car-Parrinello method [730] and conjugate gradient methods [731, 732, 669] for brevity.

action. In addition, the procedure described, requires substantial memory as both the $M \times M$ arrays $\gamma(\boldsymbol{r}\sigma, \boldsymbol{r}'\sigma')$ and $\int d^3r' \gamma(\boldsymbol{r}\sigma, \boldsymbol{r}'\sigma')\eta_n(\boldsymbol{r}')/|\boldsymbol{r}-\boldsymbol{r}'|$ have to be stored simultaneously. It is therefore preferable to use basis functions for which the matrix elements

$$(\eta_k \eta_l || \eta_m \eta_n) = \int d^3r \int d^3r' \, \frac{\eta_k^*(\boldsymbol{r})\eta_l^*(\boldsymbol{r}')\eta_m(\boldsymbol{r})\eta_n(\boldsymbol{r}')}{|\boldsymbol{r}-\boldsymbol{r}'|} \tag{C.7}$$

can be calculated analytically (like Gaussian or plane-wave basis functions). If the quantities $(\eta_k \eta_l || \eta_m \eta_n)$ are easily recalculated, it is not even necessary to store the matrix elements. Alternatively, one can store all $(\eta_k \eta_l || \eta_m \eta_n)$ before any other operation is performed. In this case the exchange term in the Hamilton matrix has the form

$$\sum_{j=1}^{N} (\eta_k \phi_j | \frac{1}{|\boldsymbol{r}-\boldsymbol{r}'|} | \phi_j \eta_n) = \sum_{l,m=1}^{M} \sum_{j=1}^{N} b_{j,l\sigma'}^* b_{j,m\sigma} (\eta_k \eta_l || \eta_m \eta_n) \,, \tag{C.8}$$

which suggests an effort proportional to NM^4. However, the evaluation of the expression (C.8) and that of the exchange energy can again be split into several independent steps, which improves the scaling behavior. One first sums up the M^2 coefficients $\sum_{j=1}^{N} b_{j,l\sigma'}^* b_{j,m\sigma}$ and stores them. This step requires M^3 operations. In the second step the resulting matrix in l,m is folded with the known matrix elements $(\eta_k \eta_l || \eta_m \eta_n)$ for each pair k,n, which requires M^4 operations. If one again stores the resulting M^2 matrix elements, the summations over i,k,n required for the calculation of the complete exchange energy are independent of the previous steps, so that the third set of summations scales as M^2 (as $\sum_{i=1}^{N} b_{i,k\sigma}^* b_{i,n\sigma'}$ is already available). Taking all steps together, one ends up with a scaling of the HF scheme proportional to M^4 in this standard implementation.

As soon as arbitrary 2-particle matrix elements

$$(\phi_i \phi_j | \frac{1}{|\boldsymbol{r}-\boldsymbol{r}'|} | \phi_k \phi_l)$$

have to be calculated, as is the case for all correlated *ab-initio* methods, the M^4-scaling can no longer be preserved by some clever sequence of operations. In addition, the scaling behavior again depends sensitively on the technical implementation. It is beyond the scope of this text to provide any details.

Appendix D
Explicit Density Functionals for the Kinetic Energy: Thomas-Fermi Models and Beyond

The theorem of Hohenberg and Kohn provides a justification of early density functional models which relied on a representation of the complete ground state energy E_0 in terms of the density,

$$E_0 = E[n_0] . \tag{D.1}$$

The first density functional of this type was the model of Thomas and Fermi (TF), which was established in the years 1927/28 [13, 14]. These authors considered a uniform gas of noninteracting electrons, the homogeneous electron gas (HEG) of Sect. 4.3, in order to derive a representation of the kinetic energy in terms of the density.

Their result can be derived by the Green's function techniques utilized in Sect. 4.3 for the discussion of the xc-energy of the HEG. In order to provide some alternative to this approach, however, a more elementary route for the derivation of the TF functional is taken in this Appendix. The Schrödinger equation for the single-particle states of the noninteracting electron gas reads

$$-\frac{\hbar^2\boldsymbol{\nabla}^2}{2m}\phi_i(\boldsymbol{r}\sigma) = \varepsilon_i\phi_i(\boldsymbol{r}\sigma) . \tag{D.2}$$

The solutions of (D.2) are given by

$$\phi_{\boldsymbol{k}s}(\boldsymbol{r}\sigma) = C\, e^{i\boldsymbol{k}\cdot\boldsymbol{r}}\chi_s(\sigma) \qquad \text{(quantum number } i \equiv \boldsymbol{k}s) , \tag{D.3}$$

with the Pauli spinors $\chi_s(\sigma)$ and the eigenvalues

$$\varepsilon_{\boldsymbol{k}} = \frac{\hbar^2\boldsymbol{k}^2}{2m} . \tag{D.4}$$

Normalizable solutions can only be obtained if $\boldsymbol{k}$ is real. However, even in the case of real $\boldsymbol{k}$ the norm of $\phi_{\boldsymbol{k}s}$ is infinite, as soon as the complete space is considered. Moreover, the differential equation (D.2) allows arbitrary real values of $\boldsymbol{k}$, so that one finds more than countably many states. It is thus necessary to regularize the problem by an additional boundary condition which ensures the normalizability of the $\phi_{\boldsymbol{k}s}$ and at the same time discretizes the spectrum. For this regularization one chooses a cubic box with sides of length L. Requiring periodic boundary conditions for all three Cartesian directions,

$$\phi_{\boldsymbol{k}s}(x+L,y,z,\sigma) = \phi_{\boldsymbol{k}s}(x,y+L,z,\sigma) = \phi_{\boldsymbol{k}s}(x,y,z+L,\sigma) = \phi_{\boldsymbol{k}s}(x,y,z,\sigma)\,, \quad \text{(D.5)}$$

leads to a quantization (i.e. discretization) of all components of $\boldsymbol{k}$,

$$k_i = \frac{2\pi}{L}\alpha_i \quad \text{with} \quad \alpha_i = 0,\pm1,\pm2,\ldots \qquad (i=1,2,3)\,. \quad \text{(D.6)}$$

Normalization to 1 inside the box is obtained for $C = 1/\sqrt{L^3}$,

$$\int_0^L dx \int_0^L dy \int_0^L dx\, e^{\frac{2\pi i}{L}(\boldsymbol{\alpha}'-\boldsymbol{\alpha})\cdot\boldsymbol{r}} \sum_{\sigma=\uparrow,\downarrow} \chi_s(\sigma)\chi_{s'}(\sigma) = L^3\,\delta_{\boldsymbol{\alpha}\boldsymbol{\alpha}'}\,\delta_{ss'}\,. \quad \text{(D.7)}$$

The single-particle states which are properly normalized within a cubic box are thus given by

$$\phi_{\boldsymbol{k}s}(\boldsymbol{r}\sigma) = \frac{e^{i\boldsymbol{k}\cdot\boldsymbol{r}}}{L^{3/2}}\chi_s(\sigma) \qquad \boldsymbol{k} = \frac{2\pi}{L}\boldsymbol{\alpha} \quad \text{with} \quad \alpha_i = 0,\pm1,\pm2,\ldots\,. \quad \text{(D.8)}$$

In the ground state of the noninteracting homogeneous electron gas each level is filled with one spin-up and one spin-down electron. The number of levels which are occupied is determined by the number of particles in the box. The eigenvalue of the energetically highest occupied state is identified with the Fermi energy ε_{F}. Consequently, the density of the system is

$$\begin{aligned} n_0 &= \sum_{i=1}^{\infty} \Theta(\varepsilon_{\mathrm{F}}-\varepsilon_i) \sum_{\sigma=\uparrow,\downarrow} \phi_i^*(\boldsymbol{r}\sigma)\phi_i(\boldsymbol{r}\sigma) \\ &= \sum_{\alpha_1,\alpha_2,\alpha_3=0}^{\infty} \Theta(\varepsilon_{\mathrm{F}}-\varepsilon_{\boldsymbol{k}}) \sum_{\sigma=\uparrow,\downarrow} \phi_{\boldsymbol{k}s}^*(\boldsymbol{r}\sigma)\phi_{\boldsymbol{k}s}(\boldsymbol{r}\sigma) \\ &= \sum_{\alpha_1,\alpha_2,\alpha_3=0}^{\infty} \Theta\left(\varepsilon_{\mathrm{F}} - \frac{\hbar^2\boldsymbol{k}^2}{2m}\right)\frac{2}{L^3}\,. \end{aligned} \quad \text{(D.9)}$$

Similarly one obtains for the kinetic energy per volume element

$$\begin{aligned} \frac{T_{\mathrm{s}}(V)}{V} &= \frac{1}{V}\sum_{i=1}^{\infty} \Theta(\varepsilon_{\mathrm{F}}-\varepsilon_i) \sum_{\sigma=\uparrow,\downarrow} \int_V d^3r\, \phi_i^*(\boldsymbol{r}\sigma)\frac{-\hbar^2\boldsymbol{\nabla}^2}{2m}\phi_i(\boldsymbol{r}\sigma) \\ &= \frac{1}{L^3}\sum_{\alpha_1,\alpha_2,\alpha_3=0}^{\infty} \Theta(\varepsilon_{\mathrm{F}}-\varepsilon_{\boldsymbol{k}}) \sum_{\sigma=\uparrow,\downarrow} \int_0^L dx \int_0^L dy \int_0^L dz\, \phi_{\boldsymbol{k}s}^*(\boldsymbol{r}\sigma)\frac{\hbar^2\boldsymbol{k}^2}{2m}\phi_{\boldsymbol{k}s}(\boldsymbol{r}\sigma) \end{aligned}$$

$$= \sum_{\alpha_1,\alpha_2,\alpha_3=0}^{\infty} \Theta\left(\varepsilon_F - \frac{\hbar^2 \boldsymbol{k}^2}{2m}\right) \frac{2}{L^3} \frac{\hbar^2 \boldsymbol{k}^2}{2m} . \tag{D.10}$$

At this point all expressions have been evaluated to a point at which the limit $L \to \infty$ can be taken, which leads back to the electron gas of infinite extension. In this limit the spacing between adjacent momenta $\boldsymbol{k}$ becomes infinitesimally small, so that the summation over all discrete values of $\boldsymbol{k}$ goes over into an integration over $\boldsymbol{k}$. The volume element of this $\boldsymbol{k}$-integration is obtained from the volume in $\boldsymbol{k}$-space which is associated with each of the discrete $\boldsymbol{k}$-values. For each of the Cartesian directions two neighboring k_i-values differ by $2\pi/L$, so that the $\boldsymbol{k}$-space volume per discrete $\boldsymbol{k}$-value is $(2\pi/L)^3$,

$$\Delta k_i = \frac{2\pi}{L} \Delta\alpha_i \quad \Longrightarrow \quad \Delta\alpha_1 \Delta\alpha_2 \Delta\alpha_3 = \left(\frac{L}{2\pi}\right)^3 \Delta^3 k$$

$$\sum_{\alpha_1,\alpha_2,\alpha_3=0}^{\infty} \quad \xrightarrow{L\to\infty} \quad \left(\frac{L}{2\pi}\right)^3 \int d^3k . \tag{D.11}$$

Introducing the Fermi momentum

$$k_F := \frac{\sqrt{2m\varepsilon_F}}{\hbar} , \tag{D.12}$$

the density and kinetic energy density are now easily evaluated using spherical coordinates,

$$\begin{aligned} n_0 &= \left(\frac{L}{2\pi}\right)^3 \int d^3k\, \Theta\left(k_F - |\boldsymbol{k}|\right) \frac{2}{L^3} \\ &= \frac{k_F^3}{3\pi^2} \end{aligned} \tag{D.13}$$

$$\begin{aligned} \frac{T_s(V)}{V} &= \left(\frac{L}{2\pi}\right)^3 \int d^3k\, \Theta\left(k_F - |\boldsymbol{k}|\right) \frac{2}{L^3} \frac{\hbar^2 \boldsymbol{k}^2}{2m} \\ &= \frac{\hbar^2 k_F^5}{10\pi^2 m} . \end{aligned} \tag{D.14}$$

Finally, one can invert the relation between n_0 and k_F,

$$k_F = \left(3\pi^2 n_0\right)^{1/3} , \tag{D.15}$$

to end up with the desired relation between the kinetic energy density t_s and the density n_0,

$$t_s \equiv \frac{T_s(V)}{V} = \frac{\hbar^2 (3\pi^2 n_0)^{5/3}}{10\pi^2 m} . \tag{D.16}$$

In order to apply this result to atoms, Thomas and Fermi (TF) relied on the local density approximation discussed in Sect. 4.3. In this approximation the energy density $t_{\rm s}(\boldsymbol{r})$ of the actual inhomogeneous system is replaced by the energy density of the electron gas, Eq. (D.16), evaluated with the local density $n(\boldsymbol{r})$. The complete kinetic energy is then given by

$$T_{\rm s}^{\rm TF} = \frac{3(3\pi^2)^{2/3}\hbar^2}{10m} \int d^3r\, n(\boldsymbol{r})^{5/3}\,. \tag{D.17}$$

This expression is manifestly a density functional. As it is derived from the non-interacting gas it represents an approximation for the Kohn-Sham kinetic energy functional $T_{\rm s}[n]$, introduced in Sect. 3.1.

The total energy functional of Thomas and Fermi neglected all exchange and correlation effects, so that only the direct Coulomb repulsion (Hartree energy) and the coupling to the external potential remain,

$$E^{\rm TF}[n] = T_{\rm s}^{\rm TF}[n] + \frac{e^2}{2}\int d^3r \int d^3r' \frac{n(\boldsymbol{r})n(\boldsymbol{r}')}{|\boldsymbol{r}-\boldsymbol{r}'|} + \int d^3r\, v_{\rm ext}(\boldsymbol{r})n(\boldsymbol{r})\,. \tag{D.18}$$

Applications can be based directly on the variational equation (2.38), reflecting the minimum principle (2.28). The welcome feature is the fact that the variational approach reduces the many-particle problem to a form which is independent of the particle number.

Considerable effort was expended in order to improve this model. The first and most important step was the inclusion of exchange by Dirac in 1930 [131]. Dirac followed the path of Thomas and Fermi and considered the exchange energy of the uniform electron gas. The exact exchange energy of the gas in the cubic box of volume $V = L^3$ has the form

$$\begin{aligned} E_{\rm x}(V) = &-\frac{e^2}{2}\sum_{i,j=1}^{\infty} \Theta(\varepsilon_{\rm F}-\varepsilon_i)\Theta(\varepsilon_{\rm F}-\varepsilon_j) \\ &\times \sum_{\sigma,\sigma'=\uparrow,\downarrow} \int_V d^3r \int d^3r' \frac{\phi_i^*(\boldsymbol{r}\sigma)\phi_j(\boldsymbol{r}\sigma)\phi_j^*(\boldsymbol{r}'\sigma')\phi_i(\boldsymbol{r}'\sigma')}{|\boldsymbol{r}-\boldsymbol{r}'|}\,. \end{aligned} \tag{D.19}$$

Insertion of the states (D.8) of the uniform gas yields for the exchange energy per volume element (after an appropriate shift of $\boldsymbol{r}'$ by $\boldsymbol{r}$)

$$e_{\rm x} \equiv \frac{E_{\rm x}(V)}{V} = -e^2 \sum_{\boldsymbol{\alpha\beta}} \Theta\left(k_{\rm F}-|\boldsymbol{k}_{\boldsymbol{\alpha}}|\right)\Theta\left(k_{\rm F}-|\boldsymbol{k}_{\boldsymbol{\beta}}|\right) \int d^3r' \frac{e^{i(\boldsymbol{k}_{\boldsymbol{\alpha}}-\boldsymbol{k}_{\boldsymbol{\beta}})\cdot\boldsymbol{r}'}}{L^6|\boldsymbol{r}'|}\,. \tag{D.20}$$

One can now use the fact that for $L\to\infty$ the summation over all integers $\boldsymbol{\alpha}$ can be replaced by an integration over $\boldsymbol{k}$, Eq. (D.11), to obtain

$$e_{\rm x} = -e^2 \int \frac{d^3k}{(2\pi)^3} \int \frac{d^3q}{(2\pi)^3} \Theta(k_{\rm F}-|\boldsymbol{k}|)\Theta(k_{\rm F}-|\boldsymbol{q}|) \int d^3r' \frac{e^{i(\boldsymbol{k}-\boldsymbol{q})\cdot\boldsymbol{r}'}}{|\boldsymbol{r}'|}\,. \tag{D.21}$$

Next, the $\boldsymbol{r}'$-integration can be carried out by introducing a suitable intermediate regularization factor $e^{-\mu|\boldsymbol{r}'|}$ in the integral and taking the limit $\mu \to 0$ after integration (compare Eq. (4.144) and the subsequent discussion in Sect. 4.4.1),

$$e_{\mathrm{X}} = -e^2 \int \frac{d^3k}{(2\pi)^3} \int \frac{d^3q}{(2\pi)^3} \Theta(k_{\mathrm{F}} - |\boldsymbol{k}|)\Theta(k_{\mathrm{F}} - |\boldsymbol{q}|) \frac{4\pi}{(\boldsymbol{k}-\boldsymbol{q})^2} \,. \tag{D.22}$$

One now first performs the $\boldsymbol{q}$-integration. Choosing the z-axis of the coordinate system for $\boldsymbol{q}$ so that it is parallel to $\boldsymbol{k}$, the $\boldsymbol{q}$-integration can be done in spherical coordinates,

$$\begin{aligned} e_{\mathrm{X}} &= -\frac{e^2}{\pi} \int \frac{d^3k}{(2\pi)^3} \Theta(k_{\mathrm{F}} - k) \int_0^{k_{\mathrm{F}}} q^2 dq \int_{-1}^{+1} d\cos(\theta) \frac{1}{k^2 + q^2 - 2kq\cos(\theta)} \\ &= \frac{e^2}{2\pi^3} \int_0^{k_{\mathrm{F}}} kdk \int_0^{k_{\mathrm{F}}} qdq \left[\ln|k-q| - \ln(k+q)\right] . \end{aligned}$$

The remaining integrations are straightforward, after splitting the range of the inner integration over q into the subregimes $[0,k]$ and $[k,k_{\mathrm{F}}]$,

$$e_{\mathrm{X}} = -\frac{e^2}{4\pi^3} k_{\mathrm{F}}^4 \,. \tag{D.23}$$

Insertion of the Fermi momentum (D.15) then leads to

$$e_{\mathrm{X}} = -\frac{e^2}{4\pi^3} \left(3\pi^2 n_0\right)^{4/3} \,. \tag{D.24}$$

Using the local density approximation, one finally arrives at the density functional

$$E_{\mathrm{X}}^{\mathrm{D}}[n] = -\frac{3(3\pi^2)^{1/3} e^2}{4\pi} \int d^3r\, n(\boldsymbol{r})^{4/3} \,. \tag{D.25}$$

$E_{\mathrm{X}}^{\mathrm{D}}[n]$ is an approximation for the exact exchange energy functional $E_{\mathrm{X}}[n]$ of DFT. As is clear from its construction, $E_{\mathrm{X}}^{\mathrm{D}}[n]$ is nothing but the LDA for exchange, Eqs. (4.99), (4.109), in modern terminology. Adding this term to the energy (D.18) constitutes the Thomas-Fermi-Dirac model.

The next step towards extending the TF model was taken by von Weizsäcker in 1935 [174]. Von Weizsäcker observed that one can express the kinetic energy of a single particle in terms of the density. In fact, if there is only one particle bound by some potential, the corresponding ground state orbital

$$\phi_i(\boldsymbol{r}\sigma) = \varphi_0(\boldsymbol{r})\chi_s(\sigma)$$

may be chosen real, so that its kinetic energy may be written as[1]

---

[1] The surface term does not contribute in the partial integration since a normalizable orbital decays sufficiently rapidly for $|\boldsymbol{r}| \to \infty$.

$$T_s = \sum_{\sigma=\uparrow,\downarrow} \int d^3r\, \phi_i^*(\boldsymbol{r}\sigma) \frac{-\hbar^2\nabla^2}{2m} \phi_i(\boldsymbol{r}\sigma) = \int d^3r \frac{[\hbar\nabla\varphi_0(\boldsymbol{r})]^2}{2m} . \tag{D.26}$$

The corresponding density is given by

$$n(\boldsymbol{r}) = \sum_{\sigma=\uparrow,\downarrow} |\phi_i(\boldsymbol{r}\sigma)|^2 = \varphi_0(\boldsymbol{r})^2 .$$

Insertion into (D.26) leads to the von Weizsäcker functional

$$T_s^{\mathrm{vW}}[n] = \frac{\hbar^2}{m} \int d^3r \frac{[\nabla n(\boldsymbol{r})]^2}{8n(\boldsymbol{r})} . \tag{D.27}$$

This density functional also represents the exact kinetic energy in the case of a noninteracting 2-particle system in which both particles occupy the same orbital φ_0, but have opposite spins. $T_s^{\mathrm{vW}}[n]$ thus agrees with the exact $T_s[n]$ of Kohn-Sham theory for a single particle and a spin-saturated pair of two particles.

The expression for $T_s^{\mathrm{vW}}[n]$ also indicates how the TF kinetic energy can be extended in order to better account for the inhomogeneity of real systems: obviously, the gradient of the density is the simplest purely density-dependent measure of the inhomogeneity in a many-particle system. The only parameter-free expression for t_s which (i) depends only locally on ∇n and (ii) does not depend on the characteristics of the external potential (as for instance on some preferred axis) is the functional (D.27). It is thus no surprise that a systematic derivation of gradient corrections for the kinetic energy, either using some form of the so-called commutator expansion [173] or following the lines of Sect. 4.4, leads to an expression which differs from $T_s^{\mathrm{vW}}[n]$ only by an overall prefactor $\lambda = 1/9$ (for all details, including higher order gradient corrections [175, 733, 194, 195, 734], see Chap. 5 of [7]). Adding $\lambda T_s^{\mathrm{vW}}[n]$ to $E^{\mathrm{TF}}[n] + E_{\mathrm{x}}^{\mathrm{D}}[n]$ constitutes the Thomas-Fermi-Dirac-Weizsäcker model.

Without going into detail, we list some further extensions of the TF-model:

- First correlation contributions were introduced by Wigner as early as 1934 [138] (see Sect. 4.3.4).
- Gradient corrections to the Dirac exchange energy were calculated subsequently, but were found to lead to a divergent behavior for small and large separations from the nucleus in atoms—compare Sect. 4.4.3.

Nonetheless, the endeavors to improve TF-type density functionals were essentially abandoned until recently, since the explicitly density-dependent representation of T_s used in these models does not allow to reproduce shell structure.

Renewed interest in functionals of the type (D.1) has been stimulated by the N^3-scaling of the Kohn-Sham approach with system size: if one wants to perform calculations for truly large quantum systems without any periodicity or other symmetry (e.g. disordered solids or huge (bio)molecules), an N^3-scaling is still prohibitive. In this case use of a *kinetic energy density functional* (KEDF) is highly attractive. In view of the limitations of the TF-type semi-local functionals a fully nonlocal ansatz is chosen for modern KEDFs [735–749]. The general form of these approximations is[2]

$$T_s^{\rm nl}[n] = T_s^{\rm TF}[n] + T_s^{\rm vW}[n] + \frac{\hbar^2 3(3\pi^2)^{2/3}}{10m} \int d^3r d^3r'\, n(\boldsymbol{r})^\alpha w_{\alpha\beta}\left(\xi_\gamma(\boldsymbol{r},\boldsymbol{r}'),\boldsymbol{r}-\boldsymbol{r}'\right) n(\boldsymbol{r}')^\beta\,, \quad \text{(D.28)}$$

with the *2-body Fermi wavevector*

$$\xi_\gamma(\boldsymbol{r},\boldsymbol{r}') = \left[\frac{(3\pi^2 n(\boldsymbol{r}))^{\gamma/3} + (3\pi^2 n(\boldsymbol{r}'))^{\gamma/3}}{2}\right]^{1/\gamma} \quad \text{(D.29)}$$

(the structure of (D.28) can be motivated by scaling arguments [750]). By construction the functional $T_s^{\rm nl}[n]$ can be exact for the electron gas with $\boldsymbol{\nabla} n = \boldsymbol{0}$ and for a 2-particle system, if the density-dependent kernel w is chosen appropriately. So, obviously one has the requirement

$$w_{\alpha\beta}\left(\xi_\gamma,\boldsymbol{r}-\boldsymbol{r}'\right) = 0$$

in the electron gas limit. Moreover, in order to recover the exact linear response result for the weakly inhomogeneous electron gas, Eq. (4.156), the kernel has to satisfy a differential equation, which allows to determine its shape. In fact, this differential equation can even be solved analytically [749], which, in spite of the nonlocality of $\xi_\gamma(\boldsymbol{r},\boldsymbol{r}')$, leads to an $N\ln(N)$ scaling of the computational effort with the system size. KEDFs can therefore provide the basis for multiscale modelling.

Selfconsistent calculations with KEDFs are usually based on pseudopotentials. The pseudopotentials have to be local, as projecting out part of the all-electron Hilbert space is not possible, if no states are involved. However, an accurate description by local pseudopotentials can only be expected for simple metals. Applications of KEDFs to bulk aluminum, aluminum surfaces and aluminum clusters [742, 743, 746, 749] demonstrated that the functional (D.28) accurately reproduces the geometry, energetics (including vacancy formation) and density profiles of the full Kohn-Sham solutions. In particular, one finds very accurate results for the relative energies of different crystal structures [746]. KEDFs perform even better for sodium [742, 743].

[2] Sometimes, even several nonlocal kernels of the form (D.28) are superposed [745],

$$\sum_{\alpha\beta} \lambda_{\alpha\beta} \, \langle n(\boldsymbol{r})^\alpha w_{\alpha\beta}\, n(\boldsymbol{r}')^\beta \rangle\,,$$

in order to allow for more flexibility. In this case $\sum_{\alpha\beta}\lambda_{\alpha\beta} = 1$ is required.

Appendix E
Asymptotic Behavior of Quasi-Particle Amplitudes

In this Appendix the asymptotic behavior of the quasi-particle amplitudes f_k, Eq. (3.104), for the case of finite systems is extracted from the differential equation (3.112). One starts by noting that a multipole expansion of the interaction $w(\boldsymbol{r},\boldsymbol{r}')$ is legitimate for large $|\boldsymbol{r}|$, as $\langle\Psi_k^{N-1}|\hat{n}(\boldsymbol{r}')|\Psi_l^{N-1}\rangle$ decays exponentially for large $|\boldsymbol{r}'|$—only bound states k are of interest, so that $|\Psi_k^{N-1}\rangle$ represents a localized wavefunction which vanishes exponentially for large $|\boldsymbol{r}|$. Restricting the discussion to the Coulomb interaction, one has

$$w(\boldsymbol{r},\boldsymbol{r}') = \frac{e^2}{|\boldsymbol{r}-\boldsymbol{r}'|} = \frac{e^2}{|\boldsymbol{r}|}\left\{1+\frac{\boldsymbol{r}\cdot\boldsymbol{r}'}{|\boldsymbol{r}|^2}+\ldots\right\}. \tag{E.1}$$

Insertion into the nonlocal term in Eq. (3.112) leads to

$$\begin{aligned}
&\sum_l \int d^3r'\, w(\boldsymbol{r},\boldsymbol{r}')\,\langle\Psi_k^{N-1}|\hat{n}(\boldsymbol{r}')|\Psi_l^{N-1}\rangle\, f_l(\boldsymbol{r}\sigma)\\
&= \frac{e^2}{|\boldsymbol{r}|}\sum_l \left\{(N-1)\delta_{kl} + \frac{\boldsymbol{r}}{|\boldsymbol{r}|^2}\cdot\langle\Psi_k^{N-1}|\int d^3r'\,\boldsymbol{r}'\hat{n}(\boldsymbol{r}')|\Psi_l^{N-1}\rangle\right\} f_l(\boldsymbol{r}\sigma)\\
&\quad +\mathcal{O}(|\boldsymbol{r}|^{-3}).
\end{aligned} \tag{E.2}$$

The kernel of the first order term is exactly the operator of the dipole moment,

$$\hat{\boldsymbol{D}} = e^2\int d^3r\,\boldsymbol{r}\,\hat{n}(\boldsymbol{r}). \tag{E.3}$$

A multipole expansion is also possible for the external potential. Using again the Coulomb form, one obtains

$$v_{\rm ext}(\boldsymbol{r}) = -\sum_\alpha \frac{Z_\alpha e^2}{|\boldsymbol{r}-\boldsymbol{R}_\alpha|} = -\frac{Ze^2}{|\boldsymbol{r}|} - \frac{\boldsymbol{r}}{|\boldsymbol{r}|^3}\cdot\sum_\alpha Z_\alpha e^2\boldsymbol{R}_\alpha + \ldots \quad ; \quad Z=\sum_\alpha Z_\alpha\,, \tag{E.4}$$

where Z_α and $\boldsymbol{R}_\alpha$ denote the charge and position of nucleus α. Insertion of (E.2)–(E.4) into (3.112) yields the asymptotic differential equation for the f_k,

$$\left\{-\frac{\hbar^2\nabla^2}{2m} - \frac{(Z-N+1)e^2}{|\boldsymbol{r}|} - \hbar\omega_k\right\} f_k(\boldsymbol{r}\sigma) + \frac{\boldsymbol{r}}{|\boldsymbol{r}|^3}\cdot\sum_l \boldsymbol{D}_{kl}\, f_l(\boldsymbol{r}\sigma) = 0. \tag{E.5}$$

All electronic and nuclear dipole contributions have been absorbed into $\boldsymbol{D}_{kl}$. The asymptotically leading term of the potential in (E.5) is the spherically symmetric monopole term. The general solution of (E.5) thus has the form

$$f_k(\boldsymbol{r}\sigma) \underset{|\boldsymbol{r}|\to\infty}{\longrightarrow} \left[\sum_{lm} c_{k\sigma}^{lm} Y_{lm}(\Omega)\right] r^{\beta_{k\sigma}-1}\, e^{-\alpha_{k\sigma} r}, \tag{E.6}$$

with coefficients $c_{k\sigma}^{lm}$, $\beta_{k\sigma}$ and $\alpha_{k\sigma}$ which remain to be determined. Insertion into (E.5) gives

$$0 = \sum_{lm} Y_{lm}(\Omega)\left\{\left[\frac{\partial^2}{\partial r^2} - \frac{l(l+1)}{r^2} + 2\frac{m}{\hbar^2}\left(\hbar\omega_k + \frac{(Z-N+1)e^2}{r}\right)\right] c_{k\sigma}^{lm}\, r^{\beta_{k\sigma}} \right.$$
$$\left. - 2\frac{m}{\hbar^2}\frac{\boldsymbol{r}}{r^3}\cdot\sum_n \boldsymbol{D}_{kn}\, c_{n\sigma}^{lm}\, r^{\beta_{n\sigma}}\, e^{(\alpha_{k\sigma}-\alpha_{n\sigma})r}\right\} e^{-\alpha_{k\sigma} r}. \tag{E.7}$$

As all f_k are coupled by the dipole moment matrix elements $\boldsymbol{D}_{kn}$ and these matrix elements do not vanish for $k \neq n$, all f_k must have the same exponential decay. This statement can be verified by reductio ad absurdum. Assume that there is one amplitude f_q which shows the weakest decay, i.e. $\alpha_{q\sigma} < \alpha_{k\sigma}$ for all $k \neq q$. Now consider the asymptotic equation for $k \neq q$. The dipole contribution of f_q dominates this asymptotic equation, i.e. the sum over n breaks down to the single term with $n = q$. Due to $\alpha_{k\sigma} - \alpha_{q\sigma} > 0$, however, this term diverges exponentially, thus requiring $c_{q\sigma}^{lm} = 0$. One ends up with a contradiction, so that all $\alpha_{k\sigma}$ must be identical. The same result is found for Hartree-Fock orbitals [751], which also satisfy coupled equations of the type (E.5).

With $\alpha_{k\sigma} \equiv \alpha_\sigma$ Eq. (E.7) reduces to

$$0 = \sum_{lm} Y_{lm}(\Omega)\left\{\left[\frac{\beta_{k\sigma}(\beta_{k\sigma}-1)}{r^2} - 2\frac{\alpha_\sigma\beta_{k\sigma}}{r} + \alpha_\sigma^2 - \frac{l(l+1)}{r^2}\right.\right.$$
$$\left. + 2\frac{m}{\hbar^2}\left(\hbar\omega_k + \frac{(Z-N+1)e^2}{r}\right)\right] c_{k\sigma}^{lm}\, r^{\beta_{k\sigma}}$$
$$\left. - 2\frac{m}{\hbar^2}\frac{\boldsymbol{r}}{r^3}\cdot\sum_n \boldsymbol{D}_{kn}\, c_{n\sigma}^{lm}\, r^{\beta_{n\sigma}}\right\}. \tag{E.8}$$

Consider now the amplitude q with the largest exponent $\beta_{k\sigma}$,

$$\beta_{q\sigma} > \beta_{k\sigma} \quad \forall\, k \neq q. \tag{E.9}$$

In the asymptotic equation for f_q the dipole term is suppressed by $1/r$ with respect to the two leading orders. Consequently one obtains

$$\alpha_\sigma^2 = -2\frac{m}{\hbar}\omega_q\,; \qquad \beta_{q\sigma} = \sqrt{\frac{m}{\hbar^2}}\frac{(Z-N+1)e^2}{\sqrt{-2\hbar\omega_q}}\,. \tag{E.10}$$

The largest exponent is therefore found for the smallest $|\omega_q|$, i.e. for $q=0$,

$$\alpha_\sigma = \sqrt{\frac{-2m\omega_0}{\hbar}} = \frac{\sqrt{2m\,\mathrm{IP}}}{\hbar}\,; \qquad \beta_{0\sigma} = \sqrt{\frac{m}{\hbar^2}}\frac{(Z-N+1)e^2}{\sqrt{-2\hbar\omega_0}}\,. \tag{E.11}$$

All other k must be suppressed relative to this leading amplitude by a factor of $1/r^2$, in order to satisfy the asymptotic equation. Only in this case can the leading contribution of the dipole term be compensated by the leading contribution of the remaining terms in the asymptotic equation (E.8) for all states $k \neq 0$,

$$0 = \sum_{lm} Y_{lm}(\Omega)\left\{2\hbar(\omega_k-\omega_0)c_{k\sigma}^{lm}\,r^{\beta_{k\sigma}} - 2\frac{\boldsymbol{r}}{r^3}\cdot\boldsymbol{D}_{k0}\,c_{0\sigma}^{lm}\,r^{\beta_{0\sigma}}\right\} \tag{E.12}$$

(as $\beta_{0\sigma} > \beta_{n\sigma}$ for $n>0$, the sum over n breaks down to a single term). One ends up with

$$\beta_{k\sigma} = \beta_{0\sigma} + 2 \quad \forall\, k>0 \tag{E.13}$$

and the coefficients $c_{k\sigma}^{lm}$ have to satisfy

$$0 = \sum_{lm} Y_{lm}(\Omega)\left\{\hbar(\omega_k-\omega_0)c_{k\sigma}^{lm} - \frac{\boldsymbol{r}}{r}\cdot\boldsymbol{D}_{k0}\,c_{0\sigma}^{lm}\right\}. \tag{E.14}$$

Equations (E.6), (E.11), (E.13), (E.14) determine the asymptotic behavior of all f_k.

Appendix F
Quantization of Noninteracting Fermions in Relativistic Quantum Field Theory

This Appendix summarizes the quantum field theoretical description of noninteracting spin-1/2 particles. In particular, the quantization procedure is reviewed, emphasizing the close relation between the minimum principle for the ground state energy and the renormalization scheme. At the same time this Appendix provides the background for the field theoretical treatment of the KS system, i.e. Eqs. (8.76)–(8.89). For brevity, we use $\hbar = c = 1$ in this Appendix.

The starting point is the *classical* field theory characterized by the Lagrangian

$$\mathscr{L}_{\rm s}(x) = \mathscr{L}_{\rm e}(x) + \mathscr{L}_{\rm ext}(x) = \overline{\psi}_{\rm s}(x)\left[i\gamma^\mu \partial_\mu - m + e\gamma^\mu V_\mu(\boldsymbol{x})\right]\psi_{\rm s}(x)\,, \qquad \text{(F.1)}$$

where $V^\mu(\boldsymbol{x})$ is a given, stationary external potential. V^μ may either represent some nuclear potential or a composite object as the total KS potential $v_{\rm s}^\mu$. The orthonormal eigenfunctions of the corresponding classical field equations will be denoted by ϕ_k, the associated single-particle energies by ε_k,

$$\left[-i\boldsymbol{\alpha}\cdot\boldsymbol{\nabla} + \beta m - e\alpha_\mu V^\mu(\boldsymbol{x})\right]\phi_k(\boldsymbol{x}) = \varepsilon_k\phi_k(\boldsymbol{x})\,. \qquad \text{(F.2)}$$

A sketch of the eigenvalue spectrum resulting in the case of an attractive V^μ is shown in Fig. F.1. It consists of a continuum of negative energy states with energies below $-m$ (i.e. $-mc^2$), a continuum of positive energy states with energies above $+m$ and a countable number of discrete levels in between (which are at least twofold degenerate in the case of time-reversal invariant systems).

In the first step one has to quantize the classical field theory. The standard canonical quantization via equal-time commutation relations for the fermion field operator $\hat{\psi}_{\rm s}$ yields

$$\hat{\psi}_{\rm s}(x) = \sum_k \hat{b}_k\phi_k(\boldsymbol{x})e^{-i\varepsilon_k t}\,, \qquad \hat{\psi}_{\rm s}^\dagger(x) = \sum_k \hat{b}_k^\dagger\phi_k^\dagger(\boldsymbol{x})e^{i\varepsilon_k t}\,, \qquad \text{(F.3)}$$

where the sums run over all negative and positive energy solutions of (F.2) and the operator-valued expansion coefficients $\hat{b}_k$ satisfy the commutation relations

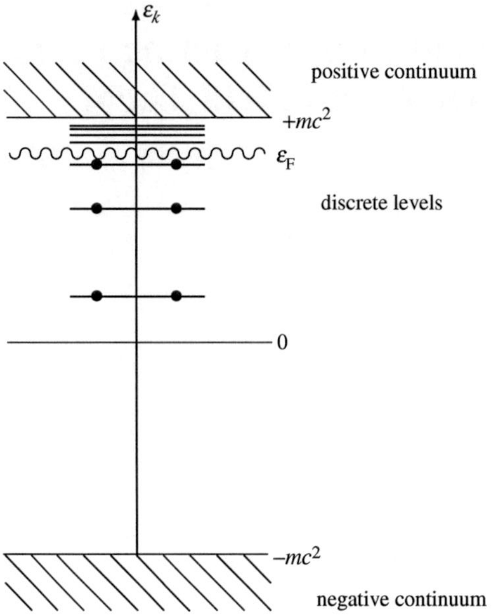

Fig. F.1 Eigenvalue spectrum of noninteracting fermions in attractive potential.

$$\left\{\hat{b}_k,\hat{b}_l\right\} = \left\{\hat{b}_k^\dagger,\hat{b}_l^\dagger\right\} = 0\,, \qquad \left\{\hat{b}_k,\hat{b}_l^\dagger\right\} = \delta_{kl}\,. \tag{F.4}$$

$\hat{b}_k$ destroys a particle in state k, so that $\hat{\psi}_s(x)$ destroys a particle at point $\boldsymbol{x}$ and time t. The canonical Hamiltonian obtained from (F.1) by insertion of (F.3) is given by

$$\begin{aligned}\hat{H}_s &= \int d^3x\, \hat{\psi}_s^\dagger(x)\Big[-i\boldsymbol{\alpha}\cdot\boldsymbol{\nabla} + \beta m - e\alpha_\mu V^\mu(\boldsymbol{x})\Big]\hat{\psi}_s(x) \\ &= \sum_k \varepsilon_k \hat{b}_k^\dagger \hat{b}_k = \sum_{\varepsilon_k\le -m} \varepsilon_k \hat{b}_k^\dagger \hat{b}_k + \sum_{-m<\varepsilon_k} \varepsilon_k \hat{b}_k^\dagger \hat{b}_k\,, \end{aligned} \tag{F.5}$$

while the canonical charge operator reads

$$\hat{Q}_s = \int d^3x\, \hat{\psi}_s^\dagger(x)\hat{\psi}_s(x) = \sum_k \hat{b}_k^\dagger \hat{b}_k = \sum_{\varepsilon_k\le -m} \hat{b}_k^\dagger \hat{b}_k + \sum_{-m<\varepsilon_k} \hat{b}_k^\dagger \hat{b}_k\,. \tag{F.6}$$

A naive application of Fermi statistics would require that all levels below the Fermi energy ε_F are occupied for the ground state $|\Phi_s\rangle$ of this system. This would imply that, in addition to a finite number of discrete levels between $-m$ and ε_F, all negative energy states are filled,

$$|\Phi_s\rangle = \prod_{\varepsilon_k\le\varepsilon_F} \hat{b}_k^\dagger |0_s'\rangle\,, \tag{F.7}$$

where $|0'_s\rangle$ denotes the vacuum with the property that $\hat{b}_k|0'_s\rangle = 0$ for all k. Obviously, $\hat{H}_s$ is not bounded from below for this kind of state and the charge $\langle\Phi_s|\hat{Q}_s|\Phi_s\rangle$ diverges.

The well-established solution to this problem is the reinterpretation of the negative energy states as unoccupied antiparticle states with positive energy $-\varepsilon_k$. The annihilation of a particle with $\varepsilon_k \leq -m$ via $\hat{b}_k$ then has to be understood as the creation of an antiparticle and vice versa, which is reflected by a redefinition of the negative energy annihilation and creation operators,

$$\hat{d}_k := \hat{b}_k^\dagger\,, \qquad \hat{d}_k^\dagger := \hat{b}_k \qquad \forall k \text{ with } \varepsilon_k \leq -m\,. \tag{F.8}$$

Equations (F.3), (F.4) then take on the forms

$$\left\{\hat{b}_k,\hat{b}_l\right\} = \left\{\hat{b}_k^\dagger,\hat{b}_l^\dagger\right\} = \left\{\hat{d}_k,\hat{d}_l\right\} = \left\{\hat{d}_k^\dagger,\hat{d}_l^\dagger\right\} = \left\{\hat{d}_k^{(\dagger)},\hat{b}_l^{(\dagger)}\right\} = 0 \tag{F.9}$$

$$\left\{\hat{b}_k,\hat{b}_l^\dagger\right\} = \left\{\hat{d}_k,\hat{d}_l^\dagger\right\} = \delta_{kl} \tag{F.10}$$

$$\hat{\psi}_s(x) = \sum_{\varepsilon_k\leq -m} \hat{d}_k^\dagger\phi_k(\boldsymbol{x})e^{-i\varepsilon_k t} + \sum_{-m<\varepsilon_k} \hat{b}_k\phi_k(\boldsymbol{x})e^{-i\varepsilon_k t}\,, \tag{F.11}$$

so that $\hat{\psi}_s(x)$ now annihilates a unit of charge at point $\boldsymbol{x}$ and time t, rather than a particle. The vacuum must be redefined accordingly,

$$\hat{b}_k|0_s\rangle = 0 \quad \forall\, \varepsilon_k > -m\,, \qquad \hat{d}_k|0_s\rangle = 0 \quad \forall\, \varepsilon_k \leq -m\,, \tag{F.12}$$

in order to ensure that neither a particle nor an antiparticle is present in the state $|0_s\rangle$. The ground state of the N-particle system is then simply given by N particles added to this vacuum,

$$|\Phi_s\rangle = \prod_{-m<\varepsilon_k\leq\varepsilon_F} \hat{b}_k^\dagger|0_s\rangle\,. \tag{F.13}$$

Insertion of (F.8) into the Hamiltonian yields

$$\begin{aligned}\hat{H}_s &= \sum_{\varepsilon_k\leq -m} \varepsilon_k\hat{d}_k\hat{d}_k^\dagger + \sum_{-m<\varepsilon_k} \varepsilon_k\hat{b}_k^\dagger\hat{b}_k \\ &= \sum_{\varepsilon_k\leq -m} (-\varepsilon_k)\hat{d}_k^\dagger\hat{d}_k + \sum_{-m<\varepsilon_k} \varepsilon_k\hat{b}_k^\dagger\hat{b}_k + \sum_{\varepsilon_k\leq -m} \varepsilon_k\,,\end{aligned} \tag{F.14}$$

so that one finds as ground state and vacuum energies,

$$\langle\Phi_s|\hat{H}_s|\Phi_s\rangle = \sum_{\varepsilon_k\leq\varepsilon_F} \varepsilon_k\,, \qquad \langle 0_s|\hat{H}_s|0_s\rangle = \sum_{\varepsilon_k\leq -m} \varepsilon_k\,. \tag{F.15}$$

Due to the simultaneous redefinition of the negative energy states and the vacuum, Eq. (F.12), the Hamiltonian is still not bounded from below. Its boundedness must be implemented by a renormalization of the energy scale, i.e. by explicit subtraction of the vacuum expectation value of $\hat{H}_s$,

$$\hat{H}'_{\mathrm{s,R}} := \hat{H}_{\mathrm{s}} - \langle 0_{\mathrm{s}}|\hat{H}_{\mathrm{s}}|0_{\mathrm{s}}\rangle \qquad \Longrightarrow \qquad \langle\Phi_{\mathrm{s}}|\hat{H}'_{\mathrm{s,R}}|\Phi_{\mathrm{s}}\rangle = \sum_{-m<\varepsilon_k\leq\varepsilon_{\mathrm{F}}} \varepsilon_k \,. \tag{F.16}$$

For this renormalized Hamiltonian one can then immediately establish a minimum principle, since any admixture of a single-particle state above ε_{F} to the N-particle state leads to a well-defined energy which is higher than $\langle\Phi_{\mathrm{s}}|\hat{H}'_{\mathrm{s,R}}|\Phi_{\mathrm{s}}\rangle$. Moreover, if one adds an antiparticle to $|\Phi_{\mathrm{s}}\rangle$, the resulting energy is at least m (i.e. mc^2) above the ground state energy (in the noninteracting theory a particle-antiparticle pair cannot annihilate). The same procedure is applied to $\hat{Q}_{\mathrm{s}}$,

$$\hat{Q}'_{\mathrm{s,R}} := \hat{Q}_{\mathrm{s}} - \langle 0_{\mathrm{s}}|\hat{Q}_{\mathrm{s}}|0_{\mathrm{s}}\rangle = -\sum_{\varepsilon_k\leq -m} \hat{d}_k^\dagger \hat{d}_k + \sum_{-m<\varepsilon_k} \hat{b}_k^\dagger \hat{b}_k \tag{F.17}$$

$$\Longrightarrow \langle\Phi_{\mathrm{s}}|\hat{Q}'_{\mathrm{s,R}}|\Phi_{\mathrm{s}}\rangle = \sum_{-m<\varepsilon_k\leq\varepsilon_{\mathrm{F}}} 1 \,, \tag{F.18}$$

which directly illustrates the opposite charges of particles and antiparticles. The subtraction of the vacuum expectation values in (F.16) and (F.17) is equivalent to a normal-ordering of the creation/annihilation operators in $\hat{H}_{\mathrm{s}}$ and $\hat{Q}_{\mathrm{s}}$.

The operators (F.16) and (F.17) are finite at this point, but they do not yet show the correct behavior under charge conjugation. Each individual field operator (F.11) transforms as [530, 531]

$$\hat{\psi}^c_{\mathrm{s}}(x) := \mathscr{C}\hat{\psi}_{\mathrm{s}}(x)\mathscr{C}^\dagger = \eta_c C \hat{\overline{\psi}}{}^T_{\mathrm{s}}(x) \,, \qquad C = i\gamma^2\gamma^0 \tag{F.19}$$

(T =transposition) with an unobservable phase η_c, so that charge conjugation reorders the field operators in the current density,

$$\begin{aligned}\mathscr{C}\hat{\overline{\psi}}_{\mathrm{s}}(x)\gamma^\mu\hat{\psi}_{\mathrm{s}}(x)\mathscr{C}^\dagger &= \sum_{a,b,c=1}^{4} \hat{\psi}_{s,a}(x)\gamma^\mu_{ba}\gamma^0_{cb}\hat{\psi}^\dagger_{s,c}(x) \\ &= \left[\gamma^0\gamma^\mu\hat{\psi}_{\mathrm{s}}(x)\right]^T\left[\hat{\psi}^\dagger_{\mathrm{s}}(x)\right]^T .\end{aligned} \tag{F.20}$$

The proper transformation behavior of the current density operator, Eq. (8.23), thus requires the presence of both possible operator orderings, which leads to the anticommutator form (8.22). For the charge operator one then obtains

$$\hat{Q}_{\mathrm{s}} = \frac{1}{2}\int d^3x\,[\hat{\psi}^\dagger_{\mathrm{s}}(x),\hat{\psi}_{\mathrm{s}}(x)] = \frac{1}{2}\left\{\sum_{\varepsilon_k\leq -m}\left[\hat{d}_k,\hat{d}_k^\dagger\right] + \sum_{-m<\varepsilon_k}\left[\hat{b}_k^\dagger,\hat{b}_k\right]\right\}. \tag{F.21}$$

This more appropriate form of $\hat{Q}_{\mathrm{s}}$ also leads to a more symmetric form of the counterterm $\langle 0_{\mathrm{s}}|\hat{Q}_{\mathrm{s}}|0_{\mathrm{s}}\rangle$ in the renormalized charge operator $\hat{Q}''_{\mathrm{s,R}}$,

$$\hat{Q}''_{\mathrm{s,R}} = \hat{Q}_{\mathrm{s}} - \langle 0_{\mathrm{s}}|\hat{Q}_{\mathrm{s}}|0_{\mathrm{s}}\rangle \qquad \langle 0_{\mathrm{s}}|\hat{Q}_{\mathrm{s}}|0_{\mathrm{s}}\rangle = \frac{1}{2}\left\{\sum_{\varepsilon_k\leq -m} 1 - \sum_{-m<\varepsilon_k} 1\right\}. \tag{F.22}$$

In the case of the homogeneous vacuum with $V^\mu = 0$ each state with energy $\varepsilon_k \geq +m$ has a unique counterpart with energy $\varepsilon_k \leq -m$, so that the sums on the right hand side of (F.22) cancel each other and $\langle 0_s|\hat{Q}_s|0_s\rangle$ vanishes.

One can proceed in a similar way for the Hamiltonian,

$$\hat{H}_s = \frac{1}{2}\int d^3x \left[\hat{\psi}_s^\dagger(x), \left(-i\boldsymbol{\alpha}\cdot\boldsymbol{\nabla} + \beta m - e\alpha_\mu V^\mu(\boldsymbol{x})\right)\hat{\psi}_s(x)\right]$$
$$= \frac{1}{2}\left\{\sum_{\varepsilon_k \leq -m} \varepsilon_k \left[\hat{d}_k, \hat{d}_k^\dagger\right] + \sum_{-m<\varepsilon_k} \varepsilon_k \left[\hat{b}_k^\dagger, \hat{b}_k\right]\right\}. \tag{F.23}$$

As the vacuum expectation value does not vanish,

$$\langle 0_s|\hat{H}_s|0_s\rangle = \frac{1}{2}\left\{\sum_{\varepsilon_k \leq -m} \varepsilon_k - \sum_{-m<\varepsilon_k} \varepsilon_k\right\}, \tag{F.24}$$

the renormalized Hamiltonian is not identical with $\hat{H}_s$,

$$\hat{H}''_{s,R} := \hat{H}_s - \langle 0_s|\hat{H}_s|0_s\rangle = -\sum_{\varepsilon_k \leq -m} \varepsilon_k \hat{d}_k^\dagger \hat{d}_k + \sum_{-m<\varepsilon_k} \varepsilon_k \hat{b}_k^\dagger \hat{b}_k . \tag{F.25}$$

The operator (F.25) measures the energy of a given state $|\Phi\rangle$ with respect to the vacuum $|0_s\rangle$ in the presence of the external potential. In the noninteracting situation these energy differences correspond directly to the "observable" ionization energies. However, the operator (F.25) does not yet reflect the fact that the vacuum energies resulting from different external potentials are not identical (Casimir effect). The differences between vacua corresponding to different V_μ are most easily seen on a local scale: the vacuum expectation value of the current density operator (8.22) reads

$$\langle 0_s|\hat{j}^\mu(x)|0_s\rangle = \frac{1}{2}\left\{\sum_{\varepsilon_k \leq -m} \overline{\phi}_k(\boldsymbol{x})\gamma^\mu \phi_k(\boldsymbol{x}) - \sum_{-m<\varepsilon_k} \overline{\phi}_k(\boldsymbol{x})\gamma^\mu \phi_k(\boldsymbol{x})\right\}. \tag{F.26}$$

While the net charge of the vacuum is zero, (F.26) shows the local polarization of the vacuum by the external potential. The corresponding energy difference becomes relevant as soon as the total energies associated with different external potentials are to be compared, as in the case of the HK theorem or the KS selfconsistency procedure. For such comparisons one needs a universal vacuum energy standard for which one chooses the vacuum $|0_0\rangle$ of the noninteracting system with $V^\mu = 0$, i.e. the homogeneous vacuum with $\langle 0_0|\hat{j}^\mu(x)|0_0\rangle = 0$,

$$\hat{b}_{0,k}|0_0\rangle = 0 \quad \forall\, \varepsilon_k > -m\,, \qquad \hat{d}_{0,k}|0_0\rangle = 0 \quad \forall\, \varepsilon_k \leq -m\,. \tag{F.27}$$

The corresponding field operator will be denoted by $\hat{\psi}_0$,

$$\hat{\psi}_0(x) = \sum_{\varepsilon_k \leq -m} \hat{d}_{0,k}^\dagger \phi_{0,k}(\boldsymbol{x})e^{-i\varepsilon_{0,k}t} + \sum_{-m<\varepsilon_k} \hat{b}_{0,k}\phi_{0,k}(\boldsymbol{x})e^{-i\varepsilon_{0,k}t}\,, \tag{F.28}$$

where the single-particle orbitals $\phi_{0,k}$ and eigenvalues $\varepsilon_{0,k}$ are the standard free plane-wave spinors and energies [531]. The final renormalized Hamiltonian is defined as

$$\hat{H}_{s,R} := \hat{H}_s - \langle 0_0|\hat{H}_0|0_0\rangle = \hat{H}''_{s,R} + \langle 0_s|\hat{H}_s|0_s\rangle - \langle 0_0|\hat{H}_0|0_0\rangle \tag{F.29}$$

$$\hat{H}_0 = \frac{1}{2}\int d^3x \left[\hat{\psi}_0^\dagger(x), \Big(-i\boldsymbol{\alpha}\cdot\boldsymbol{\nabla} + \beta m\Big)\hat{\psi}_0(x)\right]. \tag{F.30}$$

Unfortunately, there is a price to be paid for this universal definition of the energy scale. While the expectation values of (F.25) are automatically finite, the same is not true for (F.29). In order to understand the mechanism which leads to divergences, let us consider the energy of the perturbed vacuum with respect to the homogeneous vacuum (often called Casimir energy) within perturbation theory. The basic elements of the perturbation expansion are the Green's function of the perturbed vacuum,

$$iG_v^s(x,y) = \langle 0_s|T\,\hat{\psi}_s(x)\hat{\overline{\psi}}_s(y)|0_s\rangle \tag{F.31}$$

$$= \Theta(x^0-y^0)\sum_{-m<\varepsilon_k}\phi_k(\boldsymbol{x})\overline{\phi}_k(\boldsymbol{y})e^{-i\varepsilon_k(x^0-y^0)}$$

$$-\Theta(y^0-x^0)\sum_{\varepsilon_k\leq -m}\phi_k(\boldsymbol{x})\overline{\phi}_k(\boldsymbol{y})e^{-i\varepsilon_k(x^0-y^0)}, \tag{F.32}$$

and its unperturbed counterpart G_v^0,

$$iG_v^0(x,y) = \langle 0_0|T\,\hat{\psi}_0(x)\hat{\overline{\psi}}_0(y)|0_0\rangle \tag{F.33}$$

(the explicit form of G_v^0 is identical to (F.32) with ϕ_k and ε_k replaced by $\phi_{0,k}$ and $\varepsilon_{0,k}$). With these Green's functions the energy of the perturbed vacuum can be expressed as [752]

$$\langle 0_s|\hat{H}_{s,R}|0_s\rangle = -i\int d^3x \lim_{y\to x}{}_s \operatorname{tr}\Big[\Big(-i\boldsymbol{\gamma}\cdot\boldsymbol{\nabla} + m + e\not{V}(\boldsymbol{x})\Big)G_v^s(x,y)\Big]$$

$$+i\int d^3x \lim_{y\to x}{}_s \operatorname{tr}\Big[\Big(-i\boldsymbol{\gamma}\cdot\boldsymbol{\nabla} + m\Big)G_v^0(x,y)\Big], \tag{F.34}$$

where the symmetric limit,

$$\lim_{y\to x}{}_s \equiv \frac{1}{2}\left(\lim_{y\to x,y^0>x^0} + \lim_{y\to x,y^0<x^0}\right)\Bigg|_{(x-y)^2\geq 0}, \tag{F.35}$$

is a consequence of the anticommutator form of $\hat{H}_s$. Similarly, one can write the current density of the perturbed vacuum as

$$\langle 0_s|\hat{j}^\mu(x)|0_s\rangle = -i\lim_{y\to x}{}_s \operatorname{tr}[G_v^s(x,y)\gamma^\mu]. \tag{F.36}$$

For further analysis one can utilize a perturbation expansion of $G^{s}_{v}(x,y)$ in powers of the external potential, which is easily written down in terms of Feynman diagrams,

$G^{s}_{v} =$ 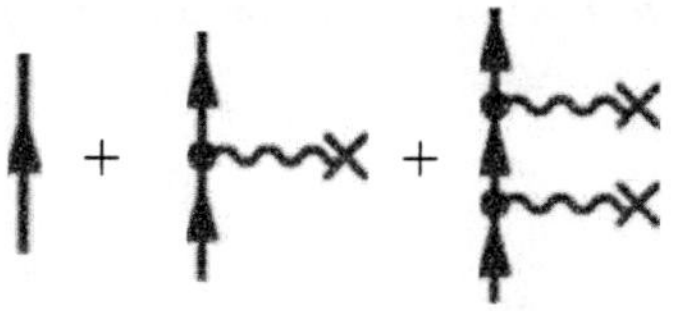 $+ \cdots .$ (F.37)

Here the solid line represents G^{0}_{v}, the wavy line the external potential and the dot denotes the vertex, i.e. in real space one has[1] (including labels)

$iG^{0}_{v,ab}(x,y) = y,b$ x,a (F.38)

$V^{\mu}(\boldsymbol{x}) = \boldsymbol{x},\mu$  (F.39)

$ie\gamma^{\mu}_{ab} =$ 

, (F.40)

where z represents the coordinates of the two Green's functions and the potential linked at the vertex (F.40), a,b are the spinor indices of the Green's functions and μ is the Minkowski index of the potential. As usual, integration over the coordinates and summation over the spinor and Minkowski indices attached to all internal vertices is implied in all composite diagrams as those of Eq. (F.37). After insertion of (F.37) into (F.34) and (F.36), one realizes that the symmetric limit induces so-called loop diagrams, as e.g.

$ie\langle 0_s|\hat{j}^{\mu}(x)|0_s\rangle =$ 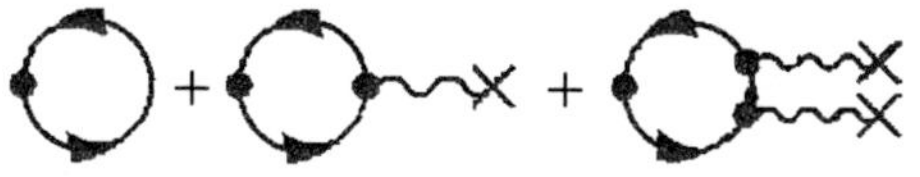 $+ \ldots .$ (F.41)

The evaluation of these expressions involves a loop-integration, either in real space or, after Fourier transformation, in momentum space. While the first and third diagram in (F.41) vanish (Furry's theorem), one identifies the second loop as the lowest order contribution to the vacuum polarization function (irreducible 2-point function) of standard vacuum QED (i.e. interacting fermions without external potential). This function is ultraviolet (UV) divergent, i.e. the loop integration diverges for large four momenta, when performed in momentum space. This introduces an UV-divergence in the current density and energy of the perturbed noninteracting vacuum: within a perturbative treatment it does not matter whether the external potential or the quantized photon field creates virtual electron–positron pairs. As a consequence, a UV renormalization procedure is required to keep $\langle 0_s|\hat{H}_{s,R}|0_s\rangle$ and $\langle 0_s|\hat{j}^{\mu}(x)|0_s\rangle$

[1] Note that these Feynman rules follow the relativistic standard [531], rather than the nonrelativistic standard. The choice (F.38)–(F.40) avoids that additional factors of i have to be assigned to a diagram by some explicit rule.

finite. Fortunately, the corresponding counterterms are completely determined by the renormalization scheme for the Green's functions of interacting vacuum QED without external potential. We are thus led to consider standard QED in some detail, which is the subject of the next Appendix.

Appendix G
Renormalization Scheme of Vacuum QED

In this Appendix we review the renormalization scheme of vacuum QED without external potential, i.e. of the Lagrangian (8.11) with

$$V^{\mu}(\boldsymbol{x}) = 0 \tag{G.1}$$

and the system being in the state with zero net charge. This summary not only serves as an introduction of the basic concepts of UV renormalization, as e.g. the counterterm technique, but also provides a number of explicit results used in the Appendices H and I. In fact, all counterterms required for inhomogeneous systems with non-vanishing current density can be extracted from the study of the vacuum Green's functions of QED without external potential: neither the presence of a perturbing external potential nor that of bound electrons introduces any new feature or new parameter. An explicit illustration of this statement has already been given in Eq. (F.41) for the case of a noninteracting inhomogeneous system. Further examples will turn up in the analysis of the homogeneous and the weakly inhomogeneous electron gas in the Appendices H and I, which also provide the background for the discussion of the existence theorem of relativistic DFT in Sect. 8.3. For brevity, we use $\hbar = c = 1$ in this and the other Appendices dealing with relativistic many-body theory.

The basic vacuum Green's functions to which we restrict the subsequent discussion are the fermion and photon propagators as well as the (reducible) vertex function,

$$G_{\mathrm{V}}(x,y) = -i\langle 0|T\hat{\psi}(x)\hat{\overline{\psi}}(y)|0\rangle \tag{G.2}$$

$$D_{\mathrm{V}}^{\mu\nu}(x,y) = -i\langle 0|T\hat{A}^{\mu}(x)\hat{A}^{\nu}(y)|0\rangle \tag{G.3}$$

$$G_{\mathrm{V}}^{(2,1)\mu}(x,y,z) = -\langle 0|T\hat{\psi}(x)\hat{\overline{\psi}}(y)\hat{A}^{\mu}(z)|0\rangle\,, \tag{G.4}$$

where $|0\rangle$ denotes the vacuum of the interacting theory. Note that we have defined the photon propagator without any prefactor of e^2, which is most suitable for the discussion of renormalization (but differs from the definitions (8.83) and (H.8) which are more adequate for the discussion of RDFT).

The standard approach to the calculation of such Green's functions is perturbation theory with respect to the electron–electron coupling constant $\alpha = e^2/(\hbar c)$. This procedure results in an expansion of the vacuum Green's functions of the interacting system in terms of vacuum expectation values of the noninteracting field operators $\hat{\psi}_0$ and $\hat{A}_0^\mu$ (see e.g. [531]). For instance, for the electron propagator one obtains

$$G_v(x-y) = -i\frac{\langle 0_0|T\,\hat{\psi}_0(x)\hat{\overline{\psi}}_0(y)\exp[ie\int d^4z\,\hat{\overline{\psi}}_0(z)\hat{\not{A}}_0(z)\hat{\psi}_0(z)]|0_0\rangle}{\langle 0_0|T\exp[ie\int d^4z\,\hat{\overline{\psi}}_0(z)\hat{\not{A}}_0(z)\hat{\psi}_0(z)]|0_0\rangle}\,, \tag{G.5}$$

where $|0_0\rangle$ is the noninteracting vacuum introduced in Eq. (F.27). In addition, the Feynman dagger notation,

$$\not{A} = A_\mu \gamma^\mu\,, \tag{G.6}$$

has been used. Analogous expressions are obtained for $D_v^{\mu\nu}$ and $G_v^{(2,1)\mu}$. The actual expansion of all vacuum expectation values of the type (G.5) in powers of e^2 is controlled by the Feynman rules resulting from the application of Wick's theorem. The basic ingredients are the noninteracting fermion propagator G_v^0, Eq. (F.33), the noninteracting photon propagator $D_{\mu\nu}^0$, Eq. (8.83), and the bare fermion-photon vertex (F.40) (together with the loop integrations and the fermion loop sign rule). It is this perturbative framework in which the concept of renormalization is usually formulated and we follow this standard.

Due to the translational invariance of QED without external potential, the situation is most conveniently analyzed in momentum space. The corresponding four-dimensional Fourier transforms can be written as

$$G_v(x-y) = \int \frac{d^4p}{(2\pi)^4}\, e^{-ip\cdot(x-y)}\, G_v(p) \tag{G.7}$$

$$D_v^{\mu\nu}(x-y) = \int \frac{d^4q}{(2\pi)^4}\, e^{-iq\cdot(x-y)}\, D_v^{\mu\nu}(q) \tag{G.8}$$

$$G_{v,\mu}^{(2,1)}(x,y,z) = \int \frac{d^4p}{(2\pi)^4}\frac{d^4k}{(2\pi)^4}\, e^{-ip\cdot(x-z)-ik\cdot(z-y)}\, G_{v,\mu}^{(2,1)}(p,k)\,. \tag{G.9}$$

This leads to loop integrations over four momenta rather than space-time coordinates (as in (G.5)), with four momentum conservation at the vertices. In momentum space the noninteracting propagator G_v^0, Eq. (F.33), is given by

$$iG_v^0(p) = i\frac{\not{p}+m}{p^2-m^2+i\eta} = \tag{G.10}$$

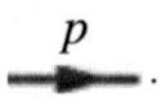

As discussed in Sect. 8.2 the form of the free photon propagator $D_{\mu\nu}^0(q)$ depends on the choice of gauge. In Sects. 8.3–8.7 Feynman gauge ($\lambda = 1$) is used, for which $D_{\mu\nu}^0$ is explicitly given by

$$D^0_{\mu\nu}(q) = D^0(q^2)\, g_{\mu\nu} \tag{G.11}$$

$$D^0(q^2) = \frac{-4\pi}{q^2 + i\eta} . \tag{G.12}$$

For the present purpose, however, Landau gauge is more adequate, which corresponds to the choice $\lambda = \infty$,

$$iD^0_{\mu\nu}(q) = i\left(g_{\mu\nu} - \frac{q_\mu q_\nu}{q^2}\right) D^0(q^2) = \tag{G.13}$$

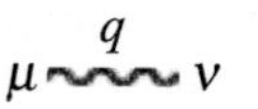

It seems worthwhile to emphasize that all covariant gauges can be handled by the same basic renormalization scheme. As $\langle 0_0|T\,\hat{\psi}_0(x)\hat{\overline{\psi}}_0(y)\hat{A}^\mu_0(z)|0_0\rangle = 0$, the first non-vanishing contribution to $G^{(2,1)\mu}_{\mathrm{v}}$ is found in first order of e,

$$G^{(2,1)}_{\mathrm{v},\mu}(p,k) = -D^0_{\mu\nu}(p-k)G^0_{\mathrm{v}}(p)e\gamma^\nu G^0_{\mathrm{v}}(k) . \tag{G.14}$$

Its core element is the simple vertex (F.40) in momentum space,

$$ie\gamma^\mu\,(2\pi)^4\,\delta^{(4)}(p-k-q) = \tag{G.15}$$

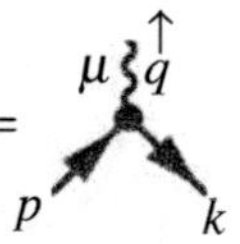

(the arrow above q indicates that this four momentum is supposed to be outgoing from the vertex, so that the argument of the δ-function is the sum over all incoming four momenta—in momentum space a direction has to be assigned to each interaction line).

In the first step of the analysis the relevant Green's functions are expressed in terms of their irreducible kernels [753], the electron self-energy $\Sigma_{\mathrm{v}}(p)$, the vacuum polarization tensor $\omega_{\mathrm{v},\mu\nu}(q)$ and the irreducible vertex function $\Gamma_{\mathrm{v},\mu}(p,k)$. The connection between these quantities is provided by Dyson equations (see e.g. [531]),

$$G_{\mathrm{v}}(p) = G^0_{\mathrm{v}}(p) + G^0_{\mathrm{v}}(p)\,\Sigma_{\mathrm{v}}(p)\,G_{\mathrm{v}}(p) \tag{G.16}$$

$$D_{\mathrm{v},\mu\nu}(q) = D^0_{\mu\nu}(q) + D^0_{\mu\rho}(q)\,\omega^{\rho\lambda}_{\mathrm{v}}(q)\,D_{\mathrm{v},\lambda\nu}(q) \tag{G.17}$$

$$G^{(2,1)}_{\mathrm{v},\mu}(p,k) = -\,e D_{\mathrm{v},\mu\nu}(p-k)\,G_{\mathrm{v}}(p)\,\left[\gamma^\nu + \Gamma^\nu_{\mathrm{v}}(p,k)\right]\,G_{\mathrm{v}}(k) . \tag{G.18}$$

The relations (G.16)–(G.18) separate the nontrivial higher order contributions in the perturbation expansions from trivial multiples of lower order terms, thus isolating the essential information contained in the Green's functions. These relations become particularly simple if (G.16) is rewritten in terms of inverse propagators,

$$G_{\mathrm{v}}(p)^{-1} = G^0_{\mathrm{v}}(p)^{-1} - \Sigma_{\mathrm{v}}(p) = \not{p} - m - \Sigma_{\mathrm{v}}(p) , \tag{G.19}$$

and if the tensor structure of $\omega^{\mu\nu}_{\mathrm{v}}(q)$,

$$\omega_{\mathrm{v}}^{\mu\nu}(q) = \left(q^2 g^{\mu\nu} - q^\mu q^\nu\right) \omega_{\mathrm{v}}(q^2) , \tag{G.20}$$

is used in (G.17),

$$D_{\mathrm{v}}^{\mu\nu}(q) = \left(g^{\mu\nu} - \frac{q^\mu q^\nu}{q^2}\right) \frac{-4\pi}{q^2[1+4\pi\omega_{\mathrm{v}}(q^2)]} . \tag{G.21}$$

The renormalization program of QED starts with and is most easily illustrated for the first order. The lowest order contributions to the three relevant irreducible 2- and 3-point functions read

$$-i\Sigma_{\mathrm{v}}^{(1)}(p) =$$

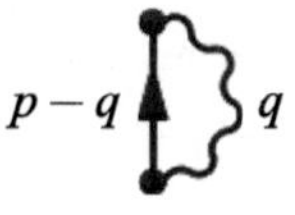

$$= e^2 \int \frac{d^4q}{(2\pi)^4} D^0_{\mu\nu}(q)\, \gamma^\mu\, G^0_{\mathrm{v}}(p-q)\, \gamma^\nu \tag{G.22}$$

$$-i\omega^{(1)}_{\mathrm{v},\mu\nu}(q) =$$

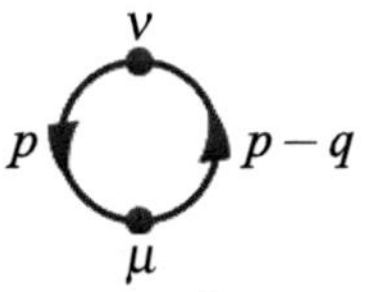

$$= -e^2 \int \frac{d^4p}{(2\pi)^4} \operatorname{tr}\left[\gamma_\mu\, G^0_{\mathrm{v}}(p)\, \gamma_\nu\, G^0_{\mathrm{v}}(p-q)\right] \tag{G.23}$$

$$ie\Gamma^{(1)}_{\mathrm{v},\mu}(p,k) =$$

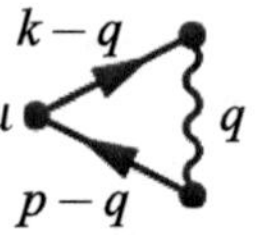

$$= -e^3 \int \frac{d^4q}{(2\pi)^4} D^0_{\rho\nu}(q) \gamma^\rho G^0_{\mathrm{v}}(p-q) \gamma_\mu G^0_{\mathrm{v}}(k-q) \gamma^\nu . \tag{G.24}$$

Insertion of (G.10), (G.13) shows that these integrals diverge for large loop (four) momentum. As these divergences result from the high energy regime they are called UV-divergences—for brevity we ignore all problems related to the infrared (low energy) regime. In order to establish a well-defined theory one first of all needs a regularization scheme which suppresses these divergences at all intermediate steps of the evaluation. Of course, this regularization must preserve the complete structure of the theory, in particular the Ward-Takahashi identities, which link the irreducible kernels (see e.g. [531]), as for example

$$(p_\mu - p'_\mu)\Gamma^\mu_{\mathrm{v}}(p,p') = \Sigma_{\mathrm{v}}(p') - \Sigma_{\mathrm{v}}(p) \tag{G.25}$$

$$\Longrightarrow \qquad \Gamma_{\mathrm{v}}^{\mu}(p,p) = -\frac{\partial}{\partial p_\mu}\Sigma_{\mathrm{v}}(p)\,. \tag{G.26}$$

For the present discussion we use dimensional regularization [754], in which all integrals of the type (G.22)–(G.24) are evaluated in a reduced number of d dimensions, rather than the 4-dimensional Minkowski space (after Wick rotation in order to obtain integrals in Euclidean space—the details of this scheme are not relevant at this point). The results evaluated for integer d are then analytically continued to non-integer d, which then allows their extension to the physically interesting limit $d \to 4$. Using the abbreviation $\Delta = (4-d)/2$ one finds for the integrals (G.22)–(G.24)

$$\Sigma_{\mathrm{v}}^{(1)}(p) = \frac{e^2}{4\pi}\Gamma(\Delta)(-\not{p}+4m) + \Sigma_{\mathrm{v,finite}}^{(1)}(p) \tag{G.27}$$

$$\omega_{\mathrm{v},\mu\nu}^{(1)}(q) = (q^2 g_{\mu\nu} - q_\mu q_\nu)\left(\frac{e^2}{12\pi^2}\Gamma(\Delta) + \omega_{\mathrm{v,finite}}^{(1)}(q^2)\right) \tag{G.28}$$

$$\Gamma_{\mathrm{v},\mu}^{(1)}(p,k) = \frac{e^2}{4\pi}\Gamma(\Delta)\gamma_\mu + \Gamma_{\mathrm{v,finite},\mu}^{(1)}(p,k)\,. \tag{G.29}$$

Here $\Gamma(\Delta)$ denotes Euler's Γ-function, in which the UV-divergences have been isolated,

$$\Gamma(\Delta) \overset{\Delta\to 0}{\longrightarrow} \frac{1}{\Delta} + \dots\;,$$

and $\Sigma_{\mathrm{v,finite}}^{(1)}$, $\omega_{\mathrm{v,finite}}^{(1)}$ and $\Gamma_{\mathrm{v,finite},\mu}^{(1)}$ represent the finite parts of the irreducible kernels in which the limit $d \to 4$ can be taken directly (the detailed form of $\Sigma_{\mathrm{v,finite}}^{(1)}$, $\omega_{\mathrm{v,finite}}^{(1)}$ and $\Gamma_{\mathrm{v,finite},\mu}^{(1)}$ is not relevant in this context). The UV-divergences now manifest themselves as simple poles in the deviation of the space-time dimensionality from $d = 4$. On the other hand, all other irreducible n-point functions are finite from the very outset (to first order).

The next step is the actual renormalization procedure. The crucial observation for both the physical interpretation as well as the technical success of this step is the fact that the divergent contributions to the three relevant functions have the same structure as the corresponding free propagators and the free vertex: the divergent part of $\Sigma_{\mathrm{v}}^{(1)}$ is just proportional to $\not{p}$ and m, but not e.g. to p^2, the divergent part of $\omega_{\mathrm{v},\mu\nu}^{(1)}$ has the same tensor structure as $D_{\mu\nu}^{0}$, Eq. (G.13), and the divergent part of $\Gamma_{\mathrm{v},\mu}^{(1)}$ is proportional to the free vertex γ_μ (but does not depend on momentum). For this reason the divergences can be absorbed into a redefinition of the constants m and e as well as a modified normalization of the field operators in the original Lagrangian. Given the form of this original, unrenormalized Lagrangian, $\mathscr{L}_{\mathrm{unren}}(\hat{\psi},\hat{A}^\mu,m,e)$, the renormalized Lagrangian is usually written as

$$\mathscr{L}_{\mathrm{R}} = \mathscr{L}_{\mathrm{unren}}\left(\sqrt{Z_2}\hat{\psi},\sqrt{Z_3}\hat{A}^\mu, m-\delta m, \frac{Z_1 e}{Z_2\sqrt{Z_3}}\right)\,. \tag{G.30}$$

The renormalization constants Z_1, Z_2, Z_3 and δm have to be interpreted as functions of the finite true physical charge e and mass m of the electrons. The relation between these quantities remains to be determined order by order in the perturbation series. In other words: the original fields and parameters in $\mathscr{L}_{\rm unren}$ are no longer interpreted as the correct physical fields and parameters, but rather as bare, unrenormalized quantities,

$$\hat{\psi}_b(x) = \sqrt{Z_2}\,\hat{\psi}(x) \tag{G.31}$$

$$\hat{A}_b^\mu(x) = \sqrt{Z_3}\,\hat{A}^\mu(x) \tag{G.32}$$

$$e_b = \frac{Z_1}{Z_2\sqrt{Z_3}}\,e \tag{G.33}$$

$$m_b = m - \delta m\,, \tag{G.34}$$

so that the renormalized Lagrangian can be reformulated in terms of the bare quantities,

$$\mathscr{L}_{\rm R} = \mathscr{L}_{\rm unren}\big(\hat{\psi}_b, \hat{A}_b^\mu, m_b, e_b\big)\,. \tag{G.35}$$

The structure of the theory, which e.g. expresses itself in Dyson equations and Ward-Takahashi identities, remains completely unchanged, due to the form invariance of the Lagrangian under the renormalization prescription. The renormalized Green's functions, i.e. the vacuum expectation values of $\hat{\psi}$ and $\hat{A}^\mu$, are now obtained as

$$\begin{aligned}
G_{\rm v,R}(x,y) &= -iZ_2^{-1}\langle 0|T\,\hat{\psi}_b(x)\hat{\overline{\psi}}_b(y)|0\rangle \\
&= Z_2^{-1}G_{\rm v}(x,y,e_b,m_b) && \text{(G.36)}\\
D_{\rm v,R}^{\mu\nu}(x,y) &= -iZ_3^{-1}\langle 0|T\hat{A}_b^\mu(x)\hat{A}_b^\nu(y)|0\rangle \\
&= Z_3^{-1}D_{\rm v}^{\mu\nu}(x,y,e_b,m_b) && \text{(G.37)}\\
G_{\rm v,R}^{(2,1)\mu}(x,y,z) &= -Z_2^{-1}Z_3^{-1/2}\langle 0|T\,\hat{\psi}_b(x)\hat{\overline{\psi}}_b(y)\hat{A}_b^\mu(z)|0\rangle \\
&= Z_2^{-1}Z_3^{-1/2}G_{\rm v}^{(2,1)\mu}(x,y,z,e_b,m_b)\,. && \text{(G.38)}
\end{aligned}$$

In these relations it has been indicated explicitly that the unrenormalized Green's functions resulting from the Lagrangian (G.35) initially depend on the bare parameters e_b and m_b.

In order to determine the unknown renormalization constants one needs some normalization conditions. These conditions result from the basic physical requirements for the Green's functions: in order to describe real fermions, which satisfy the dispersion relation $p^2 = (p^0)^2 - \boldsymbol{p}^2 = m^2$ with the finite experimental mass m in the presence of the virtual photon cloud, $G_{\rm v,R}(p)$ should reduce to the form of the free propagator $G_{\rm v}^0(p)$ with physical mass m for on-shell momentum $p^2 = m^2$, i.e. should have a simple pole with residue 1 for $p^2 = m^2$. If $\Sigma_{\rm v}$ is expressed as a function of $\not{p} - m$ (using $\not{p}^2 = p^2$) and the physical parameters,

$$\Sigma_{\mathrm{v}}(p, e_b, m_b) = \sum_{n=0}^{\infty} \Sigma_n(e,m)(\not{p} - m)^n \,, \tag{G.39}$$

the renormalized inverse propagator can be written as

$$G_{\mathrm{v,R}}^{-1}(p) = Z_2 \Bigg[\delta m - \Sigma_0(e,m) + (\not{p} - m)(1 - \Sigma_1(e,m)) \\ - \sum_{n=2}^{\infty} \Sigma_n(e,m)(\not{p} - m)^n \Bigg] \,. \tag{G.40}$$

For $G_{\mathrm{v,R}}(p)$ to have a simple pole at $p^2 = m^2$, however, one must have

$$\delta m = \Sigma_0(e,m) = \Sigma_{\mathrm{v}}(p, e_b, m_b)\Big|_{\not{p}=m} \tag{G.41}$$

$$Z_2 = (1 - \Sigma_1(e,m))^{-1} = \left(1 - \frac{d}{d\not{p}} \Sigma_{\mathrm{v}}(p, e_b, m_b)\Big|_{\not{p}=m}\right)^{-1} \,. \tag{G.42}$$

Only this choice guarantees that the higher order terms in the propagator itself are finite for $p^2 = m^2$. In fact, insertion of (G.19), (G.39), (G.41) and (G.42) into (G.36) yields

$$\begin{aligned}
G_{\mathrm{v,R}}(p) &= \left[Z_2 \left(G_{\mathrm{v}}^0(p, m_b)^{-1} - \Sigma_{\mathrm{v}}(p, e_b, m_b)\right)\right]^{-1} \\
&= \left[G_{\mathrm{v}}^0(p,m)^{-1} \left(1 - G_{\mathrm{v}}^0(p,m) Z_2 \sum_{n=2}^{\infty} \Sigma_n(e,m)(\not{p} - m)^n\right)\right]^{-1} \\
&= \sum_{k=0}^{\infty} \left(G_{\mathrm{v}}^0(p,m) Z_2 \sum_{n=2}^{\infty} \Sigma_n(e,m)(\not{p} - m)^n\right)^k G_{\mathrm{v}}^0(p,m) \\
&= \frac{\not{p} + m}{p^2 - m^2} + \frac{\not{p} + m}{p^2 - m^2} \left[Z_2 \sum_{n=2}^{\infty} \Sigma_n(e,m)(\not{p} - m)^n\right] \frac{\not{p} + m}{p^2 - m^2} + \ldots \,. \\
&= \frac{\not{p} + m}{p^2 - m^2} + \left[Z_2 \sum_{n=2}^{\infty} \Sigma_n(e,m)(\not{p} - m)^{n-2}\right] + \ldots \,.
\end{aligned}$$

Given the renormalized Green's functions, one can also define the corresponding renormalized irreducible kernels,

$$G_{\mathrm{v,R}}^{-1}(p) = \not{p} - m - \Sigma_{\mathrm{v,R}}(p) \,. \tag{G.43}$$

The additional contributions to the renormalized Green's functions resulting from renormalization are usually called counterterms. From Eqs. (G.40), (G.43) one extracts as counterterms to the self-energy,

$$\Sigma_{\mathrm{v,R}}(p) = \Sigma_{\mathrm{v}}(p) + \Delta\Sigma_{\mathrm{v}}(p) \,, \qquad \Delta\Sigma_{\mathrm{v}}(p) = -Z_2 \delta m + (1 - Z_2)(\not{p} - m) \,, \tag{G.44}$$

so that $\Sigma_{\mathrm{v,R}}$ satisfies the normalization conditions

$$\Sigma_{\mathrm{v,R}}(p)\Big|_{\not p=m}=0\,, \qquad \frac{d}{d\not p}\Sigma_{\mathrm{v,R}}(p)\Big|_{\not p=m}=0\,, \tag{G.45}$$

which should be interpreted in the spirit of the expansion (G.39) as

$$\Sigma_{0,\mathrm{R}}(e,m)=\Sigma_{1,\mathrm{R}}(e,m)=0\,.$$

One proceeds in the same fashion for the other two divergent functions. In the case of the massless photons the renormalized propagator must have a simple pole at $q^2=0$, which allows the determination of Z_3. Combination of (G.37) with (G.21) leads to

$$D^{\mu\nu}_{\mathrm{v,R}}(q) = \left(g^{\mu\nu}-\frac{q^\mu q^\nu}{q^2}\right)\frac{-4\pi}{q^2[1+4\pi\omega_{\mathrm{v,R}}(q^2)]}\,, \tag{G.46}$$

with $\omega_{\mathrm{v,R}}$ defined by

$$1+4\pi\omega_{\mathrm{v,R}}(q^2) = Z_3\left[1+4\pi\omega_{\mathrm{v}}(q^2,e_b,m_b)\right]\,. \tag{G.47}$$

The zero mass pole requirement is thus satisfied, if

$$Z_3 = \left[1+4\pi\omega_{\mathrm{v}}(q^2=0,e_b,m_b)\right]^{-1} \quad\Longleftrightarrow\quad \omega_{\mathrm{v,R}}(q^2=0)=0\,. \tag{G.48}$$

Consistent with Eq. (G.20), one then defines the renormalized irreducible polarization tensor as

$$\omega^{\mu\nu}_{\mathrm{v,R}}(q) = (q^2g^{\mu\nu}-q^\mu q^\nu)\,\omega_{\mathrm{v,R}}(q^2)\,. \tag{G.49}$$

Finally, the renormalized irreducible vertex function is defined via the Dyson equation (G.18),

$$G^{(2,1)}_{\mathrm{v,R},\mu}(p,k) = -eD_{\mathrm{R},\mu\nu}(p-k)G_{\mathrm{v,R}}(p)\left[\gamma^\nu+\Gamma^\nu_{\mathrm{v,R}}(p,k)\right]G_{\mathrm{v,R}}(k)\,, \tag{G.50}$$

using the renormalized propagators (G.36), (G.37). Combination of (G.50) with (G.38) and (G.33) then leads to

$$e\left[\gamma^\mu+\Gamma^\mu_{\mathrm{v,R}}(p,k)\right] = Z_1e\left[\gamma^\mu+\Gamma^\mu_{\mathrm{v}}(p,k,e_b,m_b)\right]\,, \tag{G.51}$$

which allows the formulation of a normalization condition for Z_1. On the mass shell, $\not p=m$, the vertex function must reduce to a pure vertex with physical charge e, in order to reproduce the Coulomb interaction for well separated electrons,

$$\gamma^\mu = Z_1\left(\gamma^\mu+\Gamma^\mu_{\mathrm{v}}(p,p,e_b,m_b)\Big|_{\not p=m}\right) \quad\Longleftrightarrow\quad \Gamma^\mu_{\mathrm{v,R}}(p,p)\Big|_{\not p=m}=0\,. \tag{G.52}$$

Using the Ward-Takahashi identity (G.26) and the expansion (G.39), one can explicitly verify that the resulting Z_1,

$$Z_1 = \left[1 + \frac{1}{4}\gamma_\mu \Gamma_{\rm v}^\mu(p,p,e_b,m_b)\Big|_{\not{p}=m}\right]^{-1}, \tag{G.53}$$

is identical with Z_2, Eq. (G.42).

As perturbation theory on the basis of (G.35) yields the irreducible functions in terms of the bare parameters, the expressions on the right-hand sides of (G.41), (G.42), (G.48) and (G.53) are obtained by use of (G.33) and (G.34), so that they themselves depend on the renormalization constants. Renormalization thus has to proceed in a recursive fashion, i.e. order by order in perturbation theory.

Explicit results are easily obtained for the first order. Use of (G.27), (G.40) gives

$$\begin{aligned} G_{\rm v,R}^{(1)}(p)^{-1} = Z_2\Bigg[\not{p}\left(1+\frac{e_b^2}{4\pi}\Gamma(\Delta)\right) - m_b\left(1+\frac{e_b^2}{\pi}\Gamma(\Delta)\right) \\ - \Sigma_{\rm v,finite}^{(1)}(p,e_b,m_b)\Bigg]. \end{aligned} \tag{G.54}$$

One now expands the right-hand side of (G.54) consistently to first order, using $Z_2 = 1 + Z_2^{(1)} + \ldots$ and $m_b = m - \delta m^{(1)} + \ldots$,

$$\begin{aligned} G_{\rm v,R}^{(1)}(p)^{-1} &= \delta m^{(1)} - \frac{3e^2}{4\pi}\Gamma(\Delta)m + (\not{p}-m)\left(1+\frac{e^2}{4\pi}\Gamma(\Delta)+Z_2^{(1)}\right) \\ &\quad - \Sigma_{\rm v,finite}^{(1)}(p,e,m). \end{aligned} \tag{G.55}$$

The conditions (G.41), (G.42) or, alternatively, (G.45), then give

$$\delta m^{(1)} = \frac{3e^2}{4\pi}\Gamma(\Delta)m + \Sigma_{\rm v,finite}^{(1)}(p,e,m)\Big|_{\not{p}=m} \tag{G.56}$$

$$Z_2^{(1)} = -\frac{e^2}{4\pi}\Gamma(\Delta) + \frac{d}{d\not{p}}\Sigma_{\rm v,finite}^{(1)}(p,e,m)\Big|_{\not{p}=m}. \tag{G.57}$$

From Eqs. (G.44), (G.56), (G.57) one extracts as first order counterterm to the self-energy,

$$\begin{aligned} \Delta\Sigma_{\rm v}^{(1)}(p) &= -\frac{3e^2}{4\pi}\Gamma(\Delta)m - \Sigma_{\rm v,finite}^{(1)}(p,e,m)\Big|_{\not{p}=m} \\ &\quad + \left[\frac{e^2}{4\pi}\Gamma(\Delta) - \frac{d}{d\not{p}}\Sigma_{\rm v,finite}^{(1)}(p,e,m)\Big|_{\not{p}=m}\right](\not{p}-m). \end{aligned} \tag{G.58}$$

Similarly, $Z_3 = 1 + Z_3^{(1)} + \ldots$ is determined by (G.28) and (G.48)

$$Z_3^{(1)} = -\frac{e^2}{3\pi}\Gamma(\Delta) \qquad \left(\omega_{\rm v,finite}^{(1)}(q^2=0)=0\right). \tag{G.59}$$

The associated counterterm for $\omega^{(1)}_{\mathrm{v},\mu\nu}$ results from (G.47) and (G.49),

$$\Delta\omega^{(1)}_{\mathrm{v},\mu\nu}(q) = -(q^2 g_{\mu\nu} - q_\mu q_\nu)\frac{e^2}{12\pi^2}\Gamma(\Delta). \tag{G.60}$$

Finally, the vertex correction is renormalized by

$$\Delta\Gamma^{(1)}_{\mathrm{v},\mu}(p,k) = Z_1^{(1)}\gamma_\mu \tag{G.61}$$

$$Z_1^{(1)} = -\frac{e^2}{4\pi}\Gamma(\Delta) - \frac{1}{4}\gamma^\mu \Gamma^{(1)}_{\mathrm{v,finite},\mu}(p,p,e,m)\Big|_{\not p=m}. \tag{G.62}$$

At this point, all Green's functions are finite to first order. At the same time, all symmetries of the theory have been preserved, which may be checked by verification of (G.26) for the renormalized functions.

This procedure can be repeated for all higher orders of perturbation theory [531], after discussion of overlapping divergences, which occur for instance in the diagram

Note that all artificial divergences resulting from use of e_b and m_b inside the finite parts of the Green's functions or from multiplication of finite terms with $Z_1^{(1)}$ etc are eliminated by higher order contributions to the renormalization constants. No further details are given at this point, as the first order provides all explicit results required for our discussion of RDFT.

The final form of the renormalized Lagrangian of QED is

$$\mathscr{L}_{\mathrm{R}} = Z_2\hat{\overline{\psi}}(x)\left(i\not\partial - m + \delta m + e\not V(\boldsymbol{x}) + e\hat{\not A}(x)\right)\hat{\psi}(x)$$
$$-\frac{Z_3}{16\pi}\hat{F}_{\mu\nu}(x)\hat{F}^{\mu\nu}(x) - \frac{Z_3\lambda}{8\pi}\left(\partial_\nu\hat{A}^\nu(x)\right)^2, \tag{G.63}$$

where we have now reintroduced the external potential V^μ in order to indicate that it has to be renormalized in the same spirit as the quantized photon field: this is immediately obvious if one analyzes the Lagrangian (G.63) in terms of perturbation theory with respect to V^μ. The renormalization constants are thus uniquely determined by vacuum QED without any external potential, so that they do not depend on the specific V^μ present. If one bases the perturbation expansion on the Lagrangian (G.63) all vacuum Green's and n-point functions of the theory (defined in terms of the physical fields $\hat{\psi}$ and $\hat{A}^\mu$) are UV finite.

Appendix H
Relativistic Homogeneous Electron Gas

In this Appendix we summarize some properties of the relativistic homogeneous electron gas (RHEG) in order to illustrate the renormalization of ground state energies (indicated in Sect. 8.2) and to provide the background for the RLDA (Sect. 8.8.3) as well as for the relativistic gradient expansion (Appendix J). For simplicity we restrict the discussion to the unpolarized RHEG (for details of the polarized RHEG see [546] and references therein).

The basic concept of the RHEG follows that of the HEG, introduced in Sect. 4.3—the RHEG consists of an infinite electron gas with density n_0 plus a neutralizing positive background charge density $n_+ = n_0$, which suppresses long-range Coulomb divergences. Now, however, the electrons and their interaction are treated on the level of the QED, i.e. the Lagrangian (8.11) with $V^\mu = 0$.

As in the preceding Appendix we use $\hbar = c = 1$.

H.1 Basic Propagators

We start by noting the basic differences between the perturbative treatment of the RHEG and that of vacuum QED, discussed in Appendix G. While the Hamiltonian of the RHEG, $\hat{H}^{\mathrm{hom}}$, is identical to that of vacuum QED, the ground state $|\Psi_0\rangle$ of the RHEG represents a gas of electrons with finite density n_0, in contrast to the ground state $|0\rangle$ of vacuum QED. As a consequence the fermion propagator,

$$G(x,y) = -i\langle\Psi_0|T\hat{\psi}(x)\hat{\overline{\psi}}(y)|\Psi_0\rangle\,, \tag{H.1}$$

differs from G_{v} already on the noninteracting level. In momentum space the noninteracting fermion propagator of the RHEG is given by

$$G^0(p) = G^0_{\mathrm{v}}(p) + G^0_{\mathrm{d}}(p) \;=\; G_-(p) + G_+(p) \tag{H.2}$$

$$G^0_{\mathrm{d}}(p) = 2\pi i\delta(p^0 - E_p)\,\frac{\not{p}+m}{2E_p}\,\Theta(k_{\mathrm{F}} - |\boldsymbol{p}|) \tag{H.3}$$

$$G_-(p) = \frac{\not{p}_- + m}{2E_p} \frac{-1}{p^0 + E_p - i\eta} \tag{H.4}$$

$$G_+(p) = \frac{\not{p}_+ + m}{2E_p} \left[\frac{\Theta(|\boldsymbol{p}| - k_F)}{p^0 - E_p + i\eta} + \frac{\Theta(k_F - |\boldsymbol{p}|)}{p^0 - E_p - i\eta} \right], \tag{H.5}$$

where $E_p = \sqrt{\boldsymbol{p}^2 + m^2}$, $p^\mu_\pm = (\pm E_p, p^i)$ and the Fermi momentum k_F is related to the electron density n_0 of the RHEG as in the nonrelativistic case,

$$n_0 = \frac{k_F^3}{3\pi^2}. \tag{H.6}$$

Two alternative forms for G^0 have been listed, the first one emphasizes its relation to the vacuum propagator $G^0_V(p)$, Eq. (G.10), the second one indicates its decomposition into positive energy (G_+) and negative energy (G_-) contributions. In the nonrelativistic limit the upper left part of the matrix $G_+(p^0 + m, \boldsymbol{p})$ goes over into the standard nonrelativistic electron gas propagator. Note that due to charge conservation the density of the RHEG is not changed by switching on the electron–electron interaction, so that n_0 also represents the density of the interacting RHEG. Equation (H.6) thus also relates the interacting density to the noninteracting k_F. Diagrammatically the full $G^0(p)$, Eq. (H.2), will be represented by

$$iG^0(p) = $$  (H.7)

in the following. The other two basic elements of perturbation theory, the noninteracting photon propagator and the simple vertex, remain unchanged. However, it seems worth pointing out that the full photon propagator

$$D_{\mu\nu}(x,y) = -ie^2 \langle \Psi_0 | T \hat{A}_\mu(x) \hat{A}_\nu(y) | \Psi_0 \rangle, \tag{H.8}$$

and the full vertex function do not: in the case of the RHEG not only virtual electron-positron pairs screen the bare interaction but also virtual electron-hole pairs. Note that we have introduced an additional factor of e^2 in the definition (H.8), as compared with the definition (G.3). This reflects the fact that it is more convenient for the subsequent discussion that the corresponding free propagator $D^0_{\mu\nu}$ approaches the Coulomb interaction in the limit $c \to \infty$.

H.2 Response Functions

Most information on the RHEG which is required in the present context is contained in the response functions of the RHEG. In our notation the time-ordered current response functions (n-point functions) are defined as

$$\chi^{(n)}_{\mu_1 \ldots \mu_n}(x_1, \ldots x_n) := (-i)^{n-1} \langle \Psi_0 | T \delta \hat{j}_{\mu_1}(x_1) \ldots \delta \hat{j}_{\mu_n}(x_n) | \Psi_0 \rangle, \tag{H.9}$$

with the operator $\delta\hat{j}_\mu$ for the induced current given by

$$\delta\hat{j}_\mu(t,\boldsymbol{r}) = \hat{j}_\mu(t,\boldsymbol{r}) - \langle\Psi_0|\hat{j}_\mu(t,\boldsymbol{r})|\Psi_0\rangle = \hat{j}_\mu(t,\boldsymbol{r}) - j_\mu(\boldsymbol{r}) . \tag{H.10}$$

For the time-independent systems of interest here a partial Fourier transformation of $\chi^{(n)}_{\mu_1\ldots\mu_n}$ is advantageous,

$$\begin{aligned}\chi^{(n)}_{\mu_1\ldots\mu_n}(t_1,\boldsymbol{r}_1;\ldots t_n,\boldsymbol{r}_n) = & \int\frac{d\omega_1}{2\pi}\cdots\int\frac{d\omega_n}{2\pi}\, e^{-i\omega_1 t_1 \ldots -i\omega_n t_n} \\ & \times 2\pi\delta(\omega_1+\ldots+\omega_n) \\ & \times\chi^{(n)}_{\mu_1\ldots\mu_n}(\omega_1,\boldsymbol{r}_1;\ldots\omega_n,\boldsymbol{r}_n) .\end{aligned} \tag{H.11}$$

The static response functions utilized in Appendix I are then obtained by taking the zero-frequency limit,

$$\chi^{(n)}_{\mu_1\ldots\mu_n}(\boldsymbol{r}_1,\ldots\boldsymbol{r}_n) \equiv \chi^{(n)}_{\mu_1\ldots\mu_n}(\omega_1=0,\boldsymbol{r}_1;\ldots\omega_n=0,\boldsymbol{r}_n) . \tag{H.12}$$

For the case of the RHEG further Fourier transformation is useful,

$$\begin{aligned}\chi^{(n)}_{\mu_1\ldots\mu_n}(q^0_1,\boldsymbol{r}_1;\ldots q^0_n,\boldsymbol{r}_n) = & \int\frac{d^3q_1}{(2\pi)^3}\cdots\int\frac{d^3q_n}{(2\pi)^3}\, e^{i\boldsymbol{r}_1\cdot\boldsymbol{q}_1+\ldots+i\boldsymbol{r}_n\cdot\boldsymbol{q}_n} \\ & \times\,(2\pi)^3\delta^{(3)}(\boldsymbol{q}_1+\ldots+\boldsymbol{q}_n) \\ & \times\,\chi^{(n)}_{\mu_1\ldots\mu_n}(q_1,\ldots q_n) .\end{aligned} \tag{H.13}$$

Current conservation then implies the transversality of $\chi^{(n)}_{\mu_1\ldots\mu_n}$ with respect to all arguments [755],

$$q_i^{\mu_i}\chi^{(n)}_{\mu_1\ldots\mu_n}(q_1,\ldots q_n) = 0 \qquad \forall\, i=1,\ldots n . \tag{H.14}$$

This relation is easily established in real space. For brevity, we only consider the linear response function explicitly,

$$\begin{aligned}\partial^\mu_x\chi^{(2)}_{\mu\nu}(x,y) = & -i\langle\Psi_0|T\left[\partial^\mu\delta\hat{j}_\mu(x)\right]\delta\hat{j}_\nu(y)|\Psi_0\rangle \\ & -i\delta(x^0-y^0)\langle\Psi_0|\delta\hat{j}_0(x)\delta\hat{j}_\nu(y)-\delta\hat{j}_\nu(y)\delta\hat{j}_0(x)|\Psi_0\rangle .\end{aligned}$$

Now the first term on the right-hand side vanishes due to current conservation (which is also valid on the level of the operator), the second term vanishes due to the vanishing equal-time commutator,

$$\left[\hat{\psi}^\dagger(t,\boldsymbol{x})\hat{\psi}(t,\boldsymbol{x}),\hat{\psi}^\dagger(t,\boldsymbol{y})\hat{\psi}(t,\boldsymbol{y})\right] = \left[\hat{\psi}^\dagger(t,\boldsymbol{x})\hat{\psi}(t,\boldsymbol{x}),\hat{\psi}^\dagger(t,\boldsymbol{y})\alpha^k\hat{\psi}(t,\boldsymbol{y})\right] = 0 .$$

The proof of (H.14) for higher order $\chi^{(n)}$ proceeds analogously, with the difference that all possible time orderings have to be taken into account in the second term. In the following the connected contributions of the $\chi^{(n)}$, for which all external vertices are linked to each other in some way, will be denoted by $\chi_c^{(n)}$, while the linear response function of the RHEG will be abbreviated by $\chi_{\mu\nu}$.

The latter function has a simple relation to the Fourier transform of the full photon propagator (H.8),

$$D_{\mu\nu}(q) = D^0_{\mu\nu}(q) + D^0_{\mu\rho}(q)\chi^{\rho\lambda}(q)D^0_{\lambda\nu}(q)\,. \tag{H.15}$$

This relation can easily be established on the basis of the equivalent of (G.5) for the photon propagator. In analogy to Eq. (G.17), one also finds a Dyson equation for the response function $\chi^{\mu\nu}$,

$$\chi^{\mu\nu}(q) = \Pi^{\mu\nu}(q) + \Pi^{\mu\rho}(q)\, D^0_{\rho\lambda}(q)\, \chi^{\lambda\nu}(q)\,. \tag{H.16}$$

Note that the present definition of the irreducible 2-point function $\Pi^{\mu\nu}$ differs from that used in Appendix G ($\omega^{\mu\nu}$) by a factor of e^2. As already indicated in the discussion of (H.8) it is more convenient for the present purpose to associate the factor of e^2 emerging from each pair of vertices in the perturbation expansion with the photon propagator than with the polarization insertion.

As a consequence of (H.14), (H.16) $\Pi^{\mu\nu}$ also satisfies the transversality relation

$$q_\mu \Pi^{\mu\nu}(q) = 0\,, \tag{H.17}$$

which determines the tensor structure of $\Pi^{\mu\nu}$. In fact, there are only two independent (4×4) polarization tensors which comply with Eq. (H.17),

$$P^{\mu\nu}_{\mathrm{L}}(q) = \frac{-1}{\boldsymbol{q}^2 q^2}\begin{pmatrix} (\boldsymbol{q}^2)^2 & \boldsymbol{q}^2 q^0 q^j \\ \boldsymbol{q}^2 q^0 q^i & (q^0)^2 q^i q^j \end{pmatrix} \tag{H.18}$$

$$P^{\mu\nu}_{\mathrm{T}}(q) = \frac{1}{\boldsymbol{q}^2}\begin{pmatrix} 0 & 0 \\ 0 & \boldsymbol{q}^2 g^{ij} + q^i q^j \end{pmatrix} \qquad (g^{ij} = -\delta_{ij}) \tag{H.19}$$

$$q_\mu P^{\mu\nu}_{\mathrm{L/T}}(q) = q_\nu P^{\mu\nu}_{\mathrm{L/T}}(q) \;=\; 0\,. \tag{H.20}$$

$\Pi^{\mu\nu}$ can therefore be written as

$$\Pi^{\mu\nu}(q) = P^{\mu\nu}_{\mathrm{L}}(q)\Pi_{\mathrm{L}}(q) - P^{\mu\nu}_{\mathrm{T}}(q)\Pi_{\mathrm{T}}(q)\,. \tag{H.21}$$

For convenience, we note some useful properties of $P^{\mu\nu}_{\mathrm{L/T}}$,

$$P_{\mathrm{L},\mu}{}^{\nu}(q)P_{\mathrm{L},\nu}{}^{\lambda}(q) = P_{\mathrm{L},\mu}{}^{\lambda}(q) \tag{H.22}$$

$$P_{\mathrm{T},\mu}{}^{\nu}(q)P_{\mathrm{T},\nu}{}^{\lambda}(q) = P_{\mathrm{T},\mu}{}^{\lambda}(q) \tag{H.23}$$

$$P_{\mathrm{L},\mu}{}^{\nu}(q)P_{\mathrm{T},\nu}{}^{\lambda}(q) = 0 \tag{H.24}$$

$$P_{\mathrm{L},\mu}{}^{0}(q)P_{\mathrm{L},0}{}^{\lambda}(q) = -\frac{\boldsymbol{q}^2}{q^2}P_{\mathrm{L},\mu}{}^{\lambda}(q) \tag{H.25}$$

$$P^{\mu\nu}_{\mathrm{L}}(q) + P^{\mu\nu}_{\mathrm{T}}(q) = g^{\mu\nu} - \frac{q^\mu q^\nu}{q^2} \tag{H.26}$$

$$P_{\mathrm{L},\mu}{}^{\mu}(q) = 1 \tag{H.27}$$

$$P_{\mathrm{T},\mu}{}^{\mu}(q) = 2\,. \tag{H.28}$$

If one decomposes $\Pi^{\mu\nu}$ into its vacuum (v) limit (obtained for $|\Psi_0\rangle \to |0\rangle$) and a remainder (the electron gas component—d),

$$\Pi^{\mu\nu}(q) = \Pi^{\mu\nu}_{\mathrm{d}}(q) + \Pi^{\mu\nu}_{\mathrm{v}}(q) \tag{H.29}$$

$$\Pi_{\mathrm{L}}(q) = \Pi_{\mathrm{L,d}}(q) + \Pi_{\mathrm{v}}(q) \tag{H.30}$$

$$\Pi_{\mathrm{T}}(q) = \Pi_{\mathrm{T,d}}(q) - \Pi_{\mathrm{v}}(q)\,, \tag{H.31}$$

the vacuum contribution can be recast in the tensor form (G.20), with the polarization function Π_{v} given by

$$\Pi_{\mathrm{v}}(q) = \frac{q^2}{e^2}\,\omega_{\mathrm{v}}(q)\,. \tag{H.32}$$

If one uses the polarization tensors (H.18), (H.19), the free photon propagator (G.13) and the longitudinal and transverse polarization functions $\Pi_{\mathrm{L/T}}$, the Dyson equation for $\chi^{\mu\nu}$ can be resolved as

$$\chi^{\mu\nu}(q) = \frac{\Pi_{\mathrm{L}}(q)}{1 - D^0(q)\Pi_{\mathrm{L}}(q)} P_{\mathrm{L}}^{\mu\nu}(q) - \frac{\Pi_{\mathrm{T}}(q)}{1 + D^0(q)\Pi_{\mathrm{T}}(q)} P_{\mathrm{T}}^{\mu\nu}(q)\,, \tag{H.33}$$

where, according to the modified definition (H.8), D^0 is given by (G.12) times an additional factor of e^2.

The full photon propagator $D^{\mu\nu}$ can now be obtained from Eq. (H.15) by insertion of (H.33). For a discussion of the renormalization of $D^{\mu\nu}$ it is instructive to rewrite the resulting expression in terms of the full vacuum photon propagator,

$$D_{\mathrm{v}}(q) = \frac{D^0(q)}{1 - D^0(q)\Pi_{\mathrm{v}}(q)}\,. \tag{H.34}$$

Insertion of (H.30) and (H.31) into (H.33) plus subsequent use of (H.34) allows a decoupling of the screening effects due to vacuum polarization from those originating from the actual electron gas,

$$D^{\mu\nu}(q) = \frac{D_{\mathrm{v}}(q)}{1 - D_{\mathrm{v}}(q)\Pi_{\mathrm{L,d}}(q)} P_{\mathrm{L}}^{\mu\nu}(q) + \frac{D_{\mathrm{v}}(q)}{1 + D_{\mathrm{v}}(q)\Pi_{\mathrm{T,d}}(q)} P_{\mathrm{T}}^{\mu\nu}(q)\,. \tag{H.35}$$

At first glance this form seems to suggest that $D^{\mu\nu}$ is UV-finite as soon as D_{v} is replaced by $D_{\mathrm{v,R}}$ defined by Eqs. (G.46)–(G.48). However, $\Pi_{\mathrm{L/T,d}}$ also contains UV-divergent subgraphs. The following 2-loop contribution may illustrate this point,

 .

If one replaces G^0 by $G^0_{\rm v}+G^0_{\rm d}$ one recognizes that besides the obvious pure vacuum loop absorbed in $D_{\rm v}$ also mixtures between the first order vertex correction loop of the vacuum, Eq. (G.24), and $G^0_{\rm d}$-type propagators occur. The counterterms required to keep such subgraphs finite are, however, completely determined by vacuum QED: similar to the renormalization of overlapping divergences, each vacuum subgraph in a (larger) non-vacuum diagram has to be supplemented individually by its associated counterterm (this also holds for multi-loop vacuum subgraphs).

For the discussion of inhomogeneity corrections to the RLDA one also needs the inverse response function $\chi^{-1}_{\mu\nu}$. However, $\chi^{-1}_{\mu\nu}(q)$ can not be an inverse of $\chi_{\mu\nu}(q)$ in the conventional understanding of an inverse matrix, as the transversality relation (H.14) requires

$$q^\mu\,\chi_{\mu\nu}(q)\,\chi^{-1,\nu\rho}(q)=0\,,$$

which is not compatible with

$$\chi_{\mu\nu}(q)\chi^{-1,\nu\rho}(q)=g_\mu{}^\rho\,.$$

In the present context $\chi^{-1,\mu\nu}(q)$ is therefore defined to satisfy

$$\chi_{\mu\nu}(q)\,\chi^{-1,\nu\rho}(q)=g_\mu{}^\rho-\frac{q_\mu q^\rho}{q^2}\,. \tag{H.36}$$

When multiplied with a quantity for which the transversality condition $q_\mu f^\mu=0$ holds, $\chi^{-1,\nu\rho}$ behaves like an ordinary inverse. For this type of inverse one obtains

$$\chi^{-1,\mu\nu}(q) = -D^0(q)\,g^{\mu\nu}+\Pi^{-1,\mu\nu}(q) \tag{H.37}$$

$$\Pi^{-1,\mu\nu}(q) = \frac{1}{\Pi_{\rm L}(q)}P_{\rm L}^{\mu\nu}-\frac{1}{\Pi_{\rm T}(q)}P_{\rm T}^{\mu\nu}\,. \tag{H.38}$$

The product of (H.37) with $\chi_{\mu\nu}$, Eq. (H.33), can be shown to satisfy Eq. (H.36) by use of (H.22)–(H.24) and (H.26).

As far as explicit approximations for the polarization functions $\Pi_{\rm L/T}$ are concerned, only very little is known even in the static limit. The complete frequency dependence is available for the noninteracting limit $\Pi^{(0)}_{\rm L/T}$, i.e. the relativistic generalization of the Lindhard function [620, 756]. In addition to its vacuum part (G.23) one has

$$\begin{aligned}\Pi^{(0)}_{{\rm d},\mu\nu}(q) = &-i\int\frac{d^4p}{(2\pi)^4}{\rm tr}\left[\gamma_\mu\,G^0_{\rm d}(p)\,\gamma_\nu\,G^0_{\rm d}(p-q)\right]\\ &-i\int\frac{d^4p}{(2\pi)^4}{\rm tr}\left[\gamma_\mu\,G^0_{\rm v}(p)\,\gamma_\nu\,G^0_{\rm d}(p-q)\right]\\ &-i\int\frac{d^4p}{(2\pi)^4}{\rm tr}\left[\gamma_\mu\,G^0_{\rm d}(p)\,\gamma_\nu\,G^0_{\rm v}(p-q)\right]\,.\end{aligned}$$

$\Pi^{(0)}_{\mathrm{d},\mu\nu}$ is sometimes called the *no-sea* limit of the full $\Pi^{(0)}_{\mu\nu}$—quite generally the no-sea approximation $\Pi_{\mathrm{ns},\mu\nu}$ is defined by neglect of all those contributions to a closed fermion loop which do not vanish in the limit $k_{\mathrm{F}} \to 0$. This *no-sea* form is not identical with the result of the more frequently applied *no-pair* approximation. The latter approximation amounts to neglecting the negative energy states completely. In the present context projecting out the negative energy states at all steps of the calculation is equivalent to a complete neglect of the negative energy component $G_{-}(p)$ of the fermion propagator,

$$\Pi^{(0)}_{\mathrm{d},\mu\nu}(q) \neq -i\int \frac{d^4p}{(2\pi)^4}\mathrm{tr}\left[\gamma_\mu G^0_+(p)\gamma_\nu G^0_+(p-q)\right] = \Pi^{(0)}_{\mathrm{np},\mu\nu}(q)\,.$$

For subsequent use we note the long-wavelength expansion of the static limit of $\Pi^{(0)}_{\mu\nu}$,

$$\Pi^{(0)}_{\mathrm{L,d}}(0,\boldsymbol{q}) = -\frac{mk_{\mathrm{F}}\eta}{\pi^2}\left\{1-\frac{1}{3}\left[1+2\frac{\beta}{\eta}\mathrm{arsinh}(\beta)\right]Q^2+\ldots\right\} \tag{H.39}$$

$$\Pi^{(0)}_{\mathrm{T,d}}(0,\boldsymbol{q}) = \frac{mk_{\mathrm{F}}\eta}{\pi^2}\left\{-\frac{2}{3}\frac{\beta}{\eta}\mathrm{arsinh}(\beta)Q^2+\ldots\right\} \tag{H.40}$$

$$\Pi^{(0)}_{\mathrm{v,R}}(0,\boldsymbol{q}) = \frac{1}{60\pi^2}\frac{\boldsymbol{q}^4}{m^2}+\ldots, \tag{H.41}$$

where $Q = |\boldsymbol{q}|/(2k_{\mathrm{F}})$ and

$$\beta = \frac{(3\pi^2 n_0)^{1/3}}{m}\,; \qquad \eta = (1+\beta^2)^{1/2}\,. \tag{H.42}$$

Beyond the noninteracting limit only the vacuum part of the 2-loop contribution to the polarization function has been evaluated [757, 758]. In addition, the screening length $\Pi_{\mathrm{L,d}}(0,\mathbf{0})$ is related to the energy density via the compressibility sum rule [759],

$$\frac{d^2}{dn_0^2}\Big[t_{\mathrm{s}}(n_0)+e_{\mathrm{xc}}(n_0)\Big] = -\frac{1}{\Pi_{\mathrm{L,d}}(0,\mathbf{0})}\,, \tag{H.43}$$

so that the long wavelength limit of higher orders of $\Pi_{\mathrm{L,d}}$ can be obtained from the associated contributions to the energy density. Finally, in the context of the quark-gluon gas the high temperature limits of certain classes of higher order diagrams have also been examined (see e.g. [760]). These results are, however, only of limited interest in the present context aiming at $T=0$ and $m \gtrsim |\boldsymbol{q}|$.

H.3 Ground State Energy

The exchange-correlation energy of the RHEG constitutes the basis for the RLDA. At the same time, it provides an instructive example for the application of the renormalization procedure described in Appendix G. We start by emphasizing that the ground state energy is defined with respect to the vacuum energy as in (8.56) with $V^\mu = 0$ (compare [621, 755]),

$$E^{\mathrm{RHEG}} = \langle \Psi_0 | \hat{H}^{\mathrm{hom}} | \Psi_0 \rangle - \langle 0 | \hat{H}^{\mathrm{hom}} | 0 \rangle + \Delta E^{\mathrm{hom}}, \tag{H.44}$$

where $\langle 0 | \hat{H}^{\mathrm{hom}} | 0 \rangle$ is the energy of the interacting, homogeneous vacuum and ΔE^{hom} represents the counterterms required to keep E^{RHEG} UV-finite. In the case of the electron gas Eq. (H.44) is applied on the level of the energy density, rather than the infinite energy itself.

The kinetic energy density t_s of the noninteracting RHEG can be evaluated without application of the UV-renormalization procedure [761],

$$\begin{aligned} t_s(n_0) &= \langle \Psi_0 | \left[\hat{\overline{\psi}}(x), \left(-i\boldsymbol{\gamma}\cdot\boldsymbol{\nabla} + (1-\gamma^0)m \right) \hat{\psi}(x) \right] | \Psi_0 \rangle \\ &\quad - \langle 0 | \left[\hat{\overline{\psi}}(x), \left(-i\boldsymbol{\gamma}\cdot\boldsymbol{\nabla} + (1-\gamma^0)m \right) \hat{\psi}(x) \right] | 0 \rangle \\ &= i \lim_{y \to x} {}_s \operatorname{tr} \left[\left(-i\boldsymbol{\gamma}\cdot\boldsymbol{\nabla} + (1-\gamma^0)m \right) G^0_{\mathrm{d}}(x-y) \right] \\ &= \frac{k_{\mathrm{F}}^5}{10\pi^2 m} \, \Phi_s(\beta) \end{aligned} \tag{H.45}$$

$$\Phi_s(\beta) = \frac{10}{\beta^5} \left[\frac{1}{8} \left(\beta\eta^3 + \beta^3\eta - \operatorname{arsinh}(\beta) \right) - \frac{1}{3}\beta^3 \right] \tag{H.46}$$

(the electron rest mass has been subtracted). The Hartree (electrostatic) energy of the RHEG vanishes, if one takes the neutralizing positive charge background into account. Following closely the derivation of Eq. (4.88), the xc-energy of the RHEG can be written in terms of a coupling constant integral over the current–current response function [618, 538],

$$e_{\mathrm{xc}}(n_0) = \frac{i}{2} \int_0^1 d\lambda \int \frac{d^4 q}{(2\pi)^4} D^0_{\mu\nu}(q) \left[\chi_\lambda^{\mu\nu}(q) - \chi_{\mathrm{v},\lambda}^{\mu\nu}(q) \right] + \Delta e^{\mathrm{hom}}. \tag{H.47}$$

$\chi_\lambda^{\mu\nu}$ is given by (H.33) with the coupling strength e^2 replaced by λe^2, $\chi_{\mathrm{v},\lambda}^{\mu\nu}$ represents its vacuum limit and Δe^{hom} is the energy density corresponding to the counterterm ΔE^{hom}.

The first order term (in e^2) in (H.47), i.e. the exchange energy of the RHEG (according to Eq. (8.92)), is the simplest energy contribution for which the UV-renormalization is nontrivial. The basic problem associated with the renormalization of energies (rather than Green's functions) is that energy expressions can not be rewritten entirely in terms of renormalized n-point functions. At least one overall loop integration remains to be treated separately (the q-integration in (H.47)). As an additional complication, this outermost loop integration often leads to overlapping divergences. An example for this statement is provided by the exchange energy, which is obtained if the full $\chi_\lambda^{\mu\nu}$ in Eq. (H.47) is replaced by its noninteracting limit $\Pi^{(0),\mu\nu}$. Visualizing the resulting integral graphically,

$-2ie_{\rm x}(n_0) =$  $-$  , (H.48)

one realizes that three divergent 1-loop subgraphs contribute to the electron gas loop,

$-i\Pi^{(0)}_{\mu\nu} =$  , $-i\Sigma^{(1)} =$  $=$  . (H.49)

While the UV-divergence of the $\Pi^{(0)}_{\mu\nu}$-subgraph is eliminated by the subtraction of the vacuum exchange energy in (H.48), the two (identical) self-energy subgraphs require additional counterterms. As one is facing overlapping divergences in (H.48) each divergent subgraph has to be renormalized individually. Of course, only the vacuum contribution to $\Sigma^{(1)}$ requires renormalization,

$$e_{\rm x}(n_0) = \frac{i}{2}\int \frac{d^4q}{(2\pi)^4} D^0_{\mu\nu}(q) \left[\Pi^{(0),\mu\nu}(q) - \Pi^{(0),\mu\nu}_{\rm v}(q)\right]$$
$$-i\int \frac{d^4p}{(2\pi)^4} \operatorname{tr}\left[G^0_{\rm d}(p)\Delta\Sigma^{(1)}_{\rm v}(p)\right] . \qquad \text{(H.50)}$$

The second line represents the lowest order contribution to the UV-counterterm $\Delta e^{\rm hom}$. As discussed in detail in Appendix G, the self-energy counterterm $\Delta\Sigma^{(1)}_{\rm v}$ is defined so that the renormalized vacuum self-energy $\Sigma_{\rm v,R}$, Eq. (G.44), satisfies the standard on-shell normalization condition (G.45), i.e. on the 1-loop level one obtains (G.58). Using the decomposition of G^0, Eq. (H.2), $e_{\rm x}(n_0)$ can thus be rewritten as

$$e_{\rm x}(n_0) = \frac{1}{2}\int \frac{d^4q}{(2\pi)^4}\int \frac{d^4p}{(2\pi)^4} D^0_{\mu\nu}(q) \operatorname{tr}\left[G^0_{\rm d}(p+q)\gamma^\mu G^0_{\rm d}(p)\gamma^\nu\right]$$
$$-i\int \frac{d^4p}{(2\pi)^4} \operatorname{tr}\left[G^0_{\rm d}(p)\Sigma^{(1)}_{\rm v,R}(p)\right] . \qquad \text{(H.51)}$$

The second term on the right-hand side vanishes according to Eqs. (H.3), (G.45),

$$\left[(\not p + m)\Sigma_{\rm v,R}(p)\right]_{p^2=m^2} = 0 .$$

Consequently, the standard renormalization scheme eliminates the vacuum corrections to $e_{\rm x}(n_0)$ completely. The first line of (H.51) can be evaluated straightforwardly [618–620],

$$e_{\rm x}(n_0) = e^{\rm NRHEG}_{\rm x}(n_0)\; \Phi_{\rm x}(\beta) \qquad \text{(H.52)}$$

$$e_{\mathrm{x}}^{\mathrm{NRHEG}}(n_0) = -\frac{e^2}{4\pi^3}\, k_{\mathrm{F}}^4 \tag{H.53}$$

$$\Phi_{\mathrm{x}}(\beta) = 1 - \frac{3}{2}\left[\frac{\eta}{\beta} - \frac{1}{\beta^2}\mathrm{arsinh}(\beta)\right]^2 . \tag{H.54}$$

Moreover, using the decomposition of the photon propagator into the Coulomb and the transverse interaction, $e_{\mathrm{x}}(n_0)$ can be split accordingly [535, 620],

$$e_{\mathrm{x}}^{\mathrm{C/T}}(n_0) = e_{\mathrm{x}}^{\mathrm{NRHEG}}(n_0)\, \Phi_{\mathrm{x}}^{\mathrm{C/T}}(\beta) \tag{H.55}$$

$$\Phi_{\mathrm{x}}^{\mathrm{C}}(\beta) = \frac{5}{6} + \frac{1}{3\beta^2} + \frac{2\eta}{3\beta}\mathrm{arsinh}(\beta) - \frac{2\eta^4}{3\beta^4}\ln(\eta) - \frac{1}{2}\left(\frac{\eta}{\beta} - \frac{\mathrm{arsinh}(\beta)}{\beta^2}\right)^2 \tag{H.56}$$

$$\Phi_{\mathrm{x}}^{\mathrm{T}}(\beta) = \frac{1}{6} - \frac{1}{3\beta^2} - \frac{2\eta}{3\beta}\mathrm{arsinh}(\beta) + \frac{2\eta^4}{3\beta^4}\ln(\eta) - \left(\frac{\eta}{\beta} - \frac{\mathrm{arsinh}(\beta)}{\beta^2}\right)^2 . \tag{H.57}$$

The UV-renormalization procedure is particularly involved for the correlation energy e_{c}, which we also discuss here for completeness. Most of the counterterms provided by Δe^{hom} are, however, included if the basic expression (H.47) is rewritten in terms of the renormalized response function $\chi_{\mathrm{R},\lambda}^{\mu\nu}$,

$$e_{\mathrm{xc}}(n_0) = \frac{i}{2}\int_0^1 d\lambda \int \frac{d^4q}{(2\pi)^4} D_{\mu\nu}^0(q)\left[\chi_{\mathrm{R},\lambda}^{\mu\nu}(q) - \chi_{\mathrm{v,R},\lambda}^{\mu\nu}(q)\right] + \Delta\tilde{e}^{\mathrm{hom}} \tag{H.58}$$

(the exchange energy has not been subtracted). The only remaining divergence (to be eliminated by $\Delta\tilde{e}^{\mathrm{hom}}$) now originates from the outermost loop integration in (H.58). It can be explicitly discussed within the so-called random phase (or ring) approximation (RPA) in which $\Pi_{\mathrm{L/T}}$ is approximated by its 1-loop contribution $\Pi_{\mathrm{L/T}}^{(0)}$ [618, 762]. Insertion of Eq. (H.33) into (H.58) then gives

$$e_{\mathrm{xc}}^{\mathrm{RPA}}(n_0) = \frac{i}{2}\int_0^1 d\lambda \int \frac{d^4q}{(2\pi)^4}\left[\frac{D^0(q)\Pi_{\mathrm{L}}^{(0)}(q)}{1-\lambda D^0(q)\Pi_{\mathrm{L}}^{(0)}(q)} - 2\frac{D^0(q)\Pi_{\mathrm{T}}^{(0)}(q)}{1+\lambda D^0(q)\Pi_{\mathrm{T}}^{(0)}(q)} - 3\frac{D^0(q)\Pi_{\mathrm{v,R}}^{(0)}(q)}{1-\lambda D^0(q)\Pi_{\mathrm{v,R}}^{(0)}(q)}\right] + \Delta\tilde{e}^{\mathrm{hom,RPA}} .$$

The coupling constant integration can be performed directly, if Eqs. (H.30), (H.31) and (H.34) are used,

$$e_{\text{xc}}^{\text{RPA}}(n_0) = -\frac{i}{2}\int \frac{d^4q}{(2\pi)^4}\left[\ln\left[1 - D_{\text{v,R}}(q)\Pi_{\text{L,d}}^{(0)}(q)\right]\right.$$
$$\left. + 2\ln\left[1 + D_{\text{v,R}}(q)\Pi_{\text{T,d}}^{(0)}(q)\right]\right] + \Delta\tilde{e}^{\text{hom,RPA}} . \quad (\text{H.59})$$

At this point it is convenient to define the vacuum-screened exchange energy,

$$e_{\text{x,s}}(n_0) = \frac{i}{2}\int \frac{d^4q}{(2\pi)^4}\left[D_{\text{v,R}}(q)\Pi_{\text{L,d}}^{(0)}(q) - 2D_{\text{v,R}}(q)\Pi_{\text{T,d}}^{(0)}(q)\right] + \Delta\tilde{e}^{\text{hom,RPA}} , \quad (\text{H.60})$$

which requires renormalization similar to its unscreened counterpart. After subtraction of $e_{\text{x,s}}$ from $e_{\text{xc}}^{\text{RPA}}$ one obtains for the correlation energy [762]

$$e_{\text{c,s}}^{\text{RPA}}(n_0) = -\frac{i}{2}\int \frac{d^4q}{(2\pi)^4}\left[\ln\left[1 - D_{\text{v,R}}(q)\Pi_{\text{L,d}}^{(0)}(q)\right]\right.$$
$$+ 2\ln\left[1 + D_{\text{v,R}}(q)\Pi_{\text{T,d}}^{(0)}(q)\right]$$
$$\left. + D_{\text{v,R}}(q)\Pi_{\text{L,d}}^{(0)}(q) - 2D_{\text{v,R}}(q)\Pi_{\text{T,d}}^{(0)}(q)\right] . \quad (\text{H.61})$$

This expression is UV-convergent as it stands, as the lowest order diagram included in (H.61) contains $\Pi_{\text{L/T,d}}^{(0)}$ already two times.[1] Finally, one can define the *no-sea* approximation of (H.61) by neglecting all screening effects due to vacuum polarization. Decomposing the result into a Coulomb and a transverse component, one ends up with

$$e_{\text{c,ns}}^{\text{C,RPA}}(n_0) = -\frac{i}{2}\int \frac{d^4q}{(2\pi)^4}\left\{\ln\left|1 - D^0(q)\Pi_{\text{L,d}}^{(0)}(q)\right| + D^0(q)\Pi_{\text{L,d}}^{(0)}(q)\right\} \quad (\text{H.62})$$

$$e_{\text{c,ns}}^{\text{T,RPA}}(n_0) = -i\int \frac{d^4q}{(2\pi)^4}\left\{\ln\left|1 + D^0(q)\Pi_{\text{T,d}}^{(0)}(q)\right| - D^0(q)\Pi_{\text{T,d}}^{(0)}(q)\right\} . \quad (\text{H.63})$$

$e_{\text{c,ns}}^{\text{C/T,RPA}}(n_0)$ has been evaluated numerically for arbitrary n_0 [538, 622]. The high-density (ultrarelativistic) limit of $e_{\text{c,ns}}^{\text{RPA}}$ is given by [618, 621]

$$e_{\text{c,ns}}^{\text{RPA}}(n_0) \xrightarrow[\beta\gg 1]{} \frac{e^4k_{\text{F}}^4}{12\pi^4}\left(\frac{3}{2}\ln\frac{\alpha}{\pi} + 1.3761 + \ldots\right) = \frac{e^4k_{\text{F}}^4}{12\pi^4}\left(-7.796 + \ldots\right) , \quad (\text{H.64})$$

where α is the fine structure constant.

In order to arrive at the RPA+, which we understand as the combination of the RPA with the remaining second order (e^4) contributions, the two second order exchange (SOX) diagrams,

[1] After Wick-rotation of q_0 in (H.61) one e.g. finds $\Pi_{\text{L,d}}^{(0)}(iq_0,\boldsymbol{q}) \sim (\boldsymbol{q}^2 + q_0^2)^{-1}$ and $\Pi_{\text{v,R}}^{(0)}(iq^0,\boldsymbol{q}) \sim (\boldsymbol{q}^2 + q_0^2)\ln|\boldsymbol{q}^2 + q_0^2|$ for large q^0 and $|\boldsymbol{q}|$ so that two factors of $\Pi_{\text{L,d}}^{(0)}(iq^0,\boldsymbol{q})$ together with the two photon propagators are sufficient to ensure UV-convergence of the outermost loop integral.

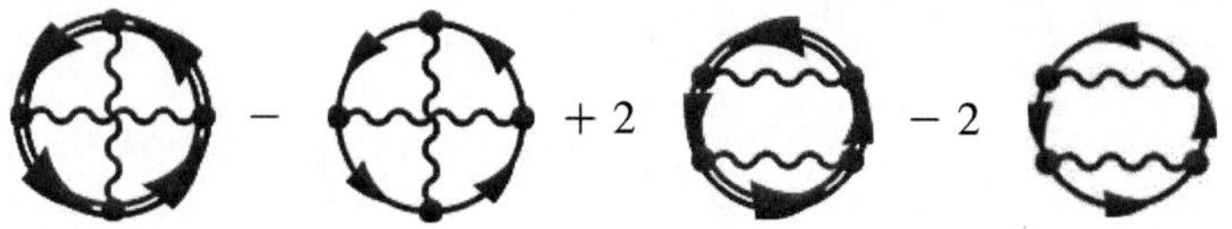

, (H.65)

have to be added to e_c^{RPA}. In contrast to the nonrelativistic situation the two rightmost diagrams do not vanish. Both types of diagrams require renormalization beyond the subtraction of their vacuum limit indicated in Eq. (H.65). The density dependence of these diagrams is not known completely. In the ultrarelativistic limit one finds for the sum of both graphs (the individual contributions are not gauge invariant) [621]

$$e_c^{SOX}(n_0) \xrightarrow[\beta\gg 1]{} \frac{e^4 k_F^4}{12\pi^4}\left(-3.18 \pm 0.12\right), \tag{H.66}$$

so that e_c^{SOX} amounts to roughly 40% of the RPA in this limit.

One can also analyze the 2-loop contribution to the screened exchange (H.60),

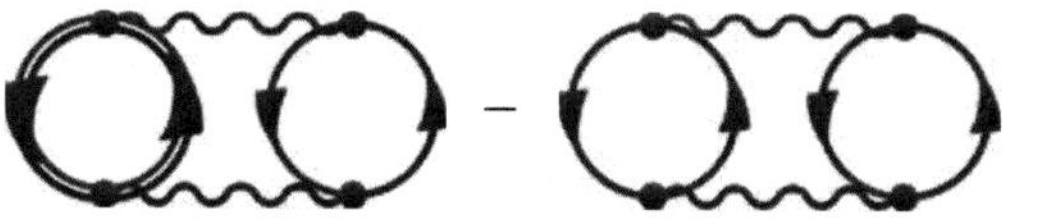

,

which (in our definition) is beyond the *no-sea* approximation. Its ultrarelativistic limit is [621],

$$e_{x,s}^{(2)}(n_0) \xrightarrow[\beta\gg 1]{} \frac{e^4 k_F^4}{12\pi^4}\left[\ln\left(2\frac{k_F}{m}\right) - \frac{11}{6}\right]. \tag{H.67}$$

In the limit of very high densities $e_{x,s}^{(2)}$ thus dominates over all other known xc-energy contributions. However, the densities required for $e_{x,s}^{(2)}$ to be of the same order of magnitude as e_c^{RPA+} are extremely high, $k_F/m \approx 10^3$, so that $e_{x,s}^{(2)}$ is not relevant for electronic structure calculations.

No calculations of e_c beyond the RPA+ are found in the literature.

H.4 Ground State Four Current

After the extensive discussion of the ground state energy little remains to be said concerning the ground state four current of the RHEG. Due to norm conservation, the interacting current must be identical with the current of the noninteracting RHEG, that is $n_0 g^{\mu 0}$, in real space. If one expresses the current expectation value in terms of the interacting propagator (H.1),

$$\langle\Psi_0|\hat{j}_\mu(x)|\Psi_0\rangle = -\, i \lim_{y\to x}{}_s \operatorname{tr}\left[G(x,y)\gamma_\mu\right],$$

one realizes that all higher order contributions resulting from the perturbation expansion of G must cancel order by order. This will be demonstrated explicitly for all first order vacuum contributions. To first order one obtains diagrammatically

$$ie\langle\Psi_0|\hat{j}_\mu(x)|\Psi_0\rangle =$$

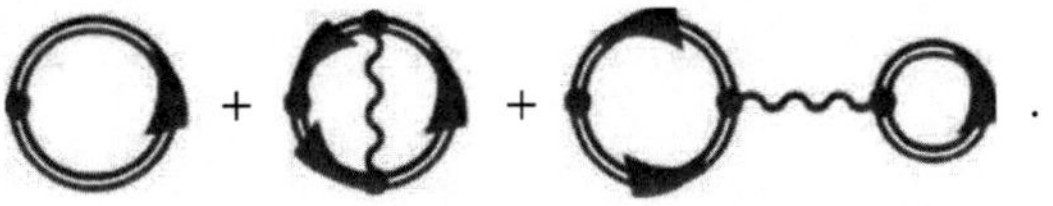

One first observes that in the pure vacuum limit, $G^0 \rightarrow G^0_{\mathrm{v}}$, all loops vanish, as at least one part of the graph is a fermion loop with an odd number of vertices (Furry's theorem)—only mixtures of vacuum subgraphs with non-vacuum components can contribute to j^μ. Moreover, the right-hand first order diagram need not be considered any further, as in addition to the electronic charge density the neutralizing positive background charge density has to be coupled to the polarization graph. Consequently, all diagrams containing tadpoles do not contribute. One is thus left with the vacuum subgraphs in the remaining first order diagram. Two vacuum subgraphs are identified, the vertex correction (left part) and the self-energy (right part). However, after renormalization the vertex correction vanishes on the mass shell due to the normalization condition (G.52). Therefore the combination of this vacuum subgraph with the remainder of the diagram, i.e. the product $\Gamma_{\mathrm{v,R}}G^0_{\mathrm{d}}$, is zero. Similarly, the vacuum self-energy is proportional to $(\not{p}-m)^2$ on the mass shell, Eq. (G.45), so that products as $G^0_{\mathrm{d}}\Sigma_{\mathrm{v,R}}G^0_{\mathrm{v}}$ vanish. Consequently, after renormalization the terms containing vacuum subgraphs give no contribution to the four current, as required. It is obvious that the argument given also applies to higher order contributions. The necessary counterterms will be denoted as $\Delta j^{\mathrm{hom}}_\mu$,

$$j_\mu = \langle\Psi_0|\hat{j}_\mu(x)|\Psi_0\rangle + \Delta j^{\mathrm{hom}}_\mu = n_0\, g_{\mu 0}\,. \tag{H.68}$$

As in the case of Δe^{hom}, $\Delta j^{\mathrm{hom}}_\mu$ is determined by the renormalization of all relevant vacuum subgraphs.

Appendix I
Renormalization of Inhomogeneous Electron Gas

In order to prepare the discussion of the relativistic generalization of the HK-theorem in Sect. 8.3 we finally consider the renormalization procedure for inhomogeneous systems, i.e. the full Lagrangian (8.11). Since the underlying renormalization program of vacuum QED is formulated within a perturbative framework (see Appendix G), we assume that the perturbing potential V^μ is sufficiently weak to allow a power series expansion of all relevant quantities with respect to V^μ. Within this approach one can explicitly derive the counterterms required for the field theoretical version of the relativistic KS equations, i.e. for the four current and kinetic energy of noninteracting particles. In this Appendix again $\hbar = c = 1$ is used.

The first quantity of interest is the four current $\delta j^\mu(\boldsymbol{r})$ induced by $V^\mu(\boldsymbol{r})$. The perturbation expansion of δj^μ with respect to V^μ can be written as

$$\delta j_\mu(\boldsymbol{r}) = \sum_{n=1}^{\infty} \frac{(-e)^n}{n!} \int d^3r_1 \dots \int d^3r_n \, \chi^{(n+1)}_{\mathrm{c},\mu\mu_1\dots\mu_n}(\boldsymbol{r}, \boldsymbol{r}_1, \dots \boldsymbol{r}_n) \times V^{\mu_1}(\boldsymbol{r}_1) \dots V^{\mu_n}(\boldsymbol{r}_n) + \Delta j^{\mathrm{inhom}}_\mu(\boldsymbol{r}), \tag{I.1}$$

where $\chi^{(n)}_{\mathrm{c},\mu_1\dots\mu_n}$ represents the static, connected response functions of the RHEG (for their precise definition see Appendix H) and $\Delta j^{\mathrm{inhom}}_\mu$ denotes the counterterms which keep δj_μ UV-finite. Of course, δj^μ satisfies current conservation,

$$\partial_\mu \delta j^\mu(\boldsymbol{r}) = \boldsymbol{\nabla} \cdot \delta \boldsymbol{j}(\boldsymbol{r}) = 0, \qquad \int d^3r \, \delta j^0(\boldsymbol{r}) = 0, \tag{I.2}$$

which is directly related to the transversality of $\chi^{(n)}_{\mathrm{c},\mu_1\dots\mu_n}$ displayed in Eq. (H.14).

The induced current (I.1) is automatically UV-finite if the expansion is based on renormalized response functions, i.e. $\Delta j^{\mathrm{inhom}}_\mu$ just sums up the terms required for the transition from the unrenormalized $\chi^{(n)}_{\mathrm{c},\mu_1\dots\mu_n}$ to their renormalized counterparts. Introducing an expansion of $\Delta j^{\mathrm{inhom}}_\mu$ in powers of V^μ one thus has

$$\Delta j_{\mu}^{\text{inhom}} = \sum_{n=1}^{\infty} \frac{(-e)^n}{n!} \int d^3 r_1 \ldots \int d^3 r_n \, \Delta \chi^{(n+1)}_{\mu\mu_1 \ldots \mu_n}(\boldsymbol{r}, \boldsymbol{r}_1, \ldots \boldsymbol{r}_n)$$
$$\times V^{\mu_1}(\boldsymbol{r}_1) \ldots V^{\mu_n}(\boldsymbol{r}_n), \qquad \text{(I.3)}$$

where $\Delta \chi^{(n)}_{\mu_1 \ldots \mu_n}$ represents the counterterms which keep the connected response function $\chi^{(n)}_{c,\mu_1 \ldots \mu_n}$ of the RHEG finite.

It is instructive to analyze $\Delta j_{\mu}^{\text{inhom}}$ for the noninteracting limit of (I.1). Using the notation of Eqs. (F.39), (F.40) and (H.7) and taking into account the fermion sign rule, the induced current is given graphically by

$ie\delta j_{\mu}^{(0)} =$  $+ \cdots$ (I.4)

(the multiplicities resulting from different ordering of vertices in case of the higher order response functions with $n \geq 2$ compensate the prefactor $1/n!$ in (I.1)). While the noninteracting 3-point function, i.e. the second graph, is UV-finite due to Furry's theorem, the noninteracting 4-point function (third diagram) is UV-finite due to its transversality and all higher order response functions are overall convergent. The only divergent term to be examined is contained in the first diagram. The counterterm for the vacuum component of $\chi^{(0)}_{\mu\nu}$, $\chi^{(0)}_{v,\mu\nu} = \Pi^{(0)}_{v,\mu\nu} = \omega^{(0)}_{v,\mu\nu}/e^2$, has been derived in Appendix G. Using dimensional regularization, one obtains Eq. (G.60) as counterterm to the lowest order 2-point function and thus after Fourier transformation,

$$\Delta j_{\mu}^{(0)}(\boldsymbol{r}) = \frac{e}{12\pi^2} \Gamma\left(\frac{4-d}{2}\right) \boldsymbol{\nabla}^2 V_{\mu}(\boldsymbol{r}), \qquad \text{(I.5)}$$

if Coulomb gauge, $\boldsymbol{\nabla} \cdot \boldsymbol{V}(\boldsymbol{r}) = 0$, is used.

The second quantity of interest is the energy shift resulting from the perturbing potential. This shift can be evaluated by use of the coupling constant integration technique with respect to V^{μ}. If one scales the associated Hamiltonian (8.43) by λ,

$$\hat{H}_{\text{ext}}(\lambda) = -\lambda e \int d^3 r \, \hat{j}^{\mu}(\boldsymbol{r}) V_{\mu}(\boldsymbol{r}), \qquad \text{(I.6)}$$

one obtains for the corresponding renormalized ground state energy

$$E(\lambda) = \langle \Psi_0(\lambda) | \hat{H}^{\text{hom}} + \hat{H}_{\text{ext}}(\lambda) | \Psi_0(\lambda) \rangle$$
$$- \langle 0 | \hat{H}^{\text{hom}} | 0 \rangle + \Delta E^{\text{hom}} + \Delta E^{\text{inhom}}(\lambda). \qquad \text{(I.7)}$$

Here $|\Psi_0(\lambda)\rangle$ denotes the ground state of the scaled Hamiltonian $\hat{H}^{\text{hom}} + \hat{H}_{\text{ext}}(\lambda)$. ΔE^{hom} provides the counterterms which, together with the vacuum expectation value $\langle 0 | \hat{H}^{\text{hom}} | 0 \rangle$, keep $E(\lambda)$ finite for $\lambda = 0$. $\Delta E^{\text{inhom}}(\lambda)$ contains all remaining counterterms. The energy of actual interest, corresponding to $\lambda = 1$, can be obtained by coupling constant integration, following the scheme in Sect. 4.2.1. Using proper normalization for all λ,

$$\langle\Psi_0(\lambda)|\Psi_0(\lambda)\rangle = 1\,,$$

one obtains by differentiation of (I.7) with respect to λ and subsequent integration from 0 to 1,

$$E(\lambda=1) = E^{\mathrm{RHEG}} - e\int_0^1 d\lambda \int d^3r\, j^\mu(\lambda,\boldsymbol{r})\, V_\mu(\boldsymbol{r}) + \Delta E^{\mathrm{inhom}}(\lambda=1)\,.$$

In this expression $j^\mu(\lambda,\boldsymbol{r})$ stands for the ground state current resulting for the coupling strength λ,

$$j^\mu(\lambda,\boldsymbol{r}) = \langle\Psi_0(\lambda)|\hat{j}^\mu(\boldsymbol{r})|\Psi_0(\lambda)\rangle\,,$$

and $E^{\mathrm{RHEG}} = E(\lambda=0)$ is to be understood as renormalized (ΔE^{hom} has been absorbed into E^{RHEG}). Insertion of (I.1) then allows to perform the λ-integration,

$$\begin{aligned} E = E^{\mathrm{RHEG}} &- e\int d^3r \langle\Psi_0(\lambda=0)|\hat{j}^\mu(\boldsymbol{r})|\Psi_0(\lambda=0)\rangle\, V_\mu(\boldsymbol{r}) \\ &+\sum_{n=2}^{\infty}\frac{(-e)^n}{n!}\int d^3r_1\ldots\int d^3r_n\,\chi^{(n)}_{\mathrm{c},\mu_1\ldots\mu_n}(\boldsymbol{r}_1,\ldots\boldsymbol{r}_n) \\ &\qquad\qquad\times V^{\mu_1}(\boldsymbol{r}_1)\ldots V^{\mu_n}(\boldsymbol{r}_n) \\ &+\Delta E^{\mathrm{inhom}}\,. \end{aligned} \tag{I.8}$$

At this point one can examine $\Delta E^{\mathrm{inhom}}$ in more detail. Its component linear in V^μ,

$$\Delta E^{\mathrm{inhom}} = -e\int d^3r\,\Delta E^{(1)}_\mu V^\mu(\boldsymbol{r}) + \mathcal{O}(V^2)\,,$$

has to keep the current expectation value of the unperturbed system, i.e. of the interacting RHEG, finite,

$$\langle\Psi_0(0)|\hat{j}_\mu(\boldsymbol{r})|\Psi_0(0)\rangle + \Delta E^{(1)}_\mu = n_0\, g_{\mu 0}\,. \tag{I.9}$$

It agrees with $\Delta j^{\mathrm{hom}}_\mu$ defined via Eq. (H.68). As discussed in Appendix F, $\Delta j^{\mathrm{hom}}_\mu$ vanishes in the noninteracting limit. All higher order ingredients of $\Delta E^{\mathrm{inhom}}$ are determined by the renormalization of the $\chi^{(n)}_{\mathrm{c},\mu_1\ldots\mu_n}$. The counterterm $\Delta E^{\mathrm{inhom}}$ is therefore closely related to $\Delta j^{\mathrm{inhom}}_\mu$, Eq. (I.3),

$$\begin{aligned} \Delta E^{\mathrm{inhom}} = \sum_{n=1}^{\infty}\frac{(-e)^n}{n!}\int d^3r_1\ldots\int d^3r_n\,\Delta\chi^{(n)}_{\mu_1\ldots\mu_n}(\boldsymbol{r}_1,\ldots\boldsymbol{r}_n) \\ \times V^{\mu_1}(\boldsymbol{r}_1)\ldots V^{\mu_n}(\boldsymbol{r}_n)\,, \end{aligned} \tag{I.10}$$

where $\Delta\chi^{(1)}_\mu \equiv \Delta j^{\mathrm{hom}}_\mu$ has been introduced for brevity. The only counterterm on the noninteracting level, corresponding to (I.5), is given by

$$\Delta E^{(0),\mathrm{inhom}} = -\frac{e^2}{24\pi^2}\Gamma\left(\frac{4-d}{2}\right)\int d^3r\, V_\mu(\boldsymbol{r})\nabla^2 V^\mu(\boldsymbol{r})\,. \tag{I.11}$$

The total energy counterterm (I.10) can be decomposed into contributions to the individual energy components. Again this can be demonstrated directly for the non-interacting case. Here the total energy is just a sum of the external potential energy,

$$E_{\text{ext}} = -e \int d^3r\, V_\mu(\boldsymbol{r}) \left[g^{\mu 0} n_0 + \delta j^\mu(\boldsymbol{r})\right] + \Delta E_{\text{ext}}^{(0),\text{inhom}}, \tag{I.12}$$

which, consistent with (I.5), requires the counterterm

$$\Delta E_{\text{ext}}^{(0),\text{inhom}} = -\frac{e^2}{12\pi^2} \Gamma\left(\frac{4-d}{2}\right) \int d^3r\, V_\mu(\boldsymbol{r}) \boldsymbol{\nabla}^2 V^\mu(\boldsymbol{r}), \tag{I.13}$$

and the noninteracting kinetic contribution T_s which absorbs the remainder of (I.11),

$$\Delta T_s^{\text{inhom}} = \frac{e^2}{24\pi^2} \Gamma\left(\frac{4-d}{2}\right) \int d^3r\, V_\mu(\boldsymbol{r}) \boldsymbol{\nabla}^2 V^\mu(\boldsymbol{r}). \tag{I.14}$$

The first order counterterms (I.5) and (I.11) are an explicit manifestation of the fact that, quite generally, Δj^μ and ΔE^{inhom} are completely determined by the external potential and the average density n_0 of the weakly inhomogeneous system. Only these two quantities enter Eqs. (I.3) and (I.10). The resulting dependence of Δj^μ and ΔE^{inhom} on V^μ is obvious, while that on n_0 results from the multi-loop contributions to the response functions.

Appendix J
Gradient Corrections to the Relativistic LDA

While the RLDA for $E_{xc}[j]$ is based on the xc-energy density of the RHEG, Eq. (H.47), the expansions (I.1) and (I.8) allow the derivation of systematic corrections to the RLDA. Restricting the discussion to the linear response contributions, Eq. (I.1) reduces to

$$\delta j^{\mu}(\boldsymbol{q}) = -e\chi^{\mu\nu}(q^0 = 0, \boldsymbol{q})\, V_{\nu}(\boldsymbol{q}), \tag{J.1}$$

with the total current given by $j^{\mu}(x) = n_0 g^{\mu 0} + \delta j^{\mu}(x)$ (response functions are always understood to be renormalized in this appendix, so that counterterms are not displayed explicitly; $\hbar = c = 1$ is again used). Using the inverse of $\chi^{\mu\nu}$, Eq. (H.36), one can rewrite (J.1) as

$$\chi^{-1}_{\rho\mu}(0,\boldsymbol{q})\delta j^{\mu}(\boldsymbol{q}) = -eV_{\rho}(\boldsymbol{q}), \tag{J.2}$$

where Coulomb gauge has been utilized. With Eqs. (J.1) and (J.2) the second order (V^2) contribution to (I.8) can be rewritten as

$$\delta E^{\mathrm{LR}} = -e\int \frac{d^3q}{(2\pi)^3}\, \delta j^{\mu}(\boldsymbol{q})\, V_{\mu}(\boldsymbol{q}) - \frac{1}{2}\int \frac{d^3q}{(2\pi)^3}\, \delta j^{\mu}(\boldsymbol{q})\, \chi^{-1}_{\mu\nu}(0,\boldsymbol{q})\, \delta j^{\nu}(-\boldsymbol{q}).$$

After insertion of the result (H.37) for the inverse response function,

$$\begin{aligned}\delta E^{\mathrm{LR}} &= -e\int \frac{d^3q}{(2\pi)^3}\, \delta j^{\mu}(\boldsymbol{q})\, V_{\mu}(\boldsymbol{q}) + \frac{1}{2}\int \frac{d^3q}{(2\pi)^3}\, \delta j^{\mu}(\boldsymbol{q})\, D^0(-\boldsymbol{q}^2)\, \delta j_{\mu}(-\boldsymbol{q}) \\ &\quad -\frac{1}{2}\int \frac{d^3q}{(2\pi)^3}\, \delta j^{\mu}(\boldsymbol{q})\, \Pi^{-1}_{\mu\nu}(0,\boldsymbol{q})\, \delta j^{\nu}(-\boldsymbol{q}),\end{aligned} \tag{J.3}$$

one can identify the first term as the linear response contribution to E_{ext}, Eq. (8.82), the second one as the induced Hartree energy (8.84). The third term represents the inhomogeneity corrections to the kinetic energy ($\delta T_{\mathrm{s}}^{\mathrm{LR}}$) and to the xc-energy ($\delta E_{\mathrm{xc}}^{\mathrm{LR}}$). $\delta T_{\mathrm{s}}^{\mathrm{LR}}$ is obtained from the noninteracting limit of $\Pi^{-1}_{\mu\nu}(\boldsymbol{q},0)$, so that the two contributions can be separated easily. Utilizing the tensor structure of $\Pi^{-1}_{\mu\nu}(\boldsymbol{q},0)$,

Eq. (H.38), as well as current conservation, $\boldsymbol{q}\cdot\boldsymbol{j}(\boldsymbol{q})=0$, one arrives at

$$\delta T_{\mathrm{s}}^{\mathrm{LR}}+\delta E_{\mathrm{xc}}^{\mathrm{LR}}=-\frac{1}{2}\int\frac{d^3q}{(2\pi)^3}\left\{\frac{\delta j^0(\boldsymbol{q})\,\delta j^0(-\boldsymbol{q})}{\Pi_{\mathrm{L}}(0,\boldsymbol{q})}+\frac{\boldsymbol{j}(\boldsymbol{q})\cdot\boldsymbol{j}(-\boldsymbol{q})}{\Pi_{\mathrm{T}}(0,\boldsymbol{q})}\right\}, \tag{J.4}$$

where $\delta\boldsymbol{j}(\boldsymbol{q})=\boldsymbol{j}(\boldsymbol{q})$ has been used.

In the next step one extracts that component of (J.4) which is part of the RLDA. This procedure involves only the δj^0-dependent term in (J.4) and follows closely the discussion of Sect. 4.4.1. For the weakly inhomogeneous gas of interest an expansion of the RLDA energy to the order $(\delta j^0)^2$ gives

$$\begin{aligned}&T_{\mathrm{s}}^{\mathrm{RLDA}}[n_0+\delta j^0]+E_{\mathrm{xc}}^{\mathrm{RLDA}}[n_0+\delta j^0]\\&=\int d^3r\,\left[t_{\mathrm{s}}(n_0+\delta j^0(\boldsymbol{r}))+e_{\mathrm{xc}}(n_0+\delta j^0(\boldsymbol{r}))\right]\\&=\int d^3r\left[t_{\mathrm{s}}(n_0)+e_{\mathrm{xc}}(n_0)+\left(\frac{dt_{\mathrm{s}}}{dn_0}(n_0)+\frac{de_{\mathrm{xc}}}{dn_0}(n_0)\right)\delta j^0(\boldsymbol{r})\right.\\&\qquad\left.+\frac{1}{2}\left(\frac{d^2t_{\mathrm{s}}}{dn_0^2}(n_0)+\frac{d^2e_{\mathrm{xc}}}{dn_0^2}(n_0)\right)\delta j^0(\boldsymbol{r})^2+\ldots\right].\end{aligned} \tag{J.5}$$

The first order term in (J.5) vanishes due to norm conservation, Eq. (I.2). The second order term can be rewritten by use of the compressibility sum rule (H.43),

$$\begin{aligned}&T_{\mathrm{s}}^{\mathrm{RLDA}}[n_0+\delta j^0]+E_{\mathrm{xc}}^{\mathrm{RLDA}}[n_0+\delta j^0]\\&=\int d^3r\,[t_{\mathrm{s}}(n_0)+e_{\mathrm{xc}}(n_0)]-\frac{1}{2}\int\frac{d^3q}{(2\pi)^3}\,\frac{\delta j^0(\boldsymbol{q})\,\delta j^0(-\boldsymbol{q})}{\Pi_{\mathrm{L}}^{(0)}(0,\boldsymbol{0})}.\end{aligned} \tag{J.6}$$

The second order term in (J.6) has to be subtracted from the complete inhomogeneity correction (J.4) as it is already contained in the RLDA,

$$\begin{aligned}\delta\tilde{T}_{\mathrm{s}}^{\mathrm{LR}}+\delta\tilde{E}_{\mathrm{xc}}^{\mathrm{LR}}=-\frac{1}{2}\int\frac{d^3q}{(2\pi)^3}&\left\{\delta j^0(\boldsymbol{q})\left[\frac{1}{\Pi_{\mathrm{L}}(0,\boldsymbol{q})}-\frac{1}{\Pi_{\mathrm{L}}(0,\boldsymbol{0})}\right]\delta j^0(-\boldsymbol{q})\right.\\&\left.+\frac{\boldsymbol{j}(\boldsymbol{q})\cdot\boldsymbol{j}(-\boldsymbol{q})}{\Pi_{\mathrm{T}}(0,\boldsymbol{q})}\right\}.\end{aligned} \tag{J.7}$$

Equation (J.7) explicitly demonstrates the current-dependence of relativistic density functionals. However, at this point $\delta\tilde{T}_{\mathrm{s}}^{\mathrm{LR}}$ and $\delta\tilde{E}_{\mathrm{xc}}^{\mathrm{LR}}$ are given as functionals of n_0 (inside $\Pi_{\mathrm{L/T}}$) and δj^0, but not yet as functionals of the complete density $j^0=n_0+\delta j^0$. Two paths can be followed towards the construction of actual density functionals: on the one hand, one can rewrite (J.7) as a fully nonlocal density functional utilizing either that $j^0(\boldsymbol{x})-j^0(\boldsymbol{y})=\delta j^0(\boldsymbol{x})-\delta j^0(\boldsymbol{y})$ [6, 85] or that $\nabla j^0(\boldsymbol{x})=\nabla\delta j^0(\boldsymbol{x})$ [158] (compare Sect. 4.4.2). On the other hand, one can restrict oneself to a long-wavelength expansion of the response kernels in (J.7), assuming $\delta j^\mu(\boldsymbol{q})$ to be strongly localized around $\boldsymbol{q}=\boldsymbol{0}$, i.e. $\delta j^\mu(\boldsymbol{x})$ to be rather delocalized. The latter approach leads to gradient corrections.

However, due to the limited information available for the relativistic polarization functions $\Pi_{L/T}$ no applications of (J.7) to E_{xc} have been reported so far. In order to illustrate the basic scheme of the gradient expansion we therefore consider T_s. After insertion of (H.39) and (H.40) into (J.7) and subsequent Fourier transformation one finds

$$\delta T_s^{[2]} = \frac{1}{72m}\int d^3x \frac{1}{n_0\eta}\left[1+2\frac{\beta}{\eta}\operatorname{arsinh}(\beta)\right]\left[\nabla\delta j^0(\boldsymbol{x})\right]^2 \\ +\frac{3\pi}{4}\int d^3x\int d^3y \frac{1}{\operatorname{arsinh}(\beta)}\frac{\boldsymbol{j}(\boldsymbol{x})\cdot\boldsymbol{j}(\boldsymbol{y})}{|\boldsymbol{x}-\boldsymbol{y}|}, \tag{J.8}$$

where the long-wavelength expansion has been taken into account to order $\boldsymbol{q}^2$ (denoted by the superscript [2]—β,η are given by Eq. (H.42)) and current conservation has been used in the second term. In the first term on the right-hand side of (J.8) one can now utilize $\nabla\delta j^0(\boldsymbol{x}) = \nabla j^0(\boldsymbol{x})$ and, correct to second order, $k_F = [3\pi^2 j^0(\boldsymbol{x})]^{1/3}$. However, the density-dependent prefactor $1/\operatorname{arsinh}(\beta)$ of the current component cannot be expressed unambiguously in terms of $j^0(\boldsymbol{x})$ as now two spatial variables are available. As in the case of the complete linear response corrections (J.7) one is left with a choice for this substitution.[1] If one abbreviates this (symmetric) function of $\boldsymbol{x}$ and $\boldsymbol{y}$ by $\bar{\beta}(\boldsymbol{x},\boldsymbol{y})$, one obtains

$$\delta T_s^{[2]}[j] = \frac{1}{72m}\int d^3x \frac{[\nabla j^0(\boldsymbol{x})]^2}{j^0(\boldsymbol{x})}\frac{1}{\eta}\left[1+2\frac{\beta}{\eta}\operatorname{arsinh}(\beta)\right] \\ +\frac{3\pi}{4}\int d^3x\int d^3y \frac{1}{\operatorname{arsinh}(\bar{\beta}(\boldsymbol{x},\boldsymbol{y}))}\frac{\boldsymbol{j}(\boldsymbol{x})\cdot\boldsymbol{j}(\boldsymbol{y})}{|\boldsymbol{x}-\boldsymbol{y}|}, \tag{J.9}$$

where β is now understood as $\beta = [3\pi^2 j^0(\boldsymbol{x})]^{1/3}/m$ ($\eta = \sqrt{1+\beta^2}$). Equation (J.9) demonstrates that current density functionals are inherently nonlocal, even in the long-wavelength limit.

One should note that the vacuum parts of $\Pi^{(0)}_{L/T}$ do not contribute to $\delta T_s^{[2]}$ as the normalization condition (G.48) together with (H.32) suppresses any vacuum contribution of the order $\boldsymbol{q}^2$. On the other hand, $\Pi^{(0)}_{v,R}$ does contribute to higher order inhomogeneity corrections (for details and a comparison with the real space gradient expansion of T_s see [532]).

In principle, this formalism can be extended to quadratic and cubic response, which allows the derivation of higher order gradient terms. In practice, however, the limited knowledge of the corresponding response functions restricts the usefulness of a first-principles determination of relativistic gradient corrections.

As a final point one should mention that gradient corrections to the relativistic $T_s[n]$ have also been derived by real-space methods [764–766, 763, 767]. These gradient terms serve as an extension of the relativistic Thomas-Fermi model [761, 768], in which the many-body problem is approached by direct solution of the basic vari-

[1] In contrast to the linear response approach the real space gradient expansion of $T_s[j]$ determines the current contribution to the second order gradient correction completely [763].

ational equation (8.72). A summary of results and details, as e.g. use of the renormalization procedure, can be found in [72, 532, 769, 770]. Expressing the noninteracting relativistic kinetic energy in terms of the density $n = j^0$,

$$T_s = \int d^3x \left\{ t_s^{[0]}[n] + t_s^{[2]}[n] + t_s^{[4]}[n] + \ldots \right\} , \tag{J.10}$$

one finds for the case of a purely electrostatic external potential (in contrast to the case of a full four potential) the expressions

$$t_s^{[0]}[n] = \frac{(3\pi^2 n)^{5/3}}{\pi^2 m} \frac{1}{\beta} \left[\frac{1}{8} \left(\beta\eta^3 + \beta^3\eta - \operatorname{arsinh}\beta \right) - \frac{\beta^3}{3} \right] \tag{J.11}$$

$$t_s^{[2]}[n] = \frac{1}{72m} \frac{(\nabla n)^2}{n\eta} \left[1 + 2\frac{\beta}{\eta} \operatorname{arsinh}\beta \right] . \tag{J.12}$$

The more complicated expression for $t_s^{[4]}[n]$ will not be given here. The agreement of (J.11) with (H.45) and of (J.12) with the density-dependent contribution to (J.9) is obvious.

Results for atoms and molecules obtained by direct application of the variational principle (without the intermediary of orbitals) are not of chemical accuracy. The functionals can, however, be useful for obtaining reasonable estimates of properties of systems that can not be investigated in such detail, for instance systems in the astrophysical field. For this purpose, it is of interest to note, that a temperature-dependent version of the relevant functionals has been derived as well [771].

Appendix K
Gordon Decomposition

The starting point for the derivation of the Gordon decomposition of the spatial components of the relativistic four current operator,

$$\hat{j}^\mu = \hat{\psi}^\dagger \alpha^\mu \hat{\psi} \qquad\qquad (\alpha^\mu = \gamma^0 \gamma^\mu), \tag{K.1}$$

is the field equation satisfied by the field operators $\hat{\psi}$, i.e. the Dirac equation,

$$\left(i\hbar c \gamma^\mu \partial_\mu - mc^2 - e\gamma^\mu A_\mu\right) \hat{\psi} = 0\,, \tag{K.2}$$

in which the potential A^μ may be operator-valued (∂_μ is defined in Eq. (8.4)). Note, however, that all subsequent steps can equally well be gone through for a current expressed in terms of single-particle orbitals,

$$j^\mu = \sum_k \Theta_k \phi_k^\dagger \alpha^\mu \phi_k\,,$$

as long as the orbitals satisfy a differential equation of the type (K.2) (as, for instance, the KS spinors).

The hermitian conjugate of (K.2) is given by

$$\hat{\psi}^\dagger \gamma^0 \left(-i\hbar c \gamma^\mu \overleftarrow{\partial_\mu} - mc^2 - e\gamma^\mu A_\mu\right) \gamma^0 = 0\,, \tag{K.3}$$

as $(\gamma^\mu)^\dagger = \gamma^0 \gamma^\mu \gamma^0$. The vector bar over $\overleftarrow{\partial_\mu}$ indicates that the derivative acts on the field operator to its left. Contraction of the field equation (K.2) with $\hat{\psi}^\dagger \alpha^k$ and of its hermitian conjugate (K.3) with $\alpha^k \hat{\psi}$ gives

$$\hat{\psi}^\dagger \alpha^k \left(i\hbar c \gamma^\mu \partial_\mu - mc^2 - e\gamma^\mu A_\mu\right) \hat{\psi} = 0$$
$$\hat{\psi}^\dagger \gamma^0 \left(-i\hbar c \gamma^\mu \overleftarrow{\partial_\mu} - mc^2 - e\gamma^\mu A_\mu\right) \gamma^0 \alpha^k \hat{\psi} = 0\,.$$

If one adds up both equations, one obtains

$$\hat{\psi}^\dagger \left[i\hbar c \gamma^0 \left(\gamma^k \gamma^\mu \partial_\mu - \gamma^\mu \gamma^k \overleftarrow{\partial_\mu} \right) - 2mc^2 \alpha^k - e\gamma^0 \left(\gamma^k \gamma^\mu + \gamma^\mu \gamma^k \right) A_\mu \right] \hat{\psi} = 0 .$$

In the next step the scalar products in Minkowski space are split into their space-time components, utilizing the commutation relations (8.7)–(8.9),

$$\begin{aligned}\hat{\psi}^\dagger \Big[-i\hbar c \left(\gamma^k \partial_0 + \gamma^k \overleftarrow{\partial_0} \right) + i\hbar c \gamma^0 \left(\gamma^k \gamma^l \partial_l - \gamma^l \gamma^k \overleftarrow{\partial_l} \right) \\ - 2mc^2 \alpha^k - 2e\gamma^0 A^k \Big] \hat{\psi} = 0 .\end{aligned}$$

At this point one can use

$$\begin{aligned}\gamma^k \gamma^l &= \frac{1}{2} \left(\left\{ \gamma^k, \gamma^l \right\} + \left[\gamma^k, \gamma^l \right] \right) \\ &= g^{kl} - i\varepsilon_{klj} \Sigma^j \qquad \text{with} \quad \Sigma^j = \begin{pmatrix} \sigma^j & 0 \\ 0 & \sigma^j \end{pmatrix}\end{aligned} \tag{K.4}$$

to obtain

$$\begin{aligned}- i\hbar c \partial_0 \left(\hat{\psi}^\dagger \gamma^k \hat{\psi} \right) + i\hbar c \hat{\psi}^\dagger \gamma^0 \left(\partial^k - \overleftarrow{\partial^k} \right) \hat{\psi} + \hbar c\, \varepsilon_{klj} \partial_l \left(\hat{\psi}^\dagger \gamma^0 \Sigma^j \hat{\psi} \right) \\ - 2mc^2 \hat{\psi}^\dagger \alpha^k \hat{\psi} - 2eA^k \hat{\psi}^\dagger \gamma^0 \hat{\psi} = 0 .\end{aligned}$$

Most of the individual terms in this equation are easily identified with established quantities. With the definitions[1]

$$\hat{j}^\mu = \left(\hat{n}, \frac{\hat{\boldsymbol{j}}}{c} \right) \tag{K.5}$$

$$\hat{\boldsymbol{j}}_\mathrm{p} = -\frac{i\hbar}{2m} \hat{\psi}^\dagger \gamma^0 \left(\overrightarrow{\boldsymbol{\nabla}} - \overleftarrow{\boldsymbol{\nabla}} \right) \hat{\psi} \tag{K.6}$$

$$\hat{\boldsymbol{m}} = \frac{e\hbar}{2mc} \hat{\psi}^\dagger \gamma^0 \boldsymbol{\Sigma} \hat{\psi} \tag{K.7}$$

$$\hat{\rho}_\mathrm{s} = \hat{\psi}^\dagger \gamma^0 \hat{\psi} \tag{K.8}$$

(note the additional factor of c which is included in the three-vector $\boldsymbol{j}$ as compared to the spatial components of j^μ!) for the paramagnetic current $\hat{\boldsymbol{j}}_\mathrm{p}$, the magnetization density $\hat{\boldsymbol{m}}$ and the scalar density $\hat{\rho}_\mathrm{s}$ one finds

$$- i\hbar c \partial_0 \left(\hat{\psi}^\dagger \boldsymbol{\gamma} \hat{\psi} \right) + 2mc \hat{\boldsymbol{j}}_\mathrm{p} + \frac{2mc^2}{e} \boldsymbol{\nabla} \times \hat{\boldsymbol{m}} - 2mc \hat{\boldsymbol{j}} - 2e\boldsymbol{A} \hat{\rho}_\mathrm{s} = 0 .$$

Extracting the spatial components of the current, one finally ends up with

---

[1] Note the relation between the gradient vector and the covariant components ∂_k,

$$\boldsymbol{\nabla} = \left(\frac{\partial}{\partial r^1}, \frac{\partial}{\partial r^2}, \frac{\partial}{\partial r^3} \right) = (\partial_1, \partial_2, \partial_3) .$$

$$\hat{\boldsymbol{j}} = -\frac{i\hbar}{2m}\partial_0\left(\hat{\psi}^\dagger \boldsymbol{\gamma}\hat{\psi}\right) + \hat{\boldsymbol{j}}_{\rm p} + \frac{c}{e}\boldsymbol{\nabla}\times\hat{\boldsymbol{m}} - \frac{e}{mc}\boldsymbol{A}\hat{\rho}_{\rm s}\,. \tag{K.9}$$

In the case of stationary systems the first operator on the right-hand side does not contribute to any expectation value of $\hat{\boldsymbol{j}}$,

$$\langle\Psi_0|\hat{\boldsymbol{j}}|\Psi_0\rangle = \langle\Psi_0|\hat{\boldsymbol{j}}_{\rm p}|\Psi_0\rangle + \frac{c}{e}\boldsymbol{\nabla}\times\langle\Psi_0|\hat{\boldsymbol{m}}|\Psi_0\rangle - \frac{e}{mc}\boldsymbol{A}\,\langle\Psi_0|\hat{\rho}_{\rm s}|\Psi_0\rangle\,. \tag{K.10}$$

Note that the precise definition of $\boldsymbol{A}$ (in particular, its sign) is determined by the differential equation (K.2).

Appendix L
Some Useful Formulae

The body of this text relies on the knowledge of a largish number of mathematical relations. A much abbreviated list is offered here.

- Laurent expansion, theorem of residues:
  A complex-valued function $f(z)$, which is analytic in the domain $\mathscr{D} \subset \mathbb{C}$ and has an isolated pole of k-th order at a point z_0 enclosed by $\mathscr{D}$, can be expanded for all $z \in \mathscr{D}$ as (Laurent expansion)

$$f(z) = \sum_{n=-\infty}^{\infty} a_n (z - z_0)^n \,, \tag{L.1}$$

  with the coefficients a_n given by

$$a_n = \oint_{\mathscr{C}} \frac{dz'}{2\pi i} \frac{f(z')}{(z' - z_0)^{n+1}} \,. \tag{L.2}$$

  Here the closed path $\mathscr{C}$ is contained fully in the domain, has a counterclockwise orientation, and encloses z_0, but no other singular point. The coefficient a_{-1}, the residue, is given by

$$a_{-1} = \oint_{\mathscr{C}} \frac{dz}{2\pi i} f(z) \,. \tag{L.3}$$

  A generalization to the case that the path encloses a set of isolated poles is the theorem of residues,

$$\oint_{\mathscr{C}} \frac{dz}{2\pi i} f(z) = \text{sum of the residues of all poles enclosed by } \mathscr{C} \,. \tag{L.4}$$

  A prominent example for the application of (L.4) is the contour integral representation of the step function,

$$\Theta(x) = \int_{-\infty}^{\infty} \frac{d\omega}{2\pi i} \frac{e^{i\omega x}}{\omega - i\eta} \,, \tag{L.5}$$

which is used to implement the time-ordering required for many response and Green's functions in frequency space.

- Dirac identity:
For integrations over frequency often the integral representation of the δ-function,

$$\delta(x) = \int_{-\infty}^{\infty} \frac{d\omega}{2\pi} e^{i\omega x}, \tag{L.6}$$

or the Dirac identity,

$$2\pi i\,\delta(x) = \frac{1}{x-i\eta} - \frac{1}{x+i\eta} \tag{L.7}$$

$$\frac{1}{x-i\eta} = \mathscr{P}\frac{1}{x} + \pi i\,\delta(x) \tag{L.8}$$

$$\frac{1}{x+i\eta} = \mathscr{P}\frac{1}{x} - \pi i\,\delta(x) \tag{L.9}$$

is used ($\mathscr{P}$ denotes the Cauchy principal value integral).

- Fourier representation of Coulomb interaction:
Whenever the Coulomb interaction has to be integrated over the complete space, use of the following regularized form is necessary

$$\int \frac{d^3q}{(2\pi)^3} \frac{e^{i\boldsymbol{q}\cdot\boldsymbol{r}}}{\boldsymbol{q}^2+\mu^2} = \frac{e^{-\mu|\boldsymbol{r}|}}{4\pi|\boldsymbol{r}|}. \tag{L.10}$$

- General identities for commutators:

$$[\hat{A}\hat{B},\hat{C}] = \hat{A}[\hat{B},\hat{C}] + [\hat{A},\hat{C}]\hat{B} \tag{L.11}$$

$$e^{\hat{A}}\,\hat{B}\,e^{-\hat{A}} = \sum_{n=0}^{\infty} \frac{1}{n!} \underbrace{[\hat{A},[\hat{A},[\ldots[\hat{A},\hat{B}]]\ldots]]}_{n \text{ times}}. \tag{L.12}$$

- Commutators involving field operators:
The following basic commutator of four field operators can be derived directly from the anticommutation rules (2.6) and (2.7),

$$\begin{aligned}
&\left[\hat{\psi}^\dagger(\boldsymbol{r}_1\sigma_1)\hat{\psi}(\boldsymbol{r}_2\sigma_2), \hat{\psi}^\dagger(\boldsymbol{r}_3\sigma_3)\hat{\psi}(\boldsymbol{r}_4\sigma_4)\right] \\
&= -\delta^{(3)}(\boldsymbol{r}_1-\boldsymbol{r}_4)\,\delta_{\sigma_1\sigma_4}\,\hat{\psi}^\dagger(\boldsymbol{r}_3\sigma_3)\hat{\psi}(\boldsymbol{r}_2\sigma_2) \\
&\quad +\delta^{(3)}(\boldsymbol{r}_2-\boldsymbol{r}_3)\,\delta_{\sigma_2\sigma_3}\,\hat{\psi}^\dagger(\boldsymbol{r}_1\sigma_1)\hat{\psi}(\boldsymbol{r}_4\sigma_4).
\end{aligned} \tag{L.13}$$

Use of this result leads to

$$0 = \left[\hat{\psi}^\dagger(\boldsymbol{r}\sigma)\hat{\psi}(\boldsymbol{r}\sigma), \hat{\psi}^\dagger(\boldsymbol{r}'\sigma')\hat{\psi}(\boldsymbol{r}'\sigma')\right] \tag{L.14}$$

$$0 = \left[\int d^3r\,\hat{\psi}^\dagger(\boldsymbol{r}\sigma)\boldsymbol{\nabla}^2\hat{\psi}(\boldsymbol{r}\sigma), \int d^3r'\,\hat{\psi}^\dagger(\boldsymbol{r}'\sigma')\hat{\psi}(\boldsymbol{r}'\sigma')\right] \tag{L.15}$$

$$\begin{aligned} 0 &= \left[\hat{\psi}^\dagger(\boldsymbol{r}\sigma)\hat{\psi}(\boldsymbol{r}\sigma)\hat{\psi}^\dagger(\boldsymbol{r}'\sigma')\hat{\psi}(\boldsymbol{r}'\sigma'),\hat{\psi}^\dagger(\boldsymbol{r}''\sigma'')\hat{\psi}(\boldsymbol{r}''\sigma'')\right] \\ &= \left[\hat{\psi}^\dagger(\boldsymbol{r}\sigma)\hat{\psi}^\dagger(\boldsymbol{r}'\sigma')\hat{\psi}(\boldsymbol{r}'\sigma')\hat{\psi}(\boldsymbol{r}\sigma),\hat{\psi}^\dagger(\boldsymbol{r}''\sigma'')\hat{\psi}(\boldsymbol{r}''\sigma'')\right]. \end{aligned} \tag{L.16}$$

With the relation (L.13) one can also evaluate the commutator of the density (2.4) and the paramagnetic current (2.158)

$$\begin{aligned} \left[\hat{\boldsymbol{j}}_{\mathrm{p}}(\boldsymbol{r}),\hat{n}(\boldsymbol{r}')\right] &= \frac{-i\hbar}{2m}\left[\boldsymbol{\nabla}\delta^{(3)}(\boldsymbol{r}-\boldsymbol{r}')\right]\sum_\sigma\left[\hat{\psi}^\dagger(\boldsymbol{r}\sigma)\hat{\psi}(\boldsymbol{r}'\sigma)+\hat{\psi}^\dagger(\boldsymbol{r}'\sigma)\hat{\psi}(\boldsymbol{r}\sigma)\right] \\ &\quad+\frac{i\hbar}{2m}\delta^{(3)}(\boldsymbol{r}-\boldsymbol{r}')\boldsymbol{\nabla}\hat{n}(\boldsymbol{r}). \end{aligned} \tag{L.17}$$

Similarly, one obtains for the commutator of the kinetic energy and density operators

$$\left[\hat{T},\hat{n}(\boldsymbol{r})\right] = i\hbar\boldsymbol{\nabla}\cdot\hat{\boldsymbol{j}}_{\mathrm{p}}(\boldsymbol{r}). \tag{L.18}$$

- Pauli matrices:
  The basic commutators (anticommutators) of the Pauli matrices,

$$[\sigma_i,\sigma_j] = 2i\sum_k\varepsilon_{ijk}\,\sigma_k \tag{L.19}$$

$$\{\sigma_i,\sigma_j\} = 2\delta_{ij}\,, \tag{L.20}$$

indicate that these operators are generators of the group $SU(2)$. The matrices are hermitian $\sigma_i^\dagger = \sigma_i$. These properties, together with a statement on the two eigenvalues of σ_z, allows the determination of an explicit representation

$$\sigma_x = \begin{pmatrix} 0 & 1 \\ 1 & 0 \end{pmatrix} \qquad \sigma_y = \begin{pmatrix} 0 & -i \\ i & 0 \end{pmatrix} \qquad \sigma_z = \begin{pmatrix} 1 & 0 \\ 0 & -1 \end{pmatrix}. \tag{L.21}$$

For handling Pauli matrices the following identity is often helpful,

$$(\boldsymbol{\sigma}\cdot\boldsymbol{a})\,(\boldsymbol{\sigma}\cdot\boldsymbol{b}) = \boldsymbol{a}\cdot\boldsymbol{b}+i\boldsymbol{\sigma}\cdot(\boldsymbol{a}\times\boldsymbol{b})\,, \tag{L.22}$$

which can be derived from the basic commutators.
The corresponding commutation relations of the relativistic (4×4) Pauli matrices (8.138) are

$$\left[\Sigma^i,\Sigma^j\right] - 2i\sum_k\varepsilon_{ijk}\,\Sigma^k \tag{L.23}$$

$$\left\{\Sigma^i,\Sigma^j\right\} = -2g^{ij}. \tag{L.24}$$

A standard representation is

$$\boldsymbol{\Sigma} = \begin{pmatrix} \boldsymbol{\sigma} & \boldsymbol{0} \\ \boldsymbol{0} & \boldsymbol{\sigma} \end{pmatrix}. \tag{L.25}$$

- Feynman rules:
These rules define the transition between the Feynman diagrams representing the individual contributions to the perturbation expansion of Green's functions (and all quantities related to them) and the corresponding analytical expressions. They are set up here in a rather general form which allows a coherent treatment of all kinds of (sub)diagrams in real space, including diagrams contributing to n-point functions such as the self-energy and the response function. The critical diagrams which require this extended form are those for which there are vertices to which only a single line (electron or interaction) is attached, as for instance the first order self-energy (3.125).
The four basic elements of Feynman diagrams representing the perturbation expansion for inhomogeneous systems are:

  - The noninteracting (as for instance the KS) Green's function (3.124),

$$G_0(\boldsymbol{r}\sigma t, \boldsymbol{r}'\sigma' t') = \int \frac{d\omega}{2\pi} e^{-i\omega(t-t')} G_0(\boldsymbol{r}\sigma, \boldsymbol{r}'\sigma', \omega) \tag{L.26}$$

$= \boldsymbol{r}'\sigma' t'$  $\boldsymbol{r}\sigma t$

$$G_0(\boldsymbol{r}\sigma, \boldsymbol{r}'\sigma', \omega) = \sum_l \left\{ (1-\Theta_l) \frac{\phi_l(\boldsymbol{r}\sigma)\phi_l^*(\boldsymbol{r}'\sigma')}{\omega - \varepsilon_l/\hbar + i\eta} + \Theta_l \frac{\phi_l(\boldsymbol{r}\sigma)\phi_l^*(\boldsymbol{r}'\sigma')}{\omega - \varepsilon_l/\hbar - i\eta} \right\}.$$

  - The Coulomb interaction, suitably extended to the time domain, in order to simplify the rules,

$w(\boldsymbol{r}-\boldsymbol{r}', t-t') = \delta(t-t') \frac{e^2}{|\boldsymbol{r}-\boldsymbol{r}'|} = \boldsymbol{r}'t'$  $\boldsymbol{r}t$. (L.27)

  - The simple vertex, drawn as a bold dot,

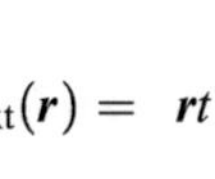
$\boldsymbol{r}\sigma t$.

    The lines attached to the dot only serve as an indication that there are (at most) two solid and one wiggly line connected to a single vertex.
  - If present, an additional perturbing external potential,

$v_{\rm ext}(\boldsymbol{r}) = \boldsymbol{r}t$ 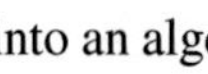 (L.28)

Any given Feynman diagram is translated into an algebraic expression according to the following rules:

1. Distinguish between simple endpoints of lines (be it solid or wiggly), i.e. endpoints not attached to a bold dot, and endpoints at vertices, characterized by the bold dot. The former endpoints will be called external points in the following, the latter internal endpoints.

2. Distinguish between internal vertices, at which one incoming and one outgoing solid (electron) line and one wiggly (interaction) line meet, and external vertices, at which one or two of these attached lines are missing.
3. Label all vertices and external points by some number i.
4. Replace each directed solid line for which the arrow points from a vertex or an external point j to a vertex or an external point i by $G_0(\boldsymbol{r}_i\sigma_i t_i, \boldsymbol{r}_j\sigma_j t_j)$.
5. Replace each wiggly line connecting the vertices or external points i and j by $w(\boldsymbol{r}_i - \boldsymbol{r}_j, t_i - t_j)$ (the direction plays no role, as the interaction is symmetric under exchange of its arguments).
6. Replace each wiggly line with a cross at its end attached to vertex i by $v_{\mathrm{ext}}(\boldsymbol{r}_i)$.
7. Replace each external vertex i with only one line attached by

$$\delta^{(3)}(\boldsymbol{r}_i - \boldsymbol{r}_{i'})\,\delta(t_i - t_{i'})\,\delta_{\sigma_i,\sigma_{i'}}\ .$$

   The space-time labels $\boldsymbol{r}_i t_i$ and $\boldsymbol{r}_{i'} t_{i'}$ are two of the arguments of the n-point function to which the diagram contributes. The same applies to both spin labels, if a wiggly line is attached to the vertex. On the other hand, if a solid line is attached to the vertex, the spin label σ_i is the spin index of this internal solid line attached, while $\sigma_{i'}$ denotes a spin argument of the n-point function.
8. Integrate over all coordinates and times associated with internal vertices,

$$\int d^3r_i\,dt_i\ ,$$

   and sum over all spins associated with internal endpoints of solid lines.
9. If the rules lead to Green's functions G_0 for which both time arguments coincide, $t_i = t_j = t$, interpret these functions as $G_0(\boldsymbol{r}_i\sigma_i t, \boldsymbol{r}_j\sigma_j t + \eta)$ and take the limit $\eta \to 0^+$ at the end of the calculation. This can only happen if the solid line ends at the same point as it starts, or if the start and end point of G_0 are connected by a single interaction line. This procedure ensures the proper operator ordering of $\hat{\psi}_0$ and $\hat{\psi}_0^\dagger$ at equal times.
10. Multiply the resulting expression by a factor of

$$(-i/\hbar)^{n+m}\, i^l\, (-1)^F$$

   for a diagram which contains n interaction, m external potential and l electron lines as well as F closed loops of solid lines.
11. Adjust the overall prefactor to the quantity (Green's or response function, density, energy, ...) which is evaluated, in accordance with the definition of this quantity. In the case of an energy (or vacuum amplitude) diagram take care of the multiplicities involved (see e.g. [95]).

For a homogeneous system, for which no external potential is present, a representation in momentum space is the appropriate choice. The three remaining basic elements after Fourier transformation are:

– The noninteracting Green's function:

$$G_0(\boldsymbol{k},\omega,\sigma\sigma') = \delta_{\sigma\sigma'}\left[\frac{\Theta(|\boldsymbol{k}|-k_{\rm F})}{\omega-\varepsilon_k/\hbar+i\eta}+\frac{\Theta(k_{\rm F}-|\boldsymbol{k}|)}{\omega-\varepsilon_k/\hbar-i\eta}\right] =$$

. (L.29)

– The Coulomb interaction (in the screened form given above):

$$w(\boldsymbol{q}) = \frac{4\pi e^2}{q^2+\mu^2} =$$

. (L.30)

– The simple vertex:

. (L.31)

The Feynman rules for a distinct diagram are in this case:

1. Assign a direction to each interaction (wiggly) line; associate energy (ω) and momentum to each line (be it solid or wiggly) and conserve energy and momentum at each vertex.
2. Replace each directed solid line for which the arrow points from vertex or external point j to vertex or external point i by $G_0(\boldsymbol{k},\omega,\sigma_i\sigma_j)$.
3. If a solid line ends at the same point as it starts, or if the start and end point of a solid line are connected by a single interaction line, interpret the associated G_0 as $e^{i\omega\eta}G_0(\boldsymbol{k},\omega,\sigma_i\sigma_j)$ and take the limit $\eta \to 0^+$ at the end of the calculation.
4. Replace each wiggly line by $w(\boldsymbol{q})$.
5. Wherever two solid lines meet at some vertex conserve the spin σ at the vertex and sum over σ.
6. Integrate over all energies and momenta which do not correspond to arguments of the Green's or n-point function.
7. Multiply the resulting expression by a factor of

$$(-i/\hbar)^{n+m}\, i^l\, (-1)^F$$

   for a diagram which contains n interaction, m external potential and l electron lines as well as F closed loops of solid lines.
8. Adjust the overall prefactor to the quantity (Green's or response function, density, energy, ...) which is evaluated, in accordance with the definition of this quantity. In the case of an energy (or vacuum amplitude) diagram take care of the multiplicities involved.

These rules are identical with those given in [94], Chaps. 9–12.

References

1. M. Born, R. Oppenheimer, Ann. Physik **84**, 457 (1927)
2. A. Szabo, N.S. Ostlund, *Modern Quantum Chemistry* (Dover, New York, NY, 1989)
3. P. Fulde, *Electron Correlations in Molecules and Solids* (Springer, Berlin, 1991)
4. W. Kohn, Rev. Mod. Phys. **71**, 1253 (1999)
5. B.G. Johnson, P.M.W. Gill, J.A. Pople, J. Chem. Phys. **97**, 7846 (1992)
6. P. Hohenberg, W. Kohn, Phys. Rev. **136 B**, 864 (1964)
7. R.M. Dreizler, E.K.U. Gross, *Density Functional Theory* (Springer, Berlin, 1990)
8. N.D. Mermin, Phys. Rev. **137 A**, 1441 (1965)
9. G.B. Bachelet, D.R. Hamann, M. Schlüter, Phys. Rev. B **26**, 4199 (1982)
10. N. Troullier, J.L. Martins, Phys. Rev. B **43**, 1993 (1991)
11. T.L. Gilbert, Phys. Rev. B **12**, 2111 (1975)
12. M. Berrondo, O. Goscinski, Int. J. Quantum Chem. **9S**, 67 (1975)
13. L.H. Thomas, Proc. Cambridge Philos. Soc. **23**, 542 (1927)
14. E. Fermi, Z. Phys. **48**, 73 (1928)
15. R.G. Parr, W. Yang, *Density-Functional Theory of Atoms and Molecules* (Oxford University Press, New York, NY, 1989)
16. W. Kohn, in *Highlights of Condensed Matter Theory*, ed. by F. Bassani, F. Fumi, M.P. Tosi (North-Holland, Amsterdam, 1985), p. 1
17. M. Levy, Phys. Rev. A **26**, 1200 (1982)
18. E.H. Lieb, in *Physics as Natural Philosophy*, ed. by A. Shimony, H. Feshbach (MIT Press, Cambridge, 1982), p. 111
19. H. Englisch, R. Englisch, Physica **121A**, 253 (1983)
20. J.T. Chayes, L. Chayes, M.B. Ruskai, J. Stat. Phys. **38**, 497 (1985)
21. M. Reed, B. Simon, *Methods of Modern Mathematical Physics, Vol.4* (Academic, New York, NY, 1978)
22. M. Levy, Proc. Natl. Acad. Sci. **76**, 6062 (1979)
23. E.H. Lieb, Int. J. Quantum Chem. **24**, 243 (1983)
24. J.E. Harriman, Phys. Rev. A **24**, 680 (1981)
25. G. Zumbach, K. Maschke, Phys. Rev. A **28**, 544 (1983)
26. G. Zumbach, K. Maschke, Phys. Rev. A **29**, E 1585 (1984)
27. R. van Leeuwen, Adv. Quantum Chem. **43**, 25 (2003)
28. H. Eschrig, *The Fundamentals of Density Functional Theory* (Edition am Gutenbergplatz, Leipzig, 2003)
29. K. Atkinson, W. Han, *Theoretical Numerical Analysis: A Functional Analysis Framework* (Springer, Berlin, 2009)
30. H. Englisch, R. Englisch, Phys. Stat. Sol. (b) **123**, 711 (1984)
31. H. Englisch, R. Englisch, Phys. Stat. Sol. (b) **124**, 373 (1984)

32. I. Ekeland, R. Teman, *Convex Analysis and Variational Problems* (North-Holland, Amsterdam, 1976)
33. J.P. Perdew, R.G. Parr, M. Levy, J.L. Balduz, Phys. Rev. Lett. **49**, 1691 (1982)
34. U. von Barth, L. Hedin, J. Phys. C **5**, 1629 (1972)
35. M.M. Pant, A.K. Rajagopal, Solid State Commun. **10**, 1157 (1972)
36. R.D. Jackson, *Classical Electrodynamics* (Wiley, New York, NY, 1975)
37. K. Capelle, G. Vignale, Phys. Rev. Lett. **86**, 5546 (2001)
38. H. Eschrig, W. Pickett, Solid State Commun. **118**, 123 (2001)
39. N.I. Gidopoulos, in *The Fundamentals of Electron Density, Density Matrix and Density Functional Theory in Atoms, Molecules and the Solid State*, ed. by N.I. Gidopoulos, S. Wilson (Kluwer, Dordrecht, 2003), p. 195
40. N.I. Gidopoulos, Phys. Rev. B **75**, 134408 (2007)
41. N. Argaman, G. Makov, Phys. Rev. B **66**, 052413 (2002)
42. Y. Tsunoda, J. Phys.: Condens. Matter **1**, 10427 (1989)
43. K. Knöpfle, L.M. Sandratskii, J. Kübler, Phys. Rev. B **62**, 5564 (2000)
44. M. Uhl, L.M. Sandratskii, J. Kübler, J. Magn. Magn. Mater. **103**, 314 (1992)
45. J. Kübler, K.H. Höck, J. Sticht, A.R. Williams, J. Phys. F **18**, 469 (1988)
46. J. Sticht, K.H. Höck, J. Kübler, J. Phys.: Condens. Matter **1**, 8155 (1989)
47. L.M. Sandratskii, Adv. Phys. **47**, 91 (1998)
48. D.M. Bylander, L. Kleinman, Phys. Rev. B **60**, R9916 (1999)
49. M. Marsman, J. Hafner, Phys. Rev. B **66**, 224409 (2002)
50. E. Sjöstedt, L. Nordström, Phys. Rev. B **66**, 014447 (2002)
51. D. Hobbs, J. Hafner, D. Spišák, Phys. Rev. B **68**, 014407 (2003)
52. M. Uchida, Y. Onose, Y. Matsui, Y. Tokura, Science **311**, 359 (2006)
53. A.M.N. Niklasson, B. Johansson, L. Nordström, Phys. Rev. Lett. **82**, 4544 (1999)
54. K. Hirai, J. Magn. Magn. Mater. **226**, 1697 (2001)
55. A.M.N. Niklasson, J.M. Wills, L. Nordström, Phys. Rev. B **63**, 104417 (2001)
56. M. Bode, M. Heide, K. von Bergmann, P. Ferriani, S. Heinze, G. Bihlmayer, A. Kubetzka, O. Pietzsch, S. Blügel, R. Wiesendanger, Nature **447**, 190 (2007)
57. T. Oda, A. Pasquarello, R. Car, Phys. Rev. Lett. **80**, 3622 (1998)
58. A. Bergman, L. Nordström, A.B. Klautau, S. Frota-Pessõa, O. Eriksson, Phys. Rev. B **75**, 224425 (2007)
59. M.B. Knickelbein, Phys. Rev. Lett. **86**, 5255 (2001)
60. D.C. Douglass, J.P. Bucher, L.A. Bloomfield, Phys. Rev. B **45**, 6341 (1992)
61. H. Eschrig, V.D.P. Servedio, J. Comput. Chem. **20**, 23 (1999)
62. L. Nordström, D.J. Singh, Phys. Rev. Lett. **76**, 4420 (1996)
63. I.V. Solovyev, A.I. Liechtenstein, V.A. Gubanov, V.P. Antropov, O.K. Andersen, Phys. Rev. B **43**, 14414 (1991)
64. E. Fawcett, Rev. Mod. Phys. **60**, 209 (1988)
65. K. Hirai, J. Phys. Soc. Jpn. **66**, 560 (1997)
66. R. Hafner, D. Spišák, R. Lorenz, J. Hafner, J. Phys.: Condens. Matter **13**, L239 (2001)
67. S. Cottenier, B.D. Vries, J. Meersschaut, M. Rots, J. Phys.: Condens. Matter **14**, 3275 (2002)
68. G. Vignale, M. Rasolt, Phys. Rev. B **37**, 10685 (1988)
69. G. Vignale, M. Rasolt, Adv. Quantum Phys. **21**, 235 (1990)
70. M. Rasolt, G. Vignale, Phys. Rev. Lett. **65**, 1498 (1990)
71. G. Vignale, in *Density Functional Theory*, *NATO ASI Series B*, vol. 337, ed. by E.K.U. Gross, R.M. Dreizler (Plenum, New York, NY, 1995), p. 485
72. E. Engel, H. Müller, C. Speicher, R.M. Dreizler, in *Density Functional Theory*, *NATO ASI Series B*, vol. 337, ed. by E.K.U. Gross, R.M. Dreizler (Plenum, New York, NY, 1995), p. 65
73. K. Capelle, E.K.U. Gross, Phys. Rev. Lett. **78**, 1872 (1997)
74. G. Vignale, M. Rasolt, Phys. Rev. Lett. **59**, 2360 (1987)
75. S. Erhard, E.K.U. Gross, Phys. Rev. A **53**, R5 (1996)
76. K. Capelle, G. Vignale, Phys. Rev. B **65**, 113106 (2002)
77. O. Gunnarsson, B.I. Lundqvist, Phys. Rev. B **13**, 4274 (1976)

78. T. Ziegler, A. Rauk, E.J. Baerends, Theor. Chim. Acta **43**, 261 (1977)
79. U. von Barth, Phys. Rev. A **20**, 1693 (1979)
80. T. Kinoshita, Phys. Rev. **105**, 1490 (1957)
81. E.R. Davidson, S.A. Hagstrom, S.J. Chakravorty, V.M. Umar, C. Froese Fischer, Phys. Rev. A **44**, 7071 (1991)
82. E.V. Ludeña, R. Lõpez-Boada, J. Maldonado, T. Koga, E.S. Kryachko, Phys. Rev. A **48**, 1937 (1993)
83. C.J. Umrigar, X. Gonze, Phys. Rev. A **50**, 3827 (1994)
84. Q. Zhao, R.G. Parr, Phys. Rev. A **46**, 2337 (1992)
85. W. Kohn, L.J. Sham, Phys. Rev. **140 A**, 1133 (1965)
86. R. Gáspár, Acta Phys. Acad. Sci. Hung. **3**, 263 (1954)
87. E.H. Lieb, in *Density Functional Methods in Physics, NATO ASI Series B*, vol. 123, ed. by R.M. Dreizler, J. da Providencia (Plenum, New York, NY, 1985), p. 31
88. J.F. Janak, Phys. Rev. B **18**, 7165 (1978)
89. J.C. Slater, J.H. Wood, Int. J. Quantum Chem. Suppl. **4**, 3 (1971)
90. J.C. Slater, Adv. Quantum Chem. **6**, 1 (1972)
91. A.K. Rajagopal, J. Callaway, Phys. Rev. B **7**, 1912 (1973)
92. C.O. Almbladh, U. von Barth, Phys. Rev. B **31**, 3231 (1985)
93. J.M. Luttinger, Phys. Rev. **119**, 1153 (1960)
94. A.L. Fetter, J.D. Walecka, *Quantum Theory of Many-Particle Systems* (McGraw-Hill, New York, NY, 1971)
95. J.W. Negele, H. Orland, *Quantum Many-Particle Systems* (Addison-Wesley, Redwood City, CA, 1988)
96. N.W. Ashcroft, N.D. Mermin, *Solid State Physics* (Saunders College, Philadelphia, PA, 1976)
97. W. Jones, N.H. March, *Theoretical Solid State Physics, Vol.1: Perfect Lattices in Equilibrium* (Wiley, New York, NY, 1973)
98. J. Callaway, *Quantum Theory of the Solid State* (Academic, New York, NY, 1974)
99. D. Mearns, Phys. Rev. B **38**, 5906 (1988)
100. D. Mearns, W. Kohn, Phys. Rev. B **39**, 10669 (1989)
101. J.P. Perdew, M. Levy, Phys. Rev. Lett. **51**, 1884 (1983)
102. L.J. Sham, M. Schlüter, Phys. Rev. Lett. **51**, 1888 (1983)
103. M. Städele, J.A. Majewski, P. Vogl, A. Görling, Phys. Rev. Lett. **79**, 2089 (1997)
104. L.D. Landau, E.M. Lifschitz, *Course of Theoretical Physics, Vol.IX* (Pergamon, New York, NY, 1980)
105. W. Koch, M.C. Holthausen, *A Chemist's Guide to Density Functional Theory* (Wiley-VCH, Weinheim, 2001)
106. K.E. Riley, B.T.O. Holt, K.M. Merz, J. Chem. Theory Comput. **3**, 407 (2007)
107. D.S. Sholl, J.A. Steckel, *Density Functional Theory: A Practical Introduction* (Wiley, New York, NY, 2009)
108. S. Kurth, J.P. Perdew, P. Blaha, Int. J. Quantum Chem. **75**, 889 (1999)
109. V.N. Staroverov, G.E. Scuseria, J. Tao, J.P. Perdew, Phys. Rev. B **69**, 075102 (2004)
110. P. Haas, F. Tran, P. Blaha, Phys. Rev. B **79**, 085104 (2009)
111. P. Haas, F. Tran, P. Blaha, K. Schwarz, R. Laskowski, Phys. Rev. B **80**, 195109 (2009)
112. P.W. Payne, J. Phys. Chem. **71**, 490 (1979)
113. V. Sahni, J. Gruenebaum, J.P. Perdew, Phys. Rev. B **26**, 4371 (1982)
114. D.C. Langreth, M.J. Mehl, Phys. Rev. B **28**, 1809 (1983)
115. S.J. Chakravorty, S.R. Gwaltney, E.R. Davidson, F.A. Parpia, C. Froese Fischer, Phys. Rev. A **47**, 3649 (1993)
116. E.K.U. Gross, M. Petersilka, T. Grabo, in *Chemical Applications of Density Functional Theory, ACS Symposium Series*, vol. 629, ed. by B.B. Laird, R.B. Ross, T. Ziegler (American Chemical Society, Washington, DC, 1996), p. 42
117. C.O. Almbladh, A.C. Pedroza, Phys. Rev. A **29**, 2322 (1984)
118. F. Aryasetiawan, M.J. Stott, Phys. Rev. B **38**, 2974 (1988)
119. A. Görling, Phys. Rev. A **46**, 3753 (1992)

120. Y. Wang, R.G. Parr, Phys. Rev. A **47**, 1591R (1993)
121. R. van Leeuwen, E.J. Baerends, Phys. Rev. A **49**, 2421 (1994)
122. R.C. Morrison, Q. Zhao, Phys. Rev. A **51**, 1980 (1995)
123. G.L. Oliver, J.P. Perdew, Phys. Rev A **20**, 397 (1979)
124. R.T. Sharp, G.K. Horton, Phys. Rev. **90**, 317 (1953)
125. J.D. Talman, W.F. Shadwick, Phys. Rev. A **14**, 36 (1976)
126. J. Harris, R.O. Jones, J. Phys. F **4**, 1170 (1974)
127. L.J. Sham, Phys. Rev. B **32**, 3876 (1985)
128. D.C. Langreth, J.P. Perdew, Phys. Rev. B **15**, 2884 (1977)
129. M. Gell-Mann, F. Low, Phys. Rev. **84**, 350 (1951)
130. J. Lindhard, Kgl. Danske Videnskab. Selskab, Mat.-Fys. Medd. **28, No.8** (1954)
131. P.A.M. Dirac, Proc. Cambridge Philos. Soc. **26**, 376 (1930)
132. W. Macke, Z. Naturforsch. **5a**, 192 (1950)
133. M. Gell-Mann, K. Brueckner, Phys. Rev. **106**, 364 (1957)
134. S.H. Vosko, L. Wilk, M. Nusair, Can. J. Phys. **58**, 1200 (1980)
135. L. Onsager, L. Mittag, M.J. Stephen, Ann. Physik **18**, 71 (1966)
136. D.F. DuBois, Ann. Phys. (N.Y.) **7**, 174 (1959)
137. W.J. Carr, A.A. Maradudin, Phys. Rev. **133**, A371 (1964)
138. E.P. Wigner, Phys. Rev. **46**, 1002 (1934)
139. E.P. Wigner, Trans. Farad. Soc. **34**, 678 (1938)
140. W.J. Carr, Phys. Rev. **122**, 1437 (1961)
141. D.M. Ceperley, B.J. Alder, Phys. Rev. Lett. **45**, 566 (1980)
142. G. Ortiz, P. Ballone, Phys. Rev. B **50**, 1391 (1994)
143. J.P. Perdew, A. Zunger, Phys. Rev. B **23**, 5048 (1981)
144. J.R. Flores, J. Chem. Phys. **98**, 5642 (1993)
145. V. Termath, W. Klopper, W. Kutzelnigg, J. Chem. Phys. **94**, 2002 (1991)
146. J.C. Slater, Phys. Rev. **81**, 385 (1951)
147. J.C. Slater, T.M. Wilson, J.H. Wood, Phys. Rev. **179**, 28 (1969)
148. R. Latter, Phys. Rev. **99**, 510 (1955)
149. J.P. Perdew, Y. Wang, Phys. Rev. B **45**, 13 244 (1992)
150. K. Brueckner, K. Sawada, Phys. Rev. **112**, 328 (1958)
151. B.S. Shastry, Phys. Rev. Lett. **38**, 449 (1977)
152. B.S. Shastry, Phys. Rev. B **17**, 385 (1978)
153. P. Nozières, *Interacting Fermi Systems* (Benjamin, Reading, MA, 1964)
154. L.J. Sham, Phys. Rev. B **7**, 4357 (1973)
155. O. Gunnarsson, M. Jonson, B.I. Lundqvist, Phys. Rev. B **20**, 3136 (1979)
156. A.H. MacDonald, K.L. Liu, S.H. Vosko, L. Wilk, Can. J. Phys. **59**, 500 (1981)
157. M.L. Plumer, D.J.W. Geldart, J. Phys. C **16**, 677 (1983)
158. E. Engel, Phys. Rev. A **51**, 1159 (1995)
159. K.L. Liu, A.H. MacDonald, S.H. Vosko, Can. J. Phys. **55**, 1991 (1977)
160. A.C. Maggs, N.W. Ashcroft, Phys. Rev. Lett. **59**, 113 (1987)
161. K. Rapcewicz, N.W. Ashcroft, Phys. Rev. B **44**, 4032 (1991)
162. Y. Andersson, D.C. Langreth, B.I. Lundqvist, Phys. Rev. Lett. **76**, 102 (1996)
163. E. Hult, Y. Andersson, B.I. Lundqvist, D.C. Langreth, Phys. Rev. Lett. **77**, 2029 (1996)
164. J.F. Dobson, B.P. Dinte, Phys. Rev. Lett. **76**, 1780 (1996)
165. W. Kohn, Y. Meir, D.E. Makarov, Phys. Rev. Lett. **80**, 4153 (1998)
166. E. Hult, H. Rydberg, B.I. Lundqvist, D.C. Langreth, Phys. Rev. B **59**, 4708 (1999)
167. H. Rydberg, B.I. Lundqvist, D.C. Langreth, M. Dion, Phys. Rev. B **62**, 6997 (2000)
168. H. Rydberg, M. Dion, N. Jacobson, S.I.S. E. Schröder P. Hyldgaard, D.C. Langreth, B.I. Lundqvist, Phys. Rev. Lett. **91**, 126402 (2003)
169. M. Dion, H. Rydberg, E. Schröder, D.C. Langreth, B.I. Lundqvist, Phys. Rev. Lett. **92**, 246401 (2004)
170. S.D. Chakarova-Käck, E. Schröder, B.I. Lundqvist, D.C. Langreth, Phys. Rev. Lett. **96**, 146107 (2006)

171. E. Engel, S.H. Vosko, Phys. Rev. B **42**, 4940 (1990)
172. S. Moroni, D.M. Ceperley, G. Senatore, Phys. Rev. Lett. **75**, 689 (1995)
173. D.A. Kirzhnits, *Field Theoretical Methods in Many-Body Systems* (Pergamon, London, 1967)
174. C.F. von Weizsäcker, Z. Phys. **96**, 431 (1935)
175. D.A. Kirzhnits, Zh. Eksp. Teor. Fiz. **32**, 115 (1957 [Sov. Phys. JETP **5**, 64 (1957)])
176. L.J. Sham, in *Computational Methods in Band Theory*, ed. by P.M. Marcus, J.F. Janak, A.R. Williams (Plenum, New York, NY, 1971), p. 458
177. E.K.U. Gross, R.M. Dreizler, Z. Phys. A **302**, 103 (1981)
178. P.R. Antoniewicz, L. Kleinman, Phys. Rev. B **31**, 6779 (1985)
179. L. Kleinman, S. Lee, Phys. Rev. B **37**, 4634 (1988)
180. J.A. Chevary, S.H. Vosko, Phys. Rev. B **42**, 5320 (1990)
181. L. Kleinman, Phys. Rev. B **10**, 2221 (1974)
182. L. Kleinman, Phys. Rev. B **12**, 3512 (1975)
183. A.K. Rajagopal, S. Ray, Phys. Rev. B **12**, 3129 (1975)
184. D.J.W. Geldart, M. Rasolt, C.O. Almbladh, Solid State Commun. **16**, 243 (1975)
185. H.H. Kranz, J.P. Löwenau, S. Schmitt-Rink, J. Phys. C **17**, 2121 (1984)
186. S.K. Ma, K.A. Brueckner, Phys. Rev. **165**, 18 (1968)
187. D.J.W. Geldart, M. Rasolt, Phys. Rev. B **13**, 1477 (1976)
188. M. Rasolt, D.J.W. Geldart, Phys. Rev. B **34**, 1325 (1986)
189. M. Rasolt, Phys. Rev. B **16**, 3234 (1977)
190. M. Rasolt, H.L. Davis, Phys. Lett. **86A**, 45 (1981)
191. Y. Wang, J.P. Perdew, Phys. Rev. B **43**, 8911 (1991)
192. C.D. Hu, D.C. Langreth, Phys. Rev. B **33**, 943 (1986)
193. Y.K. Kim, Phys. Rev. **154**, 17 (1967)
194. C.H. Hodges, Can. J. Phys. **51**, 1428 (1973)
195. M. Brack, B.K. Jennings, Y.H. Chu, Phys. Lett. **65B**, 1 (1976)
196. P.S. Svendsen, U. von Barth, Phys. Rev. B **54**, 17 402 (1996)
197. F. Herman, J.P. Van Dyke, I.B. Ortenburger, Phys. Rev. Lett. **22**, 807 (1969)
198. T. Kato, Pure Appl. Math. **10**, 151 (1957)
199. F. Herman, I.B. Ortenburger, J.P. Van Dyke, Int. J. Quantum Chem. **IIIS**, 827 (1970)
200. P. Bagno, O. Jepsen, O. Gunnarsson, Phys. Rev. B **40**, 1997 (1989)
201. D.C. Langreth, J.P. Perdew, Phys. Rev. B **21**, 5469 (1980)
202. D.C. Langreth, M.J. Mehl, Phys. Rev. Lett. **47**, 446 (1981)
203. D.C. Langreth, S.H. Vosko, Phys. Rev. Lett. **59**, 497 (1987)
204. J.P. Perdew, Phys. Rev. Lett. **55**, 1665 (1985)
205. J.P. Perdew, Y. Wang, Phys. Rev. B **33**, 8800 (1986)
206. J.P. Perdew, in *Electronic Structure of Solids 1991*, ed. by P. Ziesche, H. Eschrig (Akademie Verlag, Berlin, 1991), p. 11
207. J.P. Perdew, K. Burke, M. Ernzerhof, Phys. Rev. Lett. **77**, 3865 (1996)
208. D.C. Langreth, S.H. Vosko, in *Density Functional Theory of Many-Fermion Systems*, *Advances in Quantum Chemistry*, vol. 21, ed. by S.B. Trickey (Academic, New York, NY, 1990), p. 175
209. C.D. Hu, D.C. Langreth, Phys. Script. **32**, 391 (1985)
210. J.P. Perdew, Phys. Rev. B **33**, 8822 (1986)
211. J.P. Perdew, J.A. Chevary, S.H. Vosko, K.A. Jackson, M.R. Pederson, D.J. Singh, C. Fiolhais, Phys. Rev. B **46**, 6671 (1992)
212. E.H. Lieb, S. Oxford, Int. J. Quantum Chem. **XIX**, 427 (1981)
213. E. Engel, S.H. Vosko, Phys. Rev. B **50**, 10498 (1994)
214. I.S. Gradstein, I.M. Ryshik, *Tables* (Harri Deutsch, Frankfurt am Main, 1981)
215. A. Becke, J. Chem. Phys. **98**, 1372 (1993)
216. P.J. Stephens, F.J. Devlin, C.F. Chabalowski, M.J. Frisch, J. Phys. Chem. **98**, 11623 (1994)
217. J.P. Perdew, M. Ernzerhof, K. Burke, J. Chem. Phys. **105**, 9982 (1996)
218. A.D. Becke, J. Chem. Phys. **84**, 4524 (1986)
219. A.D. Becke, Phys. Rev. A **38**, 3098 (1988)

220. E. Engel, J.A. Chevary, L.D. Macdonald, S.H. Vosko, Z. Phys. D **23**, 7 (1992)
221. C. Lee, W. Yang, R.G. Parr, Phys. Rev. B **37**, 785 (1988)
222. R. Colle, O. Salvetti, Theoret. Chim. Acta (Berl.) **37**, 329 (1975)
223. E. Engel, S.H. Vosko, Phys. Rev. A **47**, 2800 (1993)
224. A. Becke, J. Chem. Phys. **96**, 2155 (1992)
225. B.G. Johnson, P.M. Gill, J.A. Pople, J. Chem. Phys. **98**, 5612 (1993)
226. The dissociation energy D_e in all Tables on diatomic molecules corresponds to the sum of the true dissociation energy D_0 and the zero-point energy, including the lowest order anharmonic contribution, $D_e = D_0 + \hbar\omega_e/2 - \hbar\omega_e x_e/4$. The energy D_e agrees with the well-depth of the Born-Oppenheimer surface.
227. K.P. Huber, G.L. Herzberg, *Molecular Spectra and Molecular Structure. IV. Constants of Diatomic Molecules* (Van Nostrand Reinhold, New York, NY, 1979)
228. P.J. Linstrom, W.G. Mallard (eds.), *NIST Chemistry WebBook, NIST Standard Reference Database Number 69* (National Institute of Standards and Technology, Gaithersburg, 2003)
229. D.R. Lide (ed.), *Handbook of Chemistry and Physics 1996–1997* (CRC Press, Boca Raton, FL, 1996)
230. E. Engel, U.R. Schmitt, H.J. Lüdde, A. Toepfer, E. Wüst, R.M. Dreizler, O. Knospe, R. Schmidt, P. Chattopadhyay, Phys. Rev. B **48**, 1862 (1993)
231. J.P. Perdew, S. Kurth, A. Zupan, P. Blaha, Phys. Rev. Lett. **82**, 2544 (1999)
232. J. Tao, J.P. Perdew, V.N. Staroverov, G.E. Scuseria, Phys. Rev. Lett. **91**, 146401 (2003)
233. C. Lee, D. Vanderbilt, K. Laasonen, R. Car, M. Parrinello, Phys. Rev. B **47**, 4863 (1993)
234. R.N. Barnett, U. Landman, Phys. Rev. B **48**, 2081 (1993)
235. K. Laasonen, M. Sprik, M. Parrinello, R. Car, J. Chem. Phys. **99**, 9080 (1993)
236. S. Tsuzuki, H.P. Lüthi, J. Chem. Phys. **114**, 3949 (2001)
237. P. Dufek, P. Blaha, K. Schwarz, Phys. Rev. B **50**, 7279 (1994)
238. A. Khein, D.J. Singh, C.J. Umrigar, Phys. Rev. B **51**, 4105 (1995)
239. D.J. Singh, J. Ashkenazi, Phys. Rev. B **46**, 11570 (1992)
240. F. Tran, P. Blaha, K. Schwarz, J. Phys.: Condens. Matter **19**, 196208 (2007)
241. M. Städele, M. Moukara, J.A. Majewski, P. Vogl, A. Görling, Phys. Rev. B **59**, 10031 (1999)
242. J. Heyd, J.E. Peralta, G.E. Scuseria, R.L. Martin, J. Chem. Phys. **123**, 174101 (2005)
243. E. Engel, A. Höck, R.M. Dreizler, Phys. Rev. A **62**, 042502 (2000)
244. E. Engel, S.H. Vosko, Phys. Rev. B **47**, 13164 (1993)
245. O. Gunnarsson, M. Jonson, B.I. Lundqvist, Solid State Commun. **24**, 765 (1977)
246. J.A. Alonso, L.A. Girifalco, Phys. Rev. B **17**, 3735 (1978)
247. O. Gunnarsson, M. Jonson, B.I. Lundqvist, Phys. Lett. **59A**, 177 (1976)
248. L. Fritsche, H. Gollisch, Z. Phys. B **48**, 209 (1982)
249. M.S. Hybertsen, S.G. Louie, Phys. Rev. B **30**, 5777 (1984)
250. F. Manghi, G. Riegler, C.M. Bertoni, G.B. Bachelet, Phys. Rev. B **31**, 3680 (1985)
251. D.J. Singh, Phys. Rev. B **48**, 14099 (1993)
252. O.V. Gritsenko, N.A. Cordero, A. Rubio, L.C. Balbas, J.A. Alonso, Phys. Rev. A **48**, 4197 (1993)
253. J.P.A. Charlesworth, Phys. Rev. B **53**, 12666 (1995)
254. A.R. Denton, P. Nielaba, K.J. Runge, N.W. Ashcroft, Phys. Rev. Lett. **64**, 1529 (1990)
255. S. Moroni, G. Senatore, Phys. Rev. B **44**, 9864 (1991)
256. C.N. Likos, S. Moroni, G. Senatore, Phys. Rev. B **55**, 8867 (1997)
257. W.A. Curtin, N.W. Ashcroft, Phys. Rev. A **32**, 2909 (1985)
258. A.R. Denton, N.W. Ashcroft, Phys. Rev. A **39**, 4701 (1989)
259. I. Lindgren, Int. J. Quantum Chem. Symp. **5**, 411 (1971)
260. H. Stoll, C.M.E. Pavlidou, H. Preuss, Theor. Chim. Acta **149**, 143 (1978)
261. S.H. Vosko, L. Wilk, J. Phys. B **16**, 3687 (1983)
262. M.R. Norman, D.D. Koelling, Phys. Rev. B **30**, 5530 (1984)
263. J.G. Harrison, J. Chem. Phys. **78**, 4562 (1983)
264. J.G. Harrison, J. Chem. Phys. **79**, 2265 (1983)
265. J.G. Harrison, J. Chem. Phys. **86**, 2849 (1987)

266. M.R. Pederson, C.C. Lin, J. Chem. Phys. **88**, 1807 (1988)
267. M.R. Pederson, R.A. Heaton, C.C. Lin, J. Chem. Phys. **80**, 1972 (1984)
268. M.R. Pederson, R.A. Heaton, C.C. Lin, J. Chem. Phys. **82**, 2688 (1985)
269. S. Goedecker, C.J. Umrigar, Phys. Rev. A **55**, 1765 (1997)
270. A. Svane, O. Gunnarsson, Phys. Rev. Lett. **65**, 1148 (1990)
271. Z. Szotek, W.M. Temmerman, H. Winter, Phys. Rev. B **47**, 4029R (1993)
272. Z. Szotek, W.M. Temmerman, H. Winter, Phys. Rev. Lett. **72**, 1244 (1994)
273. E. Fermi, E. Amaldi, Accad. Ital. Roma **6**, 119 (1934)
274. R. Armiento, A.E. Mattsson, Phys. Rev. B **72**, 085108 (2005)
275. M. Filatov, W. Thiel, Phys. Rev. A **57**, 189 (1998)
276. A.D. Becke, Int. J. Quantum Chem. **23**, 1915 (1983)
277. S.K. Ghosh, R.G. Parr, Phys. Rev. A **34**, 785 (1986)
278. A.D. Becke, J. Chem. Phys. **85**, 7184 (1986)
279. A.D. Becke, M.R. Roussel, Phys. Rev. A **39**, 3761 (1989)
280. A. Becke, J. Chem. Phys. **109**, 2092 (1998)
281. C. Adamo, M. Ernzerhof, G.E. Scuseria, J. Chem. Phys. **112**, 2643 (2000)
282. V.N. Staroverov, G.E. Scuseria, J. Tao, J.P. Perdew, J. Chem. Phys. **119**, 12129 (2003)
283. J.P. Perdew, J. Tao, V.N. Staroverov, G.E. Scuseria, J. Chem. Phys. **120**, 6898 (2004)
284. P.W. Anderson, Phys. Rev. **124**, 41 (1961)
285. V.I. Anisimov, J. Zaanen, O.K. Andersen, Phys. Rev. B **44**, 943 (1991)
286. V.I. Anisimov, I.V. Solovyev, M.A. Korotin, M.T. Czyzyk, G.A. Sawatzky, Phys. Rev. B **48**, 16929 (1993)
287. A.I. Liechtenstein, V.I. Anisimov, J. Zaanen, Phys. Rev. B **52**, R5467 (1995)
288. V.I. Anisimov, F. Aryasetiawan, A.I. Lichtenstein, J. Phys.: Condens. Matter **9**, 767 (1997)
289. M.E. Rose, *Elementary Theory of Angular Momentum* (Wiley, New York, NY, 1957)
290. J.C. Slater, *Quantum Theory of Molecules and Solids, Vol.4: The Self-consistent Field for Molecules and Solids* (McGraw-Hill, New York, NY, 1974), Appendix 3
291. M.T. Czyzyk, G.A. Sawatzky, Phys. Rev. B **49**, 14211 (1994)
292. A.G. Petukhov, I.I. Mazin, I. Chioncel, A.I. Lichtenstein, Phys. Rev. B **67**, 153106 (2003)
293. M. Levy, J.P. Perdew, Phys. Rev. A **32**, 2010 (1985)
294. M. Levy, W. Yang, R.G. Parr, J. Chem. Phys. **83**, 2334 (1985)
295. M. Levy, J.P. Perdew, Int. J. Quantum Chem. **28 (S19)**, 743 (1986)
296. M. Levy, in *Single-Particle Density in Physics and Chemistry*, ed. by N.H. March, B.M. Deb (Academic, London, 1987), p. 45
297. W. Yang, in *Density Matrices and Density Functionals*, ed. by R. Erdahl, V.H. Smith (Reidel, Dordrecht, 1987), p. 499
298. M. Levy, Int. J. Quantum Chem. **36 (S23)**, 617 (1989)
299. H. Ou-Yang, M. Levy, Phys. Rev. A **42**, 155 (1990)
300. H. Ou-Yang, M. Levy, Phys. Rev. A **42**, 651 (1990)
301. M. Levy, Phys. Rev. A **43**, 4637 (1991)
302. A. Görling, M. Levy, Phys. Rev. A **45**, 1509 (1992)
303. M. Levy, J.P. Perdew, Phys. Rev. B **48**, 11638 (1993)
304. M. Levy, A. Görling, Phys. Rev. A **51**, 2851 (1995)
305. M. Levy, A. Görling, Int. J. Quantum Chem. **56**, 385 (1995)
306. S. Ivanov, M. Levy, J. Phys. Chem. A **102**, 3151 (1998)
307. M. Levy, J.P. Perdew, Int. J. Quantum Chem. **49**, 539 (1994)
308. M. Levy, in *Density Functional Theory, NATO ASI Series B*, vol. 337, ed. by E.K.U. Gross, R.M. Dreizler (Plenum, New York, NY, 1995), p. 11
309. L.C. Wilson, M. Levy, Phys. Rev. B **41**, 12930 (1990)
310. Q. Zhao, M. Levy, R.G. Parr, Phys. Rev. A **47**, 918 (1993)
311. S.K. Ghosh, R.G. Parr, J. Chem. Phys. **82**, 3307 (1985)
312. F.W. Averill, G.S. Painter, Phys. Rev. B **24**, 6795 (1981)
313. H. Hellmann, *Einführung in die Quantenchemie* (Deuticke, Leipzig, 1937)
314. R.P. Feynman, Phys. Rev. **56**, 340 (1939)

315. X.J. Kong, C.T. Chan, K.M. Ho, Y.Y. Ye, Phys. Rev. B **42**, 9357 (1990)
316. P. Boschan, H. Gollisch, Z. Phys. D **17**, 127 (1990)
317. P. Mlynarski, D.R. Salahub, Phys. Rev. B **43**, 1399 (1991)
318. A. Becke, J. Chem. Phys. **97**, 9173 (1992)
319. J. Andzelm, E. Wimmer, J. Chem. Phys. **96**, 1280 (1992)
320. T. Asada, K. Terakura, Phys. Rev. B **46**, 13599 (1992)
321. M. Körling, J. Häglund, Phys. Rev. B **45**, 13293 (1992)
322. P. Söderlind, O. Eriksson, B. Johansson, J.M. Wills, Phys. Rev. B **50**, 7291 (1994)
323. A.D. Corso, A. Pasquarello, A. Baldereschi, R. Car, Phys. Rev. B **53**, 1180 (1996)
324. D.R. Hamann, Phys. Rev. B **55**, R10 157 (1997)
325. D.C. Patton, D.V. Porezag, M.R. Pederson, Phys. Rev. B **55**, 7454 (1997)
326. G. Jomard, T. Petit, A. Pasturel, L. Magaud, G. Kresse, J. Hafner, Phys. Rev. B **59**, 4044 (1999)
327. H.B. Shore, J.H. Rose, E. Zaremba, Phys. Rev. B **15**, 2858 (1977)
328. M. Weimer, F. Della Sala, A. Görling, Chem. Phys. Lett. **372**, 538 (2003)
329. R.O. Jones, O. Gunnarsson, Rev. Mod. Phys. **61**, 689 (1989)
330. D.J. Lacks, R.G. Gordon, Phys. Rev. A **47**, 4681 (1993)
331. D.C. Patton, M.R. Pederson, Phys. Rev. A **56**, R2495 (1997)
332. R.A. Aziz, M.J. Slaman, J. Chem. Phys. **94**, 8047 (1991)
333. F. Luo, G. McBane, G. Kim, C.F. Giese, W.R. Gentry, J. Chem. Phys. **98**, 3564 (1993)
334. A.D. Becke, E.R. Johnson, J. Chem. Phys. **122**, 154104 (2005)
335. A.D. Becke, E.R. Johnson, J. Chem. Phys. **123**, 154101 (2005)
336. A.D. Becke, E.R. Johnson, J. Chem. Phys. **127**, 154108 (2007)
337. F.O. Kannemann, A.D. Becke, J. Chem. Theory Comput. **5**, 719 (2009)
338. K. Terakura, T. Oguchi, A.R. Williams, J. Kübler, Phys. Rev. B **30**, 4734 (1984)
339. T.C. Leung, C.T. Chan, B.N. Harmon, Phys. Rev. B **44**, 2923 (1991)
340. F. Tran, P. Blaha, K. Schwarz, P. Novak, Phys. Rev. B **74**, 155108 (2006)
341. V.N. Glushkov, M. Levy, J. Chem. Phys. **126**, 174106 (2007)
342. C.A. Ullrich, U.J. Gossmann, E.K.U. Gross, Phys. Rev. Lett. **74**, 872 (1995)
343. A. Görling, M. Levy, Phys. Rev. A **50**, 196 (1994)
344. E. Engel, R.M. Dreizler, J. Comput. Chem. **20**, 31 (1999)
345. M.E. Casida, Phys. Rev. A **51**, 2505 (1995)
346. E. Engel, H. Jiang, A. Facco Bonetti, Phys. Rev. A **72**, 052503 (2005)
347. J.B. Krieger, Y. Li, G.J. Iafrate, Phys. Lett. A **148**, 470 (1990)
348. T. Grabo, T. Kreibich, S. Kurth, E.K.U. Gross, in *Strong Coulomb Correlations in Electronic Structure Calculations: Beyond the Local Density Approximation*, ed. by V.I. Anisimov (Gordon and Breach, Newark, NJ, 1999), p. 203
349. H. Jiang, E. Engel, J. Chem. Phys. **123**, 224102 (2005)
350. E. Engel, A. Facco Bonetti, S. Keller, I. Andrejkovics, R.M. Dreizler, Phys. Rev. A **58**, 964 (1998)
351. J.B. Krieger, Y. Li, G.J. Iafrate, Phys. Lett. A **146**, 256 (1990)
352. F. Della Sala, A. Görling, J. Chem. Phys. **115**, 5718 (2001)
353. Y. Li, J.B. Krieger, M.R. Norman, G.J. Iafrate, Phys. Rev. B **44**, 10437 (1991)
354. Y. Li, J.B. Krieger, G.J. Iafrate, Chem. Phys. Lett. **191**, 38 (1992)
355. J.B. Krieger, Y. Li, G.J. Iafrate, Phys. Rev. A **45**, 101 (1992)
356. J.B. Krieger, Y. Li, G.J. Iafrate, Phys. Rev. A **47**, 165 (1993)
357. J. Chen, J.B. Krieger, Y. Li, G.J. Iafrate, Phys. Rev. A **54**, 3939 (1996)
358. T. Kotani, Phys. Rev. B **50**, 14816 (1994)
359. T. Kotani, Phys. Rev. Lett. **74**, 2989 (1995)
360. T. Kotani, H. Akai, Phys. Rev. B **54**, 16502 (1996)
361. T. Kotani, J. Phys.: Condens. Matter **10**, 9241 (1998)
362. T. Kotani, J. Phys.: Condens. Matter **12**, 2413 (2000)
363. D.M. Bylander, L. Kleinman, Phys. Rev. Lett. **74**, 3660 (1995)
364. D.M. Bylander, L. Kleinman, Phys. Rev. B **52**, 14566 (1995)

365. D.M. Bylander, L. Kleinman, Phys. Rev. B **54**, 7891 (1996)
366. D.M. Bylander, L. Kleinman, Phys. Rev. B **55**, 9432 (1997)
367. Y.H. Kim, M. Städele, R.M. Martin, Phys. Rev. A **60**, 3633 (1999)
368. W.G. Aulbur, M. Städele, A. Görling, Phys. Rev. B **62**, 7121 (2000)
369. M. Moukara, M. Städele, J.A. Majewski, P. Vogl, A. Görling, J. Phys. C **12**, 6783 (2000)
370. A. Görling, Phys. Rev. Lett. **83**, 5459 (1999)
371. A. Heßelmann, A.W. Götz, F. Della Sala, A. Görling, J. Chem. Phys. **127**, 054102 (2007)
372. S. Ivanov, S. Hirata, R.J. Bartlett, Phys. Rev. Lett. **83**, 5455 (1999)
373. S. Hirata, S. Ivanov, I. Grabowski, R.J. Bartlett, K. Burke, J.D. Talman, J. Chem. Phys. **115**, 1635 (2001)
374. L. Fritsche, J. Yuan, Phys. Rev. A **57**, 3425 (1998)
375. R.A. Hyman, M.D. Stiles, A. Zangwill, Phys. Rev. B **62**, 15521 (2000)
376. A. Fleszar, Phys. Rev. B **64**, 245204 (2001)
377. R.J. Magyar, A. Fleszar, E.K.U. Gross, Phys. Rev. B **69**, R. J. Magyar (2004)
378. W. Yang, Q. Wu, Phys. Rev. Lett. **89**, 143002 (2002)
379. S. Kümmel, J.P. Perdew, Phys. Rev. Lett. **90**, 043004 (2003)
380. E. Engel, A. Höck, R.N. Schmid, R.M. Dreizler, N. Chetty, Phys. Rev. B **64**, 125111 (2001)
381. P. Rinke, A. Qteish, J. Neugebauer, C. Freysoldt, M. Scheffler, New J. Phys. **7**, 126 (2005)
382. A. Qteish, A.I. Al-Sharif, M. Fuchs, M. Scheffler, S. Boeck, J. Neugebauer, Comput. Phys. Commun. **169**, 28 (2005)
383. C. Froese Fischer, *The Hartree-Fock Method for Atoms* (Wiley, New York, NY, 1977)
384. L. Laaksonen, P. Pyykkö, D. Sundholm, Comput. Phys. Rep. **4**, 313 (1986)
385. T. Heaton-Burgess, F.A. Bulat, W. Yang, Phys. Rev. Lett. **98**, 256401 (2007)
386. A. Görling, A. Heßelmann, M. Jones, M. Levy, J. Chem. Phys. **128**, 104104 (2008)
387. V.N. Staroverov, G.E. Scuseria, E.R. Davidson, J. Chem. Phys. **124**, 141103 (2006)
388. A.D. McLean, B. Liu, G.S. Chandler, in *Momentum distributions, AIP Conference Proceedings*, vol. 86 (New York, NY, 1982), p. 90
389. A. Görling, Phys. Rev. B **53**, 7024 (1996)
390. E. Engel, in *A Primer in Density Functional Theory, Lecture Notes in Physics*, vol. 620, ed. by C. Fiolhais, F. Nogeira, M. Marques (Springer, Berlin, 2003), p. 56
391. E. Engel, R.N. Schmid, Phys. Rev. Lett. **103**, 036404 (2009)
392. J. van Elp, Ph.D. thesis, University of Groningen (1991)
393. F. Parmigiani, L. Sangaletti, J. Electron Spectrosc. Relat. Phenom. **98–99**, 287 (1999)
394. S. Sharma, J.K. Dewhurst, C. Ambrosch-Draxl, Phys. Rev. Lett. **95**, 136402 (2005)
395. S. Massida, M. Posternak, A. Baldereschi, Phys. Rev. B **48**, 5058 (1993)
396. C.D. Clark, P.J. Dean, P.V. Harris, Proc. Royal Soc. (London) **A277**, 312 (1964)
397. W. Bludau, A. Onton, W. Heinke, J. Appl. Phys. **45**, 1846 (1974)
398. K.H. Hellwege, O. Madelung, M. Schulz, H. Weiss (eds.), *Semiconductors, Physics of Group IV Elements and III-V Compounds*, vol. 17 (Springer, Berlin, 1982)
399. D.D. Sell, Phys. Rev. B **6**, 3750 (1972)
400. D. Pudewill, F.J. Himpsel, V. Saile, N. Schwentner, M. Skibowski, E.E. Koch, J. Jortner, J. Chem. Phys. **65**, 5226 (1977)
401. F.J. Himpsel, J.F. van der Veen, D.E. Eastman, Phys. Rev. B **22**, 1967 (1980)
402. M. Welkowsky, R. Braunstein, Phys. Rev. B **5**, 497 (1972)
403. D.E. Aspnes, Phys. Rev. B **12**, 2297 (1975)
404. T.C. Chiang, J.A. Knapp, M. Aono, D.E. Eastman, Phys. Rev. B **21**, 3513 (1980)
405. H. Armon, J.P.F. Sellschop, Phys. Rev. B **26**, 3289 (1982)
406. R.A. Roberts, W.C. Walker, Phys. Rev. **161**, 730 (1967)
407. E. Engel, Phys. Rev. B **80**, 161205(R) (2009)
408. A. Makmal, R. Armiento, E. Engel, L. Kronik, S. Kümmel, PRB **80**, 161204(R) (2009)
409. M. Grüning, A. Marini, A. Rubio, J. Chem. Phys. **124**, 154108 (2006)
410. Y.M. Niquet, X. Gonze, Phys. Rev. B **70**, 245115 (2004)
411. F. Furche, Phys. Rev. B **64**, 195120 (2001)
412. M. Fuchs, X. Gonze, Phys. Rev. B **65**, 235109 (2002)

413. N.E. Dahlen, U. von Barth, J. Chem. Phys. **120**, 6826 (2004)
414. N.E. Dahlen, U. von Barth, Phys. Rev. B **69**, 195102 (2004)
415. M. Hellgren, U. von Barth, Phys. Rev. B **76**, 075107 (2007)
416. S. Ivanov, S. Hirata, R.J. Bartlett, J. Chem. Phys. **116**, 1269 (2002)
417. I. Grabowski, S. Hirata, S. Ivanov, R.J. Bartlett, J. Chem. Phys. **116**, 4415 (2002)
418. R.J. Bartlett, I. Grabowski, S. Hirata, S. Ivanov, J. Chem. Phys. **122**, 034104 (2005)
419. R.J. Bartlett, V.F. Lotrich, I.V. Schweigert, J. Chem. Phys. **123**, 062205 (2005)
420. H. Jiang, E. Engel, J. Chem. Phys. **125**, 184108 (2006)
421. H. Jiang, E. Engel, J. Chem. Phys. **127**, 184108 (2007)
422. M. Seidl, J.P. Perdew, S. Kurth, Phys. Rev. Lett. **84**, 5070 (2000)
423. F. Furche, J. Chem. Phys. **129**, 114105 (2008)
424. J. Harl, G. Kresse, Phys. Rev. B **77**, 045136 (2008)
425. B.G. Janesko, T.M. Henderson, G.E. Scuseria, J. Chem. Phys. **130**, 081105 (2009)
426. M. Seidl, J.P. Perdew, M. Levy, Phys. Rev. A **59**, 51 (1999)
427. T. Grabo, E.K.U. Gross, Chem. Phys. Lett. **240**, 141 (1995)
428. T. Grabo, T. Kreibich, E.K.U. Gross, Mol. Engineering **7**, 27 (1997)
429. C. Filippi, C.J. Umrigar, X. Gonze, J. Chem. Phys. **107**, 9994 (1997)
430. A. Becke, J. Chem. Phys. **98**, 5648 (1993)
431. R.H. Hertwig, W. Koch, Chem. Phys. Lett. **268**, 345 (1997)
432. A.V. Arbuznikov, M. Kaupp, H. Bahmann, J. Chem. Phys. **124**, 204102 (2006)
433. M. Kaupp, H. Bahmann, A.V. Arbuznikov, J. Chem. Phys. **127**, 194102 (2007)
434. H. Jiang, E. Engel, Z. Phys. Chem. **224**, 455 (2010)
435. M. Ernzerhof, G.E. Scuseria, J. Chem. Phys. **110**, 5029 (1999)
436. X. Xu, Q. Zhang, R.P. Muller, W.A. Goddard III, J. Phys. Chem. **122**, 014105 (2005)
437. C. Adamo, V. Barone, J. Chem. Phys. **110**, 6158 (1999)
438. J. Paier, R. Hirschl, M. Marsman, G. Kresse, J. Chem. Phys. **122**, 234102 (2005)
439. J.L.F.D. Silva, M.V. Ganduglia-Pirovano, J. Sauer, V. Bayer, G. Kresse, Phys. Rev. B **75**, 045121 (2007)
440. W. Kohn, Int. J. Quantum Chem. **56**, 229 (1995)
441. J. Heyd, G.E. Scuseria, M. Ernzerhof, J. Chem. Phys. **118**, 8207 (2003)
442. J. Heyd, G.E. Scuseria, M. Ernzerhof, J. Chem. Phys. **124**, 219906 (2006)
443. A.V. Krukau, O.A. Vydrov, A.F. Izmaylov, G.E. Scuseria, J. Chem. Phys. **125**, 224106 (2006)
444. J. Heyd, G.E. Scuseria, J. Chem. Phys. **121**, 1187 (2004)
445. J. Paier, M.M. nd K. Hummer, G. Kresse, I.C. Gerber, J.G. Angyan, J. Chem. Phys. **124**, 154709 (2006)
446. E.N. Brothers, A.F. Izmaylov, J.O. Normand, V. Barone, G.E. Scuseria, J. Chem. Phys. **129**, 011102 (2008)
447. J. Jaramillo, G.E. Scuseria, M. Ernzerhof, J. Chem. Phys. **118**, 1068 (2003)
448. K. Burke, F.G. Cruz, K.C. Lam, J. Chem. Phys. **109**, 8161 (1998)
449. H. Bahmann, A. Rodenberg, A.V. Arbuznikov, M. Kaupp, J. Chem. Phys. **126**, 011103 (2007)
450. B.G. Janesko, A.V. Aliaksandr, G.E. Scuseria, J. Chem. Phys. **129**, 124110 (2008)
451. M. Lein, J.F. Dobson, E.K.U. Gross, J. Comput. Chem. **20**, 12 (1999)
452. A. Kumar, W. Meath, J. Mol. Phys. **54**, 823 (1985)
453. K.T. Tang, J.M. Norbeck, P.R. Certain, J. Chem. Phys. **64**, 3063 (1976)
454. E. Engel, A. Höck, R.M. Dreizler, Phys. Rev. A **61**, 032502 (2000)
455. D.M. Silver, Phys. Rev. A **21**, 1106 (1980)
456. F. Furche, T.V. Voorhis, J. Chem. Phys. **122**, 164106 (2005)
457. D.E. Woon, J. Chem. Phys. **100**, 2838 (1994)
458. J.G. Ángyán, I.C. Gerber, A. Savin, J. Toulouse, Phys. Rev. A **72**, 012510 (2005)
459. I.C. Gerber, J.G. Ángyán, J. Chem. Phys. **126**, 044103 (2007)
460. S. Kurth, J.P. Perdew, Phys. Rev. B **59**, 10461 (1999)
461. Z. Yan, J.P. Perdew, S. Kurth, Phys. Rev. B **61**, 16430 (2000)
462. Y. Ishikawa, K. Koc, Phys. Rev. A **50**, 4733 (1994)
463. J. J. A. Montgomery, J.W. Ochterski, G.A. Petersson, J. Chem. Phys. **101**, 5900 (1994)

464. D.E. Freund, B.D. Huxtable, J.D. Morgan III, Phys. Rev. A **29**, 980 (1984)
465. A. Facco Bonetti, E. Engel, R.N. Schmid, R.M. Dreizler, Phys. Rev. Lett. **86**, 2241 (2001)
466. A. Facco Bonetti, E. Engel, R.N. Schmid, R.M. Dreizler, Phys. Rev. Lett. **90**, 219302 (2003)
467. Y.M. Niquet, M. Fuchs, X. Gonze, J. Chem. Phys. **118**, 9504 (2003)
468. P. Mori-Sanchez, Q. Wu, W. Yang, J. Chem. Phys. **123**, 062204 (2005)
469. J. Chen, M.J. Stott, Phys. Lett. **176A**, 101 (1993)
470. J. Chen, J.B. Krieger, R.O. Esquivel, M.J. Stott, G.J. Iafrate, Phys. Rev. A **54**, 1910 (1996)
471. E. Engel, H. Jiang, Int. J. Quantum Chem. **106**, 3242 (2006)
472. E. Runge, E.K.U. Gross, Phys. Rev. Lett. **52**, 997 (1984)
473. E.K.U. Gross, W. Kohn, Phys. Rev. Lett. **55**, 2850 (1985)
474. N.D. Lang, Phys. Rev. B **52**, 5335 (1995)
475. M. Brandbyge, J.L. Mozos, P. Ordejón, J. Taylor, K. Stokbro, Phys. Rev. B **65**, 165401 (2002)
476. K. Burke, R. Car, R. Gebauer, Phys. Rev. Lett. **94**, 146803 (2005)
477. M.A.L. Marques, C.A. Ullrich, R. Nogeira, A. Rubio, K. Burke, E.K.U. Gross (eds.). *Time-Dependent Density Functional Theory*, *Lecture Notes in Physics*, vol. 706 (Springer, Berlin, 2006)
478. K. Burke, J. Werschnik, E.K.U. Gross, J. Chem. Phys. **123**, 062206 (2005)
479. S.K. Ghosh, A.K. Dhara, Phys. Rev. A **38**, 1149 (1988)
480. G. Vignale, Phys. Rev. B **70**, 201102 (2004)
481. B.X. Xu, A.K. Rajagopal, Phys. Rev. A **31**, 2682 (1985)
482. A.K. Dhara, S.K. Ghosh, Phys. Rev. A **35**, 442 (1987)
483. P.O. Löwdin, P.K. Mukherjee, Chem. Phys. Lett. **14**, 1 (1972)
484. M.Y. Amusia, V.R. Shaginyan, Phys. Lett. A **250**, 157 (1998)
485. E.K.U. Gross, C.A. Ullrich, U.J. Gossmann, in *Density Functional Theory*, *NATO ASI Series B*, vol. 337, ed. by E.K.U. Gross, R.M. Dreizler (Plenum, New York, NY, 1995), p. 149
486. E.K.U. Gross, J.F. Dobson, M. Petersilka, in *Density Functional Theory II*, *Topics in Current Chemistry*, vol. 181, ed. by R.F. Nalewajski (Springer, Berlin, 1996), p. 81
487. M.Y. Amusia, V.R. Shaginyan, Phys. Rev. A **63**, 056501 (2001)
488. R. van Leeuwen, Phys. Rev. Lett. **80**, 1280 (1998)
489. R. van Leeuwen, Int. J. Mod. Phys. B **15**, 1969 (2001)
490. M.K. Harbola, A. Banerjee, Phys. Rev. A **60**, 5101 (1999)
491. J. Schirmer, A. Dreuw, Phys. Rev. A **75**, 022513 (2007)
492. N.T. Maitra, R. van Leeuwen, K. Burke, Phys. Rev. A **78**, 056501 (2008)
493. J. Schirmer, A. Dreuw, Phys. Rev. A **78**, 056502 (2008)
494. A. Holas, M. Cinal, N.H. March, Phys. Rev. A **78**, 016501 (2008)
495. G. Vignale, Phys. Rev. A **77**, 062511 (2008)
496. H. Kohl, R.M. Dreizler, Phys. Rev. Lett. **56**, 1993 (1986)
497. S.K. Ghosh, A.K. Dhara, Phys. Rev. A **40**, 6103 (1989)
498. V. Peuckert, J. Phys. C **11**, 4945 (1978)
499. A. Zangwill, P. Soven, Phys. Rev. A **21**, 1561 (1980)
500. T. Ando, Z. Phys. B **26**, 263 (1977)
501. T. Ando, Solid State Commun. **21**, 133 (1977)
502. A. Zangwill, P. Soven, Phys. Rev. B **24**, 4121 (1981)
503. S.J.A. van Gisbergen, P.R.T. Schipper, O.V. Gritsenko, E.J. Baerends, J.G. Snijders, B. Champagne, B. Kirtman, Phys. Rev. Lett. **83**, 694 (1999)
504. J.F. Dobson, Phys. Rev. Lett. **73**, 2244 (1994)
505. W. Kohn, Phys. Rev. **123**, 1242 (1961)
506. L. Brey, N.F. Johnson, B.I. Halperin, Phys. Rev. B **40**, 10647 (1989)
507. S.K. Yip, Phys. Rev. B **43**, 1707 (1991)
508. G. Vignale, Phys. Rev. Lett. **74**, 3233 (1995)
509. G. Vignale, W. Kohn, Phys. Rev. Lett. **77**, 2037 (1996)
510. G. Vignale, C.A. Ullrich, S. Conti, Phys. Rev. Lett. **79**, 4878 (1997)
511. J.F. Dobson, M. Bünner, E.K.U. Gross, Phys. Rev. Lett. **79**, 1905 (1997)
512. I.V. Tokatly, O. Pankratov, Phys. Rev. B **67**, 201103 (2003)

513. I.V. Tokatly, Phys. Rev. B **71**, 165104 (2005)
514. I.V. Tokatly, Phys. Rev. B **71**, 165105 (2005)
515. M. Petersilka, E.K.U. Gross, Int. J. Quantum Chem. **30S**, 1393 (1996)
516. T. Grabo, M. Petersilka, E.K.U. Gross, J. Mol. Struct. (THEOCHEM) **501**, 353 (2000)
517. N.T. Maitra, F. Zhang, R.J. Cave, K. Burke, J. Chem. Phys. **120**, 5932 (2004)
518. M.E. Casida, J. Chem. Phys. **122**, 054111 (2005)
519. O. Gritsenko, E.J. Baerends, Phys. Chem. Chem. Phys. **11**, 4640 (2009)
520. M. Petersilka, U.J. Gossmann, E.K.U. Gross, Phys. Rev. Lett. **76**, 1212 (1996)
521. M. Casida, in *Recent Advances in Computational Chemistry*, vol. 1, ed. by D.P. Chong (World Scientific, Singapore, 1995), p. 155
522. M. Casida, in *Recent Developments and Applications in Density Functional Theory*, ed. by J.M. Seminario (Elsevier, Amsterdam, 1996), p. 391
523. C. Jamorski, M.E. Casida, D.R. Salahub, J. Chem. Phys. **104**, 5134 (1996)
524. J. Guan, M. Casida, D.R. Salahub, J. Mol. Struct. (THEOCHEM) **527**, 299 (2000)
525. M.A.L. Marques, E.K.U. Gross, Annu. Rev. Phys. Chem. **55**, 427 (2004)
526. A.K. Rajagopal, Phys. Rev. A **50**, 3759 (1994)
527. F.A. Parpia, W.R. Johnson, J. Phys. B **16**, L375 (1983)
528. F.A. Parpia, W.R. Johnson, J. Phys. B **17**, 531 (1984)
529. E. Engel, in *Relativistic Electronic Structure Theory, Part 1. Fundamentals*, ed. by P. Schwerdtfeger (Elsevier, Amsterdam, 2002), p. 524
530. A.O.G. Källén, in *Handbuch der Physik, Band V, Teil 1*, ed. by S. Flügge (Springer, Berlin, 1958), p. 169
531. C. Itzykson, J.B. Zuber, *Quantum Field Theory* (McGraw-Hill, New York, NY, 1980)
532. E. Engel, R.M. Dreizler, in *Density Functional Theory II, Topics in Current Chemistry*, vol. 181, ed. by R.F. Nalewajski (Springer, Berlin, 1996), p. 1
533. G. Plunien, B. Müller, W. Greiner, Phys. Rep. **134**, 87 (1986)
534. A.K. Rajagopal, J. Phys. C **11**, L943 (1978)
535. A.H. MacDonald, S.H. Vosko, J. Phys. C **12**, 2977 (1979)
536. H. Eschrig, G. Seifert, P. Ziesche, Solid State Commun. **56**, 777 (1985)
537. H.J.F. Jansen, Phys. Rev. B **38**, 8022 (1988)
538. M.V. Ramana, A.K. Rajagopal, Phys. Rev. A **24**, 1689 (1981)
539. K.N. Huang, M. Aoyagi, M.H. Chen, B. Crasemann, H. Mark, At. Data Nucl. Data Tables **18**, 243 (1976)
540. These statements should be taken with a grain of salt: In view of the prominent role of Ward identities and gauge invariance for the success of the renormalization program it seems that relativistic spin-density functional theory cannot be rigorously established within the framework of QED. On the other hand, the combination of the Hamiltonian (8.139) with projection operators onto positive energy states suffers from the same nonlinearity as the DC approach (8.73). The spin-dependent variants of RDFT are thus build on somewhat less solid ground than the four current version of RDFT. A detailed investigation of this issue is not yet available.
541. M.V. Ramana, A.K. Rajagopal, J. Phys. C **14**, 4291 (1981)
542. G. Diener, J. Gräfenstein, J. Phys. Condens. Matter **1**, 8445 (1989)
543. H.J.F. Jansen, Phys. Rev. B **59**, 4699 (1999)
544. M.V. Ramana, A.K. Rajagopal, J. Phys. C **12**, L845 (1979)
545. A.H. MacDonald, J. Phys. C **16**, 3869 (1983)
546. B.X. Xu, A.K. Rajagopal, M.V. Ramana, J. Phys. C **17**, 1339 (1984)
547. F. Wang, W.J. Liu, J. Chin. Chem. Soc. (Taipei) **50**, 597 (2003)
548. J. Anton, B. Fricke, E. Engel, Phys. Rev. A **69**, 012505 (2004)
549. R. Laskowski, G.K.H. Madsen, P. Blaha, K. Schwarz, Phys. Rev. B **69**, 140408(R) (2004)
550. P. Cortona, S. Doniach, C. Sommers, Phys. Rev. A **31**, 2842 (1985)
551. P. Cortona, Phys. Rev. B **40**, 12105 (1989)
552. H. Ebert, J. Phys.: Condens. Matter **1**, 9111 (1989)
553. J. Forstreuter, L. Steinbeck, M. Richter, H. Eschrig, Phys. Rev. B **55**, 9415 (1997)

554. H. Yamagami, A. Mavromaras, J. Kübler, J. Phys.: Condens. Matter **9**, 10881 (1997)
555. E. Engel, T. Auth, R.M. Dreizler, Phys. Rev. B **64**, 235126 (2001)
556. E. Engel, D. Ködderitzsch, H. Ebert, Phys. Rev. B **78**, 235123 (2008)
557. P.M. Morse, H. Feshbach, *Methods of Theoretical Physics* (McGraw-Hill, New York, NY, 1953)
558. H. Eschrig, M. Sargolzaei, K. Koepernik, M. Richter, Europhys. Lett. **72**, 611 (2005)
559. M.S.S. Brooks, Physica B **130**, 6 (1985)
560. J.M. Morbec, K. Capelle, Int. J. Quantum Chem. **108**, 2433 (2008)
561. D.D. Koelling, A.H. MacDonald, in *Relativistic Effects in Atoms, Molecules and Solids*, *NATO ASI Series*, vol. 87, ed. by G.L. Malli (Plenum, New York, NY, 1981), p. 227
562. J. Staunton, Rep. Prog. Phys. **57**, 1289 (1994)
563. V. Pershina, B. Fricke, in *Heavy Elements and Related New Phenomena, Vol.1*, ed. by W. Greiner, R.K. Gupta (World Scientific, Singapore, 1999), p. 194
564. H. Ebert, in *Electronic Structure and Physical Properties of Solids*, *Lecture Notes in Physics*, vol. 535, ed. by H. Dreyssé (Springer, Berlin, 2000), p. 191
565. N.E. Christensen, in *Relativistic Electronic Structure Theory, Part 1. Fundamentals*, ed. by P. Schwerdtfeger (Elsevier, Amsterdam, 2002), p. 863
566. K. Schwarz, P. Blaha, G.K.H. Madsen, Comp. Phys. Commun. **147**, 71 (2002)
567. H. Eschrig, M. Richter, I. Opahle, in *Relativistic Electronic Structure Theory, Part 2. Applications*, ed. by P. Schwerdtfeger (Elsevier, Amsterdam, 2004), p. 723
568. H. Ebert, E.K.U. Gross, in *Relativistic Effects in Heavy-Element Chemistry and Physics*, ed. by B.A. Hess (Wiley, New York, NY, 2003), p. 163
569. T.L. Loucks, *Augmented Plane Wave Method* (Benjamin, New York, NY, 1967)
570. O.K. Andersen, Phys. Rev. B **2**, 883 (1970)
571. O.K. Andersen, Phys. Rev. B **12**, 3060 (1975)
572. A.H. MacDonald, J.M. Daams, S.H. Vosko, D.D. Koelling, Phys. Rev. B **23**, 6377 (1981)
573. A.H. MacDonald, J.M. Daams, S.H. Vosko, D.D. Koelling, Phys. Rev. B **25**, 713 (1982)
574. D. Singh, S.H. Wei, H. Krakauer, Phys. Rev. Lett. **57**, 3292 (1986)
575. H. Ebert, P. Strange, , B.L. Gyorffy, J. Phys. F: Met. Phys. **18**, L135 (1988)
576. Z.W. Lu, D. Singh, H. Krakauer, Phys. Rev. B **39**, 10154 (1989)
577. P. Strange, H. Ebert, J.B. Staunton, B.L. Gyorffy, J. Phys.: Condens. Matter **1**, 2959 (1989)
578. P. Blaha, K. Schwarz, P. Sorantin, S.B. Trickey, Comput. Phys. Commun. **59**, 399 (1990)
579. H. Reinisch, H. Bross, Z. Phys. B **95**, 145 (1994)
580. P. Söderlind, J.M. Wills, B. Johansson, O. Eriksson, Phys. Rev. B **55**, 1997 (1997)
581. L. Fast, O. Eriksson, B. Johansson, J.M. Wills, G. Straub, H. Roeder, L. Nordström, Phys. Rev. Lett. **81**, 2978 (1998)
582. T. Huhne, C. Zecha, H. Ebert, P.H. Dederichs, R. Zeller, Phys. Rev. B **58**, 10236 (1998)
583. P.H.T. Philipsen, E.J. Baerends, Phys. Rev. B **61**, 1773 (2000)
584. C.J. Pickard, B. Winkler, R.K. Chen, M.C. Payne, M.H. Lee, J.S. Lin, J.A. White, V. Milman, D. Vanderbilt, Phys. Rev. Lett. **85**, 5122 (2000)
585. J. Kunes, P. Novak, R. Schmid, P. Blaha, K. Schwarz, Phys. Rev. B **64**, 153102 (2001)
586. G. Nicolay, F. Reinert, S. Hüfner, P. Blaha, Phys. Rev. B **65**, 033407 (2002)
587. C. Spiel, P. Blaha, K. Schwarz, Phys. Rev. B **79**, 115123 (2009)
588. A. Rosén, D.E. Ellis, J. Chem. Phys. **62**, 3039 (1975)
589. T. Ziegler, J.G. Snijders, E.J. Baerends, J. Chem. Phys. **74**, 1271 (1981)
590. C. van Wüllen, J. Chem. Phys. **103**, 3589 (1995)
591. E. van Lenthe, J.G. Snijders, E.J. Baerends, J. Chem. Phys. **105**, 6505 (1996)
592. W. Liu, M. Dolg, Phys. Rev. A **57**, 1721 (1998)
593. W. Liu, C. van Wüllen, J. Chem. Phys. **110**, 3730 (1999)
594. P. Belanzoni, E. van Lenthe, E.J. Baerends, J. Chem. Phys. **114**, 4421 (2001)
595. W. Liu, G. Hong, D. Dai, L. Li, M. Dolg, Theor. Chem. Acc. **96**, 75 (1997)
596. W. Liu, M. Dolg, L. Li, J. Chem. Phys. **108**, 2886 (1998)
597. P. Schwerdtfeger, M. Pernpointner, J.K. Laerdahl, J. Chem. Phys. **111**, 3357 (1999)

598. D. Geschke, S. Fritzsche, W.D. Sepp, B. Fricke, S. Varga, J. Anton, Phys. Rev. B **62**, 15439 (2000)
599. S. Varga, B. Fricke, H. Nakamatsu, T. Mukoyama, J. Anton, D. Geschke, A. Heitmann, E. Engel, T. Bastug, J. Chem. Phys. **112**, 3499 (2000)
600. T. Bastug, M. Hirata, S. Varga, B. Fricke, S. Erkoc, T. Mukoyama, Adv. Quantum Chem. **37**, 353 (2001)
601. J. Anton, T. Jacob, B. Fricke, E. Engel, Phys. Rev. Lett. **89**, 213001 (2002)
602. J. Gao, W. Liu, B. Song, C. Liu, J. Chem. Phys. **121**, 6658 (2004)
603. J. Anton, B. Fricke, P. Schwerdtfeger, Chem. Phys. **311**, 97 (2005)
604. J. Gao, W. Zou, W. Liu, Y. Xiao, D. Peng, B. Song, C. Liu, J. Chem. Phys. **123**, 054102 (2005)
605. F. Wang, T. Ziegler, E. van Lenthe, S. van Gisbergen, E.J. Baerends, J. Chem. Phys. **122**, 204103 (2005)
606. M.M. Rieger, P. Vogl, Phys. Rev. A **52**, 282 (1995)
607. T. Kreibich, E.K.U. Gross, E. Engel, Phys. Rev. A **57**, 138 (1998)
608. B.A. Shadwick, J.D. Talman, M.R. Norman, Comput. Phys. Commun. **54**, 95 (1989)
609. E. Engel, S. Keller, A. Facco Bonetti, H. Müller, R.M. Dreizler, Phys. Rev. A **52**, 2750 (1995)
610. D. Ködderitzsch, H. Ebert, E. Engel, Phys. Rev. B **77**, 045101 (2008)
611. K.G. Dyall, I.P. Grant, C.T. Johnson, F.A. Parpia, E.P. Plummer, Comput. Phys. Commun. **55**, 425 (1989)
612. E. Engel, S. Keller, R.M. Dreizler, Phys. Rev. A **53**, 1367 (1996)
613. In all our calculations the nuclei were represented by uniformly charged spheres with nuclear radii given by $R_{\rm nuc} = (1.0793A^{1/3} + 0.73587)$fm, A being the atomic mass (weighted by isotopic abundances) taken from Table III.7 of K. Hisaka, et al. (Particle Data Group), Phys. Rev. D **45**, Number 11, Part II (1992), unless explicitly stated otherwise. The speed of light had been set to $c = 137.0359895$ a.u..
614. Y. Ralchenko, A.E. Kramida, J. Reader, N.A. Team (eds.), *NIST Atomic Spectra Database (version 3.1.5), [Online]. Available: http://physics.nist.gov/asd3 [2008, September 2]* (National Institute of Standards and Technology, Gaithersburg, 2008)
615. J. Harris, R.O. Jones, J. Chem. Phys. **68**, 3316 (1978)
616. O. Gunnarsson, R.O. Jones, Phys. Rev. B **31**, 7588 (1985)
617. J.B. Lagowski, S.H. Vosko, Phys. Rev. A **39**, 4972 (1989)
618. I.A. Akhiezer, S.V. Peletminskii, Zh. Eksp. Teor. Fiz. **38**, 1829 (1960). [Sov. Phys. JETP 11, 1316 (1960)]
619. H.S. Zapolsky, Lns report, Cornell University (1960)
620. B. Jancovici, Nuovo Cim. **XXV**, 428 (1962)
621. B.A. Freedman, L.D. McLerran, Phys. Rev. D **16**, 1130 and 1147 and 1169 (1977)
622. A. Facco Bonetti, E. Engel, R.M. Dreizler, I. Andrejkovics, H. Müller, Phys. Rev. A **58**, 993 (1998)
623. E. Engel, S. Keller, R.M. Dreizler, in *Electronic Density Functional Theory: Recent Progress and New Directions*, ed. by J.F. Dobson, G. Vignale, M.P. Das (Plenum, New York, NY, 1998), p. 149
624. M. Mayer, O.D. Häberlen, N. Rösch, Phys. Rev. A **54**, 4775 (1996)
625. S. Varga, E. Engel, W.D. Sepp, B. Fricke, Phys. Rev. A **59** (1999)
626. W. Liu, C. van Wüllen, J. Chem. Phys. **113**, 2506 (2000)
627. R.N. Schmid, E. Engel, R.M. Dreizler, P. Blaha, K. Schwarz, Adv. Quantum Chem. **33**, 209 (1998)
628. P. Pyykkö, Chem. Rev. **88**, 563 (1988)
629. L. Brewer, Tech. Rep. Report LBL-3720 Rev., Lawrence Berkeley Laboratory, University of California, Berkeley (1977)
630. M.W.C. Dharma-Wardana, F. Perrot, Phys. Rev. A **26**, 2096 (1982)
631. F. Perrot, M.W.C. Dharma-Wardana, Phys. Rev. A **29**, 1378 (1984)
632. M.W.C. Dharma-Wardana, in *Density Functional Theory, NATO ASI Series B*, vol. 337, ed. by E.K.U. Gross, R.M. Dreizler (Plenum, New York, NY, 1995), p. 625

633. F. Perrot, M.W.C. Dharma-Wardana, Phys. Rev. B **52**, 5352 (1995)
634. G. Senatore, G. Pastore, Phys. Rev. Lett. **64**, 303 (1990)
635. A. Griffin, Can. J. Phys. **73**, 755 (1995)
636. G.S. Nunes, Ph.D. thesis, SUNY at Stony Brook, New York, NY (1997)
637. G.S. Nunes, J. Phys. B: At. Mol. Opt. Phys. **32**, 4293 (1999)
638. A.P. Albus, F. Illuminati, M. Wilkens, Phys. Rev. A **67**, 063606 (2003)
639. Y.E. Kim, A.L. Zubarev, Phys. Rev. A **67**, 015602 (2003)
640. C. Ebner, W.F. Saam, Phys. Rev. B **12**, 923 (1975)
641. S. Stringari, J. Treiner, Phys. Rev. B **36**, 8369 (1987)
642. F. Dalfovo, A. Lastri, L. Pricaupenko, S. Stringari, J. Treiner, Phys. Rev. B **52**, 1193 (1995)
643. B. Tanatar, K. Erkan, Phys. Rev. A **62**, 053601 (2000)
644. K.K. Rajagopal, Phys. Rev. B **76**, 054519 (2007)
645. R. Resta, Rev. Mod. Phys. **66**, 899 (1994)
646. X. Gonze, P. Ghosez, R.W. Godby, Phys. Rev. Lett. **74**, 4035 (1995)
647. G. Ortiz, R.M. Martin, Phys. Rev. B **49**, 14202 (1994)
648. R.M. Martin, G. Ortiz, Phys. Rev. B **56**, 1124 (1997)
649. D. Vanderbilt, Phys. Rev. Lett. **79**, 3966 (1997)
650. G. Ortiz, I. Souza, R.M. Martin, Phys. Rev. Lett. **80**, 353 (1998)
651. R. Resta, Phys. Rev. Lett. **77**, 2265 (1996)
652. R.M. Martin, G. Ortiz, Phys. Rev. Lett. **78**, 2028 (1997)
653. X. Gonze, P. Ghosez, R.W. Godby, Phys. Rev. Lett. **78**, 2029 (1997)
654. R. Resta, Phys. Rev. Lett. **78**, 2030 (1997)
655. B. Champagne, E.A. Perpète, S.J.A. van Gisbergen, E.J. Baerends, J.G. Snijders, C. Soubra-Ghaoui, K. Robins, B. Kirtman, J. Chem. Phys. **109**, 10489 (1998)
656. W. Yang, Phys. Rev. Lett. **66**, 1438 (1991)
657. W. Yang, Phys. Rev. A **44**, 7823 (1991)
658. Q. Zhao, W. Yang, J. Chem. Phys. **102**, 9598 (1995)
659. W. Yang, T.S. Lee, J. Chem. Phys. **103**, 5674 (1995)
660. W. Kohn, Phys. Rev. **133**, A171 (1964)
661. W. Kohn, Phys. Rev. Lett. **76**, 3168 (1996)
662. S. Goedecker, Phys. Rev. B **58**, 3501 (1998)
663. S. Ismail-Beigi, T.A. Arias, Phys. Rev. Lett. **82**, 2127 (1999)
664. S. Goedecker, Rev. Mod. Phys. **71**, 1085 (1999)
665. R.M. Martin, *Electronic Structure: Basic Theory and Practical Methods* (Cambridge University Press, Cambridge, 2004)
666. L. Seijo, Z. Barandiarán, J.M. Soler, Theor. Chem. Acc. **118**, 541 (2007)
667. T. Otsuka, T. Miyazaki, T. Ohno, D.R. Bowler, M.J. Gillan, J. Phys.: Condens. Matter **20**, 294201 (2008)
668. X. Wu, A. Selloni, R. Car, Phys. Rev. B **79**, 085102 (2009)
669. M.C. Payne, M.P. Teter, D.C. Allan, T.A. Arias, J.D. Joannopoulos, Rev. Mod. Phys. **64**, 1045 (1992)
670. L.M. Sander, H.B. Shore, L.J. Sham, Phys. Rev. Lett. **31**, 533 (1973)
671. R.K. Kalia, P. Vashishta, Phys. Rev. B **17**, 2655 (1978)
672. J.F. Capitani, R.F. Nalewajski, R.G. Parr, J. Chem. Phys. **76**, 568 (1982)
673. N. Gidopoulos, Phys. Rev. B **57**, 2146 (1998)
674. T. Kreibich, E. Gross, Phys. Rev. Lett. **86**, 2984 (2001)
675. T. Kreibich, R. van Leeuwen, E.K.U. Gross, Phys. Rev. A **78**, 022501 (2008)
676. T.C. Li, P.Q. Tong, Phys. Rev. A **34**, 529 (1986)
677. O. Butriy, H. Ebadi, P.L. de Boeij, R. van Leeuwen, E.K.U. Gross, Phys. Rev. A **76**, 052514 (2007)
678. L.N. Oliveira, E.K.U. Gross, W. Kohn, Phys. Rev. Lett. **60**, 2430 (1988)
679. K. Capelle, E.K.U. Gross, Int. J. Quantum Chem. **61**, 325 (1997)
680. O.J. Wacker, R. Kümmel, E.K.U. Gross, Phys. Rev. Lett. **73**, 2915 (1994)
681. S. Kurth, M. Marques, M. Lüders, E.K.U. Gross, Phys. Rev. Lett. **83**, 2628 (1999)

682. M. Lüders, M.A.L. Marques, N.N. Lathiotakis, A. Floris, G. Profeta, L. Fast, A. Continenza, S. Massida, E.K.U. Gross, Phys. Rev. B **72**, 024545 (2005)
683. B. Suvasini, B.L. Gyorffy, Physica C **195**, 109 (1992)
684. B. Suvasini, W.M. Temmerman, B.L. Gyorffy, Phys. Rev. B **48**, 1202 (1993)
685. W.M. Temmerman, Z. Szotek, B.L. Gyorffy, O.K. Andersen, O. Jepsen, Phys. Rev. Lett. **76**, 307 (1996)
686. B.L. Gyorffy, Z. Szotek, W.M. Temmerman, O.K. Andersen, O. Jepsen, Phys. Rev. B **58**, 1025 (1998)
687. M.A.L. Marques, M. Lüders, N.N. Lathiotakis, G. Profeta, A. Floris, L. Fast, A. Continenza, E.K.U. Gross, S. Massida, Phys. Rev. B **72**, 024546 (2005)
688. G. Profeta, C. Franchini, N.N. Lathiotakis, A. Floris, A. Sanna, M.A.L. Marques, M. Lüders, S. Massida, E.K.U. Gross, A. Continenza, Phys. Rev. Lett. **96**, 047003 (2006)
689. A. Sanna, C. Franchini, A. Floris, G. Profeta, N.N. Lathiotakis, M. Lüders, M.A.L. Marques, E.K.U. Gross, A. Continenza, S. Massida, Phys. Rev. B **73**, 144512 (2006)
690. A. Floris, A. Sanna, S. Massida, E.K.U. Gross, Phys. Rev. B **75**, 054508 (2007)
691. A. Sanna, G. Profeta, A. Floris, A. Marini, E.K.U. Gross, S. Massida, Phys. Rev. B **75**, 020511R (2007)
692. P. Cudazzo, G. Profeta, A. Sanna, A. Floris, A. Continenza, S. Massida, E.K.U. Gross, Phys. Rev. Lett. **100**, 257001 (2008)
693. C. Bersier, A. Floris, A. Sanna, G. Profeta, A. Continenza, E.K.U. Gross, S. Massida, Phys. Rev. B **79**, 104503 (2009)
694. S. Sharma, S. Shallcross, J.K. Dewhurst, A. Sanna, C. Bersier, S. Massida, E.K.U. Gross, Phys. Rev. B **80**, 184502 (2009)
695. M. Bender, P.H. Heenen, P.-G. Reinhard, Rev. Mod. Phys. **75**, 121 (2003)
696. C. Speicher, R.M. Dreizler, E. Engel, Ann. Phys. (N.Y.) **213**, 312 (1992)
697. R.N. Schmid, E. Engel, R.M. Dreizler, Phys. Rev. C **52**, 164 (1995)
698. R.N. Schmid, E. Engel, R.M. Dreizler, Phys. Rev. C **52**, 2804 (1995)
699. J. Engel, Phys. Rev. C **75**, 014306 (2007)
700. N. Barnea, Phys. Rev. C **76**, 067302 (2007)
701. J. Messud, M. Bender, E. Suraud, Phys. Rev. C **80**, 054314 (2009)
702. G.A. Lalazissis, P. Ring, D. Vretemar (eds.), *Extended Density Functionals in Nuclear Structure Physics*, *Lecture Notes in Physics*, vol. 641 (Springer, Berlin, 2004)
703. E. Drut, R.J. Furnstahl, L. Platter, Progr. Part. Nucl. Phys. **64**, 120 (2010)
704. R.A. Donnelly, R.G. Parr, J. Chem. Phys. **69**, 4431 (1978)
705. S.M. Valone, J. Chem. Phys. **73**, 4653 (1980)
706. G. Zumbach, K. Maschke, J. Chem. Phys. **82**, 5604 (1985)
707. A.M.K. Müller, Phys. Lett. A **105**, 446 (1986)
708. S. Goedecker, C.J. Umrigar, Phys. Rev. Lett. **81**, 866 (1998)
709. G. Csányi, T.A. Arias, Phys. Rev. B **61**, 7348 (2000)
710. M.A. Buijse, E.J. Baerends, Mol. Phys. **100**, 401 (2002)
711. J. Cioslowski, K. Pernal, M. Buchowiecki, J. Chem. Phys. **119**, 6443 (2003)
712. C. Kollmar, J. Chem. Phys. **121**, 11581 (2004)
713. O. Gritsenko, K. Pernal, E.J. Baerends, J. Chem. Phys. **122**, 204102 (2005)
714. N.N. Lathiotakis, N. Helbig, E.K.U. Gross, Phys. Rev. B **72**, 030501R (2005)
715. R.L. Frank, E.H. Lieb, R. Seiringer, H. Siedentop, Phys. Rev. A **76**, 052517 (2007)
716. S. Sharma, J.K. Dewhurst, N.N. Lathiotakis, E.K.U. Gross, Phys. Rev. B **78**, 201103 (2008)
717. N.N. Lathiotakis, S. Sharma, J.K. Dewhurst, F.G. Eich, M.A.L. Marques, E.K.U. Gross, Phys. Rev. A **79**, 040501R (2009)
718. K.J.H. Giesbertz, K. Pernal, O. Gritsenko, E.J. Baerends, J. Chem. Phys. **130**, 114104 (2009)
719. F. Furche, J. Chem. Phys. **114**, 5982 (2001)
720. K.J.H. Giesbertz, E.J. Baerends, O. Gritsenko, Phys. Rev. Lett. **101**, 033004 (2008)
721. P. Ziesche, Phys. Lett. A **195**, 213 (1994)
722. A. Gonis, T.C. Schulthess, J. van Ek, P.E.A. Turchi, Phys. Rev. Lett. **77**, 2981 (1996)
723. P. Ziesche, Int. J. Quantum Chem. **60**, 1361 (1996)

724. M. Levy, P. Ziesche, J. Chem. Phys. **115**, 9110 (2001)
725. F. Furche, Phys. Rev. A **70**, 022514 (2004)
726. D.A. Mazziotti, Phys. Rev. Lett. **93**, 213001 (2004)
727. D.A. Mazziotti, Phys. Rev. Lett. **97**, 143002 (2006)
728. R. Courant, D. Hilbert, *Methods of Mathematical Physics, Vol.1* (Wiley-Interscience, New York, NY, 1953)
729. Y. Saad, *Iterative Methods for Sparse Linear Systems* (PWS Publishing, Boston, MA, 1996)
730. R. Car, M. Parrinello, Phys. Rev. Lett. **55**, 2471 (1985)
731. I. Stich, R. Car, M. Parrinello, S. Baroni, Phys. Rev. B **39**, 4997 (1989)
732. M.P. Teter, M.C. Payne, D.C. Allen, Phys. Rev. B **40**, 12255 (1989)
733. A.S. Kompaneets, E.S. Pavlovskii, Zh. Eksp. Teor. Fiz. **31**, 427 (1956). [Sov. Phys. JETP **4**, 328 (1957)]
734. D.R. Murphy, Phys. Rev. A **24**, 1682 (1981)
735. E. Chaćon, J.E. Alvarellos, P. Tarazona, Phys. Rev. B **32**, 7868 (1985)
736. P. Garcia-González, J.E. Alvarellos, E. Chaćon, Phys. Rev. B **53**, 9509 (1996)
737. P. Garcia-González, J.E. Alvarellos, E. Chaćon, Phys. Rev. A **54**, 1897 (1996)
738. D. Garcia-Aldea, J.E. Alvarellos, J. Chem. Phys. **129**, 074103 (2008)
739. L.W. Wang, M.P. Teter, Phys. Rev. B **45**, 13196 (1992)
740. F. Perrot, J. Phys.: Condens. Matt. **6**, 431 (1994)
741. M. Pearson, E. Smargiassi, P.A. Madden, J. Phys.: Condens. Matt. **5**, 3221 (1993)
742. E. Smargiassi, P.A. Madden, Phys. Rev. B **49**, 5220 (1994)
743. E. Smargiassi, P.A. Madden, Phys. Rev. B **51**, 117 (1995)
744. M. Foley, P.A. Madden, Phys. Rev. B **53**, 10589 (1996)
745. Y.A. Wang, N. Govind, E.A. Carter, Phys. Rev. B **58**, 13465 (1998)
746. Y.A. Wang, N. Govind, E.A. Carter, Phys. Rev. B **60**, 16350 (1999)
747. Y.A. Wang, N. Govind, E.A. Carter, Phys. Rev. B **64**, 089903(E) (2001)
748. Y.A. Wang, N. Govind, E.A. Carter, Phys. Rev. B **64**, 129901(E) (2001)
749. G.S. Ho, V.L. Lignères, E.A. Carter, Phys. Rev. B **78**, 045105 (2008)
750. Y.A. Wang, Phys. Rev. A **55**, 4589 (1997)
751. N.C. Handy, M.T. Marron, H.J. Silverstone, Phys. Rev. **180**, 45 (1969)
752. W. Greiner, B. Müller, J. Rafelski, *Quantum Electrodynamics of Strong Fields* (Springer, Berlin, 1985)
753. n-point functions are called irreducible if their diagrammatic expansions only consist of graphs which do not split into two pieces if one internal electron or photon line is cut.
754. G. t'Hooft, M. Veltman, Nucl. Phys. B **44**, 189 (1972)
755. N.P. Landsman, C.G. van Weert, Phys. Rep. **145**, 141 (1987)
756. K. Lim, C.J. Horowitz, Nucl. Phys. A **501**, 729 (1989)
757. G. Källén, A. Sabry, Dan. Mat. Fys. Medd. **29**, No.17 (1955)
758. R. Barbieri, E. Remiddi, Nuovo Cim. A **13**, 99 (1973)
759. J.I. Kapusta, *Finite-temperature Field Theory* (Cambridge University Press, Cambridge, 1989)
760. E. Braaten, R.D. Pisarski, Nucl. Phys. B **337**, 569 (1990)
761. M.S. Vallarta, N. Rosen, Phys. Rev. **41**, 708 (1932)
762. S.A. Chin, Ann. Phys. (N.Y.) **108**, 301 (1977)
763. H. Müller, E. Engel, R.M. Dreizler, Phys. Rev. A **40**, 5542 (1989)
764. E. Engel, R.M. Dreizler, Phys. Rev. A **35**, 3607 (1987)
765. E. Engel, R.M. Dreizler, Phys. Rev. A **38**, 3909 (1988)
766. E. Engel, H. Müller, R.M. Dreizler, Phys. Rev. A **39**, 4873 (1989)
767. W.F. Pohlner, R.M. Dreizler, Phys. Rev. A **44**, 7165 (1991)
768. H. Jensen, Z. Phys. **82**, 794 (1933)
769. R.M. Dreizler, E. Engel, in *Density Functionals: Theory and Applications*, ed. by D.P. Joubert (Springer, Berlin, 1998), p. 147
770. R.M. Dreizler, in *A Primer in Density Functional Theory*, *Lecture Notes in Physics*, vol. 620, ed. by C. Fiolhais, F. Nogeira, M. Marques (Springer, Berlin, 2003), p. 123
771. H.J.A.C. Stroucken, R.M. Dreizler, Phys. Rev. A **43**, 3401 (1991)

Index

D

H

I